AF773233

# Student Solutions Manual

*to accompany*

# Chemistry

Thirteenth Edition

**Raymond Chang**
*Williams College*

**Jason Overby**
*College of Charleston*

*Prepared by*
**Jason Overby**

Mc Graw Hill Education

STUDENT SOLUTIONS MANUAL TO ACCOMPANY

CHEMISTRY

Published by McGraw-Hill Education, 2 Penn Plaza, New York, NY 10121. 

This book is printed on acid-free paper.

1 2 3 4 5 6 7 8 9 BKM 21 20 19 18 17

ISBN 978-1-260-09315-5
MHID 1-260-09315-8

Cover image: ©Shutterstock/Patrick Foto



mheducation.com/highered

# Contents

# Contents

# CHAPTER 1
# CHEMISTRY: THE STUDY OF CHANGE

## PROBLEM-SOLVING STRATEGIES AND TUTORIAL SOLUTIONS

### TYPES OF PROBLEMS

**Problem Type 1:** Density Calculations.

**Problem Type 2:** Temperature Conversions.

**(a)** °C → °F

**(b)** °F → °C

**Problem Type 3:** Scientific Notation.

**(a)** Expressing a number in scientific notation.

**(b)** Addition and subtraction.

**(c)** Multiplication and division.

**Problem Type 4:** Significant Figures.

**(a)** Addition and subtraction.

**(b)** Multiplication and division.

**Problem Type 5:** The Dimensional Analysis Method of Solving Problems.

## PROBLEM TYPE 1: DENSITY CALCULATIONS

Density is the mass of an object divided by its volume.

$$\text{density} = \frac{\text{mass}}{\text{volume}}$$

$$d = \frac{m}{V}$$

Densities of solids and liquids are typically expressed in units of grams per cubic centimeter (g/cm$^3$) or equivalently grams per milliliter (g/mL). Because gases are much less dense than solids and liquids, typical units are grams per liter (g/L).

**EXAMPLE 1.1**

**A lead brick with dimensions of 5.08 cm by 10.2 cm by 20.3 cm has a mass of 11,950 g. What is the density of lead in g/cm$^3$?**

**Strategy:** You are given the mass of the lead brick in the problem. You need to calculate the volume of the lead brick to solve for the density. The volume of a rectangular object is equal to the length × width × height.

$$\text{density} = \frac{\text{mass}}{\text{volume}}$$

**Solution:**

$$\text{Volume} = \text{length} \times \text{width} \times \text{height}$$

$$\text{Volume} = 5.08\text{ cm} \times 10.2\text{ cm} \times 20.3\text{ cm} = \mathbf{1052\text{ cm}^3}$$

Calculate the density by substituting the mass and the volume into the equation.

$$d = \frac{m}{V} = \frac{11{,}950\ \text{g}}{1052\ \text{cm}^3} = \mathbf{11.4\ g/cm^3}$$

**PRACTICE EXERCISE**

1. Platinum has a density of 21.4 g/cm$^3$. What is the mass of a small piece of platinum that has a volume of 7.50 cm$^3$?

**Text Problem: 1.24**

## PROBLEM TYPE 2: TEMPERATURE CONVERSIONS

To convert between the Fahrenheit scale and the Celsius scale, you must account for two differences between the two scales.

(1) The Fahrenheit scale defines the normal freezing point of water to be exactly 32°F, whereas the Celsius scale defines it to be exactly 0°C.

(2) A Fahrenheit degree is 5/9 the size of a Celsius degree.

### A. Converting degrees Fahrenheit to degrees Celsius

The equation needed to complete a conversion from degrees Fahrenheit to degrees Celsius is:

$$\textbf{? °C} = \textbf{(°F} - \textbf{32°F)} \times \frac{\textbf{5°C}}{\textbf{9°F}}$$

32°F is subtracted to compensate for the normal freezing point of water being 32°F, compared to 0° on the Celsius scale. We multiply by (5/9) because a Fahrenheit degree is 5/9 the size of a Celsius degree.

**EXAMPLE 1.2**

**Convert 20°F to degrees Celsius.**

$$?\text{°C} = (\text{°F} - 32\text{°F}) \times \frac{5\text{°C}}{9\text{°F}}$$

$$?\ \text{°C} = (20\text{°F} - 32\text{°F}) \times \frac{5\text{°C}}{9\text{°F}} = \mathbf{-6.7°C}$$

### B. Converting degrees Celsius to degrees Fahrenheit

The equation needed to complete a conversion from degrees Celsius to degrees Fahrenheit is:

$$\textbf{? °F} = \left(\textbf{°C} \times \frac{\textbf{9°F}}{\textbf{5°C}}\right) + \textbf{32°F}$$

°C is multiplied by (9/5) because a Celsius degree is 9/5 the size of a Fahrenheit degree. 32°F is then added to compensate for the normal freezing point of water being 32°F, compared to 0° on the Celsius scale.

**EXAMPLE 1.3**

**Normal human body temperature on the Celsius scale is 37.0°C. Convert this to the Fahrenheit scale.**

$$?\ \text{°F} = \left(\text{°C} \times \frac{9\text{°F}}{5\text{°C}}\right) + 32\text{°F}$$

$$? \,°\text{F} = \left(37.0°\text{C} \times \frac{9°\text{F}}{5°\text{C}}\right) + 32°\text{F} = \mathbf{98.6°F}$$

**PRACTICE EXERCISE**

**2.** Convert −40°F to degrees Celsius.

**Text Problems: 1.26,** 1.28

## PROBLEM TYPE 3: SCIENTIFIC NOTATION

Scientific notation is typically used when working with small or large numbers. All numbers can be expressed in the form

$$N \times 10^n$$

where $N$ is a number between 1 and 10 and $n$ is an exponent that can be a positive or negative integer, or zero.

### A. Expressing a number in scientific notation

**Strategy:** Writing scientific notation as $N \times 10^n$, we determine $n$ by counting the number of places that the decimal point must be moved to give $N$, a number between 1 and 10.

If the decimal point is moved to the left, $n$ is a positive integer, the number you are working with is larger than 10. If the decimal point is moved to the right, $n$ is a negative integer. The number you are working with is smaller than 1.

**EXAMPLE 1.4**
**Express 0.000105 in scientific notation.**

**Solution:** The decimal point must be moved four places to the right to give $N$, a number between 1 and 10. In this case,

$$N = \mathbf{1.05}$$

Since 0.000105 is a number less than one, $n$ is a negative integer. In this case, $n = -4$ (The decimal point was moved four places to the right to give $N = 1.05$).

Combining the above two steps:

$$0.000105 = \mathbf{1.05 \times 10^{-4}}$$

**Tip:** The notation $1.05 \times 10^{-4}$ means the following: Take 1.05 and multiply by $10^{-4}$ (0.0001).

$$1.05 \times 0.0001 = 0.000105$$

**EXAMPLE 1.5**
**Express 4224 in scientific notation.**

**Solution:** The decimal point must be moved three places to the left to give $N$, a number between 1 and 10. In this case,

$$N = \mathbf{4.224}$$

Since 4,224 is a number greater than one, $n$ is a positive integer. In this case, $n = 3$ (the decimal point was moved three places to the left to give $N = 4.224$).

Combining the above two steps:

$$4224 = \mathbf{4.224 \times 10^3}$$

**Tip:** The notation $4.224 \times 10^3$ means the following: Take 4.224 and multiply by $10^3$ (1000).

$$4.224 \times 1000 = 4{,}224$$

**PRACTICE EXERCISE**

3. Express the following numbers in scientific notation:

**(a)** 45,781 **(b)** 0.0000430

**Text Problem:** 1.32

## B. Addition and subtraction using scientific notation

**Strategy:** Let's express scientific notation as $N \times 10^n$. When adding or subtracting numbers using scientific notation, we must write each quantity with the same exponent, $n$. We can then add or subtract the $N$ parts of the numbers, keeping the exponent, $n$, the same.

**EXAMPLE 1.6**

**Express the answer to the following calculation in scientific notation. $(2.43 \times 10^1) + (5.955 \times 10^2) = ?$**

**Solution:** Write each quantity with the same exponent, $n$. Let's write $2.43 \times 10^1$ in such a way that $n = 2$.

**Tip:** We are *increasing* $10^n$ by a factor of 10, so we must *decrease* $N$ by a factor of 10. We move the decimal point one place to the left.

$$2.43 \times 10^1 = \mathbf{0.243 \times 10^2}$$

($n$ was increased by 1. Move the decimal point one place to the left.)

Add or subtract, as required, the $N$ parts of the numbers, keeping the exponent, $n$, the same. In this example, the process is addition.

$$\begin{array}{r} 0.243 \times 10^2 \\ +\ 5.955 \times 10^2 \\ \hline \mathbf{6.198 \times 10^2} \end{array}$$

## C. Multiplication and division using scientific notation

**Strategy:** Let's express scientific notation as $N \times 10^n$. Multiply or divide the $N$ parts of the numbers in the usual way. To come up with the correct exponent $n$, when multiplying, *add* the exponents, when dividing, *subtract* the exponents.

**EXAMPLE 1.7**

**Divide $4.2 \times 10^{-7}$ by $5.0 \times 10^{-5}$.**

**Solution:** Divide the $N$ parts of the numbers in the usual way.

$$4.2 \div 5.0 = 0.84$$

When dividing the $10^n$ parts, *subtract* the exponents.

$$0.84 \times 10^{-7-(-5)} = 0.84 \times 10^{-7+5} = \mathbf{0.84 \times 10^{-2}}$$

The usual practice is to express $N$ as a number between 1 and 10. Therefore, it is more appropriate to move the decimal point of the above number one place to the right, decreasing the exponent by 1.

$$0.84 \times 10^{-2} = \mathbf{8.4 \times 10^{-3}}$$

**Tip:** In the answer, we moved the decimal point to the right, *increasing* $N$ by a factor of 10. Therefore, we must *decrease* $10^n$ by a factor of 10. The exponent, $n$, is changed from −2 to −3.

**EXAMPLE 1.8**

**Multiply $2.2 \times 10^{-3}$ by $1.4 \times 10^6$.**

**Solution:** Multiply the $N$ parts of the numbers in the usual way.

$$2.2 \times 1.4 = 3.1$$

When multiplying the $10^n$ parts, *add* the exponents.

$$3.1 \times 10^{-3+6} = \mathbf{3.1 \times 10^3}$$

**PRACTICE EXERCISE**

4. Express the answer to the following calculations in scientific notation. Try these without using a calculator.

   (a) $2.20 \times 10^3 - 4.54 \times 10^2 =$

   (b) $4.78 \times 10^5 \div 6.332 \times 10^{-7} =$

**Text Problem: 1.34**

## PROBLEM TYPE 4: SIGNIFICANT FIGURES

See Section 1.8 of the text for guidelines for using significant figures.

### A. Addition and subtraction

**Strategy:** The number of significant figures to the right of the decimal point in the answer is determined by the lowest number of digits to the right of the decimal point in any of the original numbers.

**EXAMPLE 1.9**

**Carry out the following operations and express the answer to the correct number of significant figures.**

$$102.226 + 2.51 + 736.0 =$$

**Solution:**

$$\begin{array}{r} 102.226 \\ 2.51 \\ + \underline{736.0} \\ 840.736 \end{array} \quad \leftarrow \text{fewest digits to the right of the decimal point}$$

The 3 and 6 are nonsignificant digits, since 736.0 only has one digit to the right of the decimal point. The answer should only have one digit to the right of the decimal point.

The correct answer rounded off to the correct number of significant figures is **840.7**

**Tip:** To round off a number at a certain point, simply drop the digits that follow if the first of them is less than 5. If the first digit following the point of rounding off is equal to or greater than 5, add 1 to the preceding digit.

## B. Multiplying and dividing

**Strategy:** The number of significant figures in the answer is determined by the original number having the smallest number of significant figures.

**EXAMPLE 1.10**

**Carry out the following operations and express the answer to the correct number of significant figures.**

$$\mathbf{12 \times 2143.1 \div 3.11 = ?}$$

**Solution:**

$$12 \times 2143.1 \div 3.11 = 8269.2 = 8.2\mathbf{692} \times 10^3$$

The 6, 9, and 2 (bolded) are nonsignificant digits because the original number 12 only has two significant figures. Therefore, the answer has only two significant figures.

The correct answer rounded off to the correct number of significant figures is $\mathbf{8.3 \times 10^3}$

**PRACTICE EXERCISE**

5. Carry out the following operations and express the answer to the correct number of significant figures.

   (a) $90.25 - 83 + 1.0015 =$

   (b) $55.6 \times 3.482 \div 505.34 =$

**Text Problem: 1.38**

## PROBLEM TYPE 5: THE DIMENSIONAL ANALYSIS METHOD OF SOLVING PROBLEMS

In order to convert from one unit to another, you need to be proficient at applying dimensional analysis. See Section 1.9 of the text. Conversion factors can seem daunting, but if you keep track of the units, making sure that the appropriate units cancel, your effort will be rewarded.

*Step 1:* Map out a strategy to proceed from initial units to final units based on available conversion factors.

*Step 2:* Use the following method as many times as is necessary to ensure that you obtain the desired unit.

$$\cancel{\text{Given}}\text{ unit} \times \left(\frac{\text{desired unit}}{\cancel{\text{given}}\text{ unit}}\right) = \text{desired unit}$$

**EXAMPLE 1.11**

**How long will it take to fly from Denver to New York, a distance of 1631 miles, at a speed of 815 km/hr?**

**Strategy:** One conversion factor is given in the problem, 815 km/hr. This conversion factor can be used to convert from distance (in km) to time (in hr). If you can convert the distance of 1631 miles to km, then you can use the conversion factor (815 km/hr) to convert to time in hours. Another conversion factor that you can look up is

$$1 \text{ mi} = 1.61 \text{ km}$$

You should come up with the following strategy.

$$\text{miles} \rightarrow \text{km} \rightarrow \text{hours}$$

**Solution:** Carry out the necessary conversions, making sure that units cancel.

$$\textbf{? hours} = 1631\,\cancel{\text{mi}}\text{ (given)} \times \frac{1.61\,\cancel{\text{km}}\text{ (desired)}}{1\,\cancel{\text{mi}}\text{ (given)}} \times \frac{1\text{ h (desired)}}{815\,\cancel{\text{km}}\text{ (given)}} = \mathbf{3.22\ h}$$

**Tip**: In the first conversion factor (km/mi), km is the desired unit. When moving on to the next conversion factor (h/km), km is now given, and the desired unit is h.

**EXAMPLE 1.12**

**The *Voyager II* mission to the outer planets of our solar system transmitted by radio signals many spectacular photographs of Neptune. Radio waves, like light waves, travel at a speed of $3.00 \times 10^8$ m/s. If Neptune was 2.75 billion miles from Earth during these transmissions, how many hours were required for radio signals to travel from Neptune to Earth?**

**Strategy:** One conversion factor is given in the problem, $3.00 \times 10^8$ m/s. This conversion factor will allow you to convert from distance (in m) to time (in seconds). If you can convert the distance of 2.75 billion miles to meters, then the speed of light ($3.00 \times 10^8$ m/s) can be used to convert to time in seconds. Other conversion factors that you can look up are:

$$1 \text{ billion} = 1 \times 10^9 \qquad 60 \text{ s} = 1 \text{ min}$$

$$1 \text{ mi} = 1.61 \text{ km} \qquad 60 \text{ min} = 1 \text{ h}$$

$$1 \text{ km} = 1000 \text{ m}$$

You should come up with the following strategy.

miles → km → meters → seconds → min → hours

**Solution:** Carry out the necessary conversions, making sure that units cancel.

$$? \text{ h} = (2.75 \times 10^9 \text{ mi}) \times \frac{1.61 \text{ km}}{1 \text{ mi}} \times \frac{1000 \text{ m}}{1 \text{ km}} \times \frac{1 \text{ s}}{3.00 \times 10^8 \text{ m}} \times \frac{1 \text{ min}}{60 \text{ s}} \times \frac{1 \text{ h}}{60 \text{ min}} = \mathbf{4.10 \text{ h}}$$

**PRACTICE EXERCISES**

6. On a certain day, the concentration of carbon monoxide, CO, in the air over Denver reached $1.8 \times 10^{-5}$ g/L. Convert this concentration to mg/m$^3$.

7. Copper (Cu) is a trace element that is essential for nutrition. Newborn infants require 80 μg of Cu per kilogram of body mass per day. The Cu content of a popular baby formula is 0.48 μg of Cu per milliliter. How many milliliters should a 7.0 lb baby consume per day to obtain the minimum daily Cu requirement?

**Text Problems: 1.42, 1.44,** 1.46, 1.48, 1.50, 1.52, 1.54

## ANSWERS TO PRACTICE EXERCISES

1. 161 g Pt
2. −40°C
3. **(a)** $4.5781 \times 10^4$
   **(b)** $4.30 \times 10^{-5}$
4. **(a)** $1.75 \times 10^3$
   **(b)** $7.55 \times 10^{11}$
5. **(a)** 8
   **(b)** 0.383
6. 18 mg/m$^3$
7. 530 mL/day

## SOLUTIONS TO SELECTED TEXT PROBLEMS

**1.4** (a) hypothesis (b) law (c) theory

**1.10** (a) Cs (b) Ge (c) Ga (d) Sr
(e) U (f) Se (g) Ne (h) Cd

**1.12** (a) homogeneous mixture (b) element (c) compound
(d) homogeneous mixture (e) heterogeneous mixture (f) heterogeneous mixture
(g) element

**1.18** (a) Physical change. The helium isn't changed in any way by leaking out of the balloon.

(b) Chemical change in the battery.

(c) Physical change. The orange juice concentrate can be regenerated by evaporation of the water.

(d) Chemical change. Photosynthesis changes water, carbon dioxide, etc., into complex organic matter.

(e) Physical change. The salt can be recovered unchanged by evaporation.

**1.24** **Strategy:** We are given the density and volume of a liquid and asked to calculate the mass of the liquid. Rearrange the density equation, Equation (1.1) of the text, to solve for mass.

$$\text{density} = \frac{\text{mass}}{\text{volume}}$$

**Solution:**

$$\textbf{mass} = \text{density} \times \text{volume}$$

$$\textbf{mass of methanol} = \frac{0.7918\text{ g}}{1\text{ mL}} \times 89.9\text{ mL} = \textbf{71.2 g}$$

**1.26** **Strategy:** Find the appropriate equations for converting between Fahrenheit and Celsius and between Celsius and Fahrenheit given in Section 1.7 of the text. Substitute the temperature values given in the problem into the appropriate equation.

(a) Conversion from Fahrenheit to Celsius.

$$?\ ^\circ\text{C} = (^\circ\text{F} - 32^\circ\text{F}) \times \frac{5^\circ\text{C}}{9^\circ\text{F}}$$

$$\textbf{?}\ ^\circ\textbf{C} = (105 - 32)^\circ\text{F} \times \frac{5^\circ\text{C}}{9^\circ\text{F}} = \textbf{41}^\circ\textbf{C}$$

(b) Conversion from Celsius to Fahrenheit.

$$?\ ^\circ\text{F} = \left(^\circ\text{C} \times \frac{9^\circ\text{F}}{5^\circ\text{C}}\right) + 32^\circ\text{F}$$

$$\textbf{?}\ ^\circ\textbf{F} = \left(-11.5^\circ\text{C} \times \frac{9^\circ\text{F}}{5^\circ\text{C}}\right) + 32^\circ\text{F} = \textbf{11.3}^\circ\textbf{F}$$

**(c)** Conversion from Celsius to Fahrenheit.

$$?\,°F = \left(°C \times \frac{9°F}{5°C}\right) + 32°F$$

$$\textbf{?}\ °\textbf{F} = \left(6.3 \times 10^3\ \cancel{°C} \times \frac{9°F}{5\cancel{°C}}\right) + 32°F = \mathbf{1.1 \times 10^4\,°F}$$

**(d)** Conversion from Fahrenheit to Celsius.

$$?\,°C = (°F - 32°F) \times \frac{5°C}{9°F}$$

$$\textbf{?}\ °\textbf{C} = (451 - 32)\cancel{°F} \times \frac{5°C}{9\cancel{°F}} = \mathbf{233°C}$$

**1.28** **(a)** $K = (°C + 273°C)\dfrac{1\text{ K}}{1°C}$

$°C = K - 273 = 77\text{ K} - 273 = \mathbf{-196°C}$

**(b)** $°C = 4.2\text{ K} - 273 = \mathbf{-269°C}$

**(c)** $°C = 601\text{ K} - 273 = \mathbf{328°C}$

**1.32** **(a)** $10^{-2}$ indicates that the decimal point must be moved two places to the left.

$$1.52 \times 10^{-2} = \mathbf{0.0152}$$

**(b)** $10^{-8}$ indicates that the decimal point must be moved 8 places to the left.

$$7.78 \times 10^{-8} = \mathbf{0.0000000778}$$

**1.34** **(a)** Addition using scientific notation.

**Strategy:** Let's express scientific notation as $N \times 10^n$. When adding numbers using scientific notation, we must write each quantity with the same exponent, $n$. We can then add the $N$ parts of the numbers, keeping the exponent, $n$, the same.

**Solution:** Write each quantity with the same exponent, $n$.

Let's write 0.0095 in such a way that $n = -3$. We have decreased $10^n$ by $10^3$, so we must increase $N$ by $10^3$. Move the decimal point 3 places to the right.

$$0.0095 = 9.5 \times 10^{-3}$$

Add the $N$ parts of the numbers, keeping the exponent, $n$, the same.

$$\begin{array}{r} 9.5 \times 10^{-3} \\ +\ 8.5 \times 10^{-3} \\ \hline \mathbf{18.0 \times 10^{-3}} \end{array}$$

The usual practice is to express $N$ as a number between 1 and 10. Since we must *decrease* $N$ by a factor of 10 to express $N$ between 1 and 10 (1.8), we must *increase* $10^n$ by a factor of 10. The exponent, $n$, is increased by 1 from −3 to −2.

$$18.0 \times 10^{-3} = \mathbf{1.8 \times 10^{-2}}$$

**(b)** Division using scientific notation.

**Strategy:** Let's express scientific notation as $N \times 10^n$. When dividing numbers using scientific notation, divide the $N$ parts of the numbers in the usual way. To come up with the correct exponent, $n$, we *subtract* the exponents.

**Solution:** Make sure that all numbers are expressed in scientific notation.

$$653 = 6.53 \times 10^2$$

Divide the $N$ parts of the numbers in the usual way.

$$6.53 \div 5.75 = 1.14$$

*Subtract* the exponents, $n$.

$$1.14 \times 10^{+2-(-8)} = 1.14 \times 10^{+2+8} = \mathbf{1.14 \times 10^{10}}$$

**(c)** Subtraction using scientific notation.

**Strategy:** Let's express scientific notation as $N \times 10^n$. When subtracting numbers using scientific notation, we must write each quantity with the same exponent, $n$. We can then subtract the $N$ parts of the numbers, keeping the exponent, $n$, the same.

**Solution:** Write each quantity with the same exponent, $n$.

Let's write 850,000 in such a way that $n = 5$. This means to move the decimal point five places to the left.

$$850{,}000 = 8.5 \times 10^5$$

Subtract the $N$ parts of the numbers, keeping the exponent, $n$, the same.

$$\begin{array}{r} 8.5 \times 10^5 \\ -\ 9.0 \times 10^5 \\ \hline \mathbf{-0.5 \times 10^5} \end{array}$$

The usual practice is to express $N$ as a number between 1 and 10. Since we must *increase* $N$ by a factor of 10 to express $N$ between 1 and 10 (5), we must *decrease* $10^n$ by a factor of 10. The exponent, $n$, is decreased by 1 from 5 to 4.

$$-0.5 \times 10^5 = \mathbf{-5 \times 10^4}$$

**(d)** Multiplication using scientific notation.

**Strategy:** Let's express scientific notation as $N \times 10^n$. When multiplying numbers using scientific notation, multiply the *N* parts of the numbers in the usual way. To come up with the correct exponent, *n*, we *add* the exponents.

**Solution:** Multiply the *N* parts of the numbers in the usual way.

$$3.6 \times 3.6 = 13$$

*Add* the exponents, *n*.

$$13 \times 10^{-4+(+6)} = \mathbf{13 \times 10^2}$$

The usual practice is to express *N* as a number between 1 and 10. Since we must *decrease N* by a factor of 10 to express *N* between 1 and 10 (1.3), we must *increase* $10^n$ by a factor of 10. The exponent, *n*, is increased by 1 from 2 to 3.

$$\mathbf{13 \times 10^2 = 1.3 \times 10^3}$$

**1.36** **(a)** one **(b)** three **(c)** three **(d)** four
**(e)** two or three **(f)** one **(g)** one or two

**1.38** **(a)** Division

**Strategy:** The number of significant figures in the answer is determined by the original number having the smallest number of significant figures.

**Solution:**

$$\frac{7.310 \text{ km}}{5.70 \text{ km}} = 1.28\mathbf{3}$$

The 3 (bolded) is a nonsignificant digit because the original number 5.70 only has three significant digits. Therefore, the answer has only three significant digits.

The correct answer rounded off to the correct number of significant figures is:

**1.28** (Why are there no units?)

**(b)** Subtraction

**Strategy:** The number of significant figures to the right of the decimal point in the answer is determined by the lowest number of digits to the right of the decimal point in any of the original numbers.

**Solution:** Writing both numbers in decimal notation, we have

$$\begin{array}{r} 0.00326 \text{ mg} \\ \underline{-0.0000788 \text{ mg}} \\ 0.00318\mathbf{12} \text{ mg} \end{array}$$

The bolded numbers are nonsignificant digits because the number 0.00326 has five digits to the right of the decimal point. Therefore, we carry five digits to the right of the decimal point in our answer.

The correct answer rounded off to the correct number of significant figures is:

$$\mathbf{0.00318\ mg = 3.18 \times 10^{-3}\ mg}$$

**(c)** Addition

**Strategy:** The number of significant figures to the right of the decimal point in the answer is determined by the lowest number of digits to the right of the decimal point in any of the original numbers.

**Solution:** Writing both numbers with exponents = +7, we have

$$(0.402 \times 10^7\ \text{dm}) + (7.74 \times 10^7\ \text{dm}) = \mathbf{8.14 \times 10^7\ dm}$$

Since $7.74 \times 10^7$ has only two digits to the right of the decimal point, two digits are carried to the right of the decimal point in the final answer.

**(d)** Subtraction, addition, and division

**Strategy:** For subtraction and addition, the number of significant figures to the right of the decimal point in that part of the calculation is determined by the lowest number of digits to the right of the decimal point in any of the original numbers. For the division part of the calculation, the number of significant figures in the answer is determined by the number having the smallest number of significant figures. First, perform the subtraction and addition parts to the correct number of significant figures, and then perform the division.

**Solution:**

$$\frac{(7.8\ \text{m} - 0.34\ \text{m})}{(1.15\ \text{s} + 0.82\ \text{s})} = \frac{7.5\ \text{m}}{1.97\ \text{s}} = \mathbf{3.8\ m/s}$$

**1.40** Calculating the mean for each set of data, we find:

Tailor X: 31.5 in
Tailor Y: 32.6 in
Tailor Z: 32.1 in

From these calculations, we can conclude that the seam measurements made by Tailor Z were the most accurate of the three tailors. The precision in the measurements made by both tailors X and Z are fairly high, while the measurements made by tailor Y are less precise. In summary:

Tailor X: most precise
Tailor Y: least accurate and least precise
Tailor Z: most accurate

**1.42** (a)

**Strategy:** The problem may be stated as

$$? \text{ mg} = 242 \text{ lb}$$

A relationship between pounds and grams is given on the end sheet of your text (1 lb = 453.6 g). This relationship will allow conversion from pounds to grams. A metric conversion is then needed to convert grams to milligrams (1 mg = $1 \times 10^{-3}$ g). Arrange the appropriate conversion factors so that pounds and grams cancel, and the unit milligrams is obtained in your answer.

**Solution:** The sequence of conversions is

$$\text{lb} \rightarrow \text{grams} \rightarrow \text{mg}$$

Using the following conversion factors,

$$\frac{453.6 \text{ g}}{1 \text{ lb}} \qquad \frac{1 \text{ mg}}{1 \times 10^{-3} \text{ g}}$$

we obtain the answer in one step:

$$? \text{ mg} = 242 \cancel{\text{lb}} \times \frac{453.6 \cancel{\text{g}}}{1 \cancel{\text{lb}}} \times \frac{1 \text{ mg}}{1 \times 10^{-3} \cancel{\text{g}}} = \mathbf{1.10 \times 10^8 \text{ mg}}$$

**Check:** Does your answer seem reasonable? Should 242 lb be equivalent to 110 million mg? How many mg are in 1 lb? There are 453,600 mg in 1 lb.

(b)

**Strategy:** The problem may be stated as

$$? \text{ m}^3 = 68.3 \text{ cm}^3$$

Recall that 1 cm = $1 \times 10^{-2}$ m. We need to set up a conversion factor to convert from $\text{cm}^3$ to $\text{m}^3$.

**Solution:** We need the following conversion factor so that centimeters cancel and we end up with meters.

$$\frac{1 \times 10^{-2} \text{ m}}{1 \text{ cm}}$$

Since this conversion factor deals with length and we want volume, it must therefore be cubed to give

$$\frac{1 \times 10^{-2} \text{ m}}{1 \text{ cm}} \times \frac{1 \times 10^{-2} \text{ m}}{1 \text{ cm}} \times \frac{1 \times 10^{-2} \text{ m}}{1 \text{ cm}} = \left(\frac{1 \times 10^{-2} \text{ m}}{1 \text{ cm}}\right)^3$$

We can write

$$?\ \text{m}^3 = 68.3\ \text{cm}^3 \times \left(\frac{1 \times 10^{-2}\ \text{m}}{1\ \text{cm}}\right)^3 = \mathbf{6.83 \times 10^{-5}\ m^3}$$

**Check:** We know that 1 $\text{cm}^3 = 1 \times 10^{-6}\ \text{m}^3$. We started with $6.83 \times 10^1\ \text{cm}^3$. Multiplying this quantity by $1 \times 10^{-6}$ gives $6.83 \times 10^{-5}$.

**(c)**

**Strategy:** The problem may be stated as

$$?\ \text{L} = 7.2\ \text{m}^3$$

In Chapter 1 of the text, a conversion is given between liters and $\text{cm}^3$ (1 L = 1000 $\text{cm}^3$). If we can convert $\text{m}^3$ to $\text{cm}^3$, we can then convert to liters. Recall that 1 cm = $1 \times 10^{-2}$ m. We need to set up two conversion factors to convert from $\text{m}^3$ to L. Arrange the appropriate conversion factors so that $\text{m}^3$ and $\text{cm}^3$ cancel, and the unit liters is obtained in your answer.

**Solution:** The sequence of conversions is

$$\text{m}^3 \rightarrow \text{cm}^3 \rightarrow \text{L}$$

Using the following conversion factors,

$$\left(\frac{1\ \text{cm}}{1 \times 10^{-2}\ \text{m}}\right)^3 \qquad \frac{1\ \text{L}}{1000\ \text{cm}^3}$$

the answer is obtained in one step:

$$?\ \text{L} = 7.2\ \text{m}^3 \times \left(\frac{1\ \text{cm}}{1 \times 10^{-2}\ \text{m}}\right)^3 \times \frac{1\ \text{L}}{1000\ \text{cm}^3} = \mathbf{7.2 \times 10^3\ L}$$

**Check:** From the above conversion factors you can show that 1 $\text{m}^3 = 1 \times 10^3$ L. Therefore, 7 $\text{m}^3$ would equal $7 \times 10^3$ L, which is close to the answer.

**(d)**

**Strategy:** The problem may be stated as

$$?\ \text{lb} = 28.3\ \mu\text{g}$$

A relationship between pounds and grams is given on the end sheet of your text (1 lb = 453.6 g). This relationship will allow conversion from grams to pounds. If we can convert from μg to grams, we can then convert from grams to pounds. Recall that 1 μg = $1 \times 10^{-6}$ g. Arrange the appropriate conversion factors so that μg and grams cancel, and the unit pounds is obtained in your answer.

**Solution:** The sequence of conversions is

$$\mu g \rightarrow g \rightarrow lb$$

Using the following conversion factors,

$$\frac{1\times 10^{-6}\text{ g}}{1\ \mu\text{g}} \qquad \frac{1\text{ lb}}{453.6\text{ g}}$$

we can write

$$\textbf{? lb} = 28.3\ \cancel{\mu\text{g}} \times \frac{1\times 10^{-6}\ \cancel{\text{g}}}{1\ \cancel{\mu\text{g}}} \times \frac{1\text{ lb}}{453.6\ \cancel{\text{g}}} = \mathbf{6.24 \times 10^{-8}\ lb}$$

**Check:** Does the answer seem reasonable? What number does the prefix μ represent? Should 28.3 μg be a very small mass?

**1.44** **Strategy:** The problem may be stated as

$$?\text{ s} = 365.24\text{ days}$$

You should know conversion factors that will allow you to convert between days and hours, between hours and minutes, and between minutes and seconds. Make sure to arrange the conversion factors so that days, hours, and minutes cancel, leaving units of seconds for the answer.

**Solution:** The sequence of conversions is

$$\text{days} \rightarrow \text{hours} \rightarrow \text{minutes} \rightarrow \text{seconds}$$

Using the following conversion factors,

$$\frac{24\text{ h}}{1\text{ day}} \qquad \frac{60\text{ min}}{1\text{ h}} \qquad \frac{60\text{ s}}{1\text{ min}}$$

we can write

$$\textbf{? s} = 365.24\ \cancel{\text{day}} \times \frac{24\ \cancel{\text{h}}}{1\ \cancel{\text{day}}} \times \frac{60\ \cancel{\text{min}}}{1\ \cancel{\text{h}}} \times \frac{60\text{ s}}{1\ \cancel{\text{min}}} = \mathbf{3.1557 \times 10^{7}\ s}$$

**Check:** Does your answer seem reasonable? Should there be a very large number of seconds in 1 year?

**1.46** (a) $? \text{ in/s} = \frac{1 \text{ mi}}{8.92 \text{ min}} \times \frac{5280 \text{ ft}}{1 \text{ mi}} \times \frac{12 \text{ in}}{1 \text{ ft}} \times \frac{1 \text{ min}}{60 \text{ s}} = \mathbf{118 \text{ in/s}}$

(b) $? \text{ m/min} = \frac{1 \text{ mi}}{8.92 \text{ min}} \times \frac{1609 \text{ m}}{1 \text{ mi}} = \mathbf{1.80 \times 10^2 \text{ m/min}}$

(c) $? \text{ km/h} = \frac{1 \text{ mi}}{8.92 \text{ min}} \times \frac{1609 \text{ m}}{1 \text{ mi}} \times \frac{1 \text{ km}}{1000 \text{ m}} \times \frac{60 \text{ min}}{1 \text{ h}} = \mathbf{10.8 \text{ km/h}}$

**1.48** $? \text{ mph} = \frac{286 \text{ km}}{1 \text{ h}} \times \frac{1 \text{ mi}}{1.609 \text{ km}} = \mathbf{178 \text{ mph}}$

**1.50** $0.62 \text{ ppm Pb} = \frac{0.62 \text{ g Pb}}{1 \times 10^6 \text{ g blood}}$

$6.0 \times 10^3 \text{ g of blood} \times \frac{0.62 \text{ g Pb}}{1 \times 10^6 \text{ g blood}} = \mathbf{3.7 \times 10^{-3} \text{ g Pb}}$

**1.52** (a) $? \text{ lbs} = 70 \text{ kg} \times \frac{1 \text{ lb}}{0.4536 \text{ kg}} = \mathbf{1.5 \times 10^2 \text{ lbs}}$

(b) $? \text{ s} = 14 \times 10^9 \text{ yr} \times \frac{365 \text{ d}}{1 \text{ yr}} \times \frac{24 \text{ h}}{1 \text{ d}} \times \frac{3600 \text{ s}}{1 \text{ h}} = \mathbf{4.4 \times 10^{17} \text{ s}}$

(c) $? \text{ m} = 90 \text{ in} \times \frac{2.54 \text{ cm}}{1 \text{ in}} \times \frac{1 \times 10^{-2} \text{ m}}{1 \text{ cm}} = \mathbf{2.3 \text{ m}}$

(d) $? \text{ L} = 88.6 \text{ m}^3 \times \left(\frac{1 \text{ cm}}{1 \times 10^{-2} \text{ m}}\right)^3 \times \frac{1 \text{ L}}{1000 \text{ cm}^3} = \mathbf{8.86 \times 10^4 \text{ L}}$

**1.54** $\text{density} = \frac{0.625 \text{ g}}{1 \text{ L}} \times \frac{1 \text{ L}}{1000 \text{ mL}} \times \frac{1 \text{ mL}}{1 \text{ cm}^3} = \mathbf{6.25 \times 10^{-4} \text{ g/cm}^3}$

**1.56** See Section 1.6 of your text for a discussion of these terms.

**(a)** Chemical property. Iron has changed its composition and identity by chemically combining with oxygen and water.

**(b)** Chemical property. The water reacts with chemicals in the air (such as sulfur dioxide) to produce acids, thus changing the composition and identity of the water.

**(c)** Physical property. The color of the hemoglobin can be observed and measured without changing its composition or identity.

**(d)** Physical property. The evaporation of water does not change its chemical properties. Evaporation is a change in matter from the liquid state to the gaseous state.

**(e)** Chemical property. The carbon dioxide is chemically converted into other molecules.

**1.58** Volume of rectangular bar = length × width × height

$$\textbf{density} = \frac{m}{V} = \frac{52.7064\ \text{g}}{(8.53\ \text{cm})(2.4\ \text{cm})(1.0\ \text{cm})} = \mathbf{2.6\ g/cm^3}$$

**1.60** You are asked to solve for the inner diameter of the bottle. If we can calculate the volume that the cooking oil occupies, we can calculate the radius of the cylinder. The volume of the cylinder is, $V_{\text{cylinder}} = \pi r^2 h$ ($r$ is the inner radius of the cylinder, and $h$ is the height of the cylinder). The cylinder diameter is $2r$.

$$\text{volume of oil filling bottle} = \frac{\text{mass of oil}}{\text{density of oil}}$$

$$\text{volume of oil filling bottle} = \frac{1360\ \text{g}}{0.953\ \text{g/mL}} = 1.43 \times 10^3\ \text{mL} = 1.43 \times 10^3\ \text{cm}^3$$

Next, solve for the radius of the cylinder.

$$\text{Volume of cylinder} = \pi r^2 h$$

$$r = \sqrt{\frac{\text{volume}}{\pi \times h}}$$

$$r = \sqrt{\frac{1.43 \times 10^3\ \text{cm}^3}{\pi \times 21.5\ \text{cm}}} = 4.60\ \text{cm}$$

The inner diameter of the bottle equals $2r$.

$$\textbf{Bottle diameter} = 2r = 2(4.60\ \text{cm}) = \mathbf{9.20\ cm}$$

**1.62** $$\frac{343\ \text{m}}{1\ \text{s}} \times \frac{1\ \text{mi}}{1609\ \text{m}} \times \frac{3600\ \text{s}}{1\ \text{h}} = \mathbf{767\ mph}$$

**1.64** In order to work this problem, you need to understand the physical principles involved in the experiment in Problem 1.61. The volume of the water displaced must equal the volume of the piece of silver. If the silver did not sink, would you have been able to determine the volume of the piece of silver?

The liquid must be *less dense* than the ice in order for the ice to sink. The temperature of the experiment must be maintained at or below *0°C* to prevent the ice from melting.

**1.66** $$\text{Volume} = \frac{\text{mass}}{\text{density}}$$

$$\textbf{Volume occupied by Li} = \frac{1.20 \times 10^3\ \text{g}}{0.53\ \text{g/cm}^3} = \mathbf{2.3 \times 10^3\ cm^3}$$

**1.68** To work this problem, we need to convert from cubic feet to L. Some tables will have a conversion factor of 28.3 L = 1 ft$^3$, but we can also calculate it using the dimensional analysis method described in Section 1.9 of the text.

First, converting from cubic feet to liters:

$$(5.0 \times 10^7 \text{ ft}^3) \times \left(\frac{12 \text{ in}}{1 \text{ ft}}\right)^3 \times \left(\frac{2.54 \text{ cm}}{1 \text{ in}}\right)^3 \times \frac{1 \text{ mL}}{1 \text{ cm}^3} \times \frac{1 \times 10^{-3} \text{ L}}{1 \text{ mL}} = 1.42 \times 10^9 \text{ L}$$

The mass of vanillin (in g) is:

$$\frac{2.0 \times 10^{-11} \text{ g vanillin}}{1 \text{ L}} \times (1.42 \times 10^9 \text{ L}) = 2.84 \times 10^{-2} \text{ g vanillin}$$

The cost is:

$$(2.84 \times 10^{-2} \text{ g vanillin}) \times \frac{\$112}{50 \text{ g vanillin}} = \mathbf{\$0.064 = 6.4¢}$$

**1.70** There are 78.3 + 117.3 = 195.6 Celsius degrees between 0°S and 100°S. We can write this as a unit factor.

$$\left(\frac{195.6°\text{C}}{100°\text{S}}\right)$$

Set up the equation like a Celsius to Fahrenheit conversion. We need to subtract 117.3°C, because the zero point on the new scale is 117.3°C lower than the zero point on the Celsius scale.

$$?\ °\text{C} = \left(\frac{195.6°\text{C}}{100°\text{S}}\right)(?\ °\text{S}) - 117.3°\text{C}$$

Solving for ? °S gives: $$?\ °\text{S} = (?\ °\text{C} + 117.3°\text{C})\left(\frac{100°\text{S}}{195.6°\text{C}}\right)$$

For 25°C we have: $$?\ °\text{S} = (25 + 117.3)°\text{C}\left(\frac{100°\text{S}}{195.6°\text{C}}\right) = \mathbf{73\ °S}$$

**1.72** **(a)** $$\frac{6000 \text{ mL of inhaled air}}{1 \text{ min}} \times \frac{0.001 \text{ L}}{1 \text{ mL}} \times \frac{60 \text{ min}}{1 \text{ h}} \times \frac{24 \text{ h}}{1 \text{ day}} = \mathbf{8.6 \times 10^3 \text{ L of air/day}}$$

**(b)** $$\frac{8.6 \times 10^3 \text{ L of air}}{1 \text{ day}} \times \frac{2.1 \times 10^{-6} \text{ L CO}}{1 \text{ L of air}} = \mathbf{0.018 \text{ L CO/day}}$$

**1.74** The diameter of the basketball can be calculated from its circumference. We can then use the diameter of a ball as a conversion factor to determine the number of basketballs needed to circle the equator.

Circumference = $2\pi r$

$$d = 2r = \frac{\text{circumference}}{\pi} = \frac{29.6 \text{ in}}{\pi} = 9.42 \text{ in}$$

The circumference of the Earth is $2\pi r = (2 \times \pi \times 6400\text{ km}) = 40212\text{ km}$

$$40212\text{ km} \times \frac{1000\text{ m}}{1\text{ km}} \times \frac{1\text{ cm}}{1\times 10^{-2}\text{ m}} \times \frac{1\text{ in}}{2.54\text{ cm}} \times \frac{1\text{ ball}}{9.42\text{ in}} = 168{,}000{,}000\text{ balls}$$

We round up to an integer number of basketballs with 3 significant figures.

**1.76** Volume = surface area × depth

Recall that 1 L = 1 $dm^3$. Let's convert the surface area to units of $dm^2$ and the depth to units of dm.

$$\text{surface area} = \left(1.8 \times 10^{8}\text{ km}^2\right) \times \left(\frac{1000\text{ m}}{1\text{ km}}\right)^2 \times \left(\frac{1\text{ dm}}{0.1\text{ m}}\right)^2 = 1.8 \times 10^{16}\text{ dm}^2$$

$$\text{depth} = \left(3.9 \times 10^{3}\text{ m}\right) \times \frac{1\text{ dm}}{0.1\text{ m}} = 3.9 \times 10^{4}\text{ dm}$$

**Volume** = surface area × depth = $(1.8 \times 10^{16}\text{ dm}^2)(3.9 \times 10^{4}\text{ dm}) = 7.0 \times 10^{20}\text{ dm}^3 = \mathbf{7.0 \times 10^{20}\ L}$

**1.78** $\text{Volume of sphere} = \frac{4}{3}\pi r^3$

$$\text{Volume} = \frac{4}{3}\pi\left(\frac{15\text{ cm}}{2}\right)^3 = 1.77 \times 10^{3}\text{ cm}^3$$

$$\text{mass} = \text{volume} \times \text{density} = \left(1.77 \times 10^{3}\text{ cm}^3\right) \times \frac{22.57\text{ g Os}}{1\text{ cm}^3} \times \frac{1\text{ kg}}{1000\text{ g}} = \mathbf{4.0 \times 10^{1}\ kg\ Os}$$

$$4.0 \times 10^{1}\text{ kg Os} \times \frac{2.205\text{ lb}}{1\text{ kg}} = \mathbf{88\ lb\ Os}$$

**1.80** 62 kg = $6.2 \times 10^{4}$ g

O: $(6.2 \times 10^{4}\text{ g})(0.65) = \mathbf{4.0 \times 10^{4}\ g\ O}$

C: $(6.2 \times 10^{4}\text{ g})(0.18) = \mathbf{1.1 \times 10^{4}\ g\ C}$

H: $(6.2 \times 10^{4}\text{ g})(0.10) = \mathbf{6.2 \times 10^{3}\ g\ H}$

N: $(6.2 \times 10^{4}\text{ g})(0.03) = \mathbf{2 \times 10^{3}\ g\ N}$

Ca: $(6.2 \times 10^{4}\text{ g})(0.016) = \mathbf{9.9 \times 10^{2}\ g\ Ca}$

P: $(6.2 \times 10^{4}\text{ g})(0.012) = \mathbf{7.4 \times 10^{2}\ g\ P}$

**1.82** ? °C = $(7.3 \times 10^{2} - 273)\text{ K} = \mathbf{4.6 \times 10^{2}\,^{\circ}C}$

$$?\,^{\circ}\text{F} = \left((4.6 \times 10^{2}\,^{\circ}\text{C}) \times \frac{9^{\circ}\text{F}}{5^{\circ}\text{C}}\right) + 32^{\circ}\text{F} = \mathbf{8.6 \times 10^{2}\,^{\circ}F}$$

**1.84** $$\left(8.0\times 10^4 \text{ tons Au}\right)\times\frac{2000\text{ lb Au}}{1\text{ ton Au}}\times\frac{16\text{ oz Au}}{1\text{ lb Au}}\times\frac{\$948}{1\text{ oz Au}} = \mathbf{\$2.4\times 10^{12}}\text{ or }\mathbf{2.4\text{ trillion dollars}}$$

**1.86** $$\text{? Fe atoms} = 4.9\text{ g Fe}\times\frac{1.1\times 10^{22}\text{ Fe atoms}}{1.0\text{ g Fe}} = \mathbf{5.4\times 10^{22}\text{ Fe atoms}}$$

**1.88** 10 cm = 0.1 m. We need to find the number of times the 0.1 m wire must be cut in half until the piece left is equal to the diameter of a Cu atom, which is (2)($1.3\times 10^{-10}$ m). Let $n$ be the number of times we can cut the Cu wire in half. We can write:

$$\left(\frac{1}{2}\right)^n\times 0.1\text{ m} = 2.6\times 10^{10}\text{ m}$$

$$\left(\frac{1}{2}\right)^n = 2.6\times 10^{-9}\text{ m}$$

Taking the log of both sides of the equation:

$$n\log\left(\frac{1}{2}\right) = \log(2.6\times 10^{-9})$$

$$\mathbf{n = 29\text{ times}}$$

**1.90** Volume = area × thickness.

From the density, we can calculate the volume of the Al foil.

$$\text{Volume} = \frac{\text{mass}}{\text{density}} = \frac{3.636\text{ g}}{2.699\text{ g/cm}^3} = 1.3472\text{ cm}^3$$

Convert the unit of area from $ft^2$ to $cm^2$.

$$1.000\text{ ft}^2\times\left(\frac{12\text{ in}}{1\text{ ft}}\right)^2\times\left(\frac{2.54\text{ cm}}{1\text{ in}}\right)^2 = 929.03\text{ cm}^2$$

$$\textbf{thickness} = \frac{\text{volume}}{\text{area}} = \frac{1.3472\text{ cm}^3}{929.03\text{ cm}^2} = 1.450\times 10^{-3}\text{ cm} = \mathbf{1.450\times 10^{-2}\text{ mm}}$$

**1.92** First, let's calculate the mass (in g) of water in the pool. We perform this conversion because we know there is 1 g of chlorine needed per million grams of water.

$$\left(2.0\times 10^4\text{ gallons H}_2\text{O}\right)\times\frac{3.79\text{ L}}{1\text{ gallon}}\times\frac{1\text{ mL}}{0.001\text{ L}}\times\frac{1\text{ g}}{1\text{ mL}} = 7.58\times 10^7\text{ g H}_2\text{O}$$

Next, let's calculate the mass of chlorine that needs to be added to the pool.

$$\left(7.58 \times 10^7 \cancel{g\ H_2O}\right) \times \frac{1\text{ g chlorine}}{1 \times 10^6\ \cancel{g\ H_2O}} = 75.8\text{ g chlorine}$$

The chlorine solution is only 6 percent chlorine by mass. We can now calculate the volume of chlorine solution that must be added to the pool.

$$75.8\ \cancel{g}\text{ chlorine} \times \frac{100\%\text{ soln}}{6\%\text{ chlorine}} \times \frac{1\text{ mL soln}}{1\ \cancel{g}\text{ soln}} = \mathbf{1.3 \times 10^3\ mL\ of\ chlorine\ solution}$$

**1.94** (a) The volume of the pycnometer can be calculated by determining the mass of water that the pycnometer holds and then using the density to convert to volume.

$$(43.1195 - 32.0764)\ \cancel{g} \times \frac{1\text{ mL}}{0.99820\ \cancel{g}} = \mathbf{11.063\ mL}$$

(b) Using the volume of the pycnometer from part (a), we can calculate the density of ethanol.

$$\frac{(40.8051 - 32.0764)\text{ g}}{11.063\text{ mL}} = \mathbf{0.78900\ g/mL}$$

(c) From the volume of water added and the volume of the pycnometer, we can calculate the volume of the zinc granules by difference. Then, we can calculate the density of zinc.

$$\text{volume of water} = (62.7728 - 32.0764 - 22.8476)\ \cancel{g} \times \frac{1\text{ mL}}{0.99820\ \cancel{g}} = \mathbf{7.8630\ mL}$$

$$\text{volume of zinc granules} = 11.063\text{ mL} - 7.8630\text{ mL} = \mathbf{3.200\ mL}$$

$$\textbf{density of zinc} = \frac{22.8476\text{ g}}{3.200\text{ mL}} = \mathbf{7.140\ g/mL}$$

**1.96** First, convert 10 μm to units of cm.

$$10\ \cancel{\mu m} \times \frac{1 \times 10^{-4}\text{ cm}}{1\ \cancel{\mu m}} = 1.0 \times 10^{-3}\text{ cm}$$

Now, substitute into the given equation to solve for time.

$$t = \frac{x^2}{2D} = \frac{(1.0 \times 10^{-3}\ \cancel{cm})^2}{2(5.7 \times 10^{-7}\ \cancel{cm^2}/\text{s})} = \mathbf{0.88\ s}$$

It takes **0.88 seconds** for a glucose molecule to diffuse 10 μm.

**1.98** (a) A concentration of CO of 800 ppm in air would mean that there are 800 parts by volume of CO per 1 million parts by volume of air. Using a volume unit of liters, 800 ppm CO means that there are 800 L of CO per 1 million liters of air. The volume in liters occupied by CO in the room is:

$$17.6\ \text{m} \times 8.80\ \text{m} \times 2.64\ \text{m} = 409\ \text{m}^3 \times \left(\frac{1\ \text{cm}}{1 \times 10^{-2}\ \text{m}}\right)^3 \times \frac{1\ \text{L}}{1000\ \text{cm}^3} = 4.09 \times 10^5\ \text{L air}$$

$$4.09 \times 10^5\ \text{L air} \times \frac{8.00 \times 10^2\ \text{L CO}}{1 \times 10^6\ \text{L air}} = \mathbf{327\ L\ CO}$$

(b) $1\ \text{mg} = 1 \times 10^{-3}\ \text{g}$ and $1\ \text{L} = 1000\ \text{cm}^3$. We convert mg/m$^3$ to g/L:

$$\frac{0.050\ \text{mg}}{1\ \text{m}^3} \times \frac{1 \times 10^{-3}\ \text{g}}{1\ \text{mg}} \times \left(\frac{1 \times 10^{-2}\ \text{m}}{1\ \text{cm}}\right)^3 \times \frac{1000\ \text{cm}^3}{1\ \text{L}} = \mathbf{5.0 \times 10^{-8}\ g/L}$$

(c) $1\ \mu\text{g} = 1 \times 10^{-3}\ \text{mg}$ and $1\ \text{mL} = 1 \times 10^{-2}\ \text{dL}$. We convert mg/dL to μg/mL:

$$\frac{120\ \text{mg}}{1\ \text{dL}} \times \frac{1\ \mu\text{g}}{1 \times 10^{-3}\ \text{mg}} \times \frac{1 \times 10^{-2}\ \text{dL}}{1\ \text{mL}} = \mathbf{1.20 \times 10^3\ \mu g/mL}$$

**1.100** We wish to calculate the density and radius of the ball bearing. For both calculations, we need the volume of the ball bearing. The data from the first experiment can be used to calculate the density of the mineral oil. In the second experiment, the density of the mineral oil can then be used to determine what part of the 40.00 mL volume is due to the mineral oil and what part is due to the ball bearing. Once the volume of the ball bearing is determined, we can calculate its density and radius.

From experiment one:

$$\text{Mass of oil} = 159.446\ \text{g} - 124.966\ \text{g} = 34.480\ \text{g}$$

$$\text{Density of oil} = \frac{34.480\ \text{g}}{40.00\ \text{mL}} = 0.8620\ \text{g/mL}$$

From the second experiment:

$$\text{Mass of oil} = 50.952\ \text{g} - 18.713\ \text{g} = 32.239\ \text{g}$$

$$\text{Volume of oil} = 32.239\ \text{g} \times \frac{1\ \text{mL}}{0.8620\ \text{g}} = 37.40\ \text{mL}$$

The volume of the ball bearing is obtained by difference.

$$\text{Volume of ball bearing} = 40.00\ \text{mL} - 37.40\ \text{mL} = 2.60\ \text{mL} = 2.60\ \text{cm}^3$$

Now that we have the volume of the ball bearing, we can calculate its density and radius.

$$\text{Density of ball bearing} = \frac{18.713 \text{ g}}{2.60 \text{ cm}^3} = \mathbf{7.20 \text{ g/cm}^3}$$

Using the formula for the volume of a sphere, we can solve for the radius of the ball bearing.

$$V = \frac{4}{3}\pi r^3$$

$$2.60 \text{ cm}^3 = \frac{4}{3}\pi r^3$$

$$r^3 = 0.621 \text{ cm}^3$$

$$\boldsymbol{r} = \mathbf{0.853 \text{ cm}}$$

**1.102** We want to calculate the mass of the cylinder, which can be calculated from its volume and density. The volume of a cylinder is $\pi r^2 l$. The density of the alloy can be calculated using the mass percentages of each element and the given densities of each element.

The volume of the cylinder is:

$$V = \pi r^2 l$$

$$V = \pi(6.44 \text{ cm})^2(44.37 \text{ cm})$$

$$V = 5781 \text{ cm}^3$$

The density of the cylinder is:

$$\text{density} = (0.7942)(8.94 \text{ g/cm}^3) + (0.2058)(7.31 \text{ g/cm}^3) = 8.605 \text{ g/cm}^3$$

Now, we can calculate the mass of the cylinder.

$$\text{mass} = \text{density} \times \text{volume}$$

$$\textbf{mass} = (8.605 \text{ g/cm}^3)(5781 \text{ cm}^3) = \mathbf{4.97 \times 10^4 \text{ g}}$$

The assumption made in the calculation is that the alloy must be homogeneous in composition.

**1.104** The density of the mixed solution should be based on the percentage of each liquid and its density. Because the solid object is suspended in the mixed solution, it should have the same density as this solution. The density of the mixed solution is:

$$(0.4137)(2.0514 \text{ g/mL}) + (0.5863)(2.6678 \text{ g/mL}) = 2.413 \text{ g/mL}$$

As discussed, the density of the object should have the same density as the mixed solution (**2.413 g/mL**).

**Yes**, this procedure can be used in general to determine the densities of solids. This procedure is called the flotation method. It is based on the assumptions that the liquids are totally miscible and that the volumes of the liquids are additive.

**1.106** As water freezes, it expands. First, calculate the mass of the water at 20°C. Then, determine the volume that this mass of water would occupy at −5°C.

$$\text{Mass of water} = 242\ \cancel{\text{mL}} \times \frac{0.998\ \text{g}}{1\ \cancel{\text{mL}}} = 241.5\ \text{g}$$

$$\text{Volume of ice at } -5^\circ\text{C} = 241.5\ \cancel{\text{g}} \times \frac{1\ \text{mL}}{0.916\ \cancel{\text{g}}} = 264\ \text{mL}$$

The volume occupied by the ice is larger than the volume of the glass bottle. **The glass bottle would crack!**

# CHAPTER 2

# ATOMS, MOLECULES, AND IONS

## PROBLEM-SOLVING STRATEGIES AND TUTORIAL SOLUTIONS

**TYPES OF PROBLEMS**

**Problem Type 1:** Atomic number, Mass number, and Isotopes.

**Problem Type 2:** Empirical and Molecular Formulas.

**Problem Type 3:** Naming Compounds.
- **(a)** Ionic compounds.
- **(b)** Molecular compounds.
- **(c)** Acids.
- **(d)** Bases.

**Problem Type 4:** Formulas of Ionic Compounds.

## PROBLEM TYPE 1: ATOMIC NUMBER, MASS NUMBER, AND ISOTOPES

The **atomic number** (*Z*) is the number of protons in the nucleus of each atom of an element.

**EXAMPLE 2.1**
**What is the atomic number of an oxygen atom?**

**Solution:** The atomic number is listed above each element in the periodic table. For oxygen, the atomic number is **8**, meaning that an oxygen atom has eight protons in the nucleus.

The **mass number** (*A*) is the total number of neutrons and protons present in the nucleus of an atom of an element.

$$\textbf{mass number} = \text{number of protons} + \text{number of neutrons}$$
$$\textbf{mass number} = \text{atomic number} + \text{number of neutrons}$$

**EXAMPLE 2.2**
**A particular oxygen atom has nine neutrons in the nucleus. What is the mass number of this atom?**

**Strategy:** Looking at a periodic table, you should find that every oxygen atom has an atomic number of 8. The number of neutrons is given, so we can solve for the mass number of this atom.

**Solution:**

$$\text{mass number} = \text{atomic number} + \text{number of neutrons}$$
$$\textbf{mass number} = 8 + 9 = \textbf{17}$$

**Isotopes** are atoms that have the same atomic number, but different mass numbers. For example, there are three isotopes of oxygen found in nature, oxygen-16, oxygen-17, and oxygen-18. The accepted way to denote the atomic number and mass number of an element X is as follows:

$$^{A}_{Z}\text{X}$$

where, $A$ = mass number

$Z$ = atomic number

**EXAMPLE 2.3**

**The three isotopes of oxygen found in nature are oxygen-16, -17, and -18. Write their isotopic symbols.**

**Strategy:** The atomic number of oxygen is 8, so all isotopes of oxygen contain eight protons. The mass numbers are 16, 17, and 18, respectively.

**Solution:** $^{16}_{8}O$ $^{17}_{8}O$ $^{18}_{8}O$

The number of **electrons** in an *atom* is equal to the number of protons.

$$\textbf{number of electrons (atom)} = \text{number of protons} = \text{atomic number}$$

The number of **electrons** in an *ion* is equal to the number of protons minus the charge on the ion.

$$\textbf{number of electrons (ion)} = \text{number of protons} - \text{charge on the ion}$$

**EXAMPLE 2.4**

**What is the total number of fundamental particles (protons, neutrons, and electrons) in (a) an atom of $^{56}_{26}Fe$ and (b) an $^{56}_{26}Fe^{3+}$ ion?**

**Strategy:** Both $^{56}_{26}Fe$ and $^{56}_{26}Fe^{3+}$ have the same atomic number and mass number, but the number of electrons will be different because one species is neutral and the other has a +3 charge.

**Solution:** For both (a) and (b):

$$\textbf{number of protons} = \text{atomic number} = \mathbf{26}$$

and

$$\textbf{number of neutrons} = \text{mass number} - \text{atomic number} = 56 - 26 = \mathbf{30}$$

However, the number of electrons for the above species differ.

**(a)** $^{56}_{26}Fe$ is a neutral atom. Therefore,

$$\textbf{number of electrons} = \text{number of protons} = \mathbf{26}$$

**(b)** $^{56}_{26}Fe^{3+}$ is an ion with a +3 charge. Therefore,

$$\textbf{number of electrons} = \text{number of protons} - \text{charge} = 26 - (+3) = \mathbf{23}$$

**PRACTICE EXERCISE**

1. How many protons, neutrons, and electrons are contained in each of the following atoms or ions?

**(a)** $^{19}F$ **(b)** $^{79}Se^{2-}$ **(c)** $^{40}Ca$ **(d)** $^{48}Ti^{4+}$

**Text Problems: 2.14**, 2.16, 2.18, 2.36

## PROBLEM TYPE 2: EMPIRICAL AND MOLECULAR FORMULAS

A *molecular* formula shows the exact number of atoms of each element in the smallest unit of a substance. An *empirical formula* tells us which elements are present and the simplest whole-number ratio of their atoms. Empirical formulas are therefore the simplest chemical formulas; they are always written so that the subscripts in the molecular formulas are converted to the smallest possible whole numbers.

**EXAMPLE 2.5**

**What is the empirical formula of each of the following compounds: (a) $H_2O_2$, (b) $C_6H_8O_6$, (c) $MgCl_2$, and (d) $C_6H_6$?**

**Strategy:** An *empirical formula* tells us which elements are present and the *simplest* whole-number ratio of their atoms. Can you divide the subscripts in the formula by some factor to end up with smaller whole-number subscripts?

**Solution:**

(a) The simplest whole number ratio of the atoms in $H_2O_2$ is **HO**.

(b) The simplest whole number ratio of the atoms in $C_6H_8O_6$ is $\mathbf{C_3H_4O_3}$.

(c) The molecular formula as written contains the simplest whole number ratio of the atoms present. In this case, the molecular formula and the empirical formula are the same.

(d) The simplest whole number ratio of the atoms in $C_6H_6$ is **CH**.

**PRACTICE EXERCISE**

2. What is the empirical formula of each of the following compounds?

(a) $C_2H_6O$ (b) $C_6H_{12}O_6$ (c) $CH_3COOH$ (d) $C_{12}H_{22}O_{11}$

**Text Problems: 2.48, 2.50**

## PROBLEM TYPE 3: NAMING COMPOUNDS

### A. Naming ionic compounds

(1) **Metal cation has only one charge**. When naming ionic compounds, our reference for the names of cations and anions is Table 2.3 of the text. You should memorize the metal cations that have only one charge when they form ionic compounds. These include the alkali metals (Group 1A), which always have a +1 charge in ionic compounds, the alkaline earth metals (Group 2A), which always have a +2 charge in ionic compounds, and $Al^{3+}$, $Ag^{+}$, $Cd^{2+}$, and $Zn^{2+}$.

Since the metal cation has only one possible charge, we do not need to specify this charge in the compound. Therefore, the name of this type of ionic compound can be written simply by first naming the metal cation as it appears on the periodic table, followed by the nonmetallic anion. Anions from elements are named by changing the suffix to "-ide".

**EXAMPLE 2.6**

**Name the following ionic compounds: (a) $AlF_3$, (b) $Na_3N$, (c) $Ba(NO_3)_2$.**

**Strategy:** The metal cation in each of the compounds given has only one charge when forming ionic compounds. We do not specify this charge in naming the compound.

**Solution:**

(a) aluminum fluor*ide*

(b) sodium nitr*ide*

(c) barium nitrate

**Tip:** There are a number of ions that contain more than one atom. These ions are called **polyatomic ions**. Nitrate, $NO_3^-$, is an example. Ask your instructor which polyatomic ions you should know.

(2) **Metals that form more than one type of cation.** Transition metals typically can form more than one type of cation when forming ionic compounds. If a metal can form cations of different charges, we need to use the Stock system. In the Stock system, Roman numerals are used to specify the charge of the cation. The Roman numeral (I) is used for one positive charge, (II) for two positive charges, (III) for three positive charges, and so on.

**EXAMPLE 2.7**

**Name the following compounds: (a) FeO, (b) $Fe_2O_3$, (c) $HgSO_4$.**

**Strategy:** The metals in the compounds above can form cations of different charges. We use the Stock system to name these compounds.

**Solution:**

(a) In this compound, the iron cation has a +2 charge, since oxide has a −2 charge (see Table 2.2 of the text). Therefore, the compound is named **iron(II) oxide**.

(b) In this compound, the iron cation has a +3 charge. Therefore, the compound is named **iron(III) oxide**.

(c) This compound contains the polyatomic ion sulfate, which has a −2 charge ($SO_4^{2-}$). Thus, the charge on mercury (Hg) is +2. The compound is named **mercury(II) sulfate**.

## B. Naming molecular compounds

Unlike ionic compounds, molecular compounds contain discrete molecular units. They are usually composed of nonmetallic elements. There are two types of molecular compounds to consider.

(1) **Only one compound of the two elements exists.** If this is the case, you simply name the first element in the formula as it appears on the periodic table, followed by naming the second element with an "-ide" suffix.

**EXAMPLE 2.8**

**Name the following compounds: (a) HF, (b) SiC.**

(a) hydrogen fluoride
(b) silicon carbide

(2) **More than one compound composed of the two elements exists.** It is quite common for one pair of elements to form several different compounds. Therefore, we must be able to differentiate between the compounds. Greek prefixes are used to denote the number of atoms of each element present. See Table 2.4 in the text for the Greek prefixes used.

**EXAMPLE 2.9**

**Name the following compounds: (a) CO, (b) $CO_2$, (c) $N_2O_5$, (d) $SF_6$.**

(a) carbon *mono*xide
(b) carbon *di*oxide
(c) *di*nitrogen *pent*oxide
(d) sulfur *hexa*fluoride

> **Tip:** The prefix "mono-" may be omitted when naming the first element in a molecular compound (see carbon monoxide and carbon dioxide above). The absence of a prefix for the first element usually means that only one atom of that element is present in the molecule.

## C. Naming acids

An acid is a substance that yields hydrogen ions ($H^+$) when dissolved in water. There are two types of acids to consider.

(1) **Acids that do not contain oxygen.** This type of acid contains one or more hydrogen atoms as well as an anionic group. To name these acids, you add the prefix "hydro-" to the anion name, change the "-ide" suffix of the anion to "-ic", and then add the word "acid" at the end.

**EXAMPLE 2.10**

**Name the following binary acids: (a) HF(*aq*), (b) HCN(*aq*).**

(a) *hydro*fluoric acid

(b) $CN^-$ is a polyatomic ion called cyanide. In the acid, the "-ide" suffix is changed to "-ic". The correct name is ***hydro*cyan*ic* acid**.

> **Tip:** The (*aq*) above means that the substance is dissolved in water. HF dissolved in water is an acid and is named hydrofluoric acid. However, HF in its pure state is a molecular compound and is named hydrogen fluoride.

(2) **Oxoacids.** This type of acid contains hydrogen, oxygen, and another element. To name oxoacids, you must look carefully at the anion name. If the suffix of the anion is "-ate", change the suffix to "-ic" and add the word "acid" at the end. If the suffix of the anion is "-ite", change the suffix to "-ous" and add the word "acid" at the end.

**EXAMPLE 2.11**

**Name the following oxoacids: (a) $HNO_2$(*aq*), (b) $HClO_3$(*aq*).**

(a) The $NO_2^-$ polyatomic ion is called nitr*ite*. Simply change the suffix to "-ous" and add the word "acid". The correct name is **nitr*ous* acid**.

(b) The $ClO_3^-$ polyatomic ion is called chlor*ate*. Simply change the suffix to "-ic" and add the work "acid". The correct name is **chlor*ic* acid**.

> **Tip:** If one O atom is added to the "-ic" acid, the acid is called "per...ic" acid. For example, $HClO_4$ is named perchloric acid. Compare this acid to chloric acid above. If one O atom is removed from the "-ous" acid, the acid is called "hypo...ous" acid. For example, HClO is named hypochlorous acid. Compare this to chlorous acid, $HClO_2$.

## D. Naming Bases

A base is a substance that produces the hydroxide ion ($OH^-$) when dissolved in water. At this point, for naming purposes, we will only consider bases that contain the hydroxide ion. To name this type of base, simply name the metal cation first as it appears on the periodic table, then add "hydroxide".

**EXAMPLE 2.12**
**Name the following bases: (a) KOH, (b) $Sr(OH)_2$.**

(a) potassium hydroxide
(b) strontium hydroxide

**PRACTICE EXERCISE**

3. Name the following compounds:

(a) $MgCl_2$ (b) $CuCl_2$ (c) $HNO_3$ (d) $P_2O_5$ (e) $Ca(OH)_2$

**Text Problems: 2.60, 2.62**

## PROBLEM TYPE 4: FORMULAS OF IONIC COMPOUNDS

The formulas of ionic compounds are usually the same as their empirical formulas because ionic compounds do not consist of discrete molecular units. See Section 2.6 of the text if you need further information.

Ionic compounds are electrically neutral. In order for ionic compounds to be electrically neutral, the sum of the charges on the cation and anion in each formula unit must add up to zero. There are two possibilities to consider.

(1) If the charges on the cation and anion are numerically equal, no subscripts are necessary in the formula.
(2) If the charges on the cation and anion are numerically different, the subscript of the cation is numerically equal to the charge on the anion, and the subscript of the anion is numerically equal to the charge on the cation.

You should memorize the metal cations that have only one charge when they form ionic compounds. These include the alkali metals (Group 1A), which always have a +1 charge in ionic compounds, the alkaline earth metals (Group 2A), which always have a +2 charge in ionic compounds, and $Al^{3+}$, $Ag^{+}$, $Cd^{2+}$, and $Zn^{2+}$. All other metals can have more than one possible positive charge when forming ionic compounds. This positive charge will be specified using a Roman numeral in the name of the compound.

You should also memorize the charges of polyatomic ions (see Table 2.3 of the text) and the common charges of monatomic anions based on their positions in the periodic table (see Table 2.2 of the text).

**EXAMPLE 2.13**
**Write the formula for the ionic compound, magnesium oxide.**

**Strategy:** The magnesium cation is an alkaline earth metal cation which always has a +2 charge in an ionic compound, and the oxide anion is −2 in an ionic compound (see Table 2.2 of the text).

**Solution:** $Mg^{2+}$ and the oxide anion, $O^{2-}$, combine to form the ionic compound magnesium oxide. The sum of the charges is $+2 + (-2) = 0$, so no subscripts are necessary. The formula is **MgO**.

**EXAMPLE 2.14**

**Write the formula for the ionic compound, iron(II) chloride.**

**Strategy:** An iron cation can either have a +2 or +3 charge in an ionic compound. The Roman numeral, II, specifies that in this compound it is the +2 cation, $Fe^{2+}$. The chloride anion has a −1 charge in an ionic compound (see Table 2.2 of the text).

**Solution:** $Fe^{2+}$ and the chloride anion, $Cl^-$, combine to form the ionic compound iron(II) chloride. The charges on the cation and anion are numerically different, so make the subscript of the cation ($Fe^{2+}$) numerically equal to the charge of the anion (subscript = 1). Also, make the subscript of the anion ($Cl^-$) numerically equal to the charge of the cation (subscript = 2). The formula is **$FeCl_2$**.

> **Tip:** Check to make sure that the compound is electrically neutral by multiplying the charge of each ion by its subscript and then adding them together. The sum should equal zero.

$$(+2)(1) + (-1)(2) = 0$$

**PRACTICE EXERCISE**

4. Write the correct formulas for the following ionic compounds:

   **(a)** Sodium oxide **(b)** Copper(II) nitrate **(c)** Aluminum oxide

**Text Problems: 2.62 a, b, e, f, g, i, j**

## ANSWERS TO PRACTICE EXERCISES

**1.** **(a)** 9p, 10n, 9e
**(b)** 34p, 45n, 36e
**(c)** 20p, 20n, 20e
**(d)** 22p, 26n, 18e

**2.** **(a)** $C_2H_6O$
**(b)** $CH_2O$
**(c)** $CH_2O$
**(d)** $C_{12}H_{22}O_{11}$

**3.** **(a)** magnesium chloride
**(b)** copper(II) chloride
**(c)** nitric acid
**(d)** diphosphorus pentoxide
**(e)** calcium hydroxide

**4.** **(a)** $Na_2O$
**(b)** $Cu(NO_3)_2$
**(c)** $Al_2O_3$

## SOLUTIONS TO SELECTED TEXT PROBLEMS

**2.8** Note that you are given information to set up the unit factor relating meters and miles.

$$r_{\text{atom}} = 10^4 r_{\text{nucleus}} = 10^4 \times 2.0\ \text{cm} \times \frac{1\ \text{m}}{100\ \text{cm}} \times \frac{1\ \text{mi}}{1609\ \text{m}} = \mathbf{0.12\ mi}$$

**2.14** **Strategy:** The 239 in Pu-239 is the mass number. The **mass number** (*A*) is the total number of neutrons and protons present in the nucleus of an atom of an element. You can look up the atomic number (number of protons) on the periodic table.

**Solution:**

mass number = number of protons + number of neutrons

**number of neutrons** = mass number − number of protons = 239 − 94 = **145**

**2.16**

| Isotope | $^{15}_{7}N$ | $^{33}_{16}S$ | $^{63}_{29}Cu$ | $^{84}_{38}Sr$ | $^{130}_{56}Ba$ | $^{186}_{74}W$ | $^{202}_{80}Hg$ |
|---|---|---|---|---|---|---|---|
| No. Protons | 7 | 16 | 29 | 38 | 56 | 74 | 80 |
| No. Neutrons | 8 | 17 | 34 | 46 | 74 | 112 | 122 |
| No. Electrons | 7 | 16 | 29 | 38 | 56 | 74 | 80 |

**2.18** The accepted way to denote the atomic number and mass number of an element X is as follows:

$$^{A}_{Z}\mathbf{X}$$

where,

$A$ = mass number

$Z$ = atomic number

**(a)** $^{186}_{74}W$ **(b)** $^{201}_{80}Hg$

**2.24** **(a)** Metallic character increases as you progress down a group of the periodic table. For example, moving down Group 4A, the nonmetal carbon is at the top and the metal lead is at the bottom of the group.

**(b)** Metallic character decreases from the left side of the table (where the metals are located) to the right side of the table (where the nonmetals are located).

**2.26** F and Cl are Group 7A elements; they should have similar chemical properties. Na and K are both Group 1A elements; they should have similar chemical properties. P and N are both Group 5A elements; they should have similar chemical properties.

**2.32** **(a)** This is a diatomic molecule that is a compound.

**(b)** This is a polyatomic molecule that is a compound.

**(c)** This is a polyatomic molecule that is the elemental form of the substance. It is not a compound.

**2.34** There are more than two correct answers for each part of the problem.

**(a)** $H_2$ and $F_2$ **(b)** HCl and CO **(c)** $S_8$ and $P_4$

**(d)** $H_2O$ and $C_{12}H_{22}O_{11}$ (sucrose)

**2.36** The **atomic number (*Z*)** is the number of protons in the nucleus of each atom of an element. You can find this on a periodic table. The number of **electrons** in an *ion* is equal to the number of protons minus the charge on the ion.

**number of electrons (ion)** = number of protons − charge on the ion

| Ion | $K^+$ | $Mg^{2+}$ | $Fe^{3+}$ | $Br^-$ | $Mn^{2+}$ | $C^{4-}$ | $Cu^{2+}$ |
|---|---|---|---|---|---|---|---|
| No. protons | 19 | 12 | 26 | 35 | 25 | 6 | 29 |
| No. electrons | 18 | 10 | 23 | 36 | 23 | 10 | 27 |

**2.46** (a) The copper ion has a +1 charge and bromide has a −1 charge. The correct formula is **$CuBr$**.

(b) The manganese ion has a +3 charge and oxide has a −2 charge. The correct formula is **$Mn_2O_3$**.

(c) We have the $Hg_2^{2+}$ ion and iodide ($I^-$). The correct formula is **$Hg_2I_2$**.

(d) Magnesium ion has a +2 charge and phosphate has a −3 charge. The correct formula is **$Mg_3(PO_4)_2$**.

**2.48** **Strategy:** An *empirical formula* tells us which elements are present and the *simplest* whole-number ratio of their atoms. Can you divide the subscripts in the formula by some factor to end up with smaller whole-number subscripts?

**Solution:**

(a) Dividing both subscripts by 2, the simplest whole number ratio of the atoms in $Al_2Br_6$ is **$AlBr_3$**.

(b) Dividing all subscripts by 2, the simplest whole number ratio of the atoms in $Na_2S_2O_4$ is **$NaSO_2$**.

(c) The molecular formula as written, **$N_2O_5$**, contains the simplest whole number ratio of the atoms present. In this case, the molecular formula and the empirical formula are the same.

(d) The molecular formula as written, **$K_2Cr_2O_7$**, contains the simplest whole number ratio of the atoms present. In this case, the molecular formula and the empirical formula are the same.

**2.50** The molecular formula of ethanol is **$C_2H_6O$**.

**2.60** **Strategy:** When naming ionic compounds, our reference for the names of cations and anions is Table 2.3 of the text. Keep in mind that if a metal can form cations of different charges, we need to use the Stock system. In the Stock system, Roman numerals are used to specify the charge of the cation. The metals that have only one charge in ionic compounds are the alkali metals (+1), the alkaline earth metals (+2), $Ag^+$, $Zn^{2+}$, $Cd^{2+}$, and $Al^{3+}$.

When naming acids, binary acids are named differently than oxoacids. For binary acids, the name is based on the nonmetal. For oxoacids, the name is based on the polyatomic anion. For more detail, see Section 2.7 of the text.

**Solution:**

(a) This is an ionic compound in which the metal cation ($K^+$) has only one charge. The correct name is **potassium hypochlorite**. Hypochlorite is a polyatomic ion with one less O atom than the chlorite ion, $ClO_2^-$.

(b) **silver carbonate**

(c) This is an ionic compound in which the metal can form more than one cation. Use a Roman numeral to specify the charge of the Fe ion. Since the chloride ion has a −1 charge, the Fe ion has a +2 charge. The correct name is **iron(II) chloride.**

(d) **potassium permanganate** (e) **cesium chlorate** (f) **hypoiodous acid**

(g) This is an ionic compound in which the metal can form more than one cation. Use a Roman numeral to specify the charge of the Fe ion. Since the oxide ion has a −2 charge, the Fe ion has a +2 charge. The correct name is **iron(II) oxide.**

(h) **iron(III) oxide**

(i) This is an ionic compound in which the metal can form more than one cation. Use a Roman numeral to specify the charge of the Ti ion. Since each of the four chloride ions has a −1 charge (total of −4), the Ti ion has a +4 charge. The correct name is **titanium(IV) chloride.**

(j) **sodium hydride** (k) **lithium nitride** (l) **sodium oxide**

(m) This is an ionic compound in which the metal cation ($Na^+$) has only one charge. The $O_2^{2-}$ ion is called the peroxide ion. Each oxygen has a −1 charge. You can determine that each oxygen only has a −1 charge, because each of the two Na ions has a +1 charge. Compare this to sodium oxide in part (l). The correct name is **sodium peroxide.**

(n) **iron(III) chloride hexahydrate**

**2.62** **Strategy:** When writing formulas of molecular compounds, the prefixes specify the number of each type of atom in the compound.

When writing formulas of ionic compounds, the subscript of the cation is numerically equal to the charge of the anion, and the subscript of the anion is numerically equal to the charge on the cation. If the charges of the cation and anion are numerically equal, then no subscripts are necessary. Charges of common cations and anions are listed in Table 2.3 of the text. Keep in mind that Roman numerals specify the charge of the cation, *not* the number of metal atoms. Remember that a Roman numeral is not needed for some metal cations, because the charge is known. These metals are the alkali metals (+1), the alkaline earth metals (+2), $Ag^+$, $Zn^{2+}$, $Cd^{2+}$, and $Al^{3+}$.

When writing formulas of oxoacids, you must know the names and formulas of polyatomic anions (see Table 2.3 of the text).

**Solution:**

**(a)** The Roman numeral I tells you that the Cu cation has a +1 charge. Cyanide has a −1 charge. Since the charges are numerically equal, no subscripts are necessary in the formula. The correct formula is **CuCN**.

**(b)** Strontium is an alkaline earth metal. It only forms a +2 cation. The polyatomic ion chlorite, $ClO_2^-$, has a −1 charge. Since the charges on the cation and anion are numerically different, the subscript of the cation is numerically equal to the charge on the anion, and the subscript of the anion is numerically equal to the charge on the cation. The correct formula is $\mathbf{Sr(ClO_2)_2}$.

**(c)** Perbromic tells you that the anion of this oxoacid is perbromate, $BrO_4^-$. The correct formula is $\mathbf{HBrO_4}$***(aq)***. Remember that (*aq*) means that the substance is dissolved in water.

**(d)** Hydroiodic tells you that the anion of this binary acid is iodide, $I^-$. The correct formula is **HI(*aq*)**.

**(e)** Na is an alkali metal. It only forms a +1 cation. The polyatomic ion ammonium, $NH_4^+$, has a +1 charge and the polyatomic ion phosphate, $PO_4^{3-}$, has a −3 charge. To balance the charge, you need 2 $Na^+$ cations. The correct formula is $\mathbf{Na_2(NH_4)PO_4}$.

**(f)** The Roman numeral II tells you that the Pb cation has a +2 charge. The polyatomic ion carbonate, $CO_3^{2-}$, has a −2 charge. Since, the charges are numerically equal, no subscripts are necessary in the formula. The correct formula is $\mathbf{PbCO_3}$.

**(g)** The Roman numeral II tells you that the Sn cation has a +2 charge. Fluoride has a −1 charge. Since the charges on the cation and anion are numerically different, the subscript of the cation is numerically equal to the charge on the anion, and the subscript of the anion is numerically equal to the charge on the cation. The correct formula is $\mathbf{SnF_2}$.

**(h)** This is a molecular compound. The Greek prefixes tell you the number of each type of atom in the molecule. The correct formula is $\mathbf{P_4S_{10}}$.

**(i)** The Roman numeral II tells you that the Hg cation has a +2 charge. Oxide has a −2 charge. Since, the charges are numerically equal, no subscripts are necessary in the formula. The correct formula is **HgO**.

**(j)** The Roman numeral I tells you that the Hg cation has a +1 charge. However, this cation exists as $Hg_2^{2+}$. Iodide has a −1 charge. You need two iodide ions to balance the +2 charge of $Hg_2^{2+}$. The correct formula is $\mathbf{Hg_2I_2}$.

**(k)** This is a molecular compound. The Greek prefixes tell you the number of each type of atom in the molecule. The correct formula is $\mathbf{SeF_6}$.

**2.64** **(a)** dinitrogen pentoxide ($N_2O_5$)

**(b)** boron trifluoride ($BF_3$)

**(c)** dialuminum hexabromide ($Al_2Br_6$)

**2.66** Changing the electrical charge of an atom usually has a major effect on its chemical properties. The two electrically neutral carbon isotopes should have nearly identical chemical properties.

**2.68** Atomic number = 127 − 74 = 53. This anion has 53 protons, so it is an iodide ion. Since there is one more electron than protons, the ion has a −1 charge. The correct symbol is $\mathbf{I^-}$.

**2.70** NaCl is an ionic compound; it doesn't form molecules.

**2.72** The species and their identification are as follows:

| | | | | | |
|---|---|---|---|---|---|
| **(a)** | $SO_2$ | molecule and compound | **(g)** | $O_3$ | element and molecule |
| **(b)** | $S_8$ | element and molecule | **(h)** | $CH_4$ | molecule and compound |
| **(c)** | Cs | element | **(i)** | KBr | compound |
| **(d)** | $N_2O_5$ | molecule and compound | **(j)** | S | element |
| **(e)** | O | element | **(k)** | $P_4$ | element and molecule |
| **(f)** | $O_2$ | element and molecule | **(l)** | LiF | compound |

**2.74** **(a)** Ne, 10 p, 10 n **(b)** Cu, 29 p, 34 n **(c)** Ag, 47 p, 60 n
**(d)** W, 74 p, 108 n **(e)** Po, 84 p, 119 n **(f)** Pu, 94 p, 140 n

**2.76** **(a)** Cu **(b)** P **(c)** Kr **(d)** Cs **(e)** Al **(f)** Sb **(g)** Cl **(h)** Sr

**2.78** **(a)** Rutherford's experiment is described in detail in Section 2.2 of the text. From the average magnitude of scattering, Rutherford estimated the number of protons (based on electrostatic interactions) in the nucleus.

**(b)** Assuming that the nucleus is spherical, the volume of the nucleus is:

$$V = \frac{4}{3}\pi r^3 = \frac{4}{3}\pi(3.04 \times 10^{-13}\ \text{cm})^3 = 1.177 \times 10^{-37}\ \text{cm}^3$$

The density of the nucleus can now be calculated.

$$d = \frac{\text{m}}{\text{V}} = \frac{3.82 \times 10^{-23}\ \text{g}}{1.177 \times 10^{-37}\ \text{cm}^3} = \mathbf{3.25 \times 10^{14}\ g/cm^3}$$

To calculate the density of the space occupied by the electrons, we need both the mass of 11 electrons, and the volume occupied by these electrons.

The mass of 11 electrons is:

$$11\ \text{electrons} \times \frac{9.1095 \times 10^{-28}\ \text{g}}{1\ \text{electron}} = 1.00205 \times 10^{-26}\ \text{g}$$

The volume occupied by the electrons will be the difference between the volume of the atom and the volume of the nucleus. The volume of the nucleus was calculated above. The volume of the atom is calculated as follows:

$$186\ \cancel{\text{pm}} \times \frac{1\times10^{-12}\ \cancel{\text{m}}}{1\ \cancel{\text{pm}}} \times \frac{1\ \text{cm}}{1\times10^{-2}\ \cancel{\text{m}}} = 1.86\times10^{-8}\ \text{cm}$$

$$V_{\text{atom}} = \frac{4}{3}\pi r^3 = \frac{4}{3}\pi(1.86\times10^{-8}\ \text{cm})^3 = 2.695\times10^{-23}\ \text{cm}^3$$

$$V_{\text{electrons}} = V_{\text{atom}} - V_{\text{nucleus}} = (2.695\times10^{-23}\ \text{cm}^3) - (1.177\times10^{-37}\ \text{cm}^3) = 2.695\times10^{-23}\ \text{cm}^3$$

As you can see, the volume occupied by the nucleus is insignificant compared to the space occupied by the electrons.

The density of the space occupied by the electrons can now be calculated.

$$d = \frac{m}{V} = \frac{1.00205\times10^{-26}\ \text{g}}{2.695\times10^{-23}\ \text{cm}^3} = \mathbf{3.72\times10^{-4}\ g/cm^3}$$

The above results do support Rutherford's model. Comparing the space occupied by the electrons to the volume of the nucleus, it is clear that most of the atom is empty space. Rutherford also proposed that the nucleus was a *dense* central core with most of the mass of the atom concentrated in it. Comparing the density of the nucleus with the density of the space occupied by the electrons also supports Rutherford's model.

**2.80** The empirical and molecular formulas of acetaminophen are $\mathbf{C_8H_9NO_2}$.

**2.82** (a) The charge on the tin cation needs to be specified. The correct name is **tin(IV) chloride**.

(b) The charge on the copper ion is +1. The correct name is **copper(I) oxide**.

(c) The charge on the cobalt cation needs to be specified. The correct name is **cobalt(II) nitrate**.

(d) $Cr_2O_7^{2-}$ is the dichromate ion. The correct name is **sodium dichromate**.

**2.84** (a) Ionic compounds are typically formed between metallic and nonmetallic elements.

(b) In general the transition metals, the actinides, and the lanthanides have variable charges.

**2.86** The symbol $^{23}$Na provides more information than $_{11}$Na. The mass number plus the chemical symbol identifies a specific isotope of Na (sodium) while combining the atomic number with the chemical symbol tells you nothing new. Can other isotopes of sodium have different atomic numbers?

**2.88** Mercury (Hg) and bromine ($Br_2$)

**2.90** $H_2$, $N_2$, $O_2$, $F_2$, $Cl_2$, He, Ne, Ar, Kr, Xe, Rn

**2.92** They do not have a strong tendency to form compounds. Helium, neon, and argon are chemically inert.

**2.94** All isotopes of radium are radioactive. It is a radioactive decay product of uranium-238. Radium itself does *not* occur naturally on Earth.

**2.96** The atomic number is 77 – 43 = 34. The symbol for the anion is $\mathbf{^{77}Se^{2-}}$.

**2.98** **(a)** NaH, sodium hydride **(b)** $B_2O_3$, diboron trioxide **(c)** $Na_2S$, sodium sulfide
**(d)** $AlF_3$, aluminum fluoride **(e)** $OF_2$, oxygen difluoride **(f)** $SrCl_2$, strontium chloride

**2.100** All of these are molecular compounds. We use prefixes to express the number of each atom in the molecule. The names are nitrogen trifluoride ($NF_3$), phosphorus pentabromide ($PBr_5$), and sulfur dichloride ($SCl_2$).

**2.102**

| Cation | Anion | Formula | Name |
|---|---|---|---|
| **$Mg^{2+}$** | **$HCO_3^-$** | **$Mg(HCO_3)_2$** | Magnesium bicarbonate |
| **$Sr^{2+}$** | **$Cl^-$** | $SrCl_2$ | **Strontium chloride** |
| $Fe^{3+}$ | $NO_2^-$ | **$Fe(NO_2)_3$** | **Iron(III) nitrite** |
| **$Mn^{2+}$** | **$ClO_3^-$** | **$Mn(ClO_3)_2$** | Manganese(II) chlorate |
| **$Sn^{4+}$** | **$Br^-$** | $SnBr_4$ | **Tin(IV) bromide** |
| $Co^{2+}$ | $PO_4^{3-}$ | **$Co_3(PO_4)_2$** | **Cobalt(II) phosphate** |
| $Hg_2^{2+}$ | $I^-$ | **$Hg_2I_2$** | **Mercury(I) iodide** |
| **$Cu^+$** | **$CO_3^{2-}$** | $Cu_2CO_3$ | **Copper(I) carbonate** |
| **$Li^+$** | **$N^{3-}$** | **$Li_3N$** | Lithium nitride |
| $Al^{3+}$ | $S^{2-}$ | **$Al_2S_3$** | **Aluminum sulfide** |

**2.104** The change in energy is equal to the energy released. We call this $\Delta E$. Similarly, $\Delta m$ is the change in mass.

Because $m = \dfrac{E}{c^2}$, we have

$$\Delta m = \frac{\Delta E}{c^2} = \frac{(1.715 \times 10^3\ \text{kJ}) \times \dfrac{1000\ \text{J}}{1\ \text{kJ}}}{(3.00 \times 10^8\ \text{m/s})^2} = 1.91 \times 10^{-11}\ \text{kg} = \mathbf{1.91 \times 10^{-8}\ g}$$

Note that we need to convert kJ to J so that we end up with units of kg for the mass. $\left(1\ \text{J} = \dfrac{1\ \text{kg}\cdot\text{m}^2}{\text{s}^2}\right)$

We can add together the masses of hydrogen and oxygen to calculate the mass of water that should be formed.

$$12.096\ \text{g} + 96.000 = 108.096\ \text{g}$$

The predicted change (loss) in mass is only $1.91 \times 10^{-8}$ g which is too small a quantity to measure. Therefore, for all practical purposes, the law of conservation of mass is assumed to hold for ordinary chemical processes.

**2.106** **(a)** The volume of a sphere is

$$V = \frac{4}{3}\pi r^3$$

**Volume is proportional to the number of nucleons. Therefore,**

$$V \propto \text{A (mass number)}$$

$$r^3 \propto \text{A}$$

$$\boldsymbol{r \propto \text{A}^{1/3}}$$

**(b)** Using the equation given in the problem, we can first solve for the radius of the lithium nucleus and then solve for its volume.

$$r = r_0\text{A}^{1/3}$$

$$r = (1.2 \times 10^{-15}\ \text{m})(7)^{1/3}$$

$$r = 2.3 \times 10^{-15}\ \text{m}$$

$$V = \frac{4}{3}\pi r^3$$

$$V_{\text{nucleus}} = \frac{4}{3}\pi(2.3 \times 10^{-15}\ \text{m})^3 = \mathbf{5.1 \times 10^{-44}\ m^3}$$

**(c)** In part (b), the volume of the nucleus was calculated. Using the radius of a Li atom, the volume of a Li atom can be calculated.

$$V_{\text{atom}} = \frac{4}{3}\pi r^3 = \frac{4}{3}\pi(152 \times 10^{-12}\ \text{m})^3 = 1.47 \times 10^{-29}\ \text{m}^3$$

The fraction of the atom's volume occupied by the nucleus is:

$$\frac{V_{\text{nucleus}}}{V_{\text{atom}}} = \frac{5.1 \times 10^{-44}\ \text{m}^3}{1.47 \times 10^{-29}\ \text{m}^3} = \mathbf{3.5 \times 10^{-15}}$$

Yes, this calculation shows that the volume of the nucleus is much, much smaller than the volume of the atom, which supports Rutherford's model of an atom.

**2.108** **(a)**

| Ethane | Acetylene |
|---|---|
| 2.65 g C | 4.56 g C |
| 0.665 g H | 0.383 g H |

Let's compare the ratio of the hydrogen masses in the two compounds. To do this, we need to start with the same mass of carbon. If we were to start with 4.56 g of C in ethane, how much hydrogen would combine with 4.56 g of carbon?

$$0.665\ \text{g H} \times \frac{4.56\ \text{g C}}{2.65\ \text{g C}} = 1.14\ \text{g H}$$

We can calculate the ratio of H in the two compounds.

$$\frac{1.14\ \text{g}}{0.383\ \text{g}} \approx 3$$

This is consistent with the Law of Multiple Proportions which states that if two elements combine to form more than one compound, the masses of one element that combine with a fixed mass of the other element are in ratios of small whole numbers. In this case, the ratio of the masses of hydrogen in the two compounds is 3:1.

**(b)** For a given amount of carbon, there is 3 times the amount of hydrogen in ethane compared to acetylene. Reasonable formulas would be:

| Ethane | Acetylene |
|---|---|
| $CH_3$ | CH |
| $C_2H_6$ | $C_2H_2$ |

**2.110** The mass number is the sum of the number of protons and neutrons in the nucleus.

$$\text{Mass number} = \text{number of protons} + \text{number of neutrons}$$

Let the atomic number (number of protons) equal $A$. The number of neutrons will be $1.2A$. Plug into the above equation and solve for $A$.

$$55 = A + 1.2A$$

$$A = 25$$

The element with atomic number 25 is **manganese, Mn.**

**2.112** The acids, from left to right, are **chloric acid**, **nitrous acid**, **hydrocyanic acid**, and **sulfuric acid**.

**2.114** The formula of the ionic compound is $\mathbf{XY_2}$. Element X is most likely in Group 4B and element Y is most likely in Group 6A. A possible compound is $TiO_2$, **titanium(IV) oxide**. Other choices are elements in Group 4A: $SnO_2$ [tin(IV) oxide] and $PbO_2$ [lead(IV) oxide].

# CHAPTER 3
# MASS RELATIONSHIPS IN CHEMICAL REACTIONS

## PROBLEM-SOLVING STRATEGIES

### TYPES OF PROBLEMS

**Problem Type 1:** Calculating Average Atomic Mass.

**Problem Type 2:** Calculations Involving Molar Mass of an Element and Avogadro's Number.
- **(a)** Converting between moles of atoms and mass of atoms.
- **(b)** Calculating the mass of a single atom.
- **(c)** Converting mass in grams to number of atoms.

**Problem Type 3:** Calculations Involving Molecular Mass.
- **(a)** Calculating molecular mass.
- **(b)** Calculating the number of moles in a given amount of a compound.
- **(c)** Calculating the number of atoms in a given amount of a compound.

**Problem Type 4:** Calculations Involving Percent Composition
- **(a)** Calculating percent composition of a compound.
- **(b)** Determining empirical formula from percent composition.
- **(c)** Calculating mass from percent composition.

**Problem Type 5:** Experimental Determination of Empirical Formulas.

**Problem Type 6:** Determining the Molecular Formula of a Compound.

**Problem Type 7:** Calculating the Amounts of Reactants and Products.

**Problem Type 8:** Limiting Reagent Calculations.

**Problem Type 9:** Calculating the Percent Yield of a Reaction.

## PROBLEM TYPE 1: CALCULATING AVERAGE ATOMIC MASS

The atomic mass you look up on a periodic table is an average atomic mass. The reason for this is that most naturally occurring elements have more than one isotope. The average atomic mass can be calculated as follows:

*Step 1:* Convert the percentage of each isotope to fractions. For example, an isotope that is 69.09 percent abundant becomes 69.09/100 = 0.6909.

*Step 2:* Multiply the mass of each isotope by its abundance and add them together.

**average atomic mass** = (fraction of isotope A)(mass of isotope A) + (fraction of isotope B)(mass of isotope B) + . . . + (fraction of isotope Z)(mass of isotope Z).

**EXAMPLE 3.1**

**The element lithium has two isotopes that occur in nature: $^{6}_{3}$Li with 7.5 percent abundance and $^{7}_{3}$Li with 92.5 percent abundance. The atomic mass of $^{6}_{3}$Li is 6.01513 amu and that of $^{7}_{3}$Li is 7.01601 amu. Calculate the average atomic mass of lithium.**

**Strategy:** Each isotope contributes to the average atomic mass based on its relative abundance. Multiplying the mass of an isotope by its fractional abundance (not percent) will give the contribution to the average atomic mass of that particular isotope.

**Solution:** Convert the percentage of each isotope to fractions.

$$^{6}_{3}\mathbf{Li}: \ 7.5/100 = 0.075$$

$$^{7}_{3}\mathbf{Li}: \ 92.5/100 = 0.925$$

Multiply the mass of each isotope by its abundance and add them together.

$$\textbf{average atomic mass} = (0.075)(6.01513) + (0.925)(7.01601) = \textbf{6.94 amu}$$

**PRACTICE EXERCISE**

1. The element boron (B) consists of two stable isotopes with atomic masses of 10.0129 amu and 11.0093 amu. The average atomic mass of B is 10.81 amu. Which isotope is more abundant?

**Text Problem: 3.6**

## PROBLEM TYPE 2: CALCULATIONS INVOLVING MOLAR MASS OF AN ELEMENT AND AVOGADRO'S NUMBER

### A. Converting between moles of atoms and mass of atoms

In order to convert from one unit to another, you need to be proficient at the dimensional analysis method. See Section 1.9 of your text and Problem Type 5, Chapter 1. Unit conversions can seem daunting, but if you keep track of the units, making sure that the appropriate units cancel, your effort will be rewarded.

*Step 1:* Map out a strategy to proceed from initial units to final units based on available conversion factors.

*Step 2:* Use the following method to ensure that you obtain the desired unit.

$$\text{Given } \cancel{\text{unit}} \times \left(\frac{\text{desired unit}}{\text{given } \cancel{\text{unit}}}\right) = \text{desired unit}$$

To convert between moles and mass, you need to use the molar mass of the element as a conversion factor.

$$\cancel{\text{mol}} \times \frac{\text{g}}{\cancel{\text{mol}}} = \text{g}$$

Also, going in the opposite direction

$$\cancel{\text{g}} \times \frac{\text{mol}}{\cancel{\text{g}}} = \text{mol}$$

**Tip:** Whether you are converting from g → mol or from mol → g, you will need to use the molar mass as the conversion factor. The molar mass of an element can be found directly on the periodic table.

**EXAMPLE 3.2**

**How many grams are there in 0.130 mole of Cu?**

**Strategy:** We are given moles of copper and asked to solve for grams of copper. What conversion factor do we need to convert between moles and grams? Arrange the appropriate conversion factor so moles cancel, and the unit grams is obtained for the answer.

**Solution:** The conversion factor needed to covert between moles and grams is the molar mass. In the periodic table, we see that the molar mass of Cu is 63.55 g. This can be expressed as

$$1 \text{ mol Cu} = 63.55 \text{ g Cu}$$

From this equality, we can write two conversion factors.

$$\frac{1 \text{ mol Cu}}{63.55 \text{ g Cu}} \quad \text{and} \quad \frac{63.55 \text{ g Cu}}{1 \text{ mol Cu}}$$

The conversion factor on the right is the correct one. Moles will cancel, leaving the unit grams for the answer.

We write

$$\textbf{? g Cu} = 0.130 \cancel{\text{mol}} \text{ Cu} \times \frac{63.55 \text{ g Cu}}{1 \cancel{\text{mol}} \text{ Cu}} = \textbf{8.26 g Cu}$$

**Check:** Does a mass of 8.26 g for 0.130 mole of Cu seem reasonable? What is the mass of 1 mole of Cu?

**PRACTICE EXERCISE**

2. How many moles of Cu are in 125 g of Cu?

**Text Problem: 3.16**

## B. Calculating the mass of a single atom

To calculate the mass of a single atom, you can use Avogadro's number. The conversion factor is

$$\frac{1 \text{ mol}}{6.022 \times 10^{23} \text{ atoms}}$$

**EXAMPLE 3.3**

**Copper is a minor component of pennies minted since 1981, and it is also used in electrical cables. Calculate the mass (in grams) of a single Cu atom.**

**Strategy:** We can look up the molar mass of copper (Cu) on the periodic table (63.55 g/mol). We want to find the mass of a single atom of copper (unit of g/atom). Therefore, we need to convert from the unit mole in the denominator to the unit atom in the denominator. What conversion factor is needed to convert between moles and atoms? Arrange the appropriate conversion factor so mole in the denominator cancels, and the unit atom is obtained in the denominator.

**Solution:** The conversion factor needed is Avogadro's number. We have

$$1 \text{ mol} = 6.022 \times 10^{23} \text{ particles (atoms)}$$

From this equality, we can write two conversion factors.

$$\frac{1 \text{ mol Cu}}{6.022 \times 10^{23} \text{ Cu atoms}} \quad \text{and} \quad \frac{6.022 \times 10^{23} \text{ Cu atoms}}{1 \text{ mol Cu}}$$

The conversion factor on the left is the correct one. Moles will cancel, leaving the unit atoms in the denominator of the answer.

We write

$$? \text{ g/Cu atom} = \frac{63.55 \text{ g Cu}}{1 \text{ mol Cu}} \times \frac{1 \text{ mol Cu}}{6.022 \times 10^{23} \text{ Cu atoms}} = \mathbf{1.055 \times 10^{-22} \text{ g/Cu atom}}$$

**Check:** Should the mass of a single atom of Cu be a very small mass?

**PRACTICE EXERCISE**

3. Titanium (Ti) is a transition metal with a very high strength-to-weight ratio. For this reason, titanium is used in the construction of aircraft. What is the mass (in grams) of one Ti atom?

**Text Problem: 3.18**

## C. Converting mass in grams to number of atoms

To complete the following conversion, you need to use both molar mass and Avogadro's number as conversion factors.

**EXAMPLE 3.4**

**Zinc is the main component of pennies minted after 1981. How many zinc atoms are present in 20.0 g of Zn?**

**Strategy:** The question asks for atoms of Zu. We cannot convert directly from grams to atoms of zinc. What unit do we need to convert grams of Zn to in order to convert to atoms? What does Avogadro's number represent?

**Solution:** To calculate the number of Zn atoms, we first must convert grams of Zn to moles of Zn. We use the molar mass of zinc as a conversion factor. Once moles of Zn are obtained, we can use Avogadro's number to convert from moles of zinc to atoms of zinc.

$$1 \text{ mol Zn} = 65.39 \text{ g Zn}$$

The conversion factor needed is

$$\frac{1 \text{ mol Zn}}{65.39 \text{ g Zn}}$$

Avogadro's number is the key to the second conversion. We have

$$1 \text{ mol} = 6.022 \times 10^{23} \text{ particles (atoms)}$$

From this equality, we can write two conversion factors.

$$\frac{1 \text{ mol Zn}}{6.022 \times 10^{23} \text{ Zn atoms}} \quad \text{and} \quad \frac{6.022 \times 10^{23} \text{ Zn atoms}}{1 \text{ mol Zn}}$$

The conversion factor on the right is the one we need because it has number of Zn atoms in the numerator, which is the unit we want for the answer.

Let's complete the two conversions in one step.

grams of Zn → moles of Zn → number of Zn atoms

$$? \text{ atoms of Zn} = 20.0 \text{ g Zn} \times \frac{1 \text{ mol Zn}}{65.39 \text{ g Zn}} \times \frac{6.022 \times 10^{23} \text{ Zn atoms}}{1 \text{ mol Zn}} = \mathbf{1.84 \times 10^{23} \text{ Zn atoms}}$$

**Check:** Should 20.0 g of Zn contain fewer than Avogadro's number of atoms? What mass of Zn would contain Avogadro's number of atoms?

**PRACTICE EXERCISE**

**4.** What is the mass (in grams) of $9.09 \times 10^{23}$ atoms of Zn?

**Text Problem: 3.20**

## PROBLEM TYPE 3: CALCULATIONS INVOLVING MOLECULAR MASS

### A. Calculating molecular mass

The molecular mass is simply the sum of the atomic masses (in amu) of all the atoms in the molecule.

**EXAMPLE 3.5**

**Calculate the molecular mass of carbon tetrachloride ($CCl_4$).**

**Strategy:** How do atomic masses of different elements combine to give the molecular mass of a compound?

**Solution:** To calculate the molecular mass of a compound, we need to sum all the atomic masses of the elements in the molecule. For each element, we multiply its atomic mass by the number of atoms of that element in one molecule of the compound. We find atomic masses for the elements in the periodic table (inside front cover of the text).

$$\text{molecular mass } CCl_4 = (\text{mass of C}) + 4(\text{mass of Cl})$$

$$\textbf{molecular mass } \mathbf{CCl_4} = (12.01 \text{ amu}) + 4(35.45 \text{ amu}) = \mathbf{153.8 \text{ amu}}$$

**PRACTICE EXERCISE**

5. Bananas owe their characteristic smell and flavor to the ester, isopentyl acetate [$CH_3COOCH_2CH_2CH(CH_3)_2$]. Calculate the molecular mass of isopentyl acetate.

**Text Problem: 3.24**

## B. Calculating the number of moles in a given amount of a compound

To complete this conversion, the only conversion factor needed is the molar mass in units of g/mol. Remember, the molar mass of a compound (in grams) is numerically equal to its molecular mass (in atomic mass units). For example, the molar mass of $CCl_4$ is 153.8 g/mol, compared to its molecular mass of 153.8 amu.

**EXAMPLE 3.6**

**How many moles of ethane ($C_2H_6$) are present in 50.3 g of ethane?**

**Strategy:** First, calculate the molar mass of ethane. Then, arrange the molar mass as a conversion factor to convert from grams of ethane to moles of ethane.

**Solution:**

$$\text{molar mass of } C_2H_6 = 2(12.01\text{ g}) + 6(1.008\text{ g}) = 30.07\text{ g}$$

Hence, the conversion factor is

$$\frac{1\text{ mol } C_2H_6}{30.07\text{ g } C_2H_6}$$

Using this conversion factor, convert from grams to moles.

$$\textbf{? mol of } \mathbf{C_2H_6} = 50.3\text{ g } C_2H_6 \times \frac{1\text{ mol } C_2H_6}{30.07\text{ g } C_2H_6} = \mathbf{1.67\text{ mol}}$$

**PRACTICE EXERCISE**

6. What is the mass (in grams) of 0.436 moles of ethane ($C_2H_6$)?

**Text Problem: 3.26**

## C. Calculating the number of atoms in a given amount of a compound

Again, this is a unit conversion problem. This calculation is more difficult than the conversions above, because you must convert from *grams of compound* to *moles of compound* to *moles of a particular atom* to *number of atoms*. Sound tough? Let's try an example.

**EXAMPLE 3.7**

**How many carbon atoms are present in 50.3 g of ethane ($C_2H_6$)?**

---

**Strategy:** We started this problem in Example 3.6 when we calculated the moles of ethane in 50.3 g ethane. To continue, we need two additional conversion factors. One should represent the mole ratio between moles of C atoms and moles of ethane molecules. The other conversion factor needed is Avogadro's number.

**Solution:** The two conversion factors needed are:

$$\frac{2 \text{ mol C}}{1 \text{ mol } C_2H_6} \qquad \frac{6.022 \times 10^{23} \text{ C atoms}}{1 \text{ mol C}}$$

You should come up with the following strategy.

$$\text{grams of } C_2H_6 \rightarrow \text{moles of } C_2H_6 \rightarrow \text{moles of C} \rightarrow \text{atoms of C}$$

$$\textbf{? C atoms} = 50.3 \cancel{\text{g}} C_2H_6 \times \frac{1 \cancel{\text{mol}} C_2H_6}{30.07 \cancel{\text{g}} C_2H_6} \times \frac{2 \cancel{\text{mol}} \text{ C}}{1 \cancel{\text{mol}} C_2H_6} \times \frac{6.022 \times 10^{23} \text{ C atoms}}{1 \cancel{\text{mol}} \text{ C}}$$

$$= \mathbf{2.01 \times 10^{24} \text{ C atoms}}$$

**Check:** Does the answer seem reasonable? We have 50.3 g ethane. How many atoms of C would 30.07 g of ethane contain?

**PRACTICE EXERCISE**

7. Glucose, the sugar used by the cells of our bodies for energy, has the molecular formula, $C_6H_{12}O_6$. How many atoms of *carbon* are present in a 3.50 g sample of glucose?

**Text Problem: 3.28**

## PROBLEM TYPE 4: CALCULATIONS INVOLVING PERCENT COMPOSITION

### A. Calculating percent composition of a compound

The *percent composition by mass* is the percent by mass of each element the compound contains. Percent composition is obtained by dividing the mass of each element in 1 mole of the compound by the molar mass of the compound, then multiplying by 100 percent.

$$\textbf{percent by mass of each element} = \frac{\text{mass of element in 1 mol of compound}}{\text{molar mass of compound}} \times 100\%$$

**EXAMPLE 3.8**

**Calculate the percent composition by mass of all the elements in sodium bicarbonate, $NaHCO_3$.**

---

**Strategy:** First, calculate the molar mass of sodium bicarbonate. Then, calculate the percent by mass of each element.

**Solution:**

$$\text{molar mass sodium bicarbonate} = 22.99 \text{ g} + 1.008 \text{ g} + 12.01 \text{ g} + 3(16.00 \text{ g}) = 84.01 \text{ g}$$

$$\%\text{Na} = \frac{22.99\text{ g}}{84.01\text{ g}} \times 100\% = \mathbf{27.37\%}$$

$$\%\text{H} = \frac{1.008\text{ g}}{84.01\text{ g}} \times 100\% = \mathbf{1.200\%}$$

$$\%\text{C} = \frac{12.01\text{ g}}{84.01\text{ g}} \times 100\% = \mathbf{14.30\%}$$

$$\%\text{O} = \frac{3(16.00\text{ g})}{84.01\text{ g}} \times 100\% = \mathbf{57.14\%}$$

**Tip:** You can check your work by making sure that the mass percents of all the elements added together equals 100%. Checking above, 27.37% + 1.200% + 14.30% + 57.14% = 100.01% ≈ 100%.

**PRACTICE EXERCISE**

8. Cinnamic alcohol is used mainly in perfumes, particularly for soaps and cosmetics. Its molecular formula is $C_9H_{10}O$. Calculate the percent composition by mass of *hydrogen* in cinnamic alcohol.

**Text Problems: 3.38**, 3.40

## B. Determining empirical formula from percent composition

The procedure used above to calculate the percent composition of a compound can be reversed. Given the percent composition by mass of a compound, you can determine the empirical formula of the compound.

**EXAMPLE 3.9**

**Dieldrin, like DDT, is an insecticide that contains only C, H, Cl, and O. It is composed of 37.84 percent C, 2.12 percent H, 55.84 percent Cl, and 4.20 percent O. Determine its empirical formula.**

**Strategy:** In a chemical formula, the subscripts represent the ratio of the number of moles of each element that combine to form the compound. Therefore, we need to convert from mass percent to moles in order to determine the empirical formula. If we assume an exactly 100 g sample of the compound, do we know the mass of each element in the compound? How do we then convert from grams to moles?

**Solution:** If we have 100 g of the compound, then each percentage can be converted directly to grams. In this sample, there will be 37.84 g of C, 2.12 g of H, 55.84 g Cl, and 4.20 g of O. Because the subscripts in the formula represent a mole ratio, we need to convert the grams of each element to moles. The conversion factor needed is the molar mass of each element. Let *n* represent the number of moles of each element so that

$$n_C = 37.84\,\cancel{g}\,\text{C} \times \frac{1\text{ mol C}}{12.01\,\cancel{g}\,\text{C}} = \mathbf{3.151\text{ mol C}}$$

$$n_H = 2.12\,\cancel{g}\,\text{H} \times \frac{1\text{ mol H}}{1.008\,\cancel{g}\,\text{H}} = \mathbf{2.10\text{ mol H}}$$

$$n_{Cl} = 55.84\,\cancel{g}\,\text{Cl} \times \frac{1\text{ mol Cl}}{35.45\,\cancel{g}\,\text{Cl}} = \mathbf{1.575\text{ mol Cl}}$$

$$n_O = 4.20\ \cancel{g}\ O \times \frac{1\ \text{mol O}}{16.00\ \cancel{g}\ O} = \mathbf{0.263\ mol\ O}$$

Thus, we arrive at the formula $C_{3.151}H_{2.10}Cl_{1.575}O_{0.263}$, which gives the identity and the ratios of atoms present. However, chemical formulas are written with whole numbers.

Try to convert to whole numbers by dividing all the subscripts by the smallest subscript.

$$\textbf{C}: \frac{3.151}{0.263} = 12.0 \qquad \textbf{H}: \frac{2.10}{0.263} = 7.98 \approx 8 \qquad \textbf{Cl}: \frac{1.575}{0.263} = 5.99 \approx 6 \qquad \textbf{O}: \frac{0.263}{0.263} = 1$$

This gives us the empirical for dieldrin, $\mathbf{C_{12}H_8Cl_6O}$.

**Check:** Are the subscripts in $C_{12}H_8Cl_6O$ reduced to the smallest whole numbers?

**Tip:** It's not always this easy. Dividing by the smallest subscript often does not give all whole numbers. If this is the case, you must multiply all the subscripts by some *integer* to come up with whole number subscripts. Try the practice exercise below.

**PRACTICE EXERCISE**

9. The substance responsible for the green color on the yolk of a boiled egg is composed of 53.58 percent Fe and 46.42 percent S. Determine its empirical formula.

**Text Problems:** 3.44, **3.50**, 3.52

## C. Calculating mass from percent composition

*Step 1:* Convert the mass percentage to a fraction. For example, if the mass percent of an element in a compound were 54.73 percent, you would convert this to 54.73/100 = 0.5473.

*Step 2:* Multiply the fraction by the total mass of the compound. This gives the mass of the particular element in the compound.

**EXAMPLE 3.10**

**Calculate the mass of carbon in exactly 10 g of glucose ($C_6H_{12}O_6$).**

**Strategy:** Glucose is composed of C, H, and O. The mass due to C is based on its percentage by mass in the compound. How do we calculate mass percent of an element?

**Solution:** First, we must find the mass % of carbon in $C_6H_{12}O_6$. Then, we convert this percentage to a fraction and multiply by the mass of the compound (10 g), to find the mass of carbon in 10 g of $C_6H_{12}O_6$.

The percent by mass of carbon in glucose, is calculated as follows:

$$\text{mass \% C} = \frac{\text{mass of C in 1 mol of glucose}}{\text{molar mass of glucose}} \times 100\%$$

$$\textbf{mass \% C} = \frac{6(12.01 \text{ g})}{180.16 \text{ g}} \times 100\% = \textbf{40.00\% C}$$

Converting this percentage to a fraction, we obtain 40.00/100 = **0.4000**

Next, multiply the fraction by the total mass of the compound.

$$\textbf{? g C in 10 g glucose} = (0.4000)(10 \text{ g}) = \textbf{4.000 g C}$$

**Check:** Note that the mass percent of C is 40 percent. 40% of 10 g is 4 g.

**PRACTICE EXERCISE**

**10.** Calculate the mass of hydrogen in exactly 10 grams of glucose ($C_6H_{12}O_6$).

**Text Problem: 3.42**

## PROBLEM TYPE 5: EXPERIMENTAL DETERMINATION OF EMPIRICAL FORMULAS

See Section 3.6 of your text for a description of the experimental setup. To solve this type of problem, you must recognize that all of the carbon in the sample is converted to $CO_2$ and all the hydrogen in the sample is converted to $H_2O$. Then, you can calculate the mass of C in $CO_2$ and the mass of H in $H_2O$. Finally, you can calculate the mass of oxygen by difference, if necessary.

**EXAMPLE 3.11**

**When a 0.761-g sample of a compound containing only carbon and hydrogen is burned in an apparatus with $CO_2$ and $H_2O$ absorbers, 2.23 g $CO_2$ and 1.37 g $H_2O$ are collected. Determine the empirical formula of the compound.**

**Strategy:** Calculate the moles of C in 2.23 g $CO_2$, and the moles of H in 1.37 g $H_2O$. In this problem, we do not need to convert to grams of C and H, because there are no other elements in the compound. To calculate the moles of each component, you need the molar masses and the correct mole ratio.

You should come up with the following strategy.

$$\text{g } CO_2 \rightarrow \text{mol } CO_2 \rightarrow \text{mol C}$$

Next, determine the smallest whole number ratio in which the elements combine.

**Solution:**

$$? \text{ mol C} = 2.23 \text{ g } CO_2 \times \frac{1 \text{ mol } CO_2}{44.01 \text{ g } CO_2} \times \frac{1 \text{ mol C}}{1 \text{ mol } CO_2} = 0.507 \text{ mol C}$$

Similarly,

$$? \text{ mol H} = 1.37 \text{ g } H_2O \times \frac{1 \text{ mol } H_2O}{18.02 \text{ g } H_2O} \times \frac{2 \text{ mol H}}{1 \text{ mol } H_2O} = 0.152 \text{ mol H}$$

Thus, we arrive at the formula $C_{0.0507}H_{0.152}$, which gives the identity and the ratios of atoms present. However, chemical formulas are written with whole numbers.

Try to convert to whole numbers by dividing all the subscripts by the smallest subscript.

$$\mathbf{C:}\ \frac{0.0507}{0.0507} = 1.00 \qquad \mathbf{H:}\ \frac{0.152}{0.0507} = 3.00$$

This gives the empirical formula, $\mathbf{CH_3}$.

**PRACTICE EXERCISE**

11. Diethyl ether, commonly known as "ether", was used as an anesthetic for many years. Diethyl ether contains C, H, and O. When a 1.45 g sample of ether is burned in an apparatus such as that shown in Figure 3.6 of the text, 2.77 g of $CO_2$ and 1.70 g of $H_2O$ are collected. Determine the empirical formula of diethyl ether.

**Text Problem:** 3.148

## PROBLEM TYPE 6: DETERMINING THE MOLECULAR FORMULA OF A COMPOUND

To determine the molecular formula of a compound, we must know both the *approximate* molar mass and the empirical formula of the compound. The molecular formula will either be equal to the empirical formula or be some integral multiple of it. Thus, the molar mass divided by the empirical mass will be an integer greater than or equal to one.

$$\frac{\text{molar mass}}{\text{empirical molar mass}} \geq 1 \text{ (integer values)}$$

**EXAMPLE 3.12**

**A mass spectrum obtained on the compound in Example 3.11, shows its molecular mass to be about 31 g/mol. What is its molecular formula?**

**Strategy:** First, determine the empirical formula. Then compare the molar mass to the empirical molar mass to determine the molecular formula.

**Solution:** The empirical formula was determined in the previous example to be $\mathbf{CH_3}$.

Next, calculate the empirical molar mass.

$$\text{empirical molar mass} = 12.01\text{ g} + 3(1.008\text{ g}) = 15.03\text{ g/mol}$$

Determine the number of ($CH_3$) units present in the molecular formula. This number is found by taking the ratio

$$\frac{\text{molar mass}}{\text{empirical molar mass}} = \frac{31\text{ g}}{15.03\text{ g}} = 2.1 \approx 2$$

Thus, there are two $CH_3$ units in each molecule of the compound, so the molecular formula is $(CH_3)_2$, or $\mathbf{C_2H_6}$.

**PRACTICE EXERCISE**

12. In Example 3.9, the empirical formula of dieldrin was determined to be $C_{12}H_8Cl_6O$. If the molar mass of dieldrin is 381 ± 10 g/mol, what is the molecular formula of dieldrin?

**Text Problems:** 3.48, 3.52

## PROBLEM TYPE 7: CALCULATING THE AMOUNTS OF REACTANTS AND PRODUCTS

These types of problems are dimensional analysis problems. You must always remember to start this type of problem with a balanced chemical equation. The typical approach is given below. See Section 3.8 of your text for a step-by-step method.

Amount of reactant (expressed in grams, kg, etc.) → Moles of reactant → Moles of product → Amount of product (expressed in grams, kg, etc.)

> **Tip:** Always try to be flexible when solving problems. Most problems of this type will follow an approach similar to the one above, but you may have to modify it sometimes.

**EXAMPLE 3.13**

**Sulfur dioxide can be removed from stack gases by reaction with quicklime (CaO):**

$$SO_2(g) + CaO(s) \longrightarrow CaSO_3(s)$$

**If 975 kg of $SO_2$ are to be removed from stack gases by the above reaction, how many kilograms of CaO are required?**

**Strategy:** We compare $SO_2$ and CaO based on the *mole ratio* in the balanced equation. Before we can determine moles of CaO required, we need to convert to moles of $SO_2$. What conversion factor is needed to convert from grams of $SO_2$ to moles of $SO_2$? Once moles of CaO are obtained, another conversion factor is needed to convert from moles of CaO to grams of CaO.

**Solution:** The molar mass of $SO_2$ will allow us to convert from grams of $SO_2$ to moles of $SO_2$. The molar mass of $SO_2 = 32.07\text{ g} + 2(16.00\text{ g}) = 64.07\text{ g}$. The balanced equation is given, so the mole ratio between $SO_2$ and CaO is known, that is, 1 mole $SO_2 \simeq$ 1 mole CaO. Finally, the molar mass of CaO will convert moles of CaO to grams of CaO. This sequence of conversions is summarized as follows:

$$\text{kg } SO_2 \rightarrow \text{g } SO_2 \rightarrow \text{moles } SO_2 \rightarrow \text{moles CaO} \rightarrow \text{g CaO} \rightarrow \text{kg CaO}$$

$$\textbf{? kg CaO} = 975\text{ kg } SO_2 \times \frac{1000\text{ g } SO_2}{1\text{ kg } SO_2} \times \frac{1\text{ mol } SO_2}{64.07\text{ g } SO_2} \times \frac{1\text{ mol CaO}}{1\text{ mol } SO_2} \times \frac{56.08\text{ g CaO}}{1\text{ mol CaO}} \times \frac{1\text{ kg CaO}}{1000\text{ g CaO}}$$

$$= \textbf{853 kg CaO}$$

> **Tip:** Notice that the approach followed was a slight modification of the flow diagram given above. We went from mass of one reactant, to moles of that reactant, to moles of a second reactant, and finally to mass of second reactant.

**PRACTICE EXERCISE**

**13.** Carbon dioxide in the air of a spacecraft can be removed by its reaction with a lithium hydroxide solution.

$$CO_2(g) + 2LiOH(aq) \longrightarrow Li_2CO_3(aq) + H_2O(l)$$

On average, a person will exhale about 1 kg of $CO_2$/day. How many kilograms of LiOH are required to react with 1.0 kg of $CO_2$?

| |
|---|
| **Text Problems: 3.64**, 3.66, 3.70, **3.72**, 3.74, 3.76, 3.78 |

## PROBLEM TYPE 8: LIMITING REAGENT CALCULATIONS

When a chemist carries out a reaction, the reactants are usually not present in exact **stoichiometric amounts**, that is, in the proportions indicated by the balanced equation. The reactant used up first in a reaction is called the **limiting reagent**. When this reactant is used up, no more product can be formed.

Typically, the only difference between this type of problem and Problem Type 7, Calculating the Amounts of Reactants and Products, is that you must first determine which reactant is the limiting reagent.

**EXAMPLE 3.14**

**Phosphine ($PH_3$) burns in oxygen ($O_2$) to produce phosphorus pentoxide and water.**

$$2PH_3(g) + 4O_2(g) \longrightarrow P_2O_5(s) + 3H_2O(l)$$

**How many grams of $P_2O_5$ will be produced when 17.0 g of phosphine are reacted with 16.0 g of $O_2$?**

**Strategy:** Note that this reaction gives the amounts of both reactants, so it is likely to be a limiting reagent problem. The reactant that produces fewer moles of product is the limiting reagent because it limits the amount of product that can be produced. How do we convert from the amount of reactant to amount of product? Perform this calculation for each reactant, and then compare the moles of product, $P_2O_5$, formed by the given amounts of $PH_3$ and $O_2$ to determine which reactant is the limiting reagent.

**Solution:** We carry out two separate calculations. First, starting with 17.0 g $PH_3$, we calculate the number of moles of $P_2O_5$ that could be produced if all the $PH_3$ reacted. We complete the following conversions.

$$\text{grams of } PH_3 \rightarrow \text{moles of } PH_3 \rightarrow \text{moles of } P_2O_5$$

Combining these two conversions into one calculation, we write

$$? \text{ mol } P_2O_5 = 17.0 \text{ g } PH_3 \times \frac{1 \text{ mol } PH_3}{33.99 \text{ g } PH_3} \times \frac{1 \text{ mol } P_2O_5}{2 \text{ mol } PH_3} = 0.250 \text{ mol } P_2O_5$$

Second, starting with 16.0 g of $O_2$, we complete similar conversions.

$$\text{grams of } O_2 \rightarrow \text{moles of } O_2 \rightarrow \text{moles of } P_2O_5$$

Combining these two conversions into one calculation, we write

$$? \text{ mol } P_2O_5 = 16.0 \text{ g } O_2 \times \frac{1 \text{ mol } O_2}{32.0 \text{ g } O_2} \times \frac{1 \text{ mol } P_2O_5}{4 \text{ mol } O_2} = 0.125 \text{ mol } P_2O_5$$

The initial amount of $O_2$ limits the amount of product that can be formed; therefore, it is the limiting reagent.

The problem asks for grams of $P_2O_5$ produced. We already know the moles of $P_2O_5$ produced, 0.125 mole. Use the molar mass of $P_2O_5$ as a conversion factor to convert to grams.

$$? \text{ g } P_2O_5 = 0.125 \text{ mol } P_2O_5 \times \frac{141.94 \text{ g } P_2O_5}{1 \text{ mol } P_2O_5} = \mathbf{17.7 \text{ g } P_2O_5}$$

**Check:** Does your answer seem reasonable? 0.125 mole of product is formed. What is the mass of 1 mole of $P_2O_5$?

**PRACTICE EXERCISE**

**14.** Iron can be produced by reacting iron ore with carbon. The iron produced can then be used to make steel. The reaction is

$$2Fe_2O_3(s) + 3C(s) \xrightarrow{\text{heat}} 4Fe(l) + 3CO_2(g)$$

(a) How many grams of Fe can be produced from a mixture of 200.0 g of $Fe_2O_3$ and 300.0 g C?

(b) How many grams of excess reagent will remain after the reaction ceases?

**Text Problems:** 3.82, 3.84, 3.86

## PROBLEM TYPE 9: CALCULATING THE PERCENT YIELD OF A REACTION

The **theoretical yield** is the amount of product that would result if all the limiting reagent reacted. This is the maximum obtainable yield predicted by the balanced equation. However, the amount of product obtained is almost always less than the theoretical yield. The **actual yield** is the quantity of product that actually results from a reaction.

To determine the efficiency of a reaction, chemists often calculate the **percent yield**, which describes the proportion of the actual yield to the theoretical yield. The percent yield is calculated as follows:

$$\% \text{ yield} = \frac{\text{actual yield}}{\text{theoretical yield}} \times 100\%$$

**EXAMPLE 3.15**

**In Example 3.14, the theoretical yield of $P_2O_5$ was determined to be 17.7 g. If only 12.6 g of $P_2O_5$ are actually obtained, what is the percent yield of the reaction?**

**Solution:**

$$\% \text{ yield} = \frac{\text{actual yield}}{\text{theoretical yield}} \times 100\%$$

$$\mathbf{\% \text{ yield}} = \frac{12.6 \text{ g}}{17.7 \text{ g}} \times 100\% = \mathbf{71.2\%}$$

**PRACTICE EXERCISE**

**15.** Refer back to Practice Exercise 14 to answer this question. If the actual yield of Fe is 110 g, what is the percent yield of Fe?

**Text Problems:** 3.90, 3.92, 3.94

## ANSWERS TO PRACTICE EXERCISES

| | | |
|---|---|---|
| **1.** $^{11}B$ | **2.** 1.97 moles Cu | **3.** $7.951 \times 10^{-23}$ g/Ti atom |
| **4.** 98.7 g Zn | **5.** 130.18 amu | **6.** 13.1 g ethane |
| **7.** $7.02 \times 10^{22}$ C atoms | **8.** 7.513 percent H by mass | **9.** $Fe_2S_3$ |
| **10.** 0.67 g H | **11.** $C_2H_6O$ | **12.** $C_{12}H_8Cl_6O$ |
| **13.** 1.1 kg LiOH | **14.** **(a)** 139.9 g Fe<br>**(b)** 277 g C | **15.** 78.6 percent yield |

## SOLUTIONS TO SELECTED TEXT PROBLEMS

**3.6** **Strategy:** Each isotope contributes to the average atomic mass based on its relative abundance. Multiplying the mass of an isotope by its fractional abundance (not percent) will give the contribution to the average atomic mass of that particular isotope.

It would seem that there are two unknowns in this problem, the fractional abundance of $^6Li$ and the fractional abundance of $^7Li$. However, these two quantities are not independent of each other; they are related by the fact that they must sum to 1. Start by letting $x$ be the fractional abundance of $^6Li$. Since the sum of the two abundances must be 1, we can write

$$\text{Abundance } ^7\text{Li} = (1 - x)$$

**Solution:**

$$\textbf{Average atomic mass of Li} = 6.941 \text{ amu} = x(6.0151 \text{ amu}) + (1 - x)(7.0160 \text{ amu})$$

$$6.941 = -1.0009x + 7.0160$$

$$1.0009x = 0.075$$

$$\boldsymbol{x = 0.075}$$

$x = 0.075$ corresponds to a natural abundance of $^6Li$ of **7.5 percent**. The natural abundance of $^7Li$ is $(1 - x) = 0.925$ or **92.5 percent**.

**3.8** The unit factor required is $\left(\dfrac{6.022 \times 10^{23}\ \text{amu}}{1\ \text{g}}\right)$

$$\textbf{? amu} = 8.4\ \cancel{\text{g}} \times \frac{6.022 \times 10^{23}\ \text{amu}}{1\ \cancel{\text{g}}} = \mathbf{5.1 \times 10^{24}\ amu}$$

**3.12** The thickness of the book in miles would be:

$$\frac{0.0036\ \cancel{\text{in}}}{1\ \cancel{\text{page}}} \times \frac{1\ \cancel{\text{ft}}}{12\ \cancel{\text{in}}} \times \frac{1\ \text{mi}}{5280\ \cancel{\text{ft}}} \times (6.022 \times 10^{23}\ \cancel{\text{pages}}) = 3.42 \times 10^{16}\ \text{mi}$$

The distance, in miles, traveled by light in one year is:

$$1.00\ \cancel{\text{yr}} \times \frac{365\ \cancel{\text{day}}}{1\ \cancel{\text{yr}}} \times \frac{24\ \cancel{\text{h}}}{1\ \cancel{\text{day}}} \times \frac{3600\ \cancel{\text{s}}}{1\ \cancel{\text{h}}} \times \frac{3.00 \times 10^{8}\ \cancel{\text{m}}}{1\ \cancel{\text{s}}} \times \frac{1\ \text{mi}}{1609\ \cancel{\text{m}}} = 5.88 \times 10^{12}\ \text{mi}$$

The thickness of the book in light-years is:

$$(3.42 \times 10^{16}\ \cancel{\text{mi}}) \times \frac{1\ \text{light-yr}}{5.88 \times 10^{12}\ \cancel{\text{mi}}} = \mathbf{5.8 \times 10^{3}\ light\text{-}yr}$$

It will take light $5.8 \times 10^{3}$ years to travel from the first page to the last one!

**3.14** $$(6.00 \times 10^{9}\ \text{Co}\ \cancel{\text{atoms}}) \times \frac{1\ \text{mol Co}}{6.022 \times 10^{23}\ \text{Co}\ \cancel{\text{atoms}}} = \mathbf{9.96 \times 10^{-15}\ mol\ Co}$$

**3.16** **Strategy:** We are given moles of gold and asked to solve for grams of gold. What conversion factor do we need to convert between moles and grams? Arrange the appropriate conversion factor so moles cancel, and the unit grams is obtained for the answer.

**Solution:** The conversion factor needed to covert between moles and grams is the molar mass. In the periodic table (see inside front cover of the text), we see that the molar mass of Au is 197.0 g. This can be expressed as

$$1\ \text{mol Au} = 197.0\ \text{g Au}$$

From this equality, we can write two conversion factors.

$$\frac{1\ \text{mol Au}}{197.0\ \text{g Au}} \quad \text{and} \quad \frac{197.0\ \text{g Au}}{1\ \text{mol Au}}$$

The conversion factor on the right is the correct one. Moles will cancel, leaving the unit grams for the answer. We write

$$? \text{ g Au} = 15.3 \text{ mol Au} \times \frac{197.0 \text{ g Au}}{1 \text{ mol Au}} = \mathbf{3.01 \times 10^3 \text{ g Au}}$$

**Check:** Does a mass of 3010 g for 15.3 moles of Au seem reasonable? What is the mass of 1 mole of Au?

3.18 (a)

**Strategy:** We can look up the molar mass of arsenic (As) on the periodic table (74.92 g/mol). We want to find the mass of a single atom of arsenic (unit of g/atom). Therefore, we need to convert from the unit mole in the denominator to the unit atom in the denominator. What conversion factor is needed to convert between moles and atoms? Arrange the appropriate conversion factor so mole in the denominator cancels, and the unit atom is obtained in the denominator.

**Solution:** The conversion factor needed is Avogadro's number. We have

$$1 \text{ mol} = 6.022 \times 10^{23} \text{ particles (atoms)}$$

From this equality, we can write two conversion factors.

$$\frac{1 \text{ mol As}}{6.022 \times 10^{23} \text{ As atoms}} \quad \text{and} \quad \frac{6.022 \times 10^{23} \text{ As atoms}}{1 \text{ mol As}}$$

The conversion factor on the left is the correct one. Moles will cancel, leaving the unit atoms in the denominator of the answer.

We write

$$? \text{ g / As atom} = \frac{74.92 \text{ g As}}{1 \text{ mol As}} \times \frac{1 \text{ mol As}}{6.022 \times 10^{23} \text{ As atoms}} = \mathbf{1.244 \times 10^{-22} \text{ g/As atom}}$$

**(b)** Follow same method as part (a).

$$? \text{ g / Ni atom} = \frac{58.69 \text{ g Ni}}{1 \text{ mol Ni}} \times \frac{1 \text{ mol Ni}}{6.022 \times 10^{23} \text{ Ni atoms}} = \mathbf{9.746 \times 10^{-23} \text{ g / Ni atom}}$$

**Check:** Should the mass of a single atom of As or Ni be a very small mass?

3.20 **Strategy:** The question asks for atoms of Cu. We cannot convert directly from grams to atoms of copper. What unit do we need to convert grams of Cu to in order to convert to atoms? What does Avogadro's number represent?

**Solution:** To calculate the number of Cu atoms, we first must convert grams of Cu to moles of Cu. We use the molar mass of copper as a conversion factor. Once moles of Cu are obtained, we can use Avogadro's number to convert from moles of copper to atoms of copper.

$$1 \text{ mol Cu} = 63.55 \text{ g Cu}$$

The conversion factor needed is

$$\frac{1 \text{ mol Cu}}{63.55 \text{ g Cu}}$$

Avogadro's number is the key to the second conversion. We have

$$1 \text{ mol} = 6.022 \times 10^{23} \text{ particles (atoms)}$$

From this equality, we can write two conversion factors.

$$\frac{1 \text{ mol Cu}}{6.022 \times 10^{23} \text{ Cu atoms}} \quad \text{and} \quad \frac{6.022 \times 10^{23} \text{ Cu atoms}}{1 \text{ mol Cu}}$$

The conversion factor on the right is the one we need because it has the number of Cu atoms in the numerator, which is the unit we want for the answer.

Let's complete the two conversions in one step.

grams of Cu → moles of Cu → number of Cu atoms

$$\textbf{? atoms of Cu} = 0.063 \text{ g Cu} \times \frac{1 \text{ mol Cu}}{63.55 \text{ g Cu}} \times \frac{6.022 \times 10^{23} \text{ Cu atoms}}{1 \text{ mol Cu}} = \mathbf{6.0 \times 10^{20} \textbf{ Cu atoms}}$$

**Check:** Should 0.063 g of Cu contain fewer than Avogadro's number of atoms? What mass of Cu would contain Avogadro's number of atoms?

**3.22**

$$2 \text{ Pb atoms} \times \frac{1 \text{ mol Pb}}{6.022 \times 10^{23} \text{ Pb atoms}} \times \frac{207.2 \text{ g Pb}}{1 \text{ mol Pb}} = 6.881 \times 10^{-22} \text{ g Pb}$$

$$(5.1 \times 10^{-23} \text{ mol He}) \times \frac{4.003 \text{ g He}}{1 \text{ mol He}} = 2.0 \times 10^{-22} \text{ g He}$$

**2 atoms of lead** have a greater mass than $5.1 \times 10^{-23}$ mol of helium.

**3.24** **Strategy:** How do molar masses of different elements combine to give the molar mass of a compound?

**Solution:** To calculate the molar mass of a compound, we need to sum all the molar masses of the elements in the molecule. For each element, we multiply its molar mass by the number of moles of that element in one mole of the compound. We find molar masses for the elements in the periodic table (inside front cover of the text).

**(a)** **molar mass $Li_2CO_3$** = 2(6.941 g) + 12.01 g + 3(16.00 g) = **73.89 g**

**(b)** **molar mass $CS_2$** = 12.01 g + 2(32.07 g) = **76.15 g**

**(c)** **molar mass $CHCl_3$** = 12.01 g + 1.008 g + 3(35.45 g) = **119.37 g**

**(d)** **molar mass $C_6H_8O_6$** = 6(12.01 g) + 8(1.008 g) + 6(16.00 g) = **176.12 g**

**(e)** **molar mass $KNO_3$** = 39.10 g + 14.01 g + 3(16.00 g) = **101.11 g**

**(f)** **molar mass $Mg_3N_2$** = 3(24.31 g) + 2(14.01 g) = **100.95 g**

**3.26** **Strategy:** We are given grams of ethane and asked to solve for molecules of ethane. We cannot convert directly from grams of ethane to molecules of ethane. What unit do we need to obtain first before we can convert to molecules? How should Avogadro's number be used here?

**Solution:** To calculate number of ethane molecules, we first must convert grams of ethane to moles of ethane. We use the molar mass of ethane as a conversion factor. Once moles of ethane are obtained, we can use Avogadro's number to convert from moles of ethane to molecules of ethane.

$$\text{molar mass of } C_2H_6 = 2(12.01 \text{ g}) + 6(1.008 \text{ g}) = 30.068 \text{ g}$$

The conversion factor needed is

$$\frac{1 \text{ mol } C_2H_6}{30.068 \text{ g } C_2H_6}$$

Avogadro's number is the key to the second conversion. We have

$$1 \text{ mol} = 6.022 \times 10^{23} \text{ particles (molecules)}$$

From this equality, we can write the conversion factor:

$$\frac{6.022 \times 10^{23} \text{ ethane molecules}}{1 \text{ mol ethane}}$$

Let's complete the two conversions in one step.

grams of ethane → moles of ethane → number of ethane molecules

$$\textbf{? molecules of } \mathbf{C_2H_6} = 0.334 \text{ g } C_2H_6 \times \frac{1 \text{ mol } C_2H_6}{30.068 \text{ g } C_2H_6} \times \frac{6.022 \times 10^{23} \text{ } C_2H_6 \text{ molecules}}{1 \text{ mol } C_2H_6}$$

$$= \mathbf{6.69 \times 10^{21}\ C_2H_6\ molecules}$$

**Check:** Should 0.334 g of ethane contain fewer than Avogadro's number of molecules? What mass of ethane would contain Avogadro's number of molecules?

**3.28** **Strategy:** We are asked to solve for the number of C, S, H, and O atoms in $7.14 \times 10^3$ g of dimethyl sulfoxide (DMSO). We cannot convert directly from grams DMSO to atoms. What unit do we need to obtain first before we can convert to atoms? How should Avogadro's number be used here? How many atoms of C, S, H, or O are in 1 molecule of DMSO?

**Solution:** Let's first calculate the number of C atoms in $7.14 \times 10^3$ g of dimethyl sulfoxide. First, we must convert grams of DMSO to number of molecules of DMSO. This calculation is similar to Problem 3.26. The molecular formula of DMSO shows there are two C atoms in one DMSO molecule, which will allow us to convert to atoms of C. We need to perform three conversions:

grams of DMSO → moles of DMSO → molecules of DMSO → atoms of C

The conversion factors needed for each step are: 1) the molar mass of DMSO, 2) Avogadro's number, and 3) the number of C atoms in 1 molecule of DMSO.

We complete the three conversions in one calculation.

$$7.14 \times 10^3 \text{ g DMSO} \times \frac{1 \text{ mol DMSO}}{78.14 \text{ g DMSO}} \times \frac{6.022 \times 10^{23} \text{ DMSO molecules}}{1 \text{ mol DMSO}} \times \frac{2 \text{ C atoms}}{1 \text{ molecule DMSO}}$$

$$= \mathbf{1.10 \times 10^{26} \text{ C atoms}}$$

The above method utilizes the ratio of molecules (DMSO) to atoms (carbon). We can also solve the problem by reading the formula as the ratio of moles of DMSO to moles of carbon by using the following conversions:

grams of DMSO → moles of DMSO → moles of C → atoms of C

Try it.

**Check:** Does the answer seem reasonable? We have $7.14 \times 10^3$ g DMSO. How many atoms of C would 78.14 g of DMSO contain?

We could calculate the number of atoms of the remaining elements in the same manner, or we can use the atom ratios from the molecular formula. The sulfur atom to carbon atom ratio in a DMSO molecule is 1:2, the hydrogen atom to carbon atom ratio is 6:2 or 3:1, and the oxygen atom to carbon atom ratio is 1:2.

$$\textbf{? atoms of S} = (1.10 \times 10^{26} \text{ C atoms}) \times \frac{1 \text{ S atom}}{2 \text{ C atoms}} = \mathbf{5.50 \times 10^{25} \text{ S atoms}}$$

$$\textbf{? atoms of H} = (1.10 \times 10^{26} \text{ C atoms}) \times \frac{3 \text{ H atoms}}{1 \text{ C atom}} = \mathbf{3.30 \times 10^{26} \text{ H atoms}}$$

$$\textbf{? atoms of O} = (1.10 \times 10^{26} \text{ C atoms}) \times \frac{1 \text{ O atom}}{2 \text{ C atoms}} = \mathbf{5.50 \times 10^{25} \text{ O atoms}}$$

**3.30** Mass of water $= 2.56 \text{ mL} \times \dfrac{1.00 \text{ g}}{1.00 \text{ mL}} = 2.56 \text{ g}$

Molar mass of $H_2O$ = (16.00 g) + 2(1.008 g) = 18.016 g/mol

$$\textbf{? } \mathbf{H_2O} \textbf{ molecules} = 2.56 \text{ g } H_2O \times \frac{1 \text{ mol } H_2O}{18.016 \text{ g } H_2O} \times \frac{6.022 \times 10^{23} \text{ molecules } H_2O}{1 \text{ mol } H_2O}$$

$$= \mathbf{8.56 \times 10^{22} \text{ molecules}}$$

**3.34** Since there are two hydrogen isotopes, they can be paired in three ways: $^1H-^1H$, $^1H-^2H$, and $^2H-^2H$. There will then be three choices for each sulfur isotope. We can make a table showing all the possibilities (masses in amu):

| | $^{32}S$ | $^{33}S$ | $^{34}S$ | $^{36}S$ |
|---|---|---|---|---|
| $^1H_2$ | 34 | 35 | 36 | 38 |
| $^1H^2H$ | 35 | 36 | 37 | 39 |
| $^2H_2$ | 36 | 37 | 38 | 40 |

There will be **seven peaks** of the following mass numbers: 34, 35, 36, 37, 38, 39, and 40.

Very accurate (and expensive!) mass spectrometers can detect the mass difference between two $^{1}H$ and one $^{2}H$. How many peaks would be detected in such a "high resolution" mass spectrum?

3.38 **Strategy:** Recall the procedure for calculating a percentage. Assume that we have 1 mole of $CHCl_3$. The percent by mass of each element (C, H, and Cl) is given by the mass of that element in 1 mole of $CHCl_3$ divided by the molar mass of $CHCl_3$, then multiplied by 100 to convert from a fractional number to a percentage.

**Solution:** The molar mass of $CHCl_3$ = 12.01 g/mol + 1.008 g/mol + 3(35.45 g/mol) = 119.4 g/mol. The percent by mass of each of the elements in $CHCl_3$ is calculated as follows:

$$\%C = \frac{12.01 \text{ g/mol}}{119.4 \text{ g/mol}} \times 100\% = \mathbf{10.06\%}$$

$$\%H = \frac{1.008 \text{ g/mol}}{119.4 \text{ g/mol}} \times 100\% = \mathbf{0.8442\%}$$

$$\%Cl = \frac{3(35.45) \text{ g/mol}}{119.4 \text{ g/mol}} \times 100\% = \mathbf{89.07\%}$$

**Check:** Do the percentages add to 100%? The sum of the percentages is (10.06% + 0.8442% + 89.07%) = 99.97%. The small discrepancy from 100% is due to the way we rounded off.

3.40

| | Compound | Molar mass (g) | N% by mass |
|---|---|---|---|
| (a) | $(NH_2)_2CO$ | 60.06 | $\frac{2(14.01 \text{ g})}{60.06 \text{ g}} \times 100\% = 46.65\%$ |
| (b) | $NH_4NO_3$ | 80.05 | $\frac{2(14.01 \text{ g})}{80.05 \text{ g}} \times 100\% = 35.00\%$ |
| (c) | $HNC(NH_2)_2$ | 59.08 | $\frac{3(14.01 \text{ g})}{59.08 \text{ g}} \times 100\% = 71.14\%$ |
| (d) | $NH_3$ | 17.03 | $\frac{14.01 \text{ g}}{17.03 \text{ g}} \times 100\% = 82.27\%$ |

**Ammonia, $NH_3$,** is the richest source of nitrogen on a mass percentage basis.

3.42 **Strategy:** Tin(II) fluoride is composed of Sn and F. The mass due to F is based on its percentage by mass in the compound. How do we calculate mass percent of an element?

**Solution:** First, we must find the mass % of fluorine in $SnF_2$. Then, we convert this percentage to a fraction and multiply by the mass of the compound (24.6 g), to find the mass of fluorine in 24.6 g of $SnF_2$.

The percent by mass of fluorine in tin(II) fluoride is calculated as follows:

$$\text{mass \% F} = \frac{\text{mass of F in 1 mol } SnF_2}{\text{molar mass of } SnF_2} \times 100\%$$

$$= \frac{2(19.00\text{ g})}{156.7\text{ g}} \times 100\% = 24.25\%\text{ F}$$

Converting this percentage to a fraction, we obtain 24.25/100 = 0.2425.

Next, multiply the fraction by the total mass of the compound.

$$\textbf{? g F in 24.6 g SnF}_2 = (0.2425)(24.6\text{ g}) = \textbf{5.97 g F}$$

**Check:** As a ball-park estimate, note that the mass percent of F is roughly 25 percent, so that a quarter of the mass should be F. One quarter of approximately 24 g is 6 g, which is close to the answer.

**Note:** This problem could have been worked in a manner similar to Problem 3.46. You could complete the following conversions:

$$\text{g of SnF}_2 \rightarrow \text{mol of SnF}_2 \rightarrow \text{mol of F} \rightarrow \text{g of F}$$

**3.44** (a)

**Strategy:** In a chemical formula, the subscripts represent the ratio of the number of moles of each element that combine to form the compound. Therefore, we need to convert from mass percent to moles in order to determine the empirical formula. If we assume an exactly 100 g sample of the compound, do we know the mass of each element in the compound? How do we then convert from grams to moles?

**Solution:** If we have 100 g of the compound, then each percentage can be converted directly to grams. In this sample, there will be 40.1 g of C, 6.6 g of H, and 53.3 g of O. Because the subscripts in the formula represent a mole ratio, we need to convert the grams of each element to moles. The conversion factor needed is the molar mass of each element. Let $n$ represent the number of moles of each element so that

$$n_C = 40.1\text{ g C} \times \frac{1\text{ mol C}}{12.01\text{ g C}} = 3.339\text{ mol C}$$

$$n_H = 6.6\text{ g H} \times \frac{1\text{ mol H}}{1.008\text{ g H}} = 6.55\text{ mol H}$$

$$n_O = 53.3\text{ g O} \times \frac{1\text{ mol O}}{16.00\text{ g O}} = 3.331\text{ mol O}$$

Thus, we arrive at the formula $C_{3.339}H_{6.55}O_{3.331}$, which gives the identity and the mole ratios of atoms present. However, chemical formulas are written with whole numbers. Try to convert to whole numbers by dividing all the subscripts by the smallest subscript (3.331).

$$\textbf{C}: \frac{3.339}{3.331} \approx 1 \qquad \textbf{H}: \frac{6.55}{3.331} \approx 2 \qquad \textbf{O}: \frac{3.331}{3.331} = 1$$

This gives the empirical formula, $\mathbf{CH_2O}$.

**Check:** Are the subscripts in $CH_2O$ reduced to the smallest whole numbers?

**(b)** Following the same procedure as part (a), we find:

$$n_C = 18.4 \cancel{g}\,C \times \frac{1\ \text{mol C}}{12.01 \cancel{g}\,C} = 1.532\ \text{mol C}$$

$$n_N = 21.5 \cancel{g}\,N \times \frac{1\ \text{mol N}}{14.01 \cancel{g}\,N} = 1.535\ \text{mol N}$$

$$n_K = 60.1 \cancel{g}\,K \times \frac{1\ \text{mol K}}{39.10 \cancel{g}\,K} = 1.537\ \text{mol K}$$

Dividing by the smallest number of moles (1.532 mol) gives the empirical formula, **KCN**.

**3.48** The empirical molar mass of CH is approximately 13.018 g. Let's compare this to the molar mass to determine the molecular formula.

Recall that the molar mass divided by the empirical mass will be an integer greater than or equal to one.

$$\frac{\text{molar mass}}{\text{empirical molar mass}} \geq 1\ \text{(integer values)}$$

In this case,

$$\frac{\text{molar mass}}{\text{empirical molar mass}} = \frac{78\ \text{g}}{13.018\ \text{g}} \approx 6$$

Thus, there are six CH units in each molecule of the compound, so the molecular formula is $(CH)_6$, or $C_6H_6$.

**3.50** **METHOD 1:**

*Step 1:* Assume you have exactly 100 g of substance. 100 g is a convenient amount, because all the percentages sum to 100%. The percentage of oxygen is found by difference:

$$100\% - (19.8\% + 2.50\% + 11.6\%) = 66.1\%$$

In 100 g of PAN there will be 19.8 g C, 2.50 g H, 11.6 g N, and 66.1 g O.

*Step 2:* Calculate the number of moles of each element in the compound. Remember, an *empirical formula* tells us which elements are present and the simplest whole-number ratio of their atoms. This ratio is also a mole ratio. Use the molar masses of these elements as conversion factors to convert to moles.

$$n_C = 19.8 \cancel{g}\,C \times \frac{1\ \text{mol C}}{12.01 \cancel{g}\,C} = 1.649\ \text{mol C}$$

$$n_H = 2.50 \cancel{g}\,H \times \frac{1\ \text{mol H}}{1.008 \cancel{g}\,H} = 2.480\ \text{mol H}$$

$$n_N = 11.6 \cancel{g}\,N \times \frac{1\ \text{mol N}}{14.01 \cancel{g}\,N} = 0.8280\ \text{mol N}$$

$$n_O = 66.1 \cancel{g}\,O \times \frac{1\ \text{mol O}}{16.00 \cancel{g}\,O} = 4.131\ \text{mol O}$$

*Step 3:* Try to convert to whole numbers by dividing all the subscripts by the smallest subscript. The formula is $C_{1.649}H_{2.480}N_{0.8280}O_{4.131}$. Dividing the subscripts by 0.8280 gives the empirical formula, $\mathbf{C_2H_3NO_5}$.

To determine the molecular formula, remember that the molar mass/empirical mass will be an integer greater than or equal to one.

$$\frac{\text{molar mass}}{\text{empirical molar mass}} \geq 1 \text{ (integer values)}$$

In this case,

$$\frac{\text{molar mass}}{\text{empirical molar mass}} = \frac{120 \text{ g}}{121.05 \text{ g}} \approx 1$$

Hence, the molecular formula and the empirical formula are the same, $\mathbf{C_2H_3NO_5}$.

**METHOD 2:**

*Step 1:* Multiply the mass % (converted to a decimal) of each element by the molar mass to convert to grams of each element. Then, use the molar mass to convert to moles of each element.

$$n_C = (0.198) \times (120 \text{ g}) \times \frac{1 \text{ mol C}}{12.01 \text{ g C}} = 1.98 \text{ mol C} \approx \mathbf{2 \text{ mol C}}$$

$$n_H = (0.0250) \times (120 \text{ g}) \times \frac{1 \text{ mol H}}{1.008 \text{ g H}} = 2.98 \text{ mol H} \approx \mathbf{3 \text{ mol H}}$$

$$n_N = (0.116) \times (120 \text{ g}) \times \frac{1 \text{ mol N}}{14.01 \text{ g N}} = 0.994 \text{ mol N} \approx \mathbf{1 \text{ mol N}}$$

$$n_O = (0.661) \times (120 \text{ g}) \times \frac{1 \text{ mol O}}{16.00 \text{ g O}} = 4.96 \text{ mol O} \approx \mathbf{5 \text{ mol O}}$$

*Step 2:* Since we used the molar mass to calculate the moles of each element present in the compound, this method directly gives the molecular formula. The formula is $\mathbf{C_2H_3NO_5}$.

*Step 3:* Try to reduce the molecular formula to a simpler whole number ratio to determine the empirical formula. The formula is already in its simplest whole number ratio. The molecular and empirical formulas are the same. The empirical formula is $\mathbf{C_2H_3NO_5}$.

3.52 **METHOD 1:**

*Step 1:* Assume you have exactly 100 g of substance. 100 g is a convenient amount, because all the percentages sum to 100%. In 100 g of MSG there will be 35.51 g C, 4.77 g H, 37.85 g O, 8.29 g N, and 13.60 g Na.

*Step 2:* Calculate the number of moles of each element in the compound. Remember, an *empirical formula* tells us which elements are present and the simplest whole-number ratio of their atoms. This ratio is also a mole ratio. Let $n_C$, $n_H$, $n_O$, $n_N$, and $n_{Na}$ be the number of moles of elements present. Use the molar masses of these elements as conversion factors to convert to moles.

$$n_C = 35.51\,\cancel{g}\,C \times \frac{1 \text{ mol C}}{12.01\,\cancel{g}\,C} = 2.9567 \text{ mol C}$$

$$n_H = 4.77\,\cancel{g}\,H \times \frac{1 \text{ mol H}}{1.008\,\cancel{g}\,H} = 4.732 \text{ mol H}$$

$$n_O = 37.85\,\cancel{g}\,O \times \frac{1 \text{ mol O}}{16.00\,\cancel{g}\,O} = 2.3656 \text{ mol O}$$

$$n_N = 8.29\,\cancel{g}\,N \times \frac{1 \text{ mol N}}{14.01\,\cancel{g}\,N} = 0.5917 \text{ mol N}$$

$$n_{Na} = 13.60\,\cancel{g}\,Na \times \frac{1 \text{ mol Na}}{22.99\,\cancel{g}\,Na} = 0.59156 \text{ mol Na}$$

Thus, we arrive at the formula $C_{2.9567}H_{4.732}O_{2.3656}N_{0.5917}Na_{0.59156}$, which gives the identity and the ratios of atoms present. However, chemical formulas are written with whole numbers.

*Step 3:* Try to convert to whole numbers by dividing all the subscripts by the smallest subscript.

$$\textbf{C}: \frac{2.9567}{0.59156} = 4.9981 \approx 5 \qquad \textbf{H}: \frac{4.732}{0.59156} = 7.999 \approx 8 \qquad \textbf{O}: \frac{2.3656}{0.59156} = 3.9989 \approx 4$$

$$\textbf{N}: \frac{0.5917}{0.59156} = 1.000 \qquad \textbf{Na}: \frac{0.59156}{0.59156} = 1$$

This gives us the empirical formula for MSG, $C_5H_8O_4NNa$.

To determine the molecular formula, remember that the molar mass/empirical mass will be an integer greater than or equal to one.

$$\frac{\text{molar mass}}{\text{empirical molar mass}} \geq 1 \text{ (integer values)}$$

In this case,

$$\frac{\text{molar mass}}{\text{empirical molar mass}} = \frac{169 \text{ g}}{169.11 \text{ g}} \approx 1$$

Hence, the molecular formula and the empirical formula are the same, $\mathbf{C_5H_8O_4NNa}$. It should come as no surprise that the empirical and molecular formulas are the same since MSG stands for *monosodium*glutamate.

**METHOD 2:**

*Step 1:* Multiply the mass % (converted to a decimal) of each element by the molar mass to convert to grams of each element. Then, use the molar mass to convert to moles of each element.

$$n_C = (0.3551) \times (169\text{ g}) \times \frac{1 \text{ mol C}}{12.01\,\cancel{g}\,C} = 5.00 \text{ mol C}$$

$$n_H = (0.0477) \times (169\,\cancel{g}) \times \frac{1 \text{ mol H}}{1.008\,\cancel{g}\,H} = 8.00 \text{ mol H}$$

$$n_O = (0.3785) \times (169\,\cancel{g}) \times \frac{1 \text{ mol O}}{16.00\,\cancel{g}\,O} = 4.00 \text{ mol O}$$

$$n_N = (0.0829) \times (169\ \text{g}) \times \frac{1\ \text{mol N}}{14.01\ \text{g N}} = 1.00\ \text{mol N}$$

$$n_{Na} = (0.1360) \times (169\ \text{g}) \times \frac{1\ \text{mol Na}}{22.99\ \text{g Na}} = 1.00\ \text{mol Na}$$

*Step 2:* Since we used the molar mass to calculate the moles of each element present in the compound, this method directly gives the molecular formula. The formula is $\mathbf{C_5H_8O_4NNa}$.

**3.58** The balanced equations are as follows:

(a) $2N_2O_5 \rightarrow 2N_2O_4 + O_2$

(b) $2KNO_3 \rightarrow 2KNO_2 + O_2$

(c) $NH_4NO_3 \rightarrow N_2O + 2H_2O$

(d) $NH_4NO_2 \rightarrow N_2 + 2H_2O$

(e) $2NaHCO_3 \rightarrow Na_2CO_3 + H_2O + CO_2$

(f) $P_4O_{10} + 6H_2O \rightarrow 4H_3PO_4$

(g) $2HCl + CaCO_3 \rightarrow CaCl_2 + H_2O + CO_2$

(h) $2Al + 3H_2SO_4 \rightarrow Al_2(SO_4)_3 + 3H_2$

(i) $CO_2 + 2KOH \rightarrow K_2CO_3 + H_2O$

(j) $CH_4 + 2O_2 \rightarrow CO_2 + 2H_2O$

(k) $Be_2C + 4H_2O \rightarrow 2Be(OH)_2 + CH_4$

(l) $3Cu + 8HNO_3 \rightarrow 3Cu(NO_3)_2 + 2NO + 4H_2O$

(m) $S + 6HNO_3 \rightarrow H_2SO_4 + 6NO_2 + 2H_2O$

(n) $2NH_3 + 3CuO \rightarrow 3Cu + N_2 + 3H_2O$

**3.62** On the reactants side there are 6 A atoms and 4 B atoms. On the products side, there are 4 C atoms and 2 D atoms. Writing an equation,

$$6A + 4B \rightarrow 4C + 2D$$

Chemical equations are typically written with the smallest set of whole number coefficients. Dividing the equation by two gives,

$$\mathbf{3A + 2B \rightarrow 2C + D}$$

The correct answer is choice **(d)**.

**3.64** $Si(s) + 2Cl_2(g) \longrightarrow SiCl_4(l)$

**Strategy:** Looking at the balanced equation, how do we compare the amounts of $Cl_2$ and $SiCl_4$? We can compare them based on the mole ratio from the balanced equation.

**Solution:** Because the balanced equation is given in the problem, the mole ratio between $Cl_2$ and $SiCl_4$ is known: 2 moles $Cl_2 \simeq$ 1 mole $SiCl_4$. From this relationship, we have two conversion factors.

$$\frac{2\ \text{mol Cl}_2}{1\ \text{mol SiCl}_4} \quad \text{and} \quad \frac{1\ \text{mol SiCl}_4}{2\ \text{mol Cl}_2}$$

Which conversion factor is needed to convert from moles of $SiCl_4$ to moles of $Cl_2$? The conversion factor on the left is the correct one. Moles of $SiCl_4$ will cancel, leaving units of "mol $Cl_2$" for the answer. We calculate moles of $Cl_2$ reacted as follows:

$$\textbf{? mol Cl}_2\textbf{ reacted} = 0.507\text{ mol SiCl}_4 \times \frac{2\text{ mol Cl}_2}{1\text{ mol SiCl}_4} = \textbf{1.01 mol Cl}_2$$

**Check:** Does the answer seem reasonable? Should the moles of $Cl_2$ reacted be *double* the moles of $SiCl_4$ produced?

**3.66** Starting with the 9.8 moles of $CH_3OH$, we can use the mole ratio from the balanced equation to calculate the moles of $H_2O$ formed.

$2CH_3OH(l) + 3O_2(g) \rightarrow 2CO_2(g) + 4H_2O(l)$

$$\textbf{? mol H}_2\textbf{O} = 9.8\text{ mol CH}_3\text{OH} \times \frac{4\text{ mol H}_2\text{O}}{2\text{ mol CH}_3\text{OH}} = 20\text{ mol H}_2\text{O} = \mathbf{2.0 \times 10^1}\textbf{ mol H}_2\textbf{O}$$

**3.68** Using unit factors we convert:

g of Hg → mol Hg → mol S → g S

$$\textbf{? g S} = 246\text{ g Hg} \times \frac{1\text{ mol Hg}}{200.6\text{ g Hg}} \times \frac{1\text{ mol S}}{1\text{ mol Hg}} \times \frac{32.07\text{ g S}}{1\text{ mol S}} = \textbf{39.3 g S}$$

**3.70** **(a)** $2NaHCO_3 \longrightarrow Na_2CO_3 + H_2O + CO_2$

**(b)** Molar mass $NaHCO_3$ = 22.99 g + 1.008 g + 12.01 g + 3(16.00 g) = 84.008 g

Molar mass $CO_2$ = 12.01 g + 2(16.00 g) = 44.01 g

The balanced equation shows one mole of $CO_2$ formed from two moles of $NaHCO_3$.

$$\textbf{mass NaHCO}_3 = 20.5\text{ g CO}_2 \times \frac{1\text{ mol CO}_2}{44.01\text{ g CO}_2} \times \frac{2\text{ mol NaHCO}_3}{1\text{ mol CO}_2} \times \frac{84.008\text{ g NaHCO}_3}{1\text{ mol NaHCO}_3}$$

$$= \textbf{78.3 g NaHCO}_3$$

**3.72** $C_6H_{12}O_6 \longrightarrow 2C_2H_5OH + 2CO_2$

glucose ethanol

**Strategy:** We compare glucose and ethanol based on the *mole ratio* in the balanced equation. Before we can determine moles of ethanol produced, we need to convert to moles of glucose. What conversion factor is needed to convert from grams of glucose to moles of glucose? Once moles of ethanol are obtained, another conversion factor is needed to convert from moles of ethanol to grams of ethanol.

**Solution:** The molar mass of glucose will allow us to convert from grams of glucose to moles of glucose. The molar mass of glucose = 6(12.01 g) + 12(1.008 g) + 6(16.00 g) = 180.16 g. The balanced equation is given, so the mole ratio between glucose and ethanol is known; that is 1 mole glucose ≏ 2 moles ethanol. Finally, the molar mass of ethanol will convert moles of ethanol to grams of ethanol. This sequence of three conversions is summarized as follows:

grams of glucose → moles of glucose → moles of ethanol → grams of ethanol

$$\textbf{? g C}_2\textbf{H}_5\textbf{OH} = 500.4\ \text{g}\ C_6H_{12}O_6 \times \frac{1\ \text{mol}\ C_6H_{12}O_6}{180.16\ \text{g}\ C_6H_{12}O_6} \times \frac{2\ \text{mol}\ C_2H_5OH}{1\ \text{mol}\ C_6H_{12}O_6} \times \frac{46.068\ \text{g}\ C_2H_5OH}{1\ \text{mol}\ C_2H_5OH}$$

$$= \mathbf{255.9\ g\ C_2H_5OH}$$

**Check:** Does the answer seem reasonable? Should the mass of ethanol produced be approximately half the mass of glucose reacted? Twice as many moles of ethanol are produced compared to the moles of glucose reacted, but the molar mass of ethanol is about one-fourth that of glucose.

The liters of ethanol can be calculated from the density and the mass of ethanol.

$$\text{volume} = \frac{\text{mass}}{\text{density}}$$

$$\textbf{Volume of ethanol obtained} = \frac{255.9\ \text{g}}{0.789\ \text{g/mL}} = 324\ \text{mL} = \mathbf{0.324\ L}$$

**3.74** The balanced equation shows that eight moles of KCN are needed to combine with four moles of Au.

$$\textbf{? mol KCN} = 29.0\ \text{g Au} \times \frac{1\ \text{mol Au}}{197.0\ \text{g Au}} \times \frac{8\ \text{mol KCN}}{4\ \text{mol Au}} = \mathbf{0.294\ mol\ KCN}$$

**3.76** **(a)** $NH_4NO_3(s) \longrightarrow N_2O(g) + 2H_2O(g)$

**(b)** Starting with moles of $NH_4NO_3$, we can use the mole ratio from the balanced equation to find moles of $N_2O$. Once we have moles of $N_2O$, we can use the molar mass of $N_2O$ to convert to grams of $N_2O$. Combining the two conversions into one calculation, we have:

mol $NH_4NO_3$ → mol $N_2O$ → g $N_2O$

$$\textbf{? g N}_2\textbf{O} = 0.46\ \text{mol}\ NH_4NO_3 \times \frac{1\ \text{mol}\ N_2O}{1\ \text{mol}\ NH_4NO_3} \times \frac{44.02\ \text{g}\ N_2O}{1\ \text{mol}\ N_2O} = \mathbf{2.0 \times 10^1\ g\ N_2O}$$

**3.78** The balanced equation for the decomposition is:

$$2KClO_3(s) \longrightarrow 2KCl(s) + 3O_2(g)$$

$$\textbf{? g O}_2 = 46.0\ \text{g}\ KClO_3 \times \frac{1\ \text{mol}\ KClO_3}{122.55\ \text{g}\ KClO_3} \times \frac{3\ \text{mol}\ O_2}{2\ \text{mol}\ KClO_3} \times \frac{32.00\ \text{g}\ O_2}{1\ \text{mol}\ O_2} = \mathbf{18.0\ g\ O_2}$$

**3.82** $N_2 + 3H_2 \rightarrow 2NH_3$

9 moles of $H_2$ will react with 3 moles of $N_2$, leaving **1 mole of $H_2$ in excess**. The mole ratio between $N_2$ and $NH_3$ is 1:2. When 3 moles of $N_2$ react, **6 moles of $NH_3$ will be produced.**

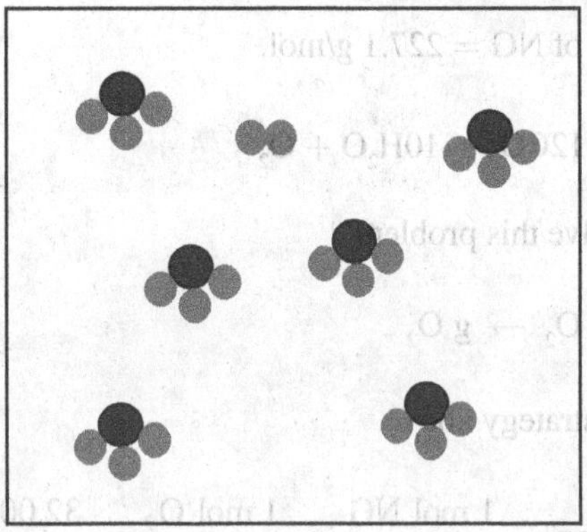


**3.84** **(a)** $2NH_3(g) + H_2SO_4(aq) \rightarrow (NH_4)_2SO_4(aq)$

**(b)** Sulfuric acid is in excess. First, let's calculate the moles of ammonia reacted to produce 20.3 g of ammonium sulfate.

$$20.3\ \text{g}\ (NH_4)_2SO_4 \times \frac{1\ \text{mol}\ (NH_4)_2SO_4}{132.15\ \text{g}\ (NH_4)_2SO_4} \times \frac{2\ \text{mol}\ NH_3}{1\ \text{mol}\ (NH_4)_2SO_4} = 0.307\ \text{mol}\ NH_3\ \text{reacted}$$

The number of moles of sulfuric acid reacted will be half the moles of ammonia reacted (see mole ratio from the balance equation).

The starting mass of each reactant is:

$$0.307\ \text{mol}\ NH_3 \times \frac{17.03\ \text{g}\ NH_3}{1\ \text{mol}\ NH_3} = \mathbf{5.23\ g\ NH_3}$$

$$0.154\ \text{mol}\ H_2SO_4 \times \frac{98.09\ \text{g}\ H_2SO_4}{1\ \text{mol}\ H_2SO_4} = 15.1\ \text{g}\ H_2SO_4\ \text{reacted}$$

$$15.1\ \text{g}\ H_2SO_4 + 5.89\ \text{g}\ H_2SO_4\ \text{unreacted} = \mathbf{21.0\ g\ H_2SO_4}$$

**3.86** This is a limiting reagent problem. Let's calculate the moles of $Cl_2$ produced assuming complete reaction for each reactant.

$$0.86\ \text{mol}\ MnO_2 \times \frac{1\ \text{mol}\ Cl_2}{1\ \text{mol}\ MnO_2} = 0.86\ \text{mol}\ Cl_2$$

$$48.2\ \text{g}\ HCl \times \frac{1\ \text{mol}\ HCl}{36.458\ \text{g}\ HCl} \times \frac{1\ \text{mol}\ Cl_2}{4\ \text{mol}\ HCl} = 0.3305\ \text{mol}\ Cl_2$$

**HCl** is the limiting reagent; it limits the amount of product produced. It will be used up first. The amount of product produced is 0.3305 mole $Cl_2$. Let's convert this to grams.

$$? \text{ g } Cl_2 = 0.3305 \text{ mol } Cl_2 \times \frac{70.90 \text{ g } Cl_2}{1 \text{ mol } Cl_2} = \mathbf{23.4 \text{ g } Cl_2}$$

**3.90** **(a)** Start with a balanced chemical equation. It's given in the problem. We use NG as an abbreviation for nitroglycerin. The molar mass of NG = 227.1 g/mol.

$$4C_3H_5N_3O_9 \longrightarrow 6N_2 + 12CO_2 + 10H_2O + O_2$$

Map out the following strategy to solve this problem.

$$\text{g NG} \rightarrow \text{mol NG} \rightarrow \text{mol } O_2 \rightarrow \text{g } O_2$$

Calculate the grams of $O_2$ using the strategy above.

$$? \text{ g } O_2 = 2.00 \times 10^2 \text{ g NG} \times \frac{1 \text{ mol NG}}{227.1 \text{ g NG}} \times \frac{1 \text{ mol } O_2}{4 \text{ mol NG}} \times \frac{32.00 \text{ g } O_2}{1 \text{ mol } O_2} = \mathbf{7.05 \text{ g } O_2}$$

**(b)** The theoretical yield was calculated in part (a), and the actual yield is given in the problem (6.55 g). The percent yield is:

$$\% \text{ yield} = \frac{\text{actual yield}}{\text{theoretical yield}} \times 100\%$$

$$\% \textbf{ yield} = \frac{6.55 \text{ g } O_2}{7.05 \text{ g } O_2} \times 100\% = \mathbf{92.9\%}$$

**3.92** The actual yield of ethylene is 481 g. Let's calculate the yield of ethylene if the reaction is 100 percent efficient. We can calculate this from the definition of percent yield. We can then calculate the mass of hexane that must be reacted.

$$\% \text{ yield} = \frac{\text{actual yield}}{\text{theoretical yield}} \times 100\%$$

$$42.5\% \text{ yield} = \frac{481 \text{ g } C_2H_4}{\text{theoretical yield}} \times 100\%$$

$$\text{theoretical yield } C_2H_4 = 1.132 \times 10^3 \text{ g } C_2H_4$$

The mass of hexane that must be reacted is:

$$(1.132 \times 10^3 \text{ g } C_2H_4) \times \frac{1 \text{ mol } C_2H_4}{28.052 \text{ g } C_2H_4} \times \frac{1 \text{ mol } C_6H_{14}}{1 \text{ mol } C_2H_4} \times \frac{86.172 \text{ g } C_6H_{14}}{1 \text{ mol } C_6H_{14}} = \mathbf{3.48 \times 10^3 \text{ g } C_6H_{14}}$$

**3.94** This is a limiting reagent problem. Let's calculate the moles of $S_2Cl_2$ produced assuming complete reaction for each reactant.

$$S_8(l) + 4Cl_2(g) \rightarrow 4S_2Cl_2(l)$$

$$4.06\ \text{g S}_8 \times \frac{1\ \text{mol S}_8}{256.56\ \text{g S}_8} \times \frac{4\ \text{mol S}_2\text{Cl}_2}{1\ \text{mol S}_8} = 0.0633\ \text{mol S}_2\text{Cl}_2$$

$$6.24\ \text{g Cl}_2 \times \frac{1\ \text{mol Cl}_2}{70.90\ \text{g Cl}_2} \times \frac{4\ \text{mol S}_2\text{Cl}_2}{4\ \text{mol Cl}_2} = 0.0880\ \text{mol S}_2\text{Cl}_2$$

$S_8$ is the limiting reagent; it limits the amount of product produced. The amount of product produced is 0.0633 mole $S_2Cl_2$. Let's convert this to grams.

$$?\ \text{g S}_2\text{Cl}_2 = 0.0633\ \text{mol S}_2\text{Cl}_2 \times \frac{135.04\ \text{g S}_2\text{Cl}_2}{1\ \text{mol S}_2\text{Cl}_2} = \mathbf{8.55\ g\ S_2Cl_2}$$

This is the theoretical yield of $S_2Cl_2$. The actual yield is given in the problem (6.55 g). The percent yield is:

$$\%\ \text{yield} = \frac{\text{actual yield}}{\text{theoretical yield}} \times 100\% = \frac{6.55\ \text{g}}{8.55\ \text{g}} \times 100\% = \mathbf{76.6\%}$$

**3.96** Start by letting $x$ be the fractional abundance of $^{85}$Rb. Since the sum of the two abundances must be 1, we can write:

$$\text{Abundance } ^{87}\text{Rb} = (1 - x)$$

$$\text{Average atomic mass of Rb} = 85.47\ \text{amu} = x(84.912\ \text{amu}) + (1 - x)(86.909\ \text{amu})$$

$$85.47 = -1.997x + 86.909$$

$$1.997x = 1.44$$

$$\boldsymbol{x = 0.721}$$

$x = 0.721$ corresponds to a natural abundance of $^{85}$Rb of **72.1 percent**. The natural abundance of $^{87}$Rb is $(1 - x) = 0.279$ or **27.9 percent**.

**3.98** $2H_2(g) + O_2(g) \rightarrow 2H_2O(g)$

We start with 8 molecules of $H_2$ and 3 molecules of $O_2$. The balanced equation shows 2 moles $H_2 \simeq$ 1 mole $O_2$. If 3 molecules of $O_2$ react, 6 molecules of $H_2$ will react, leaving 2 molecules of $H_2$ in excess. The balanced equation also shows 1 mole $O_2 \simeq$ 2 moles $H_2O$. If 3 molecules of $O_2$ react, 6 molecules of $H_2O$ will be produced.

After complete reaction, there will be **2 molecules of $H_2$** and **6 molecules of $H_2O$**. The correct diagram is choice **(b)**.

**3.100**
(a) $C_5H_{12}(l) + 8O_2(g) \rightarrow 5CO_2(g) + 6H_2O(l)$

(b) $NaHCO_3(s) + HCl(aq) \rightarrow CO_2(g) + NaCl(aq) + H_2O(l)$

(c) $6Li(s) + N_2(g) \rightarrow 2Li_3N(s)$

(d) $PCl_3(l) + 3H_2O(l) \rightarrow H_3PO_3(aq) + 3HCl(g)$

(e) $3CuO(s) + 2NH_3(g) \rightarrow 3Cu(s) + N_2(g) + 3H_2O(l)$

**3.102** We assume that all the Cl in the compound ends up as HCl and all the O ends up as $H_2O$. Therefore, we need to find the number of moles of Cl in HCl and the number of moles of O in $H_2O$.

$$\text{mol Cl} = 0.233\ \cancel{\text{g}}\ \text{HCl} \times \frac{1\ \cancel{\text{mol}}\ \text{HCl}}{36.458\ \cancel{\text{g}}\ \text{HCl}} \times \frac{1\ \text{mol Cl}}{1\ \cancel{\text{mol}}\ \text{HCl}} = 0.006391\ \text{mol Cl}$$

$$\text{mol O} = 0.403\ \cancel{\text{g}}\ \text{H}_2\text{O} \times \frac{1\ \cancel{\text{mol}}\ \text{H}_2\text{O}}{18.016\ \cancel{\text{g}}\ \text{H}_2\text{O}} \times \frac{1\ \text{mol O}}{1\ \cancel{\text{mol}}\ \text{H}_2\text{O}} = 0.02237\ \text{mol O}$$

Dividing by the smallest number of moles (0.006391 mole) gives the formula, $ClO_{3.5}$. Multiplying both subscripts by two gives the empirical formula, $\mathbf{Cl_2O_7}$.

**3.104** This problem can be solved by two different methods.

$$26.2\ \cancel{\text{g}}\ \text{H}_2\text{C}_2\text{O}_4 \cdot 2\text{H}_2\text{O} \times \frac{90.04\ \text{g H}_2\text{C}_2\text{O}_4}{126.1\ \cancel{\text{g}}\ \text{H}_2\text{C}_2\text{O}_4 \cdot 2\text{H}_2\text{O}} = \mathbf{18.7\ g\ H_2C_2O_4}$$

Or,

$$26.2\ \cancel{\text{g}}\ \text{H}_2\text{C}_2\text{O}_4 \cdot 2\text{H}_2\text{O} \times \frac{1\ \cancel{\text{mol}}\ \text{H}_2\text{C}_2\text{O}_4 \cdot 2\text{H}_2\text{O}}{126.1\ \cancel{\text{g}}\ \text{H}_2\text{C}_2\text{O}_4 \cdot 2\text{H}_2\text{O}} \times \frac{1\ \cancel{\text{mol}}\ \text{H}_2\text{C}_2\text{O}_4}{1\ \cancel{\text{mol}}\ \text{H}_2\text{C}_2\text{O}_4 \cdot 2\text{H}_2\text{O}} \times \frac{90.04\ \text{g H}_2\text{C}_2\text{O}_4}{1\ \cancel{\text{mol}}\ \text{H}_2\text{C}_2\text{O}_4}$$

$$= \mathbf{18.7\ g\ H_2C_2O_4}$$

**3.106** The symbol "O" refers to moles of oxygen atoms, not oxygen molecule ($O_2$). Look at the molecular formulas given in parts (a) and (b). What do they tell you about the relative amounts of carbon and oxygen?

**(a)** $0.212\ \cancel{\text{mol}}\ \text{C} \times \frac{1\ \text{mol O}}{1\ \cancel{\text{mol}}\ \text{C}} = \mathbf{0.212\ mol\ O}$

**(b)** $0.212\ \cancel{\text{mol}}\ \text{C} \times \frac{2\ \text{mol O}}{1\ \cancel{\text{mol}}\ \text{C}} = \mathbf{0.424\ mol\ O}$

**3.108** This is a calculation involving percent composition. Remember,

$$\text{percent by mass of each element} = \frac{\text{mass of element in 1 mol of compound}}{\text{molar mass of compound}} \times 100\%$$

The molar masses are: Al, 26.98 g/mol; $Al_2(SO_4)_3$, 342.17 g/mol; $H_2O$, 18.016 g/mol. Thus, using $x$ as the number of $H_2O$ molecules,

$$\text{mass \% Al} = \left(\frac{2(\text{molar mass of Al})}{\text{molar mass of Al}_2(\text{SO}_4)_3 + x(\text{molar mass of H}_2\text{O})}\right) \times 100\%$$

$$8.10\% = \left(\frac{2(26.98\ \text{g})}{342.17\ \text{g} + x(18.016\ \text{g})}\right) \times 100\%$$

$$(0.081)(342.17) + (0.081)(18.016)(x) = 53.96$$

$$\mathbf{x = 17.98}$$

Rounding off to a whole number of water molecules, $\mathbf{x = 18}$. Therefore, the formula is $\mathbf{Al_2(SO_4)_3 \cdot 18\ H_2O}$.

**3.110** The number of carbon atoms in a 24-carat diamond is:

$$24\ \text{carat} \times \frac{200\ \text{mg C}}{1\ \text{carat}} \times \frac{0.001\ \text{g C}}{1\ \text{mg C}} \times \frac{1\ \text{mol C}}{12.01\ \text{g C}} \times \frac{6.022 \times 10^{23}\ \text{atoms C}}{1\ \text{mol C}} = \mathbf{2.4 \times 10^{23}\ atoms\ C}$$

**3.112** The mass of oxygen in MO is 39.46 g − 31.70 g = 7.76 g O. Therefore, for every 31.70 g of M, there is 7.76 g of O in the compound MO. The molecular formula shows a mole ratio of 1 mole M : 1 mole O. First, calculate moles of M that react with 7.76 g O.

$$\text{mol M} = 7.76\ \text{g O} \times \frac{1\ \text{mol O}}{16.00\ \text{g O}} \times \frac{1\ \text{mol M}}{1\ \text{mol O}} = 0.485\ \text{mol M}$$

$$\text{molar mass M} = \frac{31.70\ \text{g M}}{0.485\ \text{mol M}} = 65.4\ \text{g/mol}$$

Thus, the atomic mass of M is **65.4 amu**. The metal is most likely **Zn**.

**3.114** The wording of the problem suggests that the actual yield is less than the theoretical yield. The percent yield will be equal to the percent purity of the iron(III) oxide. We find the theoretical yield:

$$(2.62 \times 10^3\ \text{kg } Fe_2O_3) \times \frac{1000\ \text{g } Fe_2O_3}{1\ \text{kg } Fe_2O_3} \times \frac{1\ \text{mol } Fe_2O_3}{159.7\ \text{g } Fe_2O_3} \times \frac{2\ \text{mol Fe}}{1\ \text{mol } Fe_2O_3} \times \frac{55.85\ \text{g Fe}}{1\ \text{mol Fe}} \times \frac{1\ \text{kg Fe}}{1000\ \text{g Fe}}$$

$$= 1.833 \times 10^3\ \text{kg Fe}$$

$$\text{percent yield} = \frac{\text{actual yield}}{\text{theoretical yield}} \times 100\%$$

$$\textbf{percent yield} = \frac{1.64 \times 10^3\ \text{kg Fe}}{1.833 \times 10^3\ \text{kg Fe}} \times 100\% = \mathbf{89.5\%} = \textbf{purity of } \mathbf{Fe_2O_3}$$

**3.116** The carbohydrate contains 40 percent carbon; therefore, the remaining 60 percent is hydrogen and oxygen. The problem states that the hydrogen to oxygen ratio is 2:1. We can write this 2:1 ratio as $H_2O$.

Assume 100 g of compound.

$$40.0\ \text{g C} \times \frac{1\ \text{mol C}}{12.01\ \text{g C}} = 3.331\ \text{mol C}$$

$$60.0\ \text{g } H_2O \times \frac{1\ \text{mol } H_2O}{18.016\ \text{g } H_2O} = 3.330\ \text{mol } H_2O$$

Dividing by 3.330 gives $\mathbf{CH_2O}$ for the empirical formula.

To find the molecular formula, divide the molar mass by the empirical mass.

$$\frac{\text{molar mass}}{\text{empirical mass}} = \frac{178\text{ g}}{30.026\text{ g}} \approx 6$$

Thus, there are six $CH_2O$ units in each molecule of the compound, so the molecular formula is $(CH_2O)_6$, or $\mathbf{C_6H_{12}O_6}$.

**3.118** If we assume 100 g of compound, the masses of Cl and X are 67.2 g and 32.8 g, respectively. We can calculate the moles of Cl.

$$67.2\text{ g Cl} \times \frac{1\text{ mol Cl}}{35.45\text{ g Cl}} = 1.896\text{ mol Cl}$$

Then, using the mole ratio from the chemical formula ($XCl_3$), we can calculate the moles of X contained in 32.8 g.

$$1.896\text{ mol Cl} \times \frac{1\text{ mol X}}{3\text{ mol Cl}} = 0.6320\text{ mol X}$$

0.6320 mole of X has a mass of 32.8 g. Calculating the molar mass of X:

$$\frac{32.8\text{ g X}}{0.6320\text{ mol X}} = \mathbf{51.9\text{ g/mol}}$$

The element is most likely **chromium** (molar mass = 52.00 g/mol).

**3.120** A 100 g sample of myoglobin contains 0.34 g of iron (0.34% Fe). The number of moles of Fe is:

$$0.34\text{ g Fe} \times \frac{1\text{ mol Fe}}{55.85\text{ g Fe}} = 6.09 \times 10^{-3}\text{ mol Fe}$$

Since there is one Fe atom in a molecule of myoglobin, the moles of myoglobin also equal $6.09 \times 10^{-3}$ mole. The molar mass of myoglobin can be calculated.

$$\textbf{molar mass myoglobin} = \frac{100\text{ g myoglobin}}{6.09 \times 10^{-3}\text{ mol myoglobin}} = \mathbf{1.6 \times 10^4\text{ g/mol}}$$

**3.122** If we assume 100 g of the mixture, then there are 29.96 g of Na in the mixture (29.96% Na by mass). This amount of Na is equal to the mass of Na in NaBr plus the mass of Na in $Na_2SO_4$.

$$29.96\text{ g Na} = \text{mass of Na in NaBr} + \text{mass of Na in Na}_2\text{SO}_4$$

To calculate the mass of Na in each compound, grams of compound need to be converted to grams of Na using the mass percentage of Na in the compound. If $x$ equals the mass of NaBr, then the mass of $Na_2SO_4$ is $100 - x$. Recall that we assumed 100 g of the mixture. We set up the following expression and solve for $x$.

$$29.96\text{ g Na} = \text{mass of Na in NaBr} + \text{mass of Na in Na}_2\text{SO}_4$$

$$29.96\text{ g Na} = \left[x\text{ g NaBr} \times \frac{22.99\text{ g Na}}{102.89\text{ g NaBr}}\right] + \left[(100 - x)\text{ g Na}_2\text{SO}_4 \times \frac{(2)(22.99\text{ g Na})}{142.05\text{ g Na}_2\text{SO}_4}\right]$$

$$29.96 = 0.22344x + 32.369 - 0.32369x$$

$$0.10025x = 2.409$$

$x = 24.03$ g, which equals the mass of NaBr.

The mass of $Na_2SO_4$ is $100 - x$ which equals 75.97 g.

Because we assumed 100 g of compound, the mass % of NaBr in the mixture is **24.03%** and the mass % of $Na_2SO_4$ is **75.97%**.

**3.124** $C_3H_8(g) + 5O_2(g) \rightarrow 3CO_2(g) + 4H_2O(l)$

**3.126** The mass percent of an element in a compound can be calculated as follows:

$$\text{percent by mass of each element} = \frac{\text{mass of element in 1 mol of compound}}{\text{molar mass of compound}} \times 100\%$$

The molar mass of $Ca_3(PO_4)_2 = 310.18$ g/mol

$$\%\ \mathbf{Ca} = \frac{(3)(40.08\ \text{g})}{310.18\ \text{g}} \times 100\% = \mathbf{38.76\%\ Ca}$$

$$\%\ \mathbf{P} = \frac{(2)(30.97\ \text{g})}{310.18\ \text{g}} \times 100\% = \mathbf{19.97\%\ P}$$

$$\%\ \mathbf{O} = \frac{(8)(16.00\ \text{g})}{310.18\ \text{g}} \times 100\% = \mathbf{41.27\%\ O}$$

**3.128** **Yes.** The number of hydrogen atoms in one gram of hydrogen molecules is the same as the number in one gram of hydrogen atoms. There is no difference in mass, only in the way that the particles are arranged.

Would the mass of 100 dimes be the same if they were stuck together in pairs instead of separated?

**3.130** Since we assume that water exists as either $H_2O$ or $D_2O$, the natural abundances are 99.985 percent and 0.015 percent, respectively. If we convert to molecules of water (both $H_2O$ or $D_2O$), we can calculate the molecules that are $D_2O$ from the natural abundance (0.015%).

The necessary conversions are:

mL water → g water → mol water → molecules water → molecules $D_2O$

$$400\ \text{mL water} \times \frac{1\ \text{g water}}{1\ \text{mL water}} \times \frac{1\ \text{mol water}}{18.02\ \text{g water}} \times \frac{6.022 \times 10^{23}\ \text{molecules}}{1\ \text{mol water}} \times \frac{0.015\%\ \text{molecules } D_2O}{100\%\ \text{molecules water}}$$

$$= \mathbf{2.01 \times 10^{21}\ molecules\ D_2O}$$

**3.132** First, we can calculate the moles of oxygen.

$$2.445\ \cancel{g\ C} \times \frac{1\ \cancel{mol\ C}}{12.01\ \cancel{g\ C}} \times \frac{1\ mol\ O}{1\ \cancel{mol\ C}} = 0.2036\ mol\ O$$

Next, we can calculate the molar mass of oxygen.

$$\text{molar mass O} = \frac{3.257\ g\ O}{0.2036\ mol\ O} = 16.00\ g/mol$$

If 1 mole of oxygen atoms has a mass of 16.00 g, then 1 atom of oxygen has an **atomic mass of 16.00 amu.**

**3.134** (a) The mass of chlorine is **5.0 g**.

(b) From the percent by mass of Cl, we can calculate the mass of chlorine in 60.0 g of $NaClO_3$.

$$\text{mass \% Cl} = \frac{35.45\ g\ Cl}{106.44\ g\ compound} \times 100\% = 33.31\%\ Cl$$

$$\textbf{mass Cl} = 60.0\ g \times 0.3331 = \mathbf{20.0\ g\ Cl}$$

(c) 0.10 mol of KCl contains 0.10 mol of Cl.

$$0.10\ \cancel{mol}\ Cl \times \frac{35.45\ g\ Cl}{1\ \cancel{mol}\ Cl} = \mathbf{3.5\ g\ Cl}$$

(d) From the percent by mass of Cl, we can calculate the mass of chlorine in 30.0 g of $MgCl_2$.

$$\text{mass \% Cl} = \frac{(2)(35.45\ g\ Cl)}{95.21\ g\ compound} \times 100\% = 74.47\%\ Cl$$

$$\textbf{mass Cl} = 30.0\ g \times 0.7447 = \mathbf{22.3\ g\ Cl}$$

(e) The mass of Cl can be calculated from the molar mass of $Cl_2$.

$$0.50\ \cancel{mol}\ Cl_2 \times \frac{70.90\ g\ Cl}{1\ \cancel{mol}\ Cl_2} = \mathbf{35.45\ g\ Cl}$$

Thus, (e) **0.50 mol $Cl_2$** contains the greatest mass of chlorine.

**3.136** Both compounds contain only Pt and Cl. The percent by mass of Pt can be calculated by subtracting the percent Cl from 100 percent.

**Compound A:** Assume 100 g of compound.

$$26.7\ \cancel{g}\ Cl \times \frac{1\ mol\ Cl}{35.45\ \cancel{g}\ Cl} = 0.753\ mol\ Cl$$

$$73.3\ \text{g Pt} \times \frac{1\ \text{mol Pt}}{195.1\ \text{g Pt}} = 0.376\ \text{mol Pt}$$

Dividing by the smallest number of moles (0.376 mole) gives the empirical formula, $\mathbf{PtCl_2}$.

**Compound B:** Assume 100 g of compound.

$$42.1\ \text{g Cl} \times \frac{1\ \text{mol Cl}}{35.45\ \text{g Cl}} = 1.19\ \text{mol Cl}$$

$$57.9\ \text{g Pt} \times \frac{1\ \text{mol Pt}}{195.1\ \text{g Pt}} = 0.297\ \text{mol Pt}$$

Dividing by the smallest number of moles (0.297 mole) gives the empirical formula, $\mathbf{PtCl_4}$.

**3.138** **(a)** The problem states that three molar equivalents of $C_{10}H_8N_2$ are used in the reaction.

$$6.5\ \text{g RuCl}_3\cdot 3\text{H}_2\text{O} \times \frac{1\ \text{mol RuCl}_3\cdot 3\text{H}_2\text{O}}{261.5\ \text{g RuCl}_3\cdot 3\text{H}_2\text{O}} \times \frac{3\ \text{mol C}_{10}\text{H}_8\text{N}_2}{1\ \text{mol RuCl}_3\cdot 3\text{H}_2\text{O}} \times \frac{156.2\ \text{g C}_{10}\text{H}_8\text{N}_2}{1\ \text{mol C}_{10}\text{H}_8\text{N}_2}$$

$$= \mathbf{12\ g\ C_{10}H_8N_2}$$

The problem also states that eight molar equivalents of triethylamine are used in the reaction.

$$6.5\ \text{g RuCl}_3\cdot 3\text{H}_2\text{O} \times \frac{1\ \text{mol RuCl}_3\cdot 3\text{H}_2\text{O}}{261.5\ \text{g RuCl}_3\cdot 3\text{H}_2\text{O}} \times \frac{8\ \text{mol N(CH}_2\text{CH}_3)_3}{1\ \text{mol RuCl}_3\cdot 3\text{H}_2\text{O}} \times \frac{101.2\ \text{g N(CH}_2\text{CH}_3)_3}{1\ \text{mol N(CH}_2\text{CH}_3)_3} \times \frac{1\ \text{mL}}{0.73\ \text{g}}$$

$$= \mathbf{28\ mL\ N(CH_2CH_3)_3}$$

**(b)** First, let's calculate the theoretical yield.

$$6.5\ \text{g RuCl}_3\cdot 3\text{H}_2\text{O} \times \frac{1\ \text{mol RuCl}_3\cdot 3\text{H}_2\text{O}}{261.5\ \text{g RuCl}_3\cdot 3\text{H}_2\text{O}} \times \frac{1\ \text{mol [Ru(C}_{10}\text{H}_8\text{N}_2)_3]\text{Cl}_2}{1\ \text{mol RuCl}_3\cdot 3\text{H}_2\text{O}} \times \frac{640.6\ \text{g [Ru(C}_{10}\text{H}_8\text{N}_2)_3]\text{Cl}_2}{1\ \text{mol [Ru(C}_{10}\text{H}_8\text{N}_2)_3]\text{Cl}_2}$$

$$= 16\ \text{g [Ru(C}_{10}\text{H}_8\text{N}_2)_3]\text{Cl}_2$$

The mass of product at 91% yield is: (0.91)(16 g) = $\mathbf{15\ g\ [Ru(C_{10}H_8N_2)_3]Cl_2}$.

**3.140** Both compounds contain only Mn and O. When the first compound is heated, oxygen gas is evolved. Let's calculate the empirical formulas for the two compounds, then we can write a balanced equation.

**(a) Compound X:** Assume 100 g of compound.

$$63.3\ \text{g Mn} \times \frac{1\ \text{mol Mn}}{54.94\ \text{g Mn}} = 1.15\ \text{mol Mn}$$

$$36.7\ \text{g O} \times \frac{1\ \text{mol O}}{16.00\ \text{g O}} = 2.29\ \text{mol O}$$

Dividing by the smallest number of moles (1.15 moles) gives the empirical formula, $\mathbf{MnO_2}$.

**Compound Y:** Assume 100 g of compound.

$$72.0\ \text{g Mn} \times \frac{1\ \text{mol Mn}}{54.94\ \text{g Mn}} = 1.31\ \text{mol Mn}$$

$$28.0\ \text{g O} \times \frac{1\ \text{mol O}}{16.00\ \text{g O}} = 1.75\ \text{mol O}$$

Dividing by the smallest number of moles gives $MnO_{1.33}$. Recall that an empirical formula must have whole number coefficients. Multiplying by a factor of 3 gives the empirical formula $\mathbf{Mn_3O_4}$.

**(b)** The unbalanced equation is: $MnO_2 \longrightarrow Mn_3O_4 + O_2$

Balancing by inspection gives: $\mathbf{3MnO_2 \longrightarrow Mn_3O_4 + O_2}$

**3.142** $SO_2$ is converted to $H_2SO_4$ by reaction with water. The mole ratio between $SO_2$ and $H_2SO_4$ is 1:1.

This is a unit conversion problem. You should come up with the following strategy to solve the problem.

tons $SO_2$ → ton-mol $SO_2$ → ton-mol $H_2SO_4$ → tons $H_2SO_4$

$$\textbf{? tons } \mathbf{H_2SO_4} = (4.0 \times 10^5\ \text{tons } SO_2) \times \frac{1\ \text{ton-mol } SO_2}{64.07\ \text{tons } SO_2} \times \frac{1\ \text{ton-mol } H_2SO_4}{1\ \text{ton-mol } SO_2} \times \frac{98.09\ \text{tons } H_2SO_4}{1\ \text{ton-mol } H_2SO_4}$$

$$= \mathbf{6.1 \times 10^5\ \textbf{tons}\ H_2SO_4}$$

> **Tip:** You probably won't come across a ton-mol that often in chemistry. However, it was convenient to use in this problem. We normally use a g-mol. 1 g-mol $SO_2$ has a mass of 64.07 g. In a similar manner, 1 ton-mol of $SO_2$ has a mass of 64.07 tons.

**3.144** The molecular formula of isoflurane is $\mathbf{C_3H_2ClF_5O}$. The mass percentage of each element is:

$$\%C = \frac{(3)(12.01\ \text{g})}{184.50\ \text{g}} \times 100\% = \mathbf{19.53\%}$$

$$\%H = \frac{(2)(1.008\ \text{g})}{184.50\ \text{g}} \times 100\% = \mathbf{1.093\%}$$

$$\%Cl = \frac{35.45\ \text{g}}{184.50\ \text{g}} \times 100\% = \mathbf{19.21\%}$$

$$\%F = \frac{(5)(19.00)\ \text{g}}{184.50\ \text{g}} \times 100\% = \mathbf{51.49\%}$$

$$\%O = \frac{16.00\ \text{g}}{184.50\ \text{g}} \times 100\% = \mathbf{8.672\%}$$

**Check:** 19.53% + 1.093% + 19.21% + 51.49% + 8.672% = 100.00%

**3.146** We assume that the increase in mass results from the element nitrogen. The mass of nitrogen is:

$$0.378 \text{ g} - 0.273 \text{ g} = 0.105 \text{ g N}$$

The empirical formula can now be calculated. Convert to moles of each element.

$$0.273 \text{ g Mg} \times \frac{1 \text{ mol Mg}}{24.31 \text{ g Mg}} = 0.0112 \text{ mol Mg}$$

$$0.105 \text{ g N} \times \frac{1 \text{ mol N}}{14.01 \text{ g N}} = 0.00749 \text{ mol N}$$

Dividing by the smallest number of moles gives $Mg_{1.5}N$. Recall that an empirical formula must have whole number coefficients. Multiplying by a factor of 2 gives the empirical formula $\mathbf{Mg_3N_2}$. The name of this compound is **magnesium nitride**.

**3.148** *Step 1:* Calculate the mass of C in 55.90 g $CO_2$, and the mass of H in 28.61 g $H_2O$. This is a dimensional analysis problem. To calculate the mass of each component, you need the molar masses and the correct mole ratio.

You should come up with the following strategy:

$$\text{g } CO_2 \rightarrow \text{mol } CO_2 \rightarrow \text{mol C} \rightarrow \text{g C}$$

*Step 2:*

$$? \text{ g C} = 55.90 \text{ g } CO_2 \times \frac{1 \text{ mol } CO_2}{44.01 \text{ g } CO_2} \times \frac{1 \text{ mol C}}{1 \text{ mol } CO_2} \times \frac{12.01 \text{ g C}}{1 \text{ mol C}} = 15.25 \text{ g C}$$

Similarly,

$$? \text{ g H} = 28.61 \text{ g } H_2O \times \frac{1 \text{ mol } H_2O}{18.02 \text{ g } H_2O} \times \frac{2 \text{ mol H}}{1 \text{ mol } H_2O} \times \frac{1.008 \text{ g H}}{1 \text{ mol H}} = 3.201 \text{ g H}$$

Since the compound contains C, H, and Pb, we can calculate the mass of Pb by difference.

$$51.36 \text{ g} = \text{mass C} + \text{mass H} + \text{mass Pb}$$

$$51.36 \text{ g} = 15.25 \text{ g} + 3.201 \text{ g} + \text{mass Pb}$$

$$\text{mass Pb} = 32.91 \text{ g Pb}$$

*Step 3:* Calculate the number of moles of each element present in the sample. Use molar mass as a conversion factor.

$$? \text{ mol C} = 15.25 \text{ g C} \times \frac{1 \text{ mol C}}{12.01 \text{ g C}} = 1.270 \text{ mol C}$$

Similarly,

$$? \text{ mol H} = 3.201 \text{ g H} \times \frac{1 \text{ mol H}}{1.008 \text{ g H}} = 3.176 \text{ mol H}$$

$$? \text{ mol Pb} = 32.91 \text{ g Pb} \times \frac{1 \text{ mol Pb}}{207.2 \text{ g Pb}} = 0.1588 \text{ mol Pb}$$

Thus, we arrive at the formula $Pb_{0.1588}C_{1.270}H_{3.176}$, which gives the identity and the ratios of atoms present. However, chemical formulas are written with whole numbers.

*Step 4:* Try to convert to whole numbers by dividing all the subscripts by the smallest subscript.

$$\text{Pb: } \frac{0.1588}{0.1588} = 1.00 \qquad \text{C: } \frac{1.270}{0.1588} \approx 8 \qquad \text{H: } \frac{3.176}{0.1588} \approx 20$$

This gives the empirical formula, $\mathbf{PbC_8H_{20}}$.

**3.150** (a) The following strategy can be used to convert from the volume of the Mg cube to the number of Mg atoms.

$$\text{cm}^3 \rightarrow \text{grams} \rightarrow \text{moles} \rightarrow \text{atoms}$$

$$1.0 \text{ cm}^3 \times \frac{1.74 \text{ g Mg}}{1 \text{ cm}^3} \times \frac{1 \text{ mol Mg}}{24.31 \text{ g Mg}} \times \frac{6.022 \times 10^{23} \text{ Mg atoms}}{1 \text{ mol Mg}} = \mathbf{4.3 \times 10^{22} \text{ Mg atoms}}$$

(b) Since 74 percent of the available space is taken up by Mg atoms, $4.3 \times 10^{22}$ atoms occupy the following volume:

$$0.74 \times 1.0 \text{ cm}^3 = 0.74 \text{ cm}^3$$

We are trying to calculate the radius of a single Mg atom, so we need the volume occupied by a single Mg atom.

$$\text{volume Mg atom} = \frac{0.74 \text{ cm}^3}{4.3 \times 10^{22} \text{ Mg atoms}} = 1.7 \times 10^{-23} \text{ cm}^3/\text{Mg atom}$$

The volume of a sphere is $\frac{4}{3}\pi r^3$. Solving for the radius:

$$V = 1.7 \times 10^{-23} \text{ cm}^3 = \frac{4}{3}\pi r^3$$

$$r^3 = 4.1 \times 10^{-24} \text{ cm}^3$$

$$r = 1.6 \times 10^{-8} \text{ cm}$$

Converting to picometers:

$$\textbf{radius Mg atom} = (1.6 \times 10^{-8} \text{ cm}) \times \frac{0.01 \text{ m}}{1 \text{ cm}} \times \frac{1 \text{ pm}}{1 \times 10^{-12} \text{ m}} = \mathbf{1.6 \times 10^2 \text{ pm}}$$

**3.152** The molar mass of air can be calculated by multiplying the mass of each component by its abundance and adding them together. Recall that nitrogen gas and oxygen gas are diatomic.

$$\textbf{molar mass air} = (0.7808)(28.02 \text{ g/mol}) + (0.2095)(32.00 \text{ g/mol}) + (0.0097)(39.95 \text{ g/mol}) = \mathbf{28.97 \text{ g/mol}}$$

**3.154** **(a)** $Fe_2O_3(s) + 6HCl(aq) \rightarrow 2FeCl_3(aq) + 3H_2O(l)$

**(b)** We carry out two separate calculations. First, starting with 1.22 moles of $Fe_2O_3$, we calculate the number of moles of $FeCl_3$ that could be produced if all the $Fe_2O_3$ reacted.

$$? \text{ mol FeCl}_3 = 1.22 \text{ mol Fe}_2\text{O}_3 \times \frac{2 \text{ mol FeCl}_3}{1 \text{ mol Fe}_2\text{O}_3} = 2.44 \text{ mol FeCl}_3$$

Second, starting with 289.2 g of HCl, we calculate the number of moles of $FeCl_3$ that could be produced if all the HCl reacted.

$$? \text{ mol FeCl}_3 = 289.2 \text{ g HCl} \times \frac{1 \text{ mol HCl}}{36.46 \text{ g HCl}} \times \frac{2 \text{ mol FeCl}_3}{6 \text{ mol HCl}} = 2.644 \text{ mol FeCl}_3$$

The initial amount of $Fe_2O_3$ limits the amount of product that can be formed; therefore, it is the limiting reagent.

The problem asks for grams of $FeCl_3$ produced. We already know the moles of $FeCl_3$ produced, 2.44 moles. Use the molar mass of $FeCl_3$ as a conversion factor to convert to grams.

$$\textbf{? g FeCl}_\mathbf{3} = 2.44 \text{ mol FeCl}_3 \times \frac{162.2 \text{ g FeCl}_3}{1 \text{ mol FeCl}_3} = \textbf{396 g FeCl}_\mathbf{3}$$

**3.156** **(a)** The balanced chemical equation is:

$$C_3H_8(g) + 3H_2O(g) \longrightarrow 3CO(g) + 7H_2(g)$$

**(b)** You should come up with the following strategy to solve this problem. In this problem, we use kg-mol to save a couple of steps.

$$\text{kg C}_3\text{H}_8 \rightarrow \text{mol C}_3\text{H}_8 \rightarrow \text{mol H}_2 \rightarrow \text{kg H}_2$$

$$\textbf{? kg H}_\mathbf{2} = (2.84 \times 10^3 \text{ kg C}_3\text{H}_8) \times \frac{1 \text{ kg-mol C}_3\text{H}_8}{44.09 \text{ kg C}_3\text{H}_8} \times \frac{7 \text{ kg-mol H}_2}{1 \text{ kg-mol C}_3\text{H}_8} \times \frac{2.016 \text{ kg H}_2}{1 \text{ kg-mol H}_2}$$

$$= \mathbf{9.09 \times 10^2 \text{ kg H}_2}$$

**3.158** **(a)** There is only one reactant, so when it runs out, the reaction stops. It only makes sense to discuss a limiting reagent when comparing one reactant to another reactant.

**(b)** While it is certainly possible that two reactants will be used up simultaneously, only one needs to be listed as a limiting reagent. Once that one reactant runs out, the reaction stops.

**3.160** **(a)** We need to compare the mass % of K in both KCl and $K_2SO_4$.

$$\%\text{K in KCl} = \frac{39.10 \text{ g}}{74.55 \text{ g}} \times 100\% = 52.45\% \text{ K}$$

$$\%\text{K in } K_2SO_4 = \frac{2(39.10\text{ g})}{174.27\text{ g}} \times 100\% = 44.87\%\text{ K}$$

The price is dependent on the %K.

$$\frac{\text{Price of } K_2SO_4}{\text{Price of KCl}} = \frac{\%\text{K in } K_2SO_4}{\%\text{K in KCl}}$$

$$\text{Price of } K_2SO_4 = \text{Price of KCl} \times \frac{\%\text{K in } K_2SO_4}{\%\text{K in KCl}}$$

$$\textbf{Price of } \mathbf{K_2SO_4} = \frac{\$0.55}{\text{kg}} \times \frac{44.87\%}{52.45\%} = \mathbf{\$0.47 / kg}$$

**(b)** First, calculate the number of moles of K in 1.00 kg of KCl.

$$(1.00 \times 10^3 \text{ g KCl}) \times \frac{1 \text{ mol KCl}}{74.55 \text{ g KCl}} \times \frac{1 \text{ mol K}}{1 \text{ mol KCl}} = 13.4 \text{ mol K}$$

Next, calculate the amount of $K_2O$ needed to supply 13.4 mol K.

$$13.4 \text{ mol K} \times \frac{1 \text{ mol } K_2O}{2 \text{ mol K}} \times \frac{94.20 \text{ g } K_2O}{1 \text{ mol } K_2O} \times \frac{1 \text{ kg}}{1000 \text{ g}} = \mathbf{0.631 \text{ kg } K_2O}$$

**3.162** Possible formulas for the metal bromide could be MBr, $MBr_2$, $MBr_3$, etc. Assuming 100 g of compound, the moles of Br in the compound can be determined. From the mass and moles of the metal for each possible formula, we can calculate a molar mass for the metal. The molar mass that matches a metal on the periodic table would indicate the correct formula.

Assuming 100 g of compound, we have 53.79 g Br and 46.21 g of the metal (M). The moles of Br in the compound are:

$$53.79 \text{ g Br} \times \frac{1 \text{ mol Br}}{79.90 \text{ g Br}} = 0.67322 \text{ mol Br}$$

If the formula is MBr, the moles of M are also 0.67322 mole. If the formula is $MBr_2$, the moles of M are 0.67322/2 = 0.33661 mole, and so on. For each formula (MBr, $MBr_2$, and $MBr_3$), we calculate the molar mass of the metal.

$$\text{MBr: } \frac{46.21 \text{ g M}}{0.67322 \text{ mol M}} = 68.64 \text{ g/mol (no such metal)}$$

$$MBr_2\text{: } \frac{46.21 \text{ g M}}{0.33661 \text{ mol M}} = 137.3 \text{ g/mol (The metal is Ba. The formula is } \mathbf{BaBr_2}\text{)}$$

$$MBr_3\text{: } \frac{46.21 \text{ g M}}{0.22441 \text{ mol M}} = 205.9 \text{ g/mol (no such metal)}$$

**3.164** Assume 100 g of sample. Then,

$$\text{mol Na} = 32.08\ \cancel{\text{g}}\ \text{Na} \times \frac{1\ \text{mol Na}}{22.99\ \cancel{\text{g}}\ \text{Na}} = 1.395\ \text{mol Na}$$

$$\text{mol O} = 36.01\ \cancel{\text{g}}\ \text{O} \times \frac{1\ \text{mol O}}{16.00\ \cancel{\text{g}}\ \text{O}} = 2.251\ \text{mol O}$$

$$\text{mol Cl} = 19.51\ \cancel{\text{g}}\ \text{Cl} \times \frac{1\ \text{mol Cl}}{35.45\ \cancel{\text{g}}\ \text{Cl}} = 0.5504\ \text{mol Cl}$$

Since Cl is only contained in NaCl, the moles of Cl equals the moles of Na contained in NaCl.

$$\text{mol Na (in NaCl)} = 0.5504\ \text{mol}$$

The number of moles of Na in the remaining two compounds is: 1.395 mol − 0.5504 mol = 0.8446 mol Na.

To solve for moles of the remaining two compounds, let

$$x = \text{moles of } Na_2SO_4$$
$$y = \text{moles of } NaNO_3$$

Then, from the mole ratio of Na and O in each compound, we can write

$$2x + y = \text{mol Na} = 0.8446\ \text{mol}$$
$$4x + 3y = \text{mol O} = 2.251\ \text{mol}$$

Solving two equations with two unknowns gives

$$x = 0.1414 = \text{mol } Na_2SO_4 \quad \text{and} \quad y = 0.5618 = \text{mol } NaNO_3$$

Finally, we convert to mass of each compound to calculate the mass percent of each compound in the sample. Remember, the sample size is 100 g.

$$\textbf{mass \% NaCl} = 0.5504\ \cancel{\text{mol}}\ \text{NaCl} \times \frac{58.44\ \cancel{\text{g}}\ \text{NaCl}}{1\ \cancel{\text{mol}}\ \text{NaCl}} \times \frac{1}{100\ \cancel{\text{g}}\ \text{sample}} \times 100\% = \textbf{32.17\% NaCl}$$

$$\textbf{mass \% Na}_2\textbf{SO}_4 = 0.1414\ \cancel{\text{mol}}\ Na_2SO_4 \times \frac{142.1\ \cancel{\text{g}}\ Na_2SO_4}{1\ \cancel{\text{mol}}\ Na_2SO_4} \times \frac{1}{100\ \cancel{\text{g}}\ \text{sample}} \times 100\% = \textbf{20.09\% Na}_2\textbf{SO}_4$$

$$\textbf{mass \% NaNO}_3 = 0.5618\ \cancel{\text{mol}}\ NaNO_3 \times \frac{85.00\ \cancel{\text{g}}\ NaNO_3}{1\ \cancel{\text{mol}}\ NaNO_3} \times \frac{1}{100\ \cancel{\text{g}}\ \text{sample}} \times 100\% = \textbf{47.75\% NaNO}_3$$

# CHAPTER 4
# REACTIONS IN AQUEOUS SOLUTIONS

## PROBLEM-SOLVING STRATEGIES AND TUTORIAL SOLUTIONS

### TYPES OF PROBLEMS

**Problem Type 1:** Applying Solubility Rules.

**Problem Type 2:** Writing Molecular, Ionic, and Net Ionic Equations.

**Problem Type 3:** Acid-Base Reactions.
- **(a)** Identifying Brønsted acids and bases.
- **(b)** Writing acid/base reactions.

**Problem Type 4:** Oxidation-Reduction Reactions.
- **(a)** Assigning oxidation numbers.
- **(b)** Writing oxidation/reduction half-reactions.
- **(c)** Using an activity series.

**Problem Type 5:** Concentration of Solutions.

**Problem Type 6:** Dilution of Solutions.

**Problem Type 7:** Gravimetric Analysis.

**Problem Type 8:** Acid-Base Titrations.

**Problem Type 9:** Redox Titrations.

## PROBLEM TYPE 1: APPLYING SOLUBILITY RULES

Ionic compounds are classified as "soluble", "slightly soluble", or "insoluble". Table 4.2 of your text provides solubility rules that will help you determine how a given compound behaves in aqueous solution.

**EXAMPLE 4.1**

**According to the solubility rules, which of the following compounds are soluble in water?**

**(a) $MgCO_3$ (b) $AgNO_3$ (c) $MgCl_2$ (d) $Ca_3(PO_4)_2$ (e) KOH**

**Strategy:** Although it is not necessary to memorize the solubilities of compounds, you should keep in mind the following useful rules: all ionic compounds containing alkali metal cations, the ammonium ion, and the nitrate, bicarbonate, and chlorate ions are soluble. For other compounds, refer to Table 4.2 of the text.

**Solution:**

(a) $MgCO_3$ is *insoluble* (Most ionic compounds containing carbonate ions are *insoluble*).

(b) $AgNO_3$ is *soluble* (All ionic compounds containing nitrate ions are *soluble*).

(c) $MgCl_2$ is *soluble* (Most ionic compounds containing chloride ions are *soluble*).

(d) $Ca_3(PO_4)_2$ is *insoluble* (Most ionic compounds containing phosphate ions are *insoluble*).

(e) KOH is *soluble* (All ionic compounds containing alkali metal ions are *soluble*).

PRACTICE EXERCISE

1. Predict whether the following ionic compounds are soluble or insoluble in water.

**(a)** $NaNO_3$ **(b)** AgCl **(c)** $Ba(OH)_2$ **(d)** $CaCO_3$

**Text Problems:** 4.18, **4.20**, 4.24

## PROBLEM TYPE 2: WRITING MOLECULAR, IONIC, AND NET IONIC EQUATIONS

In a *molecular equation*, the formulas are written as though all species existed as molecules or whole units. However, a molecular equation does not accurately describe what actually happens at the microscopic level. To better describe the reaction in solution, the equation should show the dissociation of dissolved ionic compounds into ions. An *ionic equation* shows dissolved ionic compounds in terms of their free ions. A *net ionic equation* shows only the species that actually take part in the reaction.

### EXAMPLE 4.2

**Write balanced molecular, ionic, and net ionic equations for the reaction that occurs when a $BaCl_2$ solution is mixed with a $Na_2SO_4$ solution.**

**Strategy:** Recall that an *ionic equation* shows dissolved ionic compounds in terms of their free ions. A *net ionic equation* shows only the species that actually take part in the reaction. What happens when ionic compounds dissolve in water? What ions are formed from the dissociation of $BaCl_2$ and $Na_2SO_4$? What happens when the cations encounter the anions in solution?

**Solution:** In solution, $BaCl_2$ dissociates into $Ba^{2+}$ and $Cl^-$ ions and $Na_2SO_4$ dissociates into $Na^+$ and $SO_4^{2-}$ ions. According to Table 4.2 of the text, barium ions ($Ba^{2+}$) and sulfate ions ($SO_4^{2-}$) will form an insoluble compound, barium sulfate ($BaSO_4$), while the other product, NaCl, is soluble and remains in solution. This is a precipitation reaction. The balanced molecular equation is:

$$BaCl_2(aq) + Na_2SO_4(aq) \longrightarrow BaSO_4(s) + 2NaCl(aq)$$

The *ionic equation* should show dissolved ionic compounds in terms of their free ions.

$$Ba^{2+}(aq) + 2Cl^-(aq) + 2Na^+(aq) + SO_4^{2-}(aq) \longrightarrow BaSO_4(s) + 2Na^+(aq) + 2Cl^-(aq)$$

As you write out the ionic equation above, you should notice that some ions ($Na^+$ and $Cl^-$) are not involved in the overall reaction. These ions are called *spectator ions*. Since the spectator ions appear on both sides of the equation and are unchanged in the chemical reaction, they can be canceled from both sides of the equation. A *net ionic equation* shows only the species that actually take part in the reaction.

Cancel the spectator ions to write the *net ionic equation*.

$$Ba^{2+}(aq) + SO_4^{2-}(aq) \longrightarrow BaSO_4(s)$$

**Check:** Note that because we balanced the molecular equation first, the net ionic equation is balanced as to the number of atoms on each side, and the number of positive and negative charges on the left-hand side of the equation is the same.

**Tip:** To help pick out the spectator ions, think about spectators at a sporting event. The spectators are at the stadium, watching the action, but they do *not* participate in the game.

**PRACTICE EXERCISE**

2. Write the balanced molecular, ionic, and net ionic equations for the following reaction:

$$CaCl_2(aq) + Na_2CO_3(aq) \longrightarrow$$

**Text Problem: 4.22**

## PROBLEM TYPE 3: ACID-BASE REACTIONS

### A. Identifying Brønsted acids and bases

A **Brønsted acid** is a proton donor, and a **Brønsted base** is a proton acceptor. To identify a Brønsted acid, you should look for a substance that contains hydrogen. The formula of inorganic acids will begin with H. For example, consider HCl (hydrochloric acid), $HNO_2$ (nitrous acid), and $H_3PO_4$ (phosphoric acid). Carboxylic acids contain the carboxyl group, −COOH. The hydrogen from the carboxyl group can be donated. Examples of carboxylic acids are $CH_3COOH$ (acetic acid) and HCOOH (formic acid).

To identify a Brønsted base, you should look for soluble hydroxide salts. The hydroxide ion ($OH^-$) will accept a proton to form $H_2O$. Also look for weak bases, which are amines. Ammonia ($NH_3$) is an example. Finally, look for anions from acids. These negatives ions can accept a proton ($H^+$). Some examples are $H_2PO_4^{2-}$, $NO_2^-$, and $HCO_3^-$.

**EXAMPLE 4.3**

**Identify each of the following species as a Brønsted acid, base, or both: (a) $HNO_3$, (b) $Ba(OH)_2$, (c) $SO_4^{2-}$, (d) $CH_3CH_2CH_2COOH$, (e) $HPO_4^{2-}$.**

**Strategy:** What are the characteristics of a Brønsted acid? Does it contain at least an H atom? With the exception of ammonia, most Brønsted bases that you will encounter at this stage are anions or soluble hydroxide salts.

**Solution:**

(a) Brønsted acid. The formula of this compound starts with H; this indicates that it is probably an acid.

(b) Brønsted base. This is a soluble hydroxide salt.

(c) Brønsted base. This negative ion can accept a proton; therefore, it is a base.

(d) Brønsted acid. This is a carboxylic acid. It contains a carboxyl group, −COOH.

(e) Both a Brønsted acid and base. This ion has a proton ($H^+$) that it can donate. It also has a negative charge and therefore can accept a proton.

**PRACTICE EXERCISE**

3. Identify each of the following species as a Brønsted acid, base, or both.

(a) $CH_3CH_2COOH$ (b) HF (c) KOH (d) $HCO_3^-$

**Text Problem: 4.32**

### B. Writing Acid-Base Reactions

An acid-base reaction is called a **neutralization reaction**. The typical products of an acid-base reaction are a salt and water.

$$\text{acid} + \text{base} \longrightarrow \text{salt} + \text{water}$$

Let's consider a generic acid, HA, reacted with a generic base, MOH.

$$HA + MOH \longrightarrow MA(salt) + H_2O$$

The $H^+$ from the acid combines with $OH^-$ from the base to produce water. The anion from the acid, $A^-$, combines with the metal cation from the base, $M^+$, to form the salt, MA.

**EXAMPLE 4.4**

**Complete and balance the following equations and write the corresponding ionic and net ionic equations:**

**(a)** $HBr(aq) + Ba(OH)_2(aq) \longrightarrow$

**(b)** $HCOOH(aq) + NaOH(aq) \longrightarrow$

**Strategy:** Recall that strong acids and strong bases are strong electrolytes. They are completely ionized in solution. An *ionic equation* will show strong acids and strong bases in terms of their free ions. Weak acids and weak bases are weak electrolytes. They only ionize to a small extent in solution. Weak acids and weak bases are shown as molecules in ionic and net ionic equations. A *net ionic equation* shows only the species that actually take part in the reaction.

**(a)**

**Solution:** HBr is a strong acid. It completely ionizes to $H^+$ and $Br^-$ ions. $Ba(OH)_2$ is a strong base. It completely ionizes to $Ba^{2+}$ and $OH^-$ ions. Since HBr is an acid, it donates an $H^+$ to the base, $OH^-$, producing water. The other product is the salt, $BaBr_2$, which is soluble and remains in solution. The balanced molecular equation is:

$$2HBr(aq) + Ba(OH)_2(aq) \longrightarrow BaBr_2(aq) + 2H_2O(l)$$

The ionic and net ionic equations are:

$$2H^+(aq) + 2Br^-(aq) + Ba^{2+}(aq) + 2OH^-(aq) \longrightarrow Ba^{2+}(aq) + 2Br^-(aq) + 2H_2O(l)$$

$$2H^+(aq) + 2OH^-(aq) \longrightarrow 2H_2O(l) \quad \text{or} \quad H^+(aq) + OH^-(aq) \longrightarrow H_2O(l)$$

**(b)**

**Solution:** HCOOH is a weak acid. It will be shown as a molecule in the ionic equation. NaOH is a strong base. It completely ionizes to $Na^+$ and $OH^-$ ions. Since HCOOH is an acid, it donates an $H^+$ to the base, $OH^-$, producing water. The other product is the salt, HCOONa, which is soluble and remains in solution. The balanced molecular equation is:

$$HCOOH(aq) + NaOH(aq) \longrightarrow HCOONa(aq) + H_2O(l)$$

The ionic and net ionic equations are:

$$HCOOH(aq) + Na^+(aq) + OH^-(aq) \longrightarrow Na^+(aq) + HCOO^-(aq) + H_2O(l)$$

$$HCOOH(aq) + OH^-(aq) \longrightarrow HCOO^-(aq) + H_2O(l)$$

**PRACTICE EXERCISE**

**4.** Write the balanced molecular, ionic, and net ionic equations for the following acid-base reaction:

$$HNO_2(aq) + Ba(OH)_2(aq) \longrightarrow$$

**Text Problem: 4.34**

## PROBLEM TYPE 4: OXIDATION-REDUCTION REACTIONS

### A. Assigning oxidation numbers

Oxidation numbers are assigned to reactants and products in oxidation-reduction (redox) reactions to keep track of electrons. An oxidation number refers to the number of charges an atom would have in a molecule (or an ionic compound) if electrons were transferred completely.

Rules for assigning oxidation numbers are in Section 4.4 of your text. These rules will be used in the following example.

To assign oxidation numbers you should refer to the following *two* steps:

*Step 1:* Use the rules in Section 4.4 to assign oxidation numbers to as many atoms as possible.

*Step 2:* Often times, one atom does not follow any rules outlined in Section 4.4. To assign an oxidation number to this atom, follow rule 6 of the text. In a neutral molecule, the sum of the oxidation numbers of all the atoms must be zero. In a polyatomic ion, the sum of the oxidation numbers of all the elements in the ion must be equal to the net charge of the ion.

**EXAMPLE 4.5**

**Assign oxidation numbers to all the atoms in the following compounds and ion:**
**(a) $Na_2SO_4$, (b) CuCl, (c) $SO_3^{2-}$**

**Strategy:** In general, we follow the rules listed in Section 4.4 of the text for assigning oxidation numbers. Remember that all alkali metals have an oxidation number of +1 in ionic compounds, and in most cases hydrogen has an oxidation number of +1 and oxygen has an oxidation number of −2 in their compounds.

**Solution:**

**(a)** Na always has an oxidation number of +1 (Rule 2). The oxidation number of oxygen in most compounds is −2 (Rule 3).

You can now assign an oxidation number to S based on Na having a +1 oxidation number and O having a −2 oxidation number. This is a neutral ionic compound, so the sum of the oxidation numbers of all the atoms must be zero.

$$2(\text{oxi. no. Na}) + (\text{oxi. no. S}) + 4(\text{oxi. no. O}) = 0$$

$$2(+1) + (\text{oxi. no. S}) + 4(-2) = 0$$

$$\mathbf{(oxi.\ no.\ S) = 8 - 2 = +6}$$

**(b)** An oxidation number of −1 can be assigned to Cl (Rule 5).

You can now assign an oxidation number to Cu. This is a neutral ionic compound.

$$(\text{oxi. no. Cu}) + (\text{oxi. no. Cl}) = 0$$
$$(\text{oxi. no. Cu}) + (-1) = 0$$

$$\mathbf{(oxi.\ no.\ Cu) = +1}$$

**(c)** An oxidation number of −2 can be assigned to oxygen (Rule 3).

You can now assign an oxidation number to S. $SO_3^{2-}$ is a polyatomic ion. The sum of the oxidation numbers of all elements in the ion must be equal to the net charge of the ion, in this case −2.

$$(\text{oxi. no. S}) + 3(\text{oxi. no. O}) = -2$$
$$(\text{oxi. no. S}) + 3(-2) = -2$$

$$\mathbf{(oxi.\ no.\ S) = -2 + 6 = +4}$$

**PRACTICE EXERCISE**

5. Assign oxidation numbers to the underlined atoms in the following molecules or ions:

(a) $\underline{C}O_3^{2-}$ (b) $\underline{Cu}Cl_2$ (c) $\underline{Ti}O_2$ (d) $\underline{N}O_3^-$

**Text Problems: 4.46**, 4.48, 4.50

## B. Writing oxidation-reduction half-reactions

**Strategy:** In order to break a redox reaction down into an oxidation half-reaction and a reduction half-reaction, you must first assign oxidation numbers to all the atoms in the reaction. In this way, you can determine which element is oxidized (loses electrons) and which element is reduced (gains electrons).

**EXAMPLE 4.6**

**For the following redox reaction, break down the reaction into its half-reactions.**

$$2Al + Fe_2O_3 \longrightarrow Al_2O_3 + 2Fe$$

**Solution: Reactants,** the oxidation number of Al is 0 (Rule 1), and the oxidation number of O in a compound is −2 (Rule 3). Solve for the oxidation number of Fe in $Fe_2O_3$. This is a neutral ionic compound.

$$2(\text{oxi. no. Fe}) + 3(\text{oxi. no. O}) = 0$$
$$2(\text{oxi. no. Fe}) + 3(-2) = 0$$
$$2(\text{oxi. no. Fe}) = +6$$

$$\textbf{(oxi. no. Fe)} = \mathbf{+3}$$

**Products,** the oxidation number of Fe is 0 (Rule 1), and the oxidation number of O in a compound is −2 (Rule 3). Solve for the oxidation number of Al in $Al_2O_3$. This is a neutral ionic compound.

$$2(\text{oxi. no. Al}) + 3(\text{oxi. no. O}) = 0$$
$$2(\text{oxi. no. Al}) + 3(-2) = 0$$
$$2(\text{oxi. no. Al}) = +6$$

$$\textbf{(oxi. no. Al)} = \mathbf{+3}$$

$$Al \longrightarrow Al^{3+} + 3e^- \quad \text{(oxidation half-reaction, } 3e^- \text{ lost)}$$
$$Fe^{3+} + 3e^- \longrightarrow Fe \quad \text{(reduction half-reaction, } 3e^- \text{ gained)}$$

**Tip:** When a species is oxidized, the oxidation number will *increase*. In this example, the oxidation number of Al *increased* from 0 to +3. When a species is reduced, the oxidation number will *decrease*. In this example, the oxidation number of Fe *decreased* from +3 to 0.

**PRACTICE EXERCISE**

6. The nickel-cadmium (nicad) battery, a popular rechargeable "dry cell" used in battery-operated tools, uses the following redox reaction to generate electricity:

$$Cd(s) + NiO_2(s) + 2H_2O(l) \longrightarrow Cd(OH)_2(s) + Ni(OH)_2(s)$$

Assign oxidation numbers to all the atoms and ions, identify the substances that are oxidized and reduced, and write oxidation and reduction half-reactions.

**Text Problem: 4.44**

### C. Using an activity series

An activity series is used to predict whether a metal or hydrogen displacement reaction will occur (see Figure 4.16 of the text). An activity series can be described as a convenient summary of the results of many possible displacement reactions.

(1) Hydrogen displacement. Any metal above hydrogen in the activity series will displace it from water or from an acid. Metals below hydrogen will *not* react with either water or an acid.

(2) Metal displacement. Any metal will react with a compound containing any metal ion listed below it.

**EXAMPLE 4.7**

**Predict the outcome of the reactions represented by the following equations by using the activity series, and balance the equations.**

**Strategy:** *Hydrogen displacement*: Any metal above hydrogen in the activity series will displace it from water or from an acid. Metals below hydrogen will *not* react with either water or an acid.

**Solution:**

**(a)** $Mg(s) + HCl(aq) \longrightarrow$

Since Mg is above hydrogen in the activity series, it will displace hydrogen from the acid.

$$Mg(s) + 2HCl(aq) \longrightarrow MgCl_2(aq) + H_2(g)$$

**(b)** $Au(s) + H_2O(l) \longrightarrow$

Since Au (gold) is below hydrogen in the activity series, it will *not* react with water. You probably already knew this.

$$Au(s) + H_2O(l) \longrightarrow \text{No reaction}$$

**Strategy:** *Metal displacement*: Any metal will react with a compound containing any metal ion listed below it.

**Solution:**

**(c)** $Cu(s) + NiCl_2(aq) \longrightarrow$

In this case $Ni^{2+}$ is listed *above* Cu in the activity series. No reaction will occur.

$$Cu(s) + NiCl_2(aq) \longrightarrow \text{No reaction}$$

**PRACTICE EXERCISE**

7. Predict the outcome of the following reactions using the activity series. If a reaction occurs, balance the equation.
   **(a)** $Al(s) + HCl(aq) \longrightarrow$
   **(b)** $Au(s) + KCl(aq) \longrightarrow$

**Text Problems: 4.52**, 4.54

## PROBLEM TYPE 5: CONCENTRATION OF SOLUTIONS

Solutions are characterized by their concentration, that is, the amount of solute dissolved in a given quantity of solvent. One of the most common units of concentration in chemistry is **molarity (*M*)**. Molarity is the number of moles of solute in 1 liter of solution:

$$M = \text{molarity} = \frac{\text{moles of solute}}{\text{liters of solution}}$$

Sometimes, it is useful to rearrange the above equation to the following form:

$$\text{moles solute} = (\text{molarity}) \times (\text{liters of solution})$$

**EXAMPLE 4.8**

**What is the molarity of a solution made by dissolving 32.1 g of $KNO_3$ in enough water to make 500 mL of solution?**

**Strategy:** Since the definition of molarity is moles solute per liters of solution, we need to convert grams of solute to moles of solute and convert mL of solution to L of solution.

**Solution:**

$$\mathcal{M}(KNO_3) = 101.1 \text{ g/mol}$$

$$? \text{ moles solute} = 32.1 \text{ g } KNO_3 \times \frac{1 \text{ mol } KNO_3}{101.1 \text{ g } KNO_3} = 0.318 \text{ mol } KNO_3$$

$$? \text{ liters of solution} = 500 \text{ mL solution} \times \frac{1 \text{ L}}{1000 \text{ mL}} = 0.500 \text{ L solution}$$

Substitute the above values into the molarity equation.

$$M = \frac{\text{moles of solute}}{\text{liters of solution}}$$

$$M = \frac{0.318 \text{ mol } KNO_3}{0.500 \text{ L solution}} = \mathbf{0.636}\ \boldsymbol{M}$$

This is normally written 0.636 $M$ $KNO_3$.

**PRACTICE EXERCISE**

8. An aqueous nutrient solution is prepared by adding 50.23 g of $KNO_3$ (molar mass = 101.1 g/mol) to enough water to fill a 40.0 L container. What is the molarity of the $KNO_3$ solution?

**EXAMPLE 4.9**

**How many moles of solute are in $2.50 \times 10^2$ mL of 0.100 $M$ KCl?**

**Strategy:** Since the problem asks for moles of solute, you must solve the equation algebraically for moles of solute.

$$\text{moles solute} = (\text{molarity}) \times (\text{liters of solution})$$

Substitute the molarity and liters of solution into the above equation to solve for moles solute.

**Solution:**

$$2.50 \times 10^2 \text{ mL} = 0.250 \text{ L}$$

$$\textbf{? moles KCl solute} = \frac{0.100 \text{ moles solute}}{1 \text{ L solution}} \times 0.250 \text{ L solution} = \mathbf{0.0250 \text{ mol}}$$

**PRACTICE EXERCISE**

9. You need to prepare 1.00 L of a 0.500 $M$ NaCl solution. What mass of NaCl (in g) must you weigh out to prepare this solution?

**Text Problems: 4.64**, 4.66, 4.68, 4.70, 4.72

## PROBLEM TYPE 6: DILUTION OF SOLUTIONS

Dilution refers to the procedure for preparing a less-concentrated solution from a more-concentrated one. The key to solving a dilution problem is to realize that

$$\text{moles of solute } \textit{before} \text{ dilution} = \text{moles of solute } \textit{after} \text{ dilution}$$

In Problem Type 5 above, we discussed how to calculate moles of solute from the molarity and the volume of solution.

$$\text{moles solute} = (\text{molarity}) \times (\text{volume of solution (in L)})$$

Thus,

$$\text{moles of solute } \textit{before} \text{ dilution (initial)} = \text{moles of solute } \textit{after} \text{ dilution (final)}$$

$$M_{\text{initial}}V_{\text{initial}} = M_{\text{final}}V_{\text{final}}$$

**EXAMPLE 4.10**

**What volume of a concentrated (12.0 *M*) hydrochloric acid stock solution is needed to prepare $8.00 \times 10^2$ mL of 0.120 *M* HCl?**

**Strategy:** Recognize that the problem asks for the initial volume of stock solution needed to prepare the dilute solution. Solve the above equation algebraically for $V_{\text{initial}}$, then substitute in the appropriate values from the problem.

**Solution:** We prepare for the calculation by tabulating our data.

$$M_i = 12.0\,M \qquad M_f = 0.120\,M$$
$$V_i = ? \qquad V_f = 8.00 \times 10^2 \text{ mL}$$

$$V_{\text{initial}} = \frac{M_{\text{final}}V_{\text{final}}}{M_{\text{initial}}}$$

$$V_{\text{initial}} = \frac{(0.120\,\cancel{M})(8.00 \times 10^2 \text{ mL})}{12.0\,\cancel{M}} = \mathbf{8.00\ mL}$$

**Tip:** The units of $V_{\text{initial}}$ and $V_{\text{final}}$ can be milliliters or liters for a dilution problem as long as they are the same. Be consistent. Also, make sure to check whether your results seem reasonable. Be sure that $M_{\text{initial}} > M_{\text{final}}$ and $V_{\text{final}} > V_{\text{initial}}$.

**PRACTICE EXERCISE**

10. A 20.0 mL sample of 0.127 *M* $Ca(NO_3)_2$ is diluted to 5.00 L. What is the molarity of the resulting solution?

**Text Problems: 4.74**, 4.76, 4.78

## PROBLEM TYPE 7: GRAVIMETRIC ANALYSIS

Gravimetric analysis is an analytical technique based on the measurement of mass. The type of gravimetric analysis discussed in your text involves the formation, isolation, and mass determination of a precipitate. This procedure is applicable only to reactions that go to completion, or have nearly a 100 percent yield. Thus, the precipitate must be insoluble rather than slightly soluble. See Section 4.6 of your text for further discussion of this technique.

The typical problem involves determining the mass percent of an element in one of the reactants. The element (ion) of interest is completely precipitated from solution. Use the following approach to solve this type of problem.

*Step 1:* From the measured mass of precipitate, calculate the mass of the element of interest in the precipitate. This is the same amount of the element that was present in the original sample.

*Step 2:* Calculate the mass percent of the element of interest in the original sample. See Problem Type 4A, Chapter 3.

> **Tip:** Try to be flexible when solving problems. Some gravimetric analysis problems may ask you to calculate the *molar concentration* of the component of interest in the sample, rather than mass percent. For this type of problem, you must modify your approach by converting grams of the component of interest to moles, then dividing by the volume of solution in liters.

**EXAMPLE 4.11**

**A 0.7469 g sample of an ionic compound containing Pb ions is dissolved in water and treated with excess $Na_2SO_4$. If the mass of $PbSO_4$ that precipitates is 0.6839 g, what is the percent by mass of Pb in the original sample?**

---

**Strategy:** We want to calculate the mass % of Pb in the original compound. Let's start with the definition of mass %.

want to calculate (mass % Pb); need to find (mass Pb); given (mass of sample)

$$\text{mass \% Pb} = \frac{\text{mass Pb}}{\text{mass of sample}} \times 100\%$$

The mass of the sample is given in the problem (0.7469 g). Therefore we need to find the mass of Pb in the original sample. We assume the precipitation is quantitative, that is, that all of the lead in the sample has been precipitated as lead sulfate. From the mass of $PbSO_4$ produced, we can calculate the mass of Pb. There is 1 mole of Pb in 1 mole of $PbSO_4$.

**Solution:** First, we calculate the mass of Pb in 0.6839 g of the $PbSO_4$ precipitate. The molar mass of $PbSO_4$ is 303.27 g/mol.

$$\textbf{? mass of Pb} = 0.6839\ \text{g PbSO}_4 \times \frac{1\ \text{mol PbSO}_4}{303.27\ \text{g PbSO}_4} \times \frac{1\ \text{mol Pb}}{1\ \text{mol PbSO}_4} \times \frac{207.2\ \text{g Pb}}{1\ \text{mol Pb}} = \textbf{0.4673 g Pb}$$

Next, we calculate the mass percent of Pb in the unknown compound.

$$\textbf{\%Pb by mass} = \frac{0.4673\ \text{g}}{0.7469\ \text{g}} \times 100\% = \textbf{62.57\%}$$

**PRACTICE EXERCISE**

11. The concentration of $Pb^{2+}$ ions in tap water could be determined by adding excess sodium sulfate solution to water. Excess sodium sulfate solution is added to 0.250 L of tap water. Write the net ionic equation and calculate the molar concentration of $Pb^{2+}$ in the water sample if 0.01685 g of solid $PbSO_4$ is formed.

**Text Problems: 4.82**, 4.84

## PROBLEM TYPE 8: ACID-BASE TITRATIONS

You must try to convince yourself that a titration problem follows the same thought process as the stoichiometry problems discussed in Chapter 3. The difference is that for acid and base solutions, you will typically be given the molarity rather than grams of substance.

Remember, you cannot directly compare grams of one substance to grams of another. Similarly, you typically cannot compare molarity or volume of one substance to that of another. Therefore, you must convert to *moles* of one substance, and then apply the correct mole ratio from the balanced chemical equation to convert to *moles* of the other substance.

A typical approach to a stoichiometry problem is outlined below.

| Amount of reactant. Expressed in grams, molarity, etc, | ⟶ | moles of reactant | ⟶ | moles of product | ⟶ | amount of product. Expressed in grams, volume, etc. |
|---|---|---|---|---|---|---|

**EXAMPLE 4.12**

**What volume of 0.900 *M* HCl is required to completely neutralize 50.0 mL of a 0.500 *M* $Ba(OH)_2$ solution?**

**Strategy:** We know the molarity of the HCl solution, and we want to calculate the volume of the HCl solution.

given ($M$ of HCl); need to find (mol HCl); want to calculate (L of HCl soln)

$$M \text{ of HCl} = \frac{\text{mol HCl}}{\text{L of HCl soln}}$$

If we can determine the moles of HCl, we can then use the definition of molarity to calculate the volume of HCl needed. From the volume and molarity of $Ba(OH)_2$, we can calculate moles of $Ba(OH)_2$. Then, using the mole ratio from the balanced equation, we can calculate moles of HCl.

**Solution:** In order to have the correct mole ratio to solve the problem, you must start with a balanced chemical equation.

$$2HCl(aq) + Ba(OH)_2(aq) \longrightarrow BaCl_2(aq) + 2H_2O(l)$$

From the molarity and volume of the $Ba(OH)_2$ solution, you can calculate moles of $Ba(OH)_2$. Then, using the mole ratio from the balanced equation above, you can calculate moles of HCl.

50.0 mL = 0.0500 L

$$? \text{ mol HCl} = 0.0500 \text{ L} \times \frac{0.500 \text{ mol Ba(OH)}_2}{1 \text{ L of solution}} \times \frac{2 \text{ mol HCl}}{1 \text{ mol Ba(OH)}_2} = 0.0500 \text{ mol HCl}$$

Thus, 0.0500 mol of HCl are required to neutralize 50.0 mL of 0.500 *M* $Ba(OH)_2$.

Solve the molarity equation algebraically for liters of solution. Then, substitute in the moles of HCl and molarity of HCl to solve for volume of HCl.

$$\text{liters of solution} = \frac{\text{moles of solute}}{M}$$

$$\textbf{volume of HCl} = \frac{0.0500\ \cancel{\text{mol}}\ \text{HCl}}{0.900\ \cancel{\text{mol}}/\text{L}} = \textbf{0.0556 L} = \textbf{55.6 mL}$$

**PRACTICE EXERCISE**

12. The distinctive odor of vinegar is due to acetic acid, $CH_3COOH$. Acetic acid reacts with sodium hydroxide in the following fashion:

$$CH_3COOH(aq) + NaOH(aq) \longrightarrow H_2O(l) + CH_3COONa(aq)$$

If 2.50 mL of vinegar requires 34.9 mL of 0.0960 *M* NaOH to reach the equivalence point in a titration, how many grams of acetic acid are in the 2.50 mL sample?

**Text Problems:** 4.90, **4.92**

## PROBLEM TYPE 9: REDOX TITRATIONS

Redox titration problems are solved in a similar manner to acid-base titration problems. A redox reaction is an *electron* transfer reaction; whereas an acid-base reaction is typically a *proton* transfer reaction. In a redox titration, an oxidizing agent is titrated against a reducing agent.

**EXAMPLE 4.13**

**A 20.32 mL volume of 0.2002 M $KMnO_4$ solution is needed to oxidize 10.00 mL of an oxalic acid ($H_2C_2O_4$) solution. What is the concentration of the oxalic acid solution? The net ionic equation is:**

$$\mathbf{2MnO_4^- + 5C_2O_4^{2-} + 16H^+ \longrightarrow 2Mn^{2+} + 10CO_2 + 8H_2O}$$

**Strategy:** We want to calculate the molarity of the oxalic acid solution. From the molarity and volume of $KMnO_4$, we can calculate moles of $KMnO_4$. Then, using the mole ratio from the balanced equation, we can calculate moles of $H_2C_2O_4$. From the moles and volume of oxalic acid, we can calculate the molarity?

**Solution:** The balanced equation is given in the problem. From the molarity and volume of the $KMnO_4$ solution, you can calculate moles of $MnO_4^-$. Then, using the mole ratio from the balanced equation above, you can calculate moles of $C_2O_4^{2-}$.

$$20.32\ \text{mL} = 0.02032\ \text{L}$$

$$0.02032\ \cancel{\text{L}}\ \text{soln} \times \frac{0.2002\ \cancel{\text{mol}}\ MnO_4^{2-}}{1\ \cancel{\text{L}}\ \text{soln}} \times \frac{5\ \text{mol}\ C_2O_4^{2-}}{2\ \cancel{\text{mol}}\ MnO_4^{2-}} = \mathbf{0.01017\ mol\ C_2O_4^{2-}}$$

The mole ratio between $C_2O_4^{2-}$ and oxalic acid, $H_2C_2O_4$, is 1:1; therefore, the number of moles of oxalic acid is 0.01017 mole.

We can now calculate the molarity of the oxalic acid solution from the moles of oxalic acid and the volume of the solution.

$$M = \frac{\text{moles of solute}}{\text{liters of solution}} = \frac{0.01017\ \text{mol oxalic acid}}{10.00 \times 10^{-3}\ \text{L soln}} = \mathbf{1.017\ \mathit{M}\ oxalic\ acid}$$

**PRACTICE EXERCISE**

**13.** Fe metal reacts with hydrochloric acid to produce $Fe^{2+}$ ions and hydrogen gas. It takes 55.6 mL of 1.15 M HCl to completely react with a piece of Fe. What is the mass of the Fe? The balanced equation is:

$$Fe(s) + 2H^+(aq) + 2Cl^-(aq) \longrightarrow Fe^{2+}(aq) + H_2(g) + 2Cl^-(aq)$$

**Text Problems: 4.96**, 4.98, 4.100, 4.102

## ANSWERS TO PRACTICE EXERCISES

**1.** **(a)** soluble **(b)** insoluble **(c)** soluble **(d)** insoluble

**2.** $CaCl_2(aq) + Na_2CO_3(aq) \longrightarrow 2NaCl(aq) + CaCO_3(s)$

$Ca^{2+}(aq) + 2Cl^-(aq) + 2Na^+(aq) + CO_3^{2-}(aq) \longrightarrow 2Na^+(aq) + 2Cl^-(aq) + CaCO_3(s)$

$Ca^{2+}(aq) + CO_3^{2-}(aq) \longrightarrow CaCO_3(s)$

**3.** **(a)** acid **(b)** acid **(c)** base **(d)** both

**4.** $2HNO_2(aq) + Ba(OH)_2(aq) \longrightarrow Ba(NO_2)_2(aq) + 2H_2O(l)$

$2HNO_2(aq) + Ba^{2+}(aq) + 2OH^-(aq) \longrightarrow Ba^{2+}(aq) + 2NO_2^-(aq) + 2H_2O(l)$

$2HNO_2(aq) + 2OH^-(aq) \longrightarrow 2NO_2^-(aq) + 2H_2O(l)$

**5.** **(a)** +4 **(b)** +2 **(c)** +4 **(d)** +5

**6.** The oxidation numbers of the atoms and ions are:

$$Cd^0(s) + Ni^{4+}O_2^{2-}(s) + 2H_2^+O^{2-}(l) \longrightarrow Cd^{2+}(OH)_2^-(s) + Ni^{2+}(OH)_2^-(s)$$

$Cd^0(s)$ is oxidized, and $Ni^{4+}$ is reduced. The oxidation and reduction half-reactions are:

$$Cd \longrightarrow Cd^{2+} + 2e^-$$
$$Ni^{4+} + 2e^- \longrightarrow Ni^{2+}$$

**7.** **(a)** $2Al(s) + 6HCl(aq) \longrightarrow 2AlCl_3(aq) + 3H_2(g)$ **(b)** No reaction

**8.** 0.0124 *M* **9.** 29.22 g NaCl **10.** $5.08 \times 10^{-4}$ *M*

**11.** $Pb^{2+}(aq) + SO_4^{2-}(aq) \longrightarrow PbSO_4(s)$

$[Pb^{2+}] = 2.222 \times 10^{-4}$ *M*

**12.** 0.201 g acetic acid **13.** 1.79 g Fe

## SOLUTIONS TO SELECTED TEXT PROBLEMS

**4.8** When NaCl dissolves in water it dissociates into $Na^+$ and $Cl^-$ ions. When the ions are hydrated, the water molecules will be oriented so that the negative end of the water dipole interacts with the positive sodium ion, and the positive end of the water dipole interacts with the negative chloride ion. The negative end of the water dipole is near the oxygen atom, and the positive end of the water dipole is near the hydrogen atoms. The diagram that best represents the hydration of NaCl when dissolved in water is choice **(c)**.

**4.10** Ionic compounds, strong acids, and strong bases (metal hydroxides) are strong electrolytes (completely broken up into ions of the compound). Weak acids and weak bases are weak electrolytes. Molecular substances other than acids or bases are nonelectrolytes.

**(a)** strong electrolyte (ionic) **(b)** nonelectrolyte

**(c)** weak electrolyte (weak base) **(d)** strong electrolyte (strong base)

**4.12** **(a)** Solid NaCl does not conduct. The ions are locked in a rigid lattice structure.

**(b)** Molten NaCl conducts. The ions can move around in the liquid state.

**(c)** Aqueous NaCl conducts. NaCl dissociates completely to $Na^+(aq)$ and $Cl^-(aq)$ in water.

**4.14** Since HCl dissolved in water conducts electricity, then HCl(*aq*) must actually exist as $H^+(aq)$ cations and $Cl^-(aq)$ anions. Since HCl dissolved in benzene solvent does not conduct electricity, then we must assume that the HCl molecules in benzene solvent do not ionize, but rather exist as un-ionized molecules.

**4.18** Refer to Table 4.2 of the text to solve this problem. $Mg(OH)_2$ is insoluble in water. It will precipitate from solution. KCl is soluble in water and will remain as $K^+$ and $Cl^-$ ions in solution. Diagram **(b)** best represents the mixture.

**4.20** **Strategy:** Although it is not necessary to memorize the solubilities of compounds, you should keep in mind the following useful rules: all ionic compounds containing alkali metal cations, the ammonium ion, and the nitrate, bicarbonate, and chlorate ions are soluble. For other compounds, refer to Table 4.2 of the text.

**Solution:**

**(a)** $CaCO_3$ is **insoluble**. Most carbonate compounds are insoluble.

**(b)** $ZnSO_4$ is **soluble**. Most sulfate compounds are soluble.

**(c)** $Hg(NO_3)_2$ is **soluble**. All nitrate compounds are soluble.

**(d)** $HgSO_4$ is **insoluble**. Most sulfate compounds are soluble, but those containing $Ag^+$, $Ca^{2+}$, $Ba^{2+}$, $Hg^{2+}$, and $Pb^{2+}$ are insoluble.

**(e)** $NH_4ClO_4$ is **soluble**. All ammonium compounds are soluble.

**4.22** **(a)**

**Strategy:** Recall that an *ionic equation* shows dissolved ionic compounds in terms of their free ions. A *net ionic equation* shows only the species that actually take part in the reaction. What happens when ionic compounds dissolve in water? What ions are formed from the dissociation of $Na_2S$ and $ZnCl_2$? What happens when the cations encounter the anions in solution?

**Solution:** In solution, $Na_2S$ dissociates into $Na^+$ and $S^{2-}$ ions and $ZnCl_2$ dissociates into $Zn^{2+}$ and $Cl^-$ ions. According to Table 4.2 of the text, zinc ions ($Zn^{2+}$) and sulfide ions ($S^{2-}$) will form an insoluble compound, zinc sulfide (ZnS), while the other product, NaCl, is soluble and remains in solution. This is a precipitation reaction. The balanced molecular equation is:

$$Na_2S(aq) + ZnCl_2(aq) \longrightarrow ZnS(s) + 2NaCl(aq)$$

The ionic and net ionic equations are:

*Ionic*: $2Na^+(aq) + S^{2-}(aq) + Zn^{2+}(aq) + 2Cl^-(aq) \longrightarrow ZnS(s) + 2Na^+(aq) + 2Cl^-(aq)$

*Net ionic*: $Zn^{2+}(aq) + S^{2-}(aq) \longrightarrow ZnS(s)$

**Check:** Note that because we balanced the molecular equation first, the net ionic equation is balanced as to the number of atoms on each side, and the number of positive and negative charges on the left-hand side of the equation is the same.

**(b)**

**Strategy:** What happens when ionic compounds dissolve in water? What ions are formed from the dissociation of $K_3PO_4$ and $Sr(NO_3)_2$? What happens when the cations encounter the anions in solution?

**Solution:** In solution, $K_3PO_4$ dissociates into $K^+$ and $PO_4^{3-}$ ions and $Sr(NO_3)_2$ dissociates into $Sr^{2+}$ and $NO_3^-$ ions. According to Table 4.2 of the text, strontium ions ($Sr^{2+}$) and phosphate ions ($PO_4^{3-}$) will form an insoluble compound, strontium phosphate [$Sr_3(PO_4)_2$], while the other product, $KNO_3$, is soluble and remains in solution. This is a precipitation reaction. The balanced molecular equation is:

$$2K_3PO_4(aq) + 3Sr(NO_3)_2(aq) \longrightarrow Sr_3(PO_4)_2(s) + 6KNO_3(aq)$$

The ionic and net ionic equations are:

*Ionic*: $6K^+(aq) + 2PO_4^{3-}(aq) + 3Sr^{2+}(aq) + 6NO_3^-(aq) \longrightarrow Sr_3(PO_4)_2(s) + 6K^+(aq) + 6NO_3^-(aq)$

*Net ionic*: $3Sr^{2+}(aq) + 2PO_4^{3-}(aq) \longrightarrow Sr_3(PO_4)_2(s)$

**Check:** Note that because we balanced the molecular equation first, the net ionic equation is balanced as to the number of atoms on each side, and the number of positive and negative charges on the left-hand side of the equation is the same.

**(c)**

**Strategy:** What happens when ionic compounds dissolve in water? What ions are formed from the dissociation of $Mg(NO_3)_2$ and NaOH? What happens when the cations encounter the anions in solution?

**Solution:** In solution, $Mg(NO_3)_2$ dissociates into $Mg^{2+}$ and $NO_3^-$ ions and NaOH dissociates into $Na^+$ and $OH^-$ ions. According to Table 4.2 of the text, magnesium ions ($Mg^{2+}$) and hydroxide ions ($OH^-$) will form an insoluble compound, magnesium hydroxide [$Mg(OH)_2$], while the other product, $NaNO_3$, is soluble and remains in solution. This is a precipitation reaction. The balanced molecular equation is:

$$Mg(NO_3)_2(aq) + 2NaOH(aq) \longrightarrow Mg(OH)_2(s) + 2NaNO_3(aq)$$

The ionic and net ionic equations are:

*Ionic*: $Mg^{2+}(aq) + 2NO_3^-(aq) + 2Na^+(aq) + 2OH^-(aq) \longrightarrow Mg(OH)_2(s) + 2Na^+(aq) + 2NO_3^-(aq)$

*Net ionic*: $Mg^{2+}(aq) + 2OH^-(aq) \longrightarrow Mg(OH)_2(s)$

**Check:** Note that because we balanced the molecular equation first, the net ionic equation is balanced as to the number of atoms on each side, and the number of positive and negative charges on the left-hand side of the equation is the same.

**4.24** **(a)** Add chloride ions. KCl is soluble, but AgCl is not.

**(b)** Add hydroxide ions. $Ba(OH)_2$ is soluble, but $Pb(OH)_2$ is insoluble.

**(c)** Add carbonate ions. $(NH_4)_2CO_3$ is soluble, but $CaCO_3$ is insoluble.

**(d)** Add sulfate ions. $CuSO_4$ is soluble, but $BaSO_4$ is insoluble.

**4.32** **Strategy:** What are the characteristics of a Brønsted acid? Does it contain at least an H atom? With the exception of ammonia, most Brønsted bases that you will encounter at this stage are anions.

**Solution:**

**(a)** $PO_4^{3-}$ in water can accept a proton to become $HPO_4^{2-}$, and is thus a **Brønsted base.**

**(b)** $ClO_2^-$ in water can accept a proton to become $HClO_2$, and is thus a **Brønsted base.**

**(c)** $NH_4^+$ dissolved in water can donate a proton $H^+$, thus behaving as a **Brønsted acid.**

**(d)** $HCO_3^-$ can either accept a proton to become $H_2CO_3$, thus behaving as a **Brønsted base**. Or, $HCO_3^-$ can donate a proton to yield $H^+$ and $CO_3^{2-}$, thus behaving as a **Brønsted acid.**

**Comment:** The $HCO_3^-$ species is said to be *amphoteric* because it possesses both acidic and basic properties.

**4.34** **Strategy:** Recall that strong acids and strong bases are strong electrolytes. They are completely ionized in solution. An *ionic equation* will show strong acids and strong bases in terms of their free ions. Weak acids and weak bases are weak electrolytes. They only ionize to a small extent in solution. Weak acids and weak bases are shown as molecules in ionic and net ionic equations. A *net ionic equation* shows only the species that actually take part in the reaction.

**(a)**

**Solution:** $CH_3COOH$ is a weak acid. It will be shown as a molecule in the ionic equation. KOH is a strong base. It completely ionizes to $K^+$ and $OH^-$ ions. Since $CH_3COOH$ is an acid, it donates an $H^+$ to the base, $OH^-$, producing water. The other product is the salt, $CH_3COOK$, which is soluble and remains in solution. The balanced molecular equation is:

$$CH_3COOH(aq) + KOH(aq) \longrightarrow CH_3COOK(aq) + H_2O(l)$$

The ionic and net ionic equations are:

*Ionic*: $CH_3COOH(aq) + K^+(aq) + OH^-(aq) \longrightarrow CH_3COO^-(aq) + K^+(aq) + H_2O(l)$

*Net ionic*: $CH_3COOH(aq) + OH^-(aq) \longrightarrow CH_3COO^-(aq) + H_2O(l)$

**(b)**

**Solution:** $H_2CO_3$ is a weak acid. It will be shown as a molecule in the ionic equation. NaOH is a strong base. It completely ionizes to $Na^+$ and $OH^-$ ions. Since $H_2CO_3$ is an acid, it donates an $H^+$ to the base, $OH^-$, producing water. The other product is the salt, $Na_2CO_3$, which is soluble and remains in solution. The balanced molecular equation is:

$$H_2CO_3(aq) + 2NaOH(aq) \longrightarrow Na_2CO_3(aq) + 2H_2O(l)$$

The ionic and net ionic equations are:

*Ionic*: $H_2CO_3(aq) + 2Na^+(aq) + 2OH^-(aq) \longrightarrow 2Na^+(aq) + CO_3^{2-}(aq) + 2H_2O(l)$

*Net ionic*: $H_2CO_3(aq) + 2OH^-(aq) \longrightarrow CO_3^{2-}(aq) + 2H_2O(l)$

**(c)**

**Solution:** $HNO_3$ is a strong acid. It completely ionizes to $H^+$ and $NO_3^-$ ions. $Ba(OH)_2$ is a strong base. It completely ionizes to $Ba^{2+}$ and $OH^-$ ions. Since $HNO_3$ is an acid, it donates an $H^+$ to the base, $OH^-$, producing water. The other product is the salt, $Ba(NO_3)_2$, which is soluble and remains in solution. The balanced molecular equation is:

$$2HNO_3(aq) + Ba(OH)_2(aq) \longrightarrow Ba(NO_3)_2(aq) + 2H_2O(l)$$

The ionic and net ionic equations are:

*Ionic*: $2H^+(aq) + 2NO_3^-(aq) + Ba^{2+}(aq) + 2OH^-(aq) \longrightarrow Ba^{2+}(aq) + 2NO_3^-(aq) + 2H_2O(l)$

*Net ionic*: $2H^+(aq) + 2OH^-(aq) \longrightarrow 2H_2O(l)$ or $H^+(aq) + OH^-(aq) \longrightarrow H_2O(l)$

**4.44** **Strategy:** In order to break a redox reaction down into an oxidation half-reaction and a reduction half-reaction, you should first assign oxidation numbers to all the atoms in the reaction. In this way, you can determine which element is oxidized (loses electrons) and which element is reduced (gains electrons).

**Solution:** In each part, the reducing agent is the reactant in the first half-reaction and the oxidizing agent is the reactant in the second half-reaction. The coefficients in each half-reaction have been reduced to smallest whole numbers.

**(a)** The product is an ionic compound whose ions are $Fe^{3+}$ and $O^{2-}$.

$$Fe \longrightarrow Fe^{3+} + 3e^-$$

$$O_2 + 4e^- \longrightarrow 2O^{2-}$$

$O_2$ is the oxidizing agent; Fe is the reducing agent.

**(b)** $Na^+$ does not change in this reaction. It is a "spectator ion."

$$2Br^- \longrightarrow Br_2 + 2e^-$$

$$Cl_2 + 2e^- \longrightarrow 2Cl^-$$

$Cl_2$ is the oxidizing agent; $Br^-$ is the reducing agent.

**(c)** Assume $SiF_4$ is made up of $Si^{4+}$ and $F^-$.

$$Si \longrightarrow Si^{4+} + 4e^-$$

$$F_2 + 2e^- \longrightarrow 2F^-$$

$F_2$ is the oxidizing agent; Si is the reducing agent.

**(d)** Assume HCl is made up of $H^+$ and $Cl^-$.

$$H_2 \longrightarrow 2H^+ + 2e^-$$

$$Cl_2 + 2e^- \longrightarrow 2Cl^-$$

$Cl_2$ is the oxidizing agent; $H_2$ is the reducing agent.

**4.46** **Strategy:** In general, we follow the rules listed in Section 4.4 of the text for assigning oxidation numbers. Remember that all alkali metals have an oxidation number of +1 in ionic compounds, and in most cases hydrogen has an oxidation number of +1 and oxygen has an oxidation number of −2 in their compounds.

**Solution:** All the compounds listed are neutral compounds, so the oxidation numbers must sum to zero (Rule 6, Section 4.4 of the text).

Let the oxidation number of P = $x$.

**(a)** $x + 1 + (3)(-2) = 0$, $\boldsymbol{x = +5}$

**(b)** $x + (3)(+1) + (2)(-2) = 0$, $\boldsymbol{x = +1}$

**(c)** $x + (3)(+1) + (3)(-2) = 0$, $\boldsymbol{x = +3}$

**(d)** $x + (3)(+1) + (4)(-2) = 0$, $\boldsymbol{x = +5}$

**(e)** $2x + (4)(+1) + (7)(-2) = 0$, $2x = 10$, $\boldsymbol{x = +5}$

**(f)** $3x + (5)(+1) + (10)(-2) = 0$, $3x = 15$, $\boldsymbol{x = +5}$

The molecules in part (a), (e), and (f) can be made by strongly heating the compound in part (d). Are these oxidation-reduction reactions?

**Check:** In each case, does the sum of the oxidation numbers of all the atoms equal the net charge on the species, in this case zero?

**4.48** All are free elements, so all have an oxidation number of **zero**.

**4.50** **(a)** N: −3 **(b)** O: −1/2 **(c)** C: −1 **(d)** C: +4

**(e)** C: +3 **(f)** O: −2 **(g)** B: +3 **(h)** W: +6

**4.52** **Strategy:** *Hydrogen displacement*: Any metal above hydrogen in the activity series will displace it from water or from an acid. Metals below hydrogen will *not* react with either water or an acid.

**Solution:** Only **(b)** Li and **(d)** Ca are above hydrogen in the activity series, so they are the only metals in this problem that will react with water.

**4.54** **(a)** $Cu(s) + HCl(aq) \rightarrow$ no reaction, since $Cu(s)$ is less reactive than the hydrogen from acids.

**(b)** $I_2(s) + NaBr(aq) \rightarrow$ no reaction, since $I_2(s)$ is less reactive than $Br_2(l)$.

**(c)** $Mg(s) + CuSO_4(aq) \rightarrow MgSO_4(aq) + Cu(s)$, since $Mg(s)$ is more reactive than Cu(s).

Net ionic equation: $Mg(s) + Cu^{2+}(aq) \rightarrow Mg^{2+}(aq) + Cu(s)$

**(d)** $Cl_2(g) + 2KBr(aq) \rightarrow Br_2(l) + 2KCl(aq)$, since $Cl_2(g)$ is more reactive than $Br_2(l)$

Net ionic equation: $Cl_2(g) + 2Br^-(aq) \rightarrow 2Cl^-(aq) + Br_2(l)$

**4.56** **(a)** Combination reaction **(b)** Decomposition reaction

**(c)** Displacement reaction **(d)** Disproportionation reaction

**4.58** The strongest oxidizing agent is the substance that is most easily reduced (i.e., the substance that can most readily gain electrons). The strongest oxidizing agent is $\mathbf{O_2^+}$.

**4.64** **Strategy:** How many moles of $NaNO_3$ does 250 mL of a 0.707 *M* solution contain? How would you convert moles to grams?

**Solution:** From the molarity (0.707 *M*), we can calculate the moles of $NaNO_3$ needed to prepare 250 mL of solution.

$$\text{Moles } NaNO_3 = \frac{0.707 \text{ mol } NaNO_3}{1000 \text{ mL soln}} \times 250 \text{ mL soln} = 0.1768 \text{ mol}$$

Next, we use the molar mass of $NaNO_3$ as a conversion factor to convert from moles to grams. $\mathcal{M}$ ($NaNO_3$) = 85.00 g/mol.

$$0.1768 \text{ mol } NaNO_3 \times \frac{85.00 \text{ g } NaNO_3}{1 \text{ mol } NaNO_3} = 15.0 \text{ g } NaNO_3$$

To make the solution, **dissolve 15.0 g of $NaNO_3$ in enough water to make 250 mL of solution.**

**Check:** As a ball-park estimate, the mass should be given by [molarity (mol/L) × volume (L) = moles × molar mass (g/mol) = grams]. Let's round the molarity to 1 *M* and the molar mass to 80 g, because we are simply making an estimate. This gives: [1 mol/L × (1/4)L × 80 g = 20 g]. This is close to our answer of 15.0 g.

**4.66** Since the problem asks for grams of solute (KOH), you should be thinking that you can calculate moles of solute from the molarity and volume of solution. Then, you can convert moles of solute to grams of solute.

$$? \text{ moles KOH solute} = \frac{5.50 \text{ moles solute}}{1000 \text{ mL solution}} \times 35.0 \text{ mL solution} = 0.1925 \text{ mol KOH}$$

The molar mass of KOH is 56.11 g/mol. Use this conversion factor to calculate grams of KOH.

$$\textbf{? grams KOH} = 0.1925 \text{ mol KOH} \times \frac{56.108 \text{ g KOH}}{1 \text{ mol KOH}} = \textbf{10.8 g KOH}$$

**4.68** (a) $$? \text{ mol } CH_3OH = 6.57 \text{ g } CH_3OH \times \frac{1 \text{ mol } CH_3OH}{32.042 \text{ g } CH_3OH} = 0.205 \text{ mol } CH_3OH$$

$$M = \frac{0.205 \text{ mol } CH_3OH}{0.150 \text{ L}} = \mathbf{1.37\ \mathit{M}}$$

(b) $$? \text{ mol } CaCl_2 = 10.4 \text{ g } CaCl_2 \times \frac{1 \text{ mol } CaCl_2}{110.98 \text{ g } CaCl_2} = 0.09371 \text{ mol } CaCl_2$$

$$M = \frac{0.09371 \text{ mol } CaCl_2}{0.220 \text{ L}} = \mathbf{0.426\ \mathit{M}}$$

(c) $$? \text{ mol } C_{10}H_8 = 7.82 \text{ g } C_{10}H_8 \times \frac{1 \text{ mol } C_{10}H_8}{128.16 \text{ g } C_{10}H_8} = 0.06102 \text{ mol } C_{10}H_8$$

$$M = \frac{0.06102 \text{ mol } C_{10}H_8}{0.0852 \text{ L}} = \mathbf{0.716\ \mathit{M}}$$

**4.70** A 250 mL sample of 0.100 *M* solution contains 0.0250 mol of solute (mol = *M* × L). The computation in each case is the same:

(a) $$0.0250\ \text{mol CsI} \times \frac{259.8\ \text{g CsI}}{1\ \text{mol CsI}} = \mathbf{6.50\ g\ CsI}$$

(b) $$0.0250\ \text{mol } H_2SO_4 \times \frac{98.086\ \text{g } H_2SO_4}{1\ \text{mol } H_2SO_4} = \mathbf{2.45\ g\ H_2SO_4}$$

(c) $$0.0250\ \text{mol } Na_2CO_3 \times \frac{105.99\ \text{g } Na_2CO_3}{1\ \text{mol } Na_2CO_3} = \mathbf{2.65\ g\ Na_2CO_3}$$

(d) $$0.0250\ \text{mol } K_2Cr_2O_7 \times \frac{294.2\ \text{g } K_2Cr_2O_7}{1\ \text{mol } K_2Cr_2O_7} = \mathbf{7.36\ g\ K_2Cr_2O_7}$$

(e) $$0.0250\ \text{mol } KMnO_4 \times \frac{158.04\ \text{g } KMnO_4}{1\ \text{mol } KMnO_4} = \mathbf{3.95\ g\ KMnO_4}$$

**4.72** The mass of $Ba(OH)_2 \times 8H_2O$ required to make 500.0 mL of a solution that is 0.1500 *M* hydroxide ions is:

$$0.5000\ \text{L} \times \frac{0.1500\ \text{mol OH}^-}{1\ \text{L soln}} \times \frac{1\ \text{mol Ba(OH)}_2 \cdot 8H_2O}{2\ \text{mol OH}^-} \times \frac{315.4\ \text{g Ba(OH)}_2 \cdot 8H_2O}{1\ \text{mol Ba(OH)}_2 \cdot 8H_2O}$$

$$= \mathbf{11.83\ g\ Ba(OH)_2 \times 8H_2O}$$

**4.74** **Strategy:** Because the volume of the final solution is greater than the original solution, this is a dilution process. Keep in mind that in a dilution, the concentration of the solution decreases, but the number of moles of the solute remains the same.

**Solution:** We prepare for the calculation by tabulating our data.

$$M_i = 0.866\ M \qquad M_f = ?$$

$$V_i = 25.0\ \text{mL} \qquad V_f = 500\ \text{mL}$$

We substitute the data into Equation (4.3) of the text.

$$M_iV_i = M_fV_f$$

$$(0.866\ M)(25.0\ \text{mL}) = M_f(500\ \text{mL})$$

$$M_f = \frac{(0.866\ M)(25.0\ \text{mL})}{500\ \text{mL}} = \mathbf{0.0433\ M}$$

**4.76** You need to calculate the final volume of the dilute solution. Then, you can subtract 505 mL from this volume to calculate the amount of water that should be added.

$$V_{\text{final}} = \frac{M_{\text{initial}}V_{\text{initial}}}{M_{\text{final}}} = \frac{(0.125\,\cancel{M})(505\text{ mL})}{(0.100\,\cancel{M})} = 631\text{ mL}$$

$$(631 - 505)\text{ mL} = \mathbf{126\text{ mL of water}}$$

**4.78** Moles of calcium nitrate in the first solution:

$$\frac{0.568\text{ mol}}{1000\text{ }\cancel{\text{mL}}\text{ soln}} \times 46.2\text{ }\cancel{\text{mL}}\text{ soln} = 0.02624\text{ mol Ca(NO}_3)_2$$

Moles of calcium nitrate in the second solution:

$$\frac{1.396\text{ mol}}{1000\text{ }\cancel{\text{mL}}\text{ soln}} \times 80.5\text{ }\cancel{\text{mL}}\text{ soln} = 0.1124\text{ mol Ca(NO}_3)_2$$

The volume of the combined solutions = 46.2 mL + 80.5 mL = 126.7 mL. The concentration of the final solution is:

$$M = \frac{(0.02624 + 0.1124)\text{mol}}{0.1267\text{ L}} = \mathbf{1.09\ M}$$

**4.82** **Strategy:** We want to calculate the mass % of Ba in the original compound. Let's start with the definition of mass %.

want to calculate → mass % Ba; need to find → mass Ba; given → mass of sample

$$\text{mass \% Ba} = \frac{\text{mass Ba}}{\text{mass of sample}} \times 100\%$$

The mass of the sample is given in the problem (0.6760 g). Therefore, we need to find the mass of Ba in the original sample. We assume the precipitation is quantitative, that is, that all of the barium in the sample has been precipitated as barium sulfate. From the mass of $BaSO_4$ produced, we can calculate the mass of Ba. There is 1 mole of Ba in 1 mole of $BaSO_4$.

**Solution:** First, we calculate the mass of Ba in 0.4105 g of the $BaSO_4$ precipitate. The molar mass of $BaSO_4$ is 233.4 g/mol.

$$?\text{ mass of Ba} = 0.4105\text{ }\cancel{\text{g}}\text{ BaSO}_4 \times \frac{1\text{ }\cancel{\text{mol}}\text{ BaSO}_4}{233.37\text{ }\cancel{\text{g}}\text{ BaSO}_4} \times \frac{1\text{ }\cancel{\text{mol}}\text{ Ba}}{1\text{ }\cancel{\text{mol}}\text{ BaSO}_4} \times \frac{137.3\text{ g Ba}}{1\text{ }\cancel{\text{mol}}\text{ Ba}}$$

$$= 0.24151\text{ g Ba}$$

Next, we calculate the mass percent of Ba in the unknown compound.

$$\mathbf{\%Ba\ by\ mass} = \frac{0.24151\text{ }\cancel{\text{g}}}{0.6760\text{ }\cancel{\text{g}}} \times 100\% = \mathbf{35.73\%}$$

**4.84** The net ionic equation is: $\mathbf{Ba^{2+}(\mathit{aq}) + SO_4^{2-}(\mathit{aq}) \longrightarrow BaSO_4(\mathit{s})}$

The answer sought is the molar concentration of $SO_4^{2-}$, that is, moles of $SO_4^{2-}$ ions per liter of solution. The dimensional analysis method is used to convert, in order:

$$\text{g of BaSO}_4 \rightarrow \text{moles BaSO}_4 \rightarrow \text{moles SO}_4^{2-} \rightarrow \text{moles SO}_4^{2-} \text{ per liter soln}$$

$$[\mathbf{SO_4^{2-}}] = 0.330 \text{ g BaSO}_4 \times \frac{1 \text{ mol BaSO}_4}{233.4 \text{ g BaSO}_4} \times \frac{1 \text{ mol SO}_4^{2-}}{1 \text{ mol BaSO}_4} \times \frac{1}{0.145 \text{ L}} = \mathbf{0.00975\ \mathit{M}}$$

**4.90** The reaction between HCl and NaOH is:

$$\text{HCl}(aq) + \text{NaOH}(aq) \rightarrow \text{H}_2\text{O}(l) + \text{NaCl}(aq)$$

We know the volume of the NaOH solution, and we want to calculate the molarity of the NaOH solution.

want to calculate ($M$ of NaOH); need to find (mol NaOH); given (L of NaOH soln)

$$M \text{ of NaOH} = \frac{\text{mol NaOH}}{\text{L of NaOH soln}}$$

If we can determine the moles of NaOH in the solution, we can then calculate the molarity of the solution. From the volume and molarity of HCl, we can calculate moles of HCl. Then, using the mole ratio from the balanced equation, we can calculate moles of NaOH.

$$? \text{ mol NaOH} = 17.4 \text{ mL HCl} \times \frac{0.312 \text{ mol HCl}}{1000 \text{ mL soln}} \times \frac{1 \text{ mol NaOH}}{1 \text{ mol HCl}} = 5.429 \times 10^{-3} \text{ mol NaOH}$$

From the moles and volume of NaOH, we calculate the molarity of the NaOH solution.

$$\mathbf{\mathit{M} \text{ of NaOH}} = \frac{\text{mol NaOH}}{\text{L of NaOH soln}} = \frac{5.429 \times 10^{-3} \text{ mol NaOH}}{25.0 \times 10^{-3} \text{ L soln}} = \mathbf{0.217\ \mathit{M}}$$

**4.92** **Strategy:** We know the molarity of the HCl solution, and we want to calculate the volume of the HCl solution.

given ($M$ of HCl); need to find (mol HCl); want to calculate (L of HCl soln)

$$M \text{ of HCl} = \frac{\text{mol HCl}}{\text{L of HCl soln}}$$

If we can determine the moles of HCl, we can then use the definition of molarity to calculate the volume of HCl needed. From the volume and molarity of NaOH or $Ba(OH)_2$, we can calculate moles of NaOH or $Ba(OH)_2$. Then, using the mole ratio from the balanced equation, we can calculate moles of HCl.

**Solution:**

**(a)** In order to have the correct mole ratio to solve the problem, you must start with a balanced chemical equation.

$$HCl(aq) + NaOH(aq) \longrightarrow NaCl(aq) + H_2O(l)$$

$$? \text{ mol HCl} = 10.0 \text{ mL} \times \frac{0.300 \text{ mol NaOH}}{1000 \text{ mL of solution}} \times \frac{1 \text{ mol HCl}}{1 \text{ mol NaOH}} = 3.00 \times 10^{-3} \text{ mol HCl}$$

From the molarity and moles of HCl, we calculate volume of HCl required to neutralize the NaOH.

$$\text{liters of solution} = \frac{\text{moles of solute}}{M}$$

$$\textbf{volume of HCl} = \frac{3.00 \times 10^{-3} \text{ mol HCl}}{0.500 \text{ mol/L}} = \mathbf{6.00 \times 10^{-3} \; L = 6.00 \; mL}$$

**(b)** This problem is similar to part (a). The difference is that the mole ratio between acid and base is 2:1.

$$2HCl(aq) + Ba(OH)_2(aq) \longrightarrow BaCl_2(aq) + 2H_2O(l)$$

$$? \text{ mol HCl} = 10.0 \text{ mL} \times \frac{0.200 \text{ mol Ba(OH)}_2}{1000 \text{ mL of solution}} \times \frac{2 \text{ mol HCl}}{1 \text{ mol Ba(OH)}_2} = 4.00 \times 10^{-3} \text{ mol HCl}$$

$$\textbf{volume of HCl} = \frac{4.00 \times 10^{-3} \text{ mol HCl}}{0.500 \text{ mol/L}} = \mathbf{8.00 \times 10^{-3} \; L = 8.00 \; mL}$$

**4.96** **Strategy:** We want to calculate the grams of $SO_2$ in the sample of air. From the molarity and volume of $KMnO_4$, we can calculate moles of $KMnO_4$. Then, using the mole ratio from the balanced equation, we can calculate moles of $SO_2$. How do we convert from moles of $SO_2$ to grams of $SO_2$?

**Solution:** The balanced equation is given in the problem.

$$5SO_2 + 2MnO_4^- + 2H_2O \longrightarrow 5SO_4^{2-} + 2Mn^{2+} + 4H^+$$

The moles of $KMnO_4$ required for the titration are:

$$\frac{0.00800 \text{ mol KMnO}_4}{1000 \text{ mL soln}} \times 7.37 \text{ mL} = 5.896 \times 10^{-5} \text{ mol KMnO}_4$$

We use the mole ratio from the balanced equation and the molar mass of $SO_2$ as conversion factors to convert to grams of $SO_2$.

$$(5.896 \times 10^{-5} \text{ mol KMnO}_4) \times \frac{5 \text{ mol SO}_2}{2 \text{ mol KMnO}_4} \times \frac{64.07 \text{ g SO}_2}{1 \text{ mol SO}_2} = \mathbf{9.44 \times 10^{-3} \; g \; SO_2}$$

**4.98** The balanced equation is given in the problem.

$$2MnO_4^- + 5H_2O_2 + 6H^+ \longrightarrow 5O_2 + 2Mn^{2+} + 8H_2O$$

First, calculate the moles of potassium permanganate in 36.44 mL of solution.

$$\frac{0.01652 \text{ mol KMnO}_4}{1000 \text{ mL soln}} \times 36.44 \text{ mL} = 6.0199 \times 10^{-4} \text{ mol KMnO}_4$$

Next, calculate the moles of hydrogen peroxide using the mole ratio from the balanced equation.

$$(6.0199 \times 10^{-4} \text{ mol KMnO}_4) \times \frac{5 \text{ mol H}_2\text{O}_2}{2 \text{ mol KMnO}_4} = 1.505 \times 10^{-3} \text{ mol H}_2\text{O}_2$$

Finally, calculate the molarity of the $H_2O_2$ solution. The volume of the solution is 0.02500 L.

$$\textbf{Molarity of H}_2\textbf{O}_2 = \frac{1.505 \times 10^{-3} \text{ mol H}_2\text{O}_2}{0.02500 \text{ L}} = \mathbf{0.06020}\ \boldsymbol{M}$$

**4.100** From the reaction of oxalic acid with NaOH, the moles of oxalic acid in 15.0 mL of solution can be determined. Then, using this number of moles and other information given, the volume of the $KMnO_4$ solution needed to react with a second sample of oxalic acid can be calculated.

First, calculate the moles of oxalic acid in the solution. $H_2C_2O_4(aq) + 2NaOH(aq) \rightarrow Na_2C_2O_4(aq) + 2H_2O(l)$

$$0.0252 \text{ L} \times \frac{0.149 \text{ mol NaOH}}{1 \text{ L soln}} \times \frac{1 \text{ mol H}_2\text{C}_2\text{O}_4}{2 \text{ mol NaOH}} = 1.877 \times 10^{-3} \text{ mol H}_2\text{C}_2\text{O}_4$$

Because we are reacting a second sample of equal volume (15.0 mL), the moles of oxalic acid will also be $1.877 \times 10^{-3}$ mole in this second sample. The balanced equation for the reaction between oxalic acid and $KMnO_4$ is:

$$2MnO_4^- + 16H^+ + 5C_2O_4^{2-} \rightarrow 2Mn^{2+} + 10CO_2 + 8H_2O$$

Let's calculate the moles of $KMnO_4$ first, and then we will determine the volume of $KMnO_4$ needed to react with the 15.0 mL sample of oxalic acid.

$$(1.877 \times 10^{-3} \text{ mol H}_2\text{C}_2\text{O}_4) \times \frac{2 \text{ mol KMnO}_4}{5 \text{ mol H}_2\text{C}_2\text{O}_4} = 7.508 \times 10^{-4} \text{ mol KMnO}_4$$

Using Equation (4.2) of the text:

$$M = \frac{n}{V}$$

$$V_{\text{KMnO}_4} = \frac{n}{M} = \frac{7.508 \times 10^{-4} \text{ mol}}{0.122 \text{ mol/L}} = \mathbf{0.00615\ L} = \mathbf{6.15\ mL}$$

**4.102** The balanced equation is:

$$2MnO_4^- + 16H^+ + 5C_2O_4^{2-} \longrightarrow 2Mn^{2+} + 10CO_2 + 8H_2O$$

$$\text{mol } MnO_4^- = \frac{9.56 \times 10^{-4} \text{ mol } MnO_4^-}{1000 \text{ mL of soln}} \times 24.2 \text{ mL} = 2.314 \times 10^{-5} \text{ mol } MnO_4^-$$

Using the mole ratio from the balanced equation, we can calculate the mass of $Ca^{2+}$ in the 10.0 mL sample of blood.

$$(2.314 \times 10^{-5} \text{ mol } MnO_4^-) \times \frac{5 \text{ mol } C_2O_4^{2-}}{2 \text{ mol } MnO_4^-} \times \frac{1 \text{ mol } Ca^{2+}}{1 \text{ mol } C_2O_4^{2-}} \times \frac{40.08 \text{ g } Ca^{2+}}{1 \text{ mol } Ca^{2+}} = 2.319 \times 10^{-3} \text{ g } Ca^{2+}$$

Converting to mg/mL:

$$\frac{2.319 \times 10^{-3} \text{ g } Ca^{2+}}{10.0 \text{ mL of blood}} \times \frac{1 \text{ mg}}{0.001 \text{ g}} = \mathbf{0.232 \text{ mg } Ca^{2+}/\text{mL of blood}}$$

**4.104** First, the gases could be tested to see if they supported combustion. $O_2$ would support combustion, $CO_2$ would not. Second, if $CO_2$ is bubbled through a solution of calcium hydroxide [$Ca(OH)_2$], a white precipitate of $CaCO_3$ forms. No reaction occurs when $O_2$ is bubbled through a calcium hydroxide solution.

**4.106** Starting with a balanced chemical equation:

$$Mg(s) + 2HCl(aq) \longrightarrow MgCl_2(aq) + H_2(g)$$

From the mass of Mg, you can calculate moles of Mg. Then, using the mole ratio from the balanced equation above, you can calculate moles of HCl reacted.

$$4.47 \text{ g Mg} \times \frac{1 \text{ mol Mg}}{24.31 \text{ g Mg}} \times \frac{2 \text{ mol HCl}}{1 \text{ mol Mg}} = 0.3677 \text{ mol HCl reacted}$$

Next we can calculate the number of moles of HCl in the original solution.

$$\frac{2.00 \text{ mol HCl}}{1000 \text{ mL soln}} \times (5.00 \times 10^2 \text{ mL}) = 1.00 \text{ mol HCl}$$

$$\text{Moles HCl remaining} = 1.00 \text{ mol} - 0.3677 \text{ mol} = 0.6323 \text{ mol HCl}$$

$$\textbf{conc. of HCl after reaction} = \frac{\text{mol HCl}}{\text{L soln}} = \frac{0.6323 \text{ mol HCl}}{0.500 \text{ L}} = 1.26 \text{ mol/L} = \mathbf{1.26}\ \boldsymbol{M}$$

**4.108** (a) The precipitation reaction is: $Al^{3+}(aq) + 3OH^-(aq) \longrightarrow Al(OH)_3(s)$. There are six 0.100 mole spheres of $OH^-(aq)$ which will combine with two 0.100 mole spheres of $Al^{3+}(aq)$, producing 0.200 mole of $Al(OH)_3(s)$. The mass of precipitate is:

$$0.200\ \text{mol Al(OH)}_3 \times \frac{78.00\ \text{g Al(OH)}_3}{1\ \text{mol Al(OH)}_3} = \mathbf{15.6\ g\ Al(OH)_3}$$

(b) The moles of ions remaining in solution are 0.100 mole $Al^{3+}$, 0.900 mole $NO_3^-$, and 0.600 mole $K^+$. The total solution volume is 400 mL. The concentration of each of these ions in the final solution is:

$$[Al^{3+}] = \frac{0.100\ \text{mol}}{0.400\ \text{L}} = \mathbf{0.250}\ \boldsymbol{M}$$

$$[NO_3^-] = \frac{0.900\ \text{mol}}{0.400\ \text{L}} = \mathbf{2.25}\ \boldsymbol{M}$$

$$[K^+] = \frac{0.600\ \text{mol}}{0.400\ \text{L}} = \mathbf{1.50}\ \boldsymbol{M}$$

There are practically no $OH^-$ ions left in solution.

**4.110** The balanced equation is:

$$2HCl(aq) + Na_2CO_3(s) \longrightarrow CO_2(g) + H_2O(l) + 2NaCl(aq)$$

The mole ratio from the balanced equation is 2 moles HCl : 1 mole $Na_2CO_3$. The moles of HCl needed to react with 0.256 g of $Na_2CO_3$ are:

$$0.256\ \text{g Na}_2\text{CO}_3 \times \frac{1\ \text{mol Na}_2\text{CO}_3}{105.99\ \text{g Na}_2\text{CO}_3} \times \frac{2\ \text{mol HCl}}{1\ \text{mol Na}_2\text{CO}_3} = 4.831 \times 10^{-3}\ \text{mol HCl}$$

$$\textbf{Molarity HCl} = \frac{\text{moles HCl}}{\text{L soln}} = \frac{4.831 \times 10^{-3}\ \text{mol HCl}}{0.0283\ \text{L soln}} = 0.171\ \text{mol/L} = \mathbf{0.171}\ \boldsymbol{M}$$

**4.112** Starting with a balanced chemical equation:

$$CH_3COOH(aq) + NaOH(aq) \longrightarrow CH_3COONa(aq) + H_2O(l)$$

From the molarity and volume of the NaOH solution, you can calculate moles of NaOH. Then, using the mole ratio from the balanced equation above, you can calculate moles of $CH_3COOH$.

$$5.75\ \text{mL solution} \times \frac{1.00\ \text{mol NaOH}}{1000\ \text{mL of solution}} \times \frac{1\ \text{mol CH}_3\text{COOH}}{1\ \text{mol NaOH}} = 5.75 \times 10^{-3}\ \text{mol CH}_3\text{COOH}$$

$$\textbf{Molarity CH}_3\textbf{COOH} = \frac{5.75 \times 10^{-3}\ \text{mol CH}_3\text{COOH}}{0.0500\ \text{L}} = \mathbf{0.115}\ \boldsymbol{M}$$

**4.114** The balanced equation is:

$$Zn(s) + 2AgNO_3(aq) \longrightarrow Zn(NO_3)_2(aq) + 2Ag(s)$$

Let $x$ = mass of Ag produced. We can find the mass of Zn reacted in terms of the amount of Ag produced.

$$x \text{ g Ag} \times \frac{1 \text{ mol Ag}}{107.9 \text{ g Ag}} \times \frac{1 \text{ mol Zn}}{2 \text{ mol Ag}} \times \frac{65.39 \text{ g Zn}}{1 \text{ mol Zn}} = 0.303x \text{ g Zn reacted}$$

The mass of Zn remaining will be:

$$2.50 \text{ g} - \text{amount of Zn reacted} = 2.50 \text{ g Zn} - 0.303x \text{ g Zn}$$

The final mass of the strip, 3.37 g, equals the mass of Ag produced + the mass of Zn remaining.

$$3.37 \text{ g} = x \text{ g Ag} + (2.50 \text{ g Zn} - 0.303\,x \text{ g Zn})$$

$$\mathbf{x = 1.25 \text{ g} = \text{mass of Ag produced}}$$

$$\textbf{mass of Zn remaining} = 3.37 \text{ g} - 1.25 \text{ g} = \mathbf{2.12 \text{ g Zn}}$$

or

$$\textbf{mass of Zn remaining} = 2.50 \text{ g Zn} - 0.303x \text{ g Zn} = 2.50 \text{ g} - (0.303)(1.25 \text{ g}) = \mathbf{2.12 \text{ g Zn}}$$

**4.116** The balanced equation is: $HNO_3(aq) + NaOH(aq) \longrightarrow NaNO_3(aq) + H_2O(l)$

$$\text{mol } HNO_3 = \frac{0.211 \text{ mol } HNO_3}{1000 \text{ mL soln}} \times 10.7 \text{ mL soln} = 2.258 \times 10^{-3} \text{ mol } HNO_3$$

$$\text{mol NaOH} = \frac{0.258 \text{ mol NaOH}}{1000 \text{ mL soln}} \times 16.3 \text{ mL soln} = 4.205 \times 10^{-3} \text{ mol NaOH}$$

Since the mole ratio from the balanced equation is 1 mole NaOH : 1 mole $HNO_3$, then $2.258 \times 10^{-3}$ mol $HNO_3$ will react with $2.258 \times 10^{-3}$ mol NaOH.

$$\text{mol NaOH remaining} = (4.205 \times 10^{-3} \text{ mol}) - (2.258 \times 10^{-3} \text{ mol}) = 1.947 \times 10^{-3} \text{ mol NaOH}$$

$$10.7 \text{ mL} + 16.3 \text{ mL} = 27.0 \text{ mL} = 0.0270 \text{ L}$$

$$\textbf{molarity NaOH} = \frac{1.947 \times 10^{-3} \text{ mol NaOH}}{0.0270 \text{ L}} = \mathbf{0.0721}\ \boldsymbol{M}$$

**4.118** The balanced equations for the two reactions are:

$$X(s) + H_2SO_4(aq) \longrightarrow XSO_4(aq) + H_2(g)$$

$$H_2SO_4(aq) + 2NaOH(aq) \longrightarrow Na_2SO_4(aq) + 2H_2O(l)$$

First, let's find the number of moles of excess acid from the reaction with NaOH.

$$0.0334\,\cancel{L} \times \frac{0.500\,\cancel{mol\,NaOH}}{1\,\cancel{L}\,soln} \times \frac{1\,mol\,H_2SO_4}{2\,\cancel{mol\,NaOH}} = 8.35 \times 10^{-3}\,mol\,H_2SO_4$$

The original number of moles of acid was:

$$0.100\,\cancel{L} \times \frac{0.500\,mol\,H_2SO_4}{1\,\cancel{L}\,soln} = 0.0500\,mol\,H_2SO_4$$

The amount of sulfuric acid that reacted with the metal, X, is

$$(0.0500\,mol\,H_2SO_4) - (8.35 \times 10^{-3}\,mol\,H_2SO_4) = 0.04165\,mol\,H_2SO_4.$$

Since the mole ratio from the balanced equation is 1 mole X : 1 mole $H_2SO_4$, then the amount of X that reacted is 0.04165 mol X.

$$\textbf{molar mass X} = \frac{1.00\,g\,X}{0.04165\,mol\,X} = \textbf{24.0 g/mol}$$

The element is **magnesium.**

**4.120** We can calculate the moles of NaOH required for neutralization. Comparing the moles of NaOH to the moles of malonic acid in 0.762 g will allow us to determine the number of ionizable H atoms in malonic acid.

$$?\,mol\,NaOH = 0.01244\,\cancel{L}\,NaOH \times \frac{1.174\,mol\,NaOH}{1\,\cancel{L}\,NaOH\,soln} = 0.01460\,mol\,NaOH$$

$$?\,mol\,C_3H_4O_4 = 0.762\,\cancel{g}\,C_3H_4O_4 \times \frac{1\,mol\,C_3H_4O_4}{104.06\,\cancel{g}\,C_3H_4O_4} = 0.00732\,mol\,C_3H_4O_4$$

Because the moles of NaOH needed for neutralization are double the number of moles of malonic acid, malonic acid must contain **two ionizable H atoms** per molecule.

**4.122** First, calculate the number of moles of glucose present.

$$\frac{0.513\,mol\,glucose}{1000\,\cancel{mL}\,soln} \times 60.0\,\cancel{mL} = 0.03078\,mol\,glucose$$

$$\frac{2.33\,mol\,glucose}{1000\,\cancel{mL}\,soln} \times 120.0\,\cancel{mL} = 0.2796\,mol\,glucose$$

Add the moles of glucose, then divide by the total volume of the combined solutions to calculate the molarity.

$$60.0\,mL + 120.0\,mL = 180.0\,mL = 0.180\,L$$

$$\textbf{Molarity of final solution} = \frac{(0.03078 + 0.2796)\,mol\,glucose}{0.180\,L} = 1.72\,mol/L = \mathbf{1.72}\,\boldsymbol{M}$$

**4.124** Iron(II) compounds can be oxidized to iron(III) compounds. The sample could be tested with a small amount of a strongly colored oxidizing agent like a $KMnO_4$ solution, which is a deep purple color. A loss of color would imply the presence of an oxidizable substance like an iron(II) salt.

**4.126** Since both of the original solutions were strong electrolytes, you would expect a mixture of the two solutions to also be a strong electrolyte. However, since the light dims, the mixture must contain fewer ions than the original solution. Indeed, $H^+$ from the sulfuric acid reacts with the $OH^-$ from the barium hydroxide to form water. The barium cations react with the sulfate anions to form insoluble barium sulfate.

$$2H^+(aq) + SO_4^{2-}(aq) + Ba^{2+}(aq) + 2OH^-(aq) \longrightarrow 2H_2O(l) + BaSO_4(s)$$

Thus, the reaction depletes the solution of ions and the conductivity decreases.

**4.128** To determine the formula of the iron chloride hydrate, we need to calculate the moles of each component in the compound (Fe, Cl, and $H_2O$). The subscripts in a chemical formula represent the mole ratio of the components in the compound.

First, we determine the moles of $H_2O$ in the hydrate.

$$\text{mass } H_2O = 5.012\text{ g} - 3.195\text{ g} = 1.817\text{ g}$$

$$\text{mol } H_2O = 1.817\text{ g } H_2O \times \frac{1\text{ mol } H_2O}{18.02\text{ g } H_2O} = 0.1008\text{ mol}$$

Next, from the mass of the AgCl precipitate, we can determine the moles of Cl and then the mass of Cl in the iron chloride hydrate.

$$7.225\text{ g AgCl} \times \frac{1\text{ mol AgCl}}{143.4\text{ g AgCl}} \times \frac{1\text{ mol Cl}}{1\text{ mol AgCl}} = 0.05038\text{ mol Cl} \times \frac{35.45\text{ g Cl}}{1\text{ mol Cl}} = 1.786\text{ g Cl}$$

Subtracting the mass of Cl from the mass of the anhydrous compound will give the mass of iron in the compound. We can then convert to moles of iron.

$$\text{mass Fe} = 3.195\text{ g} - 1.786\text{ g} = 1.409\text{ g Fe}$$

$$\text{mol Fe} = 1.409\text{ g Fe} \times \frac{1\text{ mol Fe}}{55.85\text{ g Fe}} = 0.02523\text{ mol}$$

This gives the formula, $Fe_{0.02523}Cl_{0.05038} \cdot 0.1008\,H_2O$, which gives the identity and the ratios of atoms present. However, chemical formulas are written with whole numbers. Dividing by the smallest number of moles (0.02523) gives the formula, $\mathbf{FeCl_2 \cdot 4H_2O}$.

**4.130** You could test the conductivity of the solutions. Sugar is a nonelectrolyte and an aqueous sugar solution will not conduct electricity; whereas, NaCl is a strong electrolyte when dissolved in water. Silver nitrate could be added to the solutions to see if silver chloride precipitated. In this particular case, the solutions could also be tasted.

**4.132** In a redox reaction, the oxidizing agent gains one or more electrons. In doing so, the oxidation number of the element gaining the electrons must become more negative. In the case of chlorine, the −1 oxidation number is already the most negative state possible. The chloride ion *cannot* accept any more electrons; therefore, hydrochloric acid is *not* an oxidizing agent.

**4.134** The reaction is too violent. This could cause the hydrogen gas produced to ignite, and an explosion could result.

**4.136** The solid sodium bicarbonate would be the better choice. The hydrogen carbonate ion, $HCO_3^-$, behaves as a Brønsted base to accept a proton from the acid.

$$HCO_3^-(aq) + H^+(aq) \longrightarrow H_2CO_3(aq) \longrightarrow H_2O(l) + CO_2(g)$$

The heat generated during the reaction of hydrogen carbonate with the acid causes the carbonic acid, $H_2CO_3$, that was formed to decompose to water and carbon dioxide.

The reaction of the spilled sulfuric acid with sodium hydroxide would produce sodium sulfate, $Na_2SO_4$, and water. There is a possibility that the $Na_2SO_4$ could precipitate. Also, the sulfate ion, $SO_4^{2-}$, is a weak base; therefore, the "neutralized" solution would actually be *basic*.

$$H_2SO_4(aq) + 2NaOH(aq) \longrightarrow Na_2SO_4(aq) + 2H_2O(l)$$

Also, NaOH is a caustic substance and therefore is not safe to use in this manner.

**4.138** (a) Table salt, NaCl, is very soluble in water and is a strong electrolyte. Addition of $AgNO_3$ will precipitate AgCl.

(b) Table sugar or sucrose, $C_{12}H_{22}O_{11}$, is soluble in water and is a nonelectrolyte.

(c) Aqueous acetic acid, $CH_3COOH$, the primary ingredient of vinegar, is a weak electrolyte. It exhibits all of the properties of acids (Section 4.3).

(d) Baking soda, $NaHCO_3$, is a water-soluble strong electrolyte. It reacts with acid to release $CO_2$ gas. Addition of $Ca(OH)_2$ results in the precipitation of $CaCO_3$.

(e) Washing soda, $Na_2CO_3 \cdot 10H_2O$, is a water-soluble strong electrolyte. It reacts with acids to release $CO_2$ gas. Addition of a soluble alkaline-earth salt will precipitate the alkaline-earth carbonate. Aqueous washing soda is also slightly basic (Section 4.3).

(f) Boric acid, $H_3BO_3$, is weak electrolyte and a weak acid.

(g) Epsom salt, $MgSO_4 \cdot 7H_2O$, is a water-soluble strong electrolyte. Addition of $Ba(NO_3)_2$ results in the precipitation of $BaSO_4$. Addition of hydroxide precipitates $Mg(OH)_2$.

(h) Sodium hydroxide, NaOH, is a strong electrolyte and a strong base. Addition of $Ca(NO_3)_2$ results in the precipitation of $Ca(OH)_2$.

(i) Ammonia, $NH_3$, is a sharp-odored gas that when dissolved in water is a weak electrolyte and a weak base. $NH_3$ in the gas phase reacts with HCl gas to produce solid $NH_4Cl$.

**(j)** Milk of magnesia, $Mg(OH)_2$, is an insoluble, strong base that reacts with acids. The resulting magnesium salt may be soluble or insoluble.

**(k)** $CaCO_3$ is an insoluble salt that reacts with acid to release $CO_2$ gas. $CaCO_3$ is discussed in the Chemistry in Action essays entitled, "An Undesirable Precipitation Reaction" and "Metal from the Sea" in Chapter 4.

With the exception of $NH_3$ and vinegar, all the compounds in this problem are white solids.

**4.140** The balanced equation for the reaction is:

$$\mathrm{XCl}(aq) + \mathrm{AgNO_3}(aq) \longrightarrow \mathrm{AgCl}(s) + \mathrm{XNO_3}(aq) \qquad \text{where X} = \text{Na, or K}$$

From the amount of AgCl produced, we can calculate the moles of XCl reacted (X = Na, or K).

$$1.913\text{ g AgCl} \times \frac{1\text{ mol AgCl}}{143.35\text{ g AgCl}} \times \frac{1\text{ mol XCl}}{1\text{ mol AgCl}} = 0.013345\text{ mol XCl}$$

Let $x$ = number of moles NaCl. Then, the number of moles of KCl = 0.013345 mol − $x$. The sum of the NaCl and KCl masses must equal the mass of the mixture, 0.8870 g. We can write:

$$\text{mass NaCl} + \text{mass KCl} = 0.8870\text{ g}$$

$$\left[x\text{ mol NaCl} \times \frac{58.44\text{ g NaCl}}{1\text{ mol NaCl}}\right] + \left[(0.013345 - x)\text{ mol KCl} \times \frac{74.55\text{ g KCl}}{1\text{ mol KCl}}\right] = 0.8870\text{ g}$$

$$x = 6.6958 \times 10^{-3} = \text{moles NaCl}$$

$$\text{mol KCl} = 0.013345 - x = 0.013345\text{ mol} - (6.6958 \times 10^{-3}\text{ mol}) = 6.6492 \times 10^{-3}\text{ mol KCl}$$

Converting moles to grams:

$$\text{mass NaCl} = (6.6958 \times 10^{-3}\text{ mol NaCl}) \times \frac{58.44\text{ g NaCl}}{1\text{ mol NaCl}} = 0.3913\text{ g NaCl}$$

$$\text{mass KCl} = (6.6492 \times 10^{-3}\text{ mol KCl}) \times \frac{74.55\text{ g KCl}}{1\text{ mol KCl}} = 0.4957\text{ g KCl}$$

The percentages by mass for each compound are:

$$\textbf{\% NaCl} = \frac{0.3913\text{ g}}{0.8870\text{ g}} \times 100\% = \textbf{44.11\% NaCl}$$

$$\textbf{\% KCl} = \frac{0.4957\text{ g}}{0.8870\text{ g}} \times 100\% = \textbf{55.89\% KCl}$$

**4.142** This is an acid-base reaction with $H^+$ from $HNO_3$ combining with $OH^-$ from AgOH to produce water. The other product is the salt, $AgNO_3$, which is soluble (nitrate salts are soluble, see Table 4.2 of the text).

$$AgOH(s) + HNO_3(aq) \rightarrow H_2O(l) + AgNO_3(aq)$$

Because the salt, $AgNO_3$, is soluble, it dissociates into ions in solution, $Ag^+(aq)$ and $NO_3^-(aq)$. The diagram that corresponds to this reaction is **(a)**.

**4.144** The number of moles of oxalic acid in $5.00 \times 10^2$ mL is:

$$\frac{0.100 \text{ mol } H_2C_2O_4}{1000 \text{ mL soln}} \times (5.00 \times 10^2 \text{ mL}) = 0.0500 \text{ mol } H_2C_2O_4$$

The balanced equation shows a mole ratio of 1 mol $Fe_2O_3$ : 6 mol $H_2C_2O_4$. The mass of rust that can be removed is:

$$0.0500 \text{ mol } H_2C_2O_4 \times \frac{1 \text{ mol } Fe_2O_3}{6 \text{ mol } H_2C_2O_4} \times \frac{159.7 \text{ g } Fe_2O_3}{1 \text{ mol } Fe_2O_3} = \mathbf{1.33 \text{ g } Fe_2O_3}$$

**4.146** The precipitation reaction is: $Ag^+(aq) + Br^-(aq) \longrightarrow AgBr(s)$

In this problem, the relative amounts of NaBr and $CaBr_2$ are not known. However, the total amount of $Br^-$ in the mixture can be determined from the amount of AgBr produced. Let's find the number of moles of $Br^-$.

$$1.6930 \text{ g AgBr} \times \frac{1 \text{ mol AgBr}}{187.8 \text{ g AgBr}} \times \frac{1 \text{ mol Br}^-}{1 \text{ mol AgBr}} = 9.0149 \times 10^{-3} \text{ mol Br}^-$$

The amount of $Br^-$ comes from both NaBr and $CaBr_2$. Let $x$ = number of moles NaBr. Then, the number of moles of $CaBr_2 = \frac{9.0149 \times 10^{-3} \text{ mol} - x}{2}$. The moles of $CaBr_2$ are divided by 2, because 1 mol of $CaBr_2$ produces 2 moles of $Br^-$. The sum of the NaBr and $CaBr_2$ masses must equal the mass of the mixture, 0.9157 g. We can write:

$$\text{mass NaBr} + \text{mass CaBr}_2 = 0.9157 \text{ g}$$

$$\left[x \text{ mol NaBr} \times \frac{102.89 \text{ g NaBr}}{1 \text{ mol NaBr}}\right] + \left[\left(\frac{9.0149 \times 10^{-3} - x}{2}\right) \text{mol CaBr}_2 \times \frac{199.88 \text{ g CaBr}_2}{1 \text{ mol CaBr}_2}\right] = 0.9157 \text{ g}$$

$$2.95x = 0.014751$$

$$x = 5.0003 \times 10^{-3} = \text{moles NaBr}$$

Converting moles to grams:

$$\text{mass NaBr} = (5.0003 \times 10^{-3} \text{ mol NaBr}) \times \frac{102.89 \text{ g NaBr}}{1 \text{ mol NaBr}} = 0.51448 \text{ g NaBr}$$

The percentage by mass of NaBr in the mixture is:

$$\textbf{\% NaBr} = \frac{0.51448 \text{ g}}{0.9157 \text{ g}} \times 100\% = \mathbf{56.18\%\ NaBr}$$

**4.148** There are two moles of $Cl^-$ per one mole of $CaCl_2$.

**(a)** $$25.3 \text{ g } CaCl_2 \times \frac{1 \text{ mol } CaCl_2}{110.98 \text{ g } CaCl_2} \times \frac{2 \text{ mol } Cl^-}{1 \text{ mol } CaCl_2} = 0.4559 \text{ mol } Cl^-$$

$$\textbf{Molarity } \mathbf{Cl^-} = \frac{0.4559 \text{ mol } Cl^-}{0.325 \text{ L soln}} = 1.40 \text{ mol/L} = \mathbf{1.40}\ \boldsymbol{M}$$

**(b)** We need to convert from mol/L to grams in 0.100 L.

$$\frac{1.40 \text{ mol } Cl^-}{1 \text{ L soln}} \times \frac{35.45 \text{ g Cl}}{1 \text{ mol } Cl^-} \times 0.100 \text{ L soln} = \mathbf{4.96\ g\ Cl^-}$$

**4.150** **(a)** $NH_4^+(aq) + OH^-(aq) \longrightarrow NH_3(aq) + H_2O(l)$

**(b)** From the amount of NaOH needed to neutralize the 0.2041 g sample, we can find the amount of the 0.2041 g sample that is $NH_4NO_3$.

First, calculate the moles of NaOH.

$$\frac{0.1023 \text{ mol NaOH}}{1000 \text{ mL of soln}} \times 24.42 \text{ mL soln} = 2.4982 \times 10^{-3} \text{ mol NaOH}$$

Using the mole ratio from the balanced equation, we can calculate the amount of $NH_4NO_3$ that reacted.

$$(2.4982 \times 10^{-3} \text{ mol NaOH}) \times \frac{1 \text{ mol } NH_4NO_3}{1 \text{ mol NaOH}} \times \frac{80.052 \text{ g } NH_4NO_3}{1 \text{ mol } NH_4NO_3} = 0.19999 \text{ g } NH_4NO_3$$

The purity of the $NH_4NO_3$ sample is:

$$\textbf{\% purity} = \frac{0.19999 \text{ g}}{0.2041 \text{ g}} \times 100\% = \mathbf{97.99\%}$$

**4.152** Using the rules for assigning oxidation numbers given in Section 4.4, H is +1, F is −1, so the oxidation number of O must be **zero**.

**4.154** The balanced equation is:

$$3CH_3CH_2OH + 2K_2Cr_2O_7 + 8H_2SO_4 \longrightarrow 3CH_3COOH + 2Cr_2(SO_4)_3 + 2K_2SO_4 + 11H_2O$$

From the amount of $K_2Cr_2O_7$ required to react with the blood sample, we can calculate the mass of ethanol ($CH_3CH_2OH$) in the 10.0 g sample of blood.

First, calculate the moles of $K_2Cr_2O_7$ reacted.

$$\frac{0.07654 \text{ mol } K_2Cr_2O_7}{1000 \text{ mL soln}} \times 4.23 \text{ mL} = 3.238 \times 10^{-4} \text{ mol } K_2Cr_2O_7$$

Next, using the mole ratio from the balanced equation, we can calculate the mass of ethanol that reacted.

$$3.238 \times 10^{-4} \text{ mol } K_2Cr_2O_7 \times \frac{3 \text{ mol ethanol}}{2 \text{ mol } K_2Cr_2O_7} \times \frac{46.068 \text{ g ethanol}}{1 \text{ mol ethanol}} = 0.02238 \text{ g ethanol}$$

The percent ethanol by mass is:

$$\textbf{\% by mass ethanol} = \frac{0.02238 \text{ g}}{10.0 \text{ g}} \times 100\% = \mathbf{0.224\%}$$

This is well above the legal limit of 0.08 percent by mass ethanol in the blood. The individual should be prosecuted for drunk driving.

**4.156** **(a)** $Zn(s) + H_2SO_4(aq) \longrightarrow ZnSO_4(aq) + H_2(g)$

**(b)** $2KClO_3(s) \longrightarrow 2KCl(s) + 3O_2(g)$

**(c)** $Na_2CO_3(s) + 2HCl(aq) \longrightarrow 2NaCl(aq) + CO_2(g) + H_2O(l)$

**(d)** $NH_4NO_2(s) \xrightarrow{\text{heat}} N_2(g) + 2H_2O(g)$

**4.158** $NH_4Cl$ exists as $NH_4^+$ and $Cl^-$. To form $NH_3$ and HCl, a proton ($H^+$) is transferred from $NH_4^+$ to $Cl^-$. Therefore, this is a Brønsted acid-base reaction.

**4.160** **(a)**

**First Solution:**

$$0.8214 \text{ g } KMnO_4 \times \frac{1 \text{ mol } KMnO_4}{158.04 \text{ g } KMnO_4} = 5.1974 \times 10^{-3} \text{ mol } KMnO_4$$

$$M = \frac{\text{mol solute}}{\text{L of soln}} = \frac{5.1974 \times 10^{-3} \text{ mol } KMnO_4}{0.5000 \text{ L}} = 1.0395 \times 10^{-2}\ M$$

**Second Solution:**

$$M_1V_1 = M_2V_2$$

$$(1.0395 \times 10^{-2}\ M)(2.000\ \text{mL}) = M_2(1000\ \text{mL})$$

$$M_2 = 2.079 \times 10^{-5}\ M$$

**Third Solution:**

$$M_1V_1 = M_2V_2$$

$$(2.079 \times 10^{-5}\ M)(10.00\ \text{mL}) = M_2(250.0\ \text{mL})$$

$$\mathbf{M_2 = 8.316 \times 10^{-7}\ M}$$

**(b)** From the molarity and volume of the final solution, we can calculate the moles of $KMnO_4$. Then, the mass can be calculated from the moles of $KMnO_4$.

$$\frac{8.316 \times 10^{-7}\ \text{mol KMnO}_4}{1000\ \text{mL of soln}} \times 250\ \text{mL} = 2.079 \times 10^{-7}\ \text{mol KMnO}_4$$

$$2.079 \times 10^{-7}\ \text{mol KMnO}_4 \times \frac{158.04\ \text{g KMnO}_4}{1\ \text{mol KMnO}_4} = \mathbf{3.286 \times 10^{-5}\ g\ KMnO_4}$$

This mass is too small to directly weigh accurately.

**4.162** The first titration oxidizes $Fe^{2+}$ to $Fe^{3+}$. This titration gives the amount of $Fe^{2+}$ in solution. Zn metal is added to reduce all $Fe^{3+}$ back to $Fe^{2+}$. The second titration oxidizes all the $Fe^{2+}$ back to $Fe^{3+}$. We can find the amount of $Fe^{3+}$ in the original solution by difference.

**Titration #1:** The mole ratio between $Fe^{2+}$ and $MnO_4^-$ is 5:1.

$$23.0\ \text{mL soln} \times \frac{0.0200\ \text{mol MnO}_4^-}{1000\ \text{mL soln}} \times \frac{5\ \text{mol Fe}^{2+}}{1\ \text{mol MnO}_4^-} = 2.30 \times 10^{-3}\ \text{mol Fe}^{2+}$$

$$[\mathbf{Fe^{2+}}] = \frac{\text{mol solute}}{\text{L of soln}} = \frac{2.30 \times 10^{-3}\ \text{mol Fe}^{2+}}{25.0 \times 10^{-3}\ \text{L soln}} = \mathbf{0.0920\ M}$$

**Titration #2:** The mole ratio between $Fe^{2+}$ and $MnO_4^-$ is 5:1.

$$40.0\ \text{mL soln} \times \frac{0.0200\ \text{mol MnO}_4^-}{1000\ \text{mL soln}} \times \frac{5\ \text{mol Fe}^{2+}}{1\ \text{mol MnO}_4^-} = 4.00 \times 10^{-3}\ \text{mol Fe}^{2+}$$

In this second titration, there are more moles of $Fe^{2+}$ in solution. This is due to $Fe^{3+}$ in the original solution being reduced by Zn to $Fe^{2+}$. The number of moles of $Fe^{3+}$ in solution is:

$$(4.00 \times 10^{-3}\ \text{mol}) - (2.30 \times 10^{-3}\ \text{mol}) = 1.70 \times 10^{-3}\ \text{mol Fe}^{3+}$$

$$[\mathbf{Fe^{3+}}] = \frac{\text{mol solute}}{\text{L of soln}} = \frac{1.70 \times 10^{-3}\ \text{mol Fe}^{3+}}{25.0 \times 10^{-3}\ \text{L soln}} = \mathbf{0.0680\ M}$$

**4.164** **(a)** The precipitation reaction is: $Mg^{2+}(aq) + 2OH^-(aq) \longrightarrow Mg(OH)_2(s)$

The acid-base reaction is: $Mg(OH)_2(s) + 2HCl(aq) \longrightarrow MgCl_2(aq) + 2H_2O(l)$

The redox reactions are:

$$Mg^{2+} + 2e^- \longrightarrow Mg$$
$$2Cl^- \longrightarrow Cl_2 + 2e^-$$
$$\overline{MgCl_2 \longrightarrow Mg + Cl_2}$$

**(b)** NaOH is much more expensive than CaO.

**(c)** Dolomite has the advantage of being an additional source of magnesium that can also be recovered.

**4.166** Because only B and C react with 0.5 $M$ HCl, they are more electropositive than A and D. The fact that when B is added to a solution containing the ions of the other metals, metallic A, C, and D are formed indicates that B is the most electropositive metal. Because A reacts with 6 $M$ $HNO_3$, A is more electropositive than D. The metals arranged in increasing order as reducing agents are:

$$\mathbf{D < A < C < B}$$

Examples are: D = Au, A = Cu, C = Zn, B = Mg

**4.168** **(a)** The net ionic reaction is: $\mathbf{Cu^{2+}(aq) + SO_4^{2-}(aq) + Ba^{2+}(aq) + 2OH^-(aq) \longrightarrow Cu(OH)_2(s) + BaSO_4(s)}$. See the solubility rules in Table 4.2 of the text.

**(b)** There are six 0.0500 mole spheres of $OH^-(aq)$ which will combine with three 0.0500 mole spheres of $Cu^{2+}(aq)$, producing 0.150 mole of $Cu(OH)_2(s)$. There are three 0.0500 mole spheres of $Ba^{2+}(aq)$ which will combine with three 0.0500 mole spheres of $SO_4^{2-}(aq)$, producing 0.150 mole of $BaSO_4(s)$. The mass of each precipitate formed is:

$$0.150\ \text{mol Cu(OH)}_2 \times \frac{97.57\ \text{g Cu(OH)}_2}{1\ \text{mol Cu(OH)}_2} = \mathbf{14.6\ g\ Cu(OH)_2}$$

$$0.150\ \text{mol BaSO}_4 \times \frac{233.4\ \text{g BaSO}_4}{1\ \text{mol BaSO}_4} = \mathbf{35.0\ g\ BaSO_4}$$

There is one 0.0500 mole sphere of $Cu^{2+}$ and one 0.0500 mole sphere of $SO_4^{2-}$ remaining in solution. The total solution volume is 1200 mL = 1.200 L. The concentration of each of these ions in the mixed solution is:

$$[Cu^{2+}] = [SO_4^{2-}] = \frac{0.0500\ \text{mol}}{1.200\ \text{L}} = \mathbf{0.0417\ M}$$

There are practically no $Ba^{2+}$ and $OH^-$ ions left in solution.

# CHAPTER 5
# GASES

## PROBLEM-SOLVING STRATEGIES AND TUTORIAL SOLUTIONS

### TYPES OF PROBLEMS

**Problem Type 1:** Pressure Conversions (Also see Problem Type 5, Chapter 1).

**Problem Type 2:** Calculations Using the Ideal Gas Law.

- **(a)** Given three of the four variable quantities in the equation ($P$, $V$, $n$, and $T$).
- **(b)** Only one or two variable quantities have fixed values.
- **(c)** Calculations using density or molar mass (Also see Density calculations, Problem Type 1, Chapter 1).
- **(d)** Stoichiometry involving gases (Also see Stoichiometry problems, Problem Type 7, Chapter 3).

**Problem Type 3:** Dalton's Law of Partial Pressures.

- **(a)** Introduction
- **(b)** Collecting a gas over water.

**Problem Type 4:** Root-Mean-Square (RMS) Speed.

**Problem Type 5:** Deviations from Ideal Behavior.

## PROBLEM TYPE 1: PRESSURE CONVERSIONS

(Also see Problem Type 5, Chapter 1)

In order to convert from one unit of pressure to another, you need to be proficient at the dimensional analysis method. See Section 1.9 of your text. Unit conversions can seem daunting, but if you keep track of the units, making sure that the appropriate units cancel, your effort will be rewarded.

**EXAMPLE 5.1**

**Convert 555 mmHg to kPa.**

**Strategy:** Map out a strategy to proceed from initial units to final units based on available conversion factors. Looking in Section 5.2 of your text, you should find the following conversions.

$$1 \text{ atm} = 760 \text{ mmHg}$$
$$1 \text{ atm} = 1.01325 \times 10^2 \text{ kPa}$$

You should come up with the following strategy:

$$\text{mm Hg} \rightarrow \text{atm} \rightarrow \text{kPa}$$

**Solution:** Use the following method to ensure that you obtain the desired unit.

$$\text{Given unit} \times \left(\frac{\text{desired unit}}{\text{given unit}}\right) = \text{desired unit}$$

$$\textbf{? kPa} = 555 \text{ mmHg (given)} \times \left(\frac{1 \text{ atm (desired)}}{760 \text{ mmHg (given)}}\right) \times \left(\frac{1.01325 \times 10^2 \text{ kPa (desired)}}{1 \text{ atm (given)}}\right) = \textbf{74.0 kPa}$$

> **Note**: In the first conversion factor (atm/mmHg), atm is the desired unit. When moving on to the next conversion factor (kPa/atm), atm is now given, and the desired unit is kPa.

**PRACTICE EXERCISE**

1. Convert 5.32 atm to mmHg.

**Text Problem: 5.14**

## PROBLEM TYPE 2: CALCULATIONS USING THE IDEAL GAS LAW

You will encounter five different types of problems in this chapter that use the Ideal Gas Law. Each type will be addressed individually. Remember that $R$ (the gas constant) has the units L·atm/mol·K. Therefore, in problems that contain $R$, you *must* use the following units: volume in liters (L), pressure in atmospheres (atm), and temperature in Kelvin (K).

### A. Given three of the four variable quantities in the equation (*P*, *V*, *n*, and *T*)

If you know the values of three of the variable quantities, you can solve $PV = nRT$ algebraically for the fourth one. Then, you can calculate its value by substituting in the three known quantities.

**EXAMPLE 5.2**

**Carbon monoxide gas, CO, stored in a 2.00 L container at 25.0°C exerts a pressure of 15.5 atm. How many moles of CO(g) are in the container?**

**Strategy:** This problem gives the volume, temperature, and pressure of CO gas. Is the gas undergoing a change in any of its properties? What equation should we use to solve for moles of CO? What temperature unit should be used?

**Solution:** Because no changes in gas properties occur, we can use the ideal gas equation to calculate moles. Rearranging Equation (5.8) of the text, we write:

$$n = \frac{PV}{RT}$$

Check that the three known quantities ($P$, $V$, and $T$) have the appropriate units. $P$ and $V$ have the correct units, but $T$ must be converted to units of K.

$$T(\text{K}) = {}^\circ\text{C} + 273^\circ$$
$$T(\text{K}) = 25.0^\circ + 273^\circ = 298\ \text{K}$$

Calculate the value of $n$ by substituting the three known quantities into the equation.

$$n = \frac{PV}{RT} = \frac{(15.5\ \cancel{\text{atm}})(2.00\ \cancel{\text{L}})}{298\ \cancel{\text{K}}} \times \frac{\text{mol}\cdot\cancel{\text{K}}}{0.0821\ \cancel{\text{L}}\cdot\cancel{\text{atm}}} = \mathbf{1.27\ mol\ CO}$$

**PRACTICE EXERCISE**

2. Carbon dioxide gas, $CO_2$, and water vapor are the two species primarily responsible for the greenhouse effect. Combustion of fossil fuels adds 22 billion tons of carbon dioxide to the atmosphere each year. How many liters of $CO_2$ is this at 25°C and 1.0 atm pressure?

**Text Problems: 5.32, 5.40, 5.42**

## B. Only one or two variable quantities have fixed values.

Let's look at the case where $n$ and $P$ are constant (Charles' law). We start with Equation (5.9) of the text.

$$\frac{P_1V_1}{n_1T_1} = \frac{P_2V_2}{n_2T_2} \qquad (5.9, \text{text})$$

Because $n_1 = n_2$ and $P_1 = P_2$,

$$\frac{V_1}{T_1} = \frac{V_2}{T_2}$$

**This is Charles' law.**

Solve the equation algebraically for the missing quantity. Then, you can calculate its value by substituting in the known quantities.

### EXAMPLE 5.3

**Given 10.0 L of neon gas at 5.0°C and 630 mmHg, calculate the new volume at 400°C and 2.5 atm.**

**Strategy:** The amount of gas remains constant, but the pressure, temperature, and volume change. What equation would you use to solve for the final volume?

**Solution:** We start with Equation (5.9) of the text.

$$\frac{P_1V_1}{n_1T_1} = \frac{P_2V_2}{n_2T_2}$$

Because $n_1 = n_2$,

$$\frac{P_1V_1}{T_1} = \frac{P_1V_2}{T_2}$$

Solve the equation algebraically for $V_2$, and then substitute in the known quantities to solve for $V_2$.

$$P_2 = 2.5\ \text{atm} \times \frac{760\ \text{mmHg}}{1\ \text{atm}} = 1.9 \times 10^3\ \text{mmHg}$$

$$V_2 = \frac{P_1V_1T_2}{P_2T_1}$$

$$V_2 = \frac{(630\ \text{mmHg})(10.0\ \text{L})(673\ \text{K})}{(1.9 \times 10^3\ \text{mmHg})(278\ \text{K})} = \mathbf{8.0\ L}$$

**Tip:** Since $R$ is not in this equation, the units of pressure and volume do *not* have to be atm and liters, respectively. However, temperature must *always* be in units of Kelvin for gas law calculations.

### PRACTICE EXERCISE

3. After flying in an airplane, a passenger finds that his sticky hair-gel has opened in his bag making a complete mess. This happens because the cabin is pressurized to about 8000 feet above sea level. During flight, the volume of air inside the bottle increases due to the decreased pressure. Sometimes, this can cause the flip-top cap to open. If the volume of air inside the bottle is 125 mL at sea level ($P = 760$ mmHg), what volume does the air in the bottle occupy in the airplane ($P = 595$ mmHg)? Assume that the temperature is kept constant.

**Text Problems:** 5.20, **5.22**, **5.24**, 5.34, 5.36, 5.38

## C. Calculations involving density (*d*) or molar mass (*M*)

(Also see Problem Type 1, Chapter 1)

*Step 1:* Algebraically solve the ideal gas equation for either density $\left(\frac{m}{V}\right)$ or molar mass ($M$).

$$PV = nRT$$

$$\frac{n}{V} = \frac{P}{RT} \quad \text{and} \quad n = \frac{m}{M}$$

Substituting for *n*,

$$\frac{m}{MV} = \frac{P}{RT}$$

and,

$$d = \frac{m}{V} = \frac{PM}{RT} \qquad \text{(5.11, text)}$$

Furthermore,

$$M = \frac{dRT}{P}$$

*Step 2:* Calculate the value of density or molar mass by substituting in the known quantities.

**EXAMPLE 5.4**

**What is the density of methane gas, $CH_4$, at STP.**

**Strategy:** This problem is simplified if you recall that 1 mole of an ideal gas occupies a volume of 22.4 L at STP. If we can calculate the mass of 1 mole of methane, we can solve for its density.

**Solution:** Let's assume that we have 1 mole of methane gas. At STP, 1 mole of methane will occupy a volume of 22.4 L. First, we calculate the molar mass of methane.

$$M\,(CH_4) = 12.01\text{ g} + 4(1.008\text{ g}) = 16.04\text{ g}$$

Knowing the mass of 1 mole of methane and the volume it occupies, we can calculate the density.

$$d = \frac{\text{mass}}{\text{volume}}$$

$$d = \frac{16.04\text{ g}}{22.4\text{ L}} = \mathbf{0.716\ g/L}$$

This problem could also be solved using Equation (5.11) of the text derived above.

$$d = \frac{PM}{RT}$$

Calculate the density by substituting in the known quantities.

**STP:** $P = 1.00$ atm

$T = 273$ K

$M = 16.04$ g/mol

$$d = \frac{(1.00\ \cancel{\text{atm}})\left(16.04\frac{\text{g}}{\cancel{\text{mol}}}\right)}{273\ \cancel{\text{K}}} \times \frac{\cancel{\text{mol}}\cdot\cancel{\text{K}}}{0.0821\ \text{L}\cdot\cancel{\text{atm}}} = \mathbf{0.716\ g/L}$$

**PRACTICE EXERCISE**

4. An element that exists as a diatomic gas at room temperature has a density of 1.553 g/L at 25°C and 1 atm pressure. Identify the unknown gas.

| **Text Problems: 5.44**, 5.48, 5.50, 5.52 |
|---|

## D. Stoichiometry Involving Gases

(Also see Problem Type 7, Chapter 3)

A typical gas stoichiometry problem involves the following approach:

| Amount of reactant. Expressed in grams or volume | → | moles of reactant | → | moles of product | → | amount of product. Expressed in grams, volume, etc. |
|---|---|---|---|---|---|---|

**EXAMPLE 5.5**

**Oxygen gas was discovered by decomposing mercury (II) oxide.**

$$2HgO(s) \longrightarrow 2Hg(l) + O_2(g)$$

**What volume of oxygen gas would be produced by the reaction of 35.2 g of the oxide if the gas is collected at STP?**

**Strategy:** From the moles of HgO reacted, we can calculate the moles of $O_2$ produced. From the balanced equation, we see that 2 mol HgO ≏ 1 mol $O_2$. Once moles of $O_2$ are determined, we can use the ideal gas equation to calculate the volume of $O_2$.

**Solution:** Since stoichiometry uses mole ratios, make sure that you start with a balanced equation. In this problem, you are given the balanced equation. First let's calculate moles of $O_2$ produced.

$$\text{grams HgO} \rightarrow \text{moles HgO} \rightarrow \text{moles } O_2$$

$$? \text{ mol } O_2 = 35.2 \cancel{g} \text{ HgO} \times \frac{1 \cancel{mol} \text{ HgO}}{216.6 \cancel{g} \text{ HgO}} \times \frac{1 \text{ mol } O_2}{2 \cancel{mol} \text{ HgO}} = 0.0813 \text{ mol } O_2$$

Solve the ideal gas equation algebraically for $V_{O_2}$. Then, calculate the volume by substituting the known quantities into the equation.

**STP:** $P = 1.00$ atm

$T = 273$ K

$$V_{O_2} = \frac{n_{O_2}RT}{P}$$

$$V_{O_2} = \frac{(0.0813 \cancel{mol})(273 \cancel{K})}{1.00 \cancel{atm}} \times \frac{0.0821 \text{ L} \cdot \cancel{atm}}{\cancel{mol} \cdot \cancel{K}} = \mathbf{1.82 \text{ L}}$$

**EXAMPLE 5.6**

**Calculate the volume of methane, $CH_4$, at STP required to completely consume 3.50 L of oxygen at STP.**

**Strategy:** You could follow the strategy used in the previous example to solve this problem. You could find the moles of oxygen in 3.50 L, apply the correct mole ratio to find moles of methane, and then use the ideal gas equation to find the volume of methane. However, there is a short cut if you remember Avogadro's Law. Avogadro's Law states that the volume of a gas is directly proportional to the number of moles of gas at constant temperature and pressure.

**Solution:** Write the balanced chemical equation.

$$CH_4(g) + 2O_2(g) \longrightarrow CO_2(g) + 2H_2O(l)$$

The stoichiometric ratio, $\frac{1 \text{ mol } CH_4}{2 \text{ mol } O_2}$, can be written in terms of volume, $\frac{1 \text{ L } CH_4}{2 \text{ L } O_2}$. Calculate the volume of methane using the above conversion factor.

$$\textbf{? L CH}_\mathbf{4} = 3.50 \cancel{\text{L}}\, O_2 \times \frac{1 \text{ L } CH_4}{2 \cancel{\text{L}}\, O_2} = \mathbf{1.75 \text{ L } CH_4}$$

**PRACTICE EXERCISE**

5. The decomposition of sodium azide ($NaN_3$) to sodium and nitrogen gas was one of the first reactions used to inflate air-bag systems in automobiles.

$$2NaN_3(s) \longrightarrow 2Na(l) + 3N_2(g)$$

What mass of $NaN_3$ in grams would be needed to inflate a 100 L air bag with nitrogen at 25.0°C and 755 mmHg?

**Text Problems: 5.54**, 5.56, **5.58**, 5.60, 5.62, 5.64

## PROBLEM TYPE 3: DALTON'S LAW OF PARTIAL PRESSURES

### A. Introduction

In order to understand how to approach problems with partial pressures, it is important to know (1) the derivation of Dalton's law of partial pressures and (2) the relationship between partial pressure and the total pressure.

(1) Dalton's law for two gases, **A** and **B**, in a container of volume, *V*.

$$P_{\text{Total}} = \frac{n_{\text{Total}}RT}{V}$$

$$\text{and, } n_{\text{Total}} = n_A + n_B$$

$$P_{\text{Total}} = \frac{(n_A + n_B)RT}{V}$$

$$P_{\text{Total}} = \frac{n_A RT}{V} + \frac{n_B RT}{V}$$

$$\mathbf{P_{Total} = P_A + P_B}$$

In general, $\mathbf{P_{Total} = P_1 + P_2 + P_3 + \ldots + P_n}$, where $P_1, P_2, P_3, \ldots$ are the partial pressures of components 1, 2, 3, . .

(2) The relationship between partial pressure and $P_{\text{Total}}$ for two gases, **A** and **B**, in a container of volume, *V*.

$$\frac{P_A}{P_T} = \frac{\frac{n_A RT}{V}}{\frac{(n_A + n_B)RT}{V}}$$

$$\frac{P_A}{P_T} = \frac{n_A}{(n_A + n_B)} \quad \text{and} \quad \frac{n_A}{(n_A + n_B)} = X_A$$

where $X_A$ = mole fraction of component $A$.

$$\frac{P_A}{P_T} = X_A$$

$$\mathbf{P_A = X_A P_T}$$

Similarly,

$$\mathbf{P_B = X_B P_T}$$

**In general, $P_i = X_i P_T$, where $X_i$ is the mole fraction of substance $i$.**

**EXAMPLE 5.7**

**A 2.00 L container at 22.0°C contains a mixture of 1.00 g $H_2(g)$ and 1.00 g He(g).**

**(a)** What are the partial pressures of $H_2$ and He? What is the total pressure?

**Strategy:** This is a mixture of two gases that obeys Dalton's law of partial pressures.

$$P_T = P_{H_2(g)} + P_{He(g)}$$

The partial pressure of each component can be calculated from the amount of each gas and the given conditions. The total pressure is the sum of the two partial pressures calculated.

**Solution:** First, calculate the moles of $H_2$ and He.

$$n_{H_2} = 1.00\ \text{g} \times \frac{1\ \text{mol}\ H_2}{2.016\ \text{g}\ H_2} = 0.496\ \text{mol}\ H_2$$

$$n_{He} = 1.00\ \text{g} \times \frac{1\ \text{mol He}}{4.003\ \text{g He}} = 0.250\ \text{mol He}$$

Solve the ideal gas equation algebraically for P, and then substitute in the known quantities to solve for $P_{H_2}$ or $P_{He}$.

$T(\text{K}) = 22.0° + 273° = 295\ \text{K}$

$$P_{H_2} = \frac{n_{H_2} RT}{V} \qquad P_{He} = \frac{n_{He} RT}{V}$$

$$P_{H_2} = \frac{(0.496\ \text{mol})(295\ \text{K})}{2.00\ \text{L}} \times \frac{0.0821\ \text{L} \cdot \text{atm}}{\text{mol} \cdot \text{K}} \qquad P_{He} = \frac{(0.250\ \text{mol})(295\ \text{K})}{2.00\ \text{L}} \times \frac{0.0821\ \text{L} \cdot \text{atm}}{\text{mol} \cdot \text{K}}$$

$$\mathbf{P_{H_2} = 6.01\ atm} \qquad \mathbf{P_{He} = 3.03\ atm}$$

$$P_{Total} = P_{H_2} + P_{He}$$

$$P_{Total} = 6.01\ \text{atm} + 3.03\ \text{atm} = \mathbf{9.04\ atm}$$

**(b)** What are the mole fractions of $H_2$ and He?

**Strategy:** Remember that $P_i = X_i P_T$.

**Solution:**

$$P_{H_2} = X_{H_2} P_T \qquad P_{He} = X_{He} P_T$$

$$X_{H_2} = \frac{P_{H_2}}{P_T} \qquad X_{He} = \frac{P_{He}}{P_T}$$

$$X_{H_2} = \frac{6.01 \text{ atm}}{9.04 \text{ atm}} = \mathbf{0.665} \qquad X_{He} = \frac{3.03 \text{ atm}}{9.04 \text{ atm}} = \mathbf{0.335}$$

**PRACTICE EXERCISE**

6. About two-thirds of the carbon monoxide (CO) emissions in the United States come from automobiles. CO is extremely toxic to humans because it binds 210 times more strongly to hemoglobin than does $O_2$. Hemoglobin is the iron-containing protein responsible for oxygen transport in blood. In a typical urban environment, the CO concentration is 10 parts per million (ppm). Assuming an atmospheric pressure of 750 torr, calculate the partial pressure of CO.

**Hint:** 1 ppm of a gas refers to one part by volume in 1 million volume units of the whole (volume fraction).

**Text Problems:** 5.68, 5.70, 5.72, 5.74, 5.76

## B. Collecting a gas over water

In this type of problem, you must account for the fact that

$$P_{TOTAL} = P_{SAMPLE} + P_{H_2O(g)}$$

The pressure of the gas sample of interest is:

$$P_{SAMPLE} = P_{TOTAL} - P_{H_2O(g)}$$

**EXAMPLE 5.8**

**When heated, calcium carbonate decomposes forming solid calcium oxide and carbon dioxide gas. A sample of calcium carbonate is completely decomposed by heating and 50.5 mL of gas is collected over water at 755.0 torr and 25.0°C. How much did the sample of calcium carbonate weigh? The vapor pressure of water at 25°C is 23.76 torr.**

**Strategy:** If we calculate the pressure of $CO_2$ gas collected, we can then calculate moles of $CO_2$ using the ideal gas equation. Then, using the mole ratio from the balanced equation, we can convert to moles calcium carbonate. Finally, we convert to grams of calcium carbonate.

**Solution:** We need to start with a balanced equation to have the correct mole ratio between $CO_2$ and $CaCO_3$.

$$CaCO_3(s) \longrightarrow CaO(s) + CO_2(g)$$

Since $CO_2$ is collected over water, water vapor is collected in addition to the $CO_2$ gas.

$$P_T = P_{CO_2} + P_{H_2O(g)}$$

$$P_{CO_2} = P_T - P_{H_2O(g)} = (755.0 - 23.76) \text{ torr} = 731.2 \text{ torr}$$

Solve the ideal gas equation algebraically for moles of $CO_2$. Then, substitute in the known values.

$$P_{CO_2} = 731.2\ \text{torr} \times \frac{1\ \text{atm}}{760\ \text{torr}} = 0.9621\ \text{atm}$$

$$T(\text{K}) = 25.0° + 273° = 298\ \text{K}$$

$$V = 50.5\ \text{mL} \times \frac{1\ \text{L}}{1000\ \text{mL}} = 0.0505\ \text{L}$$

$$n_{CO_2} = \frac{P_{CO_2}V}{RT}$$

$$n_{CO_2} = \frac{(0.9621\ \text{atm})(0.0505\ \text{L})}{298\ \text{K}} \times \frac{\text{mol}\cdot\text{K}}{0.0821\ \text{L}\cdot\text{atm}} = 1.99 \times 10^{-3}\ \text{mol CO}_2$$

Calculate the mass of $CaCO_3$ using the mole ratio from the balanced equation and the molar mass of $CaCO_3$.

$$?\ \mathbf{g\ CaCO_3} = (1.99 \times 10^{-3}\ \text{mol CO}_2) \times \frac{1\ \text{mol CaCO}_3}{1\ \text{mol CO}_2} \times \frac{100.09\ \text{g CaCO}_3}{1\ \text{mol CaCO}_3} = \mathbf{0.199\ g\ CaCO_3}$$

**PRACTICE EXERCISE**

7. Fermentation is one of the oldest studied biochemical processes. The fermentation process is a catabolic pathway, which involves the use of yeast to convert sugars into ethanol and carbon dioxide. In brewing beer, malt sugar is fermented to form ethanol ($CH_3CH_2OH$) and carbon dioxide ($CO_2$) according to the following overall reaction:

$$C_6H_{12}O_6(aq) \longrightarrow 2CH_3CH_2OH(aq) + 2CO_2(g)$$

A micro-brewery is making a batch of beer, and they wish to determine the mass percentage of ethanol in their brew by collecting the $CO_2(g)$ evolved. They start the process by mixing malt sugar, hops, and yeast in water. The mixture has a total mass of 800 lbs. During fermentation, 10,000 L of $CO_2$ is collected over water at 25.0°C and 750 mmHg. What is the mass percentage of ethanol in the beer? (The vapor pressure of water at 25°C is 23.76 mm Hg.)

**Text Problems:** 5.70, **5.72**

## PROBLEM TYPE 4: ROOT-MEAN-SQUARE (RMS) SPEED

This type of problem typically involves substituting known values into the following equation:

$$u_{rms} = \sqrt{\frac{3RT}{M}}$$

You must remember to use $R$ in units of $\frac{\text{J}}{\text{mol}\cdot\text{K}}$ and $M$ in units of kg/mol. 1 joule $= 1\frac{\text{kg}\cdot\text{m}^2}{\text{s}^2}$. If these units are used, the units of $u_{rms}$ are m/s.

$$u_{rms}\ (\text{units}) = \sqrt{\frac{\left(\frac{\text{J}}{\text{mol}\cdot\text{K}}\right)(\text{K})}{\left(\frac{\text{kg}}{\text{mol}}\right)}}$$

$$u_{rms}\ (\text{units}) = \sqrt{\frac{\left(\frac{\text{kg}\cdot\text{m}^2}{\text{s}^2} \times \frac{1}{\text{mol}\cdot\text{K}}\right)(\text{K})}{\left(\frac{\text{kg}}{\text{mol}}\right)}}$$

$$u_{rms}\ (\text{units}) = \sqrt{\frac{\text{m}^2}{\text{s}^2}}$$

$$u_{rms}\ (\text{units}) = \frac{\text{m}}{\text{s}}$$

**EXAMPLE 5.9**

**Which has a higher root mean square velocity, $H_2(g)$ at 150 K or He(g) at 650K?**

**Strategy:** To calculate the root-mean-square speed, we use Equation (5.16) of the text. What units should we use for *R* and *M* so the $u_{rms}$ will be expressed in units of m/s.

**Solution:** To calculate $u_{rms}$, the units of *R* should be 8.314 J/mol·K, and because 1 J = 1 kg·m²/s², the units of molar mass must be kg/mol. Let's convert the molar masses in g/mol to kg/mol.

$$M_{H_2} = \frac{2.016\ \text{g H}_2}{1\ \text{mol H}_2} \times \frac{1\ \text{kg}}{1000\ \text{g}} = 2.016 \times 10^{-3}\ \text{kg/mol}$$

$$M_{He} = \frac{4.003\ \text{g He}}{1\ \text{mol He}} \times \frac{1\ \text{kg}}{1000\ \text{g}} = 4.003 \times 10^{-3}\ \text{kg/mol}$$

Substitute the appropriate values into the equation to solve for $u_{rms}$.

$$u_{rms}(H_2) = \sqrt{\frac{(3)\left(8.314\frac{\text{J}}{\text{mol}\cdot\text{K}}\right)(150\ \text{K})}{\left(2.016 \times 10^{-3}\frac{\text{kg}}{\text{mol}}\right)}} \qquad u_{rms}(\text{He}) = \sqrt{\frac{(3)\left(8.314\frac{\text{J}}{\text{mol}\cdot\text{K}}\right)(650\ \text{K})}{\left(4.003 \times 10^{-3}\frac{\text{kg}}{\text{mol}}\right)}}$$

$$u_{rms}(H_2) = \mathbf{1.36 \times 10^3\ m/s} \qquad u_{rms}(\text{He}) = \mathbf{2.01 \times 10^3\ m/s}$$

He(g) has the higher root mean square velocity.

**PRACTICE EXERCISE**

**8.** What is the root-mean-square velocity of $F_2(g)$ at 298 K?

**Text Problems: 5.82**, 5.84

## PROBLEM TYPE 5: DEVIATIONS FROM IDEAL BEHAVIOR

Van der Waals' equation is a simple modification of the ideal gas equation. It takes into account intermolecular forces and finite molecular volumes.

Starting with the ideal gas equation:

$$P_{ideal}V_{ideal} = nRT$$

van der Waals made two corrections.

(1) $P_{ideal} = P_{real} +$ correction for attraction between particles

$$P_{ideal} = P_{real} + \frac{an^2}{V_{real}^2} \quad \text{(see Section 5.8 of the text for discussion)}$$

(2) $V_{ideal} = V_{real} -$ correction for finite volume of particles

$$V_{ideal} = V_{real} - nb \quad \text{(see Section 5.8 of the text for discussion)}$$

where,

$a$ and $b$ are constants for a particular gas (See Table 5.4 of the text)

$n$ = moles of gas

Substituting these corrections into the ideal gas equation leads to the **van der Waals equation.**

$$\left(\boldsymbol{P_{real}} + \frac{\boldsymbol{an^2}}{\boldsymbol{V_{real}^2}}\right)(\boldsymbol{V_{real}} - \boldsymbol{nb}) = \boldsymbol{nRT}$$

**EXAMPLE 5.10**

**The molar volume of isopentane ($C_5H_{12}$) is 1.00 L at 503 K and 30.0 atm.**

**(a)** Does isopentane behave like an ideal gas?

**Strategy:** In this problem we can determine if the gas deviates from ideal behavior, by comparing the ideal pressure with the actual pressure. We can calculate the ideal gas pressure using the ideal gas equation, and then compare it to the actual pressure given in the problem.

**Solution:** Solve the ideal gas equation algebraically for P, and then substitute in the known values.

$$P = \frac{nRT}{V} = \frac{(1.00\ \text{mol})(503\ \text{K})}{1.00\ \text{L}} \times \frac{0.0821\ \text{L}\cdot\text{atm}}{\text{mol}\cdot\text{K}}$$

$$P = 41.3\ \text{atm}$$

Compare the ideal pressure to the actual pressure. Calculating the percentage error would be helpful. Percentage of error is the difference between the two values divided by the actual value.

$$\%\ \text{error} = \frac{41.3\ \text{atm} - 30.0\ \text{atm}}{30.0\ \text{atm}} \times 100 = \mathbf{37.7\%}$$

Because of the large percent error, we conclude that under these conditions, $C_5H_{12}$ behaves in a nonideal manner.

**(b)** Given that $a = 17.0\ \text{L}^2\cdot\text{atm/mol}^2$ and $b = 0.136$ L/mol, calculate the pressure of isopentane as predicted by the van der Waals' equation.

**Strategy:** Calculate the pressure of 1 mol of gas at 503 K that occupies 1.00 L using van der Waals' equation.

**Solution:** Calculate the correction terms.

$$\frac{an^2}{V_{obs}^2} = \frac{\left(17.0\frac{\text{L}^2\cdot\text{atm}}{\text{mol}^2}\right)(1.00\ \text{mol})^2}{1.00\ \text{L}^2} = 17.0\ \text{atm}$$

$$nb = (1.00\ \text{mol})\left(0.136\,\frac{\text{L}}{\text{mol}}\right) = 0.136\ \text{L}$$

Substitute the values into the van der Waals' equation and solve for $P_{\text{real}}$.

$$\left(P_{\text{real}} + \frac{an^2}{V_{\text{real}}^2}\right)(V_{\text{real}} - nb) = nRT$$

$$(P_{\text{real}} + 17.0\ \text{atm})(1.00\ \text{L} - 0.136\ \text{L}) = (1.00\ \text{mol})\left(\frac{0.0821\ \text{L}\cdot\text{atm}}{\text{mol}\cdot\text{K}}\right)(503\ \text{K})$$

$$(P_{\text{real}} + 17.0\ \text{atm})(0.864) = 41.3\ \text{atm}$$

$$P_{\text{real}} + 17.0\ \text{atm} = 47.8\ \text{atm}$$

$$P_{\text{real}} = 47.8\ \text{atm} - 17.0\ \text{atm} = \mathbf{30.8\ atm}$$

The pressure calculated for 1 mole of this "real" gas at 503 K and 1.00 L using van der Waals' equation is much closer to the actual value of 30.0 atm. The percent error is only 2.7 percent.

**PRACTICE EXERCISE**

9. Using van der Waals' equation, calculate the pressure exerted by 1.00 mol of water vapor that occupies a volume of 5.55 L at 150°C.

**Text Problem: 5.94**

## ANSWERS TO PRACTICE EXERCISES

1. $4.04 \times 10^3$ mmHg
2. $V_{CO_2} = 1.1 \times 10^{16}$ L
3. $V_{\text{air}} = 160$ mL
4. The gas is fluorine, $F_2$.
5. 176 g $NaN_3$
6. $P_{CO} = 7.5 \times 10^{-3}$ torr
7. mass % ethanol = 5.0 %
8. $u_{\text{rms}} = 442$ m/s
9. $P_{\text{real}} = 6.11$ atm

## SOLUTIONS TO SELECTED TEXT PROBLEMS

5.14 **Strategy:** Because 1 atm = 760 mmHg, the following conversion factor is needed to obtain the pressure in atmospheres.

$$\frac{1 \text{ atm}}{760 \text{ mmHg}}$$

For the second conversion, 1 atm = 101.325 kPa.

**Solution:**

$$\textbf{? atm} = 606 \text{ mmHg} \times \frac{1 \text{ atm}}{760 \text{ mmHg}} = \textbf{0.797 atm}$$

$$\textbf{? kPa} = 0.797 \text{ atm} \times \frac{101.325 \text{ kPa}}{1 \text{ atm}} = \textbf{80.8 kPa}$$

5.18 (1) Recall that $V \propto \frac{1}{P}$. As the pressure is tripled, the volume will decrease to $\frac{1}{3}$ of its original volume, assuming constant $n$ and $T$. The correct choice is **(b)**.

(2) Recall that $V \propto T$. As the temperature is doubled, the volume will also double, assuming constant $n$ and $P$. The correct choice is **(a)**. The depth of color indicates the density of the gas. As the volume increases at constant moles of gas, the density of the gas will decrease. This decrease in gas density is indicated by the lighter shading.

(3) Recall that $V \propto n$. Starting with $n$ moles of gas, adding another $n$ moles of gas ($2n$ total) will double the volume. The correct choice is **(c)**. The density of the gas will remain the same as moles are doubled and volume is doubled.

(4) Recall that $V \propto T$ and $V \propto \frac{1}{P}$. Halving the temperature would decrease the volume to $\frac{1}{2}$ its original volume. However, reducing the pressure to $\frac{1}{4}$ its original value would increase the volume by a factor of 4. Combining the two changes, we have

$$\frac{1}{2} \times 4 = 2$$

The volume will double. The correct choice is **(a)**.

5.20 Temperature and amount of gas do not change in this problem ($T_1 = T_2$ and $n_1 = n_2$). Pressure and volume change; it is a Boyle's law problem.

$$\frac{P_1V_1}{n_1T_1} = \frac{P_2V_2}{n_2T_2}$$

$$P_1V_1 = P_2V_2$$

$$V_2 = 0.10\,V_1$$

$$P_2 = \frac{P_1V_1}{V_2}$$

$$P_2 = \frac{(5.3\text{ atm})V_1}{0.10V_1} = \mathbf{53\ atm}$$

**5.22** (a)

**Strategy:** The amount of gas and its temperature remain constant, but both the pressure and the volume change. What equation would you use to solve for the final volume?

**Solution:** We start with Equation (5.9) of the text.

$$\frac{P_1V_1}{n_1T_1} = \frac{P_2V_2}{n_2T_2}$$

Because $n_1 = n_2$ and $T_1 = T_2$,

$$P_1V_1 = P_2V_2$$

which is Boyle's Law. The given information is tabulated below.

| Initial conditions | Final Conditions |
|---|---|
| $P_1 = 1.2$ atm | $P_2 = 6.6$ atm |
| $V_1 = 3.8$ L | $V_2 = ?$ |

The final volume is given by:

$$V_2 = \frac{P_1V_1}{P_2}$$

$$V_2 = \frac{(1.2\text{ atm})(3.8\text{ L})}{(6.6\text{ atm})} = \mathbf{0.69\ L}$$

**Check:** When the pressure applied to the sample of air is increased from 1.2 atm to 6.6 atm, the volume occupied by the sample will decrease. Pressure and volume are inversely proportional. The final volume calculated is less than the initial volume, so the answer seems reasonable.

(b)

**Strategy:** The amount of gas and its temperature remain constant, but both the pressure and the volume change. What equation would you use to solve for the final pressure?

**Solution:** You should also come up with the equation $P_1V_1 = P_2V_2$ for this problem. The given information is tabulated below.

| Initial conditions | Final Conditions |
|---|---|
| $P_1 = 1.2$ atm | $P_2 = ?$ |
| $V_1 = 3.8$ L | $V_2 = 0.075$ L |

The final pressure is given by:

$$P_2 = \frac{P_1V_1}{V_2}$$

$$P_2 = \frac{(1.2 \text{ atm})(3.8 \not{L})}{(0.075 \not{L})} = \mathbf{61 \text{ atm}}$$

**Check:** To decrease the volume of the gas fairly dramatically from 3.8 L to 0.075 L, the pressure must be increased substantially. A final pressure of 61 atm seems reasonable.

5.24 **Strategy:** The amount of gas and its pressure remain constant, but both the temperature and the volume change. What equation would you use to solve for the final temperature? What temperature unit should we use?

**Solution:** We start with Equation (5.9) of the text.

$$\frac{P_1V_1}{n_1T_1} = \frac{P_2V_2}{n_2T_2}$$

Because $n_1 = n_2$ and $P_1 = P_2$,

$$\frac{V_1}{T_1} = \frac{V_2}{T_2}$$

which is Charles' Law. The given information is tabulated below.

| Initial conditions | Final Conditions |
|---|---|
| $T_1 = (88 + 273)\text{K} = 361$ K | $T_2 = ?$ |
| $V_1 = 9.6$ L | $V_2 = 3.4$ L |

The final temperature is given by:

$$T_2 = \frac{T_1V_2}{V_1}$$

$$T_2 = \frac{(361 \text{ K})(3.4 \not{L})}{(9.6 \not{L})} = \mathbf{1.3 \times 10^2 \text{ K}}$$

**5.26** This is a gas stoichiometry problem that requires knowledge of Avogadro's law to solve. Avogadro's law states that the volume of a gas is directly proportional to the number of moles of gas at constant temperature and pressure.

The volume ratio, 1 vol. $Cl_2$ : 3 vol. $F_2$ : 2 vol. product, can be written as a mole ratio,
1 mol $Cl_2$ : 3 mol $F_2$ : 2 mol product.

Attempt to write a balanced chemical equation. The subscript of F in the product will be three times the Cl subscript, because there are three times as many F atoms reacted as Cl atoms.

$$\mathbf{1}Cl_2(g) + \mathbf{3}F_2(g) \longrightarrow \mathbf{2}Cl_xF_{3x}(g)$$

Balance the equation. The $x$ must equal one so that there are two Cl atoms on each side of the equation. If $x = 1$, the subscript on F is 3.

$$Cl_2(g) + 3F_2(g) \longrightarrow 2ClF_3(g)$$

The formula of the product is $\mathbf{ClF_3}$.

**5.32** **Strategy:** This problem gives the amount, volume, and temperature of CO gas. Is the gas undergoing a change in any of its properties? What equation should we use to solve for the pressure? What temperature unit should be used?

**Solution:** Because no changes in gas properties occur, we can use the ideal gas equation to calculate the pressure. Rearranging Equation (5.8) of the text, we write:

$$P = \frac{nRT}{V}$$

$$P = \frac{(6.9\ \text{mol})\left(0.0821\ \frac{\text{L}\cdot\text{atm}}{\text{mol}\cdot\text{K}}\right)(62 + 273)\text{K}}{30.4\ \text{L}} = \mathbf{6.2\ atm}$$

**5.34** In this problem, the moles of gas and the volume the gas occupies are constant ($V_1 = V_2$ and $n_1 = n_2$). Temperature and pressure change.

$$\frac{P_1V_1}{n_1T_1} = \frac{P_2V_2}{n_2T_2}$$

$$\frac{P_1}{T_1} = \frac{P_2}{T_2}$$

The given information is tabulated below.

| Initial conditions | Final Conditions |
|---|---|
| $T_1 = (25 + 273)\text{K} = 298\ \text{K}$ | $T_2 = ?$ |
| $P_1 = 0.800\ \text{atm}$ | $P_2 = 2.00\ \text{atm}$ |

The final temperature is given by:

$$T_2 = \frac{T_1P_2}{P_1}$$

$$T_2 = \frac{(298\text{ K})(2.00\text{ atm})}{(0.800\text{ atm})} = \mathbf{745\text{ K}} = \mathbf{472°C}$$

**5.36** In this problem, the moles of gas and the volume the gas occupies are constant ($V_1 = V_2$ and $n_1 = n_2$). Temperature and pressure change.

$$\frac{P_1V_1}{n_1T_1} = \frac{P_2V_2}{n_2T_2}$$

$$\frac{P_1}{T_1} = \frac{P_2}{T_2}$$

The given information is tabulated below.

| Initial conditions | Final Conditions |
|---|---|
| $T_1 = 273$ K | $T_2 = (250 + 273)\text{K} = 523$ K |
| $P_1 = 1.0$ atm | $P_2 = ?$ |

The final pressure is given by:

$$P_2 = \frac{P_1T_2}{T_1}$$

$$P_2 = \frac{(1.0\text{ atm})(523\text{ K})}{273\text{ K}} = \mathbf{1.9\text{ atm}}$$

**5.38** In this problem, the moles of gas and the pressure on the gas are constant ($n_1 = n_2$ and $P_1 = P_2$). Temperature and volume are changing.

$$\frac{P_1V_1}{n_1T_1} = \frac{P_2V_2}{n_2T_2}$$

$$\frac{V_1}{T_1} = \frac{V_2}{T_2}$$

The given information is tabulated below.

| Initial conditions | Final Conditions |
|---|---|
| $T_1 = (20.1 + 273)\text{ K} = 293.1\text{ K}$ | $T_2 = (36.5 + 273)\text{K} = 309.5$ K |
| $V_1 = 0.78$ L | $V_2 = ?$ |

The final volume is given by:

$$V_2 = \frac{V_1T_2}{T_1}$$

$$V_2 = \frac{(0.78\text{ L})(309.5\,\cancel{\text{K}})}{(293.1\,\cancel{\text{K}})} = \mathbf{0.82\ L}$$

**5.40** In the problem, temperature and pressure are given. If we can determine the moles of $CO_2$, we can calculate the volume it occupies using the ideal gas equation.

$$?\text{ mol CO}_2 = 88.4\,\cancel{\text{g}}\,\text{CO}_2 \times \frac{1\text{ mol CO}_2}{44.01\,\cancel{\text{g}}\,\text{CO}_2} = 2.01\text{ mol CO}_2$$

We now substitute into the ideal gas equation to calculate volume of $CO_2$.

$$V_{\text{CO}_2} = \frac{nRT}{P} = \frac{(2.01\,\cancel{\text{mol}})\left(0.0821\frac{\text{L}\cdot\cancel{\text{atm}}}{\cancel{\text{mol}}\cdot\cancel{\text{K}}}\right)(273\,\cancel{\text{K}})}{(1\,\cancel{\text{atm}})} = \mathbf{45.1\ L}$$

Alternatively, we could use the fact that 1 mole of an ideal gas occupies a volume of 22.41 L at STP. After calculating the moles of $CO_2$, we can use this fact as a conversion factor to convert to volume of $CO_2$.

$$?\text{ L CO}_2 = 2.01\,\cancel{\text{mol}}\,\text{CO}_2 \times \frac{22.41\text{ L}}{1\,\cancel{\text{mol}}} = \mathbf{45.0\ L\ CO_2}$$

The slight difference in the results of our two calculations is due to rounding the volume occupied by 1 mole of an ideal gas to 22.41 L.

**5.42** The molar mass of $CO_2 = 44.01$ g/mol. Since $PV = nRT$, we write:

$$P = \frac{nRT}{V}$$

$$P = \frac{\left(0.050\,\cancel{\text{g}} \times \frac{1\,\cancel{\text{mol}}}{44.01\,\cancel{\text{g}}}\right)\left(0.0821\frac{\cancel{\text{L}}\cdot\text{atm}}{\cancel{\text{mol}}\cdot\cancel{\text{K}}}\right)(30 + 273)\cancel{\text{K}}}{4.6\,\cancel{\text{L}}} = \mathbf{6.1 \times 10^{-3}\ atm}$$

**5.44** **Strategy:** We can calculate the molar mass of a gas if we know its density, temperature, and pressure. What temperature and pressure units should we use?

**Solution:** We need to use Equation (5.12) of the text to calculate the molar mass of the gas.

$$\mathcal{M} = \frac{dRT}{P}$$

Before substituting into the above equation, we need to calculate the density and check that the other known quantities ($P$ and $T$) have the appropriate units.

$$d = \frac{7.10\ \text{g}}{5.40\ \text{L}} = 1.31\ \text{g/L}$$

$$T = 44^\circ + 273^\circ = 317\ \text{K}$$

$$P = 741\ \cancel{\text{torr}} \times \frac{1\ \text{atm}}{760\ \cancel{\text{torr}}} = 0.975\ \text{atm}$$

Calculate the molar mass by substituting in the known quantities.

$$\mathcal{M} = \frac{\left(1.31\ \frac{\text{g}}{\cancel{\text{L}}}\right)\left(0.0821\ \frac{\cancel{\text{L}}\cdot\cancel{\text{atm}}}{\text{mol}\cdot\cancel{\text{K}}}\right)(317\ \cancel{\text{K}})}{0.975\ \cancel{\text{atm}}} = \mathbf{35.0\ g/mol}$$

Alternatively, we can solve for the molar mass by writing:

$$\text{molar mass of compound} = \frac{\text{mass of compound}}{\text{moles of compound}}$$

Mass of compound is given in the problem (7.10 g), so we need to solve for moles of compound in order to calculate the molar mass.

$$n = \frac{PV}{RT}$$

sss

$$n = \frac{(0.975\ \cancel{\text{atm}})(5.40\ \cancel{\text{L}})}{\left(0.0821\ \frac{\cancel{\text{L}}\cdot\cancel{\text{atm}}}{\text{mol}\cdot\cancel{\text{K}}}\right)(317\ \cancel{\text{K}})} = 0.202\ \text{mol}$$

Now, we can calculate the molar mass of the gas.

$$\textbf{molar mass of compound} = \frac{\text{mass of compound}}{\text{moles of compound}} = \frac{7.10\ \text{g}}{0.202\ \text{mol}} = \mathbf{35.1\ g/mol}$$

**5.46** The number of particles in 1 L of gas at STP is:

$$\text{Number of particles} = 1.0\ \cancel{\text{L}} \times \frac{1\ \cancel{\text{mol}}}{22.414\ \cancel{\text{L}}} \times \frac{6.022 \times 10^{23}\ \text{particles}}{1\ \cancel{\text{mol}}} = 2.7 \times 10^{22}\ \text{particles}$$

$$\textbf{Number of N}_2\textbf{ molecules} = \left(\frac{78\%}{100\%}\right)(2.7 \times 10^{22}\ \text{particles}) = \mathbf{2.1 \times 10^{22}\ N_2\ molecules}$$

$$\textbf{Number of O}_2\textbf{ molecules} = \left(\frac{21\%}{100\%}\right)(2.7 \times 10^{22}\ \text{particles}) = \mathbf{5.7 \times 10^{21}\ O_2\ molecules}$$

$$\textbf{Number of Ar atoms} = \left(\frac{1\%}{100\%}\right)(2.7 \times 10^{22}\ \text{particles}) = \mathbf{3 \times 10^{20}\ Ar\ atoms}$$

**5.48** The density can be calculated from the ideal gas equation.

$$d = \frac{PM}{RT}$$

$M = 1.008 \text{ g/mol} + 79.90 \text{ g/mol} = 80.91 \text{ g/mol}$

$T = 46° + 273° = 319 \text{ K}$

$$P = 733 \text{ mmHg} \times \frac{1 \text{ atm}}{760 \text{ mmHg}} = 0.964 \text{ atm}$$

$$\mathbf{d} = \frac{(0.964 \text{ atm})\left(\dfrac{80.91 \text{ g}}{1 \text{ mol}}\right)}{319 \text{ K}} \times \frac{\text{mol}\cdot\text{K}}{0.0821 \text{ L}\cdot\text{atm}} = \mathbf{2.98 \text{ g/L}}$$

Alternatively, we can solve for the density by writing:

$$\text{density} = \frac{\text{mass}}{\text{volume}}$$

Assuming that we have 1 mole of HBr, the mass is 80.91 g. The volume of the gas can be calculated using the ideal gas equation.

$$V = \frac{nRT}{P}$$

$$V = \frac{(1 \text{ mol})\left(0.0821 \dfrac{\text{L}\cdot\text{atm}}{\text{mol}\cdot\text{K}}\right)(319 \text{ K})}{0.964 \text{ atm}} = 27.2 \text{ L}$$

Now, we can calculate the density of HBr gas.

$$\textbf{density} = \frac{\text{mass}}{\text{volume}} = \frac{80.91 \text{ g}}{27.2 \text{ L}} = \mathbf{2.97 \text{ g/L}}$$

**5.50** This is an extension of an ideal gas law calculation involving molar mass. If you determine the molar mass of the gas, you will be able to determine the molecular formula from the empirical formula.

$$M = \frac{dRT}{P}$$

Calculate the density, then substitute its value into the equation above.

$$d = \frac{0.100 \text{ g}}{22.1 \text{ mL}} \times \frac{1000 \text{ mL}}{1 \text{ L}} = 4.52 \text{ g/L}$$

$T(\text{K}) = 20° + 273° = 293 \text{ K}$

$$\mathcal{M} = \frac{\left(4.52\ \frac{\text{g}}{\text{L}}\right)\left(0.0821\ \frac{\text{L}\cdot\text{atm}}{\text{mol}\cdot\text{K}}\right)(293\ \text{K})}{1.02\ \text{atm}} = 107\ \text{g/mol}$$

Compare the empirical mass to the molar mass.

$$\text{empirical mass} = 32.07\ \text{g/mol} + 4(19.00\ \text{g/mol}) = 108.07\ \text{g/mol}$$

Remember, the molar mass will be a whole number multiple of the empirical mass. In this case, the $\frac{\text{molar mass}}{\text{empirical mass}} \approx 1$. Therefore, the molecular formula is the same as the empirical formula, $\mathbf{SF_4}$.

**5.52** Let the mole fraction of fluorine gas equal $x$ and the fraction of chlorine gas equal $1 - x$. From the molar masses of the two gases and the density equation derived from the ideal gas equation, we can solve for $x$, the fraction of the sample by mass due to fluorine gas.

$$d = \frac{P\mathcal{M}}{RT}$$

$$1.77\ \text{g/L} = \frac{(0.893\ \text{atm})\left[(x)(38.00) + (1 - x)(70.90)\right]\text{g/mol}}{\left(0.0821\ \frac{\text{L}\cdot\text{atm}}{\text{mol}\cdot\text{K}}\right)(287\ \text{K})}$$

$$41.7 = 33.9x \quad 63.3x + 63.3$$

$$x = 0.735$$

$$1 \quad x = 0.265$$

For 1 mole of the gas mixture, we have:

$$0.735\ \text{mol F}_2 \times \frac{38.00\ \text{g F}_2}{1\ \text{mol F}_2} = 27.9\ \text{g F}_2$$

$$0.265\ \text{mol Cl}_2 \times \frac{70.90\ \text{g Cl}_2}{1\ \text{mol Cl}_2} = 18.8\ \text{g Cl}_2$$

The total mass of the gases = 27.9 g + 18.8 g = 46.7 g

The percent by mass for $F_2$ and $Cl_2$ is:

$$\%\ \mathbf{F_2} = \frac{27.9\ \text{g}}{46.7\ \text{g}} \times 100\% = \mathbf{59.7\%}$$

$$\%\ \mathbf{Cl_2} = \frac{18.8\ \text{g}}{46.7\ \text{g}} \times 100\% = \mathbf{40.3\%}$$

**5.54** **Strategy:** From the moles of $CH_4$ reacted, we can calculate the moles of $CO_2$ produced. From the balanced equation, we see that 1 mol $CH_4 \simeq$ 1 mol $CO_2$. Once moles of $CO_2$ are determined, we can use the ideal gas equation to calculate the volume of $CO_2$.

**Solution:** First let's calculate moles of $CO_2$ produced.

$$? \text{ mol } CO_2 = 15.0 \text{ mol } CH_4 \times \frac{1 \text{ mol } CO_2}{1 \text{ mol } CH_4} = 15.0 \text{ mol } CO_2$$

Now, we can substitute moles, temperature, and pressure into the ideal gas equation to solve for volume of $CO_2$.

$$V = \frac{nRT}{P}$$

$$V_{CO_2} = \frac{(15.0 \text{ mol})\left(0.0821 \frac{\text{L} \cdot \text{atm}}{\text{mol} \cdot \text{K}}\right)(23 + 273)\text{K}}{0.985 \text{ atm}} = \mathbf{3.70 \times 10^2\ L}$$

**5.56** From the amount of glucose reacted (5.97 g), we can calculate the theoretical yield of $CO_2$. We can then compare the theoretical yield to the actual yield given in the problem (1.44 L) to determine the percent yield.

First, let's determine the moles of $CO_2$ that can be produced theoretically. Then, we can use the ideal gas equation to determine the volume of $CO_2$.

$$? \text{ mol } CO_2 = 5.97 \text{ g glucose} \times \frac{1 \text{ mol glucose}}{180.2 \text{ g glucose}} \times \frac{2 \text{ mol } CO_2}{1 \text{ mol glucose}} = 0.0663 \text{ mol } CO_2$$

Now, substitute moles, pressure, and temperature into the ideal gas equation to calculate the volume of $CO_2$.

$$V = \frac{nRT}{P}$$

$$V_{CO_2} = \frac{(0.0663 \text{ mol})\left(0.0821 \frac{\text{L} \cdot \text{atm}}{\text{mol} \cdot \text{K}}\right)(293 \text{ K})}{0.984 \text{ atm}} = 1.62 \text{ L}$$

This is the theoretical yield of $CO_2$. The actual yield, which is given in the problem, is 1.44 L. We can now calculate the percent yield.

$$\text{percent yield} = \frac{\text{actual yield}}{\text{theoretical yield}} \times 100\%$$

$$\textbf{percent yield} = \frac{1.44 \text{ L}}{1.62 \text{ L}} \times 100\% = \mathbf{88.9\%}$$

5.58 **Strategy:** We can calculate the moles of M reacted, and the moles of $H_2$ gas produced. By comparing the number of moles of M reacted to the number of moles $H_2$ produced, we can determine the mole ratio in the balanced equation.

**Solution:** First let's calculate the moles of the metal (M) reacted.

$$\text{mol M} = 0.225\ \cancel{\text{g}}\ \text{M} \times \frac{1\ \text{mol M}}{27.0\ \cancel{\text{g}}\ \text{M}} = 8.33 \times 10^{-3}\ \text{mol M}$$

Solve the ideal gas equation algebraically for $n_{H_2}$. Then, calculate the moles of $H_2$ by substituting the known quantities into the equation.

$$P = 741\ \cancel{\text{mmHg}} \times \frac{1\ \text{atm}}{760\ \cancel{\text{mmHg}}} = 0.975\ \text{atm}$$

$$T = 17° + 273° = 290\ \text{K}$$

$$n_{H_2} = \frac{PV_{H_2}}{RT}$$

$$n_{H_2} = \frac{(0.975\ \cancel{\text{atm}})(0.303\ \cancel{\text{L}})}{\left(0.0821\frac{\cancel{\text{L}}\cdot\cancel{\text{atm}}}{\text{mol}\cdot\cancel{\text{K}}}\right)(290\ \cancel{\text{K}})} = 1.24 \times 10^{-2}\ \text{mol}\ H_2$$

Compare the number of moles of $H_2$ produced to the number of moles of M reacted.

$$\frac{1.24 \times 10^{-2}\ \text{mol}\ H_2}{8.33 \times 10^{-3}\ \text{mol M}} \approx 1.5$$

This means that the mole ratio of $H_2$ to M is 1.5 : 1.

We can now write the balanced equation since we know the mole ratio between $H_2$ and M.

The unbalanced equation is:

$$M(s) + HCl(aq) \longrightarrow 1.5H_2(g) + M_xCl_y(aq)$$

We have 3 atoms of H on the products side of the reaction, so a 3 must be placed in front of HCl. The ratio of M to Cl on the reactants side is now 1 : 3. Therefore the formula of the metal chloride must be $MCl_3$. The balanced equation is:

$$\mathbf{M(s) + 3HCl(aq) \longrightarrow 1.5H_2(g) + MCl_3(aq)}$$

From the formula of the metal chloride, we determine that the charge of the metal is +3. Therefore, the formula of the metal oxide and the metal sulfate are $\mathbf{M_2O_3}$ and $\mathbf{M_2(SO_4)_3}$, respectively.

**5.60** **From the moles of $CO_2$ produced, we can calculate the amount of calcium carbonate that must have reacted. We can then determine the percent by mass of $CaCO_3$ in the 3.00 g sample.**

The balanced equation is:

$$CaCO_3(s) + 2HCl(aq) \longrightarrow CO_2(g) + CaCl_2(aq) + H_2O(l)$$

The moles of $CO_2$ produced can be calculated using the ideal gas equation.

$$n_{CO_2} = \frac{PV_{CO_2}}{RT}$$

$$n_{CO_2} = \frac{\left(792\ \text{mmHg} \times \frac{1\ \text{atm}}{760\ \text{mmHg}}\right)(0.656\ \text{L})}{\left(0.0821\frac{\text{L}\cdot\text{atm}}{\text{mol}\cdot\text{K}}\right)(20 + 273\ \text{K})} = \mathbf{2.84 \times 10^{-2}\ mol\ CO_2}$$

The balanced equation shows a 1:1 mole ratio between $CO_2$ and $CaCO_3$. Therefore, $2.84 \times 10^{-2}$ mole of $CaCO_3$ must have reacted.

$$?\ \text{g } CaCO_3 \text{ reacted} = (2.84 \times 10^{-2}\ \text{mol } CaCO_3) \times \frac{100.1\ \text{g } CaCO_3}{1\ \text{mol } CaCO_3} = 2.84\ \text{g } CaCO_3$$

The percent by mass of the $CaCO_3$ sample is:

$$\mathbf{\%\ CaCO_3} = \frac{2.84\ \text{g}}{3.00\ \text{g}} \times 100\% = \mathbf{94.7\%}$$

**Assumption:** The impurity (or impurities) must not react with HCl to produce $CO_2$ gas.

**5.62** The balanced equation is:

$$\mathbf{C_2H_5OH(l) + 3O_2(g) \longrightarrow 2CO_2(g) + 3H_2O(l)}$$

The moles of $O_2$ needed to react with 227 g ethanol are:

$$227\ \text{g } C_2H_5OH \times \frac{1\ \text{mol } C_2H_5OH}{46.07\ \text{g } C_2H_5OH} \times \frac{3\ \text{mol } O_2}{1\ \text{mol } C_2H_5OH} = 14.8\ \text{mol } O_2$$

14.8 moles of $O_2$ correspond to a volume of:

$$V_{O_2} = \frac{n_{O_2}RT}{P} = \frac{(14.8\ \text{mol } O_2)\left(0.0821\frac{\text{L}\cdot\text{atm}}{\text{mol}\cdot\text{K}}\right)(35 + 273\ \text{K})}{\left(790\ \text{mmHg} \times \frac{1\ \text{atm}}{760\ \text{mmHg}}\right)} = 3.60 \times 10^2\ \text{L } O_2$$

Since air is 21.0 percent $O_2$ by volume, we can write:

$$V_{\text{air}} = V_{O_2}\left(\frac{100\%\ \text{air}}{21\%\ O_2}\right) = (3.60 \times 10^2\ \text{L}\ O_2)\left(\frac{100\%\ \text{air}}{21\%\ O_2}\right) = \mathbf{1.71 \times 10^3\ L\ air}$$

**5.64** The balanced equation is: $FeS + 2HCl \rightarrow H_2S + FeCl_2$. From the moles of $H_2S$ produced, we can calculate the mass of FeS in the 4.00 g sample. The mass percent purity can then be calculated.

$$n_{H_2S} = \frac{PV}{RT} = \frac{\left(782\ \text{mmHg} \times \frac{1\ \text{atm}}{760\ \text{mmHg}}\right)(0.896\ \text{L})}{\left(0.0821\frac{\text{L}\cdot\text{atm}}{\text{mol}\cdot\text{K}}\right)(287\ \text{K})} = 0.0391\ \text{mol}\ H_2S$$

The mass of FeS in the 4.00 g sample is:

$$0.0391\ \text{mol}\ H_2S \times \frac{1\ \text{mol FeS}}{1\ \text{mol}\ H_2S} \times \frac{87.92\ \text{g FeS}}{1\ \text{mol FeS}} = 3.44\ \text{g FeS}$$

The mass percent purity is:

$$\textbf{mass percent purity} = \frac{3.44\ \text{g}}{4.00\ \text{g}} \times 100\% = \mathbf{86.0\%}$$

**5.68** Dalton's law states that the total pressure of the mixture is the sum of the partial pressures.

**(a)** $P_{\text{total}} = 0.32\ \text{atm} + 0.15\ \text{atm} + 0.42\ \text{atm} = \mathbf{0.89\ atm}$

**(b)** We know:

| Initial conditions | Final Conditions |
|---|---|
| $P_1 = (0.15 + 0.42)\text{atm} = 0.57\ \text{atm}$ | $P_2 = 1.0\ \text{atm}$ |
| $T_1 = (15 + 273)\text{K} = 288\ \text{K}$ | $T_2 = 273\ \text{K}$ |
| $V_1 = 2.5\ \text{L}$ | $V_2 = ?$ |

$$\frac{P_1V_1}{n_1T_1} = \frac{P_2V_2}{n_2T_2}$$

Because $n_1 = n_2$, we can write:

$$V_2 = \frac{P_1V_1T_2}{P_2T_1}$$

$$V_2 = \frac{(0.57\ \text{atm})(2.5\ \text{L})(273\ \text{K})}{(1.0\ \text{atm})(288\ \text{K})} = \mathbf{1.4\ L\ at\ STP}$$

**5.70** $P_{Total} = P_1 + P_2 + P_3 + \ldots + P_n$

In this case,

$$P_{Total} = P_{Ne} + P_{He} + P_{H_2O}$$

$$P_{Ne} = P_{Total} - P_{He} - P_{H_2O}$$

$$P_{Ne} = 745 \text{ mmHg} - 368 \text{ mmHg} - 28.3 \text{ mmHg} = \mathbf{349 \text{ mmHg}}$$

**5.72** **Strategy:** To solve for moles of $H_2$ generated, we must first calculate the partial pressure of $H_2$ in the mixture. What gas law do we need? How do we convert from moles of $H_2$ to amount of Zn reacted?

**Solution:** Dalton's law of partial pressure states that

$$P_{Total} = P_1 + P_2 + P_3 + \ldots + P_n$$

In this case,

$$P_{Total} = P_{H_2} + P_{H_2O}$$

$$P_{H_2} = P_{Total} - P_{H_2O}$$

$$P_{H_2} = 0.980 \text{ atm} - (23.8 \text{ mmHg})\left(\frac{1 \text{ atm}}{760 \text{ mmHg}}\right) = 0.949 \text{ atm}$$

Now that we know the pressure of $H_2$ gas, we can calculate the moles of $H_2$. Then, using the mole ratio from the balanced equation, we can calculate moles of Zn.

$$n_{H_2} = \frac{P_{H_2}V}{RT}$$

$$n_{H_2} = \frac{(0.949 \text{ atm})(7.80 \text{ L})}{(25 + 273) \text{ K}} \times \frac{\text{mol} \cdot \text{K}}{0.0821 \text{ L} \cdot \text{atm}} = 0.303 \text{ mol } H_2$$

Using the mole ratio from the balanced equation and the molar mass of zinc, we can now calculate the grams of zinc consumed in the reaction.

$$\textbf{? g Zn} = 0.303 \text{ mol } H_2 \times \frac{1 \text{ mol Zn}}{1 \text{ mol } H_2} \times \frac{65.39 \text{ g Zn}}{1 \text{ mol Zn}} = \mathbf{19.8 \text{ g Zn}}$$

**5.74** $P_i = X_i P_T$

We need to determine the mole fractions of each component in order to determine their partial pressures. To calculate mole fraction, write the balanced chemical equation to determine the correct mole ratio.

$$2NH_3(g) \longrightarrow N_2(g) + 3H_2(g)$$

The mole fractions of $H_2$ and $N_2$ are:

$$X_{H_2} = \frac{3\text{ mol}}{3\text{ mol} + 1\text{ mol}} = 0.750$$

$$X_{N_2} = \frac{1\text{ mol}}{3\text{ mol} + 1\text{ mol}} = 0.250$$

The partial pressures of $H_2$ and $N_2$ are:

$$P_{H_2} = X_{H_2}P_T = (0.750)(866\text{ mmHg}) = \mathbf{650\text{ mmHg}}$$

$$P_{N_2} = X_{N_2}P_T = (0.250)(866\text{ mmHg}) = \mathbf{217\text{ mmHg}}$$

**5.76** Let's treat each sphere as 1 mole of gas and then calculate the partial pressure of helium and hydrogen in each box. Note that the volume of the box on the right is twice that on the left.

Left container

$$P_{He} = \frac{n_{He}RT}{V} = \frac{5}{V}RT$$

$$P_{H_2} = \frac{n_{H_2}RT}{V} = \frac{4}{V}RT$$

$$P_T = P_{He} + P_{H_2} = \frac{9}{V}RT$$

Right container

$$P_{He} = \frac{11}{2V}RT = \frac{5.5}{V}RT$$

$$P_{H_2} = \frac{9}{2V}RT = \frac{4.5}{V}RT$$

$$P_T = P_{He} + P_{H_2} = \frac{10}{V}RT$$

**(a)** The **right container** has a higher total pressure.

**(b)** The **left container** has a lower partial pressure of helium.

**5.82** **Strategy:** To calculate the root-mean-square speed, we use Equation (5.16) of the text. What units should we use for $R$ and $M$ so the $u_{rms}$ will be expressed in units of m/s?

**Solution:** To calculate $u_{rms}$, the units of $R$ should be 8.314 J/mol·K, and because $1\text{ J} = 1\text{ kg}\cdot\text{m}^2/\text{s}^2$, the units of molar mass must be kg/mol.

First, let's calculate the molar masses ($M$) of $N_2$, $O_2$, and $O_3$. Remember, $M$ must be in units of kg/mol.

$$M_{N_2} = 2(14.01\text{ g/mol}) = 28.02\frac{\cancel{\text{g}}}{\text{mol}} \times \frac{1\text{ kg}}{1000\cancel{\text{g}}} = 0.02802\text{ kg/mol}$$

$$M_{O_2} = 2(16.00\text{ g/mol}) = 32.00\frac{\cancel{\text{g}}}{\text{mol}} \times \frac{1\text{ kg}}{1000\cancel{\text{g}}} = 0.03200\text{ kg/mol}$$

$$M_{O_3} = 3(16.00 \text{ g/mol}) = 48.00\frac{\cancel{\text{g}}}{\text{mol}} \times \frac{1 \text{ kg}}{1000 \cancel{\text{g}}} = 0.04800 \text{ kg/mol}$$

Now, we can substitute into Equation (5.16) of the text.

$$u_{\text{rms}} = \sqrt{\frac{3RT}{M}}$$

$$u_{\text{rms}}(N_2) = \sqrt{\frac{(3)\left(8.314\frac{\text{J}}{\text{mol}\cdot\text{K}}\right)(-23+273)\text{K}}{\left(0.02802\frac{\text{kg}}{\text{mol}}\right)}}$$

$$u_{\text{rms}}(N_2) = \mathbf{472 \text{ m/s}}$$

Similarly,

$$u_{\text{rms}}(O_2) = \mathbf{441 \text{ m/s}} \qquad u_{\text{rms}}(O_3) = \mathbf{360 \text{ m/s}}$$

**Check:** Since the molar masses of the gases increase in the order: $N_2 < O_2 < O_3$, we expect the lightest gas ($N_2$) to move the fastest on average and the heaviest gas ($O_3$) to move the slowest on average. This is confirmed in the above calculation.

**5.84** $$\textbf{RMS speed} = \sqrt{\frac{(2.0^2+2.2^2+2.6^2+2.7^2+3.3^2+3.5^2)(\text{m/s})^2}{6}} = \mathbf{2.8 \text{ m/s}}$$

$$\textbf{Average speed} = \frac{(2.0+2.2+2.6+2.7+3.3+3.5)\text{m/s}}{6} = \mathbf{2.7 \text{ m/s}}$$

The root-mean-square value is always greater than the average value, because squaring favors the larger values compared to just taking the average value.

**5.86** The separation factor is given by:

$$s = \frac{r_1}{r_2} = \sqrt{\frac{M_2}{M_1}}$$

This equation is the same as Graham's Law, Equation (5.17) of the text. For $^{235}UF_6$ and $^{238}UF_6$, we have:

$$s = \sqrt{\frac{238+(6)(19.00)}{235+(6)(19.00)}} = \mathbf{1.0043}$$

This is a very small separation factor, which is why many (thousands) stages of effusion are needed to enrich $^{235}U$.

**5.88** The rate of effusion is the number of molecules passing through a porous barrier in a given time. The molar mass of $CH_4$ is 16.04 g/mol. Using Equation (5.17) of the text, we find the molar mass of $Ni(CO)_x$.

$$\frac{r_1}{r_2} = \sqrt{\frac{\mathcal{M}_2}{\mathcal{M}_1}}$$

$$\frac{3.3}{1.0} = \sqrt{\frac{\mathcal{M}_{Ni(CO)_x}}{16.04\ \text{g/mol}}}$$

$$10.89 = \frac{\mathcal{M}_{Ni(CO)_x}}{16.04\ \text{g/mol}}$$

$$\mathcal{M}_{Ni(CO)_x} = 174.7\ \text{g/mol}$$

To find the value of $x$, we first subtract the molar mass of Ni from 174.7 g/mol.

$$174.7\ \text{g} - 58.69\ \text{g} = 116.0\ \text{g}$$

116.0 g is the mass of CO in 1 mole of the compound. The mass of 1 mole of CO is 28.01 g.

$$\frac{116.0\ \text{g}}{28.01\ \text{g}} = 4.141 \approx \mathbf{4}$$

This calculation indicates that there are 4 moles of CO in 1 mole of the compound. The value of $x$ is **4**.

**5.94** **Strategy:** In this problem we can determine if the gas deviates from ideal behavior, by comparing the ideal pressure with the actual pressure. We can calculate the ideal gas pressure using the ideal gas equation, and then compare it to the actual pressure given in the problem. What temperature unit should we use in the calculation?

**Solution:** We convert the temperature to units of Kelvin, then substitute the given quantities into the ideal gas equation.

$$T(\text{K}) = 27°\text{C} + 273° = 300\ \text{K}$$

$$P = \frac{nRT}{V} = \frac{(10.0\ \text{mol})\left(0.0821\frac{\text{L}\cdot\text{atm}}{\text{mol}\cdot\text{K}}\right)(300\ \text{K})}{1.50\ \text{L}} = 164\ \text{atm}$$

Now, we can compare the ideal pressure to the actual pressure by calculating the percent error.

$$\%\ \text{error} = \frac{164\ \text{atm} - 130\ \text{atm}}{130\ \text{atm}} \times 100\% = 26.2\%$$

Based on the large percent error, we conclude that under this condition of high pressure, the gas behaves in a **non-ideal** manner.

**5.96** When $a$ and $b$ are zero, the van der Waals equation simply becomes the ideal gas equation. In other words, an ideal gas has zero for the $a$ and $b$ values of the van der Waals equation. It therefore stands to reason that the gas with the smallest values of $a$ and $b$ will behave most like an ideal gas under a specific set of pressure and temperature conditions. Of the choices given in the problem, the gas with the smallest $a$ and $b$ values is **Ne** (see Table 5.4).

**5.98** We need to determine the molar mass of the gas. Comparing the molar mass to the empirical mass will allow us to determine the molecular formula.

$$n = \frac{PV}{RT} = \frac{(0.74\ \text{atm})\left(97.2\ \text{mL} \times \frac{0.001\ \text{L}}{1\ \text{mL}}\right)}{\left(0.0821\frac{\text{L}\cdot\text{atm}}{\text{mol}\cdot\text{K}}\right)(200 + 273)\text{K}} = 1.85 \times 10^{-3}\ \text{mol}$$

$$\text{molar mass} = \frac{0.145\ \text{g}}{1.85 \times 10^{-3}\ \text{mol}} = 78.4\ \text{g/mol}$$

The empirical mass of CH = 13.02 g/mol

Since $\frac{78.4\ \text{g/mol}}{13.02\ \text{g/mol}} = 6.02 \approx 6$, the molecular formula is $(CH)_6$ or $\mathbf{C_6H_6}$.

**5.100** The reaction is: $HCO_3^-(aq) + H^+(aq) \longrightarrow H_2O(l) + CO_2(g)$

The mass of $HCO_3^-$ reacted is:

$$3.29\ \text{g tablet} \times \frac{32.5\%\ HCO_3^-}{100\%\ \text{tablet}} = 1.07\ \text{g}\ HCO_3^-$$

$$\text{mol } CO_2 \text{ produced} = 1.07\ \text{g}\ HCO_3^- \times \frac{1\ \text{mol}\ HCO_3^-}{61.02\ \text{g}\ HCO_3^-} \times \frac{1\ \text{mol}\ CO_2}{1\ \text{mol}\ HCO_3^-} = 0.0175\ \text{mol}\ CO_2$$

$$V_{CO_2} = \frac{n_{CO_2}RT}{P} = \frac{(0.0175\ \text{mol}\ CO_2)\left(0.0821\frac{\text{L}\cdot\text{atm}}{\text{mol}\cdot\text{K}}\right)(37 + 273)\text{K}}{(1.00\ \text{atm})} = 0.445\ \text{L} = \mathbf{445\ mL}$$

**5.102** (a) The number of moles of $Ni(CO)_4$ formed is:

$$86.4\ \text{g Ni} \times \frac{1\ \text{mol Ni}}{58.69\ \text{g Ni}} \times \frac{1\ \text{mol Ni(CO)}_4}{1\ \text{mol Ni}} = 1.47\ \text{mol Ni(CO)}_4$$

The pressure of $Ni(CO)_4$ is:

$$P = \frac{nRT}{V} = \frac{(1.47\ \text{mol})\left(0.0821\frac{\text{L}\cdot\text{atm}}{\text{mol}\cdot\text{K}}\right)(43 + 273)\text{K}}{4.00\ \text{L}} = \mathbf{9.53\ atm}$$

**(b)** $Ni(CO)_4$ decomposes to produce more moles of gas (CO), which increases the pressure.

$$Ni(CO)_4(g) \longrightarrow Ni(s) + 4CO(g)$$

**5.104** Using the ideal gas equation, we can calculate the moles of gas.

$$n = \frac{PV}{RT} = \frac{(1.1\ \text{atm})\left(5.0 \times 10^2\ \text{mL} \times \dfrac{0.001\ \text{L}}{1\ \text{mL}}\right)}{\left(0.0821\dfrac{\text{L}\cdot\text{atm}}{\text{mol}\cdot\text{K}}\right)(37 + 273)\text{K}} = 0.0216\ \text{mol gas}$$

Next, use Avogadro's number to convert to molecules of gas.

$$0.0216\ \text{mol gas} \times \frac{6.022 \times 10^{23}\ \text{molecules}}{1\ \text{mol gas}} = \mathbf{1.30 \times 10^{22}\ molecules\ of\ gas}$$

The most common gases present in exhaled air are: **$CO_2$, $O_2$, $N_2$,** and **$H_2O$.**

**5.106** Mass of the Earth's atmosphere = (surface area of the earth in $cm^2$) × (mass per 1 $cm^2$ column)

Mass of a single column of air with a surface area of 1 $cm^2$ area is:

$$76.0\ \text{cm} \times 13.6\ \text{g/cm}^3 = 1.03 \times 10^3\ \text{g/cm}^2$$

The surface area of the Earth in $cm^2$ is:

$$4\pi r^2 = 4\pi(6.371 \times 10^8\ \text{cm})^2 = 5.10 \times 10^{18}\ \text{cm}^2$$

$$\textbf{Mass of atmosphere} = (5.10 \times 10^{18}\ \text{cm}^2)(1.03 \times 10^3\ \text{g/cm}^2) = 5.25 \times 10^{21}\ \text{g} = \mathbf{5.25 \times 10^{18}\ kg}$$

**5.108** To calculate the molarity of NaOH, we need moles of NaOH and volume of the NaOH solution. The volume is given in the problem; therefore, we need to calculate the moles of NaOH. The moles of NaOH can be calculated from the reaction of NaOH with HCl. The balanced equation is:

$$NaOH(aq) + HCl(aq) \longrightarrow H_2O(l) + NaCl(aq)$$

The number of moles of HCl gas is found from the ideal gas equation. $V = 0.189$ L, $T = (25 + 273)\text{K} = 298$ K, and $P = 108\ \text{mmHg} \times \dfrac{1\ \text{atm}}{760\ \text{mmHg}} = 0.142\ \text{atm}$.

$$n_{HCl} = \frac{PV_{HCl}}{RT} = \frac{(0.142\ \text{atm})(0.189\ \text{L})}{\left(0.0821\dfrac{\text{L}\cdot\text{atm}}{\text{mol}\cdot\text{K}}\right)(298\ \text{K})} = 1.10 \times 10^{-3}\ \text{mol HCl}$$

The moles of NaOH can be calculated using the mole ratio from the balanced equation.

$$(1.10 \times 10^{-3}\ \text{mol HCl}) \times \frac{1\ \text{mol NaOH}}{1\ \text{mol HCl}} = 1.10 \times 10^{-3}\ \text{mol NaOH}$$

The molarity of the NaOH solution is:

$$M = \frac{\text{mol NaOH}}{\text{L of soln}} = \frac{1.10 \times 10^{-3}\ \text{mol NaOH}}{0.0157\ \text{L soln}} = 0.0701\ \text{mol/L} = \mathbf{0.0701}\ \boldsymbol{M}$$

**5.110** To calculate the partial pressures of He and Ne, the total pressure of the mixture is needed. To calculate the total pressure of the mixture, we need the total number of moles of gas in the mixture (mol He + mol Ne).

$$n_{\text{He}} = \frac{PV}{RT} = \frac{(0.63\ \text{atm})(1.2\ \text{L})}{\left(0.0821\frac{\text{L}\cdot\text{atm}}{\text{mol}\cdot\text{K}}\right)(16 + 273)\text{K}} = 0.032\ \text{mol He}$$

$$n_{\text{Ne}} = \frac{PV}{RT} = \frac{(2.8\ \text{atm})(3.4\ \text{L})}{\left(0.0821\frac{\text{L}\cdot\text{atm}}{\text{mol}\cdot\text{K}}\right)(16 + 273)\text{K}} = 0.40\ \text{mol Ne}$$

The total pressure is:

$$P_{\text{Total}} = \frac{(n_{\text{He}} + n_{\text{Ne}})RT}{V_{\text{Total}}} = \frac{(0.032 + 0.40)\text{mol}\left(0.0821\frac{\text{L}\cdot\text{atm}}{\text{mol}\cdot\text{K}}\right)(16 + 273)\text{K}}{(1.2 + 3.4)\text{L}} = 2.2\ \text{atm}$$

$P_i = X_i P_T$. The partial pressures of He and Ne are:

$$P_{\text{He}} = \frac{0.032\ \text{mol}}{(0.032 + 0.40)\text{mol}} \times 2.2\ \text{atm} = \mathbf{0.16\ atm}$$

$$P_{\text{Ne}} = \frac{0.40\ \text{mol}}{(0.032 + 0.40)\text{mol}} \times 2.2\ \text{atm} = \mathbf{2.0\ atm}$$

**5.112** When the water enters the flask from the dropper, some hydrogen chloride dissolves, creating a partial vacuum. Pressure from the atmosphere forces more water up the vertical tube.

**5.114** Use the ideal gas equation to calculate the moles of water produced. We carry an extra significant figure in the first step of the calculation to limit rounding errors.

$$n_{\text{H}_2\text{O}} = \frac{PV}{RT} = \frac{(24.8\ \text{atm})(2.00\ \text{L})}{\left(0.0821\frac{\text{L}\cdot\text{atm}}{\text{mol}\cdot\text{K}}\right)(120 + 273)\text{K}} = 1.537\ \text{mol H}_2\text{O}$$

Next, we can determine the mass of $H_2O$ in the 54.2 g sample. Subtracting the mass of $H_2O$ from 54.2 g will give the mass of $MgSO_4$ in the sample.

$$1.537 \text{ mol H}_2\text{O} \times \frac{18.02 \text{ g H}_2\text{O}}{1 \text{ mol H}_2\text{O}} = 27.7 \text{ g H}_2\text{O}$$

$$\text{Mass MgSO}_4 = 54.2 \text{ g sample} - 27.7 \text{ g H}_2\text{O} = 26.5 \text{ g MgSO}_4$$

Finally, we can calculate the moles of $MgSO_4$ in the sample. Comparing moles of $MgSO_4$ to moles of $H_2O$ will allow us to determine the correct mole ratio in the formula.

$$26.5 \text{ g MgSO}_4 \times \frac{1 \text{ mol MgSO}_4}{120.4 \text{ g MgSO}_4} = 0.220 \text{ mol MgSO}_4$$

$$\frac{\text{mol H}_2\text{O}}{\text{mol MgSO}_4} = \frac{1.54 \text{ mol}}{0.220 \text{ mol}} = 7.00$$

Therefore, the mole ratio between $H_2O$ and $MgSO_4$ in the compound is 7 : 1. Thus, the value of $\boldsymbol{x = 7}$, and the formula is $\mathbf{MgSO_4{\cdot}7H_2O}$.

**5.116** The circumference of the cylinder is $= 2\pi r = 2\pi\left(\frac{15.0 \text{ cm}}{2}\right) = 47.1 \text{ cm}$

**(a)** The speed at which the target is moving equals:

$$\text{speed of target} = \text{circumference} \times \text{revolutions/sec}$$

$$\textbf{speed of target} = \frac{47.1 \text{ cm}}{1 \text{ revolution}} \times \frac{130 \text{ revolutions}}{1 \text{ s}} \times \frac{0.01 \text{ m}}{1 \text{ cm}} = \mathbf{61.2\ m/s}$$

**(b)** $2.80 \text{ cm} \times \frac{0.01 \text{ m}}{1 \text{ cm}} \times \frac{1 \text{ s}}{61.2 \text{ m}} = \mathbf{4.58 \times 10^{-4}\ s}$

**(c)** The Bi atoms must travel across the cylinder to hit the target. This distance is the diameter of the cylinder, which is 15.0 cm. The Bi atoms travel this distance in $4.58 \times 10^{-4}$ s.

$$\frac{15.0 \text{ cm}}{4.58 \times 10^{-4} \text{ s}} \times \frac{0.01 \text{ m}}{1 \text{ cm}} = \mathbf{328\ m/s}$$

$$u_{\text{rms}} = \sqrt{\frac{3RT}{M}} = \sqrt{\frac{3(8.314 \text{ J/K}\cdot\text{mol})(850 + 273)\text{K}}{209.0 \times 10^{-3} \text{ kg/mol}}} = \mathbf{366\ m/s}$$

The magnitudes of the speeds are comparable, but not identical. This is not surprising since 328 m/s is the velocity of a particular Bi atom, and $u_{\text{rms}}$ is an average value.

**5.118** The moles of $O_2$ can be calculated from the ideal gas equation. The mass of $O_2$ can then be calculated using the molar mass as a conversion factor.

$$n_{O_2} = \frac{PV}{RT} = \frac{(132\ \cancel{\text{atm}})(120\ \cancel{\text{L}})}{\left(0.0821\frac{\cancel{\text{L}}\cdot\cancel{\text{atm}}}{\text{mol}\cdot\cancel{\text{K}}}\right)(22 + 273)\cancel{\text{K}}} = 654\ \text{mol } O_2$$

$$\textbf{? g } \mathbf{O_2} = 654\ \cancel{\text{mol}}\ O_2 \times \frac{32.00\ \text{g } O_2}{1\ \cancel{\text{mol}}\ O_2} = \mathbf{2.09 \times 10^4\ g\ O_2}$$

The volume of $O_2$ gas under conditions of 1.00 atm pressure and a temperature of 22°C can be calculated using the ideal gas equation. The moles of $O_2$ = 654 moles.

$$V_{O_2} = \frac{n_{O_2}RT}{P} = \frac{(654\ \cancel{\text{mol}})\left(0.0821\frac{\text{L}\cdot\cancel{\text{atm}}}{\cancel{\text{mol}}\cdot\cancel{\text{K}}}\right)(22 + 273)\cancel{\text{K}}}{1.00\ \cancel{\text{atm}}} = \mathbf{1.58 \times 10^4\ L\ O_2}$$

**5.120** The fruit ripens more rapidly because the quantity (partial pressure) of ethylene gas inside the bag increases.

**5.122** As the pen is used the amount of ink decreases, increasing the volume inside the pen. As the volume increases, the pressure inside the pen decreases. The hole is needed to equalize the pressure as the volume inside the pen increases.

**5.124** **(a)** $NH_4NO_3(s) \longrightarrow N_2O(g) + 2H_2O(l)$

**(b)** $$\boldsymbol{R} = \frac{PV}{nT} = \frac{\left(718\ \cancel{\text{mmHg}} \times \frac{1\ \text{atm}}{760\ \cancel{\text{mmHg}}}\right)(0.340\ \text{L})}{\left(0.580\ \cancel{\text{g}}\ N_2O \times \frac{1\ \text{mol } N_2O}{44.02\ \cancel{\text{g}}\ N_2O}\right)(24 + 273)\text{K}} = \mathbf{0.0821\frac{L\cdot atm}{mol\cdot K}}$$

**5.126** The value of *a* indicates how strongly molecules of a given type of gas attract one anther. $\mathbf{C_6H_6}$ has the greatest intermolecular attractions due to its larger size compared to the other choices. Therefore, it has the largest *a* value.

**5.128** The gases inside the mine were a mixture of carbon dioxide, carbon monoxide, methane, and other harmful compounds. The low atmospheric pressure caused the gases to flow out of the mine (the gases in the mine were at a higher pressure), and the man suffocated.

**5.130** At the same temperature, the lighter the gas molecule, the faster it will move on average (i.e., it will have a greater average molecular speed). From left to right, the Maxwell speed distribution curves represent $\mathbf{Br_2}$ (159.8 g/mol, **red**), $\mathbf{SO_3}$ (80.07 g/mol, **yellow**), $\mathbf{N_2}$ (28.02 g/mol, **green**), and $\mathbf{CH_4}$ (16.04 g/mol, **blue**).

**5.132** **(a)** First, let's convert the concentration of hydrogen from atoms/cm$^3$ to mol/L. The concentration in mol/L can be substituted into the ideal gas equation to calculate the pressure of hydrogen.

$$\frac{1 \text{ H atom}}{1 \text{ cm}^3} \times \frac{1 \text{ mol H}}{6.022 \times 10^{23} \text{ H atoms}} \times \frac{1 \text{ cm}^3}{1 \text{ mL}} \times \frac{1 \text{ mL}}{0.001 \text{ L}} = \frac{2 \times 10^{-21} \text{ mol H}}{\text{L}}$$

The pressure of H is:

$$P = \left(\frac{n}{V}\right)RT = \left(\frac{2 \times 10^{-21} \text{ mol}}{1 \text{ L}}\right)\left(0.0821 \frac{\text{L} \cdot \text{atm}}{\text{mol} \cdot \text{K}}\right)(3 \text{ K}) = \mathbf{5 \times 10^{-22} \text{ atm}}$$

**(b)** From part (a), we know that 1 L contains 1.66 × 10$^{-21}$ mole of H atoms. We convert to the volume that contains 1.0 g of H atoms.

$$\frac{1 \text{ L}}{2 \times 10^{-21} \text{ mol H}} \times \frac{1 \text{ mol H}}{1.008 \text{ g H}} = \mathbf{5 \times 10^{20} \text{ L/g of H}}$$

**Note:** This volume is about that of all the water on Earth!

**5.134** From Table 5.3, the equilibrium vapor pressure at 30°C is 31.82 mmHg.

Converting 3.9 × 10$^3$ Pa to units of mmHg:

$$(3.9 \times 10^3 \text{ Pa}) \times \frac{760 \text{ mmHg}}{1.01325 \times 10^5 \text{ Pa}} = 29 \text{ mmHg}$$

$$\textbf{Relative Humidity} = \frac{\text{partial pressure of water vapor}}{\text{equilibrium vapor pressure}} \times 100\% = \frac{29 \text{ mmHg}}{31.82 \text{ mmHg}} \times 100\% = \mathbf{91\%}$$

**5.136** The volume of one alveoli is:

$$V = \frac{4}{3}\pi r^3 = \frac{4}{3}\pi(0.0050 \text{ cm})^3 = (5.2 \times 10^{-7} \text{ cm}^3) \times \frac{1 \text{ mL}}{1 \text{ cm}^3} \times \frac{0.001 \text{ L}}{1 \text{ mL}} = 5.2 \times 10^{-10} \text{ L}$$

The number of moles of air in one alveoli can be calculated using the ideal gas equation.

$$n = \frac{PV}{RT} = \frac{(1.0 \text{ atm})(5.2 \times 10^{-10} \text{ L})}{\left(0.0821 \frac{\text{L} \cdot \text{atm}}{\text{mol} \cdot \text{K}}\right)(37 + 273)\text{K}} = 2.0 \times 10^{-11} \text{ mol of air}$$

Since the air inside the alveoli is 14 percent oxygen, the moles of oxygen in one alveoli equals:

$$(2.0 \times 10^{-11} \text{ mol of air}) \times \frac{14\% \text{ oxygen}}{100\% \text{ air}} = 2.8 \times 10^{-12} \text{ mol } O_2$$

Converting to $O_2$ molecules:

$$(2.8 \times 10^{-12} \text{ mol } O_2) \times \frac{6.022 \times 10^{23} \; O_2 \text{ molecules}}{1 \text{ mol } O_2} = \mathbf{1.7 \times 10^{12} \; O_2 \text{ molecules}}$$

**5.138** The combined number of moles of gas in the two bulbs is 0.371 mole. From the pressure of the combined gases (1.08 atm) and the combined number of moles, we can calculate the total volume of the container. The volume of the right bulb can then be calculated.

$$V_{\text{Total}} = \frac{n_{\text{Total}}RT}{P_{\text{Total}}} = \frac{(0.371 \text{ mol})\left(0.0821 \frac{\text{L} \cdot \text{atm}}{\text{mol} \cdot \text{K}}\right)(293 \text{ K})}{1.08 \text{ atm}} = 8.26 \text{ L}$$

$$V_{\textbf{Right bulb}} = V_{\text{Total}} - V_{\text{Left bulb}} = 8.26 \text{ L} - 3.60 \text{ L} = \mathbf{4.66 \text{ L}}$$

**5.140** When calculating root-mean-square speed, remember that the molar mass must be in units of kg/mol.

$$u_{\textbf{rms}} = \sqrt{\frac{3RT}{M}} = \sqrt{\frac{3(8.314 \text{ J/mol} \cdot \text{K})(5.0 \times 10^{-8} \text{ K})}{85.47 \times 10^{-3} \text{ kg/mol}}} = \mathbf{3.8 \times 10^{-3} \text{ m/s}}$$

The mass of one Rb atom in kg is:

$$\frac{85.47 \text{ g Rb}}{1 \text{ mol Rb}} \times \frac{1 \text{ mol Rb}}{6.022 \times 10^{23} \text{ Rb atoms}} \times \frac{1 \text{ kg}}{1000 \text{ g}} = 1.419 \times 10^{-25} \text{ kg/Rb atom}$$

$$\overline{\textbf{KE}} = \frac{1}{2}m\overline{u^2} = \frac{1}{2}(1.419 \times 10^{-25} \text{ kg})(3.8 \times 10^{-3} \text{ m/s})^2 = \mathbf{1.0 \times 10^{-30} \text{ J}}$$

**5.142** The molar volume is the volume of 1 mole of gas under the specified conditions.

$$V = \frac{nRT}{P} = \frac{(1 \text{ mol})\left(0.0821 \frac{\text{L} \cdot \text{atm}}{\text{mol} \cdot \text{K}}\right)(220 \text{ K})}{\left(6.0 \text{ mmHg} \times \frac{1 \text{ atm}}{760 \text{ mmHg}}\right)} = \mathbf{2.3 \times 10^3 \text{ L}}$$

**5.144** The volume of the bulb can be calculated using the ideal gas equation. Pressure and temperature are given in the problem. Moles of air must be calculated before the volume can be determined.

$$\text{Mass of air} = 91.6843 \text{ g} - 91.4715 \text{ g} = 0.2128 \text{ g air}$$

$$\text{Molar mass of air} = (0.78 \times 28.02 \text{ g/mol}) + (0.21 \times 32.00 \text{ g/mol}) + (0.01 \times 39.95 \text{ g/mol}) = 29 \text{ g/mol}$$

$$\text{moles air} = 0.2128 \text{ g air} \times \frac{1 \text{ mol air}}{29 \text{ g air}} = 7.3 \times 10^{-3} \text{ mol air}$$

Now, we can calculate the volume of the bulb.

$$V_{\text{bulb}} = \frac{nRT}{P} = \frac{(7.3 \times 10^{-3}\ \text{mol})\left(0.0821\frac{\text{L}\cdot\text{atm}}{\text{mol}\cdot\text{K}}\right)(23 + 273)\text{K}}{\left(744\ \text{mmHg} \times \dfrac{1\ \text{atm}}{760\ \text{mmHg}}\right)} = 0.18\ \text{L} = \mathbf{1.8 \times 10^2\ mL}$$

**5.146** In Problem 5.106, the mass of the Earth's atmosphere was determined to be $5.25 \times 10^{18}$ kg. Assuming that the molar mass of air is 29.0 g/mol, we can calculate the number of molecules in the atmosphere.

**(a)** $$(5.25 \times 10^{18}\ \text{kg air}) \times \frac{1000\ \text{g}}{1\ \text{kg}} \times \frac{1\ \text{mol air}}{29.0\ \text{g air}} \times \frac{6.022 \times 10^{23}\ \text{molecules air}}{1\ \text{mol air}} = \mathbf{1.09 \times 10^{44}\ molecules}$$

**(b)** First, calculate the moles of air exhaled in every breath. (500 mL = 0.500 L)

$$n = \frac{PV}{RT} = \frac{(1\ \text{atm})(0.500\ \text{L})}{\left(0.0821\frac{\text{L}\cdot\text{atm}}{\text{mol}\cdot\text{K}}\right)(37 + 273)\text{K}} = 1.96 \times 10^{-2}\ \text{mol air/breath}$$

Next, convert to molecules of air per breath.

$$1.96 \times 10^{-2}\ \text{mol air/breath} \times \frac{6.022 \times 10^{23}\ \text{molecules air}}{1\ \text{mol air}} = \mathbf{1.18 \times 10^{22}\ molecules/breath}$$

**(c)** $$\frac{1.18 \times 10^{22}\ \text{molecules}}{1\ \text{breath}} \times \frac{12\ \text{breaths}}{1\ \text{min}} \times \frac{60\ \text{min}}{1\ \text{h}} \times \frac{24\ \text{h}}{1\ \text{day}} \times \frac{365\ \text{days}}{1\ \text{yr}} \times 35\ \text{yr} = \mathbf{2.60 \times 10^{30}\ molecules}$$

**(d)** Fraction of molecules in the atmosphere exhaled by Mozart is:

$$\frac{2.60 \times 10^{30}\ \text{molecules}}{1.09 \times 10^{44}\ \text{molecules}} = \mathbf{2.39 \times 10^{-14}}$$

Or,

$$\frac{1}{2.39 \times 10^{-14}} = 4.18 \times 10^{13}$$

Thus, about 1 molecule of air in every $4 \times 10^{13}$ molecules was exhaled by Mozart.

In a single breath containing $1.18 \times 10^{22}$ molecules, we would breathe in on average:

$$(1.18 \times 10^{22}\ \text{molecules}) \times \frac{1\ \text{Mozart air molecule}}{4 \times 10^{13}\ \text{air molecules}} = \mathbf{3 \times 10^8\ molecules\ that\ Mozart\ exhaled}$$

**(e)** We made the following assumptions:

1. Complete mixing of air in the atmosphere.
2. That no molecules escaped to the outer atmosphere.
3. That no molecules were used up during metabolism, nitrogen fixation, and so on.

**5.148** The ideal gas law can be used to calculate the moles of water vapor per liter.

$$\frac{n}{V} = \frac{P}{RT} = \frac{1.0\ \text{atm}}{\left(0.0821\frac{\text{L}\cdot\text{atm}}{\text{mol}\cdot\text{K}}\right)(100+273)\text{K}} = 0.033\ \frac{\text{mol}}{\text{L}}$$

We eventually want to find the distance between molecules. Therefore, let's convert moles to molecules, and convert liters to a volume unit that will allow us to get to distance ($m^3$).

$$\left(\frac{0.033\ \text{mol}}{1\ \text{L}}\right)\left(\frac{6.022\times 10^{23}\ \text{molecules}}{1\ \text{mol}}\right)\left(\frac{1000\ \text{L}}{1\ \text{m}^3}\right) = 2.0\times 10^{25}\ \frac{\text{molecules}}{\text{m}^3}$$

This is the number of ideal gas molecules in a cube that is 1 meter on each side. Assuming an equal distribution of molecules along the three mutually perpendicular directions defined by the cube, a linear density in one direction may be found:

$$\left(\frac{2.0\times 10^{25}\ \text{molecules}}{1\ \text{m}^3}\right)^{\frac{1}{3}} = 2.7\times 10^{8}\ \frac{\text{molecules}}{\text{m}}$$

This is the number of molecules on a line *one* meter in length. The distance between each molecule is given by:

$$\frac{1\ \text{m}}{2.70\times 10^{8}} = 3.7\times 10^{-9}\ \text{m} = \mathbf{3.7\ nm}$$

Assuming a water molecule to be a sphere with a diameter of 0.3 nm, the water molecules are separated by over 12 times their diameter: $\frac{3.7\ \text{nm}}{0.3\ \text{nm}} \approx 12$ times.

A similar calculation is done for liquid water. Starting with density, we convert to molecules per cubic meter.

$$\frac{0.96\ \text{g}}{1\ \text{cm}^3}\times\frac{1\ \text{mol H}_2\text{O}}{18.02\ \text{g H}_2\text{O}}\times\frac{6.022\times 10^{23}\ \text{molecules}}{1\ \text{mol H}_2\text{O}}\times\left(\frac{100\ \text{cm}}{1\ \text{m}}\right)^3 = 3.2\times 10^{28}\ \frac{\text{molecules}}{\text{m}^3}$$

This is the number of liquid water molecules in *one* cubic meter. From this point, the calculation is the same as that for water vapor, and the space between molecules is found using the same assumptions.

$$\left(\frac{3.2\times 10^{28}\ \text{molecules}}{1\ \text{m}^3}\right)^{\frac{1}{3}} = 3.2\times 10^{9}\ \frac{\text{molecules}}{\text{m}}$$

$$\frac{1\ \text{m}}{3.2\times 10^{9}} = 3.1\times 10^{-10}\ \text{m} = \mathbf{0.31\ nm}$$

Assuming a water molecule to be a sphere with a diameter of 0.3 nm, to one significant figure, the water molecules are touching each other in the liquid phase.

**5.150** Since the $R = 8.314$ J/mol·K and $1\text{ J} = 1\dfrac{\text{kg}\cdot\text{m}^2}{\text{s}^2}$, then the mass substituted into the equation must have units of kg and the height must have units of meters.

29 g/mol = 0.029 kg/mol

5.0 km = $5.0 \times 10^3$ m

Substituting the given quantities into the equation, we find the atmospheric pressure at 5.0 km to be:

$$P = P_0 e^{-\frac{gMh}{RT}}$$

$$P = (1.0\text{ atm})e^{-\left(\frac{(9.8\text{ m/s}^2)(0.029\text{ kg/mol})(5.0\times 10^3\text{ m})}{(8.314\text{ J/mol}\cdot\text{K})(278\text{ K})}\right)}$$

$$P = \mathbf{0.54\ atm}$$

**5.152** The relative rates of effusion of the two gases will allow the calculation of the mole fraction of each gas as it passes through the orifice. Using Graham's law of effusion,

$$\frac{r_{H_2}}{r_{D_2}} = \sqrt{\frac{M_{D_2}}{M_{H_2}}}$$

$$\frac{r_{H_2}}{r_{D_2}} = \sqrt{\frac{4.028\text{ g/mol}}{2.016\text{ g/mol}}}$$

$$\frac{r_{H_2}}{r_{D_2}} = 1.414$$

This calculation shows that $H_2$ will effuse 1.414 times faster than $D_2$. If, over a given amount of time, 1 mole of $D_2$ effuses through the orifice, 1.414 moles of $H_2$ will effuse during the same amount of time. We can now calculate the mole fraction of each gas.

$$X_{H_2} = \frac{\text{mol H}_2}{\text{total mol}} = \frac{1.414\text{ mol}}{(1+1.414)\text{mol}} = \mathbf{0.5857}$$

Because this is a two-component mixture, the mole fraction of $D_2$ is:

$$X_{D_2} = 1 - 0.5857 = \mathbf{0.4143}$$

**5.154** The reaction between Zn and HCl is: $Zn(s) + 2HCl(aq) \rightarrow H_2(g) + ZnCl_2(aq)$

From the amount of $H_2(g)$ produced, we can determine the amount of Zn reacted. Then, using the original mass of the sample, we can calculate the mass % of Zn in the sample.

$$n_{H_2} = \frac{PV_{H_2}}{RT}$$

$$n_{H_2} = \frac{\left(728\ \text{mmHg} \times \dfrac{1\ \text{atm}}{760\ \text{mmHg}}\right)(1.26\ \text{L})}{\left(0.0821\dfrac{\text{L}\cdot\text{atm}}{\text{mol}\cdot\text{K}}\right)(22+273)\text{K}} = 0.0498\ \text{mol H}_2$$

Since the mole ratio between $H_2$ and Zn is 1:1, the amount of Zn reacted is also 0.0498 mole. Converting to grams of Zn, we find:

$$0.0498\ \text{mol Zn} \times \frac{65.39\ \text{g Zn}}{1\ \text{mol Zn}} = 3.26\ \text{g Zn}$$

The mass percent of Zn in the 6.11 g sample is:

$$\textbf{mass \% Zn} = \frac{\text{mass Zn}}{\text{mass sample}} \times 100\% = \frac{3.26\ \text{g}}{6.11\ \text{g}} \times 100\% = \textbf{53.4\%}$$

**5.156** We start with Graham's Law as this problem relates to effusion of gases. Using Graham's Law, we can calculate the effective molar mass of the mixture of CO and $CO_2$. Once the effective molar mass of the mixture is known, we can determine the mole fraction of each component. Because $n \propto V$ at constant $T$ and $P$, the volume fraction = mole fraction.

$$\frac{r_{He}}{r_{mix}} = \sqrt{\frac{\mathcal{M}_{mix}}{\mathcal{M}_{He}}}$$

$$\mathcal{M}_{mix} = \left(\frac{r_{He}}{r_{mix}}\right)^2 \mathcal{M}_{He}$$

$$\mathcal{M}_{mix} = \left(\frac{\dfrac{29.7\ \text{mL}}{2.00\ \text{min}}}{\dfrac{10.0\ \text{mL}}{2.00\ \text{min}}}\right)^2 (4.003\ \text{g/mol}) = 35.31\ \text{g/mol}$$

Now that we know the molar mass of the mixture, we can calculate the mole fraction of each component.

$$X_{CO} + X_{CO_2} = 1$$

and

$$X_{CO_2} = 1 - X_{CO}$$

The mole fraction of each component multiplied by its molar mass will give the contribution of that component to the effective molar mass.

$$X_{CO}M_{CO} + X_{CO_2}M_{CO_2} = M_{mix}$$

$$X_{CO}M_{CO} + (1 - X_{CO})M_{CO_2} = M_{mix}$$

$$X_{CO}(28.01 \text{ g/mol}) + (1 - X_{CO})(44.01 \text{ g/mol}) = 35.31 \text{ g/mol}$$

$$28.01X_{CO} + 44.01 - 44.01X_{CO} = 35.31$$

$$16.00X_{CO} = 8.70$$

$$X_{CO} = 0.544$$

At constant $P$ and $T$, $n \propto V$. Therefore, volume fraction = mole fraction. As a result,

% of CO by volume = **54.4%**

% of $CO_2$ by volume = 1 − % of CO by volume = **45.6%**

**5.158** The reactions are:

$$CH_4 + 2O_2 \rightarrow CO_2 + 2H_2O$$

$$2C_2H_6 + 7O_2 \rightarrow 4CO_2 + 6H_2O$$

For a given volume and temperature, $n \propto P$. This means that the greater the pressure of reactant, the more moles of reactant, and hence the more product ($CO_2$) that will be produced. The pressure of $CO_2$ produced comes from the combustion of both methane and ethane. We set up an equation using the mole ratios from the balanced equation to convert to pressure of $CO_2$.

$$\left(P_{CH_4} \times \frac{1 \text{ mol } CO_2}{1 \text{ mol } CH_4}\right) + \left(P_{C_2H_6} \times \frac{4 \text{ mol } CO_2}{2 \text{ mol } C_2H_6}\right) = 356 \text{ mmHg } CO_2$$

$$(1) \quad P_{CH_4} + 2P_{C_2H_6} = 356 \text{ mmHg}$$

Also,

$$(2) \quad P_{CH_4} + P_{C_2H_6} = 294 \text{ mmHg}$$

Subtracting equation (2) from equation (1) gives:

$$P_{C_2H_6} = 356 - 294 = 62 \text{ mmHg}$$

$$P_{CH_4} = 294 - 62 = 232 \text{ mmHg}$$

Lastly, because $n \propto P$, we can solve for the mole fraction of each component using partial pressures.

$$X_{CH_4} = \frac{232}{294} = \mathbf{0.789} \qquad X_{C_2H_6} = \frac{62}{294} = \mathbf{0.211}$$

**5.160** **(a)** We see from the figure that two hard spheres of radius $r$ cannot approach each other more closely than $2r$ (measured from the centers). Thus, there is a sphere of radius $2r$ surrounding each hard sphere from which other hard spheres are excluded. The excluded volume/pair of molecules is:

$$V_{\text{excluded}}/\text{pair} = \frac{4}{3}\pi(2r)^3 = \frac{32}{3}\pi r^3 = \mathbf{8\left(\frac{4}{3}\pi r^3\right)}$$

This is eight times the volume of an individual molecule.

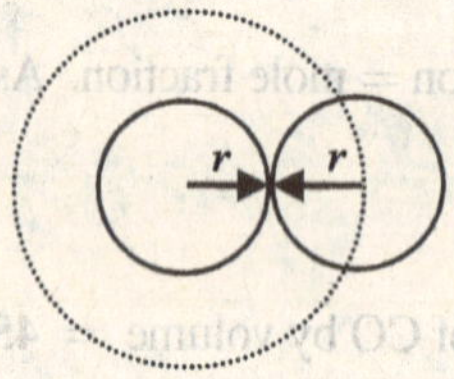


**(b)** The result in part (a) is for a pair of molecules, so the excluded volume/molecule is:

$$V_{\text{excluded}}/\text{molecule} = \frac{1}{2}\left(\frac{32}{3}\pi r^3\right) = \frac{16}{3}\pi r^3$$

To convert from excluded volume per molecule to excluded volume per mole, we need to multiply by Avogadro's number, $N_A$.

$$V_{\text{excluded}}/\text{mol} = \mathbf{\frac{16}{3}N_A\pi r^3}$$

The sum of the volumes of a mole of molecules (treated as hard spheres of radius $r$) is $\frac{4}{3}N_A\pi r^3$. The excluded volume is **four times** the volume of the spheres themselves.

**5.162** Let $u_1$ be the root-mean-square speed of the unknown gas and $u_2$ be the rms speed of HI.

$$u_1 = \sqrt{\frac{3RT}{M_1}} \qquad u_2 = \sqrt{\frac{3RT}{M_2}}$$

Dividing $u_1$ by $u_2$ gives:

$$\frac{u_1}{u_2} = \frac{\sqrt{\frac{3RT}{M_1}}}{\sqrt{\frac{3RT}{M_2}}}$$

Canceling $3RT$ from the equation gives:

$$\frac{u_1}{u_2} = \frac{\sqrt{\frac{1}{M_1}}}{\sqrt{\frac{1}{M_2}}} = \sqrt{\frac{M_2}{M_1}}$$

$$\frac{2.82}{1} = \sqrt{\frac{127.9\ \text{g/mol}}{M_1}}$$

Squaring both sides of the equation and rearranging,

$$7.95\ M_1 = 127.9\ \text{g/mol}$$

$$M_1 = 16.1\ \text{g/mol}$$

The gas is most likely **methane, $CH_4$.**

**5.164** From the root-mean-square speed, we can calculate the molar mass of the gaseous oxide.

$$u_{\text{rms}} = \sqrt{\frac{3RT}{M}}$$

$$M = \frac{3RT}{(u_{\text{rms}})^2} = \frac{3(8.314\ \text{J/mol}\cdot\text{K})(293\ \text{K})}{(493\ \text{m/s})^2} = 0.0301\ \text{kg/mol} = 30.1\ \text{g/mol}$$

The compound must be a monoxide because 2 moles of oxygen atoms would have a mass of 32.00 g. The molar mass of the other element is:

$$30.1\ \text{g/mol} - 16.00\ \text{g/mol} = 14.01\ \text{g/mol}$$

The compound is nitrogen monoxide, **NO.**

**5.166** Pressure and volume are constant. We start with Equation (5.9) of the text.

$$\frac{P_1V_1}{n_1T_1} = \frac{P_2V_2}{n_2T_2}$$

Because $P_1 = P_2$ and $V_1 = V_2$, this equation reduces to:

$$\frac{1}{n_1T_1} = \frac{1}{n_2T_2}$$

or,

$$n_1T_1 = n_2T_2$$

Because $T_1 = 2T_2$, substituting into the above equation gives:

$$2n_1T_2 = n_2T_2$$

or,

$$2n_1 = n_2$$

This equation indicates that the number of moles of gas after reaction is twice the number of moles of gas before reaction. Only reaction **(b)** fits this description.

**5.168** (i) Bulb (b) contains the same number of particles as (a), but is half the volume. The pressure will be double that of A.

$$P_B = (2)(4.0\text{ atm}) = \mathbf{8.0\ atm}$$

The volume of bulb (c) is the same as bulb (a), but there are 12 particles in (c) while there are 9 particles in (a). The pressure is directly proportional to the number of moles of gas (or particles of gas) at the same temperature and volume.

$$P_C = \left(\frac{12}{9}\right)(4.0\text{ atm}) = \mathbf{5.3\ atm}$$

(ii) When the valves are opened at constant temperature, the gases expand to fill the entire container. For example, in bulb (a), the pressure before opening the valve is 4.0 atm and the volume is 4.0 L. After the valves are opened, the volume is now 10 L (4 L + 2 L + 4 L). We use Boyle's law to calculate the partial pressure of each sample of gas after the valves are opened.

(a)
$$P_1V_1 = P_2V_2$$
$$(4.0\text{ atm})(4.0\not{L}) = P_2(10.0\not{L})$$
$$P_2 = 1.6\text{ atm}$$

(b)
$$P_1V_1 = P_2V_2$$
$$(8.0\text{ atm})(2.0\not{L}) = P_2(10.0\not{L})$$
$$P_2 = 1.6\text{ atm}$$

(c)
$$P_1V_1 = P_2V_2$$
$$(5.3\text{ atm})(4.0\not{L}) = P_2(10.0\not{L})$$
$$P_2 = 2.1\text{ atm}$$

The total pressure in the container is the sum of the partial pressures (Dalton's Law).

$$P_{\text{total}} = 1.6\text{ atm} + 1.6\text{ atm} + 2.1\text{ atm} = \mathbf{5.3\ atm}$$

There are 15 particles of gas A and 15 particles of gas B in the container. Therefore, the partial pressure of each gas will be half the total pressure of 5.3 atm.

$$P_A = P_B = \mathbf{2.65\ atm}$$

**5.170** The intensity of the peaks represents the relative abundances of the three gases. We estimate the intensity of the peaks as $CH_4$: 4.0, $C_2H_6$: 1.5, and $C_3H_8$: 2.5. The sum of the intensities is (4.0 + 1.5 + 2.5) = 8.0. Thus, from Equation (5.14) of the text, we write:

$$P_{CH_4} = \left(\frac{4.0}{8.0}\right) \times 4.50 \text{ atm} = \mathbf{2.3\ atm}$$

$$P_{C_2H_6} = \left(\frac{1.5}{8.0}\right) \times 4.50 \text{ atm} = \mathbf{0.84\ atm}$$

$$P_{C_3H_8} = \left(\frac{2.5}{8.0}\right) \times 4.50 \text{ atm} = \mathbf{1.4\ atm}$$

The sum of the partial pressures is (2.3 + 0.84 + 1.4) atm or 4.5 atm.

# CHAPTER 6
# THERMOCHEMISTRY

## PROBLEM-SOLVING STRATEGIES AND TUTORIAL SOLUTIONS

### TYPES OF PROBLEMS

**Problem Type 1:** The First Law of Thermodynamics.
- **(a)** Applying the First Law of Thermodynamics.
- **(b)** Calculating the work done in gas expansion.
- **(c)** Enthalpy and the First Law of Thermodynamics. Calculating the internal energy change of a gaseous reaction.

**Problem Type 2:** Thermochemical Equations.

**Problem Type 3:** Calculating Heat Absorbed or Released Using Specific Heat Data.

**Problem Type 4:** Calorimetry.
- **(a)** Constant-volume calorimetry.
- **(b)** Constant-pressure calorimetry.

**Problem Type 5:** Standard Enthalpy of Formation and Reaction.
- **(a)** Calculating the standard enthalpy of reaction.
- **(b)** Direct method of calculating the standard enthalpy of formation.
- **(c)** Indirect method of calculating the standard enthalpy of formation, Hess's law.

## PROBLEM TYPE 1: THE FIRST LAW OF THERMODYNAMICS

### A. Applying the First Law of Thermodynamics

The **First Law of Thermodynamics** states that energy can be converted from one form to another, but cannot be created or destroyed. Another way of stating the first law is that the energy of the universe is constant. The universe is composed of both the system and the surroundings.

$$\Delta U_{\text{sys}} + \Delta U_{\text{surr}} = 0$$

where,

the subscripts "sys" and "surr" denote system and surroundings, respectively.

However, in chemistry, we are normally interested in the changes associated with the *system* (which may be a flask containing reactants and products), not with its surroundings. Therefore, a more useful form of the first law is

$$\Delta U = q + w \qquad \text{(6.1, text)}$$

where,

$\Delta U$ is the change in the internal energy of the system
$q$ is the heat exchange between the system and surroundings
$w$ is the work done on (or by) the system

Using the sign convention for thermochemical processes (see Section 6.3 of your text for discussion), $q$ is positive for an endothermic process and negative for an exothermic process. For work, $w$ is positive for work done *on* the system *by* the surroundings and negative for work done *by* the system *on* the surroundings. Try to understand the sign convention in this manner. If a *system* loses heat to the surroundings or does work on the surroundings, we expect its internal energy to decrease since both processes are energy depleting. Conversely, if heat is added to the *system* or if work is done on the *system*, then the internal energy of the system would increase.

**EXAMPLE 6.1**

**A system does 975 kJ of work on its surroundings while at the same time it absorbs 625 kJ of heat. What is the change in energy, $\Delta U$, for the system?**

**Strategy:** The system does work on the surroundings, so what is the sign for $w$? Heat is absorbed by the gas from the surroundings. Is this an endothermic or exothermic process? What is the sign for $q$?

**Solution:** To calculate the energy change of the gas ($\Delta U$), we need Equation (6.1) of the text. To solve this problem, you must make sure to get the sign convention correct. The system does work on the surroundings; this is an energy-depleting process.

$$w = \mathbf{-975\ kJ}$$

The system absorbs 625 kJ of heat. Therefore, the internal energy of the system would increase.

$$q = \mathbf{+625\ kJ}$$

Finally,

$$\Delta U = q + w = 625\text{ kJ} + (-975\text{ kJ}) = \mathbf{-350\ kJ}$$

**PRACTICE EXERCISE**

1. The surroundings do 455 kJ of work on the system while at the same time the system releases 253 kJ of heat. What is the change in energy, $\Delta U$, for the system?

**Text Problem: 6.18**

## B. Calculating the work done in gas expansion

A useful example of mechanical work is the expansion of a gas. Picture a gas-filled cylinder that is fitted with a weightless, frictionless, movable piston, at a certain temperature, pressure, and volume. As the gas expands, it pushes the piston upward against a constant, opposing, external atmospheric pressure, $P$. The gas (system) is doing work on the surroundings. The work can be calculated as follows:

$$w = -P\Delta V \qquad (6.3,\ \text{text})$$

where,

$P$ is the external pressure
$\Delta V$ is the change in volume ($V_f - V_i$)

**Note:** The minus sign in the equation takes care of the sign convention for $w$. For gas expansion, $\Delta V > 0$, so $-P\Delta V$ is a negative quantity. When a gas expands, it's doing work on the surroundings; the internal energy of the system decreases. For gas compression, $\Delta V < 0$, so $-P\Delta V$ is a positive quantity. When a gas is compressed, the surroundings are doing work on the system, increasing the internal energy.

**EXAMPLE 6.2**

**A gas initially at a pressure of 10.0 atm and occupying a volume of 5.0 L is allowed to expand at constant temperature against a constant external pressure of 4.0 atm. After expansion, the gas occupies a volume of 12.5 L. Calculate the work done by the gas on the surroundings.**

---

**Strategy:** The work done in gas expansion is equal to the product of the external, opposing pressure and the change in volume [Equation (6.3) of the text].

$$w = -P\Delta V$$

What is the conversion factor between L·atm and J?

**Solution:** We are given the external pressure in the problem, but we must calculate $\Delta V$.

$$\Delta V = V_f - V_i = 12.5\text{ L} - 5.0\text{ L} = \mathbf{7.5\text{ L}}$$

Substitute $P$ and $\Delta V$ into Equation (6.3) of the text and solve for $w$.

$$w = -P\Delta V = -(4.0\text{ atm})(7.5\text{ L}) = \mathbf{-3.0 \times 10^1\text{ L·atm}}$$

It would be more convenient to express $w$ in units of joules. The following conversion factor can be obtained from Appendix 1 of the text:

$$1\text{ L·atm} = 101.3\text{ J}$$

Thus, we can write:

$$w = (-3.0 \times 10^1\ \cancel{\text{L·atm}}) \times \frac{101.3\text{ J}}{1\ \cancel{\text{L·atm}}} = \mathbf{-3.0 \times 10^3\text{ J}}$$

**Check:** Because this is gas expansion (work is done by the system on the surroundings), the work done has a negative sign.

**PRACTICE EXERCISE**

2. Calculate the work done on the system when 6.0 L of a gas is compressed to 1.0 L by a constant external pressure of 2.0 atm.

| **Text Problem: 6.20** |
|---|

## C. Enthalpy and the First Law of Thermodynamics—Calculating the internal energy change of a gaseous reaction

Let's return to the following form of the first law of thermodynamics.

$$\Delta U_{sys} = q + w$$

Under constant-pressure conditions we can write:

$$\Delta U = q_p + w$$

Recall that the heat evolved or absorbed ($q$) by a reaction carried out under constant-pressure conditions is equal to the enthalpy change of the system, $\Delta H$.

Thus,

$$\Delta U = \Delta H + w$$

Also, we know that for gas expansion of compression under a constant external pressure, $w = -P\Delta V$. Substituting into the above equation, we have:

$$\Delta U = \Delta H - P\Delta V$$

Also, for an ideal gas at constant pressure,

$$P\Delta V = \Delta(nRT)$$

$$\Delta V = \frac{\Delta(nRT)}{P}$$

Substituting gives:

$$\Delta U = \Delta H - \Delta(nRT)$$

Finally, at constant temperature,

$$\mathbf{\Delta U = \Delta H - RT\Delta n} \qquad \text{(6.10, text)}$$

where $\Delta n$ is defined as

$$\Delta n = \text{number of moles of product gases} - \text{number of moles of reactant gases}$$

**EXAMPLE 6.3**

**Calculate the change in internal energy when 1 mole of $H_2$ and 1/2 mole of $O_2$ are converted to 1 mole of $H_2O$ at 1 atm and 25°C.**

$$\mathbf{H_2(g) + \tfrac{1}{2}O_2(g) \longrightarrow H_2O(l) \qquad \Delta H^\circ = -286\ kJ/mol}$$

**Strategy:** Calculate the total change in the number of moles of *gas*. Note that the product is a liquid. Substitute the values of $\Delta H^\circ$ and $\Delta n$ into Equation (6.10) of the text.

**Solution:**

$$\Delta n = 0\text{ mol} - (1\text{ mol} + 1/2\text{ mol}) = -1.5\text{ mol}$$

Substitute the values for $\Delta H^\circ$ and $\Delta n$ into Equation (6.10).

$$T = 25^\circ + 273^\circ = 298\text{ K}$$

$$\Delta U^\circ = \Delta H^\circ - RT\Delta n$$

$$\mathbf{\Delta U^\circ} = -286\text{ kJ/mol} - \left(8.314\frac{\text{J}}{\text{mol}\cdot\text{K}}\right)(298\text{ K})(-1.5) \times \frac{1\text{ kJ}}{1000\text{ J}} = \mathbf{-282\ kJ/mol}$$

**PRACTICE EXERCISE**

3. Calculate the change in the internal energy when 1.0 mole of water vaporizes at 1.0 atm and 100°C. Assume that water vapor is an ideal gas and that the volume of liquid water is negligible compared with that of steam at 100°C. [$\Delta H_{vap}(H_2O) = 40.67$ kJ/mol at 100°C].

**Text Problem:** 6.28

## PROBLEM TYPE 2: THERMOCHEMICAL EQUATIONS

Equations showing both the mass and enthalpy relations are called **thermochemical equations**. The following guidelines are helpful in writing and interpreting thermochemical equations:

1. The stoichiometric coefficients always refer to the number of moles of each substance.
2. When an equation is reversed, the roles of reactants and products change. Consequently, the magnitude of $\Delta H$ for the equation remains the same, but its sign changes.
3. If both sides of a thermochemical equation are multiplied by a factor $n$, then $\Delta H$ must also be multiplied by the same factor.
4. When writing thermochemical equations, the physical states of all reactants and products must be specified, because they help determine the actual enthalpy changes.

See Section 6.4 of your text for further discussion.

**EXAMPLE 6.4**

**Given the thermochemical equation**

$$SO_2(g) + \frac{1}{2}O_2(g) \longrightarrow SO_3(g) \quad \Delta H = -99 \text{ kJ/mol}$$

**how much heat is evolved when (a) 1/2 mol of $SO_2$ reacts and (b) 3 mol of $SO_2$ reacts?**

**Strategy:** The thermochemical equation shows that for every 1 mole of $SO_2$ reacted, 99 kJ of heat are given off (note the negative sign). We can write a conversion factor from this information.

$$\frac{-99 \text{ kJ}}{1 \text{ mol } SO_2}$$

**Solution:**

(a) If 1/2 mole of $SO_2$ reacts, that means that we are multiplying the equation by 1/2. Therefore, we must multiply $\Delta H$ by 1/2.

$$\textbf{heat evolved} = 0.5 \text{ mol } SO_2 \times \frac{99 \text{ kJ}}{1 \text{ mol } SO_2} = \mathbf{5.0 \times 10^1 \text{ kJ}}$$

(b) Following the same argument as in part (a):

$$\textbf{heat evolved} = 3(99 \text{ kJ}) = \mathbf{3.0 \times 10^2 \text{ kJ}}$$

Why isn't there a negative sign in our answer? The sign convention for an exothermic reaction (energy, as heat, is released by the system) is negative ($-\Delta H$). However, in the above example, we state that heat is evolved, so a negative sign is unnecessary.

**Tip:** Remember that the heat evolved or absorbed ($q$) by a reaction carried out under constant-pressure conditions is equal to the enthalpy change of the system, $\Delta H$.

**PRACTICE EXERCISE**

4. Given the thermochemical equation

$$SO_2(g) + \frac{1}{2}O_2(g) \longrightarrow SO_3(g) \quad \Delta H = -99 \text{ kJ/mol}$$

how much heat is evolved when 75 g of $SO_2$ is combusted?

**Text Problem: 6.26**

## PROBLEM TYPE 3: CALCULATING HEAT ABSORBED OR RELEASED USING SPECIFIC HEAT DATA

If the specific heat ($s$) and the amount of substance is known, then the change in the sample's temperature ($\Delta t$) will tell us the amount of heat ($q$) that has been absorbed or released in a particular process. The equation for calculating the heat change is given by:

$$q = ms\Delta t \qquad (6.12, \text{text})$$

or

$$q = C\Delta t$$

where $m$ is the mass of the sample and $\Delta t$ is the temperature change.

$$\Delta t = t_{\text{final}} - t_{\text{initial}}$$

**EXAMPLE 6.5**

**How much heat is absorbed by 80.0 g of iron (Fe) when its temperature is raised from 25°C to 500°C? The specific heat of iron is 0.444 J/g·°C.**

**Strategy:** We know the mass, specific heat, and the change in temperature for iron. We can use Equation (6.12) of the text to solve this problem.

**Solution:** Substitute the known values into Equation (6.12).

$$q = m_{\text{Fe}}s_{\text{Fe}}\Delta t$$

$$q = (80.0\ \cancel{\text{g}})(0.444\ \text{J}/\cancel{\text{g}}\cdot\cancel{°\text{C}})(500 - 25)\cancel{°\text{C}} = \mathbf{1.69 \times 10^4\ J}$$

**PRACTICE EXERCISE**

5. A piece of iron initially at a temperature of 25°C absorbs 10.0 kJ of heat. If its mass is 50.0 g, calculate the final temperature of the piece of iron. The specific heat of iron is 0.444 J/g.°C.

**Text Problem: 6.34**

## PROBLEM TYPE 4: CALORIMETRY

### A. Constant-volume calorimetry

For a discussion of constant-volume calorimetry, see Section 6.5 of your text. Heat of combustion is usually measured in a constant-volume calorimeter. The heat released during combustion is absorbed by the calorimeter. Because no heat enters or leaves the system throughout the process, we can write:

$$q_{system} = 0 = q_{cal} + q_{rxn}$$

or,

$$q_{rxn} = -q_{cal}$$

The heat absorbed by the calorimeter can be calculated using the heat capacity of the bomb calorimeter and the temperature rise.

$$q_{cal} = C_{cal}\Delta t$$

> **Note:** The negative sign for $q_{rxn}$ indicates that heat was released during the combustion. You should expect this, because all combustion processes are exothermic. Thermal energy is transferred from the system to the surroundings.

**EXAMPLE 6.6**

**0.500 g of ethanol [$CH_3CH_2OH(l)$] was burned in a bomb calorimeter. The temperature of the water rose 1.60°C. The heat capacity of the calorimeter plus water was 9.06 kJ/°C.**

**(a) Write a balanced equation for the combustion of ethanol.**

**(b) Calculate the molar heat of combustion of ethanol.**

---

(a) Recall that a **combustion reaction** is typically a vigorous and exothermic reaction that takes place between certain substances and oxygen. If the reactant contains only C, H, and/or O, then the products are $CO_2$ and $H_2O$. Therefore, the balanced equation for the combustion of ethanol is:

$$CH_3CH_2OH(l) + 3O_2(g) \longrightarrow 2CO_2(g) + 3H_2O(g)$$

(b) **Strategy:** Knowing the heat capacity and the temperature rise, how do we calculate the heat absorbed by the calorimeter? What is the heat generated by the combustion of 0.500 g ethanol? What is the conversion factor between grams and moles of ethanol?

**Solution:** The heat absorbed by the calorimeter and water is equal to the product of the heat capacity and the temperature change. From Equation (6.16) of the text, assuming no heat is lost to the surroundings, we write

$$q_{cal} = C_{cal}\Delta t$$

$$q_{cal} = (9.06\ \text{kJ/°C})(1.60\text{°C}) = 14.5\ \text{kJ}$$

Because $q_{sys} = 0 = q_{cal} + q_{rxn}$, $q_{rxn} = -q_{cal}$. The heat change of the reaction is $-14.5$ kJ. This is the heat released by the combustion of 0.500 g of ethanol; therefore, we can write the conversion factor as

$$\frac{-14.5\ \text{kJ}}{0.500\ \text{g ethanol}}$$

From the molar mass of ethanol and the above conversion factor, the heat of combustion of 1 mole of ethanol can be calculated.

$$\mathcal{M}_{ethanol} = 2(12.01\ \text{g}) + 6(1.008\ \text{g}) + 16.00\ \text{g} = 46.07\ \text{g}$$

$$\textbf{Molar heat of combustion} = \frac{-14.5\ \text{kJ}}{0.500\ \text{g ethanol}} \times \frac{46.07\ \text{g ethanol}}{1\ \text{mol ethanol}} = \mathbf{-1.34 \times 10^3\ kJ/mol\ ethanol}$$

**PRACTICE EXERCISE**

6. The combustion of benzoic acid is often used as a standard source of heat for calibrating combustion bomb calorimeters. The heat of combustion of benzoic acid has been accurately determined to be 26.42 kJ/g. When 0.8000 g of benzoic acid was burned in a calorimeter containing water, a temperature rise of 4.08°C was observed. What is the heat capacity of the bomb calorimeter plus water?

**Text Problem: 6.108**

## B. Constant-pressure calorimetry

A simpler device than the constant-volume calorimeter is the constant-pressure calorimeter that is used to determine the heat changes for noncombustion reactions. The reactions usually occur in solution. Because the measurements are carried out under constant atmospheric pressure conditions, the heat change for the process ($q_{rxn}$) is equal to the enthalpy change ($\Delta H$).

The heat released during reaction is absorbed both by the solution in the calorimeter. We ignore the small heat capacity of the calorimeter in our calculations. Because no heat enters or leaves the system throughout the process, we can write:

$$q_{system} = 0 = q_{soln} + q_{rxn}$$

or,

$$q_{soln} = -q_{rxn}$$

The heat absorbed by the solution can be calculated using the equation

$$q_{soln} = m_{soln}s_{soln}\Delta t$$

**EXAMPLE 6.7**

**The heat of neutralization for the following reaction is –56.2 kJ/mol.**

$$\mathbf{NaOH}(aq) + \mathbf{HCl}(aq) \longrightarrow \mathbf{NaCl}(aq) + \mathbf{H_2O}(l)$$

**$1.00 \times 10^2$ mL of 1.50 *M* HCl is mixed with $1.00 \times 10^2$ mL of 1.50 *M* NaOH in a constant-pressure calorimeter. The initial temperature of the HCl and NaOH solutions is the same, 23.2°C. Calculate the final temperature of the mixed solution. Assume that the density and specific heat of the mixed solution is the same as for water (1.00 g/mL and 4.184 J/g·°C, respectively).**

**Strategy:** The neutralization reaction is exothermic. 56.2 kJ of heat are released when 1 mole of $H^+$ reacts with 1 mole of $OH^-$. Assuming no heat is lost to the surroundings, we can equate the heat lost by the reaction to the heat gained by the combined solution. How do we calculate the heat released during the reaction? Are we reacting 1 mole of $H^+$ with 1 mole of $OH^-$? How do we calculate the heat absorbed by the combined solution?

**Solution:** Assuming no heat is lost to the surroundings, we can write:

$$q_{soln} + q_{rxn} = 0$$

or

$$q_{soln} = -q_{rxn}$$

First, let's set up how we would calculate the heat gained by the solution,

$$q_{soln} = m_{soln}s_{soln}\Delta t$$

where *m* and *s* are the mass and specific heat of the solution and $\Delta t = t_f - t_i$.

We assume that the specific heat of the solution is the same as the specific heat of water, and we assume that the density of the solution is the same as the density of water (1.00 g/mL). Since the density is 1.00 g/mL, the mass of 200 mL of solution (100 mL + 100 mL) is 200 g.

Substituting into the equation above, the heat gained by the solution can be represented as:

$$q_{soln} = (2.00 \times 10^2 \text{ g})(4.184 \text{ J/g·°C})(t_f - 23.2\text{°C})$$

Next, let's calculate $q_{rxn}$, the heat released when 100 mL of 1.50 *M* HCl are mixed with 100 mL of 1.50 *M* NaOH. There is exactly enough NaOH to neutralize all the HCl. Note that 1 mole HCl ≏ 1 mole NaOH. The number of moles of HCl is:

$$(1.00 \times 10^2 \text{ mL}) \times \frac{1.50 \text{ mol HCl}}{1000 \text{ mL}} = 0.150 \text{ mol HCl}$$

The amount of heat released when 1 mole of $H^+$ is reacted is given in the problem (−56.2 kJ/mol). The amount of heat liberated when 0.150 mole of $H^+$ is reacted is:

$$q_{rxn} = 0.150 \text{ mol} \times \frac{-56.2 \times 10^3 \text{ J}}{1 \text{ mol}} = -8.43 \times 10^3 \text{ J}$$

Finally, knowing that the heat lost by the reaction equals the heat gained by the solution, we can solve for the final temperature of the mixed solution.

$$q_{soln} = -q_{rxn}$$

$$(2.00 \times 10^2 \text{ g})(4.184 \text{ J/g·°C})(t_f - 23.2\text{°C}) = -(-8.43 \times 10^3 \text{ J})$$

$$(8.37 \times 10^2)t_f - (1.94 \times 10^4) = 8.43 \times 10^3 \text{ J}$$

$$t_f = \mathbf{33.2°C}$$

**PRACTICE EXERCISE**

7. A 10.4 g sample of an unknown metal at 99.0°C was placed in a constant-pressure calorimeter containing 75.0 g of water at 23.5°C. The final temperature of the system was found to be 25.7°C. Calculate the specific heat of the metal, then use Table 6.2 of the text to predict the identity of the metal.

**Text Problem: 6.36**

## PROBLEM TYPE 5: STANDARD ENTHALPY OF FORMATION AND REACTION

The **standard enthalpy of formation** ( $\Delta H_f^\circ$ ) is defined as the heat change that results when one mole of a compound is formed from its elements at a pressure of 1 atm. The standard enthalpy of formation of any element in its most stable form is zero.

### A. Calculating the standard enthalpy of reaction

From standard enthalpies of formation, we can calculate the **standard enthalpy of reaction,** $\Delta H^\circ_{rxn}$ .

Consider the hypothetical reaction

$$aA + bB \longrightarrow cC + dD$$

where *a*, *b*, *c*, and *d* are stoichiometric coefficients.

The standard enthalpy of reaction is given by

$$\Delta H^\circ_{rxn} = [c\Delta H^\circ_f(C) + d\Delta H^\circ_f(D)] - [a\Delta H^\circ_f(A) + b\Delta H^\circ_f(B)]$$

Note that in calculations, the stoichiometric coefficients are just numbers without units.

The equation can be written in the general form:

$$\Delta H^\circ_{rxn} = \Sigma n\Delta H^\circ_f(\text{products}) - \Sigma m\Delta H^\circ_f(\text{reactants}) \quad (6.18, \text{text})$$

where *m* and *n* denote the stoichiometric coefficients for the reactants and products, and Σ (sigma) means "the sum of".

**EXAMPLE 6.8**

**A reaction used for rocket engines is**

$$N_2H_4(l) + 2H_2O_2(l) \longrightarrow N_2(g) + 4H_2O(l)$$

**What is the standard enthalpy of reaction in kilojoules? The standard enthalpies of formation are**

**$\Delta H^\circ_f [N_2H_4(l)] = 95.1$ kJ/mol, $\Delta H^\circ_f [H_2O_2(l)] = -187.8$ kJ/mol, and $\Delta H^\circ_f [H_2O(l)] = -285.8$ kJ/mol.**

**Strategy:** The enthalpy of a reaction is the difference between the sum of the enthalpies of the products and the sum of the enthalpies of the reactants. The enthalpy of each species (reactant or product) is given by the product of the stoichiometric coefficient and the standard enthalpy of formation, $\Delta H^\circ_f$, of the species.

**Solution:** We use the $\Delta H^\circ_f$ values in Appendix 3 and Equation (6.18) of the text.

$$\Delta H^\circ_{rxn} = \sum n\Delta H^\circ_f(\text{products}) - \sum m\Delta H^\circ_f(\text{reactants})$$

$$\Delta H^\circ_{rxn} = \Delta H^\circ_f[N_2(g)] + 4\Delta H^\circ_f[H_2O(l)] - \{\Delta H^\circ_f[N_2H_4(l) + 2\Delta H^\circ_f[H_2O_2(l)]\}$$

Remember, the standard enthalpy of formation of any element in its most stable form is zero. Therefore, $\Delta H^\circ_f [N_2(g)] = 0$.

$$\Delta H^\circ_{rxn} = [0 + 4(-285.8 \text{ kJ/mol})] - [95.1 \text{ kJ/mol} + 2(-187.8 \text{ kJ/mol})] = \mathbf{-862.7 \text{ kJ/mol}}$$

**PRACTICE EXERCISE**

8. The combustion of methane, the main component of natural gas, occurs according to the equation

$$CH_4(g) + 2O_2(g) \longrightarrow CO_2(g) + 2H_2O(l) \qquad \Delta H^\circ_{rxn} = -890 \text{ kJ/mol}$$

Use standard enthalpies of formation for $CO_2$ and $H_2O$ to determine the standard enthalpy of formation of methane.

**Text Problems: 6.52, 6.54, 6.56, 6.58, 6.60**

## B. Direct method of calculating the standard enthalpy of formation

This method of measuring $\Delta H^\circ_f$ applies to compounds that can be readily synthesized from their elements. The best way to describe this direct method is to look at an example.

**EXAMPLE 6.9**

**The combustion of sulfur occurs according to the following thermochemical equation:**

$$\text{S(rhombic)} + O_2(g) \longrightarrow SO_2(g) \qquad \Delta H^\circ_{\text{rxn}} = -296 \text{ kJ/mol}$$

**What is the enthalpy of formation of $SO_2$?**

---

**Strategy:** What is the $\Delta H^\circ_f$ value for an element in its standard state?

**Solution:** Knowing that the standard enthalpy of formation of any element in its most stable form is zero, and using Equation (6.18) of the text, we write:

$$\Delta H^\circ_{\text{rxn}} = \Sigma n\Delta H^\circ_f(\text{products}) - \Sigma m\Delta H^\circ_f(\text{reactants})$$

$$\Delta H^\circ_{\text{rxn}} = [\Delta H^\circ_f(SO_2)] - [\Delta H^\circ_f(S) + \Delta H^\circ_f(O_2)]$$

$$-296 \text{ kJ/mol} = [\Delta H^\circ_f(SO_2) - [0 + 0]$$

$$\Delta H^\circ_f(SO_2) = -296 \text{ kJ/mol } SO_2$$

**Note:** You should recognize that this chemical equation as written meets the definition of a *formation* reaction. Thus, $\Delta H^\circ_{\text{rxn}}$ *is* $\Delta H^\circ_f$ of $SO_2(g)$.

**PRACTICE EXERCISE**

9. Hydrogen iodide (HI) can be produced according to the following equation:

$$H_2(g) + I_2(s) \longrightarrow 2HI(g) \qquad \Delta H^\circ_{\text{rxn}} = 51.8 \text{ kJ/mol}$$

What is the enthalpy of formation ($\Delta H^\circ_f$) of HI?

**Text Problem: 6.50**

## C. Indirect method of calculating the standard enthalpy of formation, Hess's law

Many compounds cannot be directly synthesized from their elements. In these cases, $\Delta H^\circ_f$ can be determined by an indirect approach using **Hess's law**. Hess's law states that when reactants are converted to products, the change in enthalpy is the same whether the reaction takes place in one step or in a series of steps. This means that if we can break down the reaction of interest into a series of reactions for which $\Delta H^\circ_{\text{rxn}}$ can be measured, we can calculate $\Delta H^\circ_{\text{rxn}}$ for the overall reaction. Let's look at an example.

**EXAMPLE 6.10**

**From the following heats of combustion with fluorine, calculate the enthalpy of formation of methane, $CH_4$.**

(a) $CH_4(g) + 4F_2(g) \longrightarrow CF_4(g) + 4HF(g) \qquad \Delta H^\circ_{\text{rxn}} = -1942 \text{ kJ/mol}$

(b) $\text{C(graphite)} + 2F_2(g) \longrightarrow CF_4(g) \qquad \Delta H^\circ_{\text{rxn}} = -933 \text{ kJ/mol}$

(c) $H_2(g) + F_2(g) \longrightarrow 2HF(g) \qquad \Delta H^\circ_{\text{rxn}} = -542 \text{ kJ/mol}$

---

**Strategy:** Our goal is to calculate the enthalpy change for the formation of $CH_4$ from its elements C and $H_2$. This reaction does not occur directly, however, so we must use an indirect route using the information given in the three equations, which we will call equations (a), (b), and (c).

**Solution:** The enthalpy of formation of methane can be determined from the following equation.

$$C(graphite) + 2H_2(g) \longrightarrow CH_4(g) \qquad \Delta H^\circ_{rxn} = ?$$

First, we need one mole of C(graphite) as a reactant. Equation (b) has C(graphite) on the reactant side so let's keep that equation as written. Next, we need two moles of $H_2$ as a reactant. Equation (c) has 1 mole of $H_2$ as a reactant, so let's multiply this equation by 2.

$$\text{(d)} \quad 2H_2(g) + 2F_2(g) \longrightarrow 4HF(g) \qquad \Delta H^\circ_{rxn} = 2(-542\ \text{kJ/mol}) = -1084\ \text{kJ/mol}$$

Last, we need one mole of $CH_4$ as a product. Equation (a) has one mole of $CH_4$ as a reactant, so we need to reverse the equation.

$$\text{(e)} \quad CF_4(g) + 4HF(g) \longrightarrow CH_4(g) + 4F_2(g) \qquad \Delta H^\circ_{rxn} = +1942\ \text{kJ/mol}$$

| **Note:** $\Delta H^\circ_{rxn}$ changed sign when reversing the direction of the reaction. |
|---|

Adding Equations (b), (d), and (e) together, we have:

$$\text{(b)} \quad C(graphite) + \cancel{2F_2}(g) \longrightarrow \cancel{CF_4}(g) \qquad \Delta H^\circ_{rxn} = -933\ \text{kJ/mol}$$

$$\text{(d)} \quad 2H_2(g) + \cancel{2F_2}(g) \longrightarrow \cancel{4HF}(g) \qquad \Delta H^\circ_{rxn} = -1084\ \text{kJ/mol}$$

$$\text{(e)} \quad \cancel{CF_4}(g) + \cancel{4HF}(g) \longrightarrow CH_4(g) + \cancel{4F_2}(g) \qquad \Delta H^\circ_{rxn} = +1942\ \text{kJ/mol}$$

$$C(graphite) + 2H_2(g) \longrightarrow CH_4(g) \qquad \Delta H^\circ_{rxn} = -75\ \text{kJ/mol}$$

Since the above equation represents the synthesis of $CH_4$ from its elements, the $\Delta H^\circ_{rxn}$ calculated is the $\Delta H^\circ_f$ of methane.

$$\Delta H^\circ_{rxn} = \mathbf{\Delta H^\circ_f(CH_4) = -75\ kJ/mol}$$

**PRACTICE EXERCISE**

**10.** From the following enthalpies of reaction, calculate the enthalpy of combustion of methane ($CH_4$) with $F_2$:

$$CH_4(g) + 4F_2(g) \longrightarrow CF_4(g) + 4HF(g) \qquad \Delta H^\circ_{rxn} = ?$$

$$C(graphite) + 2H_2(g) \longrightarrow CH_4(g) \qquad \Delta H^\circ_{rxn} = -75\ \text{kJ/mol}$$

$$C(graphite) + 2F_2(g) \longrightarrow CF_4(g) \qquad \Delta H^\circ_{rxn} = -933\ \text{kJ/mol}$$

$$H_2(g) + F_2(g) \longrightarrow 2HF(g) \qquad \Delta H^\circ_{rxn} = -542\ \text{kJ/mol}$$

| **Text Problems: 6.62, 6.64** |
|---|

## ANSWERS TO PRACTICE EXERCISES

1. $\Delta U = 202$ kJ
2. $w = 1.0 \times 10^3$ J
3. $\Delta U = 37.57$ kJ
4. heat evolved $= 1.2 \times 10^2$ kJ
5. 475°C
6. $C_{cal} = 5.18$ kJ/°C
7. $s_{metal} = 0.906$ J/g°C. The metal is probably aluminum.
8. $\Delta H^\circ_f(CH_4) = -75.1$ kJ/mol
9. $\Delta H^\circ_f(HI) = 25.9$ kJ/mol
10. $\Delta H^\circ_{rxn} = -1942$ kJ/mol

## SOLUTIONS TO SELECTED TEXT PROBLEMS

**6.16** **(a)** Because the external pressure is zero, no work is done in the expansion.

$$w = -P\Delta V = -(0)(89.3 - 26.7)\text{mL}$$

$$\mathbf{w = 0}$$

**(b)** The external, opposing pressure is 1.5 atm, so

$$w = -P\Delta V = -(1.5 \text{ atm})(89.3 - 26.7)\text{mL}$$

$$w = -94 \text{ mL}\cdot\text{atm} \times \frac{0.001 \text{ L}}{1 \text{ mL}} = -0.094 \text{ L}\cdot\text{atm}$$

To convert the answer to joules, we write:

$$w = -0.094 \text{ L}\cdot\text{atm} \times \frac{101.3 \text{ J}}{1 \text{ L}\cdot\text{atm}} = \mathbf{-9.5 \text{ J}}$$

**(c)** The external, opposing pressure is 2.8 atm, so

$$w = -P\Delta V = -(2.8 \text{ atm})(89.3 - 26.7)\text{mL}$$

$$w = (-1.8 \times 10^2 \text{ mL}\cdot\text{atm}) \times \frac{0.001 \text{ L}}{1 \text{ mL}} = -0.18 \text{ L}\cdot\text{atm}$$

To convert the answer to joules, we write:

$$w = -0.18 \text{ L}\cdot\text{atm} \times \frac{101.3 \text{ J}}{1 \text{ L}\cdot\text{atm}} = \mathbf{-18 \text{ J}}$$

**6.18** **Strategy:** Compression is work done on the gas, so what is the sign for $w$? Heat is released by the gas to the surroundings. Is this an endothermic or exothermic process? What is the sign for $q$?

**Solution:** To calculate the energy change of the gas ($\Delta U$), we need Equation (6.1) of the text. Work of compression is positive and because heat is given off by the gas, $q$ is negative. Therefore, we have:

$$\mathbf{\Delta U} = q + w = -26 \text{ J} + 74 \text{ J} = \mathbf{48 \text{ J}}$$

As a result, the energy of the gas increases by 48 J.

**6.20** **Strategy:** The work done in gas expansion is equal to the product of the external, opposing pressure and the change in volume.

$$w = -P\Delta V$$

We assume that the volume of liquid water is zero compared to that of steam. How do we calculate the volume of the steam? What is the conversion factor between L·atm and J?

**Solution:** First, we need to calculate the volume that the water vapor will occupy ($V_f$).

Using the ideal gas equation:

$$V_{H_2O} = \frac{n_{H_2O}RT}{P} = \frac{(1 \text{ mol})\left(0.0821 \frac{\text{L}\cdot\text{atm}}{\text{mol}\cdot\text{K}}\right)(373 \text{ K})}{(1.0 \text{ atm})} = 31 \text{ L}$$

It is given that the volume occupied by liquid water is negligible. Therefore,

$$\Delta V = V_f - V_i = 31\text{ L} - 0\text{ L} = 31\text{ L}$$

Now, we substitute $P$ and $\Delta V$ into Equation (6.3) of the text to solve for $w$.

$$w = -P\Delta V = -(1.0\text{ atm})(31\text{ L}) = -31\text{ L·atm}$$

The problem asks for the work done in units of joules. The following conversion factor can be obtained from Appendix 2 of the text.

$$1\text{ L·atm} = 101.3\text{ J}$$

Thus, we can write:

$$w = -31\text{ L·atm} \times \frac{101.3\text{ J}}{1\text{ L·atm}} = \mathbf{-3.1 \times 10^3\ J}$$

**Check:** Because this is gas expansion (work is done by the system on the surroundings), the work done has a negative sign.

6.26 **Strategy:** The thermochemical equation shows that for every 2 moles of $NO_2$ produced, 114.6 kJ of heat are given off (note the negative sign of $\Delta H$). We can write a conversion factor from this information.

$$\frac{114.6\text{ kJ}}{2\text{ mol }NO_2}$$

How many moles of $NO_2$ are in $1.26 \times 10^4$ g of $NO_2$? What conversion factor is needed to convert between grams and moles?

**Solution:** We need to first calculate the number of moles of $NO_2$ in $1.26 \times 10^4$ g of the compound. Then, we can convert to the number of kilojoules produced from the exothermic reaction. The sequence of conversions is:

$$\text{grams of }NO_2 \rightarrow \text{moles of }NO_2 \rightarrow \text{kilojoules of heat generated}$$

Therefore, the heat given off is:

$$(1.26 \times 10^4\text{ g }NO_2) \times \frac{1\text{ mol }NO_2}{46.01\text{ g }NO_2} \times \frac{114.6\text{ kJ}}{2\text{ mol }NO_2} = \mathbf{1.57 \times 10^4\ kJ}$$

6.28 We initially have 6 moles of gas (3 moles of chlorine and 3 moles of hydrogen). Since our product is 6 moles of hydrogen chloride, there is no change in the number of moles of gas. Therefore there is no volume change; $\Delta V = 0$.

$$w = -P\Delta V = -(1\text{ atm})(0\text{ L}) = 0$$

$$\Delta U° = \Delta H° - P\Delta V$$

$-P\Delta V = 0$, so

$$\mathbf{\Delta U = \Delta H}$$

$$\Delta H = \Delta H^\circ_{rxn} = 3(-184.6\text{ kJ/mol}) = -553.8\text{ kJ/mol}$$

We need to multiply $\Delta H^\circ_{rxn}$ by three, because the question involves the formation of 6 moles of HCl; whereas, the equation as written only produces 2 moles of HCl.

$$\mathbf{\Delta U^\circ = \Delta H^\circ = -553.8\ kJ/mol}$$

**6.32** **Specific heat** $= \dfrac{C}{m} = \dfrac{85.7\ \text{J/°C}}{362\ \text{g}} = \mathbf{0.237\ J/g\cdot°C}$

**6.34** See Table 6.2 of the text for the specific heat of Hg.

$$q = ms\Delta t = (366\ \cancel{g})(0.139\ \text{J}/\cancel{g\cdot°C})(12.0 - 77.0)\cancel{°C} = -3.31 \times 10^3\ \text{J} = -3.31\ \text{kJ}$$

The amount of heat *liberated* is **3.31 kJ**.

**6.36** **Strategy:** We know the mass of aluminum and the initial and final temperatures of water and aluminum. We can look up the specific heats of water and aluminum in Table 6.2 of the text. Assuming no heat is lost to the surroundings, we can equate the heat lost by the aluminum to the heat gained by the water. With this information, we can solve for the mass of the water.

**Solution:** Treating the calorimeter as an isolated system (no heat lost to the surroundings), we can write:

$$q_{H_2O} + q_{Al} = 0$$

or

$$q_{H_2O} = -q_{Al}$$

The heat gained by water is given by:

$$q_{H_2O} = m_{H_2O}s_{H_2O}\Delta t = m_{H_2O}(4.184\ \text{J/g}\cdot\text{°C})(24.9 - 23.4)\text{°C}$$

where $m$ and $s$ are the mass and specific heat, and $\Delta t = t_{final} - t_{initial}$.

The heat lost by the aluminum is given by:

$$q_{Al} = m_{Al}s_{Al}\Delta t = (12.1\ \text{g})(0.900\ \text{J/g}\cdot\text{°C})(24.9 - 81.7)\text{°C}$$

Substituting into the equation derived above, we can solve for $m_{H_2O}$.

$$q_{H_2O} = -q_{Al}$$

$$m_{H_2O}(4.184\ \text{J/g}\cdot\text{°C})(24.9 - 23.4)\text{°C} = -(12.1\ \text{g})(0.900\ \text{J/g}\cdot\text{°C})(24.9 - 81.7)\text{°C}$$

$$(6.28)(m_{H_2O}) = 619$$

$$m_{H_2O} = \mathbf{98.6\ g}$$

**6.38** The heat released by the reaction heats the solution and the calorimeter: $-q_{rxn} = +(q_{soln} + q_{cal})$

The heat released by the reaction is:

$$85.0\ \cancel{mL} \times \frac{0.900\ \cancel{mol}\ \text{HCl}}{1000\ \cancel{mL}\ \text{soln}} \times \frac{-56.2\ \text{kJ}}{1\ \cancel{mol}} = -4.30\ \text{kJ}$$

We assume that the density of the solution is 1.00 g/mL, and its specific heat is 4.184 J/g·°C.

$$-q_{rxn} = +(q_{soln} + q_{cal})$$

$$4.30 \times 10^3 \text{ J} = ms\Delta t + C\Delta t$$

$$4.30 \times 10^3 \text{ J} = (170.\text{ g})(4.184 \text{ J/g·°C})(t_f - 18.24°\text{C}) + (325 \text{ J/°C})(t_f - 18.24°\text{C})$$

$$4.30 \times 10^3 = 711.28t_f - 1.2974 \times 10^4 + 325t_f - 5.928 \times 10^3$$

$$1.0363 \times 10^3 t_f = 2.3202 \times 10^4$$

$$t_f = \mathbf{22.39°C}$$

**6.46** The standard enthalpy of formation of any element in its most stable form is zero. Therefore, since $\Delta H_f^\circ(O_2) = 0$, $\mathbf{O_2}$ is the more stable form of the element oxygen at this temperature.

**6.48** (a) $\mathbf{Br_2}(\mathbf{\mathit{l}})$ is the most stable form of bromine at 25°C; therefore, $\Delta H_f^\circ[Br_2(l)] = 0$ Since $Br_2(g)$ is less stable than $Br_2(l)$, $\Delta H_f^\circ[Br_2(g)] > 0$

(b) $\mathbf{I_2}(\mathbf{\mathit{s}})$ is the most stable form of iodine at 25°C; therefore, $\Delta H_f^\circ[I_2(s)] = 0$ Since $I_2(g)$ is less stable than $I_2(s)$, $\Delta H_f^\circ[I_2(g)] > 0$.

**6.50** **Strategy:** What is the reaction for the formation of $Ag_2O$ from its elements? What is the $\Delta H_f^\circ$ value for an element in its standard state?

**Solution:** The balanced equation showing the formation of $Ag_2O(s)$ from its elements is:

$$2Ag(s) + \tfrac{1}{2}O_2(g) \longrightarrow Ag_2O(s)$$

Knowing that the standard enthalpy of formation of any element in its most stable form is zero, and using Equation (6.18) of the text, we write:

$$\Delta H^\circ_{rxn} = \sum n\Delta H_f^\circ(\text{products}) - \sum m\Delta H_f^\circ(\text{reactants})$$

$$\Delta H^\circ_{rxn} = [\Delta H_f^\circ(Ag_2O)] - \left[2\Delta H_f^\circ(Ag) + \frac{1}{2}\Delta H_f^\circ(O_2)\right]$$

$$\Delta H^\circ_{rxn} = [\Delta H_f^\circ(Ag_2O)] - [0 + 0]$$

$$\mathbf{\Delta H_f^\circ(Ag_2O) = \Delta H^\circ_{rxn}}$$

In a similar manner, you should be able to show that $\mathbf{\Delta H_f^\circ(CaCl_2) = \Delta H^\circ_{rxn}}$ for the reaction

$$Ca(s) + Cl_2(g) \longrightarrow CaCl_2(s)$$

**6.52** **Strategy:** The enthalpy of a reaction is the difference between the sum of the enthalpies of the products and the sum of the enthalpies of the reactants. The enthalpy of each species (reactant or product) is given by the product of the stoichiometric coefficient and the standard enthalpy of formation, $\Delta H_f^\circ$, of the species.

**Solution:** We use the $\Delta H_f^\circ$ values in Appendix 3 and Equation (6.18) of the text.

$$\Delta H_{rxn}^\circ = \sum n\Delta H_f^\circ(\text{products}) - \sum m\Delta H_f^\circ(\text{reactants})$$

**(a)** $HCl(g) \rightarrow H^+(aq) + Cl^-(aq)$

$$\Delta H_{rxn}^\circ = \Delta H_f^\circ(H^+) + \Delta H_f^\circ(CL^-) - \Delta H_f^\circ(HCL)$$

$$-74.9 \text{ kJ/mol} = 0 + \Delta H_f^\circ(Cl^-) - (1)(-92.3 \text{ kJ/mol})$$

$$\boldsymbol{\Delta H_f^\circ(Cl^-) = -167.2 \text{ kJ/mol}}$$

**(b)** The neutralization reaction is:

$$H^+(aq) + OH^-(aq) \rightarrow H_2O(l)$$

and,

$$\Delta H_{rxn}^\circ = \Delta H_f^\circ[(H_2O(l)] - [\Delta H_f^\circ(H^+) + \Delta H_f^\circ(OH^-)]$$

$$\Delta H_f^\circ[H_2O(l)] = -285.8 \text{ kJ/mol} \text{ (See Appendix 3 of the text.)}$$

$$\Delta H_{rxn}^\circ = (1)(-285.8 \text{ kJ/mol}) - [(1)(0 \text{ kJ/mol}) + (1)(-229.6 \text{ kJ/mol})] = \mathbf{-56.2 \text{ kJ/mol}}$$

**6.54** **(a)** $\Delta H^\circ = [2\Delta H_f^\circ(CO_2) + 2\Delta H_f^\circ(H_2O)] - [\Delta H_f^\circ(C_2H_4) + 3\Delta H_f^\circ(O_2)]$

$$\Delta H^\circ = [(2)(-393.5 \text{ kJ/mol}) + (2)(-285.8 \text{ kJ/mol})] - [(1)(52.3 \text{ kJ/mol}) + (3)(0)]$$

$$\boldsymbol{\Delta H^\circ = -1411 \text{ kJ/mol}}$$

**(b)** $\Delta H^\circ = [2\Delta H_f^\circ(H_2O) + 2\Delta H_f^\circ(SO_2)] - [2\Delta H_f^\circ(H_2S) + 3\Delta H_f^\circ(O_2)]$

$$\Delta H^\circ = [(2)(-285.8 \text{ kJ/mol}) + (2)(-296.1 \text{ kJ/mol})] - [(2)(-20.15 \text{ kJ/mol}) + (3)(0)]$$

$$\boldsymbol{\Delta H^\circ = -1124 \text{ kJ/mol}}$$

**6.56** $\Delta H_{rxn}^\circ = \sum n\Delta H_f^\circ(\text{products}) - \sum m\Delta H_f^\circ(\text{reactants})$

The reaction is:

$$H_2(g) \longrightarrow H(g) + H(g)$$

and,

$$\Delta H_{rxn}^\circ = [\Delta H_f^\circ(H) + \Delta H_f^\circ(H)] - \Delta H_f^\circ(H_2)$$

$$\Delta H_f^\circ(H_2) = 0$$

$$\Delta H_{rxn}^\circ = 436.4 \text{ kJ/mol} = 2\Delta H_f^\circ(H) - (1)(0)$$

$$\boldsymbol{\Delta H_f^\circ(H)} = \frac{436.4 \text{ kJ/mol}}{2 \text{ mol}} = \mathbf{218.2 \text{ kJ/mol}}$$

**6.58** Using the $\Delta H_f^\circ$ values in Appendix 3 and Equation (6.18) of the text, we write

$$\Delta H^\circ_{rxn} = [5\Delta H_f^\circ(B_2O_3) + 9\Delta H_f^\circ(H_2O)] - [2\Delta H_f^\circ(B_5H_9) + 12\Delta H_f^\circ(O_2)]$$

$$\Delta H^\circ = [(5)(-1263.6\ \text{kJ/mol}) + (9)(-285.8\ \text{kJ/mol})] - [(2)(73.2\ \text{kJ/mol}) + (12)(0\ \text{kJ/mol})]$$

$$\Delta H^\circ = -9036.6\ \text{kJ/mol}$$

Looking at the balanced equation, this is the amount of heat released for every 2 moles of $B_5H_9$ reacted. We can use the following ratio

$$\frac{9036.6\ \text{kJ}}{2\ \text{mol}\ B_5H_9}$$

to convert to kJ/g $B_5H_9$. The molar mass of $B_5H_9$ is 63.12 g, so

$$\text{heat } \textit{released} \text{ per gram } B_5H_9 = \frac{9036.6\ \text{kJ}}{2\ \text{mol}\ B_5H_9} \times \frac{1\ \text{mol}\ B_5H_9}{63.12\ \text{g}\ B_5H_9} = \mathbf{71.58\ kJ/g\ B_5H_9}$$

**6.60** $\Delta H^\circ_{rxn} = \sum n\Delta H_f^\circ(\text{products}) - \sum m\Delta H_f^\circ(\text{reactants})$

The balanced equation for the reaction is:

$$CaCO_3(s) \longrightarrow CaO(s) + CO_2(g)$$

$$\Delta H^\circ_{rxn} = [\Delta H_f^\circ(CaO) + \Delta H_f^\circ(CO_2)] - \Delta H_f^\circ(CaCO_3)$$

$$\Delta H^\circ_{rxn} = [(1)(-635.6\ \text{kJ/mol}) + (1)(-393.5\ \text{kJ/mol})] - (1)(-1206.9\ \text{kJ/mol}) = 177.8\ \text{kJ/mol}$$

The enthalpy change calculated above is the enthalpy change if 1 mole of $CO_2$ is produced. The problem asks for the enthalpy change if 66.8 g of $CO_2$ are produced. We need to use the molar mass of $CO_2$ as a conversion factor.

$$\mathbf{\Delta H^\circ} = 66.8\ \text{g}\ CO_2 \times \frac{1\ \text{mol}\ CO_2}{44.01\ \text{g}\ CO_2} \times \frac{177.8\ \text{kJ}}{1\ \text{mol}\ CO_2} = \mathbf{2.70 \times 10^2\ kJ}$$

**6.62** **Strategy:** Our goal is to calculate the enthalpy change for the formation of $C_2H_6$ from is elements C and $H_2$. This reaction does not occur directly, however, so we must use an indirect route using the information given in the three equations, which we will call equations (a), (b), and (c).

**Solution:** Here is the equation for the formation of $C_2H_6$ from its elements.

$$2C(\text{graphite}) + 3H_2(g) \longrightarrow C_2H_6(g) \qquad \Delta H^\circ_{rxn} = ?$$

Looking at this reaction, we need two moles of graphite as a reactant. So, we multiply Equation (a) by two to obtain:

$$\text{(d)}\quad 2C(\text{graphite}) + 2O_2(g) \longrightarrow 2CO_2(g) \qquad \Delta H^\circ_{rxn} = 2(-393.5\ \text{kJ/mol}) = -787.0\ \text{kJ/mol}$$

Next, we need three moles of $H_2$ as a reactant. So, we multiply Equation (b) by three to obtain:

$$\text{(e)}\quad 3H_2(g) + \frac{3}{2}O_2(g) \longrightarrow 3H_2O(l) \qquad \Delta H^\circ_{rxn} = 3(-285.8\ \text{kJ/mol}) = -857.4\ \text{kJ/mol}$$

Last, we need one mole of $C_2H_6$ as a product. Equation (c) has two moles of $C_2H_6$ as a reactant, so we need to reverse the equation and divide it by 2.

$$\text{(f)}\quad 2CO_2(g) + 3H_2O(l) \longrightarrow C_2H_6(g) + \frac{7}{2}O_2(g) \qquad \Delta H^\circ_{\text{rxn}} = \frac{1}{2}(3119.6\ \text{kJ/mol}) = 1559.8\ \text{kJ/mol}$$

Adding Equations (d), (e), and (f) together, we have:

| Reaction | $\Delta H^\circ$ (kJ/mol) |
|---|---|
| (d) $2C(\text{graphite}) + 2\cancel{O_2}(g) \longrightarrow 2C\cancel{O_2}(g)$ | −787.0 |
| (e) $3H_2(g) + \frac{3}{2}\cancel{O_2}(g) \longrightarrow 3H_2\cancel{O}(l)$ | −857.4 |
| (f) $2\cancel{CO_2}(g) + 3H_2\cancel{O}(l) \longrightarrow C_2H_6(g) + \frac{7}{2}\cancel{O_2}(g)$ | 1559.8 |
| $\mathbf{2C(graphite) + 3H_2(g) \longrightarrow C_2H_6(g)}$ | $\mathbf{\Delta H^\circ = -84.6\ kJ/mol}$ |

**6.64** The second and third equations can be combined to give the first equation.

$$2Al(s) + \frac{3}{2}\cancel{O_2}(g) \longrightarrow Al_2O_3(s) \qquad \Delta H^\circ = -1669.8\ \text{kJ/mol}$$

$$Fe_2O_3(s) \longrightarrow 2Fe(s) + \frac{3}{2}\cancel{O_2}(g) \qquad \Delta H^\circ = 822.2\ \text{kJ/mol}$$

$$2Al(s) + Fe_2O_3(s) \longrightarrow 2Fe(s) + Al_2O_3(s) \qquad \mathbf{\Delta H^\circ = -847.6\ kJ/mol}$$

**6.72** The gas will expand until the pressure is the same as the external pressure of 1.0 atm. We can calculate its final volume using the ideal gas equation.

$$V = \frac{nRT}{P} = \frac{(3.70\ \cancel{\text{mol}})(0.0821\ \text{L}\cdot\cancel{\text{atm}}/\cancel{\text{K}}\cdot\cancel{\text{mol}})(273 + 78.3)\cancel{\text{K}}}{1.0\ \cancel{\text{atm}}} = 1.1 \times 10^2\ \text{L}$$

We assume that the initial volume of the liquid is negligible compared to the volume of the vapor.

$$w = -P\Delta V = -(1.0\ \text{atm})(1.1 \times 10^2\ \text{L}) = -1.1 \times 10^2\ \text{L}\cdot\text{atm}$$

$$w = -1.1 \times 10^2\ \cancel{\text{L}\cdot\text{atm}} \times \frac{101.3\ \text{J}}{1\ \cancel{\text{L}\cdot\text{atm}}} = -1.1 \times 10^4\ \text{J} = -11\ \text{kJ}$$

The **expansion work** done is **11 kJ.**

**6.74**

| Reaction | $\Delta H^\circ$ (kJ/mol) |
|---|---|
| $BrF(g) \rightarrow \frac{1}{2}Br_2(l) + \frac{1}{2}F_2(g)$ | ½(188) |
| $\frac{1}{2}Br_2(l) + \frac{3}{2}F_2(g) \rightarrow BrF_3(g)$ | ½(−768) |
| $BrF(g) + F_2(g) \rightarrow BrF_3(g)$ | $\mathbf{\Delta H^\circ_{rxn} = -2.90 \times 10^2\ kJ/mol}$ |

**6.76** **(a)** $\Delta H^\circ_{\text{rxn}} = \sum n\Delta H^\circ_f(\text{products}) - \sum m\Delta H^\circ_f(\text{reactants})$

$$\Delta H^\circ_{\text{rxn}} = [4\Delta H^\circ_f(NH_3) + \Delta H^\circ_f(N_2)] - 3\Delta H^\circ_f(N_2H_4)$$

$$\mathbf{\Delta H^\circ_{rxn}} = [(4)(-46.3 \text{ kJ/mol}) + (0)] - (3)(50.42 \text{ kJ/mol}) = \mathbf{-336.5\ kJ/mol}$$

**(b)** The balanced equations are:

(1) $N_2H_4(l) + O_2(g) \longrightarrow N_2(g) + 2H_2O(l)$

(2) $4NH_3(g) + 3O_2(g) \longrightarrow 2N_2(g) + 6H_2O(l)$

The standard enthalpy change for equation (1) is:

$$\Delta H^\circ_{\text{rxn}} = \Delta H^\circ_f[N_2] + 2\Delta H^\circ_f[H_2O(l)] - \{\Delta H^\circ_f[N_2H_4(l)] + \Delta H^\circ_f[O_2]\}$$

$$\Delta H^\circ_{\text{rxn}} = [(1)(0) + (2)(-285.8 \text{ kJ/mol})] - [(1)(50.42 \text{ kJ/mol}) + (1)(0)] = -622.0 \text{ kJ/mol}$$

The standard enthalpy change for equation (2) is:

$$\Delta H^\circ_{\text{rxn}} = [2\Delta H^\circ_f(N_2) + 6\Delta H^\circ_f(H_2O)] - [4\Delta H^\circ_f(NH_3) + 3\Delta H^\circ_f(O_2)]$$

$$\Delta H^\circ_{\text{rxn}} = [(2)(0) + (6)(-285.8 \text{ kJ/mol})] - [(4)(-46.3 \text{ kJ/mol}) + (3)(0)] = -1529.6 \text{ kJ/mol}$$

We can now calculate the enthalpy change per kilogram of each substance. $\Delta H^\circ_{\text{rxn}}$ above is in units of kJ/mol. We need to convert to kJ/kg.

$$N_2H_4(l): \mathbf{\Delta H^\circ_{rxn}} = \frac{-622.0 \text{ kJ}}{1 \text{ mol } N_2H_4} \times \frac{1 \text{ mol } N_2H_4}{32.05 \text{ g } N_2H_4} \times \frac{1000 \text{ g}}{1 \text{ kg}} = \mathbf{-1.941 \times 10^4\ kJ/kg\ N_2H_4}$$

$$NH_3(g): \mathbf{\Delta H^\circ_{rxn}} = \frac{-1529.6 \text{ kJ}}{4 \text{ mol } NH_3} \times \frac{1 \text{ mol } NH_3}{17.03 \text{ g } NH_3} \times \frac{1000 \text{ g}}{1 \text{ kg}} = \mathbf{-2.245 \times 10^4\ kJ/kg\ NH_3}$$

Since **ammonia, $NH_3$**, releases more energy per kilogram of substance, it would be a better fuel.

**6.78** The heat lost by the water was absorbed by the dissolution of $NH_4NO_3$.

The heat lost by the water is:

$$q = ms\Delta t = (80.0 \text{ g})(4.184 \text{ J/g}\cdot{}^\circ\text{C})(18.1 - 21.6)\ {}^\circ\text{C}$$

$$q = -1.17 \times 10^3 \text{ J}$$

Thus, $1.17 \times 10^3$ J of heat is absorbed during the dissolution process. $\Delta H_{\text{soln}}$ is expressed in kJ/mol.

$$\mathbf{\Delta H_{soln}} = \frac{1.17 \times 10^3 \text{ J}}{3.53 \text{ g } NH_4NO_3} \times \frac{80.05 \text{ g } NH_4NO_3}{1 \text{ mol } NH_4NO_3} \times \frac{1 \text{ kJ}}{1000 \text{ J}} = \mathbf{26.5\ kJ/mol}$$

**6.80** The reaction is $2Na(s) + Cl_2(g) \rightarrow 2NaCl(s)$. First, let's calculate $\Delta H^\circ$ for this reaction using $\Delta H^\circ_f$ values in Appendix 3.

$$\Delta H^\circ_{\text{rxn}} = 2\Delta H^\circ_f(NaCl) - [2\Delta H^\circ_f(Na) + \Delta H^\circ_f(Cl_2)]$$

$$\Delta H^\circ_{\text{rxn}} = 2(-411.0 \text{ kJ/mol}) - [2(0) + 0] = -822.0 \text{ kJ/mol}$$

This is the amount of heat released when 1 mole of $Cl_2$ reacts (see balanced equation). We are not reacting 1 mole of $Cl_2$, however. From the volume and density of $Cl_2$, we can calculate grams of $Cl_2$. Then, using the molar mass of $Cl_2$ as a conversion factor, we can calculate moles of $Cl_2$. Combining these two calculations into one step, we find moles of $Cl_2$ to be:

$$2.00\ \cancel{L}\ Cl_2 \times \frac{1.88\ \cancel{g}\ Cl_2}{1\ \cancel{L}\ Cl_2} \times \frac{1\ \text{mol}\ Cl_2}{70.90\ \cancel{g}\ Cl_2} = 0.0530\ \text{mol}\ Cl_2$$

Finally, we can use the $\Delta H^\circ_{rxn}$ calculated above to find the amount of heat released when 0.0530 mole of $Cl_2$ reacts.

$$0.0530\ \cancel{mol}\ Cl_2 \times \frac{-822.0\ \text{kJ}}{1\ \cancel{mol}\ Cl_2} = -43.6\ \text{kJ}$$

The amount of heat *released* is **43.6 kJ**.

**6.82** The initial and final states of this system are identical. Since enthalpy is a state function, its value depends only upon the state of the system. The enthalpy change is **zero**.

**6.84** $H(g) + Br(g) \longrightarrow HBr(g) \quad \Delta H^\circ_{rxn} = ?$

Rearrange the equations as necessary so they can be added to yield the desired equation.

$$H(g) \longrightarrow \tfrac{1}{2}H_2(g) \qquad \Delta H^\circ_{rxn} = \tfrac{1}{2}(-436.4\ \text{kJ/mol}) = -218.2\ \text{kJ/mol}$$

$$Br(g) \longrightarrow \tfrac{1}{2}Br_2(g) \qquad \Delta H^\circ_{rxn} = \tfrac{1}{2}(-192.5\ \text{kJ/mol}) = -96.25\ \text{kJ/mol}$$

$$\tfrac{1}{2}H_2(g) + \tfrac{1}{2}Br_2(g) \longrightarrow HBr(g) \qquad \Delta H^\circ_{rxn} = \tfrac{1}{2}(-72.4\ \text{kJ/mol}) = -36.2\ \text{kJ/mol}$$

$$H(g) + Br(g) \longrightarrow HBr(g) \qquad \boldsymbol{\Delta H^\circ = -350.7\ \text{kJ/mol}}$$

**6.86** The standard enthalpy of formation is the heat change that results when 1 mole of a compound is formed from its elements at a pressure of 1 atm. The formation reaction for BaO(*s*) is:

$$Ba(s) + \tfrac{1}{2}O_2(g) \rightarrow BaO(s)$$

In the problem, we are given the heat change per 2.740 g of Ba. We convert to units of kJ/mol.

$$\Delta H^\circ_f = \frac{-11.14\ \text{kJ}}{2.740\ \cancel{g}\ Ba} \times \frac{137.3\ \cancel{g}\ Ba}{1\ \text{mol Ba}} = \mathbf{-558.2\ kJ/mol}$$

**6.88** $q_{system} = 0 = q_{metal} + q_{water} + q_{calorimeter}$

$q_{metal} + q_{water} + q_{calorimeter} = 0$

$m_{metal}s_{metal}(t_{final} - t_{initial}) + m_{water}s_{water}(t_{final} - t_{initial}) + C_{calorimeter}(t_{final} - t_{initial}) = 0$

All the needed values are given in the problem. All you need to do is plug in the values and solve for $s_{metal}$.

$$(44.0\ \text{g})(s_{metal})(28.4 - 99.0)^\circ\text{C} + (80.0\ \text{g})(4.184\ \text{J/g}\cdot^\circ\text{C})(28.4 - 24.0)^\circ\text{C} + (12.4\ \text{J/}^\circ\text{C})(28.4 - 24.0)^\circ\text{C} = 0$$

$$(-3.11 \times 10^3)s_{\text{metal}}\ (\text{g}\cdot{}^\circ\text{C}) = -1.53 \times 10^3\ \text{J}$$

$$s_{\text{metal}} = \mathbf{0.492\ J/g\cdot{}^\circ C}$$

**6.90** A good starting point would be to calculate the standard enthalpy for both reactions.

Calculate the standard enthalpy for the reaction: $C(s) + \frac{1}{2}O_2(g) \longrightarrow CO(g)$

This reaction corresponds to the standard enthalpy of formation of CO, so we use the value of −110.5 kJ/mol (see Appendix 3 of the text).

Calculate the standard enthalpy for the reaction: $C(s) + H_2O(g) \longrightarrow CO(g) + H_2(g)$

$$\Delta H^\circ_{\text{rxn}} = [\Delta H^\circ_f(CO) + \Delta H^\circ_f(H_2)] - [\Delta H^\circ_f(C) + \Delta H^\circ_f(H_2O)]$$

$$\Delta H^\circ_{\text{rxn}} = [(1)(-110.5\ \text{kJ/mol}) + (1)(0)] - [(1)(0) + (1)(-241.8\ \text{kJ/mol})] = 131.3\ \text{kJ/mol}$$

The first reaction, which is exothermic, can be used to promote the second reaction, which is endothermic. Thus, the two gases are produced alternately.

**6.92** First, calculate the energy produced by 1 mole of octane, $C_8H_{18}$.

$$C_8H_{18}(l) + \frac{25}{2}O_2(g) \longrightarrow 8CO_2(g) + 9H_2O(l)$$

$$\Delta H^\circ_{\text{rxn}} = 8\Delta H^\circ_f(CO_2) + 9\Delta H^\circ_f[H_2O(l)] - [\Delta H^\circ_f(C_8H_{18}) + \frac{25}{2}\Delta H^\circ_f(O_2)]$$

$$\Delta H^\circ_{\text{rxn}} = [(8)(-393.5\ \text{kJ/mol}) + (9)(-285.8\ \text{kJ/mol})] - [(1)(-249.9\ \text{kJ/mol}) + (\frac{25}{2})(0)]$$

$$= -5470\ \text{kJ/mol}$$

The problem asks for the energy produced by the combustion of 1 gallon of octane. $\Delta H^\circ_{\text{rxn}}$ above has units of kJ/mol octane. We need to convert from kJ/mol octane to kJ/gallon octane. The heat of combustion for 1 gallon of octane is:

$$\Delta H^\circ = \frac{-5470\ \text{kJ}}{1\ \text{mol octane}} \times \frac{1\ \text{mol octane}}{114.2\ \text{g octane}} \times \frac{2660\ \text{g}}{1\ \text{gal}} = -1.274\times10^5\ \text{kJ/gal}$$

The combustion of hydrogen corresponds to the standard heat of formation of water:

$$H_2(g) + \frac{1}{2}O_2(g) \longrightarrow H_2O(l)$$

Thus, $\Delta H^\circ_{\text{rxn}}$ is the same as $\Delta H^\circ_f$ for $H_2O(l)$, which has a value of −285.8 kJ/mol. The number of moles of hydrogen required to produce $1.274 \times 10^5$ kJ of heat is:

$$n_{H_2} = (1.274 \times 10^5\ \text{kJ}) \times \frac{1\ \text{mol}\ H_2}{285.8\ \text{kJ}} = 445.8\ \text{mol}\ H_2$$

Finally, use the ideal gas law to calculate the volume of gas corresponding to 445.8 moles of $H_2$ at 25°C and 1 atm.

$$V_{H_2} = \frac{n_{H_2}RT}{P} = \frac{(445.8\ \text{mol})\left(0.0821\ \frac{\text{L}\cdot\text{atm}}{\text{mol}\cdot\text{K}}\right)(298\ \text{K})}{(1\ \text{atm})} = \mathbf{1.09\times10^4\ L}$$

That is, the volume of hydrogen that is energy-equivalent to 1 gallon of gasoline is over **10,000 liters** at 1 atm and 25°C!

**6.94** The combustion reaction is: $C_2H_6(l) + \frac{7}{2}O_2(g) \longrightarrow 2CO_2(g) + 3H_2O(l)$

The heat released during the combustion of 1 mole of ethane is:

$$\Delta H^\circ_{\text{rxn}} = [2\Delta H^\circ_f(CO_2) + 3\Delta H^\circ_f(H_2O)] - [\Delta H^\circ_f(C_2H_6) + \frac{7}{2}\Delta H^\circ_f(O_2)]$$

$$\Delta H^\circ_{\text{rxn}} = [(2)(-393.5\ \text{kJ/mol}) + (3)(-285.8\ \text{kJ/mol})] - [(1)(-84.7\ \text{kJ/mol} + (\frac{7}{2})(0)]$$
$$= -1560\ \text{kJ/mol}$$

The heat required to raise the temperature of the water to 98°C is:

$$q = m_{H_2O}s_{H_2O}\Delta t = (855\ \text{g})(4.184\ \text{J/g}\cdot°\text{C})(98.0 - 25.0)°\text{C} = 2.61\times10^5\ \text{J} = 261\ \text{kJ}$$

The combustion of 1 mole of ethane produces 1560 kJ; the number of moles required to produce 261 kJ is:

$$261\ \text{kJ}\times\frac{1\ \text{mol ethane}}{1560\ \text{kJ}} = 0.167\ \text{mol ethane}$$

The volume of ethane is:

$$V_{\text{ethane}} = \frac{nRT}{P} = \frac{(0.167\ \text{mol})\left(0.0821\ \frac{\text{L}\cdot\text{atm}}{\text{mol}\cdot\text{K}}\right)(296\ \text{K})}{\left(752\ \text{mmHg}\times\frac{1\ \text{atm}}{760\ \text{mmHg}}\right)} = \mathbf{4.10\ L}$$

**6.96** The heat gained by the liquid nitrogen must be equal to the heat lost by the water.

$$q_{N_2} = -q_{H_2O}$$

If we can calculate the heat lost by the water, we can calculate the heat gained by 60.0 g of the nitrogen.

$$\text{Heat lost by the water} = q_{H_2O} = m_{H_2O}s_{H_2O}\Delta t$$

$$q_{H_2O} = (2.00\times10^2\ \text{g})(4.184\ \text{J/g}\cdot°\text{C})(41.0 - 55.3)°\text{C} = -1.20\times10^4\ \text{J}$$

The heat gained by 60.0 g nitrogen is the opposite sign of the heat lost by the water.

$$q_{N_2} = -q_{H_2O}$$

$$q_{N_2} = 1.20\times10^4\ \text{J}$$

The problem asks for the molar heat of vaporization of liquid nitrogen. Above, we calculated the amount of heat necessary to vaporize 60.0 g of liquid nitrogen. We need to convert from J/60.0 g $N_2$ to J/mol $N_2$.

$$\Delta H_{\text{vap}} \frac{1.20\times10^4\ \text{J}}{60.0\ \text{g}\ N_2}\times\frac{28.02\ \text{g}\ N_2}{1\ \text{mol}\ N_2} = \mathbf{5.60\times10^3\ J/mol = 5.60\ kJ/mol}$$

**6.98** Recall that the standard enthalpy of formation ( $\Delta H_f^\circ$ ) is defined as the heat change that results when 1 mole of a compound is formed from its elements at a pressure of 1 atm. Only in choice **(a)** does $\Delta H_{rxn}^\circ = \Delta H_f^\circ$. In choice (b), C(diamond) is *not* the most stable form of elemental carbon under standard conditions; C(graphite) is the most stable form.

**6.100** **(a)** No work is done by a gas expanding in a vacuum, because the pressure exerted on the gas is zero.

**(b)** $w = -P\Delta V$

$$w = -(0.20\ \text{atm})(0.50 - 0.050)\text{L} = -0.090\ \text{L·atm}$$

Converting to units of joules:

$$\boldsymbol{w} = -0.090\ \cancel{\text{L·atm}} = \frac{101.3\ \text{J}}{\cancel{\text{L·atm}}} = \mathbf{-9.1\ J}$$

**(c)** The gas will expand until the pressure is the same as the applied pressure of 0.20 atm. We can calculate its final volume using the ideal gas equation.

$$\boldsymbol{V} = \frac{nRT}{P} = \frac{(0.020\ \cancel{\text{mol}})\left(0.0821\ \frac{\text{L}\cdot\cancel{\text{atm}}}{\cancel{\text{mol}}\cdot\cancel{\text{K}}}\right)(273+20)\cancel{\text{K}}}{0.20\ \cancel{\text{atm}}} = \mathbf{2.4\ L}$$

The amount of work done is:

$$w = -P\Delta V = (0.20\ \text{atm})(2.4 - 0.050)\text{L} = -0.47\ \text{L·atm}$$

Converting to units of joules:

$$\boldsymbol{w} = -0.47\ \cancel{\text{L·atm}} \times \frac{101.3\ \text{J}}{\cancel{\text{L·atm}}} = \mathbf{-48\ J}$$

**6.102** **(a)** The more closely packed, the greater the mass of food. Heat capacity depends on both the mass and specific heat.

$$C = ms$$

The heat capacity of the food is greater than the heat capacity of air; hence, the cold in the freezer will be retained longer.

**(b)** Tea and coffee are mostly water; whereas, soup might contain vegetables and meat. Water has a higher heat capacity than the other ingredients in soup; therefore, coffee and tea retain heat longer than soup.

**6.104** $4\text{Fe}(s) + 3\text{O}_2(g) \rightarrow 2\text{Fe}_2\text{O}_3(s)$. This equation represents twice the standard enthalpy of formation of $Fe_2O_3$. From Appendix 3, the standard enthalpy of formation of $Fe_2O_3 = -822.2$ kJ/mol. So, $\Delta H^\circ$ for the given reaction is:

$$\Delta H_{rxn}^\circ = (2)(-822.2\ \text{kJ/mol}) = -1644\ \text{kJ/mol}$$

Looking at the balanced equation, this is the amount of heat released when four moles of Fe react. But, we are reacting 250 g of Fe, not 4 moles. We can convert from grams of Fe to moles of Fe, then use $\Delta H^\circ$ as a conversion factor to convert to kJ.

$$250\ \text{g Fe} \times \frac{1\ \text{mol Fe}}{55.85\ \text{g Fe}} \times \frac{-1644\ \text{kJ}}{4\ \text{mol Fe}} = -1.84 \times 10^3\ \text{kJ}$$

The amount of heat *produced* by this reaction is **1.84 × 10³ kJ**.

**6.106** The heat required to raise the temperature of 1 liter of water by 1°C is:

$$4.184\ \frac{\text{J}}{\text{g}\cdot{}^\circ\text{C}} \times \frac{1\ \text{g}}{1\ \text{mL}} \times \frac{1000\ \text{mL}}{1\ \text{L}} \times 1^\circ\text{C} = 4184\ \text{J/L}$$

Next, convert the volume of the Pacific Ocean to liters.

$$(7.2 \times 10^8\ \text{km}^3) \times \left(\frac{1000\ \text{m}}{1\ \text{km}}\right)^3 \times \left(\frac{100\ \text{cm}}{1\ \text{m}}\right)^3 \times \frac{1\ \text{L}}{1000\ \text{cm}^3} = 7.2 \times 10^{20}\ \text{L}$$

The amount of heat needed to raise the temperature of 7.2 × 10²⁰ L of water is:

$$(7.2 \times 10^{20}\ \text{L}) \times \frac{4184\ \text{J}}{1\ \text{L}} = 3.0 \times 10^{24}\ \text{J}$$

Finally, we can calculate the number of atomic bombs needed to produce this much heat.

$$(3.0 \times 10^{24}\ \text{J}) \times \frac{1\ \text{atomic bomb}}{1.0 \times 10^{15}\ \text{J}} = \mathbf{3.0 \times 10^9\ \textbf{atomic bombs} = 3.0\ \textbf{billion atomic bombs}}$$

**6.108** **Strategy:** The heat released during the reaction is absorbed by both the water and the calorimeter. How do we calculate the heat absorbed by the water? How do we calculate the heat absorbed by the calorimeter? How much heat is released when 1.9862 g of benzoic acid are reacted? The problem gives the amount of heat that is released when 1 mole of benzoic acid is reacted (−3226.7 kJ/mol).

**Solution:** The heat of the reaction (combustion) is absorbed by both the water and the calorimeter.

$$q_{\text{rxn}} = -(q_{\text{water}} + q_{\text{cal}})$$

If we can calculate both $q_{\text{water}}$ and $q_{\text{rxn}}$, then we can calculate $q_{\text{cal}}$. First, let's calculate the heat absorbed by the water.

$$q_{\text{water}} = m_{\text{water}} s_{\text{water}} \Delta t$$

$$q_{\text{water}} = (2000\ \text{g})(4.184\ \text{J/g}\cdot{}^\circ\text{C})(25.67 - 21.84)^\circ\text{C} = 3.20 \times 10^4\ \text{J} = 32.0\ \text{kJ}$$

Next, let's calculate the heat released ($q_{\text{rxn}}$) when 1.9862 g of benzoic acid are burned. $\Delta H_{\text{rxn}}$ is given in units of kJ/mol. Let's convert to $q_{\text{rxn}}$ in kJ.

$$q_{\text{rxn}} = 1.9862\ \text{g benzoic acid} \times \frac{1\ \text{mol benzoic acid}}{122.1\ \text{g benzoic acid}} \times \frac{-3226.7\ \text{kJ}}{1\ \text{mol benzoic acid}} = -52.49\ \text{kJ}$$

And,

$$q_{\text{cal}} = -q_{\text{rxn}} - q_{\text{water}}$$

$$q_{\text{cal}} = 52.49\ \text{kJ} - 32.0\ \text{kJ} = 20.5\ \text{kJ}$$

To calculate the heat capacity of the bomb calorimeter, we can use the following equation:

$$q_{cal} = C_{cal}\Delta t$$

$$C_{\text{cal}} = \frac{q_{\text{cal}}}{\Delta t} = \frac{20.5\text{ kJ}}{(25.67 - 21.84)^\circ\text{C}} = \mathbf{5.35\ kJ/^\circ C}$$

**6.110** First, let's calculate the standard enthalpy of reaction.

$$\Delta H^\circ_{\text{rxn}} = 2\Delta H^\circ_f(\text{CaSO}_4) - [2\Delta H^\circ_f(\text{CaO}) + 2\Delta H^\circ_f(\text{SO}_2) + \Delta H^\circ_f(\text{O}_2)]$$

$$= (2)(-1432.7\text{ kJ/mol}) - [(2)(-635.6\text{ kJ/mol}) + (2)(-296.1\text{ kJ/mol}) + 0]$$

$$= -1002\text{ kJ/mol}$$

This is the enthalpy change for every 2 moles of $SO_2$ that are removed. The problem asks to calculate the enthalpy change for this process if $6.6 \times 10^5$ g of $SO_2$ are removed.

$$(6.6\times10^5\text{ g SO}_2)\times\frac{1\text{ mol SO}_2}{64.07\text{ g SO}_2}\times\frac{-1002\text{ kJ}}{2\text{ mol SO}_2} = \mathbf{-5.2\times10^6\ kJ}$$

**6.112** First, we need to calculate the volume of the balloon.

$$V = \frac{4}{3}\pi r^3 = \frac{4}{3}\pi(8\text{ m})^3 = (2.1\times10^3\text{ m}^3)\times\frac{1000\text{ L}}{1\text{ m}^3} = 2.1\times10^6\text{ L}$$

**(a)** We can calculate the mass of He in the balloon using the ideal gas equation.

$$n_{\text{He}} = \frac{PV}{RT} = \frac{\left(98.7\text{ kPa}\times\frac{1\text{ atm}}{1.01325\times10^2\text{ kPa}}\right)(2.1\times10^6\text{ L})}{\left(0.0821\frac{\text{L}\cdot\text{atm}}{\text{mol}\cdot\text{K}}\right)(273+18)\text{K}} = 8.6\times10^4\text{ mol He}$$

$$\textbf{mass He} = (8.6\times10^4\text{ mol He})\times\frac{4.003\text{ g He}}{1\text{ mol He}} = \mathbf{3.4\times10^5\ g\ He}$$

**(b)** Work done $= -P\Delta V$

$$= -\left(98.7\text{ kPa}\times\frac{1\text{ atm}}{1.01325\times10^2\text{ kPa}}\right)(2.1\times10^6\text{ L})$$

$$= (-2.0\times10^6\text{ L}\cdot\text{atm})\times\frac{101.3\text{ J}}{1\text{ L}\cdot\text{atm}}$$

**Work done** $=$ $\mathbf{-2.0\times10^8\ J}$

**6.114** We use Equation (6.18) of the text.

$$\Delta H^\circ = \Delta H^\circ_f[\text{Fe(OH)}_3] - \{\Delta H^\circ_f(\text{Fe}^{3+}) + (3)[\Delta H^\circ_f(\text{OH}^-)]\}$$

$$\mathbf{\Delta H^\circ} = (-824.25\text{ kJ/mol}) - [(-47.7\text{ kJ/mol}) + (3)(-229.94\text{ kJ/mol})] = \mathbf{-86.7\ kJ/mol}$$

**6.116** (a) The heat needed to raise the temperature of the water from 3°C to 37°C can be calculated using the equation:

$$q = ms\Delta t$$

First, we need to calculate the mass of the water.

$$4 \text{ glasses of water} \times \frac{2.5\times 10^2 \text{ mL}}{1 \text{ glass}} \times \frac{1 \text{ g water}}{1 \text{ mL water}} = 1.0\times 10^3 \text{ g water}$$

The heat needed to raise the temperature of $1.0 \times 10^3$ g of water is:

$$q = ms\Delta t = (1.0\times 10^3 \text{ g})(4.184 \text{ J/g}\cdot{}^\circ\text{C})(37-3)^\circ\text{C} = 1.4\times 10^5 \text{ J} = \mathbf{1.4\times 10^2\ kJ}$$

(b) We need to calculate both the heat needed to melt the snow and also the heat needed to heat liquid water from 0°C to 37°C (normal body temperature).

The heat needed to melt the snow is:

$$(8.0\times 10^2 \text{ g})\times\frac{1 \text{ mol}}{18.02 \text{ g}}\times\frac{6.01 \text{ kJ}}{1 \text{ mol}} = 2.7\times 10^2 \text{ kJ}$$

The heat needed to raise the temperature of the water from 0°C to 37°C is:

$$q = ms\Delta t = (8.0\times 10^2 \text{ g})(4.184 \text{ J/g}\cdot{}^\circ\text{C})(37-0)^\circ\text{C} = 1.2\times 10^5 \text{ J} = 1.2\times 10^2 \text{ kJ}$$

The total heat lost by your body is:

$$(2.7\times 10^2 \text{ kJ}) + (1.2\times 10^2 \text{ kJ}) = \mathbf{3.9\times 10^2\ kJ}$$

**6.118** (a) $\Delta H^\circ = \Delta H_f^\circ(F^-) + \Delta H_f^\circ(H_2O) - [\Delta H_f^\circ(HF) + \Delta H_f^\circ(OH^-)]$

$$\Delta H^\circ = [(1)(-329.1 \text{ kJ/mol}) + (1)(-285.8 \text{ kJ/mol})] - [(1)(-320.1 \text{ kJ/mol}) + (1)(-229.6 \text{ kJ/mol})]$$

$$\mathbf{\Delta H^\circ = -65.2\ kJ/mol}$$

(b) We can add the equation given in part (a) to that given in part (b) to end up with the equation we are interested in.

$$HF(aq) + OH^-(aq) \longrightarrow F^-(aq) + H_2O(l) \qquad \Delta H^\circ = -65.2 \text{ kJ/mol}$$

$$H_2O(l) \longrightarrow H^+(aq) + OH^-(aq) \qquad \Delta H^\circ = +56.2 \text{ kJ/mol}$$

$$\mathbf{HF(aq) \longrightarrow H^+(aq) + F^-(aq) \qquad \Delta H^\circ = -9.0\ kJ/mol}$$

**6.120** The equation we are interested in is the formation of CO from its elements.

$$C(\text{graphite}) + \tfrac{1}{2}O_2(g) \longrightarrow CO(g) \qquad \Delta H^\circ = ?$$

Try to add the given equations together to end up with the equation above.

$$C(\text{graphite}) + O_2(g) \longrightarrow CO_2(g) \qquad \Delta H^\circ = -393.5 \text{ kJ/mol}$$

$$CO_2(g) \longrightarrow CO(g) + \tfrac{1}{2}O_2(g) \qquad \Delta H^\circ = +283.0 \text{ kJ/mol}$$

$$\mathbf{C(graphite) + \tfrac{1}{2}O_2(g) \longrightarrow CO(g) \qquad \Delta H^\circ = -110.5\ kJ/mol}$$

We cannot obtain $\Delta H_f^\circ$ for CO directly, because burning graphite in oxygen will form both CO and $CO_2$.

**6.122** **(a)** mass = 0.0010 kg

Potential energy $= mgh$

$= (0.0010\text{ kg})(9.8\text{ m/s}^2)(51\text{ m})$

**Potential energy = 0.50 J**

**(b)** Kinetic energy $= \frac{1}{2}mu^2 = 0.50\text{ J}$

$\frac{1}{2}(0.0010\text{ kg})u^2 = 0.50\text{ J}$

$u^2 = 1.0 \times 10^3\text{ m}^2/\text{s}^2$

$\mathbf{u = 32\text{ m/s}}$

**(c)** $q = ms\Delta t$

$0.50\text{ J} = (1.0\text{ g})(4.184\text{ J/g}\cdot{}^\circ\text{C})\Delta t$

$\mathbf{\Delta t = 0.12^\circ C}$

**6.124** The reaction we are interested in is the formation of ethanol from its elements.

$$2C(\text{graphite}) + \tfrac{1}{2}O_2(g) + 3H_2(g) \longrightarrow C_2H_5OH(l)$$

Along with the reaction for the combustion of ethanol, we can add other reactions together to end up with the above reaction.

Reversing the reaction representing the combustion of ethanol gives:

$$2CO_2(g) + 3H_2O(l) \longrightarrow C_2H_5OH(l) + 3O_2(g) \qquad \Delta H^\circ = +1367.4\text{ kJ/mol}$$

We need to add equations to add C (graphite) and remove $H_2O$ from the reactants side of the equation. We write:

$$2CO_2(g) + 3H_2O(l) \longrightarrow C_2H_5OH(l) + 3O_2(g) \qquad \Delta H^\circ = +1367.4\text{ kJ/mol}$$

$$2C(\text{graphite}) + 2O_2(g) \longrightarrow 2CO_2(g) \qquad \Delta H^\circ = 2(-393.5\text{ kJ/mol})$$

$$3H_2(g) + \tfrac{3}{2}O_2(g) \longrightarrow 3H_2O(l) \qquad \Delta H^\circ = 3(-285.8\text{ kJ/mol})$$

$$\mathbf{2C(\text{graphite}) + \tfrac{1}{2}O_2(g) + 3H_2(g) \longrightarrow C_2H_5OH(l) \qquad \Delta H^\circ_f = -277.0\text{ kJ/mol}}$$

**6.126** Heat gained by ice = Heat lost by the soft drink

$m_{\text{ice}} \times 334\text{ J/g} = -m_{\text{sd}}s_{\text{sd}}\Delta t$

$m_{\text{ice}} \times 334\text{ J/g} = -(361\text{ g})(4.184\text{ J/g}\cdot{}^\circ\text{C})(0 - 23)^\circ\text{C}$

$\mathbf{m_{\text{ice}} = 104\text{ g}}$

**6.128** The decomposition reaction is: $NH_4Cl(s) \rightarrow NH_3(g) + HCl(g)$

Using the $\Delta H^\circ_f$ values in Appendix 3 and Equation (6.18) of the text, we write:

$$\Delta H^\circ_{rxn} = [\Delta H^\circ_f(NH_3) + \Delta H^\circ_f(HCl)] - \Delta H^\circ_f(NH_4Cl)$$

$$\Delta H^\circ = [(-46.3 \text{ kJ/mol}) + (-92.3 \text{ kJ/mol})] - (-315.39 \text{ kJ/mol})$$

$$\Delta H^\circ = 176.8 \text{ kJ/mol}$$

Looking at the balanced equation, this is the amount of heat released for every 1 mole of $NH_4Cl$ decomposed. We can use the following ratio

$$\frac{176.8 \text{ kJ}}{1 \text{ mol } NH_4Cl}$$

to convert to the amount of heat required to decompose 89.7 grams of $NH_4Cl$.

$$89.7 \text{ g } NH_4Cl \times \frac{1 \text{ mol } NH_4Cl}{53.49 \text{ g } NH_4Cl} \times \frac{176.8 \text{ kJ}}{1 \text{ mol } NH_4Cl} = \mathbf{296 \text{ kJ}}$$

**6.130** From Chapter 5, we saw that the kinetic energy (or internal energy) of 1 mole of a gas is $\frac{3}{2}RT$. For 1 mole of an ideal gas, $PV = RT$. We can write:

$$\text{internal energy} = \frac{3}{2}RT = \frac{3}{2}PV$$

$$= \frac{3}{2}(1.2 \times 10^5 \text{ Pa})(5.5 \times 10^3 \text{ m}^3)$$

$$= 9.9 \times 10^8 \text{ Pa·m}^3$$

$$1 \text{ Pa·m}^3 = 1 \frac{\text{N}}{\text{m}^2}\text{m}^3 = 1 \text{ N·m} = 1 \text{ J}$$

Therefore, the internal energy is $\mathbf{9.9 \times 10^8}$ **J.**

The final temperature of the copper metal can be calculated. (10 tons = $9.07 \times 10^6$ g)

$$q = m_{Cu}s_{Cu}\Delta t$$

$$9.9 \times 10^8 \text{ J} = (9.07 \times 10^6 \text{ g})(0.385 \text{ J/g·°C})(t_f - 21°\text{C})$$

$$(3.49 \times 10^6)t_f = 1.06 \times 10^9$$

$$t_f = \mathbf{304°C}$$

**6.132** **(a)** $CaC_2(s) + 2H_2O(l) \longrightarrow Ca(OH)_2(s) + C_2H_2(g)$

**(b)** The reaction for the combustion of acetylene is:

$$2C_2H_2(g) + 5O_2(g) \longrightarrow 4CO_2(g) + 2H_2O(l)$$

We can calculate the enthalpy change for this reaction from standard enthalpy of formation values given in Appendix 3 of the text.

$$\Delta H^\circ_{rxn} = [4\Delta H^\circ_f(CO_2) + 2\Delta H^\circ_f(H_2O)] - [2\Delta H^\circ_f(C_2H_2) + 5\Delta H^\circ_f(O_2)]$$

$$\Delta H^\circ_{rxn} = [(4)(-393.5 \text{ kJ/mol}) + (2)(-285.8 \text{ kJ/mol})] - [(2)(226.6 \text{ kJ/mol}) + (5)(0)]$$

$$\Delta H^\circ_{rxn} = -2599 \text{ kJ/mol}$$

Looking at the balanced equation, this is the amount of heat released when two moles of $C_2H_2$ are reacted. The problem asks for the amount of heat that can be obtained starting with 74.6 g of $CaC_2$. From this amount of $CaC_2$, we can calculate the moles of $C_2H_2$ produced.

$$74.6\text{ g CaC}_2 \times \frac{1\text{ mol CaC}_2}{64.10\text{ g CaC}_2} \times \frac{1\text{ mol C}_2\text{H}_2}{1\text{ mol CaC}_2} = 1.16\text{ mol C}_2\text{H}_2$$

Now, we can use the $\Delta H^\circ_{rxn}$ calculated above as a conversion factor to determine the amount of heat obtained when 1.16 moles of $C_2H_2$ are burned.

$$1.16\text{ mol C}_2\text{H}_2 \times \frac{2599\text{ kJ}}{2\text{ mol C}_2\text{H}_2} = \mathbf{1.51 \times 10^3\text{ kJ}}$$

**6.134** When 1.034 g of naphthalene are burned, 41.56 kJ of heat are evolved. Let's convert this to the amount of heat evolved on a molar basis. The molar mass of naphthalene is 128.2 g/mol.

$$q = \frac{-41.56\text{ kJ}}{1.034\text{ g C}_{10}\text{H}_8} \times \frac{128.2\text{ g C}_{10}\text{H}_8}{1\text{ mol C}_{10}\text{H}_8} = -5153\text{ kJ/mol}$$

$q$ has a negative sign because this is an exothermic reaction.

This reaction is run at constant volume ($\Delta V = 0$); therefore, no work will result from the change.

$$w = -P\Delta V = 0$$

From Equation (6.4) of the text, it follows that the change in energy is equal to the heat change.

$$\mathbf{\Delta U = q + w = q_v = -5153\text{ kJ/mol}}$$

To calculate $\Delta H$, we rearrange Equation (6.10) of the text.

$$\Delta U = \Delta H - RT\Delta n$$

$$\Delta H = \Delta U + RT\Delta n$$

To calculate $\Delta H$, $\Delta n$ must be determined, which is the difference in moles of *gas* products and moles of *gas* reactants. Looking at the balanced equation for the combustion of naphthalene:

$$C_{10}H_8(s) + 12O_2(g) \rightarrow 10CO_2(g) + 4H_2O(l)$$

$$\Delta n = 10 - 12 = -2$$

$$\Delta H = \Delta U + RT\Delta n$$

$$\Delta H = -5153\text{ kJ/mol} + (8.314\text{ J/mol}\cdot\text{K})(298\text{ K})(-2) \times \frac{1\text{ kJ}}{1000\text{ J}}$$

$$\mathbf{\Delta H = -5158\text{ kJ/mol}}$$

Is $\Delta H$ equal to $q_p$ in this case?

**6.136** We know that $\Delta U = q + w$, $\Delta H = q$, and $w = -P\Delta V = -RT\Delta n$. Using thermodynamic data in Appendix 3 of the text, we can calculate $\Delta H$.

$$2H_2(g) + O_2(g) \rightarrow 2H_2O(l),\ \Delta H = 2(-285.8\text{ kJ/mol}) = -571.6\text{ kJ/mol}$$

Next, we calculate $w$. The change in moles of gas ($\Delta n$) equals $-3$.

$$w = -P\Delta V = -RT\Delta n$$

$$w = -(8.314\ \text{J/mol}\cdot\text{K})(298\ \text{K})(-3) = +7.43 \times 10^3\ \text{J/mol} = 7.43\ \text{kJ/mol}$$

$$\Delta U = q + w$$

$$\Delta U = -571.6\ \text{kJ/mol} + 7.43\ \text{kJ/mol} = \mathbf{-564.2\ kJ/mol}$$

Can you explain why $\Delta U$ is smaller (in magnitude) than $\Delta H$?

**6.138** First, we calculate $\Delta H$ for the combustion of 1 mole of glucose using data in Appendix 3 of the text. We can then calculate the heat produced in the calorimeter. Using the heat produced along with $\Delta H$ for the combustion of 1 mole of glucose will allow us to calculate the mass of glucose in the sample. Finally, the mass % of glucose in the sample can be calculated.

$$C_6H_{12}O_6(s) + 6O_2(g) \rightarrow 6CO_2(g) + 6H_2O(l)$$

$$\Delta H^\circ_{\text{rxn}} = (6)(-393.5\ \text{kJ/mol}) + (6)(-285.8\ \text{kJ/mol}) - (1)(-1274.5\ \text{kJ/mol}) = -2801.3\ \text{kJ/mol}$$

The heat produced in the calorimeter is:

$$(3.134^\circ\text{C})(19.65\ \text{kJ/}^\circ\text{C}) = 61.58\ \text{kJ}$$

Let $x$ equal the mass of glucose in the sample:

$$x\ \text{g glucose} \times \frac{1\ \text{mol glucose}}{180.2\ \text{g glucose}} \times \frac{2801.3\ \text{kJ}}{1\ \text{mol glucose}} = 61.58\ \text{kJ}$$

$$x = 3.961\ \text{g}$$

$$\%\ \text{glucose} = \frac{3.961\ \text{g}}{4.117\ \text{g}} \times 100\% = \mathbf{96.21\%}$$

**6.140** **(a)** From the mass of $CO_2$ produced, we can calculate the moles of carbon in the compound. From the mass of $H_2O$ produced, we can calculate the moles of hydrogen in the compound.

$$1.419\ \text{g}\ CO_2 \times \frac{1\ \text{mol}\ CO_2}{44.01\ \text{g}\ CO_2} \times \frac{1\ \text{mol C}}{1\ \text{mol}\ CO_2} = 0.03224\ \text{mol C}$$

$$0.290\ \text{g}\ H_2O \times \frac{1\ \text{mol}\ H_2O}{18.02\ \text{g}\ H_2O} \times \frac{2\ \text{mol H}}{1\ \text{mol}\ H_2O} = 0.03219\ \text{mol H}$$

The mole ratio between C and H is 1:1, so the empirical formula is **CH**.

**(b)** The empirical molar mass of CH is 13.02 g/mol.

$$\frac{\text{molar mass}}{\text{empirical molar mass}} = \frac{76\ \text{g}}{13.02\ \text{g}} = 5.8 \approx 6$$

Therefore, the molecular formula is $C_6H_6$, and the hydrocarbon is benzene. The combustion reaction is:

$$2C_6H_6(l) + 15O_2(g) \rightarrow 12CO_2(g) + 6H_2O(l)$$

17.55 kJ of heat is released when 0.4196 g of the hydrocarbon undergoes combustion. We can now calculate the enthalpy of combustion ( $\Delta H^\circ_{rxn}$ ) for the above reaction in units of kJ/mol. Then, from the enthalpy of combustion, we can calculate the enthalpy of formation of $C_6H_6$.

$$\frac{-17.55\text{ kJ}}{0.4196\text{ g }C_6H_6} \times \frac{78.11\text{ g }C_6H_6}{1\text{ mol }C_6H_6} \times 2\text{ mol }C_6H_6 = -6534\text{ kJ/mol}$$

$$\Delta H^\circ_{rxn} = (12)\Delta H^\circ_f(CO_2) + (6)\Delta H^\circ_f(H_2O) - (2)\Delta H^\circ_f(C_6H_6)$$

$$-6534\text{ kJ/mol} = (12)(-393.5\text{ kJ/mol}) + (6)(-285.8\text{ kJ/mol}) - (2)\Delta H^\circ_f(C_6H_6)$$

$$\Delta H^\circ_f(C_6H_6) = \mathbf{49\text{ kJ/mol}}$$

**6.142** **(a)** Heating water at room temperature to its boiling point.

**(b)** Heating water at its boiling point.

**(c)** A chemical reaction taking place in a bomb calorimeter (an isolated system) where there is no heat exchange with the surroundings.

**6.144** A → B $w = 0$, because $\Delta V = 0$

B → C $w = -P\Delta V = -(2\text{ atm})(2 - 1)\text{L} = -2\text{ L}\cdot\text{atm}$

C → D $w = 0$, because $\Delta V = 0$

D → A $w = -P\Delta V = -(1\text{ atm})(1 - 2)\text{L} = +1\text{ L}\cdot\text{atm}$

The total work done = (−2 L·atm) + (1 L·atm) = **−1 L·atm**

Converting to units of joules,

$$-1\text{ L}\cdot\text{atm} \times \frac{101.3\text{ J}}{1\text{ L}\cdot\text{atm}} = \mathbf{-101.3\text{ J}}$$

In a cyclic process, the change in a state function must be zero. We therefore conclude that work is not a state function. Note that the total work done equals the area of the enclosure.

**6.146** **(a)** **Exothermic**. Energy is released when a bond forms.

**(b)** **No clear conclusion**. Whether this process is endothermic or exothermic will depend on the amount of energy needed to break the ionic bond, and the amount of energy released during the hydration process.

**(c)** **No clear conclusion**. Whether this reaction is endothermic or exothermic will depend on the amount of energy needed to break the AB bond, and the amount of energy released when the AC bond forms.

**(d)** **Endothermic**. Energy is required to vaporize a liquid.

# CHAPTER 7
# QUANTUM THEORY AND THE ELECTRONIC STRUCTURE OF ATOMS

## PROBLEM-SOLVING STRATEGIES AND TUTORIAL SOLUTIONS

### TYPES OF PROBLEMS

**Problem Type 1:** Calculating the Frequency and Wavelength of an Electromagnetic Wave.

**Problem Type 2:** Calculating the Energy of a Photon.

**Problem Type 3:** Calculating the Energy, Wavelength, or Frequency in the Emission Spectrum of a Hydrogen Atom.

**Problem Type 4:** The de Broglie Equation: Calculating the Wavelengths of Particles.

**Problem Type 5:** Quantum Numbers.

**(a)** Labeling an atomic orbital.
**(b)** Counting the number of orbitals associated with a principal quantum number.
**(c)** Assigning quantum numbers to an electron.
**(d)** Counting the number of electrons in a principal level.

**Problem Type 6:** Writing Electron Configurations and Orbital Diagrams.

## PROBLEM TYPE 1: CALCULATING THE FREQUENCY AND WAVELENGTH OF AN ELECTROMAGNETIC WAVE

All types of *electromagnetic radiation* move through a vacuum at a speed of about $3.00 \times 10^8$ m/s, which is called the speed of light ($c$). Speed is an important property of a wave traveling through space and is equal to the product of the wavelength and the frequency of the wave. For electromagnetic waves

$$c = \lambda\nu \qquad (7.1)$$

Equation (7.1) can be rearranged as necessary to solve for either the wavelength ($\lambda$) or the frequency ($\nu$).

**EXAMPLE 7.1**

**A certain AM radio station broadcasts at a frequency of $6.00 \times 10^2$ kHz. What is the wavelength of these radio waves in meters?**

**Strategy:** We are given the frequency of an electromagnetic wave and asked to calculate the wavelength. Rearranging Equation (7.1) gives:

$$c = \lambda\nu$$

$$\lambda = \frac{c}{\nu}$$

**Solution:** Because the speed of light has units of m/s, we must convert the frequency from units of kHz to Hz ($s^{-1}$)

$$(6.00 \times 10^2 \text{ kHz}) \times \frac{1000 \text{ Hz}}{1 \text{ kHz}} = 6.00 \times 10^5 \text{ Hz} = 6.00 \times 10^5 \text{ s}^{-1}$$

Substituting in the frequency and the speed of light constant, the wavelength is:

$$\lambda = \frac{c}{\nu} = \frac{3.00 \times 10^8 \frac{\text{m}}{\cancel{\text{s}}}}{6.00 \times 10^5 \frac{1}{\cancel{\text{s}}}} = \mathbf{5.00 \times 10^2\ m}$$

**Check:** Look at Figure 7.4 of the text to confirm that this wavelength corresponds to a radio wave.

**EXAMPLE 7.2**

**What is the frequency of light that has a wavelength of 665 nm?**

---

**Strategy:** We are given the wavelength of an electromagnetic wave and asked to calculate the frequency. Rearranging Equation (7.1):

$$c = \lambda\nu$$

$$\nu = \frac{c}{\lambda}$$

**Solution:** Since the speed of light has units of m/s, we must convert the wavelength from units of nm to m.

$$665\ \cancel{\text{nm}} \times \frac{1 \times 10^{-9}\ \text{m}}{1\ \cancel{\text{nm}}} = 6.65 \times 10^{-7}\ \text{m}$$

Substituting in the wavelength and the speed of light ($3.00 \times 10^8$ m/s), the frequency is:

$$\nu = \frac{c}{\lambda} = \frac{3.00 \times 10^8 \frac{\cancel{\text{m}}}{\text{s}}}{6.65 \times 10^{-7}\ \cancel{\text{m}}} = \mathbf{4.51 \times 10^{14}\ s^{-1} = 4.51 \times 10^{14}\ Hz}$$

**PRACTICE EXERCISE**

1. Domestic microwave ovens generate microwaves with a frequency of 2.450 GHz. What is the wavelength of this microwave radiation?

**Text Problems: 7.8, 7.12, 7.16**

## PROBLEM TYPE 2: CALCULATING THE ENERGY OF A PHOTON

Max Planck said that atoms and molecules could emit (or absorb) energy only in discrete quantities. Planck gave the name *quantum* to the smallest quantity of energy that can be emitted (or absorbed) in the form of electromagnetic radiation. The energy $E$ of a single quantum of energy is given by

$$E = h\nu \qquad (7.2,\ \text{text})$$

where,

$h$ is Planck's constant = $6.63 \times 10^{-34}$ J·s
$\nu$ is the frequency of radiation

**EXAMPLE 7.3**

**The yellow light given off by a sodium vapor lamp has a wavelength of 589 nm. What is the energy of a single photon of this radiation?**

---

**Strategy:** We are given the wavelength of an electromagnetic wave and asked to calculate its energy. Equation (7.2) of the text relates the energy and frequency of an electromagnetic wave.

$$E = h\nu$$

The relationship between frequency and wavelength is:

$$\nu = \frac{c}{\lambda}$$

Substituting for the frequency gives,

$$E = \frac{hc}{\lambda}$$

**Solution:** Because the speed of light is in units of m/s, we must convert the wavelength from units of nm to m.

$$589 \text{ nm} \times \frac{1 \times 10^{-9} \text{ m}}{1 \text{ nm}} = 5.89 \times 10^{-7} \text{ m}$$

Substituting in Planck's constant, the speed of light constant, and the wavelength, the energy is:

$$E = \frac{hc}{\lambda} = \frac{(6.63 \times 10^{-34} \text{ J}\cdot\text{s})\left(3.00 \times 10^{8} \frac{\text{m}}{\text{s}}\right)}{5.89 \times 10^{-7} \text{ m}} = \mathbf{3.38 \times 10^{-19} \text{ J}}$$

**Check:** We expect the energy of a single photon to be a very small energy as calculated above, $3.38 \times 10^{-19}$ J.

**PRACTICE EXERCISE**

2. The red line in the spectrum of lithium occurs at 670.8 nm. What is the energy of a photon of this light? What is the energy of 1 mole of these photons?

3. The light-sensitive compound in most photographic films is silver bromide (AgBr). When the film is exposed, assume that the light energy absorbed dissociates the molecule into atoms. (The actual process is more complex.) If the energy of dissociation of AgBr is $1.00 \times 10^{2}$ kJ/mol, find the wavelength of light that is just able to dissociate AgBr.

**Text Problems: 7.16**, 7.18, 7.20

## PROBLEM TYPE 3: CALCULATING THE ENERGY, WAVELENGTH, OR FREQUENCY IN THE EMISSION SPECTRUM OF A HYDROGEN ATOM

Using arguments based on electrostatic interaction and Newton's laws of motion, Neils Bohr showed that the energies that the electron in the hydrogen atom can possess are given by:

$$E_n = -R_H\left(\frac{1}{n^2}\right) \qquad (7.5, \text{text})$$

where,

$R_H$ is the Rydberg constant = $2.18 \times 10^{-18}$ J

$n$ is the principal quantum number that has integer values

During the emission process in a hydrogen atom, an electron initially in an excited state characterized by the principal quantum number $n_i$ drops to a lower energy state characterized by the principal quantum number $n_f$. This lower energy state may be either another excited state or the ground state. The difference between the energies of the initial and final states is:

$$\Delta E = E_f - E_i$$

Substituting Equation (7.5) of the text into the above equation gives:

$$\Delta E = \left(\frac{-R_H}{n_f^2}\right) - \left(\frac{-R_H}{n_i^2}\right)$$

$$\Delta E = R_H\left(\frac{1}{n_i^2} - \frac{1}{n_f^2}\right)$$

Furthermore, since this transition results in the emission of a photon of frequency $\nu$ and energy $h\nu$ (See Problem Type 2), we can write:

$$\Delta E = h\nu = R_H\left(\frac{1}{n_i^2} - \frac{1}{n_f^2}\right) \quad \text{(7.6, text)}$$

**EXAMPLE 7.4**

**What wavelength of radiation will be emitted during an electron transition from the $n = 5$ state to the $n = 1$ state in the hydrogen atom? What region of the electromagnetic spectrum does this wavelength correspond to?**

**Strategy:** We are given the initial and final states in the emission process. We can calculate the energy of the emitted photon using Equation (7.6) of the text. Then, from this energy, we can solve for the wavelength. The value of Rydberg's constant is $2.18 \times 10^{-18}$ J.

**Solution:** From Equation (7.6) we write:

$$\Delta E = R_H\left(\frac{1}{n_i^2} - \frac{1}{n_f^2}\right)$$

$$\Delta E = (2.18 \times 10^{-18}\ \text{J})\left(\frac{1}{5^2} - \frac{1}{1^2}\right)$$

$$\Delta E = -2.09 \times 10^{-18}\ \text{J}$$

The negative sign for $\Delta E$ indicates that this is energy associated with an emission process. To calculate the wavelength, we will omit the minus sign for $\Delta E$ because the wavelength of the photon must be positive. We know that

$$\Delta E = h\nu$$

We also know that $\nu = \dfrac{c}{\lambda}$. Substituting into the above equation gives:

$$\Delta E = \frac{hc}{\lambda}$$

Solving the equation algebraically for the wavelength, then substituting in the known values gives:

$$\lambda = \frac{hc}{\Delta E} = \frac{(6.63 \times 10^{-34}\ \text{J}\cdot\text{s})\left(3.00 \times 10^{8}\ \frac{\text{m}}{\text{s}}\right)}{2.09 \times 10^{-18}\ \text{J}} = \mathbf{9.52 \times 10^{-8}\ m}$$

To determine the region of the electromagnetic spectrum that this wavelength corresponds to, we should convert the wavelength from units of meters to nanometers, and then compare the value to Figure 7.4 of the text.

$$(9.52 \times 10^{-8}\ \text{m}) \times \frac{1\ \text{nm}}{1 \times 10^{-9}\ \text{m}} = 95.2\ \text{nm}$$

Checking Figure 7.4, we see that the ultraviolet region of the spectrum is centered at a wavelength of 10 nm. Therefore, this emission is in the ultraviolet (UV) region of the electromagnetic spectrum.

**PRACTICE EXERCISE**

**4.** A hydrogen emission line in the ultraviolet region of the spectrum at 95.2 nm corresponds to a transition from a higher energy level $n_i$ to the $n = 1$ level. What is the value of $n_i$ for the higher energy level?

**Text Problems:** 7.30, **7.32**, 7.34

## PROBLEM TYPE 4: THE DE BROGLIE EQUATION: CALCULATING THE WAVELENGTHS OF PARTICLES

Albert Einstein showed that light (electromagnetic radiation) can possess particle like properties. De Broglie reasoned that if waves can behave like particles, then particles can exhibit wave properties. De Broglie deduced that the particle and wave properties are related by the expression:

$$\lambda = \frac{h}{mu} \qquad \text{(7.8, text)}$$

where,

$\lambda$ is the wavelength associated with the moving particle
$m$ is the mass of the particle
$u$ is the velocity of the particle

Equation (7.8) of the text implies that a particle in motion can be treated as a wave, and a wave can exhibit the properties of a particle.

**EXAMPLE 7.5**

**When an atom of Th-232 undergoes radioactive decay, an alpha particle, which has a mass of 4.0 amu, is ejected from the Th nucleus with a velocity of $1.4 \times 10^7$ m/s. What is the de Broglie wavelength of the alpha particle?**

**Strategy:** We are given the mass and the velocity of the alpha particle and asked to calculate the wavelength. We need the de Broglie equation, which is Equation (7.8) of the text. Note that because the units of Planck's constant are J·s, $m$ must be in kg and $u$ must be in m/s ($1 \text{ J} = 1 \text{ kg·m}^2/\text{s}^2$).

**Solution:** Because Planck's constant has units of J·s, and $1 \text{ J} = 1 \text{ kg·m}^2/\text{s}^2$, the mass of the alpha particle must be expressed in kilograms. Since one particle has a mass of 4.0 amu, a mole of alpha particles will have a mass of 4.0 g. A reasonable strategy to complete the conversion is:

$$\text{g/mol} \rightarrow \text{kg/mol} \rightarrow \text{kg/particle}$$

$$\frac{4.0\ \cancel{\text{g}}}{1\ \cancel{\text{mol}}} \times \frac{1\text{ kg}}{1000\ \cancel{\text{g}}} \times \frac{1\ \cancel{\text{mol}}}{6.022 \times 10^{23}\text{ particles}} = 6.6 \times 10^{-27}\text{ kg/particle}$$

Substitute the known quantities into Equation (7.8) to solve for the wavelength.

$$\lambda = \frac{h}{mu} = \frac{(6.63 \times 10^{-34}\ \cancel{\text{J·s}}) \times \left(\frac{1\ \cancel{\text{kg}}\cdot\text{m}^{\cancel{2}}/\text{s}^{\cancel{2}}}{1\ \cancel{\text{J}}}\right)}{(6.6 \times 10^{-27}\ \cancel{\text{kg}})\left(1.4 \times 10^7\ \frac{\cancel{\text{m}}}{\cancel{\text{s}}}\right)} = \mathbf{7.2 \times 10^{-15}\ m}$$

The wavelength is smaller than the diameter of the thorium nucleus, which is about $2 \times 10^{-14}$ m. Is this what you would expect?

**PRACTICE EXERCISE**

5. The average kinetic energy of a neutron at 25°C is $6.2 \times 10^{-21}$ J. What is the de Broglie wavelength of an average neutron? The mass of a neutron is 1.008 amu. (**Hint:** kinetic energy $= \frac{1}{2}mu^2$)

**Text Problems: 7.40**, 7.42

## PROBLEM TYPE 5: QUANTUM NUMBERS

See Section 7.6 of your text for a complete discussion. In quantum mechanics, three quantum numbers are required to describe the distribution of electrons in atoms.

(1) **The principal quantum number ($n$).** In a hydrogen atom, the value of $n$ determines the energy of an orbital. It can have integral values 1, 2, 3, and so forth.

(2) **The angular momentum quantum number ($l$)** tells us the "shape" of the orbitals. $l$ has possible integral values from 0 to $(n - 1)$. The value of $l$ is generally designated by the letters $s, p, d, ...,$ as follows:

| $l$ | 0 | 1 | 2 | 3 | 4 | 5 |
|---|---|---|---|---|---|---|
| **Name of orbital** | $s$ | $p$ | $d$ | $f$ | $g$ | $h$ |

(3) **The magnetic quantum number ($m_l$)** describes the orientation of the orbital in space. For a certain value of $l$, there are $(2l + 1)$ integral values of $m_l$, as follows:

$$-l, (-l + 1), ... 0, ... (+l - 1), +l$$

The number of $m_l$ values indicates the number of orbitals in a subshell with a particular $l$ value.

Finally, there is a fourth quantum number that tells us the spin of the electron. The **electron spin quantum number ($m_s$)** has values of +1/2 or −1/2, which correspond to the two spinning motions of the electron.

### A. Labeling an atomic orbital

**Strategy:** To "label" an atomic orbital, you need to specify the three quantum numbers ($n, l, m_l$) that give information about the distribution of electrons in orbitals. Remember, $m_s$ tells us the spin of the electron, which tells us nothing about the orbital.

**EXAMPLE 7.6**

**List the values of $n$, $l$, and $m_l$ for orbitals in the 2$p$ subshell.**

**Solution:** The number given in the designation of the subshell is the principal quantum number, so in this case $n = 2$. For $p$ orbitals, $l = 1$. $m_l$ can have integer values from $-l$ to $+l$. Therefore, $m_l$ can be **−1, 0,** and **+1**. (The three values for $m_l$ correspond to the three $p$ orbitals.)

**PRACTICE EXERCISE**

6. List the values of $n$, $l$, and $m_l$ for orbitals in the 4$f$ subshell.

**Text Problems: 7.52**, 7.56

### B. Counting the number of orbitals associated with a principal quantum number

**Strategy:** To work this type of problem, you must take into account the energy level ($n$), the types of orbitals in that energy level ($l$), and the number of orbitals in a subshell with a particular $l$ value ($m_l$).

**EXAMPLE 7.7**

**What is the total number of orbitals associated with the principal quantum number $n = 2$?**

**Solution:** For $n = 2$, there are only two possible values of $l$, 0 and 1. Thus, there is one 2*s* orbital, and there are three 2*p* orbitals. (For $l = 1$, there are three possible $m_l$ values, −1, 0, and +1.)

Therefore, the total number of orbitals in the $n = 2$ energy level is $1 + 3 = \mathbf{4}$.

> **Tip:** The total number of orbitals with a given $n$ value is $n^2$. For Example 7.7, the total number of orbitals in the $n = 2$ level equals $2^2 = \mathbf{4}$.

**PRACTICE EXERCISE**

7. What is the total number of orbitals associated with the principal quantum number $n = 4$?

**Text Problem:** 7.56

## C. Assigning quantum numbers to an electron

**Strategy:** In assigning quantum numbers to an electron, you need to specify all four quantum numbers. In most cases, there will be more than one possible set of quantum numbers that can designate an electron.

**EXAMPLE 7.8**

**List the different ways to write the four quantum numbers that designate an electron in a 4*s* orbital.**

**Solution:** To begin with, we know that the principal quantum number $n$ is 4, and the angular momentum quantum number $l$ is 0 (*s* orbital). For $l = 0$, there is only one possible value for $m_l$, also 0. Since the electron spin quantum number $m_s$ can be either +1/2 or −1/2, we conclude that there are two possible ways to designate the electron:

$$(4, 0, 0, +1/2) \qquad (4, 0, 0, -1/2)$$

**PRACTICE EXERCISE**

8. List the different ways to write the four quantum numbers that designate an electron in a 3*d* orbital.

**Text Problems:** 7.52, 7.54

## D. Counting the number of electrons in a principal level

**Strategy:** To work this type of problem, you need to know that the number of orbitals with a particular $l$ value is $(2l + 1)$. Also, each orbital can accommodate two electrons.

**EXAMPLE 7.9**

**What is the maximum number of electrons that can be present in the principal level for which $n = 4$?**

**Solution:** When $n = 4$, $l = 0, 1, 2,$ and 3. The number of orbitals for each $l$ value is given by

| Value of $l$ | Number of orbitals $(2l + 1)$ |
|---|---|
| 0 | 1 |
| 1 | 3 |
| 2 | 5 |
| 3 | 7 |

The total number of orbitals in the principal level $n = 4$ is sixteen. Since each orbital can accommodate two electrons, the maximum number of electrons that can reside in the orbitals is $2 \times 16 = \mathbf{32}$.

> **Tip:** The above result can be generalized by the formula $2n^2$. For Example 7.9, we have $n = 4$, so $2(4)^2 = \mathbf{32}$.

**PRACTICE EXERCISE**

**9.** What is the maximum number of electrons that can be present in the principal level for which $n = 2$?

> **Text Problem:** 7.64

## PROBLEM TYPE 6: WRITING ELECTRON CONFIGURATIONS AND ORBITAL DIAGRAMS

The electron configuration of an atom tells us how the electrons are distributed among the various atomic orbitals. To write electron configurations, you should follow the four rules or guidelines given below.

**(1)** The electron configurations of all elements except hydrogen and helium are represented by a *noble gas core*, which shows (in brackets) the noble gas element that most nearly precedes the element being considered. The noble gas core is followed by the electron configurations of filled or partially filled subshells in the outermost shells.

**(2)** *The Aufbau or "building up" principle* states that electrons are added to atomic orbitals starting with the lowest energy orbital and "building up" to higher energy orbitals.

**(3)** In many-electron atoms, the subshells are filled in the order shown in Figure 7.24 of the text.

**(4)** Each orbital can hold only *two* electrons.

The electron configuration can be represented in a more detailed manner called an ***orbital diagram*** that shows the spin of the electron. For orbital diagrams, you need to follow two additional rules given below.

**(5)** The *Pauli exclusion principle* states that no two electrons in an atom can have the same four quantum numbers. This means that electrons occupying the same orbital *cannot* have the same spin.

**(6)** *Hund's rule* states that the most stable arrangement of electrons in subshells is the one with the greatest number of parallel spins, without violating the Pauli exclusion principle.

**EXAMPLE 7.10**

**Write the ground-state electron configuration and the orbital diagram for selenium.**

**Strategy:** How many electrons are in the Se atom (Z = 34)? We start with $n = 1$ and proceed to fill orbitals in the order shown in Figure 7.23 of the text. Remember that any given orbital can hold at most 2 electrons. However, don't forget about degenerate orbitals. Starting with $n = 2$, there are three $p$ orbitals of equal energy, corresponding to $m_l = -1, 0, 1$. Starting with $n = 3$, there are five $d$ orbitals of equal energy, corresponding to $m_l = -2, -1, 0, 1, 2$. We can place electrons in the orbitals according to the Pauli exclusion principle and Hund's rule. The task is simplified if we use the noble gas core preceding Se for the inner electrons.

**Solution:** Selenium has 34 electrons. The noble gas core in this case is [Ar]. (Ar is the noble gas in the period preceding germanium.) [Ar] represents $1s^22s^22p^63s^23p^6$. This core accounts for 18 electrons, which leaves 16 electrons to place.

See Figure 7.23 of your text to check the order of filling subshells past the Ar noble gas core. You should find that the order of filling is $4s$, $3d$, $4p$. There are 16 remaining electrons to distribute among these orbitals. The $4s$ orbital can hold two electrons. Each of the five $3d$ orbitals can hold two electrons for a total of *10* electrons. This leaves four electrons to place in the $4p$ orbitals.

The electrons configuration for Se is:

$$[Ar]\,4s^2 3d^{10} 4p^4$$

To write an *orbital diagram*, we must also specify the spin of the electrons. The 4*s* and 3*d* orbitals are filled, so according to the Pauli exclusion principle, the paired electrons in the 4*s* orbital and in each of the 3*d* orbitals *must* have opposite spins.

[Ar] ↑↓ (4$s^2$) ↑↓ ↑↓ ↑↓ ↑↓ ↑↓ (3$d^{10}$)

Now, let's deal with the 4*p* electrons. Hund's rule states that the most stable arrangement of electrons in subshells is the one with the greatest number of parallel spins. In other words, we want to keep electrons unpaired if possible with parallel spins. Since there are three *p* orbitals, three of the *p* electrons can be placed individually in each of the *p* subshells with parallel spins.

↑ ↑ ↑ (4$p^3$)

Finally, the fourth *p* electron must be paired up in one of the 4*p* orbitals. The complete orbital diagram is:

[Ar] ↑↓ (4$s^2$) ↑↓ ↑↓ ↑↓ ↑↓ ↑↓ (3$d^{10}$) ↑↓ ↑ ↑ (4$p^4$)

**PRACTICE EXERCISE**

10. Write the electron configuration and the orbital diagram for iron (Fe).

**Text Problems:** 7.76, 7.78, 7.88, **7.90**, 7.92

## ANSWERS TO PRACTICE EXERCISES

1. $\lambda = 0.1224$ m

2. $E = 2.97 \times 10^{-19}$ J
   $E = 1.79 \times 10^5$ J/mol

3. $\lambda = 1.20 \times 10^{-6}$ m $= 1.20 \times 10^3$ nm

4. $n_i = 5$

5. $\lambda = 1.5 \times 10^{-10}$ m

6. $n = 4$, $l = 3$, $m_l = -3, -2, -1, 0, 1, 2, 3$

7. Total number of orbitals $= n^2 = 4^2 = 16$

8. (3, 2, −2, +1/2) (3, 2, −2, −1/2)
   (3, 2, −1, +1/2) (3, 2, −1, −1/2)
   (3, 2, 0, +1/2) (3, 2, 0, −1/2)
   (3, 2, +1, +1/2) (3, 2, +1, −1/2)
   (3, 2, +2, +1/2) (3, 2, +2, −1/2)

9. 8 electrons

10. $[Ar]4s^2 3d^6$

    [Ar] ↑↓ (4$s^2$) ↑↓ ↑ ↑ ↑ ↑ (3$d^6$)

## SOLUTIONS TO SELECTED TEXT PROBLEMS

**7.8** (a)

**Strategy:** We are given the wavelength of an electromagnetic wave and asked to calculate the frequency. Rearranging Equation (7.1) of the text and replacing *u* with *c* (the speed of light) gives:

$$\nu = \frac{c}{\lambda}$$

**Solution:** Because the speed of light is given in meters per second, it is convenient to first convert wavelength to units of meters. Recall that 1 nm = $1 \times 10^{-9}$ m (see Table 1.3 of the text). We write:

$$456 \text{ nm} \times \frac{1 \times 10^{-9}\text{ m}}{1 \text{ nm}} = 456 \times 10^{-9}\text{ m} = 4.56 \times 10^{-7}\text{ m}$$

Substituting in the wavelength and the speed of light ($3.00 \times 10^8$ m/s), the frequency is:

$$\nu = \frac{c}{\lambda} = \frac{3.00 \times 10^8 \frac{\text{m}}{\text{s}}}{4.56 \times 10^{-7}\text{ m}} = \mathbf{6.58 \times 10^{14}\ s^{-1}} \text{ or } \mathbf{6.58 \times 10^{14}\ Hz}$$

**Check:** The answer shows that $6.58 \times 10^{14}$ waves pass a fixed point every second. This very high frequency is in accordance with the very high speed of light.

(b)

**Strategy:** We are given the frequency of an electromagnetic wave and asked to calculate the wavelength. Rearranging Equation (7.1) of the text and replacing *u* with *c* (the speed of light) gives:

$$\lambda = \frac{c}{\nu}$$

**Solution:** Substituting in the frequency and the speed of light ($3.00 \times 10^8$ m/s) into the above equation, the wavelength is:

$$\lambda = \frac{c}{\nu} = \frac{3.00 \times 10^8 \frac{\text{m}}{\text{s}}}{2.45 \times 10^9 \frac{1}{\text{s}}} = 0.122 \text{ m}$$

The problem asks for the wavelength in units of nanometers. Recall that 1 nm = $1 \times 10^{-9}$ m.

$$\lambda = 0.122 \text{ m} \times \frac{1 \text{ nm}}{1 \times 10^{-9}\text{ m}} = \mathbf{1.22 \times 10^8\ nm}$$

**7.10** A radio wave is an electromagnetic wave, which travels at the speed of light. The speed of light is in units of m/s, so let's convert distance from units of miles to meters. (28 million mi = $2.8 \times 10^7$ mi)

$$? \text{ distance (m)} = (2.8 \times 10^7 \text{ mi}) \times \frac{1.61 \text{ km}}{1 \text{ mi}} \times \frac{1000 \text{ m}}{1 \text{ km}} = 4.5 \times 10^{10} \text{ m}$$

Now, we can use the speed of light as a conversion factor to convert from meters to seconds ($c = 3.00 \times 10^8$ m/s).

$$\textbf{? min} = (4.5 \times 10^{10} \text{ m}) \times \frac{1 \text{ s}}{3.00 \times 10^8 \text{ m}} = 1.5 \times 10^2 \text{ s} = \mathbf{2.5 \text{ min}}$$

**7.12** The wavelength is:

$$\lambda = \frac{1 \text{ m}}{1{,}650{,}763.73 \text{ wavelengths}} = 6.05780211 \times 10^{-7} \text{ m}$$

$$\nu = \frac{c}{\lambda} = \frac{3.00 \times 10^8 \text{ m/s}}{6.05780211 \times 10^{-7} \text{ m}} = \mathbf{4.95 \times 10^{14} \text{ s}^{-1}}$$

**7.16** **(a)**

**Strategy:** We are given the frequency of an electromagnetic wave and asked to calculate the wavelength. Rearranging Equation (7.1) of the text and replacing $u$ with $c$ (the speed of light) gives:

$$\lambda = \frac{c}{\nu}$$

**Solution:** Substituting in the frequency and the speed of light ($3.00 \times 10^8$ m/s) into the above equation, the wavelength is:

$$\lambda = \frac{3.00 \times 10^8 \frac{\text{m}}{\text{s}}}{7.5 \times 10^{14} \frac{1}{\text{s}}} = 4.0 \times 10^{-7} \text{ m} = \mathbf{4.0 \times 10^2 \text{ nm}}$$

**Check:** The wavelength of 400 nm calculated is in the blue region of the visible spectrum as expected.

**(b)**

**Strategy:** We are given the frequency of an electromagnetic wave and asked to calculate its energy. Equation (7.2) of the text relates the energy and frequency of an electromagnetic wave.

$$E = h\nu$$

**Solution:** Substituting in the frequency and Planck's constant ($6.63 \times 10^{-34}$ J·s) into the above equation, the energy of a single photon associated with this frequency is:

$$E = h\nu = (6.63 \times 10^{-34}\ \text{J}\cdot\cancel{\text{s}})\left(7.5 \times 10^{14}\ \frac{1}{\cancel{\text{s}}}\right) = \mathbf{5.0 \times 10^{-19}\ J}$$

**Check:** We expect the energy of a single photon to be a very small energy as calculated above, $5.0 \times 10^{-19}$ J.

**7.18** The energy given in this problem is for *1 mole* of photons. To apply $E = h\nu$, we must divide the energy by Avogadro's number. The energy of one photon is:

$$E = \frac{1.0 \times 10^3\ \cancel{\text{kJ}}}{1\ \cancel{\text{mol}}} \times \frac{1\ \cancel{\text{mol}}}{6.022 \times 10^{23}\ \text{photons}} \times \frac{1000\ \text{J}}{1\ \cancel{\text{kJ}}} = 1.7 \times 10^{-18}\ \text{J/photon}$$

The wavelength of this photon can be found using the relationship, $E = \dfrac{hc}{\lambda}$.

$$\lambda = \frac{hc}{E} = \frac{(6.63 \times 10^{-34}\ \cancel{\text{J}}\cdot\cancel{\text{s}})\left(3.00 \times 10^8\ \frac{\text{m}}{\cancel{\text{s}}}\right)}{1.7 \times 10^{-18}\ \cancel{\text{J}}} = 1.2 \times 10^{-7}\ \cancel{\text{m}} \times \frac{1\ \text{nm}}{1 \times 10^{-9}\ \cancel{\text{m}}} = \mathbf{1.2 \times 10^2\ nm}$$

The radiation is in the **ultraviolet** region (see Figure 7.4 of the text).

**7.20** (a) $\lambda = \dfrac{c}{\nu}$

$$\lambda = \frac{3.00 \times 10^8\ \frac{\text{m}}{\cancel{\text{s}}}}{8.11 \times 10^{14}\ \frac{1}{\cancel{\text{s}}}} = 3.70 \times 10^{-7}\ \text{m} = \mathbf{3.70 \times 10^2\ nm}$$

(b) Checking Figure 7.4 of the text, you should find that the visible region of the spectrum runs from 400 to 700 nm. 370 nm is in the **ultraviolet** region of the spectrum.

(c) $E = h\nu$. Substitute the frequency ($\nu$) into this equation to solve for the energy of one quantum associated with this frequency.

$$E = h\nu = (6.63 \times 10^{-34}\ \text{J}\cdot\cancel{\text{s}})\left(8.11 \times 10^{14}\ \frac{1}{\cancel{\text{s}}}\right) = \mathbf{5.38 \times 10^{-19}\ J}$$

**7.22** The mathematical equation for studying the photoelectric effect is

$$h\nu = W + \text{KE}$$

where $\nu$ is the frequency of light shining on the metal, $W$ is the work function, and KE is the kinetic energy of the ejected electron. We substitute h, $\nu$, and KE into the equation to solve for the work function, $W$.

$$h\nu = W + \text{KE}$$

$$(6.63 \times 10^{-34}\ \text{J}\cdot\cancel{\text{s}})(2.11 \times 10^{15}\ \cancel{\text{s}^{-1}}) = W + (5.83 \times 10^{-19}\ \text{J})$$

$$\mathbf{W = 8.16 \times 10^{-19}\ J}$$

**7.26** The emitted light could be analyzed by passing it through a prism.

**7.28** Excited atoms of the chemical elements emit the same characteristic frequencies or lines in a terrestrial laboratory, in the sun, or in a star many light-years distant from earth.

**7.30** We use more accurate values of $h$ and $c$ for this problem.

$$E = \frac{hc}{\lambda} = \frac{(6.6256 \times 10^{-34}\ \text{J}\cdot\cancel{\text{s}})(2.998 \times 10^{8}\ \cancel{\text{m/s}})}{656.3 \times 10^{-9}\ \cancel{\text{m}}} = \mathbf{3.027 \times 10^{-19}\ J}$$

**7.32** **Strategy:** We are given the initial and final states in the emission process. We can calculate the energy of the emitted photon using Equation (7.6) of the text. Then, from this energy, we can solve for the frequency of the photon, and from the frequency we can solve for the wavelength. The value of Rydberg's constant is $2.18 \times 10^{-18}$ J.

**Solution:** From Equation (7.6) we write:

$$\Delta E = R_{\text{H}}\left(\frac{1}{n_{\text{i}}^2} - \frac{1}{n_{\text{f}}^2}\right)$$

$$\Delta E = (2.18 \times 10^{-18}\ \text{J})\left(\frac{1}{4^2} - \frac{1}{2^2}\right)$$

$$\Delta E = -4.09 \times 10^{-19}\ \text{J}$$

The negative sign for $\Delta E$ indicates that this is energy associated with an emission process. To calculate the frequency, we will omit the minus sign for $\Delta E$ because the frequency of the photon must be positive. We know that

$$\Delta E = h\nu$$

Rearranging the equation and substituting in the known values,

$$\nu = \frac{\Delta E}{h} = \frac{(4.09 \times 10^{-19}\ \cancel{\text{J}})}{(6.63 \times 10^{-34}\ \cancel{\text{J}}\cdot\text{s})} = \mathbf{6.17 \times 10^{14}\ s^{-1}}\ \text{or}\ \mathbf{6.17 \times 10^{14}\ Hz}$$

We also know that $\lambda = \dfrac{c}{\nu}$. Substituting the frequency calculated above into this equation gives:

$$\lambda = \frac{3.00\times 10^{8}\ \frac{\text{m}}{\cancel{\text{s}}}}{\left(6.17\times 10^{14}\ \frac{1}{\cancel{\text{s}}}\right)} = 4.86\times 10^{-7}\ \text{m} = \mathbf{486\ nm}$$

**Check:** This wavelength is in the visible region of the electromagnetic region (see Figure 7.4 of the text). This is consistent with the fact that because $n_i = 4$ and $n_f = 2$, this transition gives rise to a spectral line in the Balmer series (see Figure 7.6 of the text).

**7.34** $$\Delta E = R_H\left(\frac{1}{n_i^2} - \frac{1}{n_f^2}\right)$$

$n_f$ is given in the problem and $R_H$ is a constant, but we need to calculate $\Delta E$. The photon energy is:

$$E = \frac{hc}{\lambda} = \frac{(6.63\times 10^{-34}\ \text{J}\cdot\cancel{\text{s}})(3.00\times 10^{8}\ \cancel{\text{m/s}})}{434\times 10^{-9}\ \cancel{\text{m}}} = 4.58\times 10^{-19}\ \text{J}$$

Since this is an emission process, the energy change $\Delta E$ must be negative, or $-4.58\times 10^{-19}$ J.

Substitute $\Delta E$ into the following equation, and solve for $n_i$.

$$\Delta E = R_H\left(\frac{1}{n_i^2} - \frac{1}{n_f^2}\right)$$

$$-4.58\times 10^{-19}\ \text{J} = (2.18\times 10^{-18}\ \text{J})\left(\frac{1}{n_i^2} - \frac{1}{2^2}\right)$$

$$\frac{1}{n_i^2} = \left(\frac{-4.58\times 10^{-19}\ \cancel{\text{J}}}{2.18\times 10^{-18}\ \cancel{\text{J}}}\right) + \frac{1}{2^2} = -0.210 + 0.250 = 0.040$$

$$n_i = \frac{1}{\sqrt{0.040}} = \mathbf{5}$$

**7.40** **Strategy:** We are given the mass and the speed of the proton and asked to calculate the wavelength. We need the de Broglie equation, which is Equation (7.8) of the text. Note that because the units of Planck's constant are J·s, $m$ must be in kg and $u$ must be in m/s (1 J = 1 kg·m$^2$/s$^2$).

**Solution:** Using Equation (7.8) we write:

$$\lambda = \frac{h}{mu}$$

$$\lambda = \frac{h}{mu} = \frac{\left(6.63 \times 10^{-34}\ \frac{\text{kg}\cdot\text{m}^2}{\text{s}^2}\cdot\text{s}\right)}{(1.673 \times 10^{-27}\ \text{kg})(2.90 \times 10^{8}\ \text{m/s})} = 1.37 \times 10^{-15}\ \text{m}$$

The problem asks to express the wavelength in nanometers.

$$\lambda = (1.37 \times 10^{-15}\ \text{m}) \times \frac{1\ \text{nm}}{1 \times 10^{-9}\ \text{m}} = \mathbf{1.37 \times 10^{-6}\ nm}$$

**7.42** First, we convert mph to m/s.

$$\frac{35\ \text{mi}}{1\ \text{h}} \times \frac{1.61\ \text{km}}{1\ \text{mi}} \times \frac{1000\ \text{m}}{1\ \text{km}} \times \frac{1\ \text{h}}{3600\ \text{s}} = 16\ \text{m/s}$$

$$\lambda = \frac{h}{mu} = \frac{\left(6.63 \times 10^{-34}\ \frac{\text{kg}\cdot\text{m}^2}{\text{s}^2}\cdot\text{s}\right)}{(2.5 \times 10^{-3}\ \text{kg})(16\ \text{m/s})} = 1.7 \times 10^{-32}\ \text{m} = \mathbf{1.7 \times 10^{-23}\ nm}$$

**7.52** **Strategy:** What are the relationships among $n$, $l$, and $m_l$?

**Solution:** We are given the principal quantum number, $n = 3$. The possible $l$ values range from 0 to $(n - 1)$. Thus, there are three possible values of $l$: 0, 1, and 2, corresponding to the $s$, $p$, and $d$ orbitals, respectively. The values of $m_l$ can vary from $-l$ to $l$. The values of $m_l$ for each $l$ value are:

$l = 0$: $m_l = 0$  $l = 1$: $m_l = -1, 0, 1$  $l = 2$: $m_l = -2, -1, 0, 1, 2$

**7.54** **(a)** The number given in the designation of the subshell is the principal quantum number, so in this case $\boldsymbol{n = 3}$. For $s$ orbitals, $\boldsymbol{l = 0}$. $m_l$ can have integer values from $-l$ to $+l$, therefore, $\boldsymbol{m_l = 0}$. The electron spin quantum number, $\boldsymbol{m_s}$, can be either **+1/2** or **−1/2**.

Following the same reasoning as part **(a)**

**(b)** $4p$: $n = 4$; $l = 1$; $m_l = -1, 0, 1$; $m_s = +1/2, -1/2$

**(c)** $3d$: $n = 3$; $l = 2$; $m_l = -2, -1, 0, 1, 2$; $m_s = +1/2, -1/2$

**7.56** For $n = 6$, the allowed values of $l$ are 0, 1, 2, 3, 4, and 5 [$l = 0$ to $(n - 1)$, integer values]. These $l$ values correspond to the $6s$, $6p$, $6d$, $6f$, $6g$, and $6h$ subshells. These subshells each have 1, 3, 5, 7, 9, and 11 orbitals, respectively (number of orbitals $= 2l + 1$).

**7.62** The two orbitals are identical in size, shape, and energy. They differ only in their orientation with respect to each other.

Can you assign a specific value of the magnetic quantum number to these orbitals? What are the allowed values of the magnetic quantum number for the 2*p* subshell?

**7.64**

| *n* value | orbital sum | total number of electrons |
|---|---|---|
| 1 | 1 | 2 |
| 2 | $1 + 3 = 4$ | 8 |
| 3 | $1 + 3 + 5 = 9$ | 18 |
| 4 | $1 + 3 + 5 + 7 = 16$ | 32 |
| 5 | $1 + 3 + 5 + 7 + 9 = 25$ | 50 |
| 6 | $1 + 3 + 5 + 7 + 9 + 11 = 36$ | 72 |

In each case the total number of orbitals is just the square of the *n* value ($n^2$). The total number of electrons is $2n^2$.

**7.66** The electron configurations for the elements are

**(a)** N: $1s^22s^22p^3$ There are three *p*-type electrons.

**(b)** Si: $1s^22s^22p^63s^23p^2$ There are six *s*-type electrons.

**(c)** S: $1s^22s^22p^63s^23p^4$ There are no *d*-type electrons.

**7.68** In the many-electron atom, the 3*p* orbital electrons are more effectively shielded by the inner electrons of the atom (that is, the 1*s*, 2*s*, and 2*p* electrons) than the 3*s* electrons. The 3*s* orbital is said to be more "penetrating" than the 3*p* and 3*d* orbitals. In the hydrogen atom there is only one electron, so the 3*s*, 3*p*, and 3*d* orbitals have the same energy.

**7.70** **(a)** $2s < 2p$ **(b)** $3p < 3d$ **(c)** $3s < 4s$ **(d)** $4d < 5f$

**7.76** For aluminum, there are not enough electrons in the 2*p* subshell. (The 2*p* subshell holds six electrons.) The number of electrons (13) is correct. The electron configuration should be $1s^22s^22p^63s^23p^1$. The configuration shown might be an excited state of an aluminum atom.

For boron, there are too many electrons. (Boron only has five electrons.) The electron configuration should be $1s^22s^22p^1$. What would be the electric charge of a boron ion with the electron arrangement given in the problem?

For fluorine, there are also too many electrons. (Fluorine only has nine electrons.) The configuration shown is that of the $F^-$ ion. The correct electron configuration is $1s^22s^22p^5$.

**7.78** You should write the electron configurations for each of these elements to answer this question. In some cases, an orbital diagram may be helpful.

B: $[He]2s^22p^1$ (1 unpaired electron)
P: $[Ne]3s^23p^3$ (3 unpaired electrons)
Mn: $[Ar]4s^23d^5$ (5 unpaired electrons)
Kr: (0 unpaired electrons)
Cd: $[Kr]5s^24d^{10}$ (0 unpaired electrons)
Pb: $[Xe]6s^24f^{14}5d^{10}6p^2$ (2 unpaired electrons)
Ne: (0 unpaired electrons, Why?)
Sc: $[Ar]4s^23d^1$ (1 unpaired electron)
Se: $[Ar]4s^23d^{10}4p^4$ (2 unpaired electrons)
Fe: $[Ar]4s^23d^6$ (4 unpaired electrons)
I: $[Kr]5s^24d^{10}5p^5$ (1 unpaired electron)

**7.88** The ground state electron configuration of Tc is: $[Kr]5s^24d^5$.

**7.90** **Strategy:** How many electrons are in the Ge atom (Z = 32)? We start with $n = 1$ and proceed to fill orbitals in the order shown in Figure 7.23 of the text. Remember that any given orbital can hold at most 2 electrons. However, don't forget about degenerate orbitals. Starting with $n = 2$, there are three *p* orbitals of equal energy, corresponding to $m_l = -1, 0, 1$. Starting with $n = 3$, there are five *d* orbitals of equal energy, corresponding to $m_l = -2, -1, 0, 1, 2$. We can place electrons in the orbitals according to the Pauli exclusion principle and Hund's rule. The task is simplified if we use the noble gas core preceding Ge for the inner electrons.

**Solution:** Germanium has 32 electrons. The noble gas core in this case is [Ar]. (Ar is the noble gas in the period preceding germanium.) [Ar] represents $1s^22s^22p^63s^23p^6$. This core accounts for 18 electrons, which leaves 14 electrons to place.

See Figure 7.23 of your text to check the order of filling subshells past the Ar noble gas core. You should find that the order of filling is 4*s*, 3*d*, 4*p*. There are 14 remaining electrons to distribute among these orbitals. The 4*s* orbital can hold two electrons. Each of the five 3*d* orbitals can hold two electrons for a total of *10* electrons. This leaves two electrons to place in the 4*p* orbitals.

The electrons configuration for Ge is:

$$[Ar]4s^23d^{10}4p^2$$

You should follow the same reasoning for the remaining atoms.

Fe: $[Ar]4s^23d^6$
Zn: $[Ar]4s^23d^{10}$
Ni: $[Ar]4s^23d^8$
W: $[Xe]6s^24f^{14}5d^4$
Tl: $[Xe]6s^24f^{14}5d^{10}6p^1$

**7.92** ↑↓ ↑ ↑ ↑ ↑↓ ↑↓ ↑ ↑ ↑↓ ↑↓ ↑↓ ↑

$3s^2$ $3p^3$ $3s^2$ $3p^4$ $3s^2$ $3p^5$

$S^+$ (5 valence electrons) | S (6 valence electrons) | $S^-$ (7 valence electrons)

3 unpaired electrons | 2 unpaired electrons | 1 unpaired electron

**$S^+$ has the most unpaired electrons**

**7.94** First, we can calculate the energy of a single photon with a wavelength of 532 nm.

$$E = \frac{hc}{\lambda} = \frac{(6.63\times10^{-34}\ \text{J}\cdot\cancel{\text{s}})(3.00\times10^{8}\ \cancel{\text{m/s}})}{532\times10^{-9}\ \cancel{\text{m}}} = 3.74\times10^{-19}\ \text{J}$$

With a power output of 25.0 mW, the energy of the laser pulse per second is 0.0250 J. The number of photons produced in a 0.0250 J pulse is:

$$0.0250\ \cancel{\text{J}} \times \frac{1\ \text{photon}}{3.74\times10^{-19}\ \cancel{\text{J}}} = \mathbf{6.68\times10^{16}\ photons}$$

**7.96** Part **(b)** is correct in the view of contemporary quantum theory. Bohr's explanation of emission and absorption line spectra appears to have universal validity. Parts **(a)** and **(c)** are artifacts of Bohr's early planetary model of the hydrogen atom and are *not* considered to be valid today.

**7.98** **(a)** With $n = 2$, there are $n^2$ orbitals $= 2^2 = 4$. $m_s = +1/2$, specifies 1 electron per orbital, for a total of **4 electrons.**

**(b)** $n = 4$ and $m_l = +1$, specifies one orbital in each subshell with $l = 1, 2,$ or $3$ (i.e., a $4p$, $4d$, and $4f$ orbital). Each of the three orbitals holds 2 electrons for a total of **6 electrons.**

**(c)** If $n = 3$ and $l = 2$, $m_l$ has the values 2, 1, 0, −1, or −2. Each of the five orbitals can hold 2 electrons for a total of **10 electrons** (2 $e^-$ in each of the five $3d$ orbitals).

**(d)** If $n = 2$ and $l = 0$, then $m_l$ can only be zero. $m_s = -1/2$ specifies 1 electron in this orbital for a total of **1 electron** (one $e^-$ in the $2s$ orbital).

**(e)** $n = 4$, $l = 3$ and $m_l = -2$, specifies one $4f$ orbital. This orbital can hold **2 electrons.**

**7.100** The wave properties of electrons are used in the operation of an electron microscope.

**7.102** (a) First convert 100 mph to units of m/s.

$$\frac{100\text{ mi}}{1\text{ h}} \times \frac{1\text{ h}}{3600\text{ s}} \times \frac{1.609\text{ km}}{1\text{ mi}} \times \frac{1000\text{ m}}{1\text{ km}} = 44.7\text{ m/s}$$

Using the de Broglie equation:

$$\lambda = \frac{h}{mu} = \frac{\left(6.63 \times 10^{-34}\ \frac{\text{kg}\cdot\text{m}^2}{\text{s}^2}\cdot\text{s}\right)}{(0.141\text{ kg})(44.7\text{ m/s})} = 1.05 \times 10^{-34}\text{ m} = \mathbf{1.05 \times 10^{-25}\ nm}$$

(b) The average mass of a hydrogen atom is:

$$\frac{1.008\text{ g}}{1\text{ mol}} \times \frac{1\text{ mol}}{6.022 \times 10^{23}\text{ atoms}} = 1.674 \times 10^{-24}\text{ g/H atom} = 1.674 \times 10^{-27}\text{ kg}$$

$$\lambda = \frac{h}{mu} = \frac{\left(6.63 \times 10^{-34}\ \frac{\text{kg}\cdot\text{m}^2}{\text{s}^2}\cdot\text{s}\right)}{(1.674 \times 10^{-27}\text{ kg})(44.7\text{ m/s})} = 8.86 \times 10^{-9}\text{ m} = \mathbf{8.86\ nm}$$

**7.104** (a) $n$ **= 2**. The possible $l$ values are from 0 up to $(n - 1)$, integer values.

(b) Possible $l$ values are **0, 1, 2,** or **3**. Possible $l$ values are from 0 up to $(n - 1)$, integer values, and possible $m_l$ values range from $-l$ to $+l$, integer values.

**7.106** (a) First, we can calculate the energy of a single photon with a wavelength of 633 nm.

$$E = \frac{hc}{\lambda} = \frac{(6.63 \times 10^{-34}\text{ J}\cdot\text{s})(3.00 \times 10^{8}\text{ m/s})}{633 \times 10^{-9}\text{ m}} = 3.14 \times 10^{-19}\text{ J}$$

The number of photons produced in a 0.376 J pulse is:

$$0.376\text{ J} \times \frac{1\text{ photon}}{3.14 \times 10^{-19}\text{ J}} = \mathbf{1.20 \times 10^{18}\ photons}$$

(b) Since a 1 W = 1 J/s, the power delivered per a 1.00 × $10^{-9}$ s pulse is:

$$\frac{0.376\text{ J}}{1.00 \times 10^{-9}\text{ s}} = 3.76 \times 10^{8}\text{ J/s} = \mathbf{3.76 \times 10^{8}\ W}$$

Compare this with the power delivered by a 100-W light bulb!

**7.108** First, let's find the energy needed to photodissociate one water molecule.

$$\frac{285.8\text{ kJ}}{1\text{ mol}} \times \frac{1\text{ mol}}{6.022 \times 10^{23}\text{ molecules}} = 4.746 \times 10^{-22}\text{ kJ/molecule} = 4.746 \times 10^{-19}\text{ J/molecule}$$

The maximum wavelength of a photon that would provide the above energy is:

$$\lambda = \frac{hc}{E} = \frac{(6.63 \times 10^{-34}\text{ J}\cdot\text{s})(3.00 \times 10^{8}\text{ m/s})}{4.746 \times 10^{-19}\text{ J}} = \mathbf{4.19 \times 10^{-7}\text{ m} = 419\text{ nm}}$$

This wavelength is in the visible region of the electromagnetic spectrum. In principle water could be dissociated into hydrogen and oxygen in this manner, but water is continuously being struck by visible radiation *without* decomposition, so the process would require a catalyst to be of practical value.

**7.110** We can rearrange the de Broglie equation to solve for velocity.

$$u = \frac{h}{m\lambda}$$

We also know that:

$$u_{\text{rms}} = \sqrt{\frac{3RT}{M}} = \sqrt{\frac{3RT}{mN_A}}$$

Note that the molar mass ($M$) equals the mass of the atom multiplied by Avogadro's number. We set the two equations equal to each other.

$$\frac{h}{m\lambda} = \sqrt{\frac{3RT}{mN_A}}$$

$$\frac{h^2}{m^2\lambda^2} = \frac{3RT}{mN_A}$$

Solving for $m$ and substituting $\text{kg}\cdot\text{m}^2/\text{s}^2$ for J gives:

$$m = \frac{h^2 N_A}{3RT\lambda^2} = \frac{\left(6.626 \times 10^{-34}\ \frac{\text{kg}\cdot\text{m}^2}{\text{s}^2}\cdot\text{s}\right)^2 (6.022 \times 10^{23}\text{ atoms/mol})}{3\left(8.314\ \frac{\text{kg}\cdot\text{m}^2}{\text{s}^2\cdot\text{mol}\cdot\text{K}}\right)(293\text{ K})(3.28 \times 10^{-11}\text{ m})^2}$$

$$m = 3.36 \times 10^{-26}\text{ kg/atom}$$

Converting to atomic mass units we find:

$$\frac{3.36 \times 10^{-26}\text{ kg}}{1\text{ atom}} \times \frac{1\text{ amu}}{1.661 \times 10^{-27}\text{ kg}} = 20.2\text{ amu}$$

The atom is **neon (Ne)**.

**7.112** The Balmer series corresponds to transitions to the $n = 2$ level.

For $He^+$:

$$\Delta E = R_{He^+}\left(\frac{1}{n_i^2} - \frac{1}{n_f^2}\right) \qquad \lambda = \frac{hc}{\Delta E} = \frac{(6.63 \times 10^{-34}\ J \cdot s)(3.00 \times 10^8\ m/s)}{\Delta E}$$

For the transition, $n = 3 \rightarrow 2$

$$\Delta E = (8.72 \times 10^{-18}\ J)\left(\frac{1}{3^2} - \frac{1}{2^2}\right) = -1.21 \times 10^{-18}\ J \qquad \lambda = \frac{1.99 \times 10^{-25}\ \cancel{J} \cdot m}{1.21 \times 10^{-18}\ \cancel{J}} = 1.64 \times 10^{-7}\ m = \mathbf{164\ nm}$$

For the transition, $n = 4 \rightarrow 2$, $\Delta E = -1.64 \times 10^{-18}$ J $\lambda =$ **121 nm**

For the transition, $n = 5 \rightarrow 2$, $\Delta E = -1.83 \times 10^{-18}$ J $\lambda =$ **109 nm**

For the transition, $n = 6 \rightarrow 2$, $\Delta E = -1.94 \times 10^{-18}$ J $\lambda =$ **103 nm**

For H, the calculations are identical to those above, except the Rydberg constant for H is $2.18 \times 10^{-18}$ J.

For the transition, $n = 3 \rightarrow 2$, $\Delta E = -3.03 \times 10^{-19}$ J $\lambda =$ **657 nm**

For the transition, $n = 4 \rightarrow 2$, $\Delta E = -4.09 \times 10^{-19}$ J $\lambda =$ **487 nm**

For the transition, $n = 5 \rightarrow 2$, $\Delta E = -4.58 \times 10^{-19}$ J $\lambda =$ **434 nm**

For the transition, $n = 6 \rightarrow 2$, $\Delta E = -4.84 \times 10^{-19}$ J $\lambda =$ **411 nm**

All the Balmer transitions for $He^+$ are in the ultraviolet region; whereas, the transitions for H are all in the visible region. Note the negative sign for energy indicating that a photon has been emitted.

**7.114** First, we need to calculate the energy of one 600 nm photon. Then, we can determine how many photons are needed to provide $4.0 \times 10^{-17}$ J of energy.

The energy of one 600 nm photon is:

$$E = \frac{hc}{\lambda} = \frac{(6.63 \times 10^{-34}\ J \cdot \cancel{s})(3.00 \times 10^8\ \cancel{m/s})}{600 \times 10^{-9}\ \cancel{m}} = 3.32 \times 10^{-19}\ J/photon$$

The number of photons needed to produce $4.0 \times 10^{-17}$ J of energy is:

$$(4.0 \times 10^{-17}\ \cancel{J}) \times \frac{1\ photon}{3.32 \times 10^{-19}\ \cancel{J}} = \mathbf{1.2 \times 10^2\ photons}$$

**7.116** The energy emitted in the 0.060 s pulse is:

$$1.6\ \cancel{W} \times \frac{1\ J/\cancel{s}}{1\ \cancel{W}} \times 0.060\ \cancel{s} = 0.096\ J$$

Next, we calculate the energy of one photon with a wavelength of 514 nm.

$$E = \frac{hc}{\lambda} = \frac{(6.63 \times 10^{-34}\ J\cdot \cancel{s})(3.00 \times 10^{8}\ \cancel{m}/\cancel{s})}{514 \times 10^{-9}\ \cancel{m}} = 3.87 \times 10^{-19}\ J$$

The number of photons emitted by the laser in the 0.060 s pulse is:

$$0.096\ \cancel{J} \times \frac{1\ \text{photon}}{3.87 \times 10^{-19}\ \cancel{J}} = \mathbf{2.5 \times 10^{17}\ photons}$$

**7.118** A "blue" photon (shorter wavelength) is higher energy than a "yellow" photon. For the same amount of energy delivered to the metal surface, there must be fewer "blue" photons than "yellow" photons. Thus, the yellow light would eject more electrons since there are more "yellow" photons. Since the "blue" photons are of higher energy, blue light will eject electrons with greater kinetic energy.

**7.120** The excited atoms are still neutral, so the total number of electrons is the same as the atomic number of the element.

**(a)** He (2 electrons), $1s^2$

**(b)** N (7 electrons), $1s^2 2s^2 2p^3$

**(c)** Na (11 electrons), $1s^2 2s^2 2p^6 3s^1$

**(d)** As (33 electrons), $[Ar]4s^2 3d^{10} 4p^3$

**(e)** Cl (17 electrons), $[Ne]3s^2 3p^5$

**7.122** Rutherford and his coworkers might have discovered the wave properties of electrons.

**7.124** The wavelength of a He atom can be calculated using the de Broglie equation. First, we need to calculate the root-mean-square speed using Equation (5.16) from the text.

$$u_{rms} = \sqrt{\frac{3\left(8.314\frac{J}{K\cdot mol}\right)(273 + 20)K}{4.003 \times 10^{-3}\ kg/mol}} = 1.35 \times 10^3\ m/s$$

To calculate the wavelength, we also need the mass of a He atom in kg.

$$\frac{4.003 \times 10^{-3}\ kg\ He}{1\ \cancel{mol}\ He} \times \frac{1\ \cancel{mol}\ He}{6.022 \times 10^{23}\ He\ atoms} = 6.647 \times 10^{-27}\ kg/atom$$

Finally, the wavelength of a He atom is:

$$\lambda = \frac{h}{mu} = \frac{(6.63 \times 10^{-34}\ J\cdot s)}{(6.647 \times 10^{-27}\ kg)(1.35 \times 10^3\ m/s)} = \mathbf{7.39 \times 10^{-11}\ m = 7.39 \times 10^{-2}\ nm}$$

**7.126** (a) **False.** $n = 2$ is the first excited state.

(b) **False.** In the $n = 4$ state, the electron is (on average) further from the nucleus and hence easier to remove.

(c) **True.**

(d) **False.** The $n = 4$ to $n = 1$ transition is a higher energy transition, which corresponds to a *shorter* wavelength.

(e) **True.**

**7.128** We use Heisenberg's uncertainty principle with the equality sign to calculate the minimum uncertainty.

$$\Delta x \Delta p = \frac{h}{4\pi}$$

The momentum ($p$) is equal to the mass times the velocity.

$$p = mu \quad \text{or} \quad \Delta p = m\Delta u$$

We can write:

$$\Delta p = m\Delta u = \frac{h}{4\pi\Delta x}$$

Finally, the uncertainty in the velocity of the oxygen molecule is:

$$\Delta u = \frac{h}{4\pi m\Delta x} = \frac{(6.63 \times 10^{-34}\ \text{J}\cdot\text{s})}{4\pi(5.3 \times 10^{-26}\ \text{kg})(5.0 \times 10^{-5}\ \text{m})} = \mathbf{2.0 \times 10^{-5}\ m/s}$$

**7.130** The Pauli exclusion principle states that no two electrons in an atom can have the same four quantum numbers. In other words, only two electrons may exist in the same atomic orbital, and these electrons must have opposite spins. **(a)** and **(f)** violate the Pauli exclusion principle.
Hund's rule states that the most stable arrangement of electrons in subshells is the one with the greatest number of parallel spins. **(b)**, **(d)**, and **(e)** violate Hund's rule.

**7.132** As an estimate, we can equate the energy for ionization ($Fe^{13+} \rightarrow Fe^{14+}$) to the average kinetic energy $\left(\frac{3}{2}RT\right)$ of the ions.

$$\frac{3.5 \times 10^4\ \text{kJ}}{1\ \text{mol}} \times \frac{1000\ \text{J}}{1\ \text{kJ}} = 3.5 \times 10^7\ \text{J}$$

$$IE = \frac{3}{2}RT$$

$$3.5 \times 10^7\ \text{J/mol} = \frac{3}{2}(8.314\ \text{J/mol}\cdot\text{K})T$$

$$\mathbf{T = 2.8 \times 10^6\ K}$$

The actual temperature can be, and most probably is, higher than this.

**7.134** Looking at the de Broglie equation $\lambda = \frac{h}{mu}$, the mass of an $N_2$ molecule (in kg) and the velocity of an $N_2$ molecule (in m/s) is needed to calculate the de Broglie wavelength of $N_2$.

First, calculate the root-mean-square velocity of $N_2$.

$\mathcal{M}(N_2) = 28.02 \text{ g/mol} = 0.02802 \text{ kg/mol}$

$$u_{\text{rms}}(N_2) = \sqrt{\frac{(3)\left(8.314\frac{\text{J}}{\text{mol}\cdot\text{K}}\right)(300\text{ K})}{\left(0.02802\frac{\text{kg}}{\text{mol}}\right)}} = 516.8 \text{ m/s}$$

Second, calculate the mass of one $N_2$ molecule in kilograms.

$$\frac{28.02\text{ g }N_2}{1\text{ mol }N_2} \times \frac{1\text{ mol }N_2}{6.022\times10^{23}\text{ }N_2\text{ molecules}} \times \frac{1\text{ kg}}{1000\text{ g}} = 4.653\times10^{-26}\text{ kg/molecule}$$

Now, substitute the mass of an $N_2$ molecule and the root-mean-square velocity into the de Broglie equation to solve for the de Broglie wavelength of an $N_2$ molecule.

$$\lambda = \frac{h}{mu} = \frac{(6.63\times10^{-34}\text{ J}\cdot\text{s})}{(4.653\times10^{-26}\text{ kg})(516.8\text{ m/s})} = \mathbf{2.76\times10^{-11}\text{ m}}$$

**7.136** The kinetic energy acquired by the electrons is equal to the voltage times the charge on the electron. After calculating the kinetic energy, we can calculate the velocity of the electrons ($\text{KE} = \frac{1}{2}mu^2$). Finally, we can calculate the wavelength associated with the electrons using the de Broglie equation.

$$\text{KE} = (5.00\times10^3\text{ V})\times\frac{1.602\times10^{-19}\text{ J}}{1\text{ V}} = 8.01\times10^{-16}\text{ J}$$

We can now calculate the velocity of the electrons.

$$\text{KE} = \frac{1}{2}mu^2$$

$$8.01\times10^{-16}\text{ J} = \frac{1}{2}(9.1094\times10^{-31}\text{ kg})u^2$$

$$u = 4.19\times10^7\text{ m/s}$$

Finally, we can calculate the wavelength associated with the electrons using the de Broglie equation.

$$\lambda = \frac{h}{mu}$$

$$\lambda = \frac{(6.63 \times 10^{-34}\ \text{J}\cdot\text{s})}{(9.1094 \times 10^{-31}\ \text{kg})(4.19 \times 10^{7}\ \text{m/s})} = \mathbf{1.74 \times 10^{-11}\ m} = \mathbf{17.4\ pm}$$

**7.138** The energy given in the problem is the energy of 1 mole of gamma rays. We need to convert this to the energy of one gamma ray, then we can calculate the wavelength and frequency of this gamma ray.

$$\frac{1.29 \times 10^{11}\ \text{J}}{1\ \text{mol}} \times \frac{1\ \text{mol}}{6.022 \times 10^{23}\ \text{gamma rays}} = 2.14 \times 10^{-13}\ \text{J/gamma ray}$$

Now, we can calculate the wavelength and frequency from this energy.

$$E = \frac{hc}{\lambda}$$

$$\lambda = \frac{hc}{E} = \frac{(6.63 \times 10^{-34}\ \text{J}\cdot\text{s})(3.00 \times 10^{8}\ \text{m/s})}{2.14 \times 10^{-13}\ \text{J}} = \mathbf{9.29 \times 10^{-13}\ m} = \mathbf{0.929\ pm}$$

and

$$E = h\nu$$

$$\nu = \frac{E}{h} = \frac{2.14 \times 10^{-13}\ \text{J}}{6.63 \times 10^{-34}\ \text{J}\cdot\text{s}} = \mathbf{3.23 \times 10^{20}\ s^{-1}}$$

**7.140** The energy of the emitted photon is:

$$E = \frac{hc}{\lambda} = \frac{(6.63 \times 10^{-34}\ \text{J}\cdot\text{s})(3.00 \times 10^{8}\ \text{m/s})}{1280 \times 10^{-9}\ \text{m}} = 1.55 \times 10^{-19}\ \text{J}$$

Since this is an emission process, the energy change, $\Delta E$, must be negative, or $-1.55 \times 10^{-19}$ J.

A wavelength of 1280 nm is in the infrared region of the electromagnetic spectrum. Transitions from higher energy levels to $n = 3$ or $n = 4$ are in the infrared region (See Figure 7.11 of the text). Let's start with an $n_f = 3$ and solve for $n_i$. If $n_i$ is a whole number, then that would be the correct transition. Substitute $\Delta E$ and $n_f$ into the Rydberg equation and solve for $n_i$.

$$\Delta E = R_H\left(\frac{1}{n_i^2} - \frac{1}{n_f^2}\right)$$

$$-1.55 \times 10^{-19}\ \text{J} = (2.18 \times 10^{-18}\ \text{J})\left(\frac{1}{n_i^2} - \frac{1}{3^2}\right)\sum_{i=1}^{n} X_i Y_i$$

$$\frac{1}{n_i^2} = \left(\frac{-1.55 \times 10^{-19}\text{ J}}{2.18 \times 10^{-18}\text{ J}}\right) + \frac{1}{3^2} = -0.0711 + 0.111 = 0.040$$

$$n_i = \frac{1}{\sqrt{0.040}} = 5$$

Therefore, the transition is from $\boldsymbol{n_i = 5}$ to $\boldsymbol{n_f = 3}$.

Another way to solve the problem is to recognize that the smallest photon energy in the Balmer series ($n_i = 3 \rightarrow n_f = 2$) is $3.0 \times 10^{-19}$ J, and the largest photon energy in the Bracket series ($n_i = \infty \rightarrow n_f = 4$) is $1.4 \times 10^{-19}$ J. Therefore, the emission must be in the Paschen series. We write:

$$-1.55 \times 10^{-19}\text{ J} = (2.18 \times 10^{-18}\text{ J})\left(\frac{1}{n_i^2} - \frac{1}{3^2}\right)$$

$$n_i = 5$$

**7.142** **(a)** Line A corresponds to the longest wavelength or lowest energy transition, which is the 3 → 2 transition. Therefore, line B corresponds to the 4 → 2 transition, and line C corresponds to the 5 → 2 transition.

**(b)** We can derive an equation for the energy change ($\Delta E$) for an electronic transition.

$$E_f = -R_H Z^2\left(\frac{1}{n_f^2}\right) \quad \text{and} \quad E_i = -R_H Z^2\left(\frac{1}{n_i^2}\right)$$

$$\Delta E = E_f - E_i = -R_H Z^2\left(\frac{1}{n_f^2}\right) - \left(-R_H Z^2\left(\frac{1}{n_i^2}\right)\right)$$

$$\Delta E = R_H Z^2\left(\frac{1}{n_i^2} - \frac{1}{n_f^2}\right)$$

Line C corresponds to the 5 → 2 transition. The energy change associated with this transition can be calculated from the wavelength (27.1 nm).

$$E = \frac{hc}{\lambda} = \frac{(6.63 \times 10^{-34}\text{ J}\cdot\cancel{\text{s}})(3.00 \times 10^8\ \cancel{\text{m/s}})}{(27.1 \times 10^{-9}\ \cancel{\text{m}})} = 7.34 \times 10^{-18}\text{ J}$$

For the 5 → 2 transition, we now know $\Delta E$, $n_i$, $n_f$, and $R_H$ ($R_H = 2.18 \times 10^{-18}$ J). Since this transition corresponds to an emission process, energy is released and $\Delta E$ is negative. ($\Delta E = -7.34 \times 10^{-18}$ J). We can now substitute these values into the equation above to solve for Z.

$$\Delta E = R_H Z^2\left(\frac{1}{n_i^2} - \frac{1}{n_f^2}\right)$$

$$-7.34 \times 10^{-18}\ \text{J} = (2.18 \times 10^{-18}\ \text{J})Z^2\left(\frac{1}{5^2} - \frac{1}{2^2}\right)$$

$$-7.34 \times 10^{-18}\ \text{J} = (-4.58 \times 10^{-19})Z^2$$

$$Z^2 = 16.0$$

$$Z = 4$$

**Z must be an integer because it represents the atomic number of the parent atom.**

**Now, knowing the value of Z, we can substitute in $n_i$ and $n_f$ for the 3 → 2 (Line A) and the 4 → 2 (Line B) transitions to solve for $\Delta E$. We can then calculate the wavelength from the energy.**

For Line A (3 → 2)

$$\Delta E = R_H Z^2\left(\frac{1}{n_i^2} - \frac{1}{n_f^2}\right) = (2.18 \times 10^{-18}\ \text{J})(4)^2\left(\frac{1}{3^2} - \frac{1}{2^2}\right)$$

$$\Delta E = -4.84 \times 10^{-18}\ \text{J}$$

$$\lambda = \frac{hc}{E} = \frac{(6.63 \times 10^{-34}\ \text{J}\cdot\text{s})(3.00 \times 10^8\ \text{m/s})}{(4.84 \times 10^{-18}\ \text{J})} = \mathbf{4.11 \times 10^{-8}\ m = 41.1\ nm}$$

For Line B (4 → 2)

$$\Delta E = R_H Z^2\left(\frac{1}{n_i^2} - \frac{1}{n_f^2}\right) = (2.18 \times 10^{-18}\ \text{J})(4)^2\left(\frac{1}{4^2} - \frac{1}{2^2}\right)$$

$$\Delta E = -6.54 \times 10^{-18}\ \text{J}$$

$$\lambda = \frac{hc}{E} = \frac{(6.63 \times 10^{-34}\ \text{J}\cdot\text{s})(3.00 \times 10^8\ \text{m/s})}{(6.54 \times 10^{-18}\ \text{J})} = \mathbf{3.04 \times 10^{-8}\ m = 30.4\ nm}$$

**(c)** The value of the final energy state is $n_f = \infty$. Use the equation derived in part (b) to solve for $\Delta E$.

$$\Delta E = R_H Z^2\left(\frac{1}{n_i^2} - \frac{1}{n_f^2}\right) = (2.18 \times 10^{-18}\ \text{J})(4)^2\left(\frac{1}{4^2} - \frac{1}{\infty^2}\right)$$

$$\mathbf{\Delta E = 2.18 \times 10^{-18}\ J}$$

**(d)** At high values of $n$, the energy levels are very closely spaced, leading to a continuum of lines. At the beginning of the continuum, the electron has been removed from the atom, and we no longer have quantized energy levels associated with the electron.

**7.144** To calculate the energy to remove an electron from the $n = 1$ state and the $n = 5$ state in the $Li^{2+}$ ion, we use the equation derived in Problem 7.142 (b).

$$\Delta E = R_H Z^2 \left(\frac{1}{n_i^2} - \frac{1}{n_f^2}\right)$$

For $n_i = 1$, $n_f = \infty$, and $Z = 3$, we have:

$$\Delta E = (2.18 \times 10^{-18}\ \text{J})(3)^2 \left(\frac{1}{1^2} - \frac{1}{\infty^2}\right) = \mathbf{1.96 \times 10^{-17}\ J}$$

For $n_i = 5$, $n_f = \infty$, and $Z = 3$, we have:

$$\Delta E = (2.18 \times 10^{-18}\ \text{J})(3)^2 \left(\frac{1}{5^2} - \frac{1}{\infty^2}\right) = \mathbf{7.85 \times 10^{-19}\ J}$$

To calculate the wavelength of the emitted photon in the electronic transition from $n = 5$ to $n = 1$, we first calculate $\Delta E$ and then calculate the wavelength.

$$\Delta E = R_H Z^2 \left(\frac{1}{n_i^2} - \frac{1}{n_f^2}\right) = (2.18 \times 10^{-18}\ \text{J})(3)^2 \left(\frac{1}{5^2} - \frac{1}{1^2}\right) = -1.88 \times 10^{-17}\ \text{J}$$

We ignore the minus sign for $\Delta E$ in calculating $\lambda$.

$$\lambda = \frac{hc}{\Delta E} = \frac{(6.63 \times 10^{-34}\ \text{J}\cdot\text{s})(3.00 \times 10^8\ \text{m/s})}{1.88 \times 10^{-17}\ \text{J}}$$

$$\boldsymbol{\lambda = 1.06 \times 10^{-8}\ \text{m} = 10.6\ \text{nm}}$$

**7.146** The uncertainty in the position of the particle is $\Delta x$ in Heisenberg's uncertainty principle. The de Broglie wavelength, $h/mu$, equals the minimum uncertainty in the position of this particle. We use the equal sign in the uncertainty equation to calculate the *minimum* uncertainty value. $\Delta p = m\Delta u$, which gives:

$$\Delta x \Delta p = \frac{h}{4\pi}$$

$$\frac{h}{mu}(m\Delta u) = \frac{h}{4\pi}$$

$$\Delta u = \frac{u}{4\pi} = \frac{1.2 \times 10^5\ \text{m/s}}{4\pi} = \mathbf{9.5 \times 10^3\ m/s}$$

**7.148** We calculate $W$ (the work function) at a wavelength of 351 nm. Once $W$ is known, we can then calculate the velocity of an ejected electron using light with a wavelength of 313 nm.

First, we convert wavelength to frequency.

$$\nu = \frac{c}{\lambda} = \frac{3.00 \times 10^8 \text{ m/s}}{351 \times 10^{-9} \text{ m}} = 8.55 \times 10^{14} \text{ s}^{-1}$$

$$h\nu = W + \frac{1}{2}m_e u^2$$

$$(6.63 \times 10^{-34} \text{ J}\cdot\text{s})(8.55 \times 10^{14} \text{ s}^{-1}) = W + \frac{1}{2}(9.1094 \times 10^{-31} \text{ kg})(0 \text{ m/s})^2$$

$$W = 5.67 \times 10^{-19} \text{ J}$$

Next, we convert a wavelength of 313 nm to frequency, and then calculate the velocity of the ejected electron.

$$\nu = \frac{c}{\lambda} = \frac{3.00 \times 10^8 \text{ m/s}}{313 \times 10^{-9} \text{ m}} = 9.58 \times 10^{14} \text{ s}^{-1}$$

$$h\nu = W + \frac{1}{2}m_e u^2$$

$$(6.63 \times 10^{-34} \text{ J}\cdot\text{s})(9.58 \times 10^{14} \text{ s}^{-1}) = (5.67 \times 10^{-19} \text{ J}) + \frac{1}{2}(9.1094 \times 10^{-31} \text{ kg})u^2$$

$$6.82 \times 10^{-20} = (4.5547 \times 10^{-31})u^2$$

$$u = \mathbf{3.87 \times 10^5 \text{ m/s}}$$

**7.150** We note that the maximum solar radiation centers around 500 nm. Thus, over billions of years, organisms have adjusted their development to capture energy at or near this wavelength. The two most notable cases are photosynthesis and vision.

**7.152** At a node, the wave function is zero. This indicates that there is zero probability of finding an electron at this distance from the nucleus.

$$\psi_{2s} = \frac{1}{\sqrt{2a_0^3}}\left(1 - \frac{\rho}{2}\right)e^{-\rho/2}$$

$$0 = \frac{1}{\sqrt{2a_0^3}}\left(1 - \frac{\rho}{2}\right)e^{-\rho/2}$$

The right hand side of the equation will equal zero when $\frac{\rho}{2} = 1$. This is the location of the node.

$$\frac{\rho}{2} = 1$$

$$\rho = 2 = Z\left(\frac{r}{a_0}\right)$$

For a hydrogen atom, the atomic number, Z, is 1. The location of the node as a distance from the nucleus, r, is

$$2 = \frac{r}{a_0}$$

$$r = 2a_0$$

$$r = (2)(0.529\text{ nm}) = \mathbf{1.06\text{ nm}}$$

**7.154** (a) $$\text{m} = 5672\text{ amu} \times \frac{1\text{ g}}{6.022 \times 10^{23}\text{ amu}} \times \frac{1\text{ kg}}{1000\text{ g}} = 9.419 \times 10^{-24}\text{ kg}$$

$$\lambda = \frac{\text{h}}{\text{mu}} = \frac{6.63 \times 10^{-34}\text{ J}\cdot\text{s}}{9.419 \times 10^{-24}\text{ kg} \times 63\text{ m/s}} = 1.1 \times 10^{-12}\text{ m}$$

$$1.1 \times 10^{-12}\text{ m} \times \frac{1 \times 10^{12}\text{ pm}}{1\text{ m}} = \mathbf{1.1\text{ pm}}$$

(b) The wavelength is over three orders-of-magnitude shorter that the diameter of the molecule – long enough to give detectable wave properties, but too short to cause significant deviations from the predictions of classical mechanics.

# CHAPTER 8
# PERIODIC RELATIONSHIPS AMONG THE ELEMENTS

## PROBLEM-SOLVING STRATEGIES AND TUTORIAL SOLUTIONS

### TYPES OF PROBLEMS

**Problem Type 1:** Writing an Electron Configuration and Identifying an Element (see Problem Type 7.6).

**Problem Type 2:** Electron Configurations of Cations and Anions.

**Problem Type 3:** Comparing the Sizes of Atoms.

**Problem Type 4:** Comparing the Sizes of Ions.

**Problem Type 5:** Comparing the Ionization Energies of Elements.

**Problem Type 6:** Electron Affinity.

## PROBLEM TYPE 1: WRITING AN ELECTRON CONFIGURATION AND IDENTIFYING AN ELEMENT

See Problem Type 7.6 for information on writing electron configurations. When examining the electron configurations of the elements in a particular group, a clear pattern emerges. Elements in the same group have the same electron configuration of their *outer* electrons. The outer electrons of an atom, which are those involved in chemical bonding, are called **valence electrons**. Having the same number of valence electrons accounts for the similarities in chemical behavior among the elements within each of these groups.

In regards to identifying an element, the text considers a number of items. First, according to the type of subshell being filled, the elements can be divided into categories—the representative elements, the noble gases, the transition metals, the lanthanides, and the actinides.

> The **representative elements** are the elements in Groups 1A through 7A, all of which have *incompletely* filled *s* or *p* subshells of the highest principal quantum number.
>
> The **noble gases**, with the exception of helium, all have a *completely* filled *p* subshell.
>
> The **transition metals** are the elements in Groups 1B and 3B through 8B that have *incompletely* filled *d* subshells, or readily produce cations with *incompletely* filled *d* subshells.
>
> The **lanthanides** and **actinides** are sometimes called *f*-block transition elements because they have *incompletely* filled *f* subshells.

Second, you might be able to classify the element as a metal, nonmetal, or metalloid. However, sometimes you need to be careful in making this classification. For example in Group 4A, carbon is a nonmetal, silicon and germanium are metalloids, and tin and lead are metals.

Third, the problem might ask whether the element is paramagnetic or diamagnetic. In Chapter 7, you learned that an element that contains unpaired electrons is *paramagnetic*, and an element in which all electrons are paired is *diamagnetic*.

**EXAMPLE 8.1**

**A neutral atom of a certain element has 19 electrons. Without consulting a periodic table, answer the following questions: (a) What is the electron configuration of the element? (b) How should the element be classified? (c) Are the atoms of this element diamagnetic or paramagnetic?**

**Strategy:** (a) We refer to the building-up principle discussed in Section 7.9 of the text. We start writing the electron configuration with principal quantum number $n = 1$ and continue upward in energy until all electrons are accounted for. (b) What are the electron configuration characteristics of representative elements, transition elements, and noble gases? (c) Examine the pairing scheme of the electrons in the outermost shell. What determines whether an element is diamagnetic or paramagnetic?

**Solution:**

(a) We know that for $n = 1$, we have a 1*s* orbital (2 electrons). For $n = 2$, we have a 2*s* orbital (2 electrons) and three 2*p* orbitals (6 electrons). For $n = 3$, we have a 3*s* orbital (2 electrons) and three 3*p* orbitals (6 electrons). The number of electrons left to place is $19 - 18 = 1$. This electron is placed in the 4*s* orbital. The electron configuration is $1s^2 2s^2 2p^6 3s^2 3p^6 4s^1$ or $[Ar]4s^1$.

(b) Because the 4*s* subshell is not completely filled, this is a *representative element.* Without consulting a periodic table, you might know that the alkali metal family has one valence electron in the *s* subshell. You could then further classify this element as an *alkali metal.*

(c) There is *one* unpaired electron in the *s* subshell. Therefore, the atoms of this element are paramagnetic.

**Check:** For (b), note that a transition metal possesses an incompletely filled *d* subshell, and a noble gas has a completely filled outer-shell. For (c), recall that if the atoms of an element contain an odd number of electrons, the element must be paramagnetic.

**PRACTICE EXERCISE**

1. A neutral atom of a certain element has 36 electrons. Without consulting a periodic table, answer the following questions: (a) What is the electron configuration of the element? (b) How should the element be classified? (c) Are the atoms of this element diamagnetic or paramagnetic?

**Text Problems: 8.20**, 8.22, 8.24

## PROBLEM TYPE 2: ELECTRON CONFIGURATIONS OF CATIONS AND ANIONS

Writing an electron configuration for a cation or anion requires only a slight extension of the method used for neutral atoms. Let's group the ions in two categories.

**(1) Ions Derived from Representative Elements**

**(a)** In the formation of a **cation** from the neutral atom of a representative element, one or more electrons are removed from the highest occupied *n* shell. The number of electrons removed is equal to the charge of the ion.

Examples: Following are the electron configurations of some neutral atoms and their corresponding cations.

Mg: $[Ne]3s^2$ $Mg^{2+}$: [Ne]

K: $[Ar]4s^1$ $K^+$: [Ar]

**Note:** Each ion has a stable noble gas electron configuration.

**(b)** In the formation of an **anion** from the neutral atom of a representative element, one or more electrons are added to the highest partially filled *n* shell. The number of electrons added is equal to the charge of the ion.

Examples: Following are the electron configurations of some neutral atoms and their corresponding anions.

O: $[He]2s^22p^4$ $O^{2-}$: [Ne]

Cl: $[Ne]3s^23p^5$ $Cl^-$: [Ar]

**Note:** Each ion has a stable noble gas electron configuration.

**(2) Cations Derived from Transition Metals**

In a neutral transition metal atom, the n*s* orbital is filled before the $(n-1)d$ orbitals. See Figure 7.23 of the text. However, in a transition metal ion, the $(n-1)d$ orbitals are more stable than the *ns* orbital. Hence, when a cation is formed from an atom of a transition metal, electrons are ***always*** removed first from the *ns* orbital and then from the $(n-1)d$ orbitals if necessary. See Section 8.2 of the text for a more complete discussion.

Examples: Following are the electron configurations of some neutral transition metals and their corresponding cations.

Fe: $[Ar]4s^23d^6$ $Fe^{2+}$: $[Ar]3d^6$

Fe: $[Ar]4s^23d^6$ $Fe^{3+}$: $[Ar]3d^5$

Mn: $[Ar]4s^23d^5$ $Mn^{7+}$: [Ar]

**PRACTICE EXERCISE**

2. Write electron configurations for the following ions: $N^{3-}$, $Ba^{2+}$, $Zn^{2+}$, and, $V^{5+}$.

**Text Problems:** 8.26, **8.28**, 8.30, 8.32

## PROBLEM TYPE 3: COMPARING THE SIZES OF ATOMS

The general periodic trends in atomic size are:

(1) Moving from left to right across a row (period) of the periodic table, the atomic radius ***decreases*** due to an increase in effective nuclear charge.

(2) Moving down a column (group) of the periodic table, the atomic radius ***increases*** since the orbital size increases with increasing principal quantum number.

For a more detailed discussion, see Section 8.3 of the text.

**EXAMPLE 8.2**

**Which one of the following has the smallest atomic radius? (a) Li, (b) Na, (c) Be, (d) Mg**

**Strategy:** What are the trends in atomic radii in a periodic group and in a particular period. Which of the above elements are in the same group and which are in the same period?

**Solution:** Recall that the general periodic trends in atomic size are:

(1) Moving from left to right across a row (period) of the periodic table, the atomic radius ***decreases*** due to an increase in effective nuclear charge.

(2) Moving down a column (group) of the periodic table, the atomic radius ***increases*** since the orbital size increases with increasing principal quantum number.

Atomic radii *increase* going down a group of the periodic table; therefore, Li atoms have a smaller radius than Na atoms, and Be atoms have a smaller radius than Mg atoms. We have narrowed our choices to Li and Be. Atomic radii decrease when moving from left to right across a row of the periodic table. Thus, **Be atoms** are smaller than Li atoms as well as the other choices given.

**Check:** See Figure 8.5 of the text to confirm that the answer is correct.

**PRACTICE EXERCISE**

3. Which atom should have the largest atomic radius? (a) Br, (b) Cl, (c) Se, (d) Ge, (e) C.

**Text Problems: 8.38**, 8.40, 8.42

## PROBLEM TYPE 4: COMPARING THE SIZES OF IONS

When a neutral atom is converted to an ion, we expect a change in size. When forming an *anion* from an atom, its size (or radius) *increases*, because the nuclear charge remains the same but the repulsion resulting from the additional electron(s) enlarges the domain of the electron cloud. Conversely, when forming a *cation* from an atom, its size *decreases*, because removing one or more electrons reduces electron-electron repulsion but the nuclear charge remains the same. Thus, the electron cloud shrinks.

The general periodic trends in ionic size are:

(1) Similar to atomic size, ionic radii increase when moving down a column (group) of the periodic table.

(2) The next trend only applies to **isoelectronic ions**. Isoelectronic ions have the same number of electrons and the same electron configuration.

(a) Cations are smaller than anions. For example, $Mg^{2+}$ is smaller than $O^{2-}$. Both ions have the same number of electrons (10), but Mg (Z = 12) has more protons then O (Z = 8). The larger effective nuclear charge of $Mg^{2+}$ results in a smaller radius.

(b) Considering only *isoelectronic cations*, the radius of a tripositive ion is smaller than the radius of a dipositive ion, which is smaller than the radius of a unipositive ion. For example, $Al^{3+}$ is smaller than $Mg^{2+}$, which is smaller than $Na^+$. Each of the cations has 10 electrons, but $Al^{3+}$ has 13 protons, $Mg^{2+}$ has 12 protons, and $Na^+$ has 11 protons. As the effective nuclear charge increases, the ionic radius decreases.

(c) Considering only *isoelectronic anions*, the radius increases as we go from an ion with a uninegative charge, to one with a dinegative charge, and so on. For example, $N^{3-}$ is larger than $O^{2-}$, which is larger than $F^-$. Each of the anions has 10 electrons, but $N^{3-}$ has 7 protons, $O^{2-}$ has 8 protons, and $F^-$ has 9 protons. As the effective nuclear charge increases, ionic radius decreases.

**EXAMPLE 8.3**

**In each of the following pairs, choose the ion with the *larger* ionic radius: (a) $K^+$ and $Na^+$; (b) $K^+$ and $Ca^{2+}$; (c) $K^+$ and $Cl^-$.**

**Strategy:** In comparing ionic radii, it is useful to classify the ions into three categories: (1) isoelectronic ions, (2) ions that carry the same charges and are generated from atoms of the same periodic group, and (3) ions that carry different charges but are generated from the same atom. In case (1), ions carrying a greater negative charge are always larger; in case (2), ions from atoms having a greater atomic number are always larger; in case (3), ions have a smaller positive charge are always larger.

**Solution:**

**(a)** $K^+$ and $Na^+$ are in the same group of the periodic table (Group 1A, alkali metals). As you proceed down a group of the periodic table, ionic radii increase. This occurs because orbital size increases with increasing principal quantum number. Therefore, $\mathbf{K^+}$ has the larger ionic radius.

**(b)** $K^+$ and $Ca^{2+}$ are isoelectronic cations. Both ions have 18 electrons, but $Ca^{2+}$ has one more proton than $K^+$. The greater effective nuclear charge of $Ca^{2+}$, pulls the 18 electrons more strongly toward the nucleus. Therefore, $\mathbf{K^+}$ has the larger ionic radius. For isoelectronic cations, the radius of a dipositive ion is smaller than the radius of a unipositive ion.

**(c)** $K^+$ and $Cl^-$ are isoelectronic species. Both ions have 18 electrons, but $K^+$ has the greater effective nuclear charge, with 19 protons compared to 17 protons for $Cl^-$. Therefore, $\mathbf{Cl^-}$ has the larger ionic radius.

**PRACTICE EXERCISE**

4. Which of the following has the largest radius: $Na^+$, $Mg^{2+}$, $Al^{3+}$, $S^{2-}$, or $Cl^-$?

**Text Problems: 8.44**, 8.46

## PROBLEM TYPE 5: COMPARING THE IONIZATION ENERGIES OF ELEMENTS

**Ionization energy** is the minimum energy required to remove an electron from a gaseous atom in its ground state. An equation that represents this process is

$$\text{energy} + X(g) \longrightarrow X^+(g) + e^-$$

where,

X represents a gaseous atom of any element

$e^-$ is an electron

The more "tightly" the electron is held in the atom, the more difficult it will be to remove the electron. Hence, the more "tightly" the electron is held, the higher the ionization energy.

The general periodic trends for ionization energy are:

(1) Moving from left to right across a row of the periodic table, the ionization energy ***increases*** due to an increase in effective nuclear charge. As the effective nuclear charge increases, the electrons will be held more "tightly" by the nucleus, making them more difficult to remove.

(2) Moving down a column (group) of the periodic table, the ionization energy ***decreases***. As the principal quantum number $n$ increases, so does the average distance of a valence electron from the nucleus. A greater separation between the electron and the nucleus results in a weaker attraction; hence, a valence electron becomes increasingly easier to remove as we proceed down a group of the periodic table.

As with most periodic trends, there are exceptions. For a more detailed discussion, see Section 8.4 of the text.

**EXAMPLE 8.4**
**Which of the following has the highest ionization energy: K, Br, Cl, or S?**

**Strategy:** The ionization energy increases from left to right across a row, and it decreases moving down a column or family of the periodic table.

**Solution:** Moving across rows of the periodic table, the ionization energy of Cl is greater than for S, and the ionization energy of Br is greater than for K. We have narrowed our choices to Cl and Br. Moving down a group of the periodic table, the ionization decreases. Therefore, **Cl** has a higher ionization energy than Br as well as the other choices given.

**PRACTICE EXERCISE**

5. Based on periodic trends, which of the following elements has the greatest ionization energy: Cl, K, S, Se, or Br?

**Text Problems:** 8.52, 8.54, **8.56**

## PROBLEM TYPE 6: ELECTRON AFFINITY

**Electron affinity** is the energy change when an electron is accepted by an atom in the gaseous state. An equation that represents this process is:

$$X(g) + e^- \longrightarrow X^-(g)$$

where,

X represents a gaseous atom of any element

$e^-$ is an electron

A positive electron affinity signifies that energy is liberated when an electron is added to an atom. The more positive the electron affinity, the greater the tendency of the atom to accept an electron. Electron affinity is positive if the reaction is exothermic and negative if the reaction is endothermic.

The general periodic trends for electron affinity are:

(1) The tendency to accept electrons increases (that is, electron affinity values become more positive) as we move from left to right across a period.

(2) The electron affinities of metals are generally less than those of nonmetals.

(3) Electron affinity values vary little within a given group.

**EXAMPLE 8.5**
**Explain why the electron affinities of the halogens are all positive.**

The outer-shell electron configuration of the halogens is $ns^2np^5$. For the process

$$X(g) + e^- \longrightarrow X^-(g)$$

where X denotes a member of the halogen family, the accepted electron would fill the outer shell, giving the halogen ion a stable noble gas electron configuration. This is a favorable process; consequently, halogens have a strong tendency to accept an extra electron (i.e., a highly positive electron affinity).

**Text Problems:** **8.62**, 8.64

## ANSWERS TO PRACTICE EXERCISES

**1.** **(a)** $1s^22s^22p^63s^23p^64s^23d^{10}4p^6$

**(b)** Since the 4*p* subshell is completely filled, this is a *noble gas*. All noble gases are *nonmetals*.

**(c)** Since the outer shell is completely filled, there are *no* unpaired electrons. Therefore, the atoms of this element are diamagnetic.

**2.** $N^{3-}$: [Ne] $Ba^{2+}$: [Xe] $Zn^{2+}$: $[Ar]3d^{10}$ $V^{5+}$: [Ar]

**3.** **Ge** has the largest atomic radius of the group.

**4.** The ions are isoelectronic. $S^{2-}$ has the largest radius.

**5.** Cl

## SOLUTIONS TO SELECTED TEXT PROBLEMS

**8.20** **Strategy:** **(a)** We refer to the building-up principle discussed in Section 7.9 of the text. We start writing the electron configuration with principal quantum number $n = 1$ and continue upward in energy until all electrons are accounted for. **(b)** What are the electron configuration characteristics of representative elements, transition elements, and noble gases? **(c)** Examine the pairing scheme of the electrons in the outermost shell. What determines whether an element is diamagnetic or paramagnetic?

**Solution:**

**(a)** We know that for $n = 1$, we have a 1*s* orbital (2 electrons). For $n = 2$, we have a 2*s* orbital (2 electrons) and three 2*p* orbitals (6 electrons). For $n = 3$, we have a 3*s* orbital (2 electrons). The number of electrons left to place is $17 - 12 = 5$. These five electrons are placed in the 3*p* orbitals. The electron configuration is $\mathbf{1s^22s^22p^63s^23p^5}$ or $\mathbf{[Ne]3s^23p^5}$.

**(b)** Because the 3*p* subshell is not completely filled, this is a **representative element**. Without consulting a periodic table, you might know that the halogen family has seven valence electrons. You could then further classify this element as a *halogen*. In addition, all halogens are *nonmetals*.

**(c)** If you were to write an orbital diagram for this electron configuration, you would see that there is *one* unpaired electron in the *p* subshell. Remember, the three 3*p* orbitals can hold a total of six electrons. Therefore, the atoms of this element are **paramagnetic**.

**Check:** For (b), note that a transition metal possesses an incompletely filled *d* subshell, and a noble gas has a completely filled outer-shell. For (c), recall that if the atoms of an element contain an odd number of electrons, the element must be paramagnetic.

**8.22** Elements that have the same number of valence electrons will have similarities in chemical behavior. Looking at the periodic table, elements with the same number of valence electrons are in the same group. Therefore, the pairs that would represent similar chemical properties of their atoms are:

**(a)** and **(d)** **(b)** and **(e)** **(c)** and **(f)**.

**8.24** **(a)** Group 1A **(b)** Group 5A **(c)** Group 8A **(d)** Group 8B

Identify the elements.

**8.26** You should realize that the metal ion in question is a transition metal ion because it has five electrons in the 3*d* subshell. Remember that in a transition metal ion, the $(n-1)d$ orbitals are more stable than the *ns* orbital. Hence, when a cation is formed from an atom of a transition metal, electrons are *always* removed first from the *ns* orbital and then from the $(n-1)d$ orbitals if necessary. Since the metal ion has a +3 charge, three electrons have been removed. Since the 4*s* subshell is less stable than the 3*d*, two electrons would have been lost from the 4*s* and one electron from the 3*d*. Therefore, the electron configuration of the neutral atom is $[Ar]4s^2 3d^6$. This is the electron configuration of iron. Thus, the metal is **iron**.

**8.28** **Strategy:** In the formation of a **cation** from the neutral atom of a representative element, one or more electrons are *removed* from the highest occupied *n* shell. In the formation of an **anion** from the neutral atom of a representative element, one or more electrons are *added* to the highest partially filled *n* shell. Representative elements typically gain or lose electrons to achieve a stable noble gas electron configuration. When a cation is formed from an atom of a transition metal, electrons are *always* removed first from the *ns* orbital and then from the $(n-1)d$ orbitals if necessary.

**Solution:**

(a) [Ne]

(b) same as (a). Do you see why?

(c) [Ar]

(d) Same as (c). Do you see why?

(e) Same as (c)

(f) $[Ar]3d^6$. Why isn't it $[Ar]4s^2 3d^4$?

(g) $[Ar]3d^9$. Why not $[Ar]4s^2 3d^7$?

(h) $[Ar]3d^{10}$. Why not $[Ar]4s^2 3d^8$?

**8.30** (a) $Cr^{3+}$ (b) $Sc^{3+}$ (c) $Rh^{3+}$ (d) $Ir^{3+}$

**8.32** Isoelectronic means that the species have the same number of electrons and the same electron configuration.

$Be^{2+}$ and He (2 $e^-$) $F^-$ and $N^{3-}$ (10 $e^-$) $Fe^{2+}$ and $Co^{3+}$ (24 $e^-$) $S^{2-}$ and Ar (18 $e^-$)

**8.38** **Strategy:** What are the trends in atomic radii in a periodic group and in a particular period? Which of the above elements are in the same group and which are in the same period?

**Solution:** Recall that the general periodic trends in atomic size are:

(1) Moving from left to right across a row (period) of the periodic table, the atomic radius ***decreases*** due to an increase in effective nuclear charge.

(2) Moving down a column (group) of the periodic table, the atomic radius ***increases*** since the orbital size increases with increasing principal quantum number.

The atoms that we are considering are all in the same period of the periodic table. Hence, the atom furthest to the left in the row will have the largest atomic radius, and the atom furthest to the right in the row will have the smallest atomic radius. Arranged in order of decreasing atomic radius, we have:

$$\textbf{Na} > \textbf{Mg} > \textbf{Al} > \textbf{P} > \textbf{Cl}$$

**Check:** See Figure 8.5 of the text to confirm that the above is the correct order of decreasing atomic radius.

**8.40** **Fluorine** is the smallest atom in Group 7A. Atomic radius increases moving down a group since the orbital size increases with increasing principal quantum number, *n*.

**8.42** The atomic radius is largely determined by how strongly the outer-shell electrons are held by the nucleus. The larger the effective nuclear charge, the more strongly the electrons are held and the smaller the atomic radius. For the second period, the atomic radius of Li is largest because the 2*s* electron is well shielded by the filled 1*s* shell. The effective nuclear charge that the outermost electrons feel increases across the period as a result of incomplete shielding by electrons in the same shell. Consequently, the orbital containing the electrons is compressed and the atomic radius decreases.

**8.44** **Strategy:** In comparing ionic radii, it is useful to classify the ions into three categories: (1) isoelectronic ions, (2) ions that carry the same charges and are generated from atoms of the same periodic group, and (3) ions that carry different charges but are generated from the same atom. In case (1), ions carrying a greater negative charge are always larger; in case (2), ions from atoms having a greater atomic number are always larger; in case (3), ions that have a smaller positive charge are always larger.

**Solution:** The ions listed are all isoelectronic. They each have ten electrons. The ion with the fewest protons will have the largest ionic radius, and the ion with the most protons will have the smallest ionic radius. The effective nuclear charge increases with increasing number of protons. The electrons are attracted more strongly by the nucleus, decreasing the ionic radius. $N^{3-}$ has only 7 protons resulting in the smallest attraction exerted by the nucleus on the 10 electrons. $N^{3-}$ is the largest ion of the group. $Mg^{2+}$ has 12 protons resulting in the largest attraction exerted by the nucleus on the 10 electrons. $Mg^{2+}$ is the smallest ion of the group. The order of increasing atomic radius is:

$$\mathbf{Mg^{2+} < Na^{+} < F^{-} < O^{2-} < N^{3-}}$$

**8.46** Both selenium and tellurium are Group 6A elements. Since atomic radius increases going down a column in the periodic table, it follows that $\mathbf{Te^{2-}}$ must be larger than $Se^{2-}$.

**8.48** $\mathbf{H^-}$ is larger. Both $H^-$ and He have 2 electrons (they are isoelectronic), but they have different numbers of protons in the nucleus. Since $H^-$ has only 1 proton, the two electrons will experience less nuclear charge, and $H^-$ will be larger than He.

**8.52** The general periodic trend for first ionization energy is that it increases across a period (row) of the periodic table and it decreases down a group (column). Of the choices, K will have the smallest ionization energy. Ca, just to the right of K, will have a higher first ionization energy. Moving to the right across the periodic table, the ionization energies will continue to increase as we move to P. Continuing across to Cl and moving up the halogen group, F will have a higher ionization energy than P. Finally, Ne is to the right of F in period two, thus it will have a higher ionization energy. The correct order of increasing first ionization energy is:

$$\mathbf{K < Ca < P < F < Ne}$$

You can check the above answer by looking up the first ionization energies for these elements in Table 8.2 of the text.

**8.54** The Group 3A elements (such as Al) all have a single electron in the outermost *p* subshell, which is well shielded from the nuclear charge by the inner electrons and the $ns^2$ electrons. Therefore, less energy is needed to remove a single *p* electron than to remove a paired *s* electron from the same principal energy level (such as for Mg).

**8.56** **Strategy:** Removal of the outermost electron requires less energy if it is shielded by a filled inner shell.

**Solution:** The lone electron in the 3*s* orbital will be much easier to remove. This lone electron is shielded from the nuclear charge by the filled inner shell. Therefore, the ionization energy of 496 kJ/mol is paired with the electron configuration $1s^2 2s^2 2p^6 3s^1$.

A noble gas electron configuration, such as $1s^2 2s^2 2p^6$, is a very stable configuration, making it extremely difficult to remove an electron. The 2*p* electron is not as effectively shielded by electrons in the same energy level. The high ionization energy of 2080 kJ/mol would be associated with the element having this noble gas electron configuration.

**Check:** Compare this answer to the data in Table 8.2. The electron configuration of $1s^2 2s^2 2p^6 3s^1$ corresponds to a Na atom, and the electron configuration of $1s^2 2s^2 2p^6$ corresponds to a Ne atom.

**8.58** The atomic number of mercury is 80. We carry an extra significant figure throughout this calculation to avoid rounding errors.

$$\Delta E = (2.18 \times 10^{-18}\ \text{J})(80^2)\left(\frac{1}{1^2} - \frac{1}{\infty^2}\right) = 1.395 \times 10^{-14}\ \text{J/ion}$$

$$\boldsymbol{\Delta E} = \frac{1.395 \times 10^{-14}\ \cancel{\text{J}}}{1\ \cancel{\text{ion}}} \times \frac{6.022 \times 10^{23}\ \cancel{\text{ions}}}{1\ \text{mol}} \times \frac{1\ \text{kJ}}{1000\ \cancel{\text{J}}} = \mathbf{8.40 \times 10^6\ kJ/mol}$$

**8.62** **Strategy:** What are the trends in electron affinity in a periodic group and in a particular period? Which of the above elements are in the same group and which are in the same period?

**Solution:** One of the general periodic trends for electron affinity is that the tendency to accept electrons increases (that is, electron affinity values become more positive) as we move from left to right across a period. However, this trend does not include the noble gases. We know that noble gases are extremely stable, and they do not want to gain or lose electrons. Therefore, **helium, He,** would have the lowest electron affinity.

Based on the periodic trend discussed above, **Cl** would be expected to have the highest electron affinity. Addition of an electron to Cl forms $Cl^-$, which has a stable noble gas electron configuration.

**8.64** Alkali metals have a valence electron configuration of $ns^1$ so they can accept another electron in the ns orbital. On the other hand, alkaline earth metals have a valence electron configuration of $ns^2$. Alkaline earth metals have little tendency to accept another electron, as it would have to go into a higher energy *p* orbital.

**8.68** Since ionization energies decrease going down a column in the periodic table, francium should have the lowest first ionization energy of all the alkali metals. As a result, Fr should be the most reactive of all the

Group 1A elements toward water and oxygen. The reaction with oxygen would probably be similar to that of K, Rb, or Cs.

What would you expect the formula of the oxide to be? The chloride?

**8.70** The Group 1B elements are much less reactive than the Group 1A elements. The 1B elements are more stable because they have much higher ionization energies resulting from incomplete shielding of the nuclear charge by the inner *d* electrons. The $ns^1$ electron of a Group 1A element is shielded from the nucleus more effectively by the completely filled noble gas core. Consequently, the outer *s* electrons of 1B elements are more strongly attracted by the nucleus.

**8.72** (a) Lithium oxide is a basic oxide. It reacts with water to form the metal hydroxide:

$$Li_2O(s) + H_2O(l) \longrightarrow 2LiOH(aq)$$

(b) Calcium oxide is a basic oxide. It reacts with water to form the metal hydroxide:

$$CaO(s) + H_2O(l) \longrightarrow Ca(OH)_2(aq)$$

(c) Sulfur trioxide is an acidic oxide. It reacts with water to form sulfuric acid:

$$SO_3(g) + H_2O(l) \longrightarrow H_2SO_4(aq)$$

**8.74** As we move down a column, the metallic character of the elements increases. Since magnesium and barium are both Group 2A elements, we expect barium to be more metallic than magnesium and **BaO** to be more basic than MgO.

**8.76** (a) bromine (b) nitrogen (c) rubidium (d) magnesium

**8.78** $P^{3-}$, $S^{2-}$, $Cl^-$, $K^+$, $Ca^{2+}$, $Sc^{3+}$, $Ti^{4+}$, $V^{5+}$, $Cr^{6+}$, $Mn^{7+}$

**8.80** According to the *Handbook of Chemistry and Physics* (1966-67 edition), potassium metal has a melting point of 63.6°C, bromine is a reddish brown liquid with a melting point of −7.2°C, and potassium bromide (KBr) is a colorless solid with a melting point of 730°C. **M** is **potassium** (K) and **X** is **bromine** (Br).

**8.82** $O^+$ and N    Ar and $S^{2-}$    Ne and $N^{3-}$    Zn and $As^{3+}$    $Cs^+$ and Xe

**8.84** (a) and (d)

**8.86** Fluorine is a yellow-green gas that attacks glass; chlorine is a pale yellow gas; bromine is a fuming red liquid; and iodine is a dark, metallic-looking solid.

**8.88** **(a)** This reaction represents the first and second ionizations of magnesium (Mg) and the electron affinity of fluorine (F).

$$Mg(g) \rightarrow Mg^+(g) + e^- \qquad \Delta H = 738.1 \text{ kJ/mol}$$
$$Mg^+(g) \rightarrow Mg^{2+}(g) + e^- \qquad \Delta H = 1450 \text{ kJ/mol}$$
$$2F(g) + 2e^- \rightarrow 2F^-(g) \qquad \Delta H = 2(-328 \text{ kJ/mol})$$
$$Mg(g) + 2F(g) \rightarrow Mg^{2+}(g) + 2F^-(g) \qquad \mathbf{\Delta H = 1532\ kJ/mol}$$

**(b)** This reaction represents the first, second, and third ionizations of aluminum (Al) and the first and second electron affinities of oxygen (O).

$$2(Al(g) \rightarrow Al^+(g) + e^-) \qquad \Delta H = 2(577.9 \text{ kJ/mol})$$
$$2(Al^+(g) \rightarrow Al^{2+}(g) + e^-) \qquad \Delta H = 2(1820 \text{ kJ/mol})$$
$$2(Al^{2+}(g) \rightarrow Al^{3+}(g) + e^-) \qquad \Delta H = 2(2750 \text{ kJ/mol})$$
$$3(O(g) + e^- \rightarrow O^-(g)) \qquad \Delta H = 3(-141 \text{ kJ/mol})$$
$$3(O^-(g) + e^- \rightarrow O^{2-}(g)) \qquad \Delta H = 3(844 \text{ kJ/mol})$$
$$2Al(g) + 3O(g) \rightarrow 2Al^{3+}(g) + 3O^{2-}(g) \qquad \mathbf{\Delta H = 12{,}405\ kJ/mol}$$

**8.90** Fluorine

**8.92** $H^-$ and He are isoelectronic species with two electrons. Since $H^-$ has only one proton compared to two protons for He, the nucleus of $H^-$ will attract the two electrons less strongly compared to He. Therefore, $H^-$ is larger.

**8.94**

| Oxide | Name | Property |
|---|---|---|
| $Li_2O$ | lithium oxide | basic |
| BeO | beryllium oxide | amphoteric |
| $B_2O_3$ | boron oxide | acidic |
| $CO_2$ | carbon dioxide | acidic |
| $N_2O_5$ | dinitrogen pentoxide | acidic |

Note that only the highest oxidation states are considered.

**8.96** In its chemistry, hydrogen can behave like an alkali metal ($H^+$) and like a halogen ($H^-$). $H^+$ is a single proton.

**8.98** Replacing Z in the equation given in Problem 8.57 with (Z − σ) gives:

$$E_n = (2.18 \times 10^{-18}\ \text{J})(Z - \sigma)^2\left(\frac{1}{n^2}\right)$$

For helium, the atomic number (Z) is 2, and in the ground state, its two electrons are in the first energy level, so $n = 1$. Substitute Z, $n$, and the first ionization energy into the above equation to solve for $\sigma$.

$$E_1 = 3.94 \times 10^{-18}\text{ J} = (2.18 \times 10^{-18}\text{ J})(2 - \sigma)^2\left(\frac{1}{1^2}\right)$$

$$(2 - \sigma)^2 = \frac{3.94 \times 10^{-18}\text{ J}}{2.18 \times 10^{-18}\text{ J}}$$

$$2 - \sigma = \sqrt{1.81}$$

$$\sigma = 2 - 1.35 = \mathbf{0.65}$$

**8.100** The volume of a sphere is $\frac{4}{3}\pi r^3$.

The percentage of volume occupied by $K^+$ compared to K is:

$$\frac{\text{volume of K}^+\text{ ion}}{\text{volume of K atom}} \times 100\% = \frac{\frac{4}{3}\pi(133\text{ pm})^3}{\frac{4}{3}\pi(227\text{ pm})^3} \times 100\% = 20.1\%$$

Therefore, there is a decrease in volume of (100 – 20.1)% = **79.9%** when $K^+$ is formed from K.

**8.102** Rearrange the given equation to solve for ionization energy.

$$IE = h\nu - \frac{1}{2}mu^2$$

or,

$$IE = \frac{hc}{\lambda} - KE$$

The kinetic energy of the ejected electron is given in the problem. Substitute $h$, $c$, and $\lambda$ into the above equation to solve for the ionization energy.

$$\textbf{IE} = \frac{(6.63 \times 10^{-34}\text{ J}\cdot\text{s})(3.00 \times 10^8\text{ m/s})}{162 \times 10^{-9}\text{ m}} - (5.34 \times 10^{-19}\text{ J})$$

$$IE = \mathbf{6.94 \times 10^{-19}\text{ J}}$$

We might also want to express the ionization energy in kJ/mol.

$$\frac{6.94 \times 10^{-19}\text{ J}}{1\text{ photon}} \times \frac{6.022 \times 10^{23}\text{ photons}}{1\text{ mol}} \times \frac{1\text{ kJ}}{1000\text{ J}} = \mathbf{418\text{ kJ/mol}}$$

To ensure that the ejected electron is the valence electron, UV light of the *longest* wavelength (lowest energy) should be used that can still eject electrons.

**8.104** We want to determine the second ionization energy of lithium.

$$Li^{+} \longrightarrow Li^{2+} + e^{-} \qquad IE_2 = ?$$

The equation given in Problem 8.57 allows us to determine the third ionization energy for Li. Knowing the total energy needed to remove all three electrons from Li, we can calculate the second ionization energy by difference.

$$\text{Energy needed to remove three electrons} = IE_1 + IE_2 + IE_3$$

First, let's calculate $IE_3$. For Li, Z = 3, and $n = 1$ because the third electron will come from the 1$s$ orbital.

$$IE_3 = \Delta E = E_\infty - E_3$$

$$\mathbf{IE_3} = -(2.18 \times 10^{-18}\text{ J})(3)^2\left(\frac{1}{\infty^2}\right) + (2.18 \times 10^{-18}\text{ J})(3)^2\left(\frac{1}{1^2}\right)$$

$$IE_3 = +1.96 \times 10^{-17}\text{ J}$$

Converting to units of kJ/mol:

$$\mathbf{IE_3} = (1.96 \times 10^{-17}\text{ J}) \times \frac{6.022 \times 10^{23}\text{ ions}}{1\text{ mol}} = 1.18 \times 10^{7}\text{ J/mol} = 1.18 \times 10^{4}\text{ kJ/mol}$$

$$\text{Energy needed to remove three electrons} = IE_1 + IE_2 + IE_3$$

$$1.96 \times 10^{4}\text{ kJ/mol} = 520\text{ kJ/mol} + IE_2 + (1.18 \times 10^{4}\text{ kJ/mol})$$

$$\boldsymbol{IE_2 = 7.28 \times 10^{3}\text{ kJ/mol}}$$

**8.106** X must belong to Group 4A; it is probably **Sn** or **Pb** because it is not a very reactive metal (it is certainly not reactive like an alkali metal).

Y is a nonmetal since it does *not* conduct electricity. Since it is a light yellow solid, it is probably **phosphorus** (Group 5A).

Z is an **alkali metal** since it reacts with air to form a basic oxide or peroxide.

**8.108** $Na \longrightarrow Na^{+} + e^{-} \qquad IE_1 = 495.9\text{ kJ/mol}$

This equation is the reverse of the electron affinity for $Na^+$. Therefore, the electron affinity of $Na^+$ is **+495.9 kJ/mol.** Note that the electron affinity is positive, indicating that energy is liberated when an electron is added to an atom or ion. You should expect this since we are adding an electron to a positive ion.

**8.110** The reaction representing the electron affinity of chlorine is:

$$Cl(g) + e^{-} \longrightarrow Cl^{-}(g) \qquad \Delta H^\circ = +349\text{ kJ/mol}$$

It follows that the energy needed for the reverse process is also +349 kJ/mol.

$$Cl^{-}(g) + h\nu \longrightarrow Cl(g) + e^{-} \qquad \Delta H^\circ = +349\text{ kJ/mol}$$

The energy above is the energy of one mole of photons. We need to convert to the energy of one photon in order to calculate the wavelength of the photon.

$$\frac{349\ \text{kJ}}{1\ \text{mol photons}} \times \frac{1\ \text{mol photons}}{6.022 \times 10^{23}\ \text{photons}} \times \frac{1000\ \text{J}}{1\ \text{kJ}} = 5.80 \times 10^{-19}\ \text{J/photon}$$

Now, we can calculate the wavelength of a photon with this energy.

$$\lambda = \frac{hc}{E} = \frac{(6.63 \times 10^{-34}\ \text{J}\cdot\text{s})(3.00 \times 10^{8}\ \text{m/s})}{5.80 \times 10^{-19}\ \text{J}} = 3.43 \times 10^{-7}\ \text{m} = \mathbf{343\ nm}$$

The radiation is in the **ultraviolet** region of the electromagnetic spectrum.

**8.112** The equation that we want to calculate the energy change for is:

$$Na(s) \longrightarrow Na^{+}(g) + e^{-} \qquad \Delta H° = ?$$

Can we take information given in the problem and other knowledge to end up with the above equation? This is a Hess's law problem (see Chapter 6).

In the problem we are given: $Na(s) \longrightarrow Na(g) \qquad \Delta H° = 108.4\ \text{kJ/mol}$

We also know the ionization energy of Na (*g*). $Na(g) \longrightarrow Na^{+}(g) + e^{-} \qquad \Delta H° = 495.9\ \text{kJ/mol}$

---

Adding the two equations: $Na(s) \longrightarrow Na^{+}(g) + e^{-} \qquad \mathbf{\Delta H° = 604.3\ kJ/mol}$

**8.114** The electron configuration of titanium is: $[Ar]4s^23d^2$. Titanium has four valence electrons, so the maximum oxidation number it is likely to have in a compound is +4. The compounds followed by the oxidation state of titanium are: $K_3TiF_6$, +3; $K_2Ti_2O_5$, +4; $TiCl_3$, +3; $K_2TiO_4$, +6; and $K_2TiF_6$, +4. $\mathbf{K_2TiO_4}$ is unlikely to exist because of the oxidation state of Ti of +6. Titanium in an oxidation state greater than +4 is unlikely because of the very high ionization energies needed to remove the fifth and sixth electrons.

**8.116** The unbalanced ionic equation is: $MnF_6^{2-} + SbF_5 \longrightarrow SbF_6^{-} + MnF_3 + F_2$

In this redox reaction, $Mn^{4+}$ is reduced to $Mn^{3+}$, and $F^-$ from both $MnF_6^{2-}$ and $SbF_5$ is oxidized to $F_2$.

We can simplify the half-reactions. $Mn^{4+} \xrightarrow{\text{reduction}} Mn^{3+}$

$$F^- \xrightarrow{\text{oxidation}} F_2$$

Balancing the two half-reactions: $Mn^{4+} + e^- \longrightarrow Mn^{3+}$

$$2F^- \longrightarrow F_2 + 2e^-$$

Adding the two half-reactions: $2Mn^{4+} + 2F^- \longrightarrow 2Mn^{3+} + F_2$

We can now reconstruct the complete balanced equation. In the balanced equation, we have 2 moles of Mn ions and 1 mole of $F_2$ on the products side.

$$\mathbf{2K_2MnF_6 + SbF_5 \longrightarrow KSbF_6 + 2MnF_3 + 1F_2}$$

We can now balance the remainder of the equation by inspection. Notice that there are 4 moles of $K^+$ on the left, but only 1 mole of $K^+$ on the right. The balanced equation is:

$$\mathbf{2K_2MnF_6 + 4SbF_5 \longrightarrow 4KSbF_6 + 2MnF_3 + F_2}$$

**8.118** To work this problem, assume that the oxidation number of oxygen is −2.

| Oxidation number | Chemical formula |
|---|---|
| +1 | $N_2O$ |
| +2 | NO |
| +3 | $N_2O_3$ |
| +4 | $NO_2$, $N_2O_4$ |
| +5 | $N_2O_5$ |

**8.120** The larger the effective nuclear charge, the more tightly held are the electrons. Thus, the atomic radius will be small, and the ionization energy will be large. The quantities show an opposite periodic trend.

**8.122** We assume that the m.p. and b.p. of bromine will be between those of chlorine and iodine.

Taking the average of the melting points and boiling points:

$$\textbf{m.p.} = \frac{-101.0°\text{C} + 113.5°\text{C}}{2} = \mathbf{6.3°C} \qquad \text{(Handbook: } -7.2°\text{C)}$$

$$\textbf{b.p.} = \frac{-34.6°\text{C} + 184.4°\text{C}}{2} = \mathbf{74.9°C} \qquad \text{(Handbook: } 58.8\ °\text{C)}$$

The estimated values do not agree very closely with the actual values because $Cl_2(g)$, $Br_2(l)$, and $I_2(s)$ are in different physical states. If you were to perform the same calculations for the noble gases, your calculations would be much closer to the actual values.

**8.124** There is a large jump in energy between $IE_2$ and $IE_3$. This indicates that the element has two valence electrons. This representative element is an **alkaline earth metal.**

**8.126** **(a)** It was determined that the periodic table was based on atomic number, not atomic mass.

**(b)** Argon:

$$(0.00337 \times 35.9675 \text{ amu}) + (0.00063 \times 37.9627 \text{ amu}) + (0.9960 \times 39.9624 \text{ amu}) = \mathbf{39.95\ amu}$$

Potassium:

$$(0.93258 \times 38.9637 \text{ amu}) + (0.000117 \times 39.9640 \text{ amu}) + (0.0673 \times 40.9618 \text{ amu}) = \mathbf{39.10\ amu}$$

**8.128** $Z = 119$

Electron configuration: $[Rn]7s^2 5f^{14} 6d^{10} 7p^6 8s^1$

**8.130** There is a large jump from the third to the fourth ionization energy, indicating a change in the principal quantum number *n*. In other words, the fourth electron removed is an inner, noble gas core electron, which is difficult to remove. Therefore, the element is in **Group 3A**.

**8.132** **(a)** $SiH_4$, $GeH_4$, $SnH_4$, $PbH_4$

**(b)** Metallic character increases going down a family of the periodic table. Therefore, RbH would be more ionic than NaH.

**(c)** Since Ra is in Group 2A, we would expect the reaction to be the same as other alkaline earth metals with water.

$$\text{Ra}(s) + 2\text{H}_2\text{O}(l) \rightarrow \text{Ra(OH)}_2(aq) + \text{H}_2(g)$$

**(d)** Beryllium (diagonal relationship)

**8.134** Coulomb's Law states that the potential energy (*E*) between two ions is directly proportional to the product of their charges and inversely proportional to the distance of separation between them.

$$E = k\frac{Q_{\text{cation}}Q_{\text{anion}}}{r}$$

In this case the charges of the ions are the same, so any difference in binding energy is due to the distance of separation (*r*) between the ions. $Mg^{2+}$ is the smallest ion of the group and thus will have the greatest tendency for binding to phosphate. $Ba^{2+}$ is the largest ion of the group and thus will have the lowest tendency for binding to phosphate.

**8.136** The importance and usefulness of the periodic table lie in the fact that we can use our understanding of the general properties and trends within a group or a period to predict with considerable accuracy the properties of any element, even though the element may be unfamiliar to us. For example, elements in the same group or family have the same valence electron configurations. Due to the same number of valence electrons occupying similar orbitals, elements in the same family have similar chemical properties. In addition, trends in properties such as ionization energy, atomic radius, electron affinity, and metallic character can be predicted based on an element's position in the periodic table. Ionization energy typically increases across a period of the period table and decreases down a group. Atomic radius typically decreases across a period and increases down a group. Electron affinity typically increases across a period and decreases down a group. Metallic character typically decreases across a period and increases down a group. The periodic table is an extremely useful tool for a scientist. Without having to look in a reference book for a particular element's properties, one can look at its position in the periodic table and make educated predictions as to its many properties such as those mentioned above.

**8.138** The first statement that an allotropic form of the element is a colorless crystalline solid, might lead you to think about diamond, a form of carbon. When carbon is reacted with excess oxygen, the colorless gas, carbon dioxide is produced.

$$C(s) + O_2(g) \rightarrow CO_2(g)$$

When $CO_2(g)$ is dissolved in water, carbonic acid is produced.

$$CO_2(g) + H_2O(l) \rightarrow H_2CO_3(aq)$$

Element X is most likely carbon, choice **(c)**.

**8.140** The ionization energy of 412 kJ/mol represents the energy difference between the ground state and the dissociation limit, whereas the ionization energy of 126 kJ/mol represents the energy difference between the first excited state and the dissociation limit. Therefore, the energy difference between the ground state and the excited state is:

$$\Delta E = (412 - 126)\text{ kJ/mol} = 286\text{ kJ/mol}$$

The energy of light emitted in a transition from the first excited state to the ground state is therefore 286 kJ/mol. We first convert this energy to units of J/photon, and then we can calculate the wavelength of light emitted in this electronic transition.

$$E = \frac{286 \times 10^3\text{ J}}{1\text{ mol}} \times \frac{1\text{ mol}}{6.022 \times 10^{23}\text{ photons}} = 4.75 \times 10^{-19}\text{ J/photon}$$

$$\lambda = \frac{hc}{\mathbf{E}} = \frac{(6.63 \times 10^{-34}\text{ J}\cdot\text{s})(3.00 \times 10^8\text{ m/s})}{4.75 \times 10^{-19}\text{ J}} = \mathbf{4.19 \times 10^{-7}\ m} = \mathbf{419\ nm}$$

**8.142** In He, $r$ is greater than that in H. Also, the shielding in He makes $Z_{eff}$ less than 2. Therefore, $I_1(\text{He}) < 2I(\text{H})$. In $He^+$, there is only one electron so there is no shielding. The greater attraction between the nucleus and the lone electron reduces $r$ to less than the $r$ of hydrogen. Therefore, $I_2(\text{He}) > 2I(\text{H})$.

**8.144** We rearrange the equation given in the problem to solve for $Z_{eff}$.

$$Z_{eff} = n\sqrt{\frac{IE_1}{1312\text{ kJ/mol}}}$$

Li: $$Z_{eff} = (2)\sqrt{\frac{520\text{ kJ/mol}}{1312\text{ kJ/mol}}} = \mathbf{1.26}$$

Na: $$Z_{eff} = (3)\sqrt{\frac{495.9\text{ kJ/mol}}{1312\text{ kJ/mol}}} = \mathbf{1.84}$$

K: $$Z_{eff} = (4)\sqrt{\frac{418.7\text{ kJ/mol}}{1312\text{ kJ/mol}}} = \mathbf{2.26}$$

As we move down a group, $Z_{eff}$ increases. This is what we would expect because shells with larger $n$ values are less effective at shielding the outer electrons from the nuclear charge.

$$\text{Li:} \quad \frac{Z_{eff}}{n} = \frac{1.26}{2} = \mathbf{0.630}$$

$$\text{Na:} \quad \frac{Z_{eff}}{n} = \frac{1.84}{3} = \mathbf{0.613}$$

$$\text{K:} \quad \frac{Z_{eff}}{n} = \frac{2.26}{4} = \mathbf{0.565}$$

The $Z_{eff}/n$ values are fairly constant, meaning that the screening per shell is about the same.

**8.146** Please see the website, www.webelements.com. Click on the "Biology" tab above the periodic table, and then click on each of the listed elements. A brief summary of the biological role of each element is provided.

# CHAPTER 9
# CHEMICAL BONDING I: BASIC CONCEPTS

## PROBLEM-SOLVING STRATEGIES AND TUTORIAL SOLUTIONS

### TYPES OF PROBLEMS

**Problem Type 1:** Classifying Chemical Bonds.

**Problem Type 2:** Calculating the Lattice Energy of Ionic Compounds.

**Problem Type 3:** Writing Lewis Structures.

**Problem Type 4:** Formal Charges.

- **(a)** Assigning formal charges.
- **(b)** Choosing the most plausible Lewis structure based on formal charges.

**Problem Type 5:** Drawing Resonance Structures.

**Problem Type 6:** Exceptions to the Octet Rule.

- **(a)** The incomplete octet.
- **(b)** Odd-electron molecules.
- **(c)** The expanded octet.

**Problem Type 7:** Using Bond enthalpies to Estimate the Enthalpy of a Reaction.

## PROBLEM TYPE 1: CLASSIFYING CHEMICAL BONDS

We can classify bonds as three different types: ionic, polar covalent, or covalent.

In a **covalent bond**, the electron pair of the bond is shared *equally* by the two atoms.

In a **polar covalent bond**, the electron pair of the bond is shared *unequally* by the two atoms. The electrons spend more time in the vicinity of one atom than the other.

Covalent bonds (polar or nonpolar) are typically formed between two nonmetals.

In an **ionic bond**, the electron or electrons are nearly completely transferred from one atom to another. Ionic bonds are typically formed between a metal cation and a nonmetal anion or a metal cation and a polyatomic anion.

A property that helps us distinguish a nonpolar covalent bond from a polar covalent bond is **electronegativity**, the ability of an atom to attract toward itself the electrons in a chemical bond. Elements with high electronegativities have a greater tendency to attract electrons than elements with low electronegativities. Linus Pauling devised a method for calculating *relative* electronegativities of most elements. These values are shown in Figure 9.5 of the text.

Only atoms of the same element, which have the same electronegativity, can be joined by a pure *covalent bond*. Atoms of elements with similar electronegativities tend to form *polar covalent bonds* with each other because the shift in electron density is usually small. There is no sharp distinction between a polar bond and an ionic bond, but the following rule is helpful in distinguishing them. An *ionic bond* forms when the electronegativity difference between the two bonding atoms is 2.0 or more. This rule applies to most but not all ionic compounds.

**EXAMPLE 9.1**

**Classify the following bonds as ionic, polar covalent, or covalent: (a) the bond in KCl, (b) the OH bond in $H_2O$, and (c) the OO bond in oxygen gas ($O_2$).**

**Strategy:** We can look up electronegativity values in Figure 9.5 of the text. The amount of ionic character is based on the electronegativity difference between the two atoms. The larger the electronegativity difference, the greater the ionic character.

**Solution:**

(a) In Figure 9.5 of the text, we see that the electronegativity difference between K and Cl is 2.2, above the 2.0 guideline. Therefore, the bond between K and Cl is ionic. Remember that an ionic bond is typically formed between a metal cation and a nonmetal anion or a metal cation and a polyatomic anion.

(b) The electronegativity difference between O and H is 1.4, which is appreciable but not large enough to qualify $H_2O$ as an ionic compound. Therefore, the bond between O and H is polar covalent. Recall that polar covalent compounds are typically formed between two different nonmetals.

(c) The two O atoms are identical in every respect. Therefore, the bond between them is purely covalent.

**PRACTICE EXERCISE**

1. Classify the following bonds as ionic, polar covalent, or covalent: (a) the NH bond in $NH_3$, (b) the OO bond in hydrogen peroxide ($H_2O_2$), and (c) the bond in NaF.

**Text Problems:** 9.20, **9.36**, 9.38, 9.40

## PROBLEM TYPE 2: CALCULATING THE LATTICE ENERGY OF IONIC COMPOUNDS

See Section 9.3 of the text for a complete discussion of lattice energy. Example 9.2 below will illustrate the method used to calculate the lattice energy of an ionic compound.

**EXAMPLE 9.2**

**Calculate the lattice energy of magnesium oxide, MgO, given that the enthalpy of sublimation of Mg is 150 kJ/mol and the electron affinity of $O^-$ is −780 kJ/mol.**

**Strategy:** We want to calculate the enthalpy change corresponding to the following reaction:

$$Mg^{2+}(g) + O^{2-}(g) \longrightarrow MgO(s) \qquad \Delta H^\circ = ?$$

This reaction is the reverse of the reaction for the lattice energy. If we can calculate $\Delta H^\circ$ for the above reaction, we can determine the *lattice energy*. Starting with Mg(*s*) and $O_2(g)$, we can follow two different pathways to the product, MgO(*s*). The enthalpy changes for the two pathways will be equal.

Pathway 1: This is the overall reaction, which is the $\Delta H_f^\circ$ of MgO(*s*). You can look up the appropriate value in Appendix 3 of the text.

$$Mg(s) + O_2(g) \longrightarrow MgO(s) \qquad \Delta H^\circ = -601.8 \text{ kJ/mol}$$

Pathway 2: This is the indirect pathway. Using a series of steps, we can form the product $MgO(s)$. The last step in the process will be:

$$Mg^{2+}(g) + O^{2-}(g) \longrightarrow MgO(s)$$

This reaction is the reverse of the reaction for the lattice energy. Knowing the enthalpy changes for the other steps, we can calculate the enthalpy change for the step above.

**Solution:**

*Step 1*: Convert solid magnesium to magnesium vapor.

$$Mg(s) \longrightarrow Mg(g) \qquad \Delta H_1^\circ = 150 \text{ kJ/mol}$$

This is the energy needed to sublime $Mg(s)$.

*Step 2*: Dissociate ½ mole of $O_2$ gas into separate gaseous O atoms.

$$\frac{1}{2}O_2(g) \longrightarrow O(g) \qquad \Delta H_2^\circ = \left(\frac{1}{2}\right)(498.7 \text{ kJ/mol}) = 249.4 \text{ kJ/mol}$$

498.7 kJ is the amount of energy needed to break a mole of O=O bonds. Here we are breaking the bonds in half a mole of $O_2$. You can find this bond enthalpy in Table 9.4 of the text.

*Step 3*: Ionize 1 mole of gaseous Mg atoms.

$$Mg(g) \longrightarrow Mg^{2+}(g) + 2e^- \qquad \Delta H_3^\circ = 738.1 \text{ kJ/mol} + 1450 \text{ kJ/mol} = 2188 \text{ kJ/mol}$$

This process corresponds to the first and second ionization energies of Mg. See Table 8.2 of the text for the ionization energies.

*Step 4*: Add 2 moles of electrons to 1 mole of gaseous O atoms.

$$O(g) + e^- \longrightarrow O^-(g) \qquad \Delta H_{4a}^\circ = -142 \text{ kJ/mol}$$

$$O^-(g) + e^- \longrightarrow O^{2-}(aq) \qquad \Delta H_{4b}^\circ = 780 \text{ kJ/mol}$$

This process corresponds to the opposite of the electron affinity value of O and the electron affinity value of $O^-$. The electron affinity of O is given in Table 8.3 of the text.

*Step 5*: Combine 1 mole of gaseous $Mg^{2+}$ and 1 mole of gaseous $O^{2-}$ to form 1 mole of solid MgO.

$$Mg^{2+}(g) + O^{2-}(g) \longrightarrow MgO(s) \qquad \Delta H_5^\circ = ?$$

This is the enthalpy change that we wish to calculate because it is the opposite of the lattice energy. Changing the sign of the value that we calculate will give us the lattice energy.

*Step 6*: The enthalpy change for the two pathways is the same. We can write:

$$\Delta H^\circ(\text{pathway 1}) = \Delta H^\circ(\text{pathway 2})$$

$$\Delta H_f^\circ[MgO(s)] = \Delta H_1^\circ + \Delta H_2^\circ + \Delta H_3^\circ + \Delta H_{4a}^\circ + \Delta H_{4b}^\circ + \Delta H_5^\circ$$

Substituting the values from above:

$$-601.8 \text{ kJ/mol} = (150 + 249.4 + 2188 - 142 + 780)\text{kJ/mol} + \Delta H_5^\circ$$

$$\Delta H_5^\circ = -3827 \text{ kJ/mol}$$

This enthalpy change corresponds to the reaction that is the reverse of the lattice energy. Therefore, the lattice energy will have the opposite sign of $\Delta H_5^\circ$.

$$\mathrm{MgO}(s) \longrightarrow \mathrm{Mg}^{2+}(g) + \mathrm{O}^{2-}(g) \qquad \textbf{lattice energy} = \textbf{+3827 kJ/mol}$$

**PRACTICE EXERCISE**

**2.** Given that the enthalpy of sublimation of sodium (Na) is 108 kJ/mol and $\Delta H_f^\circ$ [NaF(*s*)] = −570 kJ/mol, calculate the lattice energy of NaF(*s*). See Tables 8.2, 8.3, and 9.4 of the text for other data.

**Text Problem:** 9.26

## PROBLEM TYPE 3: WRITING LEWIS STRUCTURES

The general rules for writing Lewis structures are given below. For more detailed rules, see Section 9.6 of the text.

1. Write the skeletal structure of the compound, using chemical symbols and placing bonded atoms next to one another. In general, the least electronegative atom occupies the central position. Hydrogen and fluorine usually occupy the terminal (end) positions in the Lewis structure.
2. Count the number of valence electrons present. For polyatomic anions, add the number of negative charges to that total. For polyatomic cations, subtract the number of positive charges from the number of valence electrons.
3. Draw a single covalent bond between the central atom and each of the surrounding atoms. Complete the octets of the atoms bonded to the central atom with lone pairs. The total number of electrons to be used is that determined in step 2.
4. Sometimes there will not be enough electrons to satisfy the octet rule of the central atom by placing lone pairs. Try adding double or triple bonds between the surrounding atoms and the central atom, using the lone pairs from the surrounding atoms.

**EXAMPLE 9.3**

**Write the Lewis structure for $SO_2$.**

**Strategy:** We follow the procedure for drawing Lewis structures outlined in Section 9.6 of the text.

**Solution:**

*Step 1:* Sulfur is less electronegative than oxygen, so it occupies the central position. The skeletal structure is:

O S O

*Step 2:* The outer-shell electron configurations of O and S are $2s^2 2p^4$ and $3s^2 3p^4$, respectively. Thus, there are

$$6 + (2 \times 6) = 18 \text{ valence electrons}$$

> **Tip:** For the representative elements (Group 1A – 7A), the number of valence electrons is equal to the group number. For example, both O and S are in Group 6A; thus, they both have 6 valence electrons.

*Step 3:* We draw a single covalent bond between S and each O, and then complete the octets for the O atoms. We place the remaining two electrons on S.

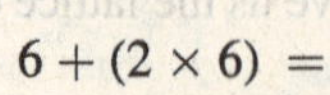

*Step 4:* The octet rule is satisfied for the oxygen atoms; however, the S atom does *not* satisfy the octet rule. Let's try making a sulfur-to-oxygen double bond by moving a lone pair from one of the O atoms to form another bond with S.

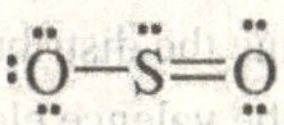

Now, the octet rule is also satisfied for the S atom.

**Check:** As a final check, we verify that there are 18 valence electrons in the Lewis structure of $SO_2$.

**EXAMPLE 9.4**

**Write the Lewis structure for $NO_3^-$.**

**Strategy:** We follow the procedure for drawing Lewis structures outlined in Section 9.6 of the text.

**Solution:**

*Step 1:* Nitrogen is less electronegative than oxygen, so it occupies the central position. The skeletal structure is:

O

O N O

*Step 2:* Nitrate is a polyatomic anion, so we must add the negative charge to the number of valence electrons. The outer-shell electron configurations of O (Group 4A) and N (Group 5A) are $2s^22p^4$ and $2s^22p^3$, respectively. Thus, there are

$$5 + (3 \times 6) + 1 = 24 \text{ valence electrons}$$

*Step 3:* We draw a single covalent bond between N and each O, and then complete the octets for the O atoms.

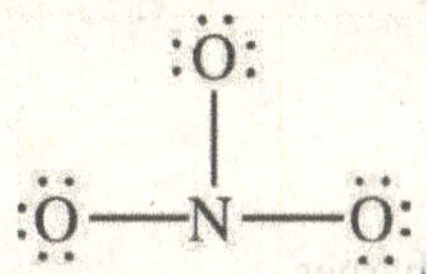

*Step 4:* The octet rule is satisfied for the oxygen atoms; however, the N atom does *not* satisfy the octet rule. Let's try making a nitrogen-to-oxygen double bond by moving a lone pair from one of the O atoms to form another bond with N.

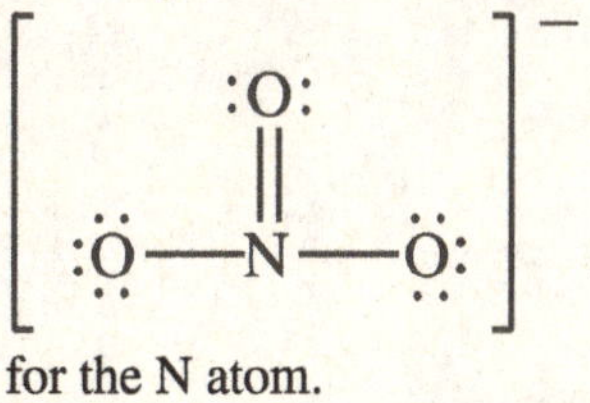

Now, the octet rule is also satisfied for the N atom.

**Check:** As a final check, we verify that there are 24 valence electrons in the Lewis structure of $NO_3^-$.

Also notice that we draw a bracket with the charge of the polyatomic ion around the Lewis structure. This is to distinguish an ion from a neutral molecule.

**PRACTICE EXERCISES**

3. Write the Lewis structure for $AsCl_3$.
4. Write the Lewis structure for cyanide ion, $CN^-$.

**Text Problems:** 9.44, **9.48**, 9.52

## PROBLEM TYPE 4: FORMAL CHARGES

**By comparing the number of electrons in an isolated atom (valence electrons) with the number of electrons that are associated with the same atom in a Lewis structure, we can determine the distribution of electrons in the molecule and draw the most plausible Lewis structure. This difference between the valence electrons in an isolated atom and the number of electrons assigned to that atom in a Lewis structure is called the atom's formal charge.**

**To assign the number of electrons on an atom in a Lewis structure, we proceed as:**

- **All the atom's nonbonding electrons are assigned to the atom.**
- **Half of the bonding electrons are assigned to the atom. For example, a single bond is a two-electron bond. One electron from the bond would be assigned to the given atom.**

**When you write formal charges, the following rules are helpful.**

- **For neutral molecules, the sum of the formal charges must add up to zero.**
- **For cations, the sum of the formal charges must equal the positive charge.**
- **For anions, the sum of the formal charges must equal the negative charge.**

### A. Assigning formal charges

**EXAMPLE 9.5**

**Assign formal charges to the atoms in the following Lewis structures.**

(a) :C≡O:

(b) :Ö—S̈=Ö (with lone pairs as drawn)

**Strategy:** Use the approach discussed above to assign formal charges.

**Solution:**

**(a)** The formal charge on each atom can be calculated using the following approach:

:C≡O:

| | C | O |
|---|---|---|
| Valence $e^-$ | 4 | 6 |
| $e^-$ assigned to atom | 5 | 5 |
| Difference (formal charge) | 1− | 1+ |

Some chemists do not approve of this structure for CO because it places a positive formal charge on the more electronegative oxygen atom.

**(b)** The formal charge on each atom can be calculated using the following approach:

:Ö—S̈=Ö

| | O | S | O |
|---|---|---|---|
| Valence $e^-$ | 6 | 6 | 6 |
| $e^-$ assigned to atom | 7 | 5 | 6 |
| Difference (formal charge) | 1− | 1+ | 0 |

**PRACTICE EXERCISE**

**5.** Assign formal charges to the atoms in the following Lewis structures.

(a) 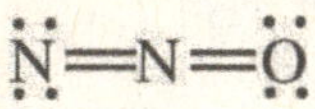

(b) 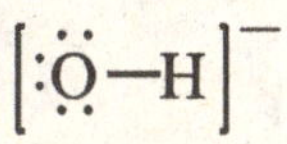

**Text Problems:** 9.58, 9.60

## B. Choosing the most plausible Lewis structure based on formal charges

The following guidelines show how to use formal charges to select a plausible Lewis structure for a given compound.

- For neutral molecules, a Lewis structure in which there are no formal charges is preferable to one in which formal charges are present.
- Lewis structures with large formal charges are less plausible than those with small formal charges.
- When comparing two structures with similar magnitudes of formal charges, the most plausible structure is the one in which negative formal charges are placed on the more electronegative atoms.

**EXAMPLE 9.6**

**Two possible Lewis structures for $BF_3$ are shown below. Which is the more reasonable structure in terms of formal charges?**

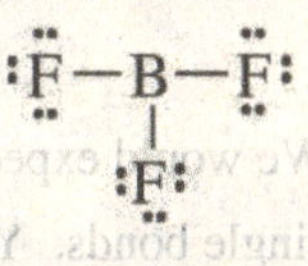

(a)

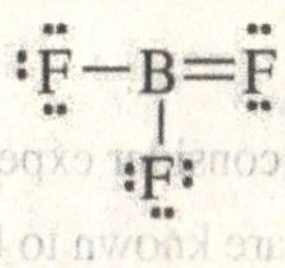

(b)

The formal charge on each atom in (a) can be calculated using the following approach:

B atom: formal charge = 3 valence $e^-$ − 3 $e^-$ assigned to the atom = 0

F atoms: formal charge = 7 valence $e^-$ − 7 $e^-$ assigned to the atom = 0

The formal charge on each atom in (b) can be calculated using the following approach:

B atom: formal charge = 3 valence $e^-$ − 4 $e^-$ assigned to the atom = 1−

F atom (double bond): formal charge = 7 valence $e^-$ − 6 $e^-$ assigned to the atom = 1+

F atoms (single bond): formal charge = 7 valence $e^-$ − 7 $e^-$ assigned to the atom = 0

The rule used to establish the most plausible structure is: For neutral molecules, a Lewis structure in which there are no formal charges is preferable to one in which formal charges are present. Thus, structure (a) is preferred over structure (b). We could also rule out structure (b) as the most plausible structure because there is a positive formal charge on the very electronegative F atom.

**PRACTICE EXERCISE**

6. Consider the following Lewis structures for the sulfate ion. Assign formal charges to each atom in the structure, and then determine which structure is more reasonable based on formal charges?

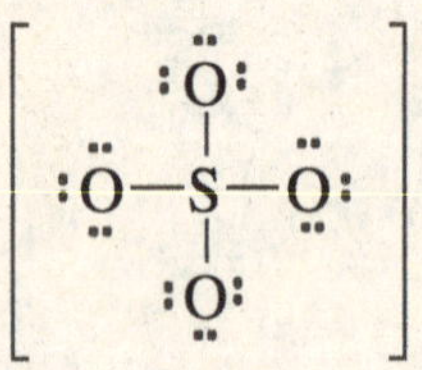

(a)

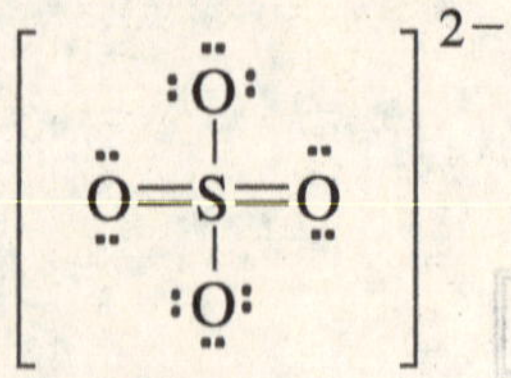

(b)

**Text Problem: 9.66**

## PROBLEM TYPE 5: DRAWING RESONANCE STRUCTURES

The Lewis structure for $SO_3$ is shown below.

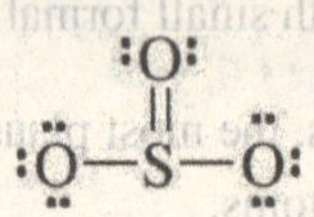

We can draw two more equivalent Lewis structures with the double bond between S and a different oxygen atom.

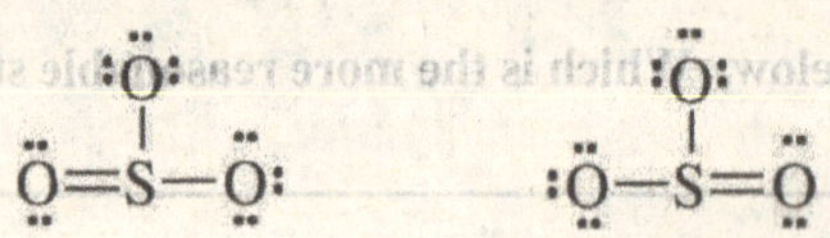

Which is the correct structure? Let's consider experimental data. We would expect the S–O bond to be longer than the S=O bond because double bonds are known to be shorter than single bonds. Yet experimental evidence shows that all three sulfur-to-oxygen bonds are equal in length. Therefore, none of the three structures shown accurately represents the molecule. We resolve this conflict by using all *three* Lewis structures to represent $SO_3$.

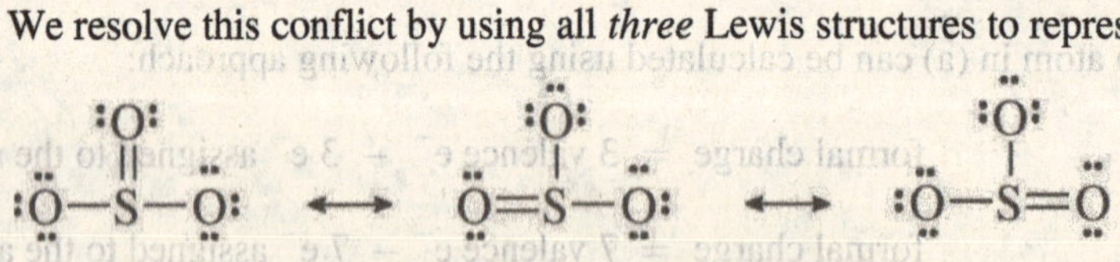

Each of the three structures is called a **resonance structure**. A resonance structure is one of two or more Lewis structures for a single molecule that cannot be described fully with only one Lewis structure. The symbol ⟷ indicates that the structures shown are resonance structures.

**EXAMPLE 9.7**

**We drew the Lewis structure for nitrate ion, $NO_3^-$, in Example 9.4. However, experimental evidence shows that all N–O bonds are equivalent. Draw resonance structures to indicate the equivalence of the N–O bonds.**

**Strategy:** We follow the procedure for drawing Lewis structures outlined in Section 9.6 of the text. After we complete the Lewis structure, we draw the resonance structures.

**Solution:** The Lewis structure drawn in Example 9.4 shown with formal charges is:

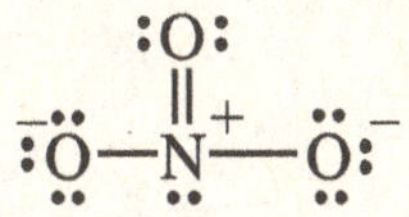

This structure, while a correct Lewis structure, does not show the equivalence of all three N–O bonds. Three contributing resonance structures can be drawn.

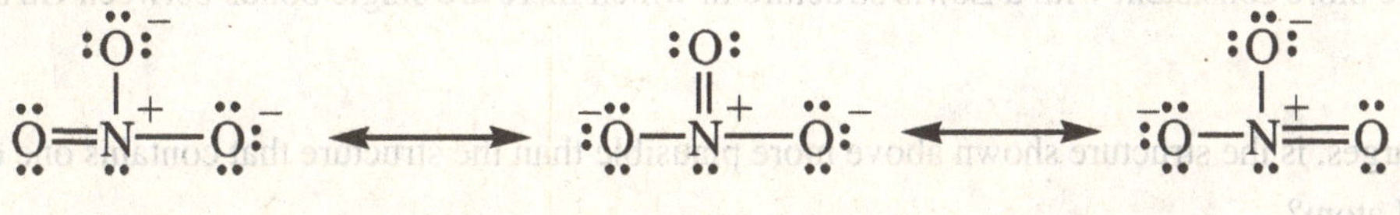

Resonance does not mean that the nitrate ion shifts quickly back and forth from one resonance structure to the other. Keep in mind that *none* of the resonance structures adequately represents the actual molecule, which has its own unique, stable structure. The actual structure is an average or hybrid of the above three structures. Resonance is a human invention, designed to address the limitations in these simple bonding models.

**PRACTICE EXERCISE**

**7.** Draw all the resonance structures for $N_2O$. The skeletal structure is N–N–O.

| Text Problems: 9.56, 9.58, 9.60 |
|---|

## PROBLEM TYPE 6: EXCEPTIONS TO THE OCTET RULE

### A. The incomplete octet

In some compounds the number of electrons surrounding the central atom in a stable molecule is fewer than eight. **Be**, **B**, and **Al** tend to form compounds in which they are surrounded by fewer than eight electrons.

**EXAMPLE 9.8**

**Draw the Lewis structure for $GaI_3$.**

**Strategy:** We follow the procedure outlined in Section 9.6 of the text for drawing Lewis structures.

**Solution:**

*Step 1:* Gallium is less electronegative than iodine, so it occupies the central position. The skeletal structure is:

I
I Ga I

*Step 2:* The outer-shell electron configurations of Ga (Group IIIA) and I (Group VIIA) are $4s^2 4p^1$ and $5s^2 5p^5$, respectively. Thus, there are

$$3 + (3 \times 7) = 24 \text{ valence electrons}$$

*Step 3:* We draw a single covalent bond between Ga and each I, and then complete the octets for the I atoms.

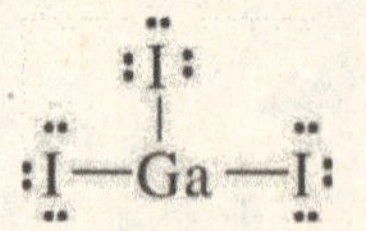

The octet rule is satisfied for the iodine atoms; however, the Ga atom does *not* satisfy the octet rule. A resonance structure with a double bond between Ga and I can be drawn that satisfies the octet rule for Ga. However, the properties of $GaI_3$ are more consistent with a Lewis structure in which there are single bonds between Ga and each I, as shown above.

Based on formal charges, is the structure shown above more plausible than the structure that contains one double bond between Ga and an I atom?

**PRACTICE EXERCISE**

8. Write the Lewis structure for $BCl_3$.

**Text Problems: 9.66,** 9.70

## B. Odd-electron molecules

Some molecules contain an **odd** number of electrons. To satisfy the octet rule, we need an even number of electrons. Therefore, the octet rule cannot be satisfied in a molecule that has an odd number of electrons. When drawing a Lewis structure, if an atom has fewer than eight electrons, make sure that it is the least electronegative atom in the compound.

**EXAMPLE 9.9**

**Draw the Lewis structure for $NO_2$ (all bonds are equivalent).**

**Strategy:** We follow the procedure outlined in Section 9.6 of the text for drawing Lewis structures.

**Solution:**

*Step 1:* Nitrogen is less electronegative than oxygen, so it occupies the central position. The skeletal structure is:

O N O

*Step 2:* The outer-shell electron configurations of N (Group 5A) and O (Group 6A) are $2s^22p^3$ and $2s^22p^4$, respectively. Thus, there are

$$5 + (2 \times 6) = 17 \text{ valence electrons}$$

*Step 3:* We have an odd number of valence electrons, so the octet rule cannot be satisfied for at least one of the atoms in the molecule. Either nitrogen or oxygen in the structure will have fewer than eight electrons, because a second-row element cannot exceed an octet of electrons. Since N is less electronegative than O, nitrogen should be electron deficient. We draw a single covalent bond between N and each O, and then complete the octets for the O atoms.

*Step 4:* We have one electron left to place on the molecule. Placing the electron on N only gives five electrons around N. However, if we make a nitrogen-to-oxygen double bond by moving a lone pair from one of the O atoms to form another bond with N, nitrogen will be surrounded by seven electrons.

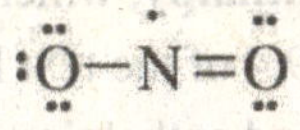

We could also draw resonance structures in which oxygen is electron deficient with seven electrons. However, the above structure is the most plausible since the least electronegative element is electron deficient.

## C. The expanded octet

A number of compounds contain more than eight electrons around an atom. These **expanded octets** only occur for atoms of elements in and beyond the third period of the periodic table. Atoms from the third period on can accommodate more than eight electrons because they also have *d* orbitals that can be used in bonding.

**EXAMPLE 9.10**

**Draw the Lewis structure for $ClF_3$.**

**Strategy:** We follow the procedure outlined in Section 9.6 of the text for drawing Lewis structures.

**Solution:**

*Step 1:* Chlorine is less electronegative than fluorine, so it occupies the central position. The skeletal structure is:

F
F Cl F

*Step 2:* The outer-shell electron configurations of F and Cl are $2s^22p^5$ and $3s^23p^5$, respectively. Thus, there are

$$7 + (3 \times 7) = 28 \text{ valence electrons}$$

*Step 3:* We draw a single covalent bond between Cl and each F, and then complete the octets for the F atoms.

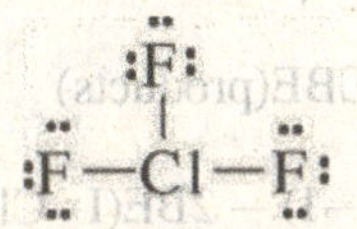

*Step 4:* At this point, we still have two electron pairs (4 $e^-$) to place. Fluorine cannot exceed an octet of electrons, so the electrons must be placed as lone pairs on chlorine. The correct Lewis structure is:

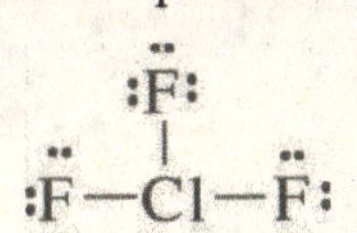

Chlorine can exceed an octet of electrons. An expanded octet can occur for atoms of elements in and beyond the third period of the periodic table.

**Check:** As a final check, we verify that there are 28 valence electrons in the Lewis structure of $ClF_3$.

**PRACTICE EXERCISE**

9. Write the Lewis structure for $PCl_5$.

**Text Problem:** 9.68

## PROBLEM TYPE 7: USING BOND ENTHALPIES TO ESTIMATE THE ENTHALPY OF A REACTION

A quantitative measure of the stability of a molecule is its **bond enthalpy**, which is the enthalpy change required to break a particular bond in one mole of gaseous molecules. Table 9.4 of the text lists the average bond enthalpies of a number of bonds found in polyatomic molecules, as well as the bond enthalpies of several diatomic molecules.

In many cases, it is possible to predict the approximate enthalpy of reaction by using the average bond enthalpies. Energy *is always required* to break chemical bonds and chemical bond formation is always accompanied by a *release of energy*. We can estimate the enthalpy of reaction by counting the total number of bonds broken and formed in the reaction and recording all the corresponding energy changes. The enthalpy of reaction in the *gas phase* is given by:

$$\Delta H^\circ = \text{total energy input} - \text{total energy released}$$

$$\Delta H^\circ = \Sigma\text{BE(reactants)} - \Sigma\text{BE(products)}$$

Where,

BE is the average bond enthalpy

$\Sigma$ represents summation

If the total energy input is greater than the total energy released, $\Delta H^\circ$ *is positive* and the reaction is *endothermic*. On the other hand, if more energy is released than absorbed, $\Delta H^\circ$ *is negative* and the reaction is *exothermic*.

**EXAMPLE 9.11**

**Use average bond enthalpies to estimate $\Delta H^\circ$ for the following reaction:**

$$Cl_2(g) + I_2(g) \longrightarrow 2ICl(g)$$

**The average I–Cl bond enthalpy is 210 kJ.**

**Strategy:** Keep in mind that bond breaking is an energy absorbing (endothermic) process and bond making is an energy releasing (exothermic) process. Therefore, the overall energy change is the difference between these two opposing processes, as described in Equation (9.3) of the text.

**Solution:**

$$\Delta H^\circ = \Sigma\text{BE(reactants)} - \Sigma\text{BE(products)}$$

$$\Delta H^\circ = \text{BE(Cl–Cl)} + \text{BE(I–I)} - 2\text{BE(I–Cl)}$$

$$= 242.7 \text{ kJ/mol} + 151.0 \text{ kJ/mol} - 2(210 \text{ kJ/mol})$$

$$\boldsymbol{\Delta H^\circ = -26 \text{ kJ/mol}}$$

Is the reaction exothermic or endothermic?

**PRACTICE EXERCISE**

10. Estimate the enthalpy change for the following reaction:

$$N_2(g) + O_2(g) \longrightarrow 2NO(g)$$

**Hint:** NO has a double bond [BE(N=O) = 630 kJ/mol]

**Text Problems: 9.74**, 9.76

## ANSWERS TO PRACTICE EXERCISES

1. (a) polar covalent (b) covalent (c) ionic

2. lattice energy (NaF) = 919 kJ/mol

3. 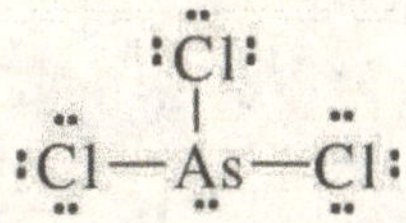

4. 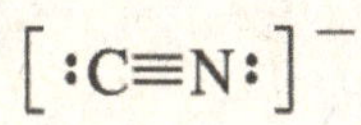

5. (a) Formal charge (left N) = 5 valence $e^-$ − 6 $e^-$ assigned to the atom = 1−

   Formal charge (middle N) = 5 valence $e^-$ − 4 $e^-$ assigned to the atom = 1+

   Formal charge (O) = 6 valence $e^-$ − 6 $e^-$ assigned to the atom = 0

   (b) Formal charge (O) = 6 valence $e^-$ − 7 $e^-$ assigned to the atom = 1−

   Formal charge (H) = 1 valence electron − 1 $e^-$ assigned to the atom = 0

6. (b) is the more plausible structure based on formal charges. There is a large positive (+2) formal charge on the S atom in structure (a). The formal charge on the S atom in structure (b) is zero.

7. 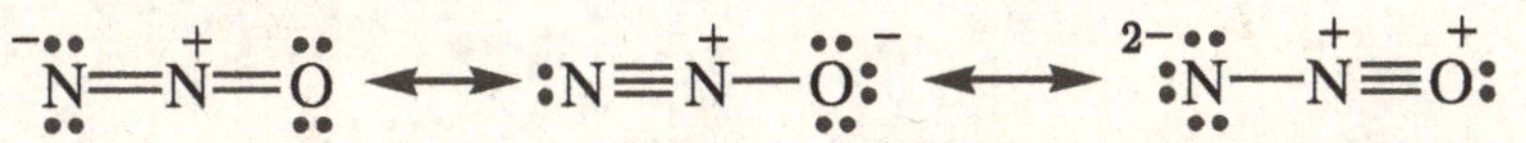

8. 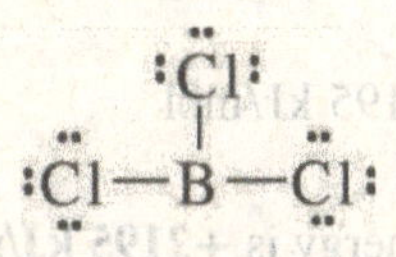

9. The lone pairs have been left off the chlorine atoms.

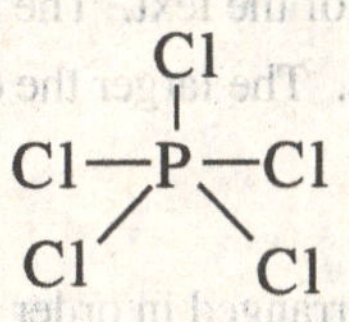

10. $\Delta H° = 180$ kJ/mol

## SOLUTIONS TO SELECTED TEXT PROBLEMS

**9.16** (a) RbI, rubidium iodide (b) $Cs_2SO_4$, cesium sulfate

(c) $Sr_3N_2$, strontium nitride (d) $Al_2S_3$, aluminum sulfide

**9.18** The Lewis representations for the reactions are:

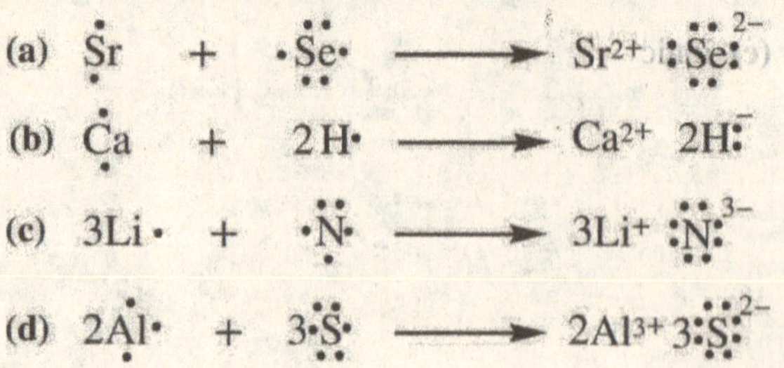


**9.20** **(a)** Covalent ($BF_3$, boron trifluoride) **(b)** ionic (KBr, potassium bromide)

**9.26**

| | | |
|---|---|---|
| (1) | $Ca(s) \rightarrow \cancel{Ca(g)}$ | $\Delta H_1^\circ = 121$ kJ/mol |
| (2) | $Cl_2(g) \rightarrow 2\cancel{Cl(g)}$ | $\Delta H_2^\circ = 242.8$ kJ/mol |
| (3) | $\cancel{Ca(g)} \rightarrow \cancel{Ca^+(g)} + \cancel{e^-}$ | $\Delta H_3^{\circ\prime} = 589.5$ kJ/mol |
| | $\cancel{Ca^+(g)} \rightarrow \cancel{Ca^{2+}(g)} + \cancel{e^-}$ | $\Delta H_3^{\circ\prime\prime} = 1145$ kJ/mol |
| (4) | $2[\cancel{Cl(g)} + \cancel{e^-} \rightarrow \cancel{Cl^-(g)}]$ | $\Delta H_4^\circ = 2(-349\text{ kJ/mol}) = -698$ kJ/mol |
| (5) | $\cancel{Ca^{2+}(g)} + 2\cancel{Cl^-(g)} \rightarrow CaCl_2(s)$ | $\Delta H_5^\circ = ?$ |
| | $Ca(s) + Cl_2(g) \rightarrow CaCl_2(s)$ | $\Delta H_{\text{overall}}^\circ = -795$ kJ/mol |

Thus we write:

$$\Delta H_{\text{overall}}^\circ = \Delta H_1^\circ + \Delta H_2^\circ + \Delta H_3^{\circ\prime} + \Delta H_3^{\circ\prime\prime} + \Delta H_4^\circ + \Delta H_5^\circ$$

$$\Delta H_5^\circ = (-795 - 121 - 242.8 - 589.5 - 1145 + 698)\text{kJ/mol} = -2195\text{ kJ/mol}$$

The lattice energy is represented by the reverse of equation (5); therefore, the lattice energy is **+2195 kJ/mol**.

**9.36** **Strategy:** We can look up electronegativity values in Figure 9.5 of the text. The amount of ionic character is based on the electronegativity difference between the two atoms. The larger the electronegativity difference, the greater the ionic character.

**Solution:** Let ΔEN = electronegativity difference. The bonds arranged in order of increasing ionic character are:

C−H (ΔEN = 0.4) < Br−H (ΔEN = 0.7) < F−H (ΔEN = 1.9) < Li−Cl (ΔEN = 2.0)

< Na−Cl (ΔEN = 2.1) < K−F (ΔEN = 3.2)

**9.38** The order of increasing ionic character is:

Cl−Cl (zero difference in electronegativity) < Br−Cl (difference 0.2) < Si−C (difference 0.7)

< Cs−F (difference 3.3).

**9.40** (a) The two silicon atoms are the same. The bond is covalent.

(b) The electronegativity difference between Cl and Si is 3.0 − 1.8 = 1.2. The bond is polar covalent.

(c) The electronegativity difference between F and Ca is 4.0 − 1.0 = 3.0. The bond is ionic.

(d) The electronegativity difference between N and H is 3.0 − 2.1 = 0.9. The bond is polar covalent.

**9.44**

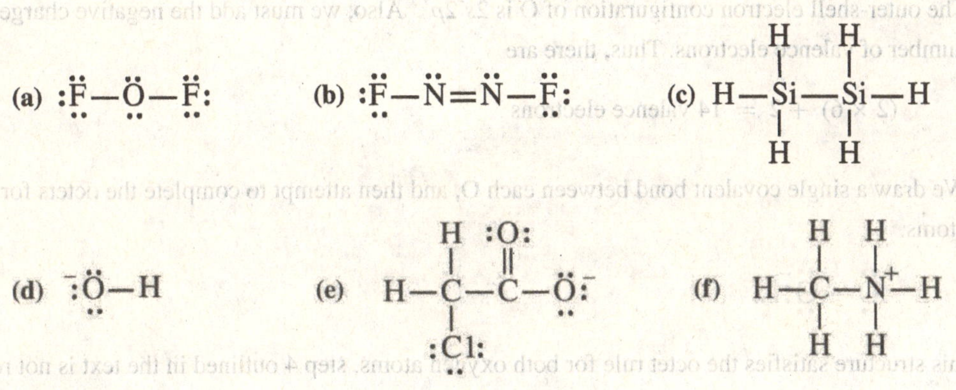

**9.46** The Lewis dot structures are:

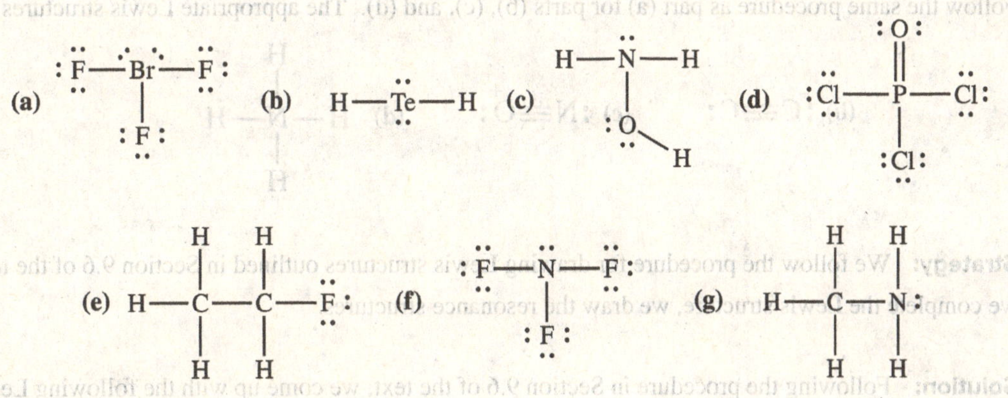

The octet rule is exceeded in each case.

**9.48** (a) Neither oxygen atom has a complete octet. The left-most hydrogen atom is forming two bonds ($4\ e^-$). Hydrogen can only be surrounded by at most two electrons.

(b) The correct structure is:

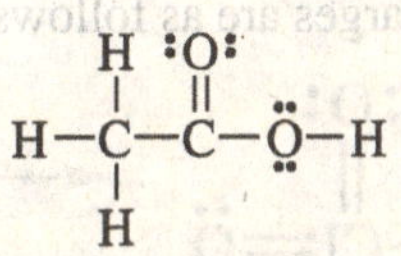

Do the two structures have the same number of electrons? Is the octet rule satisfied for all atoms other than hydrogen, which should have a duet of electrons?

**9.52** **Strategy:** We follow the procedure for drawing Lewis structures outlined in Section 9.6 of the text.

**Solution:**

**(a)**

*Step 1:* It is obvious that the skeletal structure is: O O

*Step 2:* The outer-shell electron configuration of O is $2s^22p^4$. Also, we must add the negative charges to the number of valence electrons. Thus, there are

$$(2 \times 6) + 2 = 14 \text{ valence electrons}$$

*Step 3:* We draw a single covalent bond between each O, and then attempt to complete the octets for the O atoms.

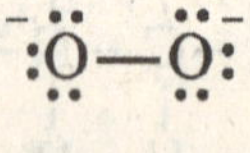

Because this structure satisfies the octet rule for both oxygen atoms, step 4 outlined in the text is not required.

**Check:** As a final check, we verify that there are 14 valence electrons in the Lewis structure of $O_2^-$.

Follow the same procedure as part (a) for parts (b), (c), and (d). The appropriate Lewis structures are:

**(b)** 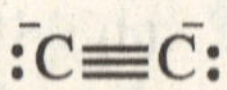 **(c)** 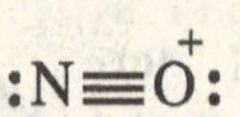 **(d)** 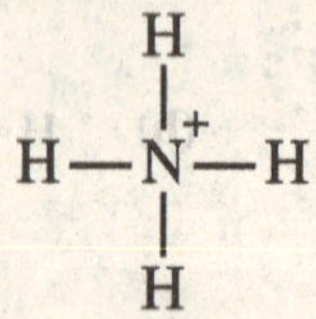

**9.56** **Strategy:** We follow the procedure for drawing Lewis structures outlined in Section 9.6 of the text. After we complete the Lewis structure, we draw the resonance structures.

**Solution:** Following the procedure in Section 9.6 of the text, we come up with the following Lewis structure for $ClO_3^-$.

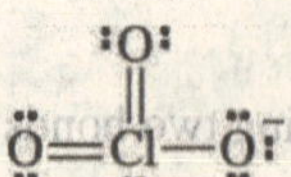

We can draw two more equivalent Lewis structures with the double bond between Cl and a different oxygen atom.

The resonance structures with formal charges are as follows:

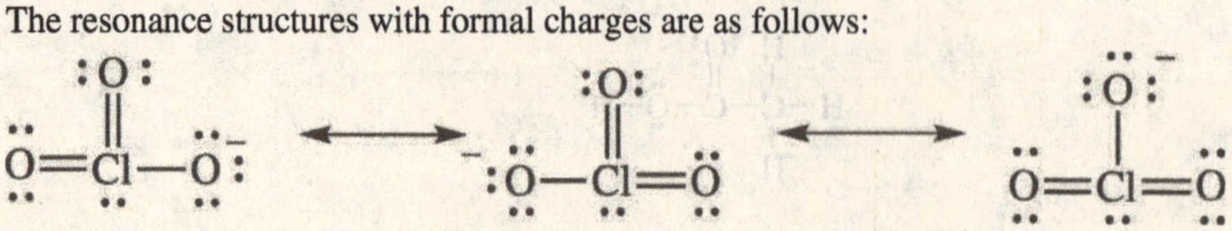

**9.58** The structures of the most important resonance forms are:

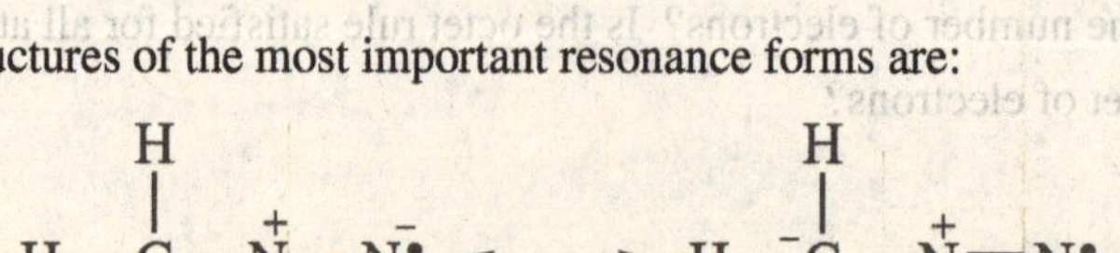

**9.60** Three reasonable resonance structures for $OCN^-$ are:

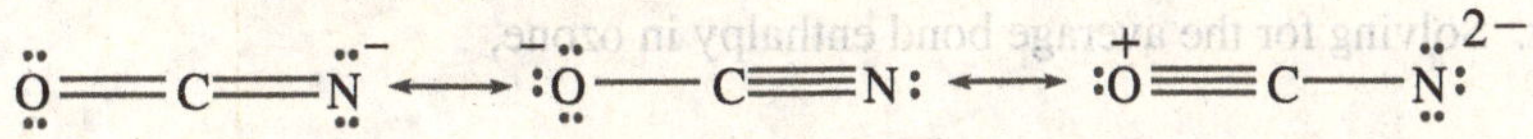

**9.66** **Strategy:** We follow the procedure outlined in Section 9.6 of the text for drawing Lewis structures. We assign formal charges as discussed in Section 9.7 of the text.

**Solution:** Drawing the structure with single bonds between Be and each of the Cl atoms, the octet rule for Be is *not* satisfied. The Lewis structure is:

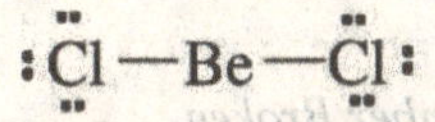

An octet of electrons on Be can only be formed by making two double bonds as shown below:

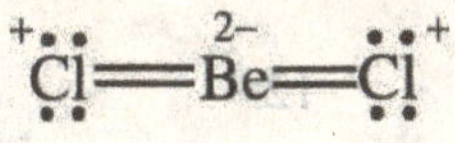

This places a high negative formal charge on Be and positive formal charges on the Cl atoms. This structure distributes the formal charges counter to the electronegativities of the elements. It is not a plausible Lewis structure.

**9.68** The outer electron configuration of antimony is $5s^2 5p^3$. The Lewis structure is shown below. All five valence electrons are shared in the five covalent bonds. The octet rule is not obeyed. (The electrons on the chlorine atoms have been omitted for clarity.)

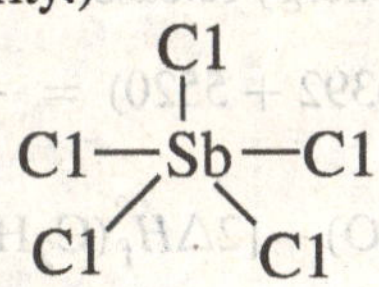

Can Sb have an expanded octet?

**9.70** The reaction can be represented as:

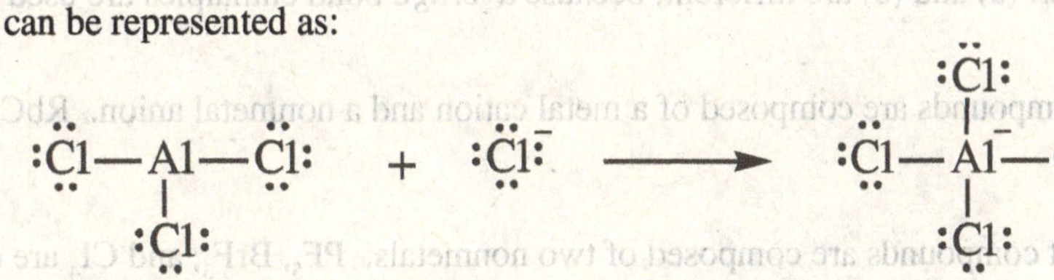

The new bond formed is called a **coordinate covalent bond.**

**9.74** **Strategy:** Keep in mind that bond breaking is an energy absorbing (endothermic) process and bond making is an energy releasing (exothermic) process. Therefore, the overall energy change is the difference between these two opposing processes, as described in Equation (9.3) of the text.

**Solution:** There are two oxygen-to-oxygen bonds in ozone. We will represent these bonds as O–O. However, these bonds might not be true oxygen-to-oxygen single bonds. Using Equation (9.3) of the text, we write:

$$\Delta H^\circ = \sum BE(\text{reactants}) - \sum BE(\text{products})$$

$$\Delta H^\circ = BE(O{=}O) - 2BE(O{-}O)$$

In the problem, we are given $\Delta H°$ for the reaction, and we can look up the O=O bond enthalpy in Table 9.4 of the text. Solving for the average bond enthalpy in ozone,

$$-2\text{BE(O–O)} = \Delta H° - \text{BE(O=O)}$$

$$\text{BE(O}-\text{O)} = \frac{\text{BE(O=O)} - \Delta H°}{2} = \frac{498.7\text{ kJ/mol} + 107.2\text{ kJ/mol}}{2} = \mathbf{303.0\text{ kJ/mol}}$$

Considering the resonance structures for ozone, is it expected that the O–O bond enthalpy in ozone is between the single O–O bond enthalpy (142 kJ) and the double O=O bond enthalpy (498.7 kJ)?

**9.76** (a)

| *Bonds Broken* | *Number Broken* | *Bond Enthalpy* (kJ/mol) | *Enthalpy Change* (kJ) |
|---|---|---|---|
| C – H | 12 | 414 | 4968 |
| C – C | 2 | 347 | 694 |
| O = O | 7 | 498.7 | 3491 |

| *Bonds Formed* | *Number Formed* | *Bond Enthalpy* (kJ/mol) | *Enthalpy Change* (kJ) |
|---|---|---|---|
| C = O | 8 | 799 | 6392 |
| O – H | 12 | 460 | 5520 |

$$\boldsymbol{\Delta H°} = \text{total energy input} - \text{total energy released}$$

$$= (4968 + 694 + 3491) - (6392 + 5520) = \mathbf{-2759\text{ kJ/mol}}$$

(b) $\Delta H° = 4\Delta H_f°(CO_2) + 6\Delta H_f°(H_2O) - [2\Delta H_f°(C_2H_6) + 7\Delta H_f°(O_2)]$

$$\boldsymbol{\Delta H°} = (4)(-393.5\text{ kJ/mol}) + (6)(-241.8\text{ kJ/mol}) - [(2)(-84.7\text{ kJ/mol}) + (7)(0)] = \mathbf{-2855\text{ kJ/mol}}$$

The answers for part (a) and (b) are different, because *average* bond enthalpies are used for part (a).

**9.78** Typically, ionic compounds are composed of a metal cation and a nonmetal anion. RbCl and $KO_2$ are ionic compounds.

Typically, covalent compounds are composed of two nonmetals. $PF_5$, $BrF_3$, and $CI_4$ are covalent compounds.

**9.80** Recall that you can classify bonds as ionic or covalent based on electronegativity difference.

The melting points (°C) are shown in parentheses following the formulas.

| | | | | |
|---|---|---|---|---|
| **Ionic:** | NaF (993) | $MgF_2$ (1261) | $AlF_3$ (1291) | |
| **Covalent:** | $SiF_4$ (−90.2) | $PF_5$ (−83) | $SF_6$ (−121) | $ClF_3$ (−83) |

Is there any correlation between ionic character and melting point?

**9.82** KF is an ionic compound. It is a solid at room temperature made up of $K^+$ and $F^-$ ions. It has a high melting point, and it is a strong electrolyte. Benzene, $C_6H_6$, is a covalent compound that exists as discrete molecules. It is a liquid at room temperature. It has a low melting point, is insoluble in water, and is a nonelectrolyte.

**9.84** The resonance structures are:

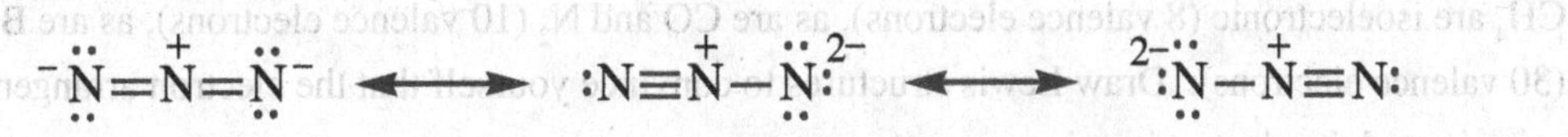

Which is the most plausible structure based on a formal charge argument?

**9.86** (a) An example of an aluminum species that satisfies the octet rule is the anion $AlCl_4^-$. The Lewis dot structure is drawn in Problem 9.66.

(b) An example of an aluminum species containing an expanded octet is anion $AlF_6^{3-}$. (How many pairs of electrons surround the central atom?)

(c) An aluminum species that has an incomplete octet is the compound $AlCl_3$. The dot structure is given in Problem 9.66.

**9.88** $CF_2$ would be very unstable because carbon does not have an octet. (How many electrons does it have?)

$LiO_2$ would not be stable because the lattice energy between $Li^+$ and superoxide $O_2^-$ would be too low to stabilize the solid.

$CsCl_2$ requires a $Cs^{2+}$ cation. The second ionization energy is too large to be compensated by the increase in lattice energy.

$PI_5$ appears to be a reasonable species (compared to $PF_5$ in Example 9.10 of the text). However, the iodine atoms are too large to have five of them "fit" around a single P atom.

**9.90** (a) false (b) true (c) false (d) false

For question (c), what is an example of a second-period species that violates the octet rule?

**9.92** The formation of $CH_4$ from its elements is:

$$C(s) + 2H_2(g) \longrightarrow CH_4(g)$$

The reaction could take place in two steps:

Step 1: $C(s) + 2H_2(g) \longrightarrow \cancel{C}(g) + \cancel{4H}(g)$ $\quad \Delta H^\circ_{rxn} = (716 + 872.8)\text{kJ/mol} = 1589\text{ kJ/mol}$

Step 2: $\cancel{C}(g) + \cancel{4H}(g) \longrightarrow CH_4(g)$ $\quad \Delta H^\circ_{rxn} \approx -4 \times (\text{bond energy of C} - \text{H bond})$

$$\approx -4 \times 414\text{ kJ/mol} = -1656\text{ kJ/mol}$$

Therefore, $\Delta H^\circ_f(CH_4)$ would be approximately the sum of the enthalpy changes for the two steps. See Section 6.6 of the text (Hess's law).

$$\Delta H^\circ_f(CH_4) = \Delta H^\circ_{rxn}(1) + \Delta H^\circ_{rxn}(2)$$

$$\Delta H^\circ_f(CH_4) = (1589 - 1656)\text{kJ/mol} = \mathbf{-67\ kJ/mol}$$

The actual value of $\Delta H^\circ_f(CH_4) = -74.85$ kJ/mol.

**9.94** Only $N_2$ has a triple bond. Therefore, it has the shortest bond length.

**9.96** To be isoelectronic, molecules must have the same number and arrangement of valence electrons. $NH_4^+$ and $CH_4$ are isoelectronic (8 valence electrons), as are CO and $N_2$ (10 valence electrons), as are $B_3N_3H_6$ and $C_6H_6$ (30 valence electrons). Draw Lewis structures to convince yourself that the electron arrangements are the same in each isoelectronic pair.

**9.98** The reaction can be represented as:

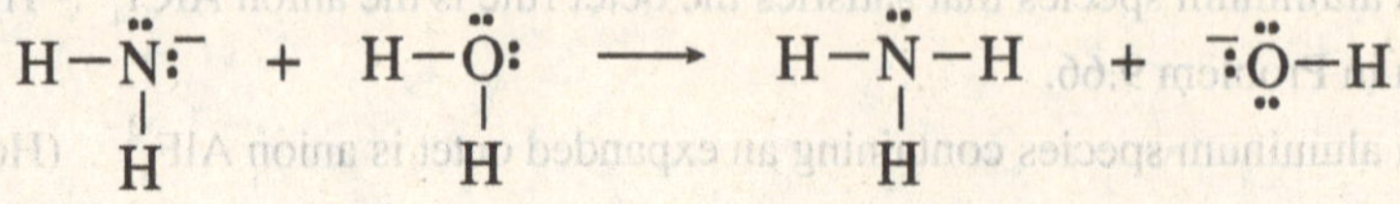

**9.100** The central iodine atom in $I_3^-$ has *ten* electrons surrounding it: two bonding pairs and three lone pairs. The central iodine has an expanded octet. Elements in the second period such as fluorine cannot have an expanded octet as would be required for $F_3^-$.

**9.102** The skeletal structure is:

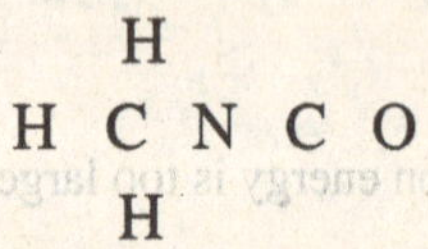

The number of valence electron is: $(1 \times 3) + (2 \times 4) + 5 + 6 = 22$ valence electrons

We can draw two resonance structures for methyl isocyanate.

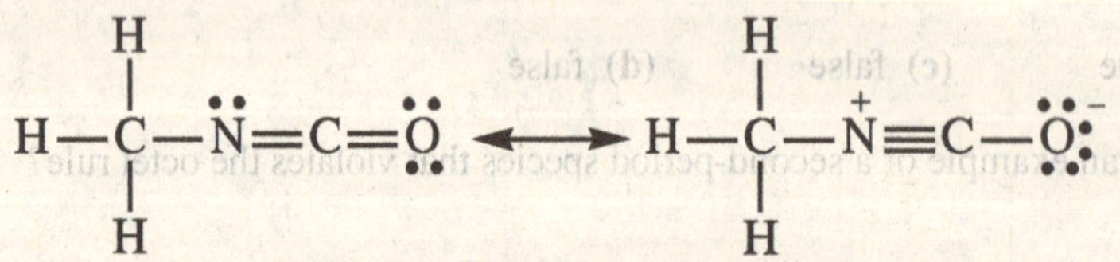

**9.104** **(a)** This is a very good resonance form; there are no formal charges and each atom satisfies the octet rule.

**(b)** This is a second choice after (a) because of the positive formal charge on the oxygen (high electronegativity).

**(c)** This is a poor choice for several reasons. The formal charges are placed counter to the electronegativities of C and O, the oxygen atom does not have an octet, and there is no bond between that oxygen and carbon!

**(d)** This is a mediocre choice because of the large formal charge and lack of an octet on carbon.

**9.106** The nonbonding electron pairs around Cl and F are omitted for simplicity.

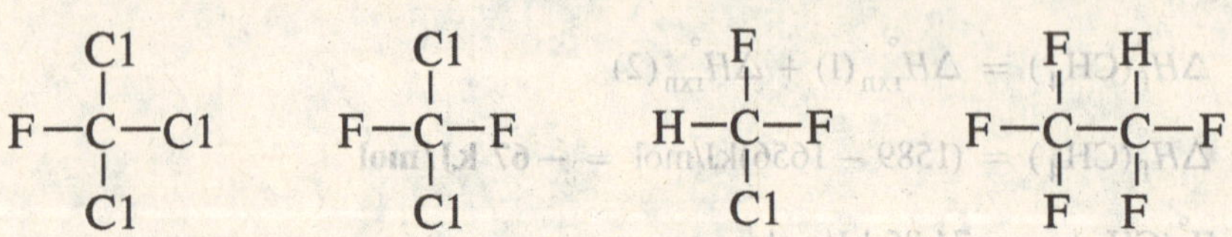

**9.108** **(a)** Using Equation (9.3) of the text,

$$\Delta H = \sum \text{BE(reactants)} - \sum \text{BE(products)}$$

$$\boldsymbol{\Delta H} = [(436.4 + 151.0) - 2(298.3)] = \mathbf{-9.2\ kJ/mol}$$

**(b)** Using Equation (6.18) of the text,

$$\Delta H° = 2\Delta H_f^\circ[\text{HI}(g)] - \{\Delta H_f^\circ[\text{H}_2(g)] + \Delta H_f^\circ[\text{I}_2(g)]\}$$

$$\boldsymbol{\Delta H°} = (2)(25.9\ \text{kJ/mol}) - [(0) + (1)(61.0\ \text{kJ/mol})] = \mathbf{-9.2\ kJ/mol}$$

**9.110** The Lewis structures are:

**(a)** $:\bar{C}\equiv O:^+$ **(b)** $:N\equiv O:^+$ **(c)** $:\bar{C}\equiv N:$ **(d)** $:N\equiv N:$

**9.112** True. Each noble gas atom already has completely filled *ns* and *np* subshells.

**9.114** **(a)** The bond enthalpy of $F_2^-$ is the energy required to break up $F_2^-$ into an F atom and an $F^-$ ion.

$$F_2^-(g) \longrightarrow F(g) + F^-(g)$$

We can arrange the equations given in the problem so that they add up to the above equation. See Section 6.6 of the text (Hess's law).

$$\begin{array}{llll} F_2^-(g) & \longrightarrow & \cancel{F_2}(g) + \cancel{e^-} & \Delta H° = 290\ \text{kJ/mol} \\ \cancel{F_2}(g) & \longrightarrow & \cancel{2F}(g) & \Delta H° = 156.9\ \text{kJ/mol} \\ \cancel{F}(g) + \cancel{e^-} & \longrightarrow & F^-(g) & \Delta H° = -333\ \text{kJ/mol} \\ \hline F_2^-(g) & \longrightarrow & F(g) + F^-(g) & \end{array}$$

The bond enthalpy of $F_2^-$ is the sum of the enthalpies of reaction.

$$\text{BE}(\mathbf{F_2^-}) = [290 + 156.9 + (-333\ \text{kJ})]\text{kJ/mol} = \mathbf{114\ kJ/mol}$$

**(b)** The bond in $F_2^-$ is weaker (114 kJ/mol) than the bond in $F_2$ (156.9 kJ/mol), because the extra electron increases repulsion between the F atoms.

**9.116** In **(a)** there is a lone pair on the C atom and the negative formal charge is on the less electronegative C atom.

**9.118** **(a)** $:\dot{N}=\ddot{O} \longleftrightarrow {}^-:\ddot{N}=\dot{O}^+$

The first structure is the most important. Both N and O have formal charges of zero. In the second structure, the more electronegative oxygen atom has a formal charge of +1. Having a positive formal charge on an highly electronegative atom is not favorable. In addition, both structures leave one atom with an incomplete octet. This cannot be avoided due to the odd number of electrons.

**(b)** It is not possible to draw a structure with a triple bond between N and O.

$$:N\equiv\dot{O}$$

Any structure drawn with a triple bond will lead to an expanded octet. Elements in the second row of the period table cannot exceed the octet rule.

**9.120**

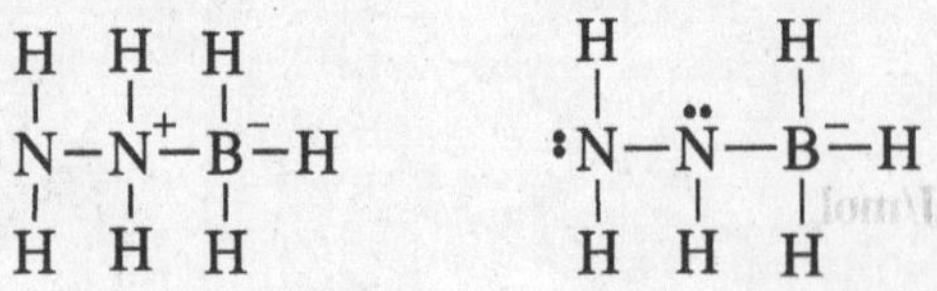

**9.122** The following Lewis structure obeys the octet rule, but places a lone pair and a negative charge on the C atom in $CO_3^{2-}$. It is not a stable structure.

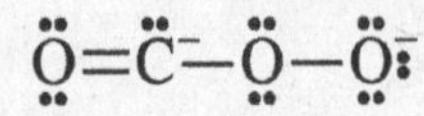

**9.124**

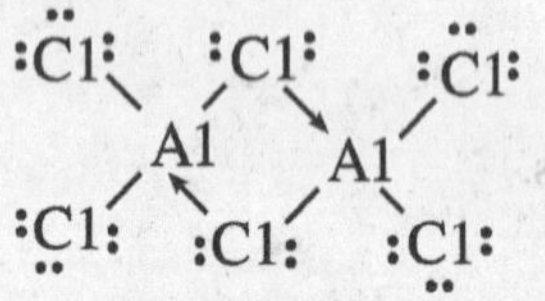

The arrows indicate coordinate covalent bonds.

**9.126** There are four C–H bonds in $CH_4$, so the average bond enthalpy of a C–H bond is:

$$\frac{1656 \text{ kJ/mol}}{4} = 414 \text{ kJ/mol}$$

The Lewis structure of propane is:

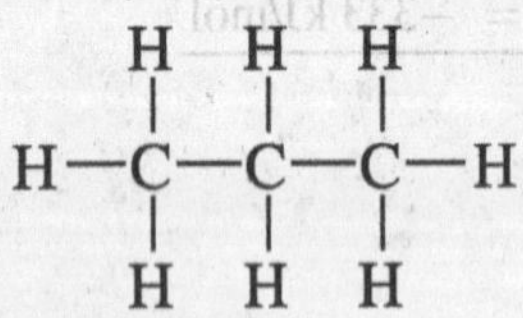

There are eight C–H bonds and two C–C bonds. We write:

$$8(\text{C–H}) + 2(\text{C–C}) = 4006 \text{ kJ/mol}$$

$$8(414 \text{ kJ/mol}) + 2(\text{C–C}) = 4006 \text{ kJ/mol}$$

$$2(\text{C–C}) = 694 \text{ kJ/mol}$$

So, the average bond enthalpy of a C–C bond is: $\frac{694}{2}$ kJ/mol = **347 kJ/mol**

**9.128**

| *Bonds Broken* | *Number Broken* | *Bond Enthalpy* (kJ/mol) | *Enthalpy Change* (kJ) |
|---|---|---|---|
| C = C | 1 | 620 | 620 |
| C – H | 4 | 414 | 1656 |
| Cl – Cl | 1 | 242.7 | 242.7 |

| *Bonds Formed* | *Number Formed* | *Bond Enthalpy* (kJ/mol) | *Enthalpy Change* (kJ) |
|---|---|---|---|
| C – C | 1 | 347 | 347 |
| C – H | 4 | 414 | 1656 |
| C – Cl | 2 | 328 | 656 |

$$\boldsymbol{\Delta H°} = \text{total energy input} - \text{total energy released}$$

$$= (620 + 1656 + 242.7) - (347 + 1656 + 656) = \mathbf{-140\ kJ/mol}$$

We next calculate $\Delta H$ from $\Delta H_f^\circ$ values.

$$\Delta H° = \Delta H_f^\circ(C_2H_4Cl_2) - [\Delta H_f^\circ(C_2H_4) + \Delta H_f^\circ(Cl_2)]$$

$$\boldsymbol{\Delta H°} = (-132\ \text{kJ/mol}) - [52.3\ \text{kJ/mol} + 0] = \mathbf{-184\ kJ/mol}$$

**9.130**

(a) 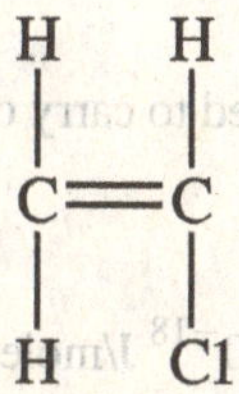

(b) 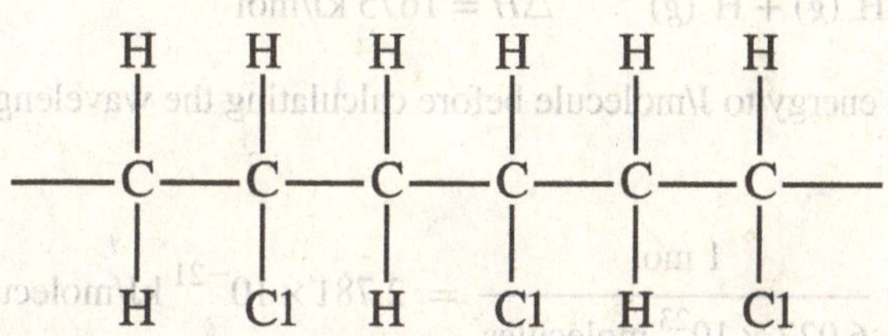

(c) In the formation of poly(vinyl chloride) from vinyl chloride, for every C=C double bond broken, 2 C–C single bonds are formed. No other bonds are broken or formed. The energy changes for 1 mole of vinyl chloride reacted are:

$$\text{total energy input (breaking C=C bonds)} = 620\ \text{kJ}$$

$$\text{total energy released (forming C–C bonds)} = 2 \times 347\ \text{kJ} = 694\ \text{kJ}$$

$$\Delta H° = 620\ \text{kJ} - 694\ \text{kJ} = -74\ \text{kJ}$$

The negative sign shows that this is an exothermic reaction. To find the enthalpy change when $1.0 \times 10^3$ kg of vinyl chloride react, we proceed as follows:

$$\boldsymbol{\Delta H} = (1.0 \times 10^6\ \text{g}\ C_2H_3Cl) \times \frac{1\ \text{mol}\ C_2H_3Cl}{62.49\ \text{g}\ C_2H_3Cl} \times \frac{-74\ \text{kJ}}{1\ \text{mol}\ C_2H_3Cl} = \mathbf{-1.2 \times 10^6\ kJ}$$

**9.132** $\text{EN(O)} = \dfrac{1314 + 141}{2} = 727.5$ $\quad\text{EN(F)} = \dfrac{1680 + 328}{2} = 1004$

$$\text{EN(Cl)} = \frac{1251 + 349}{2} = 800$$

Using Mulliken's definition, the electronegativity of chlorine is greater than that of oxygen, and fluorine is still the most electronegative element. We can convert to the Pauling scale by dividing each of the above by 230 kJ/mol.

$\text{EN(O)} = \dfrac{727.5}{230} = \mathbf{3.16}$ $\quad\text{EN(F)} = \dfrac{1004}{230} = \mathbf{4.37}$ $\quad\text{EN(Cl)} = \dfrac{800}{230} = \mathbf{3.48}$

These values compare to the Pauling values for oxygen of 3.5, fluorine of 4.0, and chlorine of 3.0.

**9.134** (1) You could determine the magnetic properties of the solid. An $Mg^+O^-$ solid would be paramagnetic while $Mg^{2+}O^{2-}$ solid is diamagnetic.

(2) You could determine the lattice energy of the solid. $Mg^+O^-$ would have a lattice energy similar to $Na^+Cl^-$. This lattice energy is much lower than the lattice energy of $Mg^{2+}O^{2-}$.

**9.136** The bond enthalpy of $H_2$ is 436.4 kJ/mol (See Table 9.4 of the text). This represents the process:

$$H_2(g) \rightarrow H(g) + H(g)$$

To arrive at the reaction shown in the problem, we need both the ionization energy and the electron affinity for H(*g*).

$$\begin{array}{ll} H_2(g) \rightarrow H(g) + H(g) & \Delta H = 436.4\text{ kJ/mol} \\ H(g) \rightarrow H^+(g) & \Delta H = 1312\text{ kJ/mol} \\ H(g) \rightarrow H^-(g) & \Delta H = -73\text{ kJ/mol} \\ \hline H_2(g) \rightarrow H^+(g) + H^-(g) & \Delta H = 1675\text{ kJ/mol} \end{array}$$

We convert this energy to J/molecule before calculating the wavelength of light needed to carry out the reaction.

$$\frac{1675\text{ kJ}}{1\text{ mol}} \times \frac{1\text{ mol}}{6.022 \times 10^{23}\text{ molecules}} = 2.781 \times 10^{-21}\text{ kJ/molecule} = 2.781 \times 10^{-18}\text{ J/molecule}$$

$$\lambda = \frac{hc}{E} = \frac{(6.63 \times 10^{-34}\text{ J}\cdot\text{s})(3.00 \times 10^8\text{ m/s})}{2.781 \times 10^{-18}\text{ J}} = \mathbf{7.15 \times 10^{-8}\text{ m} = 71.5\text{ nm}}$$

**9.138** We can arrange the equations for the lattice energy of KCl, ionization energy of K, and electron affinity of Cl, to end up with the desired equation.

$$\begin{array}{ll} K^+(g) + Cl^-(g) \rightarrow KCl(s) & \Delta H° = -699\text{ kJ/mol (equation for lattice energy of KCl, reversed)} \\ K(g) \rightarrow K^+(g) + e^- & \Delta H° = 418.7\text{ kJ/mol (ionization energy of K)} \\ Cl(g) + e^- \rightarrow Cl^-(g) & \Delta H° = -349\text{ kJ/mol (electron affinity of Cl)} \\ \hline K(g) + Cl(g) \rightarrow KCl(s) & \Delta H° = (-699 + 418.7 + -349)\text{ kJ/mol} = \mathbf{-629\text{ kJ/mol}} \end{array}$$

**9.140** From Table 9.4 of the text, we can find the bond enthalpies of C–N and C=N. The average can be calculated, and then the maximum wavelength associated with this enthalpy can be calculated.

The average bond enthalpy for C–N and C=N is:

$$\frac{(276 + 615)\text{ kJ/mol}}{2} = 446\text{ kJ/mol}$$

We need to convert this to units of J/bond before the maximum wavelength to break the bond can be calculated. Because there is only 1 CN bond per molecule, there is Avogadro's number of bonds in 1 mole of the amide group.

$$\frac{446\ \text{kJ}}{1\ \text{mol}} \times \frac{1\ \text{mol}}{6.022 \times 10^{23}\ \text{bonds}} \times \frac{1000\ \text{J}}{1\ \text{kJ}} = 7.41 \times 10^{-19}\ \text{J/bond}$$

The maximum wavelength of light needed to break the bond is:

$$\lambda_{\text{max}} = \frac{hc}{E} = \frac{(6.63 \times 10^{-34}\ \text{J}\cdot\text{s})(3.00 \times 10^{8}\ \text{m/s})}{7.41 \times 10^{-19}\ \text{J}} = \mathbf{2.68 \times 10^{-7}\ m = 268\ nm}$$

**9.142** (a) We divide the equation given in the problem by 4 to come up with the equation for the decomposition of *1 mole* of nitroglycerin.

$$C_3H_5N_3O_9(l) \rightarrow 3CO_2(g) + \tfrac{5}{2}H_2O(g) + \tfrac{3}{2}N_2(g) + \tfrac{1}{4}O_2(g)$$

We calculate $\Delta H°$ using Equation (6.18) and the enthalpy of formation values from Appendix 3 of the text.

$$\Delta H^{\circ}_{\text{rxn}} = \sum n\Delta H^{\circ}_{\text{f}}(\text{products}) - \sum m\Delta H^{\circ}_{\text{f}}(\text{reactants})$$

$$\Delta H^{\circ}_{\text{rxn}} = (3)(-395.5\ \text{kJ/mol}) + \left(\frac{5}{2}\right)(-241.8\ \text{kJ/mol}) - (1)(-371.1\ \text{kJ/mol}) = \mathbf{-1413.9\ kJ/mol}$$

Next, we calculate $\Delta H°$ using bond enthalpy values from Table 9.4 of the text.

| *Bonds Broken* | *Number Broken* | *Bond Enthalpy* (kJ/mol) | *Enthalpy Change* (kJ/mol) |
|---|---|---|---|
| C–H | 5 | 414 | 2070 |
| C–C | 2 | 347 | 694 |
| C–O | 3 | 351 | 1053 |
| N–O | 6 | 176 | 1056 |
| N=O | 3 | 607 | 1821 |
| *Bonds Formed* | *Number Formed* | *Bond Enthalpy* (kJ/mol) | *Enthalpy Change* (kJ/mol) |
| C=O | 6 | 799 | 4794 |
| O–H | (5/2)(2) = 5 | 460 | 2300 |
| N≡N | 1.5 | 941.4 | 1412.1 |
| O=O | 0.25 | 498.7 | 124.7 |

From Equation (9.3) of the text:

$$\Delta H° = \Sigma\text{BE(reactants)} - \Sigma\text{BE(products)}$$

$$\mathbf{\Delta H°} = (6694 - 8630.8)\text{kJ/mol} = \mathbf{-1937\ kJ/mol}$$

The $\Delta H°$ values do not agree exactly because average bond enthalpies are used, and nitroglycerin is a liquid (strictly, the bond enthalpy values are for gases).

**(b)** One mole of nitroglycerin generates, $(3 + 2.5 + 1.5 + 0.25) = 7.25$ moles of gas. One mole of an ideal gas occupies a volume of 22.41 L at STP.

$$7.25\ \text{mol} \times \frac{22.41\ \text{L}}{1\ \text{mol}} = \mathbf{162\ L}$$

**(c)** We calculate the pressure exerted by 7.25 moles of gas occupying a volume of 162 L at a temperature of 3000 K.

$$P = \frac{nRT}{V} = \frac{(7.25\ \text{mol})(0.0821\ \text{L}\cdot\text{atm}/\text{mol}\cdot\text{K})(3000\ \text{K})}{162\ \text{L}} = \mathbf{11.0\ atm}$$

**9.144** There are no lone pairs on adjacent atoms in $C_2H_6$, there is one lone pair on each nitrogen atom in $N_2H_4$, and there are two lone pairs on each oxygen atom in $H_2O_2$. Draw Lewis structures to determine the number of lone pairs in each molecule. Looking at Table 9.4 of the text which lists bond enthalpies of diatomic molecules and average bond enthalpies of polyatomic molecules, we can estimate the bond enthalpies of C–C in $C_2H_6$, N–N in $N_2H_4$, and O–O in $H_2O_2$ by looking up the average bond enthalpy values. These will not be exact bond enthalpy values for the given molecules, but the values will be approximate and give us a measure of what effect lone pairs on adjacent atoms have on the strength of the particular bonds. The values given in Table 9.4 are:

| | |
|---|---|
| C–C | 347 kJ/mol |
| N–N | 193 kJ/mol |
| O–O | 142 kJ/mol |

Comparing these values, it is clear that lone pairs on adjacent atoms weaken the particular bond. The C–C bond in $C_2H_6$, with no adjacent lone pairs is the strongest bond, the N–N bond in $N_2H_4$, with one lone pair on each nitrogen atom is a weaker bond, and the O–O bond in $H_2O_2$, with two lone pairs on each oxygen atom is the weakest bond.

**9.146** The energy needed to dissociate the chlorine molecule can be calculated from the wavelength.

$$E = \frac{hc}{\lambda} = \frac{(6.63 \times 10^{-34}\ \text{J}\cdot\text{s})(3.00 \times 10^{8}\ \text{m/s})}{471.7 \times 10^{-9}\ \text{m}} = \frac{4.22 \times 10^{-19}\ \text{J}}{1\ \text{molecule}} \times \frac{6.022 \times 10^{23}\ \text{molecules}}{1\ \text{mole}} \times \frac{1\ \text{kJ}}{1000\ \text{J}}$$

$$= 254\ \text{kJ/mol}$$

Because one chlorine atom is in an excited state that is 10.5 kJ/mol above the ground state, we subtract this value from 254 kJ/mol to calculate the bond enthalpy of the $Cl_2$ molecule.

**Bond enthalpy $Cl_2$** = 254 kJ/mol – 10.5 kJ/mol = **244 kJ/mol**

**9.148** From Table 9.5 we see that the C–F bond (450 kJ/mol) is stronger than the C–C bond (347 kJ/mol). Thus, the reaction between $C_2H_6$ and $F_2$ favors the formation of the C–F bond over the C–C bond.

# CHAPTER 10
# CHEMICAL BONDING II: MOLECULAR GEOMETRY AND HYBRIDIZATION OF ATOMIC ORBITALS

## PROBLEM-SOLVING STRATEGIES AND TUTORIAL SOLUTIONS

### TYPES OF PROBLEMS

**Problem Type 1:** Molecular Geometry.

- **(a)** Molecules in which the central atom has *no* lone pairs.
- **(b)** Molecules in which the central atom has one or more lone pairs.
- **(c)** Geometry of molecules with more than one central atom.

**Problem Type 2:** Predicting Dipole Moments.

**Problem Type 3:** Hybridization of Atomic Orbitals.

- **(a)** Hybridization of *s* and *p* orbitals.
- **(b)** Hybridization of *s*, *p*, and *d* orbitals.
- **(c)** Hybridization in molecules containing double and triple bonds.

**Problem Type 4:** Molecular Orbital Diagrams.

## PROBLEM TYPE 1: MOLECULAR GEOMETRY

The model that we are going to use to study molecular geometry is called the **valence-shell electron-pair repulsion (VSEPR) model**. It accounts for the geometric arrangement of electron pairs around a central atom in terms of the repulsion between electron pairs. The geometry that a molecule ultimately adopts *minimizes* electron-pair repulsion.

**Guidelines for Applying the VSEPR Model**

1. Write the Lewis structure of the molecule.
2. Only consider the electron pairs around the *central atom* (the atom that is bonded to more than one other atom). Count the number of electron pairs around the central atom (bonding pairs and lone pairs). For counting purposes, treat double and triple bonds as though they were single bonds. Refer to Table 10.1 of the text to predict the overall arrangement of the electron pairs.
3. Use Tables 10.1 and 10.2 of the text to predict the *geometry* of the molecule.
4. In predicting bond angles, note that a lone pair repels another lone pair or a bonding pair more strongly than a bonding pair repels another bonding pair. There is no easy way to predict bond angles accurately when the central atom possesses one or more lone pairs.

### A. Molecules in which the central atom has *no* lone pairs

For simplicity, we will only consider molecules that contain atoms of two elements, A and B, where A is the central atom. These molecules have the general formula $AB_x$, where $x$ is an integer 2, 3, . . . . In most cases, $x$ is between 2 and 6.

Table 10.1 of the text shows five possible arrangements of electron pairs around the central atom A. Remember, these arrangements are adopted because electron-pair repulsions are minimized.

Below is a condensed version of Table 10.1 from the text.

**Arrangement of electron pairs around a central atom (A) in a molecule, and geometry of molecules if the central atom has *no* lone pairs.**

| Number of electron pairs around central atom | Arrangement of electron pairs | Molecular geometry |
|---|---|---|
| 2 | Linear, 180° | Linear, $AB_2$ |
| 3 | Trigonal planar, 120° | Trigonal planar, $AB_3$ |
| 4 | Tetrahedral, 109.5° | Tetrahedral, $AB_4$ |
| 5 | Trigonal bipyramidal, 120°, 90° | Trigonal bipyramidal, $AB_5$ |
| 6 | Octahedral, 90° | Octahedral, $AB_6$ |

> **Note:** Since there are *no* lone pairs around the central atom in all the examples shown above, the molecular geometry is *always* the same as the arrangement of electron pairs around the central atom.

**EXAMPLE 10.1**

**Use the VSEPR model to predict the geometry of the following molecules and ions: (a) $HgCl_2$, (b) $SnCl_4$, (c) $NO_3^-$, (d) $PF_5$.**

**Strategy:** The sequence of steps in determining molecular geometry is as follows:

draw Lewis structure ⟶ find arrangement of electrons pairs ⟶ find arrangement of bonding pairs ⟶ determine geometry based on bonding pairs

**Solution:**

**(a)**

*Step 1:* Write the Lewis structure of the molecule (see Chapter 9).

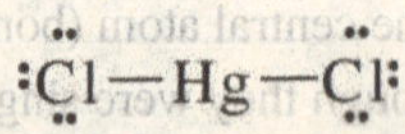

*Step 2:* Count the number of electron pairs around the central atom. There are two electron pairs around Hg.

*Step 3:* Since there are two electron pairs around Hg, the electron-pair arrangement that minimizes repulsion is **linear**. In addition, since there are no lone pairs around the central atom, the geometry is also **linear** ($AB_2$).

**(b)**

*Step 1:* Write the Lewis structure of the molecule.

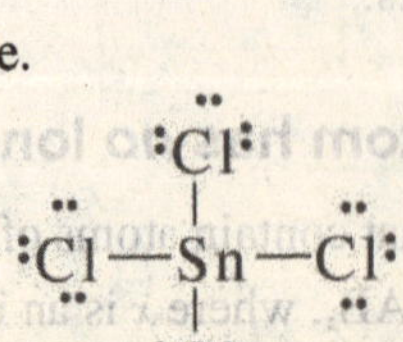

*Step 2:* Count the number of electron pairs around the central atom. There are four electron pairs around Sn.

*Step 3:* Since there are four electron pairs around Sn, the electron-pair arrangement that minimizes repulsion is **tetrahedral.**

In addition, since there are no lone pairs around the central atom, the geometry is also **tetrahedral** ($AB_4$).

**(c)**

*Step 1:* Write the Lewis structure of the molecule.

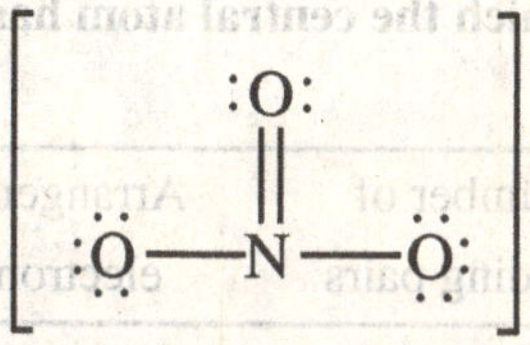

*Step 2:* Count the number of electron pairs around the central atom. There are three electron pairs around N. Remember, for VSEPR purposes, a double or triple bond counts the same as a single bond.

*Step 3:* Since there are three electron pairs around N, the electron-pair arrangement that minimizes repulsion is **trigonal planar**.

In addition, since there are no lone pairs around the central atom, the geometry is also **trigonal planar** ($AB_3$).

**(d)**

*Step 1:* Write the Lewis structure of the molecule.

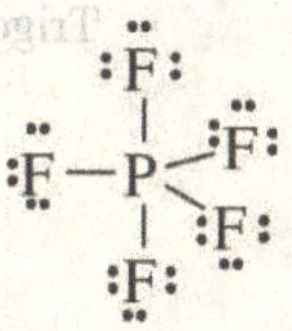

*Step 2:* Count the number of electron pairs around the central atom. There are five electron pairs around P.

Since there are five electron pairs around P, the electron-pair arrangement that minimizes repulsion is **trigonal bipyramidal.**

In addition, since there are no lone pairs around the central atom, the geometry is also **trigonal bipyramidal** ($AB_5$).

**PRACTICE EXERCISE**

1. Use the VSEPR model to predict the geometry of the following molecules: (a) $CO_2$, (b) $CCl_4$,(c) $SO_3$, (d) $SF_6$.

**Text Problems: 10.8**, 10.10, 10.12, 10.14

## B. Molecules in which the central atom has one or more lone pairs

If lone pairs are present on the central atom, the overall arrangement of electron pairs is ***not*** the same as the geometry of the molecule.

To keep track of the total number of bonding pairs and lone pairs, we will designate molecules with lone pairs as $AB_xE_y$, where A is the central atom, B is a surrounding atom, and E is a lone pair on the central atom, A. Both $x$ and $y$ are integers: $x = 2, 3, \ldots,$ and $y = 1, 2, \ldots.$

When working with molecules that have lone pairs on the central atom, remember to count all electron pairs on the central atom, both bonding pairs and lone pairs. The total number of electron pairs around the central atom determines the electron arrangement around the central atom. See Table 10.1. However, the molecular geometry will not be the same as this electron arrangement. Use Table 10.2 of the text to determine the molecular geometry, or a better option is to build a model. Gum drops and toothpicks make an effective and inexpensive model kit. And besides, you can eat your model when you are finished. (Not the toothpicks!)

A condensed version of Table 10.2 of the text follows.

**Geometry of simple molecules and ions in which the central atom has one or more lone pairs.**

| Class of molecule | Total # of electron pairs on central atom | Number of bonding pairs | Arrangement of electron pairs | Number of lone pairs | Molecular geometry |
|---|---|---|---|---|---|
| $AB_2E$ | 3 | 2 | Trigonal planar | 1 | Bent |
| $AB_3E$ | 4 | 3 | Tetrahedral | 1 | Trigonal pyramid |
| $AB_2E_2$ | 4 | 2 | Tetrahedral | 2 | Bent |
| $AB_4E$ | 5 | 4 | Trigonal bipyramidal | 1 | Distorted tetrahedron |
| $AB_3E_2$ | 5 | 3 | Trigonal bipyramidal | 2 | T-shaped |
| $AB_2E_3$ | 5 | 2 | Trigonal bipyramidal | 3 | Linear |
| $AB_5E$ | 6 | 5 | Octahedral | 1 | Square pyramidal |
| $AB_4E_2$ | 6 | 4 | Octahedral | 2 | Square planar |

**Note:** The arrangement of 3 electron-pairs is always trigonal planar; the arrangement of 4 electron-pairs is tetrahedral; the arrangement of 5 electron-pairs is trigonal bipyramidal; the arrangement of 6 electron-pairs is octahedral. However, if there are any lone pairs on the central atom, the molecular geometry will be different from the electron arrangement. **Build models!**

**EXAMPLE 10.2**

**Use the VSEPR model to predict the geometry of the following molecules: (a) $O_3$, (b) $XeF_2$, (c) $IF_5$.**

**Strategy:** The sequence of steps in determining molecular geometry is as follows:

draw Lewis structure ⟶ find arrangement of electrons pairs ⟶ find arrangement of bonding pairs ⟶ determine geometry based on bonding pairs

**Solution:**

(a)

*Step 1:* Write the Lewis structure of the molecule (see Chapter 9).

*Step 2:* Count the number of electron pairs around the central atom. There are three electron pairs around the central oxygen. Remember, for VSEPR purposes, a double or triple bond counts the same as a single bond.

*Step 3:* Since there are three electron pairs around the central O, the electron-pair arrangement that minimizes repulsion is **trigonal planar**.

However, there is one lone pair of electrons around the central atom. Consulting a model or Table 10.2 of the text, you should find that the geometry of the molecule is **bent** ($AB_2E$).

**(b)**

*Step 1:* Write the Lewis structure of the molecule.

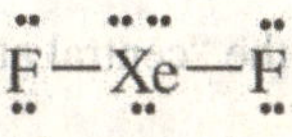

*Step 2:* Count the number of electron pairs around the central atom. There are five electron pairs around xenon.

*Step 3:* Since there are five electron pairs around xenon, the electron-pair arrangement that minimizes repulsion is **trigonal bipyramidal.**

However, there are three lone pairs of electrons around the central atom. Consulting a model or Table 10.2 of the text, you should find that the geometry of the molecule is **linear** ($AB_2E_3$).

**(c)**

*Step 1:* Write the Lewis structure of the molecule.

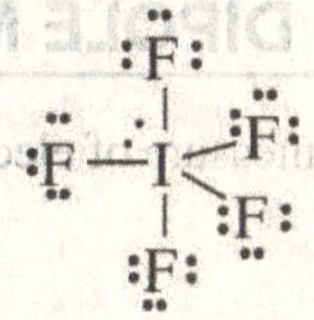

*Step 2:* Count the number of electron pairs around the central atom. There are six electron pairs around iodine.

*Step 3:* Since there are six electron pairs around iodine, the electron-pair arrangement that minimizes repulsion is **octahedral**.

However, there is one lone pair of electrons around the central atom. Consulting a model or Table 10.2 of the text, you should find that the geometry of the molecule is **square pyramidal** ($AB_5E$).

**PRACTICE EXERCISE**

2. Use the VSEPR model to predict the geometry of the following molecules: (a) $SO_2$, (b) $XeF_4$, (c) $SF_4$.

| **Text Problems:** 10.10, 10.12, 10.14 |
|---|

## C. Geometry of molecules with more than one central atom

So far, we have discussed the geometry of molecules having only one central atom. (The term "central atom" means an atom that is not a terminal atom in a polyatomic molecule.) Many molecules will have more than one "central atom". In these cases, you have to apply the VSEPR method presented in parts A and B above to each of the central atoms.

**EXAMPLE 10.3**

**Use the VSEPR model to predict the geometry of $C_2H_6$.**

*Step 1:* Write the Lewis structure of the molecule. Hydrogens must be in terminal positions. The two carbons must be the "central atoms".

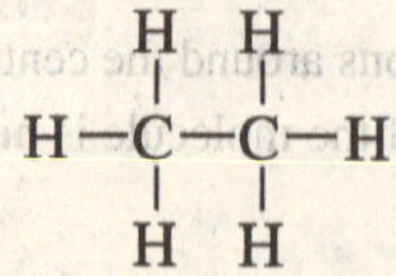

*Step 2:* Count the number of electron pairs around the "central atoms". There are four electron pairs around each carbon.

*Step 3:* Since there are four electron pairs around each carbon, the electron-pair arrangement around each carbon that minimizes repulsion is **tetrahedral**.

In addition, since there are no lone pairs around the central atoms, their geometries are also **tetrahedral**.

**PRACTICE EXERCISE**

3. Use the VSEPR model to predict the geometry of $C_2H_2$.

## PROBLEM TYPE 2: PREDICTING DIPOLE MOMENTS

To determine if a molecule has a dipole moment (a measure of electrical charge separation in a molecule) you must consider two factors.

1. Are the *bonds* in the molecule polar?

A bond will be polar if there is a difference in electronegativity between the two atoms comprising the bond (see Section 9.5 of the text). For example, in an O–H bond, there is a shift in electron density from H to O because O is more electronegative than H. The shift in electron density is symbolized by placing a crossed arrow ( +⟶ ) above the bond to indicate the direction of the shift in electron density. This is called a *bond* moment. For example:

⟵+

O – H

The consequent charge separation can be represented as:

$\delta^-$ $\delta^+$

O – H

where δ (delta) represents a partial charge.

2. Is the *molecule* polar?

Diatomic molecules containing atoms of the same element do not have dipole moments and so are **nonpolar molecules**. There is no difference in electronegativity since the two elements are the same.

However, most molecules will have polar bonds (bond moments). If a molecule has polar bonds, does this mean that it is a polar molecule? Not necessarily. The *bond moment* is a vector quantity, which means that it has both a magnitude and direction. The measured *dipole moment* of the molecule is equal to the vector sum of the bond moments. For example, in $CO_2$, the two bond moments are equal in magnitude and opposite in direction. The sum or resultant dipole moment will be *zero*. Hence, $CO_2$ is a **nonpolar molecule**.

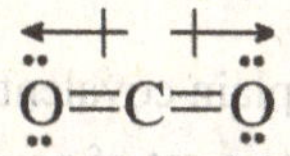

In summary, to determine if a molecule is **polar** (i.e., does the molecule have a dipole moment), you must sum the individual bond moments to determine if there is a resultant dipole moment.

**EXAMPLE 10.4**

**Predict whether each of the following molecules has a dipole moment: (a) $CCl_4$ and (b) $CHCl_3$.**

**Strategy:** Keep in mind that the dipole moment of a molecule depends on both the difference in electronegativities of the elements present and its geometry. A molecule can have polar bonds (if the bonded atoms have different electronegativities), but it may not possess a dipole moment if it has a highly symmetrical geometry.

**Solution:**

(a) Write the Lewis structure for the molecule. The lone pairs on Cl have been omitted.

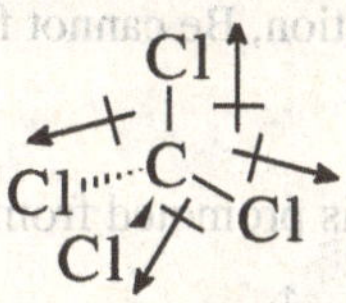

Shown on the Lewis structure are the bond moments. Chlorine is more electronegative than C, so the arrows indicate the shift in electron density toward Cl. However, these polar bonds are arranged in a symmetric tetrahedral fashion about the central C atom. The sum or resultant dipole moment is *zero*. $CCl_4$ is a **nonpolar molecule**.

(b) Write the Lewis structure for the molecule. The lone pairs on Cl have been omitted.

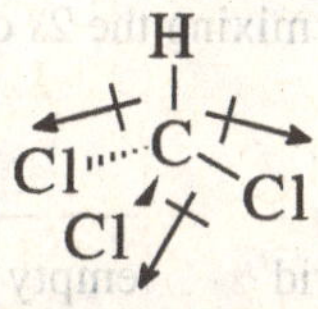

Shown on the Lewis structure are the bond moments. Chlorine is more electronegative than C, so the arrows indicate the shift in electron density toward Cl. The electronegativity difference between C and H is very small, so a C–H bond is essentially nonpolar. In this case, the three bond moments partially reinforce each other. Thus, $CHCl_3$ has a dipole moment and is therefore a **polar molecule**.

**PRACTICE EXERCISE**

4. Predict whether each of the following molecules has a dipole moment: (a) CO, (b) $BCl_3$, and (c) $XeF_4$.

**Text Problems:** 10.20, 10.22, **10.24**

## PROBLEM TYPE 3: HYBRIDIZATION OF ATOMIC ORBITALS

The **VSEPR** model is very powerful considering its simplicity; however, it does not explain chemical bond formation in any detail. In the 1930s, **valence bond (VB) theory** was introduced to account for chemical bond formation. **VB** theory describes covalent bonding as the overlapping of atomic orbitals. This means that the orbitals share a common region in space.

**VB** theory uses a concept called **hybridization** to explain covalent bonding. Hybridization is the mixing of atomic orbitals in an atom (usually a central atom) to generate a set of new atomic orbitals, called *hybrid orbitals*. Hybrid orbitals are atomic orbitals obtained when two or more nonequivalent orbitals of the same atom combine. The hybrid orbitals are used to form covalent bonds.

### A. Hybridization of *s* and *p* orbitals

**1. *sp* hybrid orbitals**

Let's consider the central atom Be in the $BeH_2$ molecule. Be has a ground state electron configuration of $1s^22s^2$. Only valence electrons are involved in bonding, so an orbital diagram of the valence electrons is:

$\uparrow\downarrow$    __ __ __
2*s*    2*p*

With this ground-state electron configuration, Be cannot form bonds with H because its valence electrons are paired in a 2*s* orbital.

To explain the bonding, first an electron is promoted from the 2*s* orbital to a 2*p* orbital.

$\uparrow$    $\uparrow$ __ __
2*s*    2*p*

Now, we have two different orbitals that can bond to the two hydrogens, which would result in two nonequivalent Be–H bonds. However, experimental evidence shows that there are two equivalent Be–H bonds.

This is where hybridization comes in. By mixing the 2*s* orbital with one of the 2*p* orbitals, we can generate two equivalent *sp* hybrid orbitals.

$\uparrow$ $\uparrow$    __ __
*sp* hybrid orbitals    empty *p* orbitals

Figure 10.10 of the text shows the shape and orientation of the *sp* hybrid orbitals. These two hybrid orbitals lie along the same line so that the angle between them is 180° (linear arrangement). Each of the BeH bonds is formed by the overlap of a Be *sp* hybrid and a H 1*s* orbital. The resulting $BeH_2$ molecule has a linear geometry.

**2. $sp^2$ hybrid orbitals**

Following the same type of argument for *sp* hybrids, $sp^2$ hybrid orbitals are formed by mixing an *s* orbital with two *p* orbitals. Three equivalent $sp^2$ hybrid orbitals are formed.

$\uparrow$    $\uparrow$ $\uparrow$ __
*s*    *p*

Mixing an *s* with two *p* orbitals gives:

↑ ↑ ↑ ___

$sp^2$ hybrid orbitals — empty *p* orbital

Figure 10.12 of the text shows the shape and orientation of the $sp^2$ hybrid orbitals. These three hybrid orbitals lie in a plane with an angle between any two hybrids of 120°. The three $sp^2$ hybrid orbitals are arranged in a trigonal planar fashion.

3. **$sp^3$ hybrid orbitals**

Again, following the same argument as above, $sp^3$ hybrid orbitals are formed by mixing an *s* orbital with three *p* orbitals. Four equivalent $sp^3$ hybrid orbitals are formed.

↑ (*s*) ↑ ↑ ↑ (*p*)

Mixing an *s* with three *p* orbitals gives:

↑ ↑ ↑ ↑

$sp^3$ hybrid orbitals

Figure 10.7 of the text shows the shape and orientation of the $sp^3$ hybrid orbitals. These four equivalent hybrid orbitals are directed toward the four corners of a regular tetrahedron with 109.5° bond angles.

Summarizing,

| Type of hybrid | No. of equivalent hybrid orbitals | Arrangement of hybrid orbitals | No. of empty *p* orbitals |
|---|---|---|---|
| *sp* | 2 | linear, 180° | 2 |
| $sp^2$ | 3 | trigonal planar, 120° | 1 |
| $sp^3$ | 4 | tetrahedral, 109.5° | 0 |

**EXAMPLE 10.5**

**Determine the hybridization of the central (underlined) atom in each of the following molecules: (a) $\underline{C}Cl_4$ and (b) $\underline{B}Cl_3$.**

**Strategy:** The steps for determining the hybridization of the central atom in a molecule are:

draw Lewis Structure of the molecule ⟶ use VSEPR to determine the electron pair arrangement surrounding the central atom (Table 10.1 of the text) ⟶ use Table 10.4 of the text to determine the hybridization state of the central atom

**Solution:**

**(a)** Write the Lewis structure of the molecule. The lone pairs on the chlorines have been omitted.

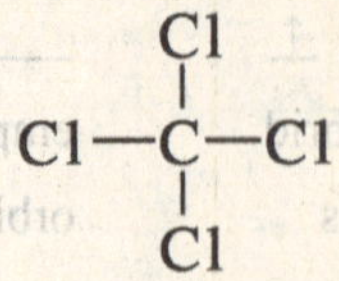

Count the number of electron pairs around the central atom. Since there are four electron pairs around C, the electron arrangement that minimizes electron-pair repulsion is **tetrahedral.**

We conclude that C is ***sp*$^3$ hybridized** because it has the electron arrangement of four *sp*$^3$ hybrid orbitals.

**(b)** Write the Lewis structure of the molecule. The lone pairs on the chlorines have been omitted.

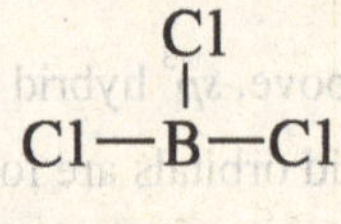

Count the number of electron pairs around the central atom. Since there are three electron pairs around B, the electron arrangement that minimizes electron-pair repulsion is **trigonal planar**.

We conclude that B is ***sp*$^2$ hybridized** because it has the electron arrangement of three *sp*$^2$ hybrid orbitals.

**PRACTICE EXERCISE**

**5.** Determine the hybridization of the central (underlined) atom in each of the following molecules or ions:

(a) $\underline{C}O_2$ and (b) $\underline{C}O_3^{2-}$.

| **Text Problems: 10.32**, 10.34, 10.40, **10.42** |
|---|

## B. Hybridization of *s*, *p*, and *d* orbitals

### 1. *sp*$^3$*d* hybrid orbitals

We use the same approach that we used for hybridizing *s* and *p* orbitals, but now we are also mixing in a *d* orbital. The *sp*$^3$*d* hybrid orbitals are formed by mixing an *s* orbital, three *p* orbitals, and a *d* orbital. Five equivalent *sp*$^3$*d* hybrid orbitals are formed.

Mixing an *s* orbital, three *p* orbitals, and a *d* orbital gives:

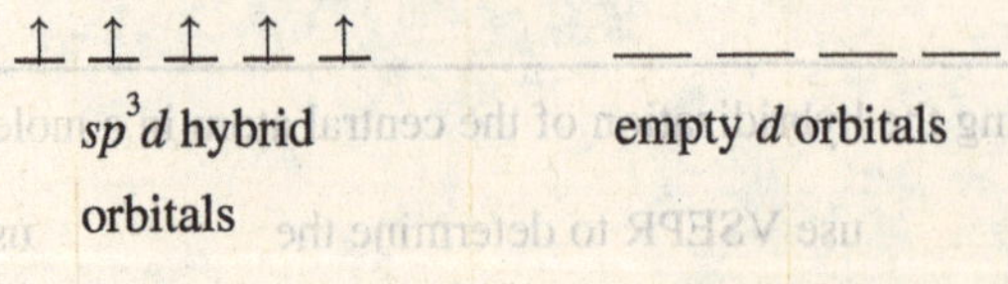

Table 10.4 of the text shows the shape and orientation of the $sp^3d$ hybrid orbitals. These five equivalent hybrid orbitals are directed toward the five corners of a trigonal bipyramid with bond angles of 120° and 90°.

**2. $sp^3d^2$ hybrid orbitals**

We use the same approach that we used above, but now we are mixing in two *d* orbitals. The $sp^3d^2$ hybrid orbitals are formed by mixing an *s* orbital, three *p* orbitals, and two *d* orbitals. Six equivalent $sp^3d^2$ hybrid orbitals are formed.

↑ ↑ ↑ ↑ ↑ ↑ ___ ___ ___
*s* *p* *d*

Mixing an *s* orbital, three *p* orbitals, and two *d* orbitals gives:

↑ ↑ ↑ ↑ ↑ ↑ ___ ___ ___
$sp^3d^2$ hybrid orbitals empty *d* orbitals

Table 10.4 of the text shows the shape and orientation of the $sp^3d^2$ hybrid orbitals. These six equivalent hybrid orbitals are directed toward the six corners of an octahedron with bond angles of 90°.

Summarizing,

| Type of hybrid | No. of equivalent hybrid orbitals | Arrangement of hybrid orbitals |
|---|---|---|
| $sp^3d$ | 5 | trigonal bipyramid, 120°, 90° |
| $sp^3d^2$ | 6 | octahedral, 90° |

**EXAMPLE 10.6**

**Describe the hybridization of xenon in xenon tetrafluoride ($XeF_4$).**

**Strategy:** The steps for determining the hybridization of the central atom in a molecule are:

draw Lewis Structure of the molecule ⟶ use VSEPR to determine the electron pair arrangement surrounding the central atom (Table 10.1 of the text) ⟶ use Table 10.4 of the text to determine the hybridization state of the central atom

**Solution:** Write the Lewis structure for $XeF_4$. The lone pairs of electrons on F have been omitted.

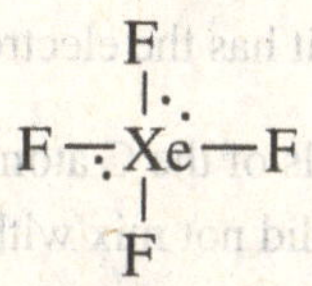

Count the number of electron pairs around the central atom. Since there are six electron pairs around Xe, the electron arrangement that minimizes electron-pair repulsions is **octahedral.**

We conclude that Xe is **$sp^3d^2$ hybridized** because it has the electron arrangement of six $sp^3d^2$ hybrid orbitals.

**Tip:** It is important to use the electron arrangement to determine the hybridization of the central atom and not the geometry. The two lone pairs on Xe are occupying two of the hybrid orbitals, so the lone pairs must be included to determine the correct hybridization.

**PRACTICE EXERCISE**

6. Describe the hybridization of sulfur in sulfur tetrafluoride, $SF_4$.

**Text Problem: 10.36**

## C. Hybridization in molecules containing double or triple bonds

We can determine the hybridization of molecules containing double or triple bonds in the same manner as molecules with single bonds. Furthermore, we would like to determine which orbitals overlap to form the double or triple bond.

In a double or triple bond, there are two types of covalent bonds formed. One involves end-to-end overlap of orbitals in which the electron density is concentrated between the nuclei of the bonding atoms. A bond of this type is called a **sigma bond ($\sigma$ bond)**. The second type involves side-to-side overlap of orbitals in which electron density is concentrated above and below the plane of the nuclei of the bonding atoms. This type of bond is called a **pi bond ($\pi$ bond)**. For the molecules we will be considering, a $\pi$ bond is formed from the side-to-side overlap of two *p* orbitals.

**EXAMPLE 10.7**

**Describe the bonding in carbon dioxide, $CO_2$.**

**Strategy:** The steps for determining the hybridization of the central atom in a molecule are:

draw Lewis Structure of the molecule → use VSEPR to determine the electron pair arrangement surrounding the central atom (Table 10.1 of the text) → use Table 10.4 of the text to determine the hybridization state of the central atom

**Solution:** Write the Lewis structure for $CO_2$.

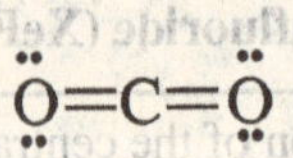

Count the number of electron pairs around the central atom. Since there are two electron pairs around C, the electron arrangement that minimizes electron-pair repulsions is **linear.**

We conclude that C is ***sp* hybridized** because it has the electron arrangement of two *sp* hybrid orbitals.

Next, count the number of electron pairs around each oxygen atom. Since there are three electron pairs around each O, the electron arrangement that minimizes repulsions is **trigonal planar**.

We conclude that each O is **$sp^2$ hybridized** because it has the electron arrangement of three $sp^2$ hybrid orbitals.

Describing the bonding in $CO_2$, each of the *sp* orbitals of the C atom forms a sigma bond with an $sp^2$ hybrid on each of the O atoms. Carbon has two "pure" *p* orbitals that did not mix with the *s* orbital. Each of these *p* orbitals of the C atom forms a pi bond by overlapping in a side-to-side fashion with a *p* orbital on each of the oxygen atoms. Each double bond is composed of one $\sigma$ bond and one $\pi$ bond. Finally, the two lone pairs on each O atom are placed in its two remaining $sp^2$ orbitals.

> **Tip:** A double bond is typically composed of one σ bond and one π bond. A triple bond is composed of one σ bond and two π bonds.

**PRACTICE EXERCISE**

**7.** Describe the bonding in a nitrogen molecule, $N_2$.

> **Text Problems:** 10.36, **10.38**

## PROBLEM TYPE 4: MOLECULAR ORBITAL DIAGRAMS

When working molecular orbital problems, realize that this is another bonding model that is different from the other models we have encountered. So far, we have looked at the Lewis model and valence bond theory. We will focus on molecular orbital diagrams for homonuclear diatomic molecules of second-period elements. Examples include $N_2$, $O_2$, and $F_2$.

### A. Writing electron configurations

For homonuclear diatomic molecules of second-period elements, the types of molecular orbitals used in bonding will be similar. The order of filling molecular orbitals for $Li_2$, $B_2$, $C_2$, and $N_2$ is:

$\sigma^\star_{2p_x}$ ——

$\pi^\star_{2p_y}, \pi^\star_{2p_z}$ —— ——

$\sigma_{2p_x}$ ——

$\pi_{2p_y}, \pi_{2p_z}$ —— ——

$\sigma^\star_{2s}$ ——

$\sigma_{2s}$ ——

$\sigma^\star_{1s}$ ——

$\sigma_{1s}$ ——

The order of filling molecular orbitals for $O_2$ and $F_2$ is:

$\sigma^\star_{2p_x}$ ——

$\pi^\star_{2p_y}, \pi^\star_{2p_z}$ —— ——

$\pi_{2p_y}, \pi_{2p_z}$ —— ——

$\sigma_{2p_x}$ ——

$\sigma^\star_{2s}$ ——

$\sigma_{2s}$ ——

$\sigma^\star_{1s}$ ——

$\sigma_{1s}$ ——

Note that for $O_2$ and $F_2$, the $\sigma_{2p_x}$ is lower in energy than the $\pi_{2p_y}, \pi_{2p_z}$ molecular orbitals.

**EXAMPLE 10.8**

**Write the electron configuration for the ion, $O_2^-$.**

---

**Strategy:** Count the number of electrons in the ion.

Each oxygen has 8 electrons, plus we need to add one electron for the negative charge. The total number of electrons in the ion is 17.

Place the electrons in molecular orbitals following this convention.

1. Build up from the lowest energy molecular orbital to higher energy orbitals.
2. Each molecular orbital can hold a maximum of two electrons.
3. Follow Hund's rule: the most stable arrangement of electrons in molecular orbitals with equal energy is the one with the greatest number of parallel spins.
4. When electrons are paired in a molecular orbital, they must have opposite spin.

**Solution:** Placing 17 electrons following the above rules gives:

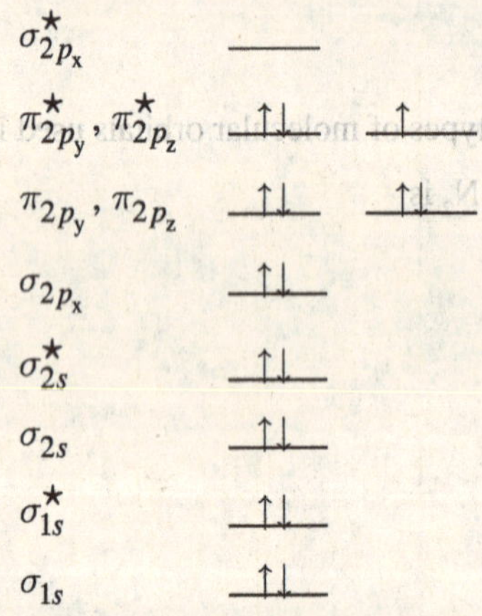


The electron configuration is: $(\sigma_{1s})^2(\sigma^\star_{1s})^2(\sigma_{2s})^2(\sigma^\star_{2s})^2(\sigma_{2p_x})^2(\pi_{2p_y})^2(\pi_{2p_z})^2(\pi^\star_{2p_y})^2(\pi^\star_{2p_z})^1$

**PRACTICE EXERCISE**

8. Write the electron configuration for $C_2^+$.

| **Text Problems:** 10.52, 10.54, 10.56, 10.58, 10.60 |
|---|

## B. Calculating bond order

Placing electrons in a "bonding" molecular orbital yields a stable covalent bond, whereas placing electrons in an "antibonding" molecular orbital results in an unstable bond. We can evaluate the stability of molecules or ions by calculating their **bond order**. The bond order indicates the strength of the bond; the greater the bond order, the stronger the bond. We can calculate bond order as follows:

$$\text{bond order} = \frac{1}{2}\left(\begin{matrix}\text{number of electrons} \\ \text{in bonding MOs}\end{matrix} - \begin{matrix}\text{number of electrons} \\ \text{in antibonding MOs}\end{matrix}\right)$$

**EXAMPLE 10.9**

**Determine the bond order of $O_2^-$.**

---

Referring to Example 10.8, the bond order is:

$$\textbf{bond order} = \frac{1}{2}(10 - 7) = \mathbf{1.5}$$

The bond order of $O_2$ is 2. This indicates that $O_2$ is more stable than $O_2^-$. Is this what you would expect?

**PRACTICE EXERCISE**

9. Determine the bond order of $C_2^+$.

| **Text Problems:** 10.52, 10.54, 10.56, 10.58 |
|---|

## C. Determining the magnetic character of a molecule or ion

In Section 7.8 of the text, the terms "paramagnetic" and "diamagnetic" were discussed. A *paramagnetic* substance contains unpaired electrons and is *attracted* by an external magnetic field. Any substance with an *odd* number of electrons must be paramagnetic, because we need an even number of electrons for complete pairing. In a *diamagnetic* substance, all the electron spins are paired. Diamagnetic substances are slightly *repelled* by an external magnetic field.

Substances containing an even number of electrons may be either diamagnetic or paramagnetic. A molecular orbital diagram can be helpful in determining the magnetic character of a molecule or ion with an even number of electrons. $O_2$ is an example of a molecule with an even number of electrons (16) that is paramagnetic (see Table 10.5 of the text).

**EXAMPLE 10.10**

**Determine the magnetic character of $O_2^-$.**

---

Since $O_2^-$ has an odd number of electrons (17), it is paramagnetic. We could also write the molecular orbital diagram for $O_2^-$ to determine its magnetic character. Looking at the MO diagram for $O_2^-$ in Example 10.8, we see that there is a single unpaired electron in the $\pi^{\star}_{2p_z}$ molecular orbital. Hence, $O_2^-$ is paramagnetic.

**PRACTICE EXERCISE**

10. Determine the magnetic character of $C_2^+$.

| **Text Problems:** 10.56, 10.58 |
|---|

## ANSWERS TO PRACTICE EXERCISES

1. (a) linear (b) tetrahedral (c) trigonal planar (d) octahedral
2. (a) bent (b) square planar (c) distorted tetrahedron (seesaw)

3. The electron arrangement around each C that minimizes electron-pair repulsion is linear. Since there are no lone pairs around each C, the geometry is also **linear.**

4. **(a)** Yes, the molecule is polar. **(b)** No, the molecule is nonpolar. **(c)** No, the molecule is nonpolar.

5. **(a)** *sp* **(b)** $sp^2$

6. When you draw the Lewis structure for $SF_4$, you will find five electron pairs around the central atom, S (four bonding pairs and one lone pair). The electron arrangement that minimizes electron-pair repulsion is trigonal bipyramidal. You should conclude that S is $dsp^3$ hybridized because it has the electron arrangement of five $dsp^3$ hybrid orbitals.

7. :N≡N:

The structure of $N_2$ is **linear**. We conclude that each N is ***sp* hybridized** because it has the electron arrangement of two *sp* hybrid orbitals. An *sp* orbital of one N atom overlaps with an *sp* orbital on the other N to form a sigma bond. The other *sp* orbital of each N contains the lone pair of electrons. Each nitrogen atom has two "pure" *p* orbitals that did not mix with the *s* orbital. The two *p* orbitals on one N can form two pi bonds by overlapping side-to-side with the two *p* orbitals on the other N atom. The triple bond is composed of one σ bond and two π bonds.

8. $(\sigma_{1s})^2(\sigma_{1s}^\star)^2(\sigma_{2s})^2(\sigma_{2s}^\star)^2(\pi_{2p_y})^2(\pi_{2p_z})^1$

9. bond order = 1.5

10. $C_2^+$ is paramagnetic.

## SOLUTIONS TO SELECTED TEXT PROBLEMS

**10.8** **Strategy:** The sequence of steps in determining molecular geometry is as follows:

draw Lewis structure ⟶ find arrangement of electrons pairs ⟶ find arrangement of bonding pairs ⟶ determine geometry based on bonding pairs

**Solution:**

| Lewis structure | Electron pairs on central atom | Electron arrangement | Lone pairs | Geometry |
|---|---|---|---|---|
| (a) 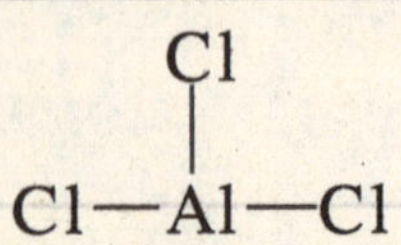 | 3 | trigonal planar | 0 | trigonal planar, $AB_3$ |
| (b) Cl—Zn—Cl | 2 | linear | 0 | linear, $AB_2$ |

(c) 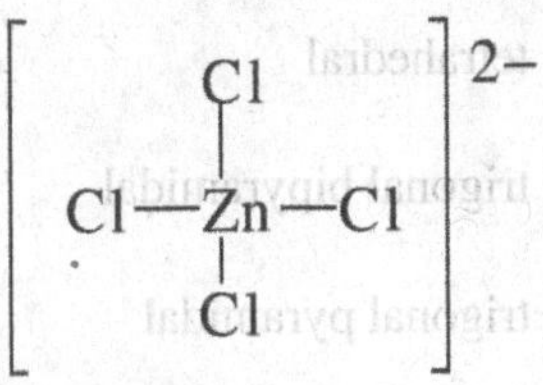 4 tetrahedral 0 tetrahedral, $AB_4$

**10.10** We use the following sequence of steps to determine the geometry of the molecules.

draw Lewis structure ⟶ find arrangement of electrons pairs ⟶ find arrangement of bonding pairs ⟶ determine geometry based on bonding pairs

**(a)** Looking at the Lewis structure we find 4 pairs of electrons around the central atom. The electron pair arrangement is tetrahedral. Since there are no lone pairs on the central atom, the geometry is also **tetrahedral**.

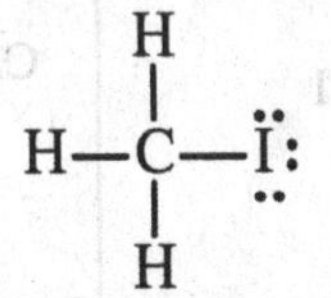

**(b)** Looking at the Lewis structure we find 5 pairs of electrons around the central atom. The electron pair arrangement is trigonal bipyramidal. There are two lone pairs on the central atom, which occupy positions in the trigonal plane. The geometry is **t-shaped**.

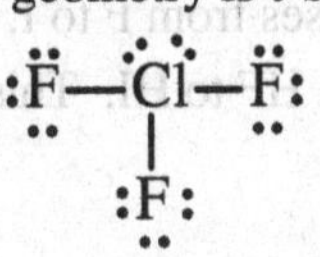

**(c)** Looking at the Lewis structure we find 4 pairs of electrons around the central atom. The electron pair arrangement is tetrahedral. There are two lone pairs on the central atom. The geometry is **bent**.

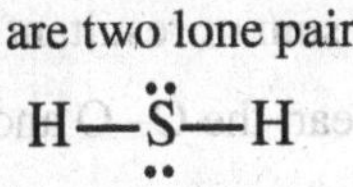

**(d)** Looking at the Lewis structure, there are 3 VSEPR pairs of electrons around the central atom. Recall that a double bond counts as one VSEPR pair. The electron pair arrangement is trigonal planar. Since there are no lone pairs on the central atom, the geometry is also **trigonal planar**.

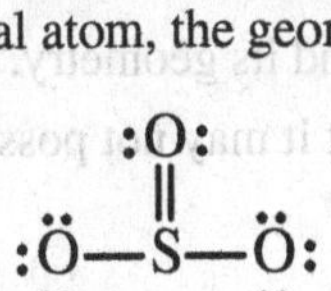

**(e)** Looking at the Lewis structure, there are 4 pairs of electrons around the central atom. The electron pair arrangement is tetrahedral. Since there are no lone pairs on the central atom, the geometry is also **tetrahedral**.

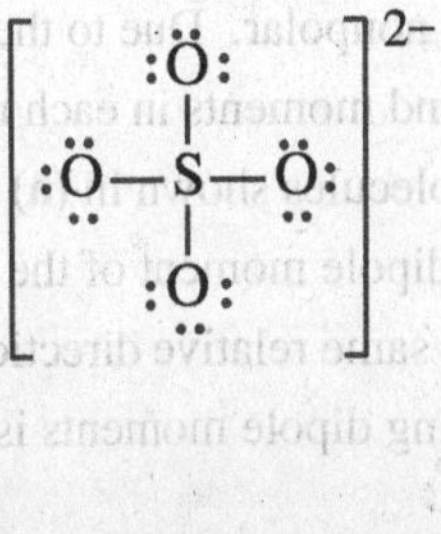

**10.12** (a) $AB_4$ tetrahedral

(b) $AB_2E_2$ bent

(c) $AB_3$ trigonal planar

(d) $AB_2E_3$ linear

(e) $AB_4E_2$ square planar

(f) $AB_4$ tetrahedral

(g) $AB_5$ trigonal bipyramidal

(h) $AB_3E$ trigonal pyramidal

(i) $AB_4$ tetrahedral

**10.14** Only molecules with four bonds to the central atom and no lone pairs are tetrahedral ($AB_4$).

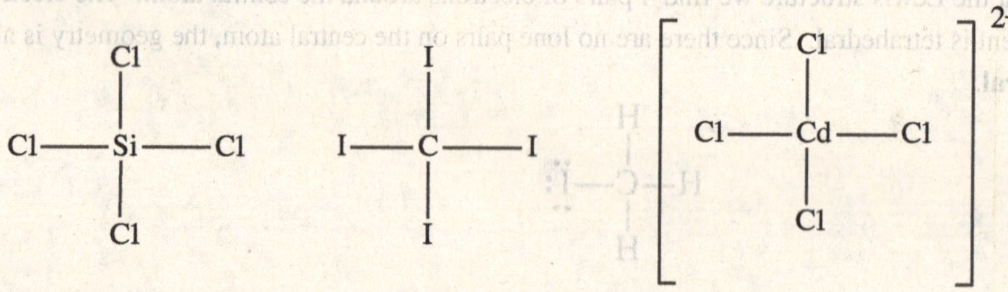


What are the Lewis structures and shapes for $XeF_4$ and $SeF_4$?

**10.20** The electronegativity of the halogens decreases from F to I. Thus, the polarity of the H−X bond (where X denotes a halogen atom) also decreases from HF to HI. This difference in electronegativity accounts for the decrease in dipole moment.

**10.22** Draw the Lewis structures. Both molecules are linear ($AB_2$). In $CS_2$, the two C−S bond moments are equal in magnitude and opposite in direction. The sum or resultant dipole moment will be *zero*. Hence, $CS_2$ is a nonpolar molecule. Even though OCS is linear, the C−O and C−S bond moments are not exactly equal, and there will be a small net dipole moment. Hence, OCS has a **larger** dipole moment than $CS_2$ (zero).

**10.24** **Strategy:** Keep in mind that the dipole moment of a molecule depends on both the difference in electronegativities of the elements present and its geometry. A molecule can have polar bonds (if the bonded atoms have different electronegativities), but it may not possess a dipole moment if it has a highly symmetrical geometry.

**Solution:** Each vertex of the hexagonal structure of benzene represents the location of a C atom. Around the ring, there is no difference in electronegativity between C atoms, so the only bonds we need to consider are the polar C−Cl bonds.

The molecules shown in **(b)** and **(d)** are nonpolar. Due to the high symmetry of the molecules and the equal magnitude of the bond moments, the bond moments in each molecule cancel one another. The resultant dipole moment will be *zero*. For the molecules shown in **(a)** and **(c),** the bond moments do not cancel and there will be net dipole moments. The dipole moment of the molecule in **(a)** is larger than that in **(c)**, because in **(a)** all the bond moments point in the same relative direction, reinforcing each other (see Lewis structure below). Therefore, the order of increasing dipole moments is:

**(b) = (d) < (c) < (a).**

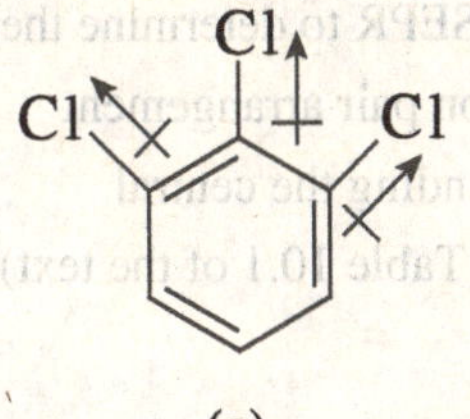

**(a)**

**10.32** **Strategy:** The steps for determining the hybridization of the central atom in a molecule are:

draw Lewis Structure of the molecule → use VSEPR to determine the electron pair arrangement surrounding the central atom (Table 10.1 of the text) → use Table 10.4 of the text to determine the hybridization state of the central atom

**Solution:**

**(a)** Write the Lewis structure of the molecule.

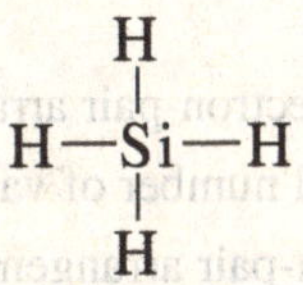

Count the number of electron pairs around the central atom. Since there are four electron pairs around Si, the electron arrangement that minimizes electron-pair repulsion is **tetrahedral.**

We conclude that Si is ***sp*$^3$ hybridized** because it has the electron arrangement of four *sp*$^3$ hybrid orbitals.

**(b)** Write the Lewis structure of the molecule.

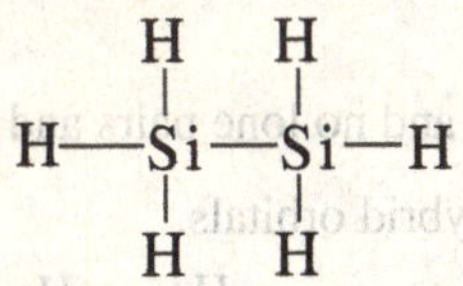

Count the number of electron pairs around the "central atoms". Since there are four electron pairs around each Si, the electron arrangement that minimizes electron-pair repulsion for each Si is **tetrahedral.**

We conclude that each Si is ***sp*$^3$ hybridized** because it has the electron arrangement of four *sp*$^3$ hybrid orbitals.

**10.34** Draw the Lewis structures. Before the reaction, boron is *sp*$^2$ hybridized (trigonal planar electron arrangement) in $BF_3$ and nitrogen is *sp*$^3$ hybridized (tetrahedral electron arrangement) in $NH_3$. After the reaction, boron and nitrogen are both *sp*$^3$ hybridized (tetrahedral electron arrangement).

**10.36** **Strategy:** The steps for determining the hybridization of the central atom in a molecule are:

draw Lewis Structure of the molecule → use VSEPR to determine the electron pair arrangement surrounding the central atom (Table 10.1 of the text) → use Table 10.4 of the text to determine the hybridization state of the central atom

**Solution:**

Write the Lewis structure of the molecule.

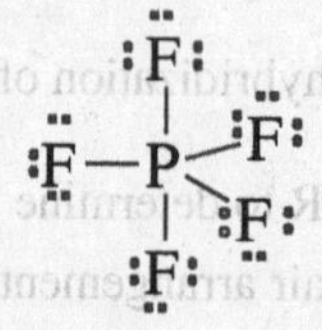

Count the number of electron pairs around the central atom. Since there are five electron pairs around P, the electron arrangement that minimizes electron-pair repulsion is **trigonal bipyramidal** ($AB_5$).

We conclude that P is ***sp*$^3$*d* hybridized** because it has the electron arrangement of five *sp*$^3$*d* hybrid orbitals.

**10.38** An *sp*$^3$*d*$^2$ hybridization indicates that the electron-pair arrangement about iodine is octahedral. If four fluorines are placed around iodine, the total number of valence electrons is 35. Thirty-six electrons are required to complete an octahedral electron-pair arrangement with four bonds and two lone pairs of electrons. Adding one valence electron gives the anion, $\mathbf{IF_4^-}$.

**10.40** (a) Each carbon has four bond pairs and no lone pairs and therefore has a tetrahedral electron pair arrangement. This implies *sp*$^3$ hybrid orbitals.

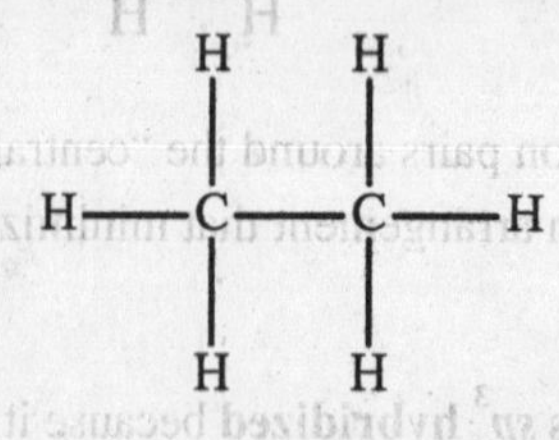

(b) The left-most carbon is tetrahedral and therefore has *sp*$^3$ hybrid orbitals. The two carbon atoms connected by the double bond are trigonal planar with *sp*$^2$ hybrid orbitals.

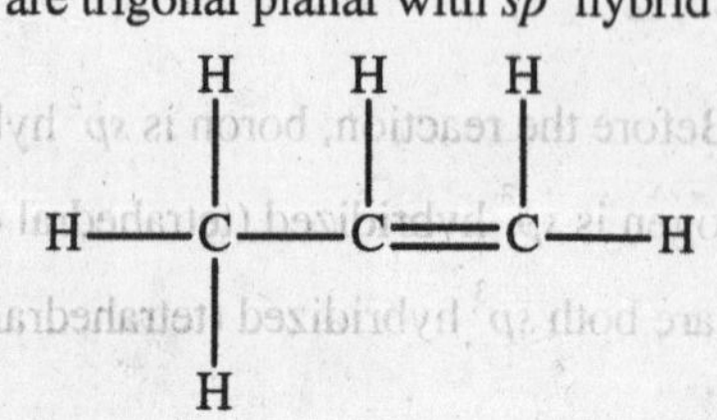

**(c)** Carbons 1 and 4 have $sp^3$ hybrid orbitals. Carbons 2 and 3 have $sp$ hybrid orbitals.

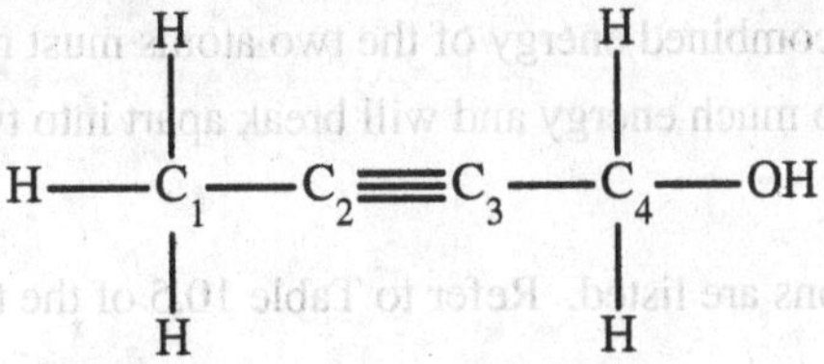

**(d)** The left-most carbon is tetrahedral ($sp^3$ hybrid orbitals). The carbon connected to oxygen is trigonal planar (why?) and has $sp^2$ hybrid orbitals.

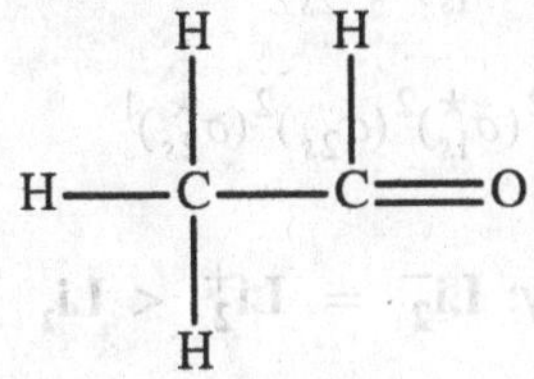

**(e)** The left-most carbon is tetrahedral ($sp^3$ hybrid orbitals). The other carbon is trigonal planar with $sp^2$ hybridized orbitals.

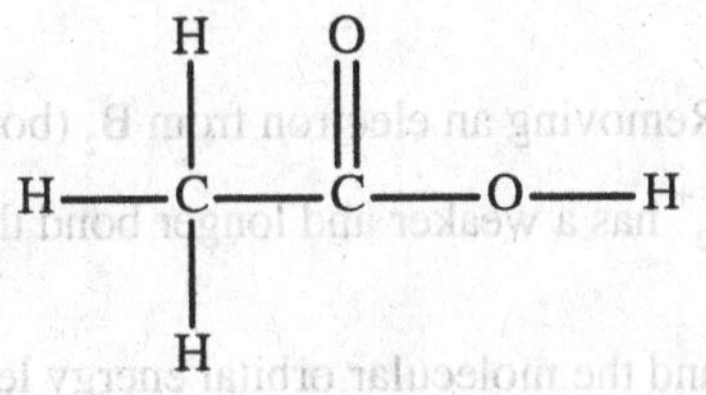

**10.42** **Strategy:** The steps for determining the hybridization of the central atom in a molecule are:

draw Lewis Structure of the molecule ⟶ use VSEPR to determine the electron pair arrangement surrounding the central atom (Table 10.1 of the text) ⟶ use Table 10.4 of the text to determine the hybridization state of the central atom

**Solution:**

Write the Lewis structure of the molecule. Several resonance forms with formal charges are shown.

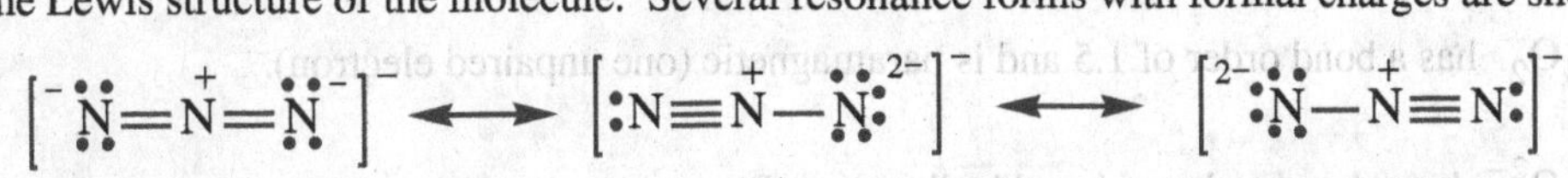

Count the number of electron pairs around the central atom. Since there are two electron pairs around N, the electron arrangement that minimizes electron-pair repulsion is **linear** ($AB_2$). Remember, for VSEPR purposes a multiple bond counts the same as a single bond.

We conclude that N is ***sp* hybridized** because it has the electron arrangement of two *sp* hybrid orbitals.

**10.44** A single bond is usually a sigma bond, a double bond is usually a sigma bond and a pi bond, and a triple bond is always a sigma bond and two pi bonds. Therefore, there are **nine pi bonds** and **nine sigma bonds** in the molecule.

**10.50** In order for the two hydrogen atoms to combine to form a $H_2$ molecule, the electrons must have opposite spins. Furthermore, the combined energy of the two atoms must not be too great. Otherwise, the $H_2$ molecule will possess too much energy and will break apart into two hydrogen atoms.

**10.52** The electron configurations are listed. Refer to Table 10.5 of the text for the molecular orbital diagram.

$Li_2$: $(\sigma_{1s})^2(\sigma_{1s}^\star)^2(\sigma_{2s})^2$ bond order $= 1$

$Li_2^+$: $(\sigma_{1s})^2(\sigma_{1s}^\star)^2(\sigma_{2s})^1$ bond order $= \frac{1}{2}$

$Li_2^-$: $(\sigma_{1s})^2(\sigma_{1s}^\star)^2(\sigma_{2s})^2(\sigma_{2s}^\star)^1$ bond order $= \frac{1}{2}$

Order of increasing stability: $\mathbf{Li_2^-} = \mathbf{Li_2^+} < \mathbf{Li_2}$

In reality, $Li_2^+$ is more stable than $Li_2^-$ because there is less electrostatic repulsion in $Li_2^+$.

**10.54** See Table 10.5 of the text. Removing an electron from $B_2$ (bond order = 1) gives $B_2^+$, which has a bond order of (1/2). Therefore, $\mathbf{B_2^+}$ has a weaker and longer bond than $B_2$.

**10.56** In both the Lewis structure and the molecular orbital energy level diagram (Table 10.5 of the text), the oxygen molecule has a double bond (bond order = 2). The principal difference is that the molecular orbital treatment predicts that the molecule will have two unpaired electrons (paramagnetic). Experimentally this is found to be true.

**10.58** We refer to Table 10.5 of the text.

$O_2$ has a bond order of 2 and is paramagnetic (two unpaired electrons).

$O_2^+$ has a bond order of 2.5 and is paramagnetic (one unpaired electron).

$O_2^-$ has a bond order of 1.5 and is paramagnetic (one unpaired electron).

$O_2^{2-}$ has a bond order of 1 and is diamagnetic.

Based on molecular orbital theory, the stability of these molecules increases as follows:

$$\mathbf{O_2^{2-} < O_2^- < O_2 < O_2^+}$$

**10.60** As discussed in the text (see Table 10.5), the single bond in $B_2$ is a pi bond (the electrons are in a pi *bonding* molecular orbital) and the double bond in $C_2$ is made up of two pi bonds (the electrons are in the pi *bonding* molecular orbitals).

**10.62** 1) Atoms are far apart. There is no interaction.

2) Atoms approach each other. Attractive forces are stronger than repulsive forces, so the potential energy of the system decreases. The 2*p* orbitals on F begin to overlap.

3) The system is most stable; potential energy reaches a minimum. This point represents the equilibrium bond length of $F_2$. There is significant orbital overlap, and the electrons spend time in the region between nuclei where they can interact with both nuclei.

4) As the distance between nuclei continues to decrease, nuclear-nuclear and electron-electron repulsions increase leading to an increase in potential energy.

5) If the distance between nuclei were to decrease further, the potential energy would continue to rise until it becomes positive. The $F_2$ molecule is no longer stable.

**10.66** The symbol on the left shows the pi bond delocalized over the entire molecule. The symbol on the right shows only one of the two resonance structures of benzene; it is an incomplete representation.

**10.68** (a) Two Lewis resonance forms are shown below. Formal charges different than zero are indicated.

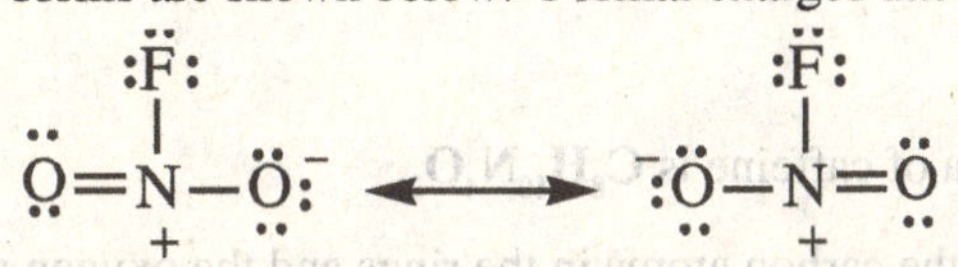

(b) There are no lone pairs on the nitrogen atom; it should have a trigonal planar electron pair arrangement and therefore use $sp^2$ hybrid orbitals.

(c) The bonding consists of sigma bonds joining the nitrogen atom to the fluorine and oxygen atoms. In addition there is a pi molecular orbital delocalized over the N and O atoms. Is nitryl fluoride isoelectronic with the carbonate ion?

**10.70** The Lewis structures of ozone are:

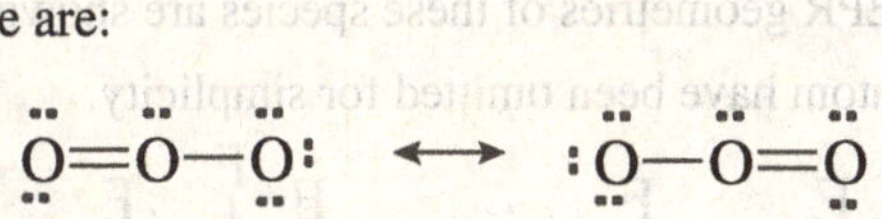

The central oxygen atom is $sp^2$ hybridized ($AB_2E$). The unhybridized $2p_z$ orbital on the central oxygen overlaps with the $2p_z$ orbitals on the two end atoms.

**10.72** **Strategy:** The sequence of steps in determining molecular geometry is as follows:

draw Lewis structure $\longrightarrow$ find arrangement of electrons pairs $\longrightarrow$ find arrangement of bonding pairs $\longrightarrow$ determine geometry based on bonding pairs

**Solution:**

Write the Lewis structure of the molecule.

Count the number of electron pairs around the central atom. There are two electron pairs around Hg.

Since there are two electron pairs around Hg, the electron-pair arrangement that minimizes electron-pair repulsion is **linear**.

In addition, since there are no lone pairs around the central atom, the geometry is also **linear** ($AB_2$).

You could establish the geometry of $HgBr_2$ by measuring its dipole moment. If mercury(II) bromide were bent, it would have a measurable dipole moment. Experimentally, it has no dipole moment and therefore must be linear.

**10.74** According to valence bond theory, a pi bond is formed through the side-to-side overlap of a pair of *p* orbitals. As atomic size increases, the distance between atoms is too large for *p* orbitals to overlap effectively in a side-to-side fashion. If two orbitals overlap poorly, that is, they share very little space in common, then the resulting bond will be very weak. This situation applies in the case of pi bonds between silicon atoms as well as between any other elements not found in the second period. It is usually far more energetically favorable for silicon, or any other heavy element, to form two single (sigma) bonds to two other atoms than to form a double bond (sigma + pi) to only one other atom.

**10.76** **(a)** The molecular formula of caffeine is $\mathbf{C_8H_{10}N_4O_2}$.

**(b)** The hybridizations of the carbon atoms in the rings and the oxygen atoms are $sp^2$. The carbon atoms of the methyl groups ($-CH_3$) are $sp^3$ hybridized. The three nitrogen atoms with bonded methyl groups ($-CH_3$) are $sp^3$ hybridized. The fourth nitrogen is $sp^2$ hybridized.

**(c)** The geometry about the $sp^2$ hybridized carbons is **trigonal planar**, and the geometry about the $sp^3$ hybridized carbons is **tetrahedral**. The geometry about the $sp^2$ hybridized nitrogen is **bent**, and the geometry about the $sp^3$ hybridized nitrogen atoms is **trigonal pyramidal**.

**10.78** The Lewis structures and VSEPR geometries of these species are shown below. The three nonbonding pairs of electrons on each fluorine atom have been omitted for simplicity.

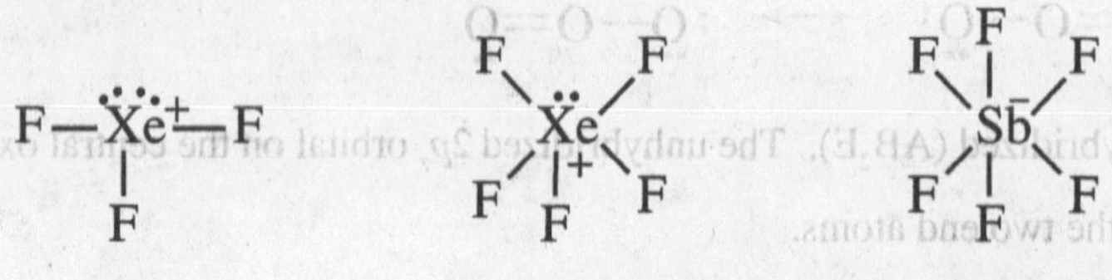

| $AB_3E_2$ | $AB_5E$ | $AB_6$ |
| --- | --- | --- |
| T-shaped | Square Pyramidal | Octahedral |

**10.80** To predict the bond angles for the molecules, you would have to draw the Lewis structure and determine the geometry using the VSEPR model. From the geometry, you can predict the bond angles.

**(a)** $BeCl_2$: $AB_2$ type, 180° (linear).

**(b)** $BCl_3$: $AB_3$ type, 120° (trigonal planar).

(c) $CCl_4$: $AB_4$ type, 109.5° (tetrahedral).

(d) $CH_3Cl$: $AB_4$ type, 109.5° (tetrahedral with a possible slight distortion resulting from the different sizes of the chlorine and hydrogen atoms).

(e) $Hg_2Cl_2$: Each mercury atom is of the $AB_2$ type. The entire molecule is linear, 180° bond angles.

(f) $SnCl_2$: $AB_2E$ type, roughly 120° (bent).

(g) $H_2O_2$: The atom arrangement is HOOH. Each oxygen atom is of the $AB_2E_2$ type and the H−O−O angles will be roughly 109.5°.

(h) $SnH_4$: $AB_4$ type, 109.5° (tetrahedral).

**10.82** Since arsenic and phosphorus are both in the same group of the periodic table, this problem is exactly like Problem 10.40. $AsF_5$ is an $AB_5$ type molecule, so the geometry is trigonal bipyramidal. We conclude that As is ***sp*$^3$*d* hybridized** because it has the electron arrangement of five $sp^3d$ hybrid orbitals.

**10.84** Only $ICl_2^-$ and $CdBr_2$ will be linear. The rest are bent.

**10.86** **(a)**

**Strategy:** The steps for determining the hybridization of the central atom in a molecule are:

| | | | | |
|---|---|---|---|---|
| draw Lewis Structure of the molecule | ⟶ | use VSEPR to determine the electron pair arrangement surrounding the central atom (Table 10.1 of the text) | ⟶ | use Table 10.4 of the text to determine the hybridization state of the central atom |

**Solution:**

The geometry around each nitrogen is identical. To complete an octet of electrons around N, you must add a lone pair of electrons. Count the number of electron pairs around N. There are three electron pairs around each N.

Since there are three electron pairs around N, the electron-pair arrangement that minimizes electron-pair repulsion is **trigonal planar**.

We conclude that each N is ***sp*$^2$ hybridized** because it has the electron arrangement of three $sp^2$ hybrid orbitals.

**(b)**

**Strategy:** Keep in mind that the dipole moment of a molecule depends on both the difference in electronegativities of the elements present and its geometry. A molecule can have polar bonds (if the bonded atoms have different electronegativities), but it may not possess a dipole moment if it has a highly symmetrical geometry.

**Solution:** An N–F bond is polar because F is more electronegative than N. The structure on the right has a dipole moment because the two N–F bond moments do not cancel each other out and so the molecule has a net dipole moment. On the other hand, the two N–F bond moments in the left-hand structure cancel. The sum or resultant dipole moment will be *zero*.

**10.88** In 1,2-dichloroethane, the two C atoms are joined by a sigma bond. Rotation about a sigma bond does not destroy the bond, and the bond is therefore free (or relatively free) to rotate. Thus, all angles are permitted and the molecule is nonpolar because the C–Cl bond moments cancel each other because of the averaging effect brought about by rotation. In *cis*-dichloroethylene the two C–Cl bonds are locked in position. The π bond between the C atoms prevents rotation (in order to rotate, the π bond must be broken, using an energy source such as light or heat). Therefore, there is no rotation about the C=C in *cis*-dichloroethylene, and the molecule is polar.

**10.90** $O_3$, CO, $CO_2$, $NO_2$, $N_2O$, $CH_4$, and $CFCl_3$ are greenhouse gases.

**10.92** The Lewis structure is:

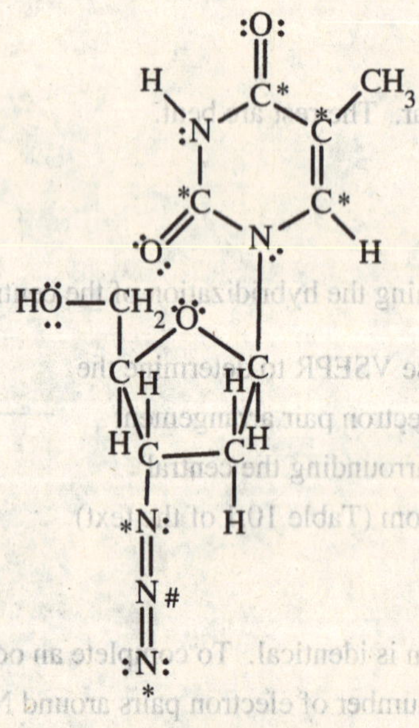

The carbon atoms and nitrogen atoms marked with an asterisk (C∗ and N∗) are $sp^2$ hybridized; unmarked carbon atoms and nitrogen atoms are $sp^3$ hybridized; and the nitrogen atom marked with (#) is *sp* hybridized.

**10.94** C has no *d* orbitals but Si does (3*d*). Thus, $H_2O$ molecules can add to Si in hydrolysis (valence-shell expansion).

**10.96** The carbons are in $sp^2$ hybridization states. The nitrogens are in the $sp^3$ hybridization state, except for the ring nitrogen double-bonded to a carbon that is $sp^2$ hybridized. The oxygen atom is $sp^2$ hybridized.

**10.98** **(a)** Use a conventional oven. A microwave oven would not cook the meat from the outside toward the center (it penetrates).

**(b)** Polar molecules absorb microwaves and would interfere with the operation of radar.

**(c)** Too much water vapor (polar molecules) absorbed the microwaves, interfering with the operation of radar.

**10.100** Molecules (a) and (b) are polar.

**10.102** The smaller size of F compared to Cl results in a shorter F–F bond than a Cl–Cl bond. The closer proximity of the lone pairs of electrons on the F atoms results in greater electron-electron repulsions that weaken the bond.

**10.104** $1\text{ D} = 3.336 \times 10^{-30}\text{ C}\cdot\text{m}$

electronic charge $(e) = 1.6022 \times 10^{-19}\text{ C}$

$$\frac{\mu}{ed} \times 100\% = \frac{1.92\,\cancel{\text{D}} \times \dfrac{3.336 \times 10^{-30}\,\cancel{\text{C}}\cdot\cancel{\text{m}}}{1\,\cancel{\text{D}}}}{(1.6022 \times 10^{-19}\,\cancel{\text{C}}) \times (91.7 \times 10^{-12}\,\cancel{\text{m}})} \times 100\% = \mathbf{43.6\%}\text{ ionic character}$$

**10.106** The second and third vibrational motions are responsible for $CO_2$ to behave as a greenhouse gas. $CO_2$ is a nonpolar molecule. The second and third vibrational motions create a changing dipole moment. The first vibration, a symmetric stretch, does *not* create a dipole moment. Since CO, $NO_2$, and $N_2O$ are all polar molecules, they will also act as greenhouse gases.

**10.108** **(a)** A σ bond is formed by orbitals overlapping end-to-end. Rotation will not break this end-to-end overlap. A π bond is formed by the sideways overlapping of orbitals. The two 90° rotations (180° total) will break and then reform the pi bond, thereby converting *cis*-dichloroethylene to *trans*-dichloroethylene.

**(b)** The pi bond is weaker because of the lesser extent of sideways orbital overlap, compared to the end-to-end overlap in a sigma bond.

**(c)** The bond enthalpy is given in the unit, kJ/mol. To find the longest wavelength of light needed to bring about the conversion from *cis* to *trans*, we need the energy to break a pi bond in a single molecule. We convert from kJ/mol to J/molecule.

$$\frac{270\text{ kJ}}{1\,\cancel{\text{mol}}} \times \frac{1\,\cancel{\text{mol}}}{6.022 \times 10^{23}\text{ molecules}} = 4.48 \times 10^{-22}\text{ kJ/molecule} = 4.48 \times 10^{-19}\text{ J/molecule}$$

Now that we have the energy needed to cause the conversion from *cis* to *trans* in one molecule, we can calculate the wavelength from this energy.

$$E = \frac{hc}{\lambda}$$

$$\lambda = \frac{hc}{E} = \frac{(6.63 \times 10^{-34}\ \text{J}\cdot\text{s})(3.00 \times 10^{8}\ \text{m/s})}{4.48 \times 10^{-9}\ \text{J}}$$

$$\lambda = \mathbf{4.44 \times 10^{-7}\ m = 444\ nm}$$

**10.110** In each case, we examine the molecular orbital that is occupied by the valence electrons of the molecule to see if it is a bonding or antibonding molecular orbital. If the electron is in a bonding molecular orbital, it is more stable than an electron in an atomic orbital (1*s* or 2*p* atomic orbital) and thus will have a higher ionization energy compared to the lone atom. On the other hand, if the electron is in an antibonding molecular orbital, it is less stable than an electron in an atomic orbital (1*s* or 2*p* atomic orbital) and thus will have a lower ionization energy compared to the lone atom. Refer to Table 10.5 of the text.

**(a)** $H_2$ **(b)** $N_2$ **(c)** O **(d)** F

**10.112** **(a)** Looking at the electronic configuration for $N_2$ shown in Table 10.5 of the text, we write the electronic configuration for $P_2$.

$$[Ne_2](\sigma_{3s})^2(\sigma^\star_{3s})^2(\pi_{3p_y})^2(\pi_{3p_z})^2(\sigma_{3p_x})^2$$

**(b)** Past the $Ne_2$ core configuration, there are 8 bonding electrons and 2 antibonding electrons. The bond order is:

$$\textbf{bond order} = \tfrac{1}{2}(8 - 2) = \mathbf{3}$$

**(c)** All the electrons in the electronic configuration are paired. $P_2$ is **diamagnetic.**

**10.114** The Lewis structure shows 4 pairs of electrons on the two oxygen atoms. From Table 10.5 of the text, we see that these 8 valence electrons are placed in the $\sigma_{2p_x}$, $\pi_{2p_y}$, $\pi_{2p_z}$, $\pi^\star_{2p_y}$, and $\pi^\star_{2p_z}$ orbitals. For all the electrons to be paired, energy is needed to flip the spin in one of the antibonding molecular orbitals ($\pi^\star_{2p_y}$ or $\pi^\star_{2p_z}$). According to Hund's rule, this arrangement is less stable than the ground-state configuration shown in Table 10.5, and hence the Lewis structure shown actually corresponds to an excited state of the oxygen molecule.

**10.116** You should draw the correct Lewis structures for each species. $ClF_3$ has a T-shaped geometry and is $sp^3d$ hybridized. $AsF_5$ has a trigonal bipyramidal geometry and is $sp^3d$ hybridized. $ClF_2^+$ has a bent or v-shaped geometry and is $sp^3$ hybridized. $AsF_6^-$ has an octahedral geometry and is $sp^3d^2$ hybridized.

**10.118** **(a)** Although the O atoms are $sp^3$ hybridized, they are locked in a planar structure by the benzene rings. The molecule is symmetrical and therefore does not possess a dipole moment.

**(b)** 20 σ bonds and 6 π bonds.

**10.120** **(a)** $:\overset{-}{C}\equiv\overset{+}{O}:$

This is the only reasonable Lewis structure for CO. The electronegativity difference suggests that electron density should concentrate on the O atom, but assigning formal charges places a negative charge on the C atom. Therefore, we expect CO to have a small dipole moment.

**(b)** The Lewis structure shows a triple bond. The molecular orbital description gives a bond order of 3, just like $N_2$ (see Table 10.5 of the text).

**(c)** Normally, we would expect the more electronegative atom to bond with the metal ion. In this case, however, the small dipole moment suggests that the C atom may form a stronger bond with $Fe^{2+}$ than O. More elaborate analysis of the orbitals involved shows that this is indeed the case.

**10.122** The skeletal structure of carbon suboxide is:

$$O{=}C{=}C{=}C{=}O$$

The C=O bond moments on each end of the molecule are equal and opposite and will cancel. Carbon suboxide is linear (the C atoms are all *sp* hybridized) and does **not possess a dipole moment**.

**10.124** We refer to Table 10.5 of the text.

$NO^{2-}$ has a bond order of 1.5.

$NO^-$ has a bond order of 2. ($NO^-$ is isoelectronic with $O_2$.)

NO has a bond order of 2.5.

$NO^+$ has a bond order of 3. ($NO^+$ is isoelectronic with CO.)

$NO^{2+}$ has a bond order of 2.5.

Arranged in order of increasing bond order, we have:

$$\mathbf{NO^{2-} < NO^- < NO = NO^{2+} < NO^+}$$

# CHAPTER 11
# INTERMOLECULAR FORCES AND LIQUIDS AND SOLIDS

## PROBLEM-SOLVING STRATEGIES AND TUTORIAL SOLUTIONS

### TYPES OF PROBLEMS

**Problem Type 1:** Identifying Intermolecular Forces.

**Problem Type 2:** Identifying Hydrogen Bonds.

**Problem Type 3:** Counting the Number of Atoms in a Unit Cell.

**Problem Type 4:** Calculating Density from Crystal Structure and Atomic Radius.

## PROBLEM TYPE 1: IDENTIFYING INTERMOLECULAR FORCES

**Intermolecular forces** are attractive forces between molecules. Intermolecular forces account for the existence of the condensed states of matter--liquids and solids. To understand the properties of condensed matter, you must understand the different types of intermolecular forces.

(1) **Dipole-dipole forces:** These are attractive forces that act between polar molecules, that is, between molecules that possess dipole moments (see Section 11.2 of the text). The larger the dipole moments, the greater the attractive force.

(2) **Ion-dipole forces:** These are attractive forces that occur between an ion (either a cation or an anion) and a polar molecule (see Figure 11.2 of the text). The strength of this interaction depends on the charge and size of the ion and on the magnitude of the dipole moment and the size of the molecule.

(3) **Dispersion forces:** These are attractive forces that arise as a result of temporary dipoles induced in the atoms or molecules. This is the only type of intermolecular force that exists between nonpolar atoms or molecules. The likelihood of a dipole moment being induced depends on the polarizability of the atom or molecule. *Polarizability* is the ease with which the electron distribution in the atom or molecule can be distorted. Dispersion forces usually increase with molar mass.

For a complete discussion of intermolecular forces, see Section 11.2 of the text.

**EXAMPLE 11.1**

**Indicate all the different types of intermolecular forces that exist in each of the following substances: (a) $CCl_4(l)$ and (b) $HBr(l)$.**

---

**Strategy:** Classify the species into three categories: ionic, polar (possessing a dipole moment), and nonpolar. Keep in mind that dispersion forces exist between *all* species.

**Solution:**

(a) $CCl_4$ is nonpolar. The only type of intermolecular forces present in nonpolar molecules is *dispersion forces*.

**(b)** HBr is a polar molecule. The types of intermolecular forces present are *dipole-dipole* and *dispersion forces*. There is no hydrogen bonding in HBr. The Br atom is too large and is not electronegative enough.

**PRACTICE EXERCISE**

1. Which of the following substances should have the strongest intermolecular attractive forces: $N_2$, Ar, $F_2$, or $Cl_2$?
2. The dipole moment (μ) in HCl is 1.03 D, and in HCN it is 2.99 D. Which substance should have the higher boiling point?

**Text Problems: 11.8**, 11.10, 11.16, **11.18**

## PROBLEM TYPE 2: IDENTIFYING HYDROGEN BONDS

The boiling points of $NH_3$, $H_2O$, and HF are much higher than expected if the boiling point is solely based on molar mass (see Figure 11.6 of the text). The high boiling points in these compounds are due to extensive *hydrogen bonding*. The hydrogen bond is a special type of dipole-dipole interaction between the hydrogen atom in a polar bond, such as N–H, O–H, or F–H, and an electronegative O, N, or F atom. This interaction is written

A–H···B or A–H···A

where A and B represent O, N, or F. A–H is one molecule or part of a molecule and B is a part of another molecule. The dotted line represents the hydrogen bond.

The average energy of a hydrogen bond is quite large for a dipole-dipole interaction (up to 40 kJ/mol). Thus, hydrogen bonds are a powerful force in determining the structures and properties of many compounds.

**EXAMPLE 11.2**

**Predict whether hydrogen bonding intermolecular attractions are present in the following substances: (a) $CH_3OH(l)$ and (b) $CH_3CH_2OCH_2CH_3(l)$.**

**Strategy:** If a molecule contains an N–H, O–H, or F–H bond it can form intermolecular hydrogen bonds. A hydrogen bond is a particularly strong dipole-dipole intermolecular attraction.

**Solution:**

**(a)** $CH_3OH$ (methanol) is polar and has a hydrogen atom bound to an oxygen atom. Hydrogen bonding intermolecular forces are present between $CH_3OH$ molecules. Dipole-dipole and dispersion forces are also present.

**(b)** $CH_3CH_2OCH_2CH_3$ (diethyl ether) is polar and does contain both oxygen atoms and hydrogen atoms. However, all the hydrogen atoms are bonded to carbon, not oxygen. There is no hydrogen bonding in diethyl ether, because carbon is not electronegative enough. The intermolecular attractive forces present are dipole-dipole and dispersion forces.

**PRACTICE EXERCISE**

3. Which member of each pair has the stronger intermolecular forces of attraction: (a) $H_2O$ or $H_2S$, (b) HCl or HF, and (c) $NH_3$ or $PH_3$?

**Text Problems:** 11.12, **11.14, 11.18**, 11.20

## PROBLEM TYPE 3: COUNTING THE NUMBER OF ATOMS IN A UNIT CELL

To solve this problem, you must realize that because every unit cell in a crystalline solid is adjacent to other unit cells, most of a cell's atoms are shared by neighboring cells. In cubic cells, each corner atom is shared by eight unit cells [see Figure 11.19(a) of the text]. A face-centered atom is shared by two unit cells [See Figure 11.19(c) of the text]. A center atom in a unit cell is solely contained by that unit cell and is not shared. An edge atom is shared by four unit cells. Table 11.1 summarizes this information.

**TABLE 11.1**

| Type of atom | Amount of atom contained in unit cell |
|---|---|
| Corner | 1/8 |
| Edge | 1/4 |
| Face-centered | 1/2 |
| Center | 1 |

**EXAMPLE 11.3**

**If atoms of a solid occupy a face-centered cubic lattice, how many atoms are there per unit cell?**

**Strategy:** Recall that a corner atom is shared with 8 unit cells and therefore only 1/8 of corner atom is within a given unit cell. Also recall that a face atom is shared with 2 unit cells and therefore 1/2 of a face atom is within a given unit cell. See Figure 11.19 of the text.

**Solution:** In a face-centered cubic unit cell, there are atoms at each of the eight corners, and there is one atom in each of the six faces.

(8 corner atoms)(1/8 atom per corner) + (6 face-centered atoms)(1/2 atom per face) = **4 atoms/unit cell**

**PRACTICE EXERCISE**

4. Atoms of polonium (Po) occupy a simple cubic lattice. How many Po atoms are there per unit cell?

**Text Problems:** 11.38, **11.44**

## PROBLEM TYPE 4: CALCULATING DENSITY FROM CRYSTAL STRUCTURE AND ATOMIC RADIUS

*Step 1:* To solve this type of problem, you must know how many atoms are contained in the different types of cubic unit cells. Table 11.2 summarizes the number of atoms per unit cell.

**TABLE 11.2**

| Type of cubic unit cell | Number of atoms/unit cell |
|---|---|
| Simple cubic | 1 |
| Body-centered cubic | 2 |
| Face-centered cubic | 4 |

*Step 2:* You can look up the relationship between edge length ($a$) and atomic radius ($r$). See Figure 11.22 of the text. Table 11.3 summarizes these relationships.

**TABLE 11.3**

| Type of cubic unit cell | Relationship between a and r |
|---|---|
| Simple cubic | $a = 2r$ |
| Body-centered cubic | $a = \frac{4}{\sqrt{3}}r$ |
| Face-centered cubic | $a = \sqrt{8}\,r$ |

*Step 3:* Density = mass/volume. The volume of a cube is equal to the edge length cubed.

$$V = a^3$$

You will know how many atoms are in a unit cell (*Step 1* above). To calculate the mass, you need to convert from atoms/unit cell to grams/unit cell. A reasonable strategy would be

$$\frac{\text{atoms}}{\text{unit cell}} \rightarrow \frac{\text{mol}}{\text{unit cell}} \rightarrow \frac{\text{grams}}{\text{unit cell}}$$

*Step 4:* Substitute the mass and volume into the density equation to solve for the density.

**EXAMPLE 11.4**

**Nickel crystallizes in a face-centered cubic lattice with an edge length of 352 pm. Calculate the density of nickel.**

**Strategy:** To calculate the density, we need to find the mass and the volume of the unit cell. The volume of the unit cell can be calculated from the edge length. Because nickel crystallizes in a face-centered cubic lattice, there are four Ni atoms per unit cell. We can convert from 4 atoms/unit cell to grams/unit cell.

**Solution:** Density = mass/volume. The volume of a cube is equal to the edge length cubed.

$$V = a^3$$

$$V = (325\text{ pm})^3 = 3.43 \times 10^7\text{ pm}^3$$

We convert $pm^3$ to $cm^3$ because the density of a solid is typically expressed in units of $g/cm^3$.

$$(3.43 \times 10^7\text{ pm}^3) \times \left(\frac{1 \times 10^{-12}\text{ m}}{1\text{ pm}}\right)^3 \times \left(\frac{1\text{ cm}}{1 \times 10^{-2}\text{ m}}\right)^3 = 3.43 \times 10^{-23}\text{ cm}^3$$

There are four atoms/unit cell. To calculate the mass, you need to convert from atoms/unit cell to grams/unit cell. A reasonable strategy would be:

$$\text{atoms/unit cell} \rightarrow \text{mol/unit cell} \rightarrow \text{grams/unit cell}$$

$$\frac{4\text{ Ni atoms}}{\text{unit cell}} \times \frac{1\text{ mol Ni}}{6.022 \times 10^{23}\text{ Ni atoms}} \times \frac{58.71\text{ g Ni}}{1\text{ mol Ni}} = \frac{3.900 \times 10^{-22}\text{ g Ni}}{1\text{ unit cell}}$$

Substitute the mass and volume into the density equation to solve for the density.

$$d = \frac{m}{V} = \frac{3.900 \times 10^{-22} \text{ g Ni}}{3.43 \times 10^{-23} \text{ cm}^3} = \mathbf{11.4 \text{ g/cm}^3}$$

**PRACTICE EXERCISE**

5. Potassium crystallizes in a body-centered cubic lattice and has a density of 0.856 g/cm$^3$ at 25°C.
   (a) How many atoms are there per unit cell?
   (b) What is the length of an edge of the unit cell?

**Text Problems: 11.40,** 11.42

## ANSWERS TO PRACTICE EXERCISES

1. $Cl_2$ 2. HCN 3. **(a)** $H_2O$ **(b)** HF **(c)** $NH_3$

4. 1 atom/unit cell 5. **(a)** 2 atoms/unit cell **(b)** $a = 5.34 \times 10^{-8}$ cm $= 534$ pm

## SOLUTIONS TO SELECTED TEXT PROBLEMS

**11.8** **Strategy:** Classify the species into three categories: ionic, polar (possessing a dipole moment), and nonpolar. Keep in mind that dispersion forces exist between *all* species.

**Solution:** The three molecules are essentially nonpolar. There is little difference in electronegativity between carbon and hydrogen. Thus, the only type of intermolecular attraction in these molecules is dispersion forces. Other factors being equal, the molecule with the greater number of electrons will exert greater intermolecular attractions. By looking at the molecular formulas you can predict that the order of increasing boiling points will be $CH_4 < C_3H_8 < C_4H_{10}$.

On a very cold day, propane and butane would be liquids (boiling points −44.5°C and −0.5°C, respectively); only **methane** would still be a gas (boiling point −161.6°C).

**11.10** **(a)** Benzene ($C_6H_6$) molecules are nonpolar. Only **dispersion** forces will be present.

**(b)** Chloroform ($CH_3Cl$) molecules are polar (why?). **Dispersion** and **dipole-dipole** forces will be present.

**(c)** Phosphorus trifluoride ($PF_3$) molecules are polar. **Dispersion** and **dipole-dipole** forces will be present.

**(d)** Sodium chloride (NaCl) is an ionic compound. **Ion-ion** and **dispersion** forces will be present.

**(e)** Carbon disulfide ($CS_2$) molecules are nonpolar. Only **dispersion** forces will be present.

**11.12** In this problem you must identify the species capable of hydrogen bonding among themselves, not with water. In order for a molecule to be capable of hydrogen bonding with another molecule like itself, it must have at least one hydrogen atom bonded to N, O, or F. Of the choices, only **(e)** $CH_3COOH$ (acetic acid) shows this structural feature. The others cannot form hydrogen bonds among themselves.

**11.14** **Strategy:** The molecule with the stronger intermolecular forces will have the higher boiling point. If a molecule contains an N−H, O−H, or F−H bond it can form intermolecular hydrogen bonds. A hydrogen bond is a particularly strong dipole-dipole intermolecular attraction.

**Solution:** 1-butanol has the higher boiling point because the molecules can form hydrogen bonds with each other (it contains an O−H bond). Diethyl ether molecules do contain both oxygen atoms and hydrogen atoms. However, all the hydrogen atoms are bonded to carbon, not oxygen. There is no hydrogen bonding in diethyl ether, because carbon is not electronegative enough.

**11.16** (a) Xe: it has more electrons and therefore stronger dispersion forces.

(b) $CS_2$: it has more electrons (both molecules nonpolar) and therefore stronger dispersion forces.

(c) $Cl_2$: it has more electrons (both molecules nonpolar) and therefore stronger dispersion forces.

(d) LiF: it is an ionic compound, and the ion-ion attractions are much stronger than the dispersion forces between $F_2$ molecules.

(e) $NH_3$: it can form hydrogen bonds and $PH_3$ cannot.

**11.18** **Strategy:** Classify the species into three categories: ionic, polar (possessing a dipole moment), and nonpolar. Also look for molecules that contain an N−H, O−H, or F−H bond, which are capable of forming intermolecular hydrogen bonds. Keep in mind that dispersion forces exist between *all* species.

**Solution:**

(a) Water has O−H bonds. Therefore, water molecules can form hydrogen bonds. The attractive forces that must be overcome are hydrogen bonding and dispersion forces.

(b) Bromine ($Br_2$) molecules are nonpolar. Only dispersion forces must be overcome.

(c) Iodine ($I_2$) molecules are nonpolar. Only dispersion forces must be overcome.

(d) In this case, the F−F bond must be broken. This is an *intra*molecular force between two F atoms, not an *inter*molecular force between $F_2$ molecules. The attractive forces of the covalent bond must be overcome.

**11.20** The lower melting compound (shown below) can form hydrogen bonds only with itself (*intra*molecular hydrogen bonds), as shown in the figure. Such bonds do not contribute to *inter*molecular attraction and do not help raise the melting point of the compound. The other compound can form *inter*molecular hydrogen bonds; therefore, it will take a higher temperature to provide molecules of the liquid with enough kinetic energy to overcome these attractive forces to escape into the gas phase.

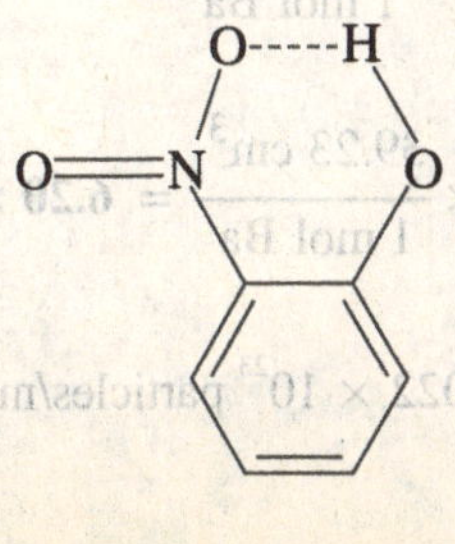

**11.32** Ethylene glycol has two –OH groups, allowing it to exert strong intermolecular forces through hydrogen bonding. Its viscosity should fall between ethanol (1 OH group) and glycerol (3 OH groups).

**11.38** A corner sphere is shared equally among eight unit cells, so only one-eighth of each corner sphere "belongs" to any one unit cell. A face-centered sphere is divided equally between the two unit cells sharing the face. A body-centered sphere belongs entirely to its own unit cell.

In a *simple cubic cell* there are eight corner spheres. One-eighth of each belongs to the individual cell giving a total of **one** whole sphere per cell. In a *body-centered cubic cell,* there are eight corner spheres and one body-center sphere giving a total of **two** spheres per unit cell (one from the corners and one from the body-center). In a *face-center* sphere, there are eight corner spheres and six face-centered spheres (six faces). The total number of spheres would be **four**: one from the corners and three from the faces.

**11.40** **Strategy:** The problem gives a generous hint. First, we need to calculate the volume (in cm$^3$) occupied by 1 mole of Ba atoms. Next, we calculate the volume that a Ba atom occupies. Once we have these two pieces of information, we can multiply them together to end up with the number of Ba atoms per mole of Ba.

$$\frac{\text{number of Ba atoms}}{\cancel{\text{cm}^3}} \times \frac{\cancel{\text{cm}^3}}{1 \text{ mol Ba}} = \frac{\text{number of Ba atoms}}{1 \text{ mol Ba}}$$

**Solution:** The volume that contains one mole of barium atoms can be calculated from the density using the following strategy:

$$\frac{\text{volume}}{\text{mass of Ba}} \rightarrow \frac{\text{volume}}{\text{mol Ba}}$$

$$\frac{1 \text{ cm}^3}{3.50 \cancel{\text{g}} \text{ Ba}} \times \frac{137.3 \cancel{\text{g}} \text{ Ba}}{1 \text{ mol Ba}} = \frac{39.23 \text{ cm}^3}{1 \text{ mol Ba}}$$

We carry an extra significant figure in this calculation to limit rounding errors. Next, the volume that contains two barium atoms is the volume of the body-centered cubic unit cell. Some of this volume is empty space because packing is only 68.0 percent efficient. But, this will not affect our calculation.

$$V = a^3$$

Let's also convert to cm$^3$.

$$V = (502 \cancel{\text{pm}})^3 \times \left(\frac{1 \times 10^{-12} \cancel{\text{m}}}{1 \cancel{\text{pm}}}\right)^3 \times \left(\frac{1 \text{ cm}}{0.01 \cancel{\text{m}}}\right)^3 = \frac{1.265 \times 10^{-22} \text{ cm}^3}{2 \text{ Ba atoms}}$$

We can now calculate the number of barium atoms in one mole using the strategy presented above.

$$\frac{\text{number of Ba atoms}}{\cancel{\text{cm}^3}} \times \frac{\cancel{\text{cm}^3}}{1 \text{ mol Ba}} = \frac{\text{number of Ba atoms}}{1 \text{ mol Ba}}$$

$$\frac{2 \text{ Ba atoms}}{1.265 \times 10^{-22} \cancel{\text{cm}^3}} \times \frac{39.23 \cancel{\text{cm}^3}}{1 \text{ mol Ba}} = \mathbf{6.20 \times 10^{23} \text{ atoms/mol}}$$

This is close to Avogadro's number, $6.022 \times 10^{23}$ particles/mol.

**11.42** The mass of the unit cell is the mass in grams of two europium atoms.

$$m = \frac{2 \text{ Eu atoms}}{1 \text{ unit cell}} \times \frac{1 \text{ mol Eu}}{6.022 \times 10^{23} \text{ Eu atoms}} \times \frac{152.0 \text{ g Eu}}{1 \text{ mol Eu}} = 5.048 \times 10^{-22} \text{ g Eu/unit cell}$$

$$V = \frac{5.048 \times 10^{-22} \text{ g}}{1 \text{ unit cell}} \times \frac{1 \text{ cm}^3}{5.26 \text{ g}} = 9.60 \times 10^{-23} \text{ cm}^3\text{/unit cell}$$

The edge length ($a$) is:

$$a = V^{1/3} = (9.60 \times 10^{-23} \text{ cm}^3)^{1/3} = 4.58 \times 10^{-8} \text{ cm} = \mathbf{458 \text{ pm}}$$

**11.44** **Strategy:** Recall that a corner atom is shared with 8 unit cells and therefore only 1/8 of a corner atom is within a given unit cell. Also recall that a face atom is shared with 2 unit cells and therefore 1/2 of a face atom is within a given unit cell. See Figure 11.19 of the text.

**Solution:** In a face-centered cubic unit cell, there are atoms at each of the eight corners, and there is one atom in each of the six faces. Only one-half of each face-centered atom and one-eighth of each corner atom belongs to the unit cell.

$$\text{X atoms/unit cell} = \text{(8 corner atoms)(1/8 atom per corner)} = 1 \text{ X atom/unit cell}$$

$$\text{Y atoms/unit cell} = \text{(6 face-centered atoms)(1/2 atom per face)} = 3 \text{ Y atoms/unit cell}$$

The unit cell is the smallest repeating unit in the crystal; therefore, the empirical formula is $\mathbf{XY_3}$.

**11.48** Rearranging the Bragg equation, we have:

$$\lambda = \frac{2d\sin\theta}{n} = \frac{2(282 \text{ pm})(\sin 23.0^\circ)}{1} = 220 \text{ pm} = \mathbf{0.220 \text{ nm}}$$

**11.52** See Table 11.4 of the text. The properties listed are those of a **molecular solid**.

**11.54** In a molecular crystal the lattice points are occupied by molecules. Of the solids listed, the ones that are composed of molecules are $Se_8$, HBr, $CO_2$, $P_4O_6$, and $SiH_4$. In covalent crystals, atoms are held together in an extensive three-dimensional network entirely by covalent bonds. Of the solids listed, the ones that are composed of atoms held together by covalent bonds are Si and C.

**11.56** In diamond, each carbon atom is covalently bonded to four other carbon atoms. Because these bonds are strong and uniform, diamond is a very hard substance. In graphite, the carbon atoms in each layer are linked by strong bonds, but the layers are bound by weak dispersion forces. As a result, graphite may be cleaved easily between layers and is not hard.

In graphite, all atoms are $sp^2$ hybridized; each atom is covalently bonded to three other atoms. The remaining unhybridized $2p$ orbital is used in pi bonding forming a delocalized molecular orbital. The electrons are free to move around in this extensively delocalized molecular orbital making graphite a good conductor of electricity in directions along the planes of carbon atoms.

**11.76** *Step 1:* Warming ice to the melting point.

$$q_1 = ms\Delta t = (866\text{ g H}_2\text{O})(2.03\text{ J/g}\cdot{}^\circ\text{C})[0-(-10)]^\circ\text{C} = 17.6\text{ kJ}$$

*Step 2:* Converting ice at the melting point to liquid water at 0°C. (See Table 11.8 of the text for the heat of fusion of water.)

$$q_2 = 866\text{ g H}_2\text{O} \times \frac{1\text{ mol}}{18.02\text{ g H}_2\text{O}} \times \frac{6.01\text{ kJ}}{1\text{ mol}} = 289\text{ kJ}$$

*Step 3:* Heating water from 0°C to 100°C.

$$q_3 = ms\Delta t = (866\text{ g H}_2\text{O})(4.184\text{ J/g}\cdot{}^\circ\text{C})[(100-0)^\circ\text{C}] = 362\text{ kJ}$$

*Step 4:* Converting water at 100°C to steam at 100°C. (See Table 11.6 of the text for the heat of vaporization of water.)

$$q_4 = 866\text{ g H}_2\text{O} \times \frac{1\text{ mol}}{18.02\text{ g H}_2\text{O}} \times \frac{40.79\text{ kJ}}{1\text{ mol}} = 1.96\times 10^3\text{ kJ}$$

*Step 5:* Heating steam from 100°C to 126°C.

$$q_5 = ms\Delta t = (866\text{ g H}_2\text{O})(1.99\text{ J/g}\cdot{}^\circ\text{C})[(126-100)^\circ\text{C}] = 44.8\text{ kJ}$$

$$q_{\text{total}} = q_1 + q_2 + q_3 + q_4 + q_5 = \mathbf{2.67\times 10^3\ kJ}$$

How would you set up and work this problem if you were computing the heat lost in cooling steam from 126°C to ice at −10°C?

**11.78** $\Delta H_{\text{vap}} = \Delta H_{\text{sub}} - \Delta H_{\text{fus}} = 62.30\text{ kJ/mol} - 15.27\text{ kJ/mol} = \mathbf{47.03\ kJ/mol}$

**11.80** Two phase changes occur in this process. First, the liquid is turned to solid (freezing), then the solid ice is turned to gas (sublimation).

**11.82** When steam condenses to liquid water at 100°C, it releases a large amount of heat equal to the enthalpy of vaporization. Thus steam at 100°C exposes one to more heat than an equal amount of water at 100°C.

**11.84** We can use a modified form of the Clausius-Clapeyron equation to solve this problem. See Equation (11.5) in the text.

$P_1 = 40.1$ mmHg $\quad\quad P_2 = ?$

$T_1 = 7.6°\text{C} = 280.6\text{ K} \quad\quad T_2 = 60.6°\text{C} = 333.6\text{ K}$

$$\ln\frac{P_1}{P_2} = \frac{\Delta H_{vap}}{R}\left(\frac{1}{T_2} - \frac{1}{T_1}\right)$$

$$\ln\frac{40.1}{P_2} = \frac{31000\text{ J/mol}}{8.314\text{ J/K}\cdot\text{mol}}\left(\frac{1}{333.6\text{ K}} - \frac{1}{280.6\text{ K}}\right)$$

$$\ln\frac{40.1}{P_2} = -2.11$$

Taking the antilog of both sides, we have:

$$\frac{40.1}{P_2} = 0.121$$

$$\boldsymbol{P_2 = 331\text{ mmHg}}$$

**11.86** The liquid nitrogen (boiling point = 77 K = −196°C) will vaporize very quickly, absorbing energy from your skin and the surrounding air molecules. The small amount of liquid nitrogen will only be in contact with your skin for a short time. A few drops of boiling water (boiling point = 100°C) on your skin will release energy as it cools. Water has a high specific heat and therefore will releases a large amount of energy per amount of substance as it cools.

**11.90** Initially, the ice melts because of the increase in pressure. As the wire sinks into the ice, the water above the wire refreezes. Eventually the wire actually moves completely through the ice block without cutting it in half.

**11.92** Region labels: The region containing point A is the solid region. The region containing point B is the liquid region. The region containing point C is the gas region.

**(a)** Raising the temperature at constant pressure beginning at A implies starting with solid ice and warming until melting occurs. If the warming continued, the liquid water would eventually boil and change to steam. Further warming would increase the temperature of the steam.

**(b)** At point C water is in the gas phase. Cooling without changing the pressure would eventually result in the formation of solid ice. Liquid water would never form.

**(c)** At B the water is in the liquid phase. Lowering the pressure without changing the temperature would eventually result in boiling and conversion to water in the gas phase.

**11.94** **(a)** A low surface tension means the attraction between molecules making up the surface is weak. Water has a high surface tension; water bugs could not "walk" on the surface of a liquid with a low surface tension.

**(b)** A low critical temperature means a gas is very difficult to liquefy by cooling. This is the result of weak intermolecular attractions. Helium has the lowest known critical temperature (5.3 K).

**(c)** A low boiling point means weak intermolecular attractions. It takes little energy to separate the particles. All ionic compounds have extremely high boiling points.

**(d)** A low vapor pressure means it is difficult to remove molecules from the liquid phase because of high intermolecular attractions. Substances with low vapor pressures have high boiling points (why?).

Thus, only choice **(d)** indicates strong intermolecular forces in a liquid. The other choices indicate weak intermolecular forces in a liquid.

**11.96** The properties of hardness, high melting point, poor conductivity, and so on, could place boron in either the ionic or covalent categories. However, boron atoms will not alternately form positive and negative ions to achieve an ionic crystal. The structure is **covalent** because the units are single boron atoms.

**11.98** The unit cell of iodine is **orthorhombic**. See Figure 11.15 of the text. Note that all angles of the unit cell are 90° and that the edges all have different lengths ($a \neq b \neq c$).

**11.100** The vapor pressure of mercury (as well as all other substances) is 760 mmHg at its normal boiling point.

**11.102** It has reached the critical point; the point of critical temperature ($T_c$) and critical pressure ($P_c$).

**11.104** Crystalline $SiO_2$. Its regular structure results in a more efficient packing.

**11.106** Assuming that liquid remains in the container after equilibrium has been established, the vapor pressure of a liquid in a closed container depends on (c) temperature and (d) the intermolecular forces between the molecules in the liquid.

**11.108** (a) **False.** Permanent dipoles are usually much stronger than temporary dipoles.

(b) **False.** The hydrogen atom must be bonded to N, O, or F.

(c) **True.**

(d) **False.** The magnitude of the attraction depends on both the ion charge and the polarizability of the neutral atom or molecule.

**11.110** Sublimation temperature is −78°C or 195 K at a pressure of 1 atm.

$$\ln\frac{P_1}{P_2} = \frac{\Delta H_{sub}}{R}\left(\frac{1}{T_2} - \frac{1}{T_1}\right)$$

$$\ln\frac{1}{P_2} = \frac{25.9 \times 10^3\ \text{J/mol}}{8.314\ \text{J/mol}\cdot\text{K}}\left(\frac{1}{150\ \text{K}} - \frac{1}{195\ \text{K}}\right)$$

$$\ln\frac{1}{P_2} = 4.79$$

Taking the antiln of both sides gives:

$$P_2 = 8.3 \times 10^{-3}\ \text{atm}$$

**11.112** (a) **$K_2S$**: Ionic forces are much stronger than the dipole-dipole forces in $(CH_3)_3N$.

(b) **$Br_2$**: Both molecules are nonpolar; but $Br_2$ has more electrons. (The boiling point of $Br_2$ is 50°C and that of $C_4H_{10}$ is −0.5°C.)

**11.114** $CH_4$ is a tetrahedral, nonpolar molecule that can only exert weak dispersion type attractive forces. $SO_2$ is bent (why?) and possesses a dipole moment, which gives rise to stronger dipole-dipole attractions. **Sulfur dioxide** will have a larger value of "*a*" in the van der Waals equation (*a* is a measure of the strength of the interparticle attraction) and will behave less like an ideal gas than methane.

**11.116** The standard enthalpy change for the formation of gaseous iodine from solid iodine is simply the difference between the standard enthalpies of formation of the products and the reactants in the equation:

$$I_2(s) \rightarrow I_2(g)$$

$$\Delta H_{vap} = \Delta H_f^\circ[I_2(g)] - \Delta H_f^\circ[I_2(s)] = 62.4\text{ kJ/mol} - 0\text{ kJ/mol} = \mathbf{62.4\ kJ/mol}$$

**11.118** The amount of heat needed to raise the temperature of 50.0 g of water from 20°C to its boiling point (100°C) is:

$$q_1 = ms\Delta t = (50.0\ \text{g H}_2\text{O})(4.184\ \text{J/g}\cdot{}^\circ\text{C})[(100-20)^\circ\text{C}] = 16.7\text{ kJ}$$

Next, let's calculate the amount of heat needed to convert water at 100°C to steam at 100°C. (See Table 11.6 of the text for the heat of vaporization of water.)

$$q_2 = 50.0\ \text{g H}_2\text{O} \times \frac{1\text{ mol}}{18.02\ \text{g H}_2\text{O}} \times \frac{40.79\text{ kJ}}{1\text{ mol}} = 113\text{ kJ}$$

The amount of heat left to heat the steam is:

$$150.0\text{ kJ} - 16.7\text{ kJ} - 113\text{ kJ} = 20.3\text{ kJ}$$

Let $t_f$ be the final temperature of steam. We write;

$$q = ms\Delta t$$

$$20.3 \times 10^3\text{ J} = (50.0\text{ g})(1.99\text{ J/g}\cdot{}^\circ\text{C})(t_f - 100)^\circ\text{C}$$

$$2.03 \times 10^4 = 99.5t_f - 9.95 \times 10^3$$

$$t_f = \mathbf{304^\circ C}$$

**11.120** Smaller ions have more concentrated charges (charge densities) and are more effective in ion-dipole interaction. The greater the ion-dipole interaction, the larger is the heat of hydration.

**11.122** **(a)** For the process: $Br_2(l) \rightarrow Br_2(g)$

$$\Delta H^\circ = \Delta H^\circ_f[Br_2(g)] - \Delta H^\circ_f[Br_2(l)] = (1)(30.7\text{ kJ/mol}) - 0 = \mathbf{30.7\ kJ/mol}$$

**(b)** For the process: $Br_2(g) \rightarrow 2Br(g)$

$\Delta H^\circ$ = **192.5 kJ/mol** (from Table 9.4 of the text)

As expected, the bond enthalpy represented in part (b) is much greater than the energy of vaporization represented in part (a). It requires more energy to break the bond than to vaporize the molecule.

**11.124** **(a)** Decreases **(b)** No change **(c)** No change

**11.126** Recall that a corner atom/ion is shared with 8 unit cells and therefore only 1/8 of a corner atom is within a given unit cell. Also recall that a face atom/ion is shared with 2 unit cells and therefore 1/2 of a face atom is within a given unit cell. See Figure 11.19 of the text. In Figure 11.26(a), there is **one $Cs^+$ ion** in the center of the unit cell and 8 $Cl^-$ ions at the corners. Since 1/8 of a corner ion is within a given unit cell, there is **one $Cl^-$ ion** within the unit cell. In Figure 11.26(b), there are **four $Zn^{2+}$ ions** within the unit cell, and there are 8 $S^{2-}$ corner ions and 6 $S^{2-}$ face ions. Since 1/8 of a corner ion and 1/2 of a face ion is within a given unit cell, there are **four $S^{2-}$ ions** within the unit cell. In Figure 11.26(c), there are **eight $F^-$ ions** within the unit cell, and there are 8 $Ca^{2+}$ corner ions and 6 $Ca^{2+}$ face ions. Since 1/8 of a corner ion and 1/2 of a face ion is within a given unit cell, there are **four $Ca^{2+}$ ions** within the unit cell.

**11.128** $CaCO_3(s) \rightarrow CaO(s) + CO_2(g)$

Three phases (two solid and one gas). $CaCO_3$ and CaO constitute two separate solid phases because they are separated by well-defined boundaries.

**11.130** $SiO_2$ has an extensive three-dimensional structure. $CO_2$ exists as discrete molecules. It will take much more energy to break the strong network covalent bonds of $SiO_2$; therefore, $SiO_2$ has a much higher boiling point than $CO_2$.

**11.132** The moles of water vapor can be calculated using the ideal gas equation.

$$n = \frac{PV}{RT} = \frac{\left(187.5\text{ mmHg} \times \dfrac{1\text{ atm}}{760\text{ mmHg}}\right)(5.00\text{ L})}{\left(0.0821\dfrac{\text{L}\cdot\text{atm}}{\text{mol}\cdot\text{K}}\right)(338\text{ K})} = 0.0445\text{ mol}$$

$$\text{mass of water vapor} = 0.0445\text{ mol} \times 18.02\text{ g/mol} = 0.802\text{ g}$$

Now, we can calculate the percentage of the 1.20 g sample of water that is vapor.

$$\textbf{\% of H}_2\textbf{O vaporized} = \frac{0.802\text{ g}}{1.20\text{ g}} \times 100\% = \mathbf{66.8\%}$$

**11.134** The packing efficiency is: $\dfrac{\text{volume of atoms in unit cell}}{\text{volume of unit cell}} \times 100\%$

An atom is assumed to be spherical, so the volume of an atom is $(4/3)\pi r^3$. The volume of a cubic unit cell is $a^3$ ($a$ is the length of the cube edge). The packing efficiencies are calculated below:

**(a)** Simple cubic cell: cell edge ($a$) = $2r$

$$\text{Packing efficiency} = \frac{\left(\dfrac{4\pi r^3}{3}\right) \times 100\%}{(2r)^3} = \frac{4\pi r^3 \times 100\%}{24r^3} = \frac{\pi}{6} \times 100\% = \mathbf{52.4\%}$$

**(b)** Body-centered cubic cell: cell edge $= \dfrac{4r}{\sqrt{3}}$

$$\text{Packing efficiency} = \frac{2 \times \left(\dfrac{4\pi r^3}{3}\right) \times 100\%}{\left(\dfrac{4r}{\sqrt{3}}\right)^3} = \frac{2 \times \left(\dfrac{4\pi r^3}{3}\right) \times 100\%}{\left(\dfrac{64r^3}{3\sqrt{3}}\right)} = \frac{2\pi\sqrt{3}}{16} \times 100\% = \mathbf{68.0\%}$$

Remember, there are two atoms per body-centered cubic unit cell.

**(c)** Face-centered cubic cell: cell edge $= \sqrt{8}r$

$$\text{Packing efficiency} = \frac{4 \times \left(\dfrac{4\pi r^3}{3}\right) \times 100\%}{\left(\sqrt{8}r\right)^3} = \frac{\left(\dfrac{16\pi r^3}{3}\right) \times 100\%}{8r^3\sqrt{8}} = \frac{2\pi}{3\sqrt{8}} \times 100\% = \mathbf{74.0\%}$$

Remember, there are four atoms per face-centered cubic unit cell.

**11.136** For a face-centered cubic unit cell, the length of an edge ($a$) is given by:

$$a = \sqrt{8}r$$

$$a = \sqrt{8}\,(191\text{ pm}) = 5.40 \times 10^2\text{ pm}$$

The volume of a cube equals the edge length cubed ($a^3$).

$$V = a^3 = (5.40 \times 10^2\text{ pm})^3 \times \left(\frac{1 \times 10^{-12}\text{ m}}{1\text{ pm}}\right)^3 \times \left(\frac{1\text{ cm}}{1 \times 10^{-2}\text{ m}}\right)^3 = 1.57 \times 10^{-22}\text{ cm}^3$$

Now that we have the volume of the unit cell, we need to calculate the mass of the unit cell in order to calculate the density of Ar. The number of atoms in one face-centered cubic unit cell is four.

$$m = \frac{4\text{ atoms}}{1\text{ unit cell}} \times \frac{1\text{ mol}}{6.022 \times 10^{23}\text{ atoms}} \times \frac{39.95\text{ g}}{1\text{ mol}} = \frac{2.65 \times 10^{-22}\text{ g}}{1\text{ unit cell}}$$

$$d = \frac{m}{V} = \frac{2.65 \times 10^{-22}\text{ g}}{1.57 \times 10^{-22}\text{ cm}^3} = \mathbf{1.69\ g/cm^3}$$

**11.138** **(a)** Two triple points: Diamond/graphite/liquid and graphite/liquid/vapor.

**(b)** Diamond.

**(c)** Apply high pressure at high temperature.

**11.140** The cane is made of many molecules held together by intermolecular forces. The forces are strong and the molecules are packed tightly. Thus, when the handle is raised, all the molecules are raised because they are held together.

**11.142** When the tungsten filament inside the bulb is heated to a high temperature (about 3000°C), the tungsten sublimes (solid → gas phase transition) and then it condenses on the inside walls of the bulb. The inert, pressurized Ar gas retards sublimation and oxidation of the tungsten filament.

**11.144** The fuel source for the Bunsen burner is most likely methane gas. When methane burns in air, carbon dioxide and water are produced.

$$CH_4(g) + 2O_2(g) \rightarrow CO_2(g) + 2H_2O(g)$$

The water vapor produced during the combustion condenses to liquid water when it comes in contact with the outside of the cold beaker.

**11.146** First, we need to calculate the volume (in $cm^3$) occupied by 1 mole of Fe atoms. Next, we calculate the volume that a Fe atom occupies. Once we have these two pieces of information, we can multiply them together to end up with the number of Fe atoms per mole of Fe.

$$\frac{\text{number of Fe atoms}}{\cancel{cm^3}} \times \frac{\cancel{cm^3}}{1 \text{ mol Fe}} = \frac{\text{number of Fe atoms}}{1 \text{ mol Fe}}$$

The volume that contains one mole of iron atoms can be calculated from the density using the following strategy:

$$\frac{\text{volume}}{\text{mass of Fe}} \rightarrow \frac{\text{volume}}{\text{mol Fe}}$$

$$\frac{1 \text{ cm}^3}{7.874 \cancel{g} \text{ Fe}} \times \frac{55.85 \cancel{g} \text{ Fe}}{1 \text{ mol Fe}} = \frac{7.093 \text{ cm}^3}{1 \text{ mol Fe}}$$

Next, the volume that contains two iron atoms is the volume of the body-centered cubic unit cell. Some of this volume is empty space because packing is only 68.0 percent efficient. But, this will not affect our calculation.

$$V = a^3$$

Let's also convert to $cm^3$.

$$V = (286.7 \cancel{pm})^3 \times \left(\frac{1 \times 10^{-12} \cancel{m}}{1 \cancel{pm}}\right)^3 \times \left(\frac{1 \text{ cm}}{0.01 \cancel{m}}\right)^3 = \frac{2.357 \times 10^{-23} \text{ cm}^3}{2 \text{ Fe atoms}}$$

We can now calculate the number of iron atoms in one mole using the strategy presented above.

$$\frac{\text{number of Fe atoms}}{\cancel{cm^3}} \times \frac{\cancel{cm^3}}{1 \text{ mol Fe}} = \frac{\text{number of Fe atoms}}{1 \text{ mol Fe}}$$

$$\frac{2\text{ Fe atoms}}{2.357\times10^{-23}\text{ cm}^3}\times\frac{7.093\text{ cm}^3}{1\text{ mol Ba}} = \mathbf{6.019\times10^{23}\text{ Fe atoms/mol}}$$

The small difference between the above number and $6.022 \times 10^{23}$ is the result of rounding off and using rounded values for density and other constants.

**11.148** Figure 11.29 of the text shows that all alkali metals have a body-centered cubic structure. Figure 11.22 of the text gives the equation for the radius of a body-centered cubic unit cell.

$$r = \frac{\sqrt{3}a}{4}\text{, where } a \text{ is the edge length.}$$

Because $V = a^3$, if we can determine the volume of the unit cell ($V$), then we can calculate $a$ and $r$.

Using the ideal gas equation, we can determine the moles of metal in the sample.

$$n = \frac{PV}{RT} = \frac{\left(19.2\text{ mmHg}\times\frac{1\text{ atm}}{760\text{ mmHg}}\right)(0.843\text{ L})}{\left(0.0821\frac{\text{L}\cdot\text{atm}}{\text{mol}\cdot\text{K}}\right)(1235\text{ K})} = 2.10\times10^{-4}\text{ mol}$$

Next, we calculate the volume of the cube, and then convert to the volume of one unit cell.

$$\text{Vol. of cube} = (0.171\text{ cm})^3 = 5.00\times10^{-3}\text{ cm}^3$$

This is the volume of $2.10 \times 10^{-4}$ mole. We convert from volume/mole to volume/unit cell.

$$\text{Vol. of unit cell} = \frac{5.00\times10^{-3}\text{ cm}^3}{2.10\times10^{-4}\text{ mol}}\times\frac{1\text{ mol}}{6.022\times10^{23}\text{ atoms}}\times\frac{2\text{ atoms}}{1\text{ unit cell}} = 7.91\times10^{-23}\text{ cm}^3\text{/unit cell}$$

Recall that there are 2 atoms in a body-centered cubic unit cell.

Next, we can calculate the edge length ($a$) from the volume of the unit cell.

$$a = \sqrt[3]{V} = \sqrt[3]{7.91\times10^{-23}\text{ cm}^3} = 4.29\times10^{-8}\text{ cm}$$

Finally, we can calculate the radius of the alkali metal.

$$r = \frac{\sqrt{3}a}{4} = \frac{\sqrt{3}(4.29\times10^{-8}\text{ cm})}{4} = 1.86\times10^{-8}\text{ cm} = \mathbf{186\text{ pm}}$$

Checking Figure 8.5 of the text, we conclude that the metal is **sodium, Na.**

To calculate the density of the metal, we need the mass and volume of the unit cell. The volume of the unit cell has been calculated ($7.91 \times 10^{-23}$ cm$^3$/unit cell). The mass of the unit cell is

$$2\text{ Na atoms}\times\frac{22.99\text{ amu}}{1\text{ Na atom}}\times\frac{1\text{ g}}{6.022\times10^{23}\text{ amu}} = 7.635\times10^{-23}\text{ g}$$

$$d = \frac{m}{V} = \frac{7.635\times10^{-23}\text{ g}}{7.91\times10^{-23}\text{ cm}^3} = \mathbf{0.965\text{ g/cm}^3}$$

The density value also matches that of sodium.

**11.150** The original diagram shows that as heat is supplied to the water, its temperature rises. At the boiling point (represented by the horizontal line), water is converted to steam. Beyond this point the temperature of the steam rises above 100°C.

Choice (a) is eliminated because it shows no change from the original diagram even though the mass of water is doubled.

Choice (b) is eliminated because the rate of heating is greater than that for the original system. Also, it shows water boiling at a higher temperature, which is not possible.

Choice (c) is eliminated because it shows that water now boils at a temperature below 100°C, which is not possible.

Choice **(d)** therefore represents what actually happens. The heat supplied is enough to bring the water to its boiling point, but not raise the temperature of the steam.

**11.152** At the normal boiling point, the pressure of HF is 1 atm. We use Equation (5.11) of the text to calculate the density of HF.

$$d = \frac{PM}{RT} = \frac{(1\ \cancel{\text{atm}})(20.01\frac{\text{g}}{\cancel{\text{mol}}})}{\left(0.0821\frac{\text{L}\cdot\cancel{\text{atm}}}{\cancel{\text{mol}}\cdot\cancel{\text{K}}}\right)(273+19.5)\cancel{\text{K}}} = \mathbf{0.833\ g/L}$$

The fact that the measured density is larger suggests that HF molecules must be associated to some extent in the gas phase. This is not surprising considering that HF molecules form strong intermolecular hydrogen bonds.

# CHAPTER 12
# PHYSICAL PROPERTIES OF SOLUTIONS

## PROBLEM-SOLVING STRATEGIES AND TUTORIAL SOLUTIONS

### TYPES OF PROBLEMS

**Problem Type 1:** Predicting Solubility Based on Intermolecular Forces.

**Problem Type 2:** Types of Concentration Units.
- **(a)** Percent by mass.
- **(b)** Molarity.
- **(c)** Molality.

**Problem Type 3:** Converting between Concentration Units.
- **(a)** Converting molality to molarity.
- **(b)** Converting molarity to molality.
- **(c)** Converting percent by mass to molality.

**Problem Type 4:** Effect of Pressure on Solubility: Henry's Law.

**Problem Type 5:** Colligative Properties of Nonelectrolytes.
- **(a)** Vapor-pressure lowering, Raoult's law.
- **(b)** Boiling-point elevation.
- **(c)** Freezing-point depression.
- **(d)** Osmotic pressure.

**Problem Type 6:** Determining Molar Mass Using Colligative Properties.
- **(a)** Calculating molar mass from freezing-point depression.
- **(b)** Calculating molar mass from osmotic pressure.

**Problem Type 7:** Colligative Properties of Electrolytes.

## PROBLEM TYPE 1: PREDICTING SOLUBILITY BASED ON INTERMOLECULAR FORCES

Two substances with intermolecular forces of similar type and magnitude are likely to be soluble in each other. The saying "*like dissolves like*" will help you predict the solubility of a substance in a solvent.

**EXAMPLE 12.1**
**Which of the following would be a better solvent for molecular $I_2(s)$: $CCl_4$ or $H_2O$?**

**Strategy:** In predicting solubility, remember the saying: Like dissolves like. A nonpolar solute will dissolve in a nonpolar solvent; ionic compounds will generally dissolve in polar solvents due to favorable ion-dipole interactions; solutes that can form hydrogen bonds with a solvent will have high solubility in the solvent.

**Solution:** $I_2$ is a nonpolar molecule (see Chapter 10). Using the like-dissolves-like rule, $I_2$ will be more soluble in the nonpolar solvent, $CCl_4$, than in the polar solvent, $H_2O$.

**PRACTICE EXERCISE**

1. In which solvent will NaBr be more soluble, benzene ($C_6H_6$) or water?

| **Text Problems: 12.10**, 12.12 |
|---|

## PROBLEM TYPE 2: TYPES OF CONCENTRATION UNITS

We will focus on three of the most common units of concentration: percent by mass, molarity, and molality.

### A. Percent by mass

The percent by mass is defined as:

$$\textbf{percent by mass of solute} = \frac{\text{mass of solute}}{\text{mass of solute} + \text{mass of solvent}} \times 100\% \quad \text{(12.1, text)}$$

$$= \frac{\text{mass of solute}}{\text{mass of soln}} \times 100\%$$

The percent by mass has no units because it is a ratio of two similar quantities.

**EXAMPLE 12.2**

**The dehydrated form of Epsom salt is magnesium sulfate. What is the percent $MgSO_4$ by mass in a solution made from 16.0 g $MgSO_4$ and 100 mL of $H_2O$ at 25°C? The density of water at 25°C is 0.997 g/mL.**

**Strategy:** We are given the mass of the solute, and the volume and density of the solvent. The mass of solvent can be calculated from its density and volume, and then we can use Equation (12.1) of the text to solve for the mass percent of $MgSO_4$ in the solution.

**Solution:**
Calculate the mass of 100 mL of water using the density of water as a conversion factor.

$$?\text{ g of water} = 100\ \cancel{\text{mL}}\ H_2O \times \frac{0.997\text{ g }H_2O}{1\ \cancel{\text{mL}}\ H_2O} = 99.7\text{ g }H_2O$$

Substitute the mass of solute and the mass of solvent into Equation (12.1) to calculate the percent by mass of $MgSO_4$.

$$\textbf{percent by mass MgSO}_\mathbf{4} = \frac{\text{mass of solute}}{\text{mass of solute} + \text{mass of solvent}} \times 100\%$$

$$= \frac{16.0\text{ g}}{16.0\text{ g} + 99.7\text{ g}} \times 100\% = \mathbf{13.8\%}$$

**PRACTICE EXERCISE**

2. An aqueous solution contains 167 g $CuSO_4$ in 820 mL of solution. The density of the solution is 1.195 g/mL. Calculate the percent $CuSO_4$ by mass in the solution.

| **Text Problem: 12.16** |
|---|

### B. Molarity (*M*)

We have already defined molarity in Chapter 4. Molarity is defined as

$$\text{molarity} = \frac{\text{moles of solute}}{\text{liters of soln}}$$

Molarity has units of mol/L.

**EXAMPLE 12.3**

**What is the molarity of the $MgSO_4$ solution made in Example 12.2? Assume that the density remains unchanged upon addition of $MgSO_4$ to the water.**

**Strategy:** To calculate molarity, we need moles of solute and volume of solution. Moles of solute can be calculated from the mass of solute. To find the volume of the solution, we add together the mass of the solute and solvent and then use the density to convert to volume.

**Solution:** Calculate the number of moles $MgSO_4$ in 16.0 g. Use the molar mass of $MgSO_4$ as a conversion factor.

$$? \text{ mol MgSO}_4 = 16.0 \cancel{\text{g}} \text{ MgSO}_4 \times \frac{1 \text{ mol MgSO}_4}{120.38 \cancel{\text{g}} \text{ MgSO}_4} = 0.133 \text{ mol MgSO}_4$$

Calculate the volume of the solution. The total mass of the solution is:

$$16.0 \text{ g MgSO}_4 + \left(100 \cancel{\text{mL}} \text{ H}_2\text{O} \times \frac{0.997 \text{ g H}_2\text{O}}{1 \cancel{\text{mL}} \text{ H}_2\text{O}}\right) = 115.7 \text{ g}$$

Assuming that the density of the solution is the same as that of water, we can calculate the volume of the solution as follows:

$$\text{Volume of solution} = 115.7 \cancel{\text{g}} \times \frac{1 \text{ mL}}{0.997 \cancel{\text{g}}} = 116.0 \text{ mL soln}$$

We calculate the molarity of the solution by dividing the moles of solute by volume of solution (in L).

$$\textbf{molarity} = \frac{\text{moles of solute}}{\text{liters of soln}} = \frac{0.133 \text{ mol}}{0.116 \text{ L}} = \mathbf{1.15}\ \boldsymbol{M}$$

**PRACTICE EXERCISE**

3. An aqueous solution contains 167 g $CuSO_4$ in 820 mL of solution. The density of the solution is 1.195 g/mL. Calculate the molarity of the solution.

## C. Molality (*m*)

Molality is defined as the number of moles of solute per mass of solvent (in kg).

$$\text{molality} = \frac{\text{moles of solute}}{\text{mass of solvent (kg)}} \qquad (12.2, \text{text})$$

**EXAMPLE 12.4**

**What is the molality of the $MgSO_4$ solution made in Example 12.2?**

**Strategy:** To calculate molality, we need moles of solute and mass of solvent (water) in kg. From Example 12.3, we know the moles of solute (0.133 mole $MgSO_4$), and from Example 12.2, we know the mass of water (99.7 g). We will repeat these calculations below, and then solve for molality.

**Solution:** Calculate the number of moles $MgSO_4$ in 16.0 g. Use the molar mass of $MgSO_4$ as a conversion factor.

$$? \text{ mol MgSO}_4 = 16.0 \cancel{\text{g}} \text{ MgSO}_4 \times \frac{1 \text{ mol MgSO}_4}{120.38 \cancel{\text{g}} \text{ MgSO}_4} = 0.133 \text{ mol MgSO}_4$$

Calculate the mass (in kg) of 100 mL of $H_2O$. Use the density of water as a conversion factor.

$$? \text{ mass of } H_2O = 100 \text{ mL } H_2O \times \frac{0.997 \text{ g } H_2O}{1 \text{ mL } H_2O} \times \frac{1 \text{ kg}}{1000 \text{ g}} = 0.0997 \text{ kg } H_2O$$

Calculate the molality by substituting the mol of solute and the mass of solvent (in kg) into Equation (12.2) of the text.

$$\textbf{molality} = \frac{\text{moles of solute}}{\text{mass of solvent (kg)}} = \frac{0.133 \text{ mol}}{0.0997 \text{ kg}} = \mathbf{1.33}\ \boldsymbol{m}$$

**PRACTICE EXERCISE**

4. An aqueous solution contains 167 g $CuSO_4$ in 820 mL of solution. The density of the solution is 1.195 g/mL. Calculate the molality of the solution.

**Text Problems:** 12.18, 12.20

## PROBLEM TYPE 3: CONVERTING BETWEEN CONCENTRATION UNITS

There are advantages and disadvantages to each type of concentration unit. An advantage of molality and percent by mass is that the concentration is temperature independent. On the other hand, molarity changes with temperature, because solution volume typically increases with increasing temperature. However, the advantage of molarity is that it is generally easier to measure the volume of solution than to weigh the solvent or solution.

### A. Converting molality to molarity

**EXAMPLE 12.5**

**Calculate the molarity of a 2.44 *m* NaCl solution given that its density is 1.089 g/mL.**

**Strategy:** To calculate molarity, we need moles of solute and volume of solution. The moles of solute can be obtained directly from the molality. 2.44 *m* means 2.44 moles of solute per 1 kg (1000 g) of solvent. Next, we need to determine the mass of the solution, and then we can use the density to convert to volume of solution. To calculate the mass of the solution, you must calculate the mass of the solute and then add that to the mass of water (1 kg = 1000 g).

**Solution:** Use the molar mass of the solute as a conversion factor to convert from moles of solute to mass of solute.

$$? \text{ mass of solute} = 2.44 \text{ mol NaCl} \times \frac{58.44 \text{ g NaCl}}{1 \text{ mol NaCl}} = 143 \text{ g NaCl}$$

$$? \text{ mass of solution} = \text{mass of solute} + \text{mass of solvent}$$

$$= 143 \text{ g} + 1000 \text{ g} = 1143 \text{ g}$$

From the mass of the solution, we can calculate the volume of solution using the solution density as a conversion factor.

$$? \text{ volume of solution} = 1143 \text{ g} \times \frac{1 \text{ mL}}{1.089 \text{ g}} = 1.050 \times 10^3 \text{ mL} = 1.050 \text{ L}$$

Dividing moles of solute (2.44 moles) by liters of solution gives the molarity of the solution.

$$\textbf{molarity} = \frac{\text{mol of solute}}{\text{L of soln}} = \frac{2.44 \text{ mol}}{1.050 \text{ L}} = \mathbf{2.32}\ \boldsymbol{M}$$

**PRACTICE EXERCISE**

5. Concentrated hydrochloric acid is 15.7 *m*. Calculate the molarity of concentrated HCl given that its density is 1.18 g/mL.

**Text Problem: 12.22**

## B. Converting molarity to molality

**EXAMPLE 12.6**

**Calculate the molality of a 2.55 *M* NaCl solution given that its density is 1.089 g/mL.**

**Strategy:** To calculate molality, we need moles of solute and mass of solvent (in kg). The moles of solute can be obtained directly from the molarity. 2.55 *M* means 2.55 moles of solute per 1 L of solution. Next, we can determine the mass of the solution from the density and volume of solution. Subtracting off the mass of the solute from the mass of the solution will give the mass of solvent.

**Solution:** From the volume of solution, we can calculate the mass of the solution using the solution density as a conversion factor. Remember that 2.55 *M* means 2.55 moles of solute per 1 L (1000 mL) of solution.

$$? \text{ mass of solution} = 1000 \cancel{\text{mL}} \text{ soln} \times \frac{1.089 \text{ g}}{1 \cancel{\text{mL}} \text{ soln}} = 1089 \text{ g}$$

To calculate the mass of the *solvent* (water), you need to subtract the mass due to the *solute* from the mass of solution calculated. You can calculate the mass of the solute from the moles of solute using molar mass as a conversion factor.

$$? \text{ mass of solute} = 2.55 \cancel{\text{mol}} \text{ NaCl} \times \frac{58.44 \text{ g NaCl}}{1 \cancel{\text{mol}} \text{ NaCl}} = 149 \text{ g NaCl}$$

$$? \text{ mass of solvent} = \text{mass of soln} - \text{mass of solute}$$

$$= 1089 \text{ g} - 149 \text{ g} = 9.40 \times 10^2 \text{ g} = 0.940 \text{ kg solvent}$$

The moles of solute are given in the molarity (2.55 mol). Divide moles of solute by mass of solvent in kg to calculate the molality of the solution.

$$\textbf{molality} = \frac{\text{mol of solute}}{\text{kg of solvent}} = \frac{2.55 \text{ mol}}{0.940 \text{ kg}} = \mathbf{2.71}\ \boldsymbol{m}$$

**PRACTICE EXERCISE**

6. Concentrated sulfuric acid, $H_2SO_4$ is 18.0 *M*. Calculate the molality of concentrated $H_2SO_4$ given that its density is 1.83 g/mL.

## C. Converting percent by mass to molality

**EXAMPLE 12.7**

**Concentrated hydrochloric acid is 36.5 percent HCl by mass. Its density is 1.18 g/mL. Calculate the molality of concentrated HCl.**

**Strategy:** In solving this type of problem, it is convenient to assume that we start with 100.0 grams of the solution. If the mass of hydrochloric acid is 36.5% of 100.0 g, or 36.5 g, the percent by mass of water must be 100.0% – 36.5% = 63.5%. The mass of water in 100.0 g of solution would be 63.5 g. From the definition of molality, we need to find moles of solute (hydrochloric acid) and kilograms of solvent (water).

**Solution:** Since the definition of molality is

$$\text{molality} = \frac{\text{moles of solute}}{\text{mass of solvent (kg)}}$$

we first convert 36.5 g HCl to moles of HCl using its molar mass, then we convert 63.5 g of $H_2O$ to units of kilograms.

$$36.5\ \cancel{g}\ \text{HCl} \times \frac{1\ \text{mol HCl}}{36.46\ \cancel{g}\ \text{HCl}} = 1.00\ \text{mol HCl}$$

$$63.5\ \cancel{g}\ H_2O \times \frac{1\ \text{kg}}{1000\ \cancel{g}} = 0.0635\ \text{kg}\ H_2O$$

Lastly, we divide moles of solute by mass of solvent in kg to calculate the molality of the solution.

$$\textbf{molality} = \frac{\text{mol of solute}}{\text{kg of solvent}} = \frac{1.00\ \text{mol}}{0.0635\ \text{kg}} = \mathbf{15.7}\ \boldsymbol{m}$$

**PRACTICE EXERCISE**

7. What is the molality of a 3.0 percent hydrogen peroxide ($H_2O_2$) aqueous solution? The density of the solution is 1.0 g/mL.

**Text Problem: 12.22**

## PROBLEM TYPE 4: EFFECT OF PRESSURE ON SOLUBILITY: HENRY'S LAW

Solubility is defined as the maximum amount of a solute that will dissolve in a given quantity of solvent at a specific temperature. For all practical purposes, external pressure has no influence on the solubilities of liquids and solids, but it does greatly affect the solubility of gases. There is a quantitative relationship between gas solubility and pressure called **Henry's law**, which states that the solubility of a gas in a liquid is proportional to the pressure of the gas over the solution:

$$c = kP$$

where,

$c$ is the molar concentration (mol/L) of the dissolved gas.
$P$ is the pressure, in atmospheres, of the gas over the solution.
$k$ is a constant for a given gas that depends only on the temperature. $k$ has units of mol/L·atm.

As the pressure of the gas over the solution increases, more gas molecules strike the surface of the liquid increasing the number of gas molecules that dissolve in the liquid. For further discussion, see Section 12.5 of the text.

**EXAMPLE 12.8**

**What is the concentration of $O_2$ at 25°C in water that is saturated with *air* at an atmospheric pressure of 645 mmHg? The Henry's law constant (*k*) for oxygen is $3.5 \times 10^{-4}$ mol/L·atm. Assume that the mole fraction of oxygen in air is 0.209.**

**Strategy:** We want to calculate the molar concentration ($c$) of $O_2$ in water. We can use Henry's law, $c_{O_2} = kP_{O_2}$. $k$ is given in the problem. To calculate the molar concentration, we must first calculate the partial pressure of oxygen in air. The partial pressure of $O_2$ can be found using Dalton's law of partial pressures (see Chapter 5).

**Solution:**

$$P_{O_2} = X_{O_2}P_T = (0.209)(645 \text{ mmHg}) \times \frac{1 \text{ atm}}{760 \text{ mmHg}} = 0.177 \text{ atm}$$

Substitute the partial pressure of $O_2$ and $k$ into the Henry's law expression to solve for the molar concentration of $O_2$ in water.

$$c_{O_2} = kP_{O_2} = (3.5 \times 10^{-4} \text{ mol/L}\cdot\text{atm})(0.177 \text{ atm}) = \mathbf{6.2 \times 10^{-5} \text{ mol/L}}$$

**PRACTICE EXERCISE**

8. The Henry's law constant for CO is $9.73 \times 10^{-4}$ mol/L·atm at 25°C. What is the concentration of dissolved CO in water if the partial pressure of CO in the air is 0.015 mmHg?

**Text Problems:** 12.36, **12.38**

## PROBLEM TYPE 5: COLLIGATIVE PROPERTIES OF NONELECTROLYTES

Several important properties of solutions depend on the number of solute particles in solution and not on the nature of the solute particles. These properties are called **colligative properties.**

### A. Vapor pressure lowering, Raoult's law

If a solute is nonvolatile, the vapor pressure of its solution is always less than that of the pure solvent. Raoult's law quantifies this relationship by stating that the partial pressure of a solvent over a solution, $P_1$, is given by the vapor pressure of the pure solvent, $P_1^\circ$, times the mole fraction of the solvent in the solution, $X_1$:

$$P_1 = X_1P_1^\circ \qquad (12.4, \text{ text})$$

If the solution contains only one solute, $X_1 = 1 - X_2$, where $X_2$ is the mole fraction of the solute. Substituting for $X_1$ in Equation (12.4) of the text gives:

$$P_1 = (1 - X_2)P_1^\circ$$

$$P_1 = P_1^\circ - X_2P_1^\circ$$

$$P_1^\circ - P_1 = \Delta P = X_2P_1^\circ \qquad (12.5, \text{ text})$$

Equation (12.5) of the text shows that a decrease in vapor pressure, $\Delta P$, is directly proportional to the concentration of the solute in solution, $X_2$.

**EXAMPLE 12.9**

**Calculate the vapor pressure of an aqueous solution at 30°C made from $1.00 \times 10^2$ g of sucrose ($C_{12}H_{22}O_{11}$) and $1.00 \times 10^2$ g of water. The vapor pressure of pure water at 30°C is 31.8 mmHg.**

**Strategy:** Equation (12.4) of the text gives a relationship between the vapor pressure of the solvent over a solution and the mole fraction of the solvent. If we can calculate the mole fraction of the solvent, we can calculate the vapor pressure of the solution.

**Solution:**

$$\text{mol water} = (1.00 \times 10^2 \text{ g water}) \times \frac{1 \text{ mol water}}{18.02 \text{ g water}} = 5.55 \text{ mol}$$

$$\text{mol sucrose} = (1.00 \times 10^2 \text{ g sucrose}) \times \frac{1 \text{ mol sucrose}}{342.3 \text{ g sucrose}} = 0.292 \text{ mol}$$

$$X_{\text{water}} = \frac{\text{mol}_{\text{water}}}{\text{mol}_{\text{water}} + \text{mol}_{\text{sucrose}}} = \frac{5.55 \text{ mol}}{5.55 \text{ mol} + 0.292 \text{ mol}} = 0.950$$

Substitute $X_{\text{water}}$ and the vapor pressure of pure water into Equation (12.4) to solve for the vapor pressure of the solution.

$$P_{\text{soln}} = X_{\text{water}} P^{\circ}_{\text{water}} = (0.950)(31.8 \text{ mmHg}) = \mathbf{30.2 \text{ mmHg}}$$

**Tip:** You could also have solved this problem using Equation (12.5). You could calculate the change in vapor pressure ($\Delta P$) from the mole fraction of solute. Then, you could calculate the vapor pressure of the solution by subtracting $\Delta P$ from the vapor pressure of the pure solvent (water).

**PRACTICE EXERCISE**

**9.** At 25°C, the vapor pressure of pure water is 23.76 mmHg and that of an aqueous sucrose ($C_{12}H_{22}O_{11}$) is 23.28 mmHg. Calculate the molality of the solution.

**Text Problems: 12.50**, 12.52, 12.54

## B. Boiling-point elevation

The **boiling point** of a solution is the temperature at which its vapor pressure equals the external atmospheric pressure (see Section 11.8 of the text). We just saw in Part (a) that a nonvolatile solute always decreases the vapor pressure of the solution relative to the pure solvent. Consequently, the boiling point of the solution is *higher* than the pure solvent, because more energy in the form of heat must be added to raise the vapor pressure of the solution to the external atmospheric pressure. The change in boiling point is proportional to the concentration of solute.

$$\Delta T_b = K_b m \quad \text{(12.6, text)}$$

where,

$\Delta T_b = T_b - T_b^{\circ}$ (where $T_b$ is the boiling point of the solution and $T_b^{\circ}$ is the boiling point of the pure solvent)

$m$ is the molal concentration of the solute

$K_b$ is the molal boiling-point elevation constant of the solvent with units of °C/$m$

Do you know why molality is used for the concentration instead of molarity? Because we are dealing with a system that is not kept at constant temperature, we cannot express the concentration in molarity because molarity changes with temperature.

**EXAMPLE 12.10**

**What is the boiling point of an "antifreeze/coolant" solution made from a 50-50 mixture (by volume) of ethylene glycol, $C_2H_6O_2$ and water? Assume the density of water is 1.00 g/mL and the density of ethylene glycol is 1.11 g/mL.**

**Strategy:** Using Equation (12.6) of the text, we can calculate the change in boiling point, $\Delta T_b$, by first calculating the molality of the solution and then multiplying by $K_b$ for water (see Table 12.2 of the text).ss

**Solution:** For simplicity, assume that you have 100.0 mL of solution. Since the mixture is 50-50 by volume, there are 50.0 mL of ethylene glycol and 50.0 mL of water. To calculate the molality of the solution, you need the moles of solute (ethylene glycol) and the mass of solvent (water) in kg.

$$\text{mol of ethylene glycol} = 50.0\ \text{mL } C_2H_6O_2 \times \frac{1.11\ \text{g } C_2H_6O_2}{1\ \text{mL } C_2H_6O_2} \times \frac{1\ \text{mol } C_2H_6O_2}{62.07\ \text{g } C_2H_6O_2} = 0.894\ \text{mol}$$

$$\text{mass of water} = 50.0\ \text{mL } H_2O \times \frac{1.00\ \text{g } H_2O}{1\ \text{mL } H_2O} \times \frac{1\ \text{kg}}{1000\ \text{g}} = 0.0500\ \text{kg}$$

$$\text{molality} = \frac{\text{mol solute}}{\text{kg solvent}} = \frac{0.894\ \text{mol}}{0.0500\ \text{kg}} = 17.9\ m$$

Substitute the molality of the solution and $K_b$ into Equation (12.6) to solve for the change in boiling point. Then, add the change in boiling point to the normal boiling point of water (100.0°C) to calculate the boiling point of the solution.

$$\Delta T_b = K_b m = (0.52°\text{C}/m)(17.9\ m) = 9.3°\text{C}$$

$$\textbf{b.p. of soln} = 100.0°\text{C} + 9.3°\text{C} = \mathbf{109.3°C}$$

**PRACTICE EXERCISE**

**10.** What is the boiling point of an aqueous solution of a nonvolatile solute that freezes at −3.0°C?

## C. Freezing-point depression

The **freezing point** of a liquid (or the melting point of a solid) is the temperature at which the solid and liquid phases coexist in equilibrium. It might be easier to understand freezing-point depression by looking at the opposite of freezing, melting. To melt a solid, the intermolecular forces holding the solid molecules together must be overcome. Adding another solid substance to a pure solid disrupts the intermolecular forces of the formerly pure solid. Hence, it is easier to overcome the intermolecular forces and the mixture melts at a lower temperature than the pure solid. The melting point (or the freezing point) is depressed. The depression of freezing point can be represented by the following equation.

$$\Delta T_f = K_f m \qquad (12.7, \text{text})$$

where,

$\Delta T_f = T_f^\circ - T_f$ (where $T_f$ is the freezing point of the solution and $T_f^\circ$ is the freezing point of the pure solvent)
$m$ is the molal concentration of the solute
$K_f$ is the molal freezing-point depression constant with units of °C/$m$

**EXAMPLE 12.11**

**How many grams of isopropyl alcohol, $C_3H_7OH$, should be added to 1.0 L of water to give a solution that will not freeze above −16°C? Assume the density of water is 1.00 g/mL.**

**Strategy:** We want to lower the freezing point of 1.0 L of water by 16°C. Using Equation (12.7), we can calculate the molality of the solution needed. Then, from the molality, we can calculate the grams of isopropyl alcohol needed.

**Solution:** Rearrange Equation (12.7) of the text to solve for the molality. You can look up $K_f$ in Table 12.2 of the text.

$$m = \frac{\Delta T_f}{K_f} = \frac{16°\text{C}}{1.86°\text{C}/m} = 8.6\ m$$

8.6 $m$ means that the solution contains 8.6 mol of solute per 1 kg of solvent. Assume that the density of water is 1.0 g/mL; thus, 1 kg (1000 g) of water has a volume of 1 L. Convert 8.6 mol of isopropyl alcohol to grams of isopropyl alcohol using the molar mass as a conversion factor.

$$\textbf{? g isopropyl alcohol} = 8.6\ \text{mol } C_3H_7OH \times \frac{60.09\ \text{g } C_3H_7OH}{1\ \text{mol } C_3H_7OH} = \mathbf{5.2 \times 10^2\ g}$$

**PRACTICE EXERCISE**

11. Benzene melts at 5.50°C. When 2.11 g of naphthalene, $C_{10}H_8$, is added to 100 g of benzene, the solution freezes at 4.65°C. Calculate the freezing-point depression constant ($K_f$) for benzene.

| **Text Problems:** 12.56, 12.60 |
|---|

## D. Osmotic pressure

**Osmosis** is the net movement of solvent molecules through a semipermeable membrane from a pure solvent or from a dilute solution to a more concentrated solution. The **osmotic pressure** ($\pi$) of a solution is the pressure required to stop osmosis. The osmotic pressure of a solution is given by:

$$\pi = MRT \qquad (12.8,\ \text{text})$$

where,

$M$ is the molarity of the solution
$R$ is the gas constant (0.0821 L·atm/K·mol)
$T$ is the absolute temperature in K

**EXAMPLE 12.12**

**The average osmotic pressure of seawater is about 30.0 atm at 25°C. Calculate the molar concentration of an aqueous solution of urea ($NH_2CONH_2$) that is isotonic with seawater.**

**Strategy:** A solution of urea that is isotonic with seawater must have the same osmotic pressure, 30.0 atm. Solve Equation (12.8) of the text algebraically for the molar concentration. Then, substitute $\pi$, $R$, and $T$ (in K) into the equation to solve for the molar concentration of the urea solution.

**Solution:**

$$\textbf{molarity} = \frac{\pi}{RT} = \frac{30.0\ \cancel{\text{atm}}}{298\ \cancel{\text{K}}} \times \frac{\text{mol}\cdot\cancel{\text{K}}}{0.0821\ \text{L}\cdot\cancel{\text{atm}}} = \mathbf{1.23}\ \boldsymbol{M}$$

**PRACTICE EXERCISE**

12. The walls of red blood cells are semipermeable membranes, and the solution of NaCl within those walls exerts an osmotic pressure of 7.82 atm at 37°C. What concentration of NaCl must a *surrounding* solution have so that this pressure is balanced and cell rupture (hemolysis) is prevented?

# PROBLEM TYPE 6: DETERMINING MOLAR MASS USING COLLIGATIVE PROPERTIES

Any of the four colligative properties discussed in Problem Type 5 can be used to calculate the molar mass of the solute. However, in practice, only freezing-point depression and osmotic pressure are used because they show the most pronounced changes.

## A. Calculating molar mass from freezing-point depression

You can solve this type of problem by first calculating the molality of the solute using Equation (12.7) of the text and then calculating the molar mass from the molality.

**EXAMPLE 12.13**

**Benzene has a normal freezing point of 5.51°C. The addition of 1.25 g of an unknown compound to 85.0 g of benzene produces a solution with a freezing point of 4.52°C. What is the molar mass of the unknown compound?**

**Strategy:** We are asked to calculate the molar mass of the unknown. Grams of the compound are given in the problem, so we need to solve for moles of compound.

want to calculate

given

$$\text{molar mass of unknown} = \frac{\text{grams of unknown}}{\text{moles of unknown}}$$

need to find

From the freezing point depression, we can calculate the molality of the solution. Then, from the molality, we can determine the number of moles in 1.25 g of the unknown compound.

**Solution:** Solve Equation (12.7) algebraically for molality ($m$), then substitute $\Delta T_f$ and $K_f$ into the equation to calculate the molality.

$$\Delta T_f = 5.51°C - 4.52°C = 0.99°C$$

$$m = \frac{\Delta T_f}{K_f} = \frac{0.99\cancel{°C}}{5.12\cancel{°C}/m} = 0.19\ m$$

Multiplying the molality by the mass of solvent (in kg) gives moles of unknown solute. Then, dividing the mass of solute (in g) by the moles of solute, gives the molar mass of the unknown solute.

$$\text{? mol of unknown solute} = \frac{0.19 \text{ mol solute}}{1\ \cancel{\text{kg}} \text{ benzene}} \times 0.085\ \cancel{\text{kg}} \text{ benzene} = 0.016 \text{ mol solute}$$

$$\textbf{molar mass of unknown} = \frac{1.25 \text{ g}}{0.016 \text{ mol}} = \textbf{78 g/mol}$$

**PRACTICE EXERCISE**

13. When 48 g of glucose (a nonelectrolyte) is dissolved in 500 g of $H_2O$, the solution has a freezing point of −0.94°C. What is the molar mass of glucose?

14. In the course of research, a chemist isolates a new compound. An elemental analysis shows the following: C, 50.7 percent; H, 4.25 percent; O, 45.1 percent. If 5.01 g of the compound is dissolved in 100 g of water, a solution with a freezing point of −0.65°C is produced. What is the molecular formula of the compound?

**Text Problems: 12.58, 12.62**

## B. Calculating the molar mass from osmotic pressure

You can solve this type of problem by first calculating the molarity of the solution using Equation (12.8) of the text and then calculating the molar mass from the molarity.

**EXAMPLE 12.14**

**30.0 g of sucrose is dissolved in water making $1.00 \times 10^2$ mL of solution. The solution has an osmotic pressure of 20.8 atm at 16.0°C. What is the molar mass of sucrose?**

**Strategy:** We are asked to calculate the molar mass of the sucrose. Grams of the sucrose are given in the problem, so we need to solve for moles of sucrose.

want to calculate → molar mass of sucrose; given → grams of sucrose; need to find → moles of sucrose

$$\text{molar mass of sucrose} = \frac{\text{grams of sucrose}}{\text{moles of sucrose}}$$

From the osmotic pressure of the solution, we can calculate the molarity of the solution. Then, from the molarity, we can determine the number of moles in 30.0 g of sucrose. What units should we use for π and temperature?

**Solution:** First, we calculate the molarity using Equation (12.8) of the text.

$$\pi = MRT$$

$$\text{molarity} = \frac{\pi}{RT} = \frac{20.8 \text{ atm}}{289 \text{ K}} \times \frac{\text{mol} \cdot \text{K}}{0.0821 \text{ L} \cdot \text{atm}} = 0.877\ M$$

Multiplying the molarity by the volume of solution (in L) gives moles of solute (sucrose).

$$? \text{ mol of sucrose} = (0.877 \text{ mol/L})(0.100 \text{ L}) = 0.0877 \text{ mol sucrose}$$

Lastly, dividing the mass of sucrose (in g) by the moles of sucrose, gives the molar mass of sucrose.

$$\textbf{molar mass of sucrose} = \frac{30.0 \text{ g sucrose}}{0.0877 \text{ mol sucrose}} = \textbf{342 g/mol}$$

**PRACTICE EXERCISE**

15. Peruvian Indians use a dart poison from root extracts called curare. It is a nonelectrolyte. The osmotic pressure at 20.0°C of an aqueous solution containing 0.200 g of curare in $1.00 \times 10^2$ mL of solution is 56.2 mmHg. Calculate the molar mass of curare.

**Text Problems: 12.64**, 12.66

## PROBLEM TYPE 7: COLLIGATIVE PROPERTIES OF ELECTROLYTES

Electrolytes dissociate into ions in solution, so this requires us to take a slightly different approach than that used for the colligative properties of nonelectrolytes. Remember, it is the number of solute particles that determines the colligative properties of a solution. To account for the dissociation of an electrolyte into ions, the equations for colligative properties must be modified as follows:

$$\Delta T_b = iK_bm \qquad (12.10, \text{text})$$

$$\Delta T_f = iK_fm \qquad (12.11, \text{text})$$

$$\pi = iMRT \qquad (12.12, \text{text})$$

where,

*i* is the van't Hoff factor which is defined as:

$$i = \frac{\text{actual number of particles in soln after dissociation}}{\text{number of formula units initially dissolved in soln}}$$

Thus, $i$ should be 1 for all nonelectrolytes. For strong electrolytes such as KCl and $BaSO_4$, $i$ should be 2, and for strong electrolytes such as $CaCl_2$ and $Na_2CO_3$, $i$ should be 3. In reality, the colligative properties of *electrolyte* solutions are usually smaller than anticipated. At higher concentrations, electrostatic forces come into play, drawing cations and anions together. The ion pairs that are formed decrease the number of solute particles in solution. Table 12.3 of the text lists the van't Hoff factor ($i$) for various solutes.

**EXAMPLE 12.15**

**Calculate the value of *i* for an electrolyte that should dissociate into 2 ions, if a 1.0 *m* aqueous solution of the electrolyte freezes at −3.28°C. Why is the value of *i* less than 2?**

**Strategy:** Solve Equation (12.11) of the text algebraically for $i$, then substitute the values of $\Delta T_f$, $K_f$, and $m$ into the equation to solve for $i$.

**Solution:**

$\Delta T_f = 3.28°C$

$$i = \frac{\Delta T_f}{K_f m} = \frac{3.28\cancel{°C}}{(1.86\cancel{°C}/\cancel{m})(1.0\ \cancel{m})} = \mathbf{1.76}$$

The value if $i$ is less than 2 because ion pairing reduces the number of particles in solution. Remember, colligative properties depend only on the number of solute particles in solution and *not* on the nature of the solute particles.

**PRACTICE EXERCISE**

**16.** Arrange the following aqueous solutions in order of increasing boiling points: 0.100 $m$ ethanol, 0.050 $m$ $Ca(NO_3)_2$, 0.100 $m$ NaBr, 0.050 $m$ HCl.

**Text Problems:** 12.70, 12.72, 12.74, **12.76**, 12.78

## ANSWERS TO PRACTICE EXERCISES

1. Remember that "like dissolves like". Therefore, the ionic solid NaBr will be more soluble in the polar solvent, $H_2O$.

2. 17.0 percent $CuSO_4$

3. 1.28 $M$ $CuSO_4$

4. 1.29 $m$ $CuSO_4$

5. 11.8 $M$ HCl

6. $m = 3 \times 10^2\ m$

7. 0.91 $m$ $H_2O_2$

8. $1.9 \times 10^{-8}\ M$ CO

9. $m = 1.1\ m$

10. b.p. of soln = 100.84°C

11. $K_f$ (benzene) = 5.15°C/$m$

12. 0.15 $M$ NaCl

13. molar mass = $1.9 \times 10^2$ g/mol

14. $C_6H_6O_4$

15. molar mass of curare = $6.50 \times 10^2$ g/mol

16. ethanol ≈ HCl < $Ca(NO_3)_2$ < NaBr

## SOLUTIONS TO SELECTED TEXT PROBLEMS

**12.10** **Strategy:** In predicting solubility, remember the saying: Like dissolves like. A nonpolar solute will dissolve in a nonpolar solvent; ionic compounds will generally dissolve in polar solvents due to favorable ion-dipole interactions; solutes that can form hydrogen bonds with a solvent will have high solubility in the solvent.

**Solution:** Strong hydrogen bonding (dipole-dipole attraction) is the principal intermolecular attraction in liquid ethanol, but in liquid cyclohexane the intermolecular forces are dispersion forces because cyclohexane is nonpolar. Cyclohexane cannot form hydrogen bonds with ethanol, and therefore cannot attract ethanol molecules strongly enough to form a solution.

**12.12** The longer the C–C chain, the more the molecule "looks like" a hydrocarbon and the less important the –OH group becomes. Hence, as the C–C chain length increases, the molecule becomes less polar. Since "like dissolves like," as the molecules become more nonpolar, the solubility in polar water decreases. The –OH group of the alcohols can form strong hydrogen bonds with water molecules, but this property decreases as the chain length increases.

**12.16** **Strategy:** We are given the percent by mass of the solute and the mass of the solute. We can use Equation (12.1) of the text to solve for the mass of the solvent (water).

**Solution:**

(a) The percent by mass is defined as

$$\text{percent by mass of solute} = \frac{\text{mass of solute}}{\text{mass of solute} + \text{mass of solvent}} \times 100\%$$

Substituting in the percent by mass of solute and the mass of solute, we can solve for the mass of solvent (water).

$$16.2\% = \frac{5.00 \text{ g urea}}{5.00 \text{ g urea} + \text{mass of water}} \times 100\%$$

$$(0.162)(\text{mass of water}) = 5.00 \text{ g} - (0.162)(5.00\text{g})$$

$$\textbf{mass of water} = \textbf{25.9 g}$$

(b) Similar to part (a),

$$1.5\% = \frac{26.2 \text{ g MgCl}_2}{26.2 \text{ g MgCl}_2 + \text{mass of water}} \times 100\%$$

$$\textbf{mass of water} = \mathbf{1.72 \times 10^3}\textbf{ g}$$

**12.18** $$\text{molality} = \frac{\text{moles of solute}}{\text{mass of solvent (kg)}}$$

**(a)** $$\text{mass of 1 L soln} = 1000\ \text{mL} \times \frac{1.08\ \text{g}}{1\ \text{mL}} = 1080\ \text{g}$$

$$\text{mass of water} = 1080\ \text{g} - \left(2.50\ \text{mol NaCl} \times \frac{58.44\ \text{g NaCl}}{1\ \text{mol NaCl}}\right) = 934\ \text{g} = 0.934\ \text{kg}$$

$$m = \frac{2.50\ \text{mol NaCl}}{0.934\ \text{kg H}_2\text{O}} = \mathbf{2.68}\ \boldsymbol{m}$$

**(b)** 100 g of the solution contains 48.2 g KBr and 51.8 g $H_2O$.

$$\text{mol of KBr} = 48.2\ \text{g KBr} \times \frac{1\ \text{mol KBr}}{119.0\ \text{g KBr}} = 0.405\ \text{mol KBr}$$

$$\text{mass of H}_2\text{O (in kg)} = 51.8\ \text{g H}_2\text{O} \times \frac{1\ \text{kg}}{1000\ \text{g}} = 0.0518\ \text{kg H}_2\text{O}$$

$$m = \frac{0.405\ \text{mol KBr}}{0.0518\ \text{kg H}_2\text{O}} = \mathbf{7.82}\ \boldsymbol{m}$$

**12.20** Let's assume that we have 1.0 L of a 0.010 *M* solution.

Assuming a solution density of 1.0 g/mL, the mass of 1.0 L (1000 mL) of the solution is 1000 g or $1.0 \times 10^3$ g.

The mass of 0.010 mole of urea is:

$$0.010\ \text{mol urea} \times \frac{60.06\ \text{g urea}}{1\ \text{mol urea}} = 0.60\ \text{g urea}$$

The mass of the solvent is:

$$(\text{solution mass}) - (\text{solute mass}) = (1.0 \times 10^3\ \text{g}) - (0.60\ \text{g}) = 1.0 \times 10^3\ \text{g} = 1.0\ \text{kg}$$

$$m = \frac{\text{moles solute}}{\text{mass solvent}} = \frac{0.010\ \text{mol}}{1.0\ \text{kg}} = \mathbf{0.010}\ \boldsymbol{m}$$

**12.22** **(a)** Converting mass percent to molality.

**Strategy:** In solving this type of problem, it is convenient to assume that we start with 100.0 grams of the solution. If the mass of sulfuric acid is 98.0% of 100.0 g, or 98.0 g, the percent by mass of water must be 100.0% − 98.0% = 2.0%. The mass of water in 100.0 g of solution would be 2.0 g. From the definition of molality, we need to find moles of solute (sulfuric acid) and kilograms of solvent (water).

**Solution:** Since the definition of molality is

$$\text{molality} = \frac{\text{moles of solute}}{\text{mass of solvent (kg)}}$$

we first convert 98.0 g $H_2SO_4$ to moles of $H_2SO_4$ using its molar mass, then we convert 2.0 g of $H_2O$ to units of kilograms.

$$98.0\,\text{g}\,H_2SO_4 \times \frac{1\text{ mol }H_2SO_4}{98.09\,\text{g}\,H_2SO_4} = 0.999\text{ mol }H_2SO_4$$

$$2.0\,\text{g}\,H_2O \times \frac{1\text{ kg}}{1000\text{ g}} = 2.0 \times 10^{-3}\text{ kg }H_2O$$

Lastly, we divide moles of solute by mass of solvent in kg to calculate the molality of the solution.

$$m = \frac{\text{mol of solute}}{\text{kg of solvent}} = \frac{0.999\text{ mol}}{2.0 \times 10^{-3}\text{ kg}} = \mathbf{5.0 \times 10^2}\ \boldsymbol{m}$$

**(b)** Converting molality to molarity.

**Strategy:** From part (a), we know the moles of solute (0.999 mole $H_2SO_4$) and the mass of the solution (100.0 g). To solve for molarity, we need the volume of the solution, which we can calculate from its mass and density.

**Solution:** First, we use the solution density as a conversion factor to convert to volume of solution.

$$?\text{ volume of solution} = 100.0\text{ g} \times \frac{1\text{ mL}}{1.83\text{ g}} = 54.6\text{ mL} = 0.0546\text{ L}$$

Since we already know moles of solute from part (a), 0.999 mole $H_2SO_4$, we divide moles of solute by liters of solution to calculate the molarity of the solution.

$$M = \frac{\text{mol of solute}}{\text{L of soln}} = \frac{0.999\text{ mol}}{0.0546\text{ L}} = \mathbf{18.3}\ \boldsymbol{M}$$

**12.24** Assume 100.0 g of solution.

**(a)** The mass of ethanol in the solution is $0.100 \times 100.0\text{ g} = 10.0\text{ g}$. The mass of the water is $100.0\text{ g} - 10.0\text{ g} = 90.0\text{ g} = 0.0900\text{ kg}$. The amount of ethanol in moles is:

$$10.0\text{ g ethanol} \times \frac{1\text{ mol}}{46.07\text{ g}} = 0.217\text{ mol ethanol}$$

$$m = \frac{\text{mol solute}}{\text{kg solvent}} = \frac{0.217 \text{ mol}}{0.0900 \text{ kg}} = \mathbf{2.41}\ \boldsymbol{m}$$

**(b)** The volume of the solution is:

$$100.0 \text{ g} \times \frac{1 \text{ mL}}{0.984 \text{ g}} = 102 \text{ mL} = 0.102 \text{ L}$$

The amount of ethanol in moles is 0.217 mole [part (a)].

$$M = \frac{\text{mol solute}}{\text{liters of soln}} = \frac{0.217 \text{ mol}}{0.102 \text{ L}} = \mathbf{2.13}\ \boldsymbol{M}$$

**(c)** $$\textbf{Solution volume} = 0.125 \text{ mol} \times \frac{1 \text{ L}}{2.13 \text{ mol}} = \mathbf{0.0587\ L = 58.7\ mL}$$

**12.28** At 75°C, 155 g of $KNO_3$ dissolves in 100 g of water to form 255 g of solution. When cooled to 25°C, only 38.0 g of $KNO_3$ remain dissolved. This means that (155 − 38.0) g = 117 g of $KNO_3$ will crystallize. The amount of $KNO_3$ formed when 100 g of saturated solution at 75°C is cooled to 25°C can be found by a simple unit conversion.

$$100 \text{ g saturated soln} \times \frac{117 \text{ g KNO}_3 \text{ crystallized}}{255 \text{ g saturated soln}} = \mathbf{45.9\ g\ KNO_3}$$

**12.36** According to Henry's law, the solubility of a gas in a liquid increases as the pressure increases ($c = kP$). The soft drink tastes flat at the bottom of the mine because the carbon dioxide pressure is greater and the dissolved gas is not released from the solution. As the miner goes up in the elevator, the atmospheric carbon dioxide pressure decreases and dissolved gas is released from his stomach.

**12.38** **Strategy:** The given solubility allows us to calculate Henry's law constant ($k$), which can then be used to determine the concentration of $N_2$ at 4.0 atm. We can then compare the solubilities of $N_2$ in blood under normal pressure (0.80 atm) and under a greater pressure that a deep-sea diver might experience (4.0 atm) to determine the moles of $N_2$ released when the diver returns to the surface. From the moles of $N_2$ released, we can calculate the volume of $N_2$ released.

**Solution:** First, calculate the Henry's law constant, $k$, using the concentration of $N_2$ in blood at 0.80 atm.

$$k = \frac{c}{P}$$

$$k = \frac{5.6 \times 10^{-4} \text{ mol/L}}{0.80 \text{ atm}} = 7.0 \times 10^{-4} \text{ mol/L} \cdot \text{atm}$$

Next, we can calculate the concentration of $N_2$ in blood at 4.0 atm using $k$ calculated above.

$$c = kP$$

$$c = (7.0 \times 10^{-4}\ \text{mol/L·atm})(4.0\ \text{atm}) = 2.8 \times 10^{-3}\ \text{mol/L}$$

From each of the concentrations of $N_2$ in blood, we can calculate the number of moles of $N_2$ dissolved by multiplying by the total blood volume of 5.0 L. Then, we can calculate the number of moles of $N_2$ released when the diver returns to the surface.

The number of moles of $N_2$ in 5.0 L of blood at 0.80 atm is:

$$(5.6 \times 10^{-4}\ \text{mol/L})(5.0\ \text{L}) = 2.8 \times 10^{-3}\ \text{mol}$$

The number of moles of $N_2$ in 5.0 L of blood at 4.0 atm is:

$$(2.8 \times 10^{-3}\ \text{mol/L})(5.0\ \text{L}) = 1.4 \times 10^{-2}\ \text{mol}$$

The amount of $N_2$ released in moles when the diver returns to the surface is:

$$(1.4 \times 10^{-2}\ \text{mol}) - (2.8 \times 10^{-3}\ \text{mol}) = 1.1 \times 10^{-2}\ \text{mol}$$

Finally, we can now calculate the volume of $N_2$ released using the ideal gas equation. The total pressure pushing on the $N_2$ that is released is atmospheric pressure (1 atm).

The volume of $N_2$ released is:

$$V_{N_2} = \frac{nRT}{P}$$

$$V_{N_2} = \frac{(1.1 \times 10^{-2}\ \text{mol})(273 + 37)\text{K}}{(1.0\ \text{atm})} \times \frac{0.0821\ \text{L·atm}}{\text{mol·K}} = \mathbf{0.28\ L}$$

**12.50** **Strategy:** From the vapor pressure of water at 20°C and the change in vapor pressure for the solution (2.0 mmHg), we can solve for the mole fraction of sucrose using Equation (12.5) of the text. From the mole fraction of sucrose, we can solve for moles of sucrose. Lastly, we convert form moles to grams of sucrose.

**Solution:** Using Equation (12.5) of the text, we can calculate the mole fraction of sucrose that causes a 2.0 mmHg drop in vapor pressure.

$$\Delta P = X_2 P_1^{\text{o}}$$

$$\Delta P = X_{\text{sucrose}} P_{\text{water}}^{\text{o}}$$

$$X_{\text{sucrose}} = \frac{\Delta P}{P_{\text{water}}^{\text{o}}} = \frac{2.0\ \text{mmHg}}{17.5\ \text{mmHg}} = 0.11$$

From the definition of mole fraction, we can calculate moles of sucrose.

$$X_{sucrose} = \frac{n_{sucrose}}{n_{water} + n_{sucrose}}$$

$$\text{moles of water} = 552\ \text{g} \times \frac{1\ \text{mol}}{18.02\ \text{g}} = 30.6\ \text{mol}\ H_2O$$

$$X_{sucrose} = 0.11 = \frac{n_{sucrose}}{30.6 + n_{sucrose}}$$

$$n_{sucrose} = 3.8\ \text{mol sucrose}$$

Using the molar mass of sucrose as a conversion factor, we can calculate the mass of sucrose.

$$\textbf{mass of sucrose} = 3.8\ \text{mol sucrose} \times \frac{342.3\ \text{g sucrose}}{1\ \text{mol sucrose}} = \mathbf{1.3 \times 10^3\ g\ sucrose}$$

**12.52** For any solution the sum of the mole fractions of the components is always 1.00, so the mole fraction of 1−propanol is 0.700. The partial pressures are:

$$P_{ethanol} = X_{ethanol} \times P^{o}_{ethanol} = (0.300)(100\ \text{mmHg}) = \mathbf{30.0\ mmHg}$$

$$P_{1-propanol} = X_{1-propanol} \times P^{o}_{1-propanol} = (0.700)(37.6\ \text{mmHg}) = \mathbf{26.3\ mmHg}$$

Is the vapor phase richer in one of the components than the solution? Which component? Should this always be true for ideal solutions?

**12.54** This problem is very similar to Problem 12.50.

$$\Delta P = X_{urea}P^{o}_{water}$$

$$2.50\ \text{mmHg} = X_{urea}(31.8\ \text{mmHg})$$

$$X_{urea} = 0.0786$$

The number of moles of water is:

$$n_{water} = 450\ \text{g}\ H_2O \times \frac{1\ \text{mol}\ H_2O}{18.02\ \text{g}\ H_2O} = 25.0\ \text{mol}\ H_2O$$

$$X_{urea} = \frac{n_{urea}}{n_{water} + n_{urea}}$$

$$0.0786 = \frac{n_{\text{urea}}}{25.0 + n_{\text{urea}}}$$

$$n_{\text{urea}} = 2.13 \text{ mol}$$

$$\textbf{mass of urea} = 2.13 \text{ mol urea} \times \frac{60.06 \text{ g urea}}{1 \text{ mol urea}} = \textbf{128 g of urea}$$

**12.56** $$m = \frac{\Delta T_f}{K_f} = \frac{1.1°C}{1.86°C/m} = \mathbf{0.59\ m}$$

**12.58 METHOD 1:**

**Strategy:** First, we can determine the empirical formula from mass percent data. Then, we can determine the molar mass from the freezing-point depression. Finally, from the empirical formula and the molar mass, we can find the molecular formula.

**Solution:** If we assume that we have 100 g of the compound, then each percentage can be converted directly to grams. In this sample, there will be 40.0 g of C, 6.7 g of H, and 53.3 g of O. Because the subscripts in the formula represent a mole ratio, we need to convert the grams of each element to moles. The conversion factor needed is the molar mass of each element. Let $n$ represent the number of moles of each element so that

$$n_C = 40.0 \text{ g C} \times \frac{1 \text{ mol C}}{12.01 \text{ g C}} = 3.33 \text{ mol C}$$

$$n_H = 6.7 \text{ g H} \times \frac{1 \text{ mol H}}{1.008 \text{ g H}} = 6.6 \text{ mol H}$$

$$n_O = 53.3 \text{ g O} \times \frac{1 \text{ mol O}}{16.00 \text{ g O}} = 3.33 \text{ mol O}$$

Thus, we arrive at the formula $C_{3.33}H_{6.6}O_{3.3}$, which gives the identity and the ratios of atoms present. However, chemical formulas are written with whole numbers. Try to convert to whole numbers by dividing all the subscripts by the smallest subscript.

$$\text{C:}\ \frac{3.33}{3.33} = 1.00 \qquad \text{H:}\ \frac{6.6}{3.33} = 2.0 \qquad \text{O:}\ \frac{3.33}{3.33} = 1.00$$

This gives us the empirical, $CH_2O$.

Now, we can use the freezing point data to determine the molar mass. First, calculate the molality of the solution.

$$m = \frac{\Delta T_f}{K_f} = \frac{1.56°C}{8.00°C/m} = 0.195\ m$$

Multiplying the molality by the mass of solvent (in kg) gives moles of unknown solute. Then, dividing the mass of solute (in g) by the moles of solute gives the molar mass of the unknown solute.

$$? \text{ mol of unknown solute} = \frac{0.195 \text{ mol solute}}{1 \text{ kg diphenyl}} \times 0.0278 \text{ kg diphenyl}$$

$$= 0.00542 \text{ mol solute}$$

$$\textbf{molar mass of unknown} = \frac{0.650 \text{ g}}{0.00542 \text{ mol}} = \mathbf{1.20 \times 10^2 \text{ g/mol}}$$

Finally, we compare the empirical molar mass to the molar mass above.

$$\text{empirical molar mass} = 12.01 \text{ g} + 2(1.008 \text{ g}) + 16.00 \text{ g} = 30.03 \text{ g/mol}$$

The number of ($CH_2O$) units present in the molecular formula is:

$$\frac{\text{molar mass}}{\text{empirical molar mass}} = \frac{1.20 \times 10^2 \text{ g}}{30.03 \text{ g}} = 4.00$$

Thus, there are four $CH_2O$ units in each molecule of the compound, so the molecular formula is $(CH_2O)_4$, or **$C_4H_8O_4$**.

**METHOD 2:**

**Strategy:** As in Method 1, we determine the molar mass of the unknown from the freezing point data. Once the molar mass is known, we can multiply the mass % of each element (converted to a decimal) by the molar mass to convert to grams of each element. From the grams of each element, the moles of each element can be determined and hence the mole ratio in which the elements combine.

**Solution:** We use the freezing point data to determine the molar mass. First, calculate the molality of the solution.

$$m = \frac{\Delta T_f}{K_f} = \frac{1.56°C}{8.00°C/m} = 0.195\ m$$

Multiplying the molality by the mass of solvent (in kg) gives moles of unknown solute. Then, dividing the mass of solute (in g) by the moles of solute, gives the molar mass of the unknown solute.

$$? \text{ mol of unknown solute} = \frac{0.195 \text{ mol solute}}{1 \text{ kg diphenyl}} \times 0.0278 \text{ kg diphenyl}$$

$$= 0.00542 \text{ mol solute}$$

$$\textbf{molar mass of unknown} = \frac{0.650 \text{ g}}{0.00542 \text{ mol}} = \mathbf{1.20 \times 10^2 \text{ g/mol}}$$

Next, we multiply the mass % (converted to a decimal) of each element by the molar mass to convert to grams of each element. Then, we use the molar mass to convert to moles of each element.

$$n_C = (0.400) \times (1.20 \times 10^2 \text{ g}) \times \frac{1 \text{ mol C}}{12.01 \text{ g C}} = 4.00 \text{ mol C}$$

$$n_H = (0.067) \times (1.20 \times 10^2 \text{ g}) \times \frac{1 \text{ mol H}}{1.008 \text{ g H}} = 7.98 \text{ mol H}$$

$$n_O = (0.533) \times (1.20 \times 10^2 \text{ g}) \times \frac{1 \text{ mol O}}{16.00 \text{ g O}} = 4.00 \text{ mol O}$$

Since we used the molar mass to calculate the moles of each element present in the compound, this method directly gives the molecular formula. The formula is $\mathbf{C_4H_8O_4}$.

**12.60** We first find the number of moles of gas using the ideal gas equation.

$$n = \frac{PV}{RT} = \frac{\left(748 \text{ mmHg} \times \frac{1 \text{ atm}}{760 \text{ mmHg}}\right)(4.00 \text{ L})}{(27 + 273)\text{K}} \times \frac{\text{mol} \cdot \text{K}}{0.0821 \text{ L} \cdot \text{atm}} = 0.160 \text{ mol}$$

$$\text{molality} = \frac{0.160 \text{ mol}}{0.0580 \text{ kg benzene}} = 2.76\ m$$

$$\Delta T_f = K_f m = (5.12°\text{C}/m)(2.76\ m) = 14.1°\text{C}$$

$$\textbf{freezing point} = 5.5°\text{C} - 14.1°\text{C} = \mathbf{-8.6°C}$$

**12.62** First, from the freezing point depression we can calculate the molality of the solution. See Table 12.2 of the text for the normal freezing point and $K_f$ value for benzene.

$$\Delta T_f = (5.5 - 4.3)°\text{C} = 1.2°\text{C}$$

$$m = \frac{\Delta T_f}{K_f} = \frac{1.2°\text{C}}{5.12°\text{C}/m} = 0.23\ m$$

Multiplying the molality by the mass of solvent (in kg) gives moles of unknown solute. Then, dividing the mass of solute (in g) by the moles of solute, gives the molar mass of the unknown solute.

$$\text{? mol of unknown solute} = \frac{0.23 \text{ mol solute}}{1 \text{ kg benzene}} \times 0.0250 \text{ kg benzene}$$

$$= 0.0058 \text{ mol solute}$$

$$\textbf{molar mass of unknown} = \frac{2.50 \text{ g}}{0.0058 \text{ mol}} = \mathbf{4.3 \times 10^2\ g/mol}$$

The empirical molar mass of $C_6H_5P$ is 108.1 g/mol. Therefore, the molecular formula is $(C_6H_5P)_4$ or $\mathbf{C_{24}H_{20}P_4}$.

**12.64** **Strategy:** We are asked to calculate the molar mass of the polymer. Grams of the polymer are given in the problem, so we need to solve for moles of polymer.

want to calculate → molar mass of polymer; given → grams of polymer; need to find → moles of polymer

$$\text{molar mass of polymer} = \frac{\text{grams of polymer}}{\text{moles of polymer}}$$

From the osmotic pressure of the solution, we can calculate the molarity of the solution. Then, from the molarity, we can determine the number of moles in 0.8330 g of the polymer. What units should we use for π and temperature?

**Solution:** First, we calculate the molarity using Equation (12.8) of the text.

$$\pi = MRT$$

$$M = \frac{\pi}{RT} = \frac{\left(5.20\ \text{mmHg} \times \frac{1\ \text{atm}}{760\ \text{mmHg}}\right)}{298\ \text{K}} \times \frac{\text{mol}\cdot\text{K}}{0.0821\ \text{L}\cdot\text{atm}} = 2.80 \times 10^{-4}\ M$$

Multiplying the molarity by the volume of solution (in L) gives moles of solute (polymer).

$$?\ \text{mol of polymer} = (2.80 \times 10^{-4}\ \text{mol/L})(0.170\ \text{L}) = 4.76 \times 10^{-5}\ \text{mol polymer}$$

Lastly, dividing the mass of polymer (in g) by the moles of polymer, gives the molar mass of the polymer.

$$\textbf{molar mass of polymer} = \frac{0.8330\ \text{g polymer}}{4.76 \times 10^{-5}\ \text{mol polymer}} = \mathbf{1.75 \times 10^4\ g/mol}$$

**12.66** We use the osmotic pressure data to determine the molarity.

$$M = \frac{\pi}{RT} = \frac{4.61\ \text{atm}}{(20 + 273)\ \text{K}} \times \frac{\text{mol}\cdot\text{K}}{0.0821\ \text{L}\cdot\text{atm}} = 0.192\ \text{mol/L}$$

Next we use the density and the solution mass to find the volume of the solution.

$$\text{mass of soln} = 6.85\ \text{g} + 100.0\ \text{g} = 106.9\ \text{g soln}$$

$$\text{volume of soln} = 106.9\ \text{g soln} \times \frac{1\ \text{mL}}{1.024\ \text{g}} = 104.4\ \text{mL} = 0.1044\ \text{L}$$

Multiplying the molarity by the volume (in L) gives moles of solute (carbohydrate).

$$\text{mol of solute} = M \times \text{L} = (0.192\ \text{mol/}\cancel{\text{L}})(0.1044\ \cancel{\text{L}}) = 0.0200\ \text{mol solute}$$

Finally, dividing mass of carbohydrate by moles of carbohydrate gives the molar mass of the carbohydrate.

$$\textbf{molar mass} = \frac{6.85\ \text{g carbohydrate}}{0.0200\ \text{mol carbohydrate}} = \mathbf{343\ g/mol}$$

12.70 Boiling point, vapor pressure, and osmotic pressure all depend on particle concentration. Therefore, these solutions also have the same **boiling point, osmotic pressure**, and **vapor pressure**.

12.72 The freezing point will be depressed most by the solution that contains the most solute particles. You should try to classify each solute as a strong electrolyte, a weak electrolyte, or a nonelectrolyte. All three solutions have the same concentration, so comparing the solutions is straightforward. HCl is a strong electrolyte, so under ideal conditions it will completely dissociate into two particles per molecule. The concentration of particles will be 1.00 *m*. Acetic acid is a weak electrolyte, so it will only dissociate to a small extent. The concentration of particles will be greater than 0.50 *m*, but less than 1.00 *m*. Glucose is a nonelectrolyte, so glucose molecules remain as glucose molecules in solution. The concentration of particles will be 0.50 *m*. For these solutions, the order in which the freezing points become *lower* is:

**0.50 *m* glucose > 0.50 *m* acetic acid > 0.50 *m* HCl**

In other words, the HCl solution will have the lowest freezing point (greatest freezing point depression).

12.74 Using Equation (12.5) of the text, we can find the mole fraction of the NaCl. We use subscript 1 for $H_2O$ and subscript 2 for NaCl.

$$\Delta P = X_2 P_1^\circ$$

$$X_2 = \frac{\Delta P}{P_1^\circ}$$

$$X_2 = \frac{23.76\ \text{mmHg} - 22.98\ \text{mmHg}}{23.76\ \text{mmHg}} = 0.03283$$

Let's assume that we have 1000 g (1 kg) of water as the solvent, because the definition of molality is moles of solute per kg of solvent. We can find the number of moles of particles dissolved in the water using the definition of mole fraction.

$$X_2 = \frac{n_2}{n_1 + n_2}$$

$$n_1 = 1000\ \cancel{\text{g}}\ H_2O \times \frac{1\ \text{mol}\ H_2O}{18.02\ \cancel{\text{g}}\ H_2O} = 55.49\ \text{mol}\ H_2O$$

$$\frac{n_2}{55.49 + n_2} = 0.03283$$

$$n_2 = 1.884 \text{ mol}$$

Since NaCl dissociates to form two particles (ions), the number of moles of NaCl is half of the above result.

$$\text{Moles NaCl} = 1.884 \text{ mol particles} \times \frac{1 \text{ mol NaCl}}{2 \text{ mol particles}} = 0.9420 \text{ mol}$$

The molality of the solution is:

$$\frac{0.9420 \text{ mol}}{1.000 \text{ kg}} = \mathbf{0.9420}\ \boldsymbol{m}$$

12.76 **Strategy:** We want to calculate the osmotic pressure of a NaCl solution. Since NaCl is a strong electrolyte, $i$ in the van't Hoff equation is 2.

$$\pi = iMRT$$

Since, $R$ is a constant and $T$ is given, we need to first solve for the molarity of the solution in order to calculate the osmotic pressure ($\pi$). If we assume a given volume of solution, we can then use the density of the solution to determine the mass of the solution. The solution is 0.86% by mass NaCl, so we can find grams of NaCl in the solution.

**Solution:** To calculate molarity, let's assume that we have 1.000 L of solution ($1.000 \times 10^3$ mL). We can use the solution density as a conversion factor to calculate the mass of $1.000 \times 10^3$ mL of solution.

$$(1.000 \times 10^3 \text{ mL soln}) \times \frac{1.005 \text{ g soln}}{1 \text{ mL soln}} = 1005 \text{ g of soln}$$

Since the solution is 0.86% by mass NaCl, the mass of NaCl in the solution is:

$$1005 \text{ g} \times \frac{0.86\%}{100\%} = 8.6 \text{ g NaCl}$$

The molarity of the solution is:

$$\frac{8.6 \text{ g NaCl}}{1.000 \text{ L}} \times \frac{1 \text{ mol NaCl}}{58.44 \text{ g NaCl}} = 0.15\ M$$

Since NaCl is a strong electrolyte, we assume that the van't Hoff factor is 2. Substituting $i$, $M$, $R$, and $T$ into the equation for osmotic pressure gives:

$$\pi = iMRT = (2)\left(\frac{0.15 \text{ mol}}{\text{L}}\right)\left(\frac{0.0821 \text{ L}\cdot\text{atm}}{\text{mol}\cdot\text{K}}\right)(310 \text{ K}) = \mathbf{7.6\ atm}$$

**12.78** From Table 12.3 of the text, $i = 1.3$

$$\pi = iMRT$$

$$\pi = (1.3)\left(\frac{0.0500\ \text{mol}}{\text{L}}\right)\left(\frac{0.0821\ \text{L}\cdot\text{atm}}{\text{mol}\cdot\text{K}}\right)(298\ \text{K})$$

$$\boldsymbol{\pi = 1.6\ \text{atm}}$$

**12.82** Note that octane has the lowest density and is immiscible with both water and methanol. Octane will layer on top of the miscible mixture of water and methanol. The correct picture is **(c)**.

**12.84** At constant temperature, the osmotic pressure of a solution is proportional to the molarity. When equal volumes of the two solutions are mixed, the molarity will just be the mean of the molarities of the two solutions (assuming additive volumes). Since the osmotic pressure is proportional to the molarity, the osmotic pressure of the solution will be the mean of the osmotic pressure of the two solutions.

$$\pi = \frac{2.4\ \text{atm} + 4.6\ \text{atm}}{2} = \mathbf{3.5\ atm}$$

**12.86** **(a)** We use Equation (12.4) of the text to calculate the vapor pressure of each component.

$$P_1 = X_1 P_1^{\text{o}}$$

First, you must calculate the mole fraction of each component.

$$X_{\text{A}} = \frac{n_{\text{A}}}{n_{\text{A}} + n_{\text{B}}} = \frac{1.00\ \text{mol}}{1.00\ \text{mol} + 1.00\ \text{mol}} = 0.500$$

Similarly,

$$X_{\text{B}} = 0.500$$

Substitute the mole fraction calculated above and the vapor pressure of the pure solvent into Equation (12.4) to calculate the vapor pressure of each component of the solution.

$$P_{\text{A}} = X_{\text{A}} P_{\text{A}}^{\text{o}} = (0.500)(76\ \text{mmHg}) = 38\ \text{mmHg}$$

$$P_{\text{B}} = X_{\text{B}} P_{\text{B}}^{\text{o}} = (0.500)(132\ \text{mmHg}) = 66\ \text{mmHg}$$

The total vapor pressure is the sum of the vapor pressures of the two components.

$$P_{\text{Total}} = P_{\text{A}} + P_{\text{B}} = 38\ \text{mmHg} + 66\ \text{mmHg} = \mathbf{104\ mmHg}$$

**(b)** This problem is solved similarly to part (a).

$$X_{\text{A}} = \frac{n_{\text{A}}}{n_{\text{A}} + n_{\text{B}}} = \frac{2.00\ \text{mol}}{2.00\ \text{mol} + 5.00\ \text{mol}} = 0.286$$

**Similarly,**

$$X_B = 0.714$$

$$P_A = X_A P_A^{\circ} = (0.286)(76 \text{ mmHg}) = 22 \text{ mmHg}$$

$$P_B = X_B P_B^{\circ} = (0.714)(132 \text{ mmHg}) = 94 \text{ mmHg}$$

$$\mathbf{P_{Total}} = P_A + P_B = 22 \text{ mmHg} + 94 \text{ mmHg} = \mathbf{116 \text{ mmHg}}$$

**12.88** From the osmotic pressure, you can calculate the molarity of the solution.

$$M = \frac{\pi}{RT} = \frac{\left(30.3 \text{ mmHg} \times \dfrac{1 \text{ atm}}{760 \text{ mmHg}}\right)}{308 \text{ K}} \times \frac{\text{mol} \cdot \text{K}}{0.0821 \text{ L} \cdot \text{atm}} = 1.58 \times 10^{-3} \text{ mol/L}$$

Multiplying molarity by the volume of solution in liters gives the moles of solute.

$$(1.58 \times 10^{-3} \text{ mol solute/L soln}) \times (0.262 \text{ L soln}) = 4.14 \times 10^{-4} \text{ mol solute}$$

Divide the grams of solute by the moles of solute to calculate the molar mass.

$$\textbf{molar mass of solute} = \frac{1.22 \text{ g}}{4.14 \times 10^{-4} \text{ mol}} = \mathbf{2.95 \times 10^3 \text{ g/mol}}$$

**12.90** From the freezing point depression, we can calculate the molality of the solution. From the molality, the moles of solute (naphthalene) and then the mass can be determined. Solve Equation (12.7) of the text algebraically for molality ($m$), then substitute $\Delta T_f$ and $K_f$ into the equation to calculate the molality. You can find the $K_f$ value for benzene in Table 12.2 of the text.

$$m = \frac{\Delta T_f}{K_f} = \frac{2.00°\text{C}}{5.12°\text{C}/m} = 0.391\ m$$

The moles of naphthalene is:

$$\text{moles solute} = (m)(\text{kg solvent}) = (0.391 \text{ mol/kg})(0.250 \text{ kg}) = 0.0978 \text{ mol}$$

The mass of naphthalene that must be added to lower the freezing point of 250 g of benzene by 2.00°C is:

$$\textbf{mass naphthalene} = 0.0978 \text{ mol} \times \frac{128.2 \text{ g}}{1 \text{ mol}} = \mathbf{12.5 \text{ g}}$$

**12.92** Solve Equation (12.7) of the text algebraically for molality ($m$), then substitute $\Delta T_f$ and $K_f$ into the equation to calculate the molality. You can find the normal freezing point for benzene and $K_f$ for benzene in Table 12.2 of the text.

$$\Delta T_f = 5.5°C - 3.9°C = 1.6°C$$

$$m = \frac{\Delta T_f}{K_f} = \frac{1.6°\cancel{C}}{5.12°\cancel{C}/m} = 0.31\ m$$

Multiplying the molality by the mass of solvent (in kg) gives moles of unknown solute. Then, dividing the mass of solute (in g) by the moles of solute, gives the molar mass of the unknown solute.

$$?\text{ mol of unknown solute} = \frac{0.31\text{ mol solute}}{1\ \cancel{kg}\text{ benzene}} \times (8.0 \times 10^{-3}\ \cancel{kg}\text{ benzene})$$

$$= 2.5 \times 10^{-3}\text{ mol solute}$$

$$\text{molar mass of unknown} = \frac{0.50\text{ g}}{2.5 \times 10^{-3}\text{ mol}} = 2.0 \times 10^{2}\text{ g/mol}$$

The molar mass of cocaine $C_{17}H_{21}NO_4$ = 303 g/mol, so **the compound is not cocaine**. We assume in our analysis that the compound is a pure, monomeric, nonelectrolyte.

**12.94** The molality of the solution assuming $AlCl_3$ to be a nonelectrolyte is:

$$\text{mol } AlCl_3 = 1.00\ \cancel{g}\ AlCl_3 \times \frac{1\text{ mol } AlCl_3}{133.3\ \cancel{g}\ AlCl_3} = 0.00750\text{ mol } AlCl_3$$

$$m = \frac{0.00750\text{ mol}}{0.0500\text{ kg}} = 0.150\ m$$

The molality calculated with Equation (12.7) of the text is:

$$m = \frac{\Delta T_f}{K_f} = \frac{1.11°\cancel{C}}{1.86°\cancel{C}/m} = 0.597\ m$$

The ratio $\frac{0.597\ m}{0.150\ m}$ is 4. Thus each $AlCl_3$ dissociates as follows:

$$AlCl_3(s) \rightarrow Al^{3+}(aq) + 3Cl^-(aq)$$

**12.96** Henry's law states that the solubility of a gas in a liquid is proportional to the pressure of the gas over the solution. See Example 12.6 of the text for the Henry's law constants for $O_2$ and $N_2$ at 25°C.

$$c = kP$$

$$c_{O_2} = k_{O_2}P_{O_2} = \left(1.3 \times 10^{-3} \frac{\text{mol}}{\text{L}\cdot\cancel{\text{atm}}}\right)(0.20\ \cancel{\text{atm}}) = 2.6 \times 10^{-4}\ M$$

$$c_{N_2} = k_{N_2}P_{N_2} = \left(6.8 \times 10^{-4} \frac{\text{mol}}{\text{L}\cdot\cancel{\text{atm}}}\right)(0.80\ \cancel{\text{atm}}) = 5.4 \times 10^{-4}\ M$$

Assuming the density of water to be 1.00 g/mL, the moles of water in 1 L is:

$$1000\ \cancel{\text{g}}\ H_2O \times \frac{1\ \text{mol}\ H_2O}{18.02\ \cancel{\text{g}}\ H_2O} = 55.5\ \text{mol}\ H_2O$$

The mole fractions of $O_2$ and $N_2$ in water at 25°C are:

$$X_{O_2} = \frac{2.6 \times 10^{-4}\ \text{mol}\ O_2}{55.5\ \text{mol}\ H_2O} = \mathbf{4.7 \times 10^{-6}}$$

$$X_{N_2} = \frac{5.4 \times 10^{-4}\ \text{mol}\ N_2}{55.5\ \text{mol}\ H_2O} = \mathbf{9.7 \times 10^{-6}}$$

Let's compare the mole fractions of $O_2$ and $N_2$ in both air and water.

$$\text{In air:}\quad \frac{X_{O_2}}{X_{N_2}} = \frac{0.20}{0.80} = 0.25$$

$$\text{In water:}\quad \frac{X_{O_2}}{X_{N_2}} = \frac{4.7 \times 10^{-6}}{9.7 \times 10^{-6}} = 0.48$$

The mole fraction of $O_2$ compared to the mole fraction of $N_2$ in water is greater compared to its mole fraction in air. This difference is because $O_2$ is more soluble in water than $N_2$. Note the small magnitude of the mole fraction of $O_2$ in water. Fish must be very efficient in extracting $O_2$ from the very small concentration of $O_2$ found in water.

**12.98** Solution A: Let molar mass be $M$.

$$\Delta P = X_A P_A^\circ$$

$$(760 - 754.5) = X_A(760)$$

$$X_A = 7.237 \times 10^{-3}$$

$$n = \frac{\text{mass}}{\text{molar mass}}$$

$$X_A = \frac{n_A}{n_A + n_{\text{water}}} = \frac{5.00/M}{5.00/M + 100/18.02} = 7.237 \times 10^{-3}$$

$$M = 124 \text{ g/mol}$$

Solution B: Let molar mass be $M$

$$\Delta P = X_B P_B^o$$

$$X_B = 7.237 \times 10^{-3}$$

$$n = \frac{\text{mass}}{\text{molar mass}}$$

$$X_B = \frac{n_B}{n_B + n_{\text{benzene}}} = \frac{2.31/M}{2.31/M + 100/78.11} = 7.237 \times 10^{-3}$$

$$M = 248 \text{ g/mol}$$

The molar mass in benzene is about twice that in water. This suggests some sort of dimerization is occurring in a nonpolar solvent such as benzene.

**12.100** As the chain becomes longer, the alcohols become more like hydrocarbons (nonpolar) in their properties. The alcohol with five carbons (*n*-pentanol) would be the best solvent for iodine (a) and *n*-pentane (c) (why?). Methanol ($CH_3OH$) is the most water like and is the best solvent for an ionic solid like KBr.

**12.102** $I_2 - H_2O$: Dipole - induced dipole.

$I_3^- - H_2O$: Ion - dipole. Stronger interaction causes more $I_2$ to be converted to $I_3^-$.

**12.104** **(a)** If the membrane is permeable to all the ions and to the water, the result will be the same as just removing the membrane. You will have two solutions of equal NaCl concentration.

**(b)** This part is tricky. The movement of one ion but not the other would result in one side of the apparatus acquiring a positive electric charge and the other side becoming equally negative. This has never been known to happen, so we must conclude that migrating ions always drag other ions of the opposite charge with them. In this hypothetical situation only water would move through the membrane from the dilute to the more concentrated side.

**(c)** This is the classic osmosis situation. Water would move through the membrane from the dilute to the concentrated side.

**12.106** First, we calculate the number of moles of HCl in 100 g of solution.

$$n_{\text{HCl}} = 100 \text{ g soln} \times \frac{37.7 \text{ g HCl}}{100 \text{ g soln}} \times \frac{1 \text{ mol HCl}}{36.46 \text{ g HCl}} = 1.03 \text{ mol HCl}$$

Next, we calculate the volume of 100 g of solution.

$$V = 100 \text{ g} \times \frac{1 \text{ mL}}{1.19 \text{ g}} \times \frac{1 \text{ L}}{1000 \text{ mL}} = 0.0840 \text{ L}$$

Finally, the molarity of the solution is:

$$\frac{1.03\text{ mol}}{0.0840\text{ L}} = \mathbf{12.3}\ \boldsymbol{M}$$

**12.108** Let the mass of NaCl be $x$ g. Then, the mass of sucrose is $(10.2 - x)$g.

We know that the equation representing the osmotic pressure is:

$$\pi = MRT$$

$\pi$, $R$, and $T$ are given. Using this equation and the definition of molarity, we can calculate the percentage of NaCl in the mixture.

$$\text{molarity} = \frac{\text{mol solute}}{\text{L soln}}$$

Remember that NaCl dissociates into two ions in solution; therefore, we multiply the moles of NaCl by two.

$$\text{mol solute} = 2\left(x\text{ g NaCl} \times \frac{1\text{ mol NaCl}}{58.44\text{ g NaCl}}\right) + \left((10.2 - x)\text{g sucrose} \times \frac{1\text{ mol sucrose}}{342.3\text{ g sucrose}}\right)$$

$$\text{mol solute} = 0.03422x + 0.02980 - 0.002921x$$

$$\text{mol solute} = 0.03130x + 0.02980$$

$$\text{Molarity of solution} = \frac{\text{mol solute}}{\text{L soln}} = \frac{(0.03130x + 0.02980)\text{mol}}{0.250\text{ L}}$$

Substitute molarity into the equation for osmotic pressure to solve for $x$.

$$\pi = MRT$$

$$7.32\text{ atm} = \left(\frac{(0.03130x + 0.02980)\text{mol}}{0.250\text{ L}}\right)\left(0.0821\frac{\text{L}\cdot\text{atm}}{\text{mol}\cdot\text{K}}\right)(296\text{ K})$$

$$0.0753 = 0.03130x + 0.02980$$

$$x = 1.45\text{ g} = \text{mass of NaCl}$$

$$\textbf{Mass \% NaCl} = \frac{1.45\text{ g}}{10.2\text{ g}} \times 100\% = \mathbf{14.2\%}$$

**12.110** See Figure 12.8 of the text. The graph in the problem shows a positive deviation from Raoult's law. This means that the intermolecular forces between A and B molecules are weaker than those between A molecules and between B molecules, and therefore there is a greater tendency for these molecules to leave the solution than in the case of an ideal solution. Consequently, the vapor pressure of the solution is greater than the sum of the vapor pressures as predicted by Raoult's law for the same concentration. Also, the heat of solution is positive (that is, mixing is an endothermic process).

At $X_A = 0.20$, the vapor pressure of the solution is less than the vapor pressure of pure A and higher than the vapor pressure of pure B. This means that the solution has a higher boiling point than liquid A and a lower boiling point than liquid B.

The false statements are **(a)** and **(d)**.

**12.112** **(a)** Solubility decreases with increasing lattice energy.

**(b)** Ionic compounds are more soluble in a polar solvent.

**(c)** Solubility increases with enthalpy of hydration of the cation and anion.

**12.114** $$\text{molality} = \frac{98.0\text{ g } H_2SO_4 \times \dfrac{1\text{ mol } H_2SO_4}{98.09\text{ g } H_2SO_4}}{2.0\text{ g } H_2O \times \dfrac{1\text{ kg } H_2O}{1000\text{ g } H_2O}} = \mathbf{5.0 \times 10^2}\ \boldsymbol{m}$$

We can calculate the density of sulfuric acid from the molarity.

$$\text{molarity} = 18\ M = \frac{18\text{ mol } H_2SO_4}{1\text{ L soln}}$$

The 18 mol of $H_2SO_4$ has a mass of:

$$18\text{ mol } H_2SO_4 \times \frac{98.0\text{ g } H_2SO_4}{1\text{ mol } H_2SO_4} = 1.8 \times 10^3\text{ g } H_2SO_4$$

$$1\text{ L} = 1000\text{ mL}$$

$$\textbf{density} = \frac{\text{mass } H_2SO_4}{\text{volume}} = \frac{1.8 \times 10^3\text{ g}}{1000\text{ mL}} = \mathbf{1.80\ g/mL}$$

**12.116** $P_A = X_A P_A^\circ$

$$P_{\text{ethanol}} = (0.62)(108\text{ mmHg}) = 67.0\text{ mmHg}$$

$$P_{\text{1-propanol}} = (0.38)(40.0\text{ mmHg}) = 15.2\text{ mmHg}$$

In the vapor phase:

$$X_{\text{ethanol}} = \frac{67.0}{67.0 + 15.2} = \mathbf{0.815}$$

**12.118** $NH_3$ can form hydrogen bonds with water; $NCl_3$ cannot. (Like dissolves like.)

**12.120** We can calculate the molality of the solution from the freezing point depression.

$$\Delta T_f = K_f m$$

$$0.203 = 1.86\ m$$

$$m = \frac{0.203}{1.86} = 0.109\ m$$

The molality of the original solution was 0.106 *m*. Some of the solution has ionized to $H^+$ and $CH_3COO^-$.

| | $CH_3COOH$ $\rightleftharpoons$ | $CH_3COO^-$ + | $H^+$ |
|---|---|---|---|
| Initial | 0.106 *m* | 0 | 0 |
| Change | $-x$ | $+x$ | $+x$ |
| Equil. | $0.106\ m - x$ | $x$ | $x$ |

At equilibrium, the total concentration of species in solution is 0.109 *m*.

$$(0.106 - x) + 2x = 0.109\ m$$

$$x = 0.003\ m$$

The percentage of acid that has undergone ionization is:

$$\frac{0.003\ m}{0.106\ m} \times 100\% = \mathbf{3\%}$$

**12.122** First, we can calculate the molality of the solution from the freezing point depression.

$$\Delta T_f = (5.12)m$$

$$(5.5 - 3.5) = (5.12)m$$

$$m = 0.39$$

Next, from the definition of molality, we can calculate the moles of solute.

$$m = \frac{\text{mol solute}}{\text{kg solvent}}$$

$$0.39\ m = \frac{\text{mol solute}}{80 \times 10^{-3}\ \text{kg benzene}}$$

$$\text{mol solute} = 0.031\ \text{mol}$$

The molar mass (Mss) of the solute is:

$$\frac{3.8 \text{ g}}{0.031 \text{ mol}} = \mathbf{1.2 \times 10^2 \text{ g/mol}}$$

The molar mass of $CH_3COOH$ is 60.05 g/mol. Since the molar mass of the solute calculated from the freezing point depression is twice this value, the structure of the solute most likely is a dimer that is held together by hydrogen bonds.

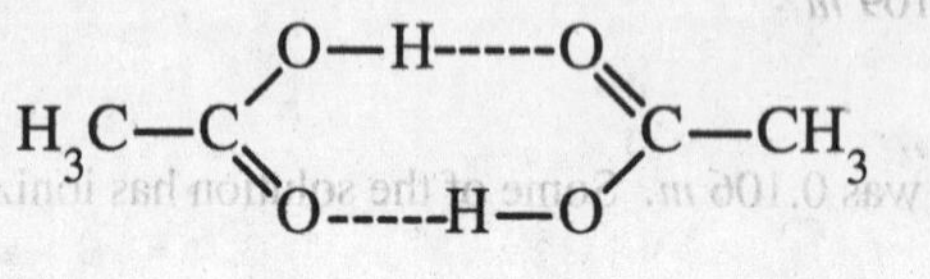

A dimer

**12.124** (a) $\Delta T_f = K_f m$

$2 = (1.86)(m)$

**molality = 1.1 *m***

This concentration is too high and is *not* a reasonable physiological concentration.

(b) Although the protein is present in low concentrations, it can prevent the formation of ice crystals.

**12.126** As the water freezes, dissolved minerals in the water precipitate from solution. The minerals refract light and create an opaque appearance.

**12.128** To solve for the molality of the solution, we need the moles of solute (urea) and the kilograms of solvent (water). If we assume that we have 1 mole of water, we know the mass of water. Using the change in vapor pressure, we can solve for the mole fraction of urea and then the moles of urea.

Using Equation (12.5) of the text, we solve for the mole fraction of urea.

$\Delta P = 23.76 \text{ mmHg} - 22.98 \text{ mmHg} = 0.78 \text{ mmHg}$

$$\Delta P = X_2 P_1^\circ = X_{\text{urea}} P_{\text{water}}^\circ$$

$$X_{\text{urea}} = \frac{\Delta P}{P_{\text{water}}^\circ} = \frac{0.78 \text{ mmHg}}{23.76 \text{ mmHg}} = 0.033$$

Assuming that we have 1 mole of water, we can now solve for moles of urea.

$$X_{\text{urea}} = \frac{\text{mol urea}}{\text{mol urea} + \text{mol water}}$$

$$0.033 = \frac{n_{\text{urea}}}{n_{\text{urea}} + 1}$$

$$0.033n_{\text{urea}} + 0.033 = n_{\text{urea}}$$

$$0.033 = 0.967n_{\text{urea}}$$

$$n_{\text{urea}} = 0.034 \text{ mol}$$

1 mole of water has a mass of 18.02 g or 0.01802 kg. We now know the moles of solute (urea) and the kilograms of solvent (water), so we can solve for the molality of the solution.

$$m = \frac{\text{mol solute}}{\text{kg solvent}} = \frac{0.034 \text{ mol}}{0.01802 \text{ kg}} = \mathbf{1.9\ m}$$

**12.130** **(a)** The solution is prepared by mixing equal masses of A and B. Let's assume that we have 100 grams of each component. We can convert to moles of each substance and then solve for the mole fraction of each component.

Since the molar mass of A is 100 g/mol, we have 1.00 mole of A. The moles of B are:

$$100 \text{ g B} \times \frac{1 \text{ mol B}}{110 \text{ g B}} = 0.909 \text{ mol B}$$

The mole fraction of A is:

$$X_A = \frac{n_A}{n_A + n_B} = \frac{1}{1 + 0.909} = \mathbf{0.524}$$

Since this is a two component solution, the mole fraction of B is: $X_B = 1 - 0.524 = \mathbf{0.476}$

**(b)** We can use Equation (12.4) of the text and the mole fractions calculated in part (a) to calculate the partial pressures of A and B over the solution.

$$P_A = X_A P_A^{\circ} = (0.524)(95 \text{ mmHg}) = \mathbf{50\ mmHg}$$

$$P_B = X_B P_B^{\circ} = (0.476)(42 \text{ mmHg}) = \mathbf{20\ mmHg}$$

**(c)** Recall that pressure of a gas is directly proportional to moles of gas ($P \propto n$). The ratio of the partial pressures calculated in part (b) is 50 : 20, and therefore the ratio of moles will also be 50 : 20. Let's assume that we have 50 moles of A and 20 moles of B. We can solve for the mole fraction of each component and then solve for the vapor pressures using Equation (12.4) of the text.

The mole fraction of A is:

$$X_A = \frac{n_A}{n_A + n_B} = \frac{50}{50 + 20} = \mathbf{0.71}$$

Since this is a two component solution, the mole fraction of B is: $X_B = 1 - 0.71 = \mathbf{0.29}$

The vapor pressures of each component above the solution are:

$$P_A = X_A P_A^\circ = (0.71)(95\ \text{mmHg}) = \mathbf{67\ mmHg}$$

$$P_B = X_B P_B^\circ = (0.29)(42\ \text{mmHg}) = \mathbf{12\ mmHg}$$

**12.132** To calculate the mole fraction of urea in the solutions, we need the moles of urea and the moles of water. The number of moles of urea in each beaker is:

$$\text{moles urea (1)} = \frac{0.10\ \text{mol}}{1\ \cancel{\text{L}}} \times 0.050\ \cancel{\text{L}} = 0.0050\ \text{mol}$$

$$\text{moles urea (2)} = \frac{0.20\ \text{mol}}{1\ \cancel{\text{L}}} \times 0.050\ \cancel{\text{L}} = 0.010\ \text{mol}$$

The number of moles of water in each beaker initially is:

$$\text{moles water} = 50\ \cancel{\text{mL}} \times \frac{1\ \cancel{\text{g}}}{1\ \cancel{\text{mL}}} \times \frac{1\ \text{mol}}{18.02\ \cancel{\text{g}}} = 2.8\ \text{mol}$$

The mole fraction of urea in each beaker initially is:

$$X_1 = \frac{0.0050\ \text{mol}}{0.0050\ \text{mol} + 2.8\ \text{mol}} = 1.8 \times 10^{-3}$$

$$X_2 = \frac{0.010\ \text{mol}}{0.010\ \text{mol} + 2.8\ \text{mol}} = 3.6 \times 10^{-3}$$

Equilibrium is attained by the transfer of water (via water vapor) from the less concentrated solution to the more concentrated one until the mole fractions of urea are equal. At this point, the mole fractions of water in each beaker are also equal, and Raoult's law implies that the vapor pressures of the water over each beaker are the same. Thus, there is no more net transfer of solvent between beakers. Let $y$ be the number of moles of water transferred to reach equilibrium.

$$X_1\ \text{(equil.)} = X_2\ \text{(equil.)}$$

$$\frac{0.0050\ \text{mol}}{0.0050\ \text{mol} + 2.8\ \text{mol} - y} = \frac{0.010\ \text{mol}}{0.010\ \text{mol} + 2.8\ \text{mol} + y}$$

$$0.014 + 0.0050y = 0.028 - 0.010y$$

$$y = 0.93$$

The mole fraction of urea at equilibrium is:

$$\frac{0.010 \text{ mol}}{0.010 \text{ mol} + 2.8 \text{ mol} + 0.93 \text{ mol}} = \mathbf{2.7 \times 10^{-3}}$$

This solution to the problem assumes that the volume of water left in the bell jar as vapor is negligible compared to the volumes of the solutions. It is interesting to note that at equilibrium, 16.8 mL of water has been transferred from one beaker to the other.

**12.134** Starting with $n = kP$ and substituting into the ideal gas equation ($PV = nRT$), we find:

$$PV = (kP)RT$$

$$V = kRT$$

This equation shows that the volume of a gas that dissolves in a given amount of solvent is dependent on the *temperature*, not the pressure of the gas.

**12.136** To calculate the freezing point of the solution, we need the solution molality and the freezing-point depression constant for water (see Table 12.2 of the text). We can first calculate the molarity of the solution using Equation (12.8) of the text: $\pi = MRT$. The solution molality can then be determined from the molarity.

$$M = \frac{\pi}{RT} = \frac{10.50 \text{ atm}}{(0.0821 \text{ L}\cdot\text{atm/mol}\cdot\text{K})(298 \text{ K})} = 0.429\ M$$

Let's assume that we have 1 L (1000 mL) of solution. The mass of 1000 mL of solution is:

$$\frac{1.16 \text{ g}}{1 \text{ mL}} \times 1000 \text{ mL} = 1160 \text{ g soln}$$

The mass of the solvent ($H_2O$) is:

$$\text{mass H}_2\text{O} = \text{mass soln} - \text{mass solute}$$

$$\text{mass H}_2\text{O} = 1160 \text{ g} - \left(0.429 \text{ mol glucose} \times \frac{180.2 \text{ g glucose}}{1 \text{ mol glucose}}\right) = 1083 \text{ g} = 1.083 \text{ kg}$$

The molality of the solution is:

$$\text{molality} = \frac{\text{mol solute}}{\text{kg solvent}} = \frac{0.429 \text{ mol}}{1.083 \text{ kg}} = 0.396\ m$$

The freezing point depression is:

$$\Delta T_f = K_f m = (1.86°\text{C}/m)(0.396\ m) = 0.737°\text{C}$$

The solution will freeze at 0°C − 0.737°C = **−0.737°C**

**12.138** Valinomycin contains both polar and nonpolar groups. The polar groups bind the $K^+$ ions and the nonpolar $-CH_3$ groups allow the valinomycin molecule to dissolve in the the nonpolar lipid barrier of the cell. Once dissolved in the lipid barrier, the $K^+$ ions transport across the membrane into the cell to offset the ionic balance.

**12.140** To lift the ice cube, the string is wetted and laid on top of the ice cube. Salt is shaken onto the top of the ice cube and the moistened string. The presence of salt lowers the freezing point of the ice resulting in the melting of the ice on the surface. Melting is an endothermic process. The water in the moist string freezes, and the string becomes attached to the ice cube. The ice cube can now be lifted out of the glass.

# CHAPTER 13
# CHEMICAL KINETICS

## PROBLEM-SOLVING STRATEGIES AND TUTORIAL SOLUTIONS

### TYPES OF PROBLEMS

**Problem Type 1:** Writing Rate Expressions.

**Problem Type 2:** Determining the Rate Law of a Reaction.

**Problem Type 3:** First-Order Reactions.
- **(a)** Analyzing a first-order reaction.
- **(b)** Determining the half-life of a first-order reaction.
- **(c)** Analyzing first-order kinetics.

**Problem Type 4:** Second-Order Reactions.
- **(a)** Analyzing a second-order reaction.
- **(b)** Determining the half-life of a second-order reaction.
- **(c)** Analyzing second-order kinetics.

**Problem Type 5:** The Arrhenius Equation.
- **(a)** Applying the Arrhenius equation.
- **(b)** Applying a modified form of the Arrhenius equation that relates the rate constants at two different temperatures.

**Problem Type 6:** Studying Reaction Mechanisms.

## PROBLEM TYPE 1: WRITING RATE EXPRESSIONS

As a chemical reaction proceeds, the concentrations of reactants and products change with time. As the reaction

$$A + B \longrightarrow C$$

progresses, the concentration of C increases. The rate can be expressed as the change in concentration of C during the time interval $\Delta t$.

$$\text{rate} = \frac{\Delta[C]}{\Delta t}$$

For a specific reaction, we need to take into account the stoichiometry of the balanced equation. For example, let's express the rate of the following reaction in terms of the concentrations of the individual reactants and products.

$$2NO(g) + O_2(g) \longrightarrow 2NO_2(g)$$

Notice from the balanced equation that the concentration of NO will decrease twice as fast as that of $O_2$. We can write the rate as

$$\text{rate} = -\frac{1}{2}\frac{\Delta[NO]}{\Delta t} \quad \text{or} \quad \text{rate} = -\frac{\Delta[O_2]}{\Delta t} \quad \text{or} \quad \text{rate} = \frac{1}{2}\frac{\Delta[NO_2]}{\Delta t}$$

Division of each concentration by the coefficient from the balanced equation makes all of the above rates equal. Notice also that a negative sign is inserted before terms involving reactants. The $\Delta$[NO] is a negative quantity because the concentration of NO *decreases* with time. Therefore, multiplying Δ[NO] by a negative sign makes the rate of reaction a positive quantity.

For a general reaction:

$$a\text{A} + b\text{B} \longrightarrow c\text{C} + d\text{D}$$

the rate of reaction can be expressed in terms of any reactant or product.

$$\text{rate} = -\frac{1}{a}\frac{\Delta[\text{A}]}{\Delta t} = -\frac{1}{b}\frac{\Delta[\text{B}]}{\Delta t} = \frac{1}{c}\frac{\Delta[\text{C}]}{\Delta t} = \frac{1}{d}\frac{\Delta[\text{D}]}{\Delta t}$$

**EXAMPLE 13.1**

**Write expressions for the rate of the following reaction in terms of each of the reactants and products.**

$$2N_2O_5(g) \longrightarrow 4NO_2(g) + O_2(g)$$

**Strategy:** The rate is defined as the change in concentration of a reactant or product with time. Each "change in concentration" term is divided by the corresponding stoichiometric coefficient. Terms involving reactants are preceded by a minus sign.

**Solution:** **Rate** $= -\frac{1}{2}\frac{\Delta[N_2O_5]}{\Delta t} = \frac{1}{4}\frac{\Delta[NO_2]}{\Delta t} = \frac{\Delta[O_2]}{\Delta t}$

**PRACTICE EXERCISE**

1. Oxygen gas can be formed by the decomposition of nitrogen monoxide (nitric oxide).

$$2NO(g) \longrightarrow O_2(g) + N_2(g)$$

If the rate of formation of $O_2$ is 0.054 *M*/s, what is the rate of change of the NO concentration?

**Text Problems:** 13.6, **13.8**

## PROBLEM TYPE 2: DETERMINING THE RATE LAW OF A REACTION

The **rate law** is an expression relating the rate of a reaction to the rate constant and the concentrations of reactants. For a general reaction of the type

$$a\text{A} + b\text{B} \longrightarrow c\text{C} + d\text{D}$$

the rate law takes the form

$$\text{rate} = k[\text{A}]^x[\text{B}]^y$$

The term *k* is the **rate constant**, a constant of proportionality between the reaction rate and the concentrations of the reactants. The sum of the powers to which all reactant concentrations appearing in the rate law are raised is called the overall **reaction order**. In the rate law expression shown above, the overall reaction order is given by $x + y$.

*k, x, and y must be determined experimentally*. One method to determine *x* and *y* is to keep the concentrations of all but one reactant fixed. Then, the rate of reaction is measured as a function of the concentration of the one reactant whose concentration is varied. Any variation in rate is due solely to the variation in this reactant's concentration.

The example below will show you how to determine the rate law and how to calculate the rate constant.

**EXAMPLE 13.2**

**The following rate data were collected for the reaction**

$$2NO + 2H_2 \longrightarrow N_2 + 2H_2O$$

**(a) Determine the rate law.**

**(b) Calculate the rate constant.**

| Experiment | [NO] ($M$) | $[H_2]$ ($M$) | $\Delta[N_2]/\Delta t$ ($M$/h) |
|---|---|---|---|
| 1 | 0.60 | 0.15 | 0.076 |
| 2 | 0.60 | 0.30 | 0.15 |
| 3 | 0.60 | 0.60 | 0.30 |
| 4 | 1.20 | 0.60 | 1.21 |

**Strategy:** We are given a set of concentrations and rate data and asked to determine the rate law and the rate constant. We assume that the rate law takes the form

$$\text{rate} = k[NO]^x[H_2]^y$$

How do we use the data to determine $x$ and $y$? Once the orders of the reactants are known, we can calculate $k$ for any set of rate and concentrations.

**Solution:**

**(a)** Experiments 1 and 2 show that when we double the concentration of $H_2$ at constant concentration of NO, the rate doubles. Taking the ratio of the rates from these two experiments:

$$\frac{\text{rate}_2}{\text{rate}_1} = \frac{0.15\ M/\text{h}}{0.076\ M/\text{h}} \approx 2 = \frac{k(0.60)^x(0.30)^y}{k(0.60)^x(0.15)^y}$$

Therefore,

$$\frac{(0.30)^y}{(0.15)^y} = 2^y = 2$$

or, $y = 1$. That is, the reaction is first-order in $H_2$. Experiments 3 and 4 indicate that doubling [NO] at constant $[H_2]$ quadruples the rate. Here we write the ratio as:

$$\frac{\text{rate}_4}{\text{rate}_3} = \frac{1.21\ M/\text{h}}{0.30\ M/\text{h}} \approx 4 = \frac{k(1.20)^x(0.60)^y}{k(0.60)^x(0.60)^y}$$

Therefore,

$$\frac{(1.20)^x}{(0.60)^x} = 2^x = 4$$

or, $x = 2$. That is, the reaction is second-order in NO. Hence, the rate law is given by:

$$\textbf{rate} = \boldsymbol{k[NO]^2[H_2]}$$

**(b)** The rate constant $k$ can be calculated using the data from any one of the experiments. Rearranging the rate law and using the first set of data, we find:

$$k = \frac{\text{rate}}{[NO]^2[H_2]}$$

$$k = \frac{0.076\ \cancel{M}/\text{h}}{(0.60\ M)^2(0.15\ \cancel{M})} = \mathbf{1.4\ M^{-2}h^{-1}}$$

**PRACTICE EXERCISE**

2. The following experimental data were obtained for the reaction

$$2A + B \longrightarrow \text{products}$$

What is the rate law for this reaction?

| Experiment | $[A]_0$ ($M$) | $[B]_0$ ($M$) | Rate ($M$/s) |
|---|---|---|---|
| 1 | 0.40 | 0.20 | $5.6 \times 10^{-3}$ |
| 2 | 0.80 | 0.20 | $5.5 \times 10^{-3}$ |
| 3 | 0.40 | 0.40 | $22.3 \times 10^{-3}$ |

**Text Problems:** 13.14, **13.16**, 13.18, **13.56 a,b**

## PROBLEM TYPE 3: FIRST-ORDER REACTIONS

One of the most widely encountered kinetic forms is the first-order rate equation. Consider the reaction,

$$A \longrightarrow \text{products}$$

For a first-order reaction, the exponent of [A] in the rate law is 1. We can write:

$$\text{rate} = k[A]$$

We also know that

$$\text{rate} = \frac{-\Delta[A]}{\Delta t}$$

Combining the two equations gives:

$$k[A] = \frac{-\Delta[A]}{\Delta t}$$

Using calculus, we can integrate both sides of the above equation to give:

$$\ln\frac{[A]_t}{[A]_0} = -kt \qquad (13.3, \text{text})$$

where ln is the natural logarithm, and $[A]_0$ and $[A]_t$ are the concentrations of A at times $t = 0$ and $t = t$, respectively.

### A. Analyzing a first-order reaction

**EXAMPLE 13.3**

**Methyl isocyanide undergoes a first-order isomerization to from methyl cyanide.**

$$CH_3NC(g) \longrightarrow CH_3CN(g)$$

**The reaction was studied at 199° C. The initial concentration of $CH_3NC$ was 0.0258 mol/L and after 11.4 min, analysis showed the concentration of the product to be $1.30 \times 10^{-3}$ mol/L.**

**(a) What is the first-order rate constant?**

**(b) How long will it take for 90.0 percent of the $CH_3NC$ to react?**

**(a)**

**Strategy:** The relationship between the concentration of a reactant at different times in a first-order reaction is given by Equations (13.3) and (13.4) of the text. From the initial concentration and the concentration at time = 11.4 min, the first-order rate constant can be determined.

**Solution:** We can calculate $[CH_3NC]_t$ by realizing that the amount of product formed equals the amount of reactant lost due to the 1:1 mole ratio between reactant and product. Thus,

$$[CH_3NC]_t = [CH_3NC]_0 - (1.30 \times 10^{-3}\ M)$$

$$= 0.0258\ M - (1.30 \times 10^{-3}\ M) = 0.0245\ M$$

Using Equation (13.3) of the text, we plug in the concentrations and time to solve for $k$.

$$\ln\frac{[CH_3NC]_t}{[CH_3NC]_0} = -kt$$

$$\ln\frac{0.0245\ M}{0.0258\ M} = -k(11.4\ \text{min})$$

$$k = -\frac{\ln(0.9496)}{11.4\ \text{min}} = \mathbf{4.54 \times 10^{-3}\ min^{-1}}$$

**(b)**

**Strategy:** The relationship between the concentration of a reactant at different times in a first-order reaction is given by Equations (13.3) and (13.4) of the text. We are asked to determine the time required for 90% of $CH_3NC$ to react. If we initially have 100% of the compound and 90% has reacted, then what is left must be (100% – 90%), or 10%. Thus, the ratio of the percentages will be equal to the ratio of the actual concentrations; that is, $[A]_t/[A]_0$ = 10%/100%, or 0.10/1.00.

**Solution:** The time required for 90% of $CH_3NC$ to react can be found using Equation (13.3) of the text.

$$\ln\frac{[CH_3NC]_t}{[CH_3NC]_0} = -kt$$

$$\ln\frac{(0.10)}{(1.00)} = -(4.54 \times 10^{-3}\ \text{min}^{-1})t$$

$$t = -\frac{\ln(0.10)}{4.54 \times 10^{-3}\ \text{min}^{-1}} = \mathbf{507\ min}$$

**PRACTICE EXERCISE**

3. The hydrolysis of sucrose ($C_{12}H_{22}O_{11}$) yields the simple sugars glucose ($C_6H_{12}O_6$) and fructose ($C_6H_{12}O_6$), which happen to be isomers.

$$C_{12}H_{22}O_{11} + H_2O \longrightarrow C_6H_{12}O_6 + C_6H_{12}O_6$$

The reaction is first-order in glucose concentration.

$$\text{rate} = k[C_{12}H_{22}O_{11}]$$

At 27°C, the rate constant is $2.1 \times 10^{-6}\ s^{-1}$. Starting with a sucrose solution with a concentration of 0.10 $M$ at 27°C, what would the concentration of sucrose be 24 hours later? (The solution is maintained at 27°C.)

**Text Problem: 13.26**

## B. Determining the half-life of a first-order reaction

The **half-life** of a reaction, $t_{\frac{1}{2}}$, is the time required for the concentration of a reactant to decrease to half of its initial concentration. From Equation (13.3) of the text, we can write

$$\ln\frac{[\mathrm{A}]_0}{[\mathrm{A}]_t} = kt$$

From the definition of half-life, when $t = t_{\frac{1}{2}}$,

$$[\mathrm{A}]_t = \frac{[\mathrm{A}]_0}{2}$$

Substituting gives:

$$\ln\frac{[\mathrm{A}]_0}{\frac{[\mathrm{A}]_0}{2}} = kt_{\frac{1}{2}}$$

$$\ln 2 = kt_{\frac{1}{2}}$$

$$t_{\frac{1}{2}} = \frac{\ln 2}{k} = \frac{0.693}{k} \qquad \text{(13.6, text)}$$

Equation (13.6) of the text shows that the half-life of a first-order reaction is *independent* of the initial concentration of the reactant. Thus, it would take the same time for the concentration of the reactant to decrease from 1.0 *M* to 0.50 *M* as it would to decrease from 0.10 *M* to 0.050 *M*.

The half-life can also be used to determine the rate constant of a first-order reaction.

**EXAMPLE 13.4**

**Methyl isocyanide undergoes a first-order isomerization to form methyl cyanide.**

$$\mathbf{CH_3NC(}g\mathbf{)} \longrightarrow \mathbf{CH_3CN(}g\mathbf{)}$$

**The reaction was studied at 199°C. The initial concentration of $CH_3NC$ was 0.0258 mol/L and after 11.4 min, analysis showed the concentration of the product to be $1.30 \times 10^{-3}$ mol/L. Using the rate constant calculated in Example 13.3, calculate the half-life of methyl isocyanide.**

---

**Strategy:** To calculate the half-life of a first-order reaction, we use Equation (13.6) of the text. The rate constant for this reaction was determined in Example 13.3 ($k = 4.54 \times 10^{-3}\ \mathrm{min}^{-1}$).

**Solution:** For a first-order reaction, we only need the rate constant to calculate the half-life. From Equation (13.6),

$$t_{\frac{1}{2}} = \frac{0.693}{k} = \frac{0.693}{4.54 \times 10^{-3}\ \mathrm{min}^{-1}} = \mathbf{153\ min}$$

**PRACTICE EXERCISE**

4. The hydrolysis of sucrose ($C_{12}H_{22}O_{11}$) yields the simple sugars glucose ($C_6H_{12}O_6$) and fructose ($C_6H_{12}O_6$), which happen to be isomers.

$$C_{12}H_{22}O_{11} + H_2O \longrightarrow C_6H_{12}O_6 + C_6H_{12}O_6$$

The reaction is first-order in sucrose concentration.

$$\text{rate} = k[C_{12}H_{22}O_{11}]$$

At 27°C, the rate constant is $2.1 \times 10^{-6}\ s^{-1}$. What is the half-life of sucrose at 27°C?

**Text Problem: 13.26**

## C. Analyzing first-order kinetics

From Equation (13.3),

$$\ln\frac{[A]_t}{[A]_0} = -kt \qquad (13.3, \text{text})$$

$$\ln[A]_t - \ln[A]_0 = -kt$$

or

$$\ln[A]_t = -kt + \ln[A]_0 \qquad (13.4, \text{text})$$

Equation (13.4) has the form of a linear equation.

$$y = mx + b$$

where,

*m* is the slope of the line

*b* is the *y*-intercept

A plot of ln[A] versus *t* (*y* vs. *x*) gives a straight line with a slope of $-k$ (*m*). Thus, we can calculate the rate constant *k* from the slope of the plot.

**EXAMPLE 13.5**

**At 500 K, butadiene gas converts to cyclobutene gas:**

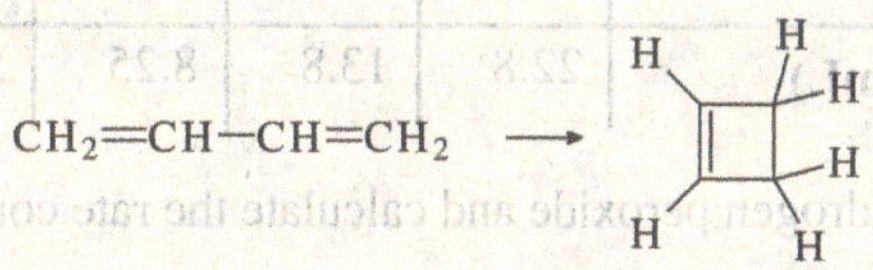

**Given the following data, is the reaction first-order in butadiene concentration?**

| Time from start (s) | Concentration of butadiene (*M*) |
|---|---|
| 195 | 0.0162 |
| 604 | 0.0147 |
| 1246 | 0.0129 |
| 2180 | 0.0110 |
| 4140 | 0.0084 |
| 8135 | 0.0057 |

**Strategy:** If the reaction is first-order in butadiene, then a plot of ln[butadiene] versus *t* will be a straight line.

**Solution:**

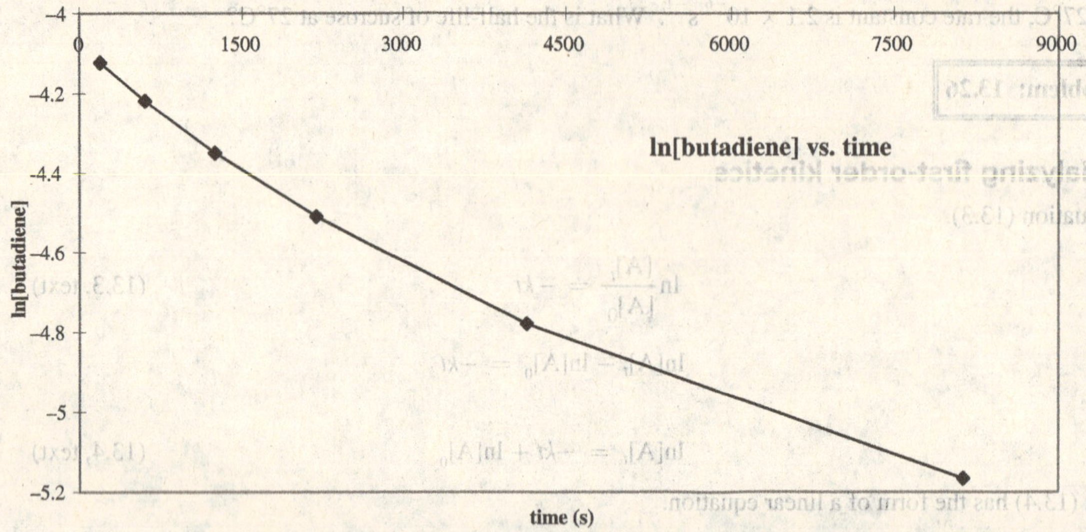


As you can see, a plot of ln[butadiene] versus $t$ does *not* give a straight line. Hence, the reaction is *not* first-order in butadiene.

**PRACTICE EXERCISE**

**5.** In a certain experiment, the rate of hydrogen peroxide decomposition,

$$2H_2O_2 \rightarrow 2H_2O + O_2$$

is followed by titration against a potassium permanganate solution. At regular intervals, an equal volume of $H_2O_2$ is withdrawn to give the following data:

| **Time (min)** | 0 | 10.0 | 20.0 | 30.0 |
|---|---|---|---|---|
| **Volume of $KMnO_4$ used (mL)** | 22.8 | 13.8 | 8.25 | 5.00 |

Confirm that the reaction is first-order in hydrogen peroxide and calculate the rate constant.

**Hint:** The concentration of hydrogen peroxide is directly proportional to the volume (in mL) of $KMnO_4$ used in each titration.

**Text Problem:** 13.20

## PROBLEM TYPE 4: SECOND-ORDER REACTIONS

Second-order reactions are also encountered quite often. Consider the reaction,

$$A \longrightarrow \text{products}$$

For a second-order reaction, the exponent of [A] in the rate law is 2. We can write:

$$\text{rate} = k[A]^2$$

We also know that

$$\text{rate} = \frac{-\Delta[A]}{\Delta t}$$

Combining the two equations gives:

$$k[A]^2 = \frac{-\Delta[A]}{\Delta t}$$

Using calculus, we can integrate both sides of the above equation to give:

$$\frac{1}{[A]_t} = kt + \frac{1}{[A]_0} \qquad (13.7, \text{text})$$

where $[A]_0$ and $[A]_t$ are the concentrations of A at times $t = 0$ and $t = t$, respectively.

## A. Analyzing a second-order reaction

**EXAMPLE 13.6**
**At 500 K, butadiene gas converts to cyclobutene gas:**

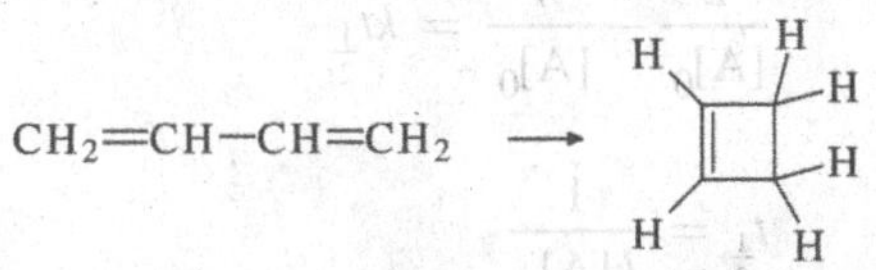

**At 500 K, the rate constant for the reaction is 0.0143/*M*·s. If the initial concentration of butadiene is 0.272 *M*, calculate the concentration of butadiene after 30.0 min.**

---

**Strategy:** The relationship between the concentration of a reactant at different times in a second-order reaction is given by Equation (13.7) of the text. From the initial concentration and the rate constant, the concentration at $t = 30$ min can be determined.

**Solution:** Using Equation (13.7) of the text, we plug in the initial concentration, *k*, and *t*, to solve for the concentration at $t = 30$ min.

$$\frac{1}{[\text{butadiene}]_t} = kt + \frac{1}{[\text{butadiene}]_0}$$

$$\frac{1}{[\text{butadiene}]_t} = (0.0143\ /M\cdot\text{s})(1.80 \times 10^3\ \text{s}) + \frac{1}{0.272\ M}$$

$$\frac{1}{[\text{butadiene}]_t} = 29.4\ M^{-1}$$

$$\mathbf{[butadiene] = 0.0340}\ \boldsymbol{M}$$

**PRACTICE EXERCISE**

6. For the reaction shown in Example 13.6 above, how long will it take for the concentration of butadiene to decrease from its initial concentration of 0.272 *M* to 0.100 *M*?

**Text Problem:** 13.28

## B. Determining the half-life of a second-order reaction

The **half-life** of a reaction, $t_{\frac{1}{2}}$, is the time required for the concentration of a reactant to decrease to half of its initial concentration. Starting with Equation (13.7) of the text,

$$\frac{1}{[A]_t} = kt + \frac{1}{[A]_0} \qquad (13.7,\ \text{text})$$

and the definition of half-life, $t = t_{\frac{1}{2}}$, we write:

$$[A]_t = \frac{[A]_0}{2}$$

Substituting into Equation (13.7) gives:

$$\frac{1}{\frac{[A]_0}{2}} = kt_{\frac{1}{2}} + \frac{1}{[A]_0}$$

$$\frac{2}{[A]_0} - \frac{1}{[A]_0} = kt_{\frac{1}{2}}$$

$$t_{\frac{1}{2}} = \frac{1}{k[A]_0} \qquad (13.8,\ \text{text})$$

Equation (13.8) of the text shows that the half-life of a second-order reaction is *dependent* on the initial concentration of the reactant, unlike the half-life of a first-order reaction.

**EXAMPLE 13.7**
**The following reaction follows second-order kinetics:**

$CH_2{=}CH{-}CH{=}CH_2 \longrightarrow$ 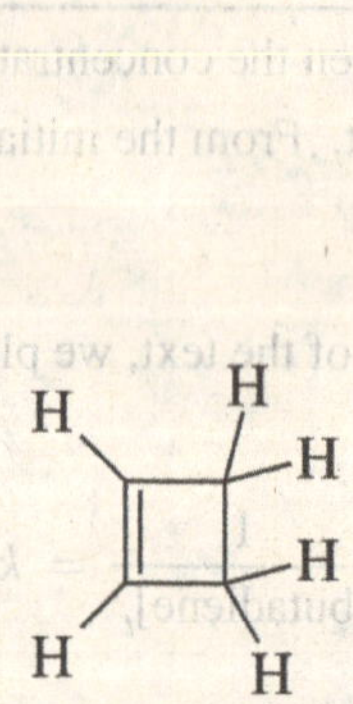

butadiene(*g*) cyclobutene(*g*)

**At 500 K, the rate constant is 0.0143/*M*·s. The initial concentration of butadiene is 0.272 *M*. What is the half-life for this reaction?**

---

**Strategy:** To calculate the half-life of a second-order reaction, we use Equation (13.8) of the text. The initial concentration and the rate constant are needed to solve for the half-life.

**Solution:** For a second-order reaction, we need the rate constant and the initial concentration to calculate the half-life. From Equation (13.8),

$$t_{\frac{1}{2}} = \frac{1}{k[A]_0} = \frac{1}{(0.0143\ /M\cdot s)(0.272\ M)} = \mathbf{257\ s}$$

**PRACTICE EXERCISE**

7. For the reaction shown in Example 13.7 above, the half-life of the reaction is determined to be 66.6 s at 500 K. What is the initial concentration of butadiene? ($k = 0.0143$ /$M$·s at 500 K)

## C. Analyzing second-order kinetics

Equation (13.7) of the text has the form of a linear equation.

$$\frac{1}{[A]_t} = kt + \frac{1}{[A]_0} \qquad (13.7, \text{text})$$

$$y = mx + b$$

where,

$m$ is the slope of the line
$b$ is the $y$-intercept

A plot of $\frac{1}{[A]}$ versus $t$ ($y$ vs. $x$) gives a straight line with a slope of $k$ ($m$). Thus, we can calculate the rate constant $k$ from the slope of the plot.

**EXAMPLE 13.8**

**At 500 K, butadiene gas converts to cyclobutene gas:**

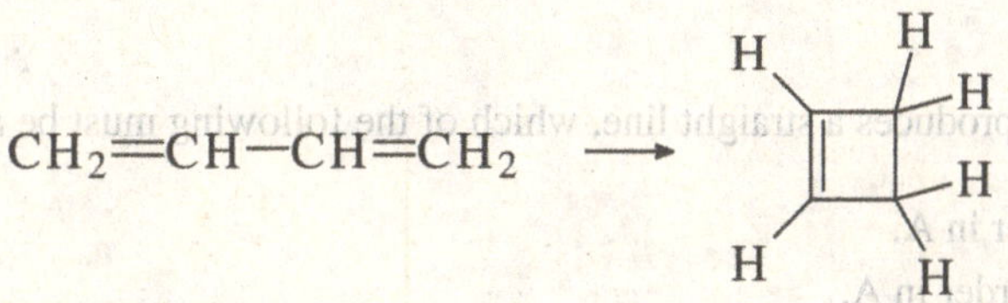

**Given the following data, is the reaction second-order in butadiene concentration?**

| Time from start (s) | Concentration of butadiene ($M$) |
|---|---|
| 195 | 0.0162 |
| 604 | 0.0147 |
| 1246 | 0.0129 |
| 2180 | 0.0110 |
| 4140 | 0.0084 |
| 8135 | 0.0057 |

**Strategy:** If the reaction is second-order in butadiene, then a plot of $\frac{1}{[\text{butadiene}]}$ versus $t$ will be a straight line.

**Solution:**

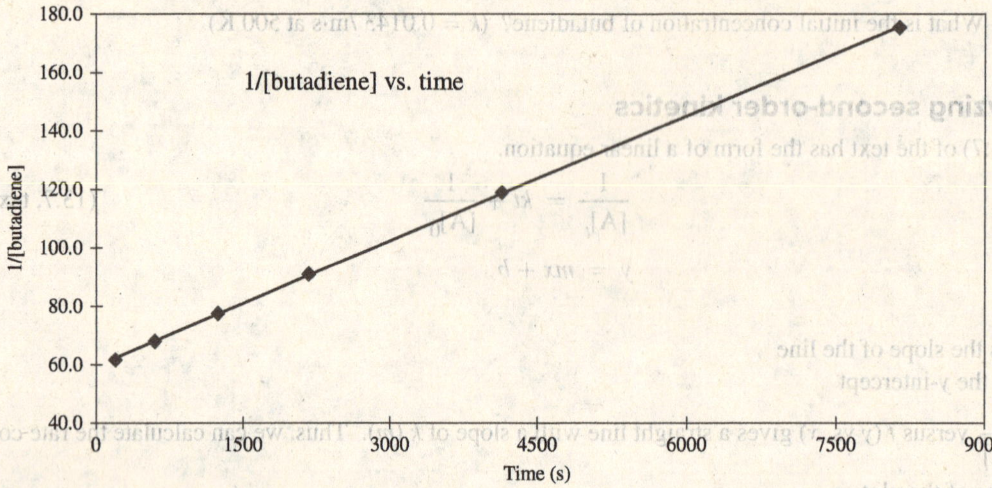


As you can see, a plot of $\frac{1}{[\text{butadiene}]}$ versus *t does* give a straight line. Hence, the reaction is second-order in butadiene.

**PRACTICE EXERCISE**

**8.** If a plot of $\frac{1}{[\text{A}]}$ versus time produces a straight line, which of the following must be *true*?

a. The reaction is first-order in A.
b. The reaction is second-order in A.
c. The reaction is first-order in two reactants.
d. The rate of the reaction does not depend on the concentration of A.
e. None of the above

**Text Problem:** 13.20

## PROBLEM TYPE 5: THE ARRHENIUS EQUATION

The dependence of the rate constant of a reaction on temperature can be expressed by the following equation, called the **Arrhenius equation.**

$$k = Ae^{-E_a/RT}$$

where, $E_a$ is the activation energy of the reaction (in kJ/mol)
$R$ is the gas constant (8.314 J/mol·K)
$T$ is the absolute temperature (in K)
$e$ is the base of the natural logarithm scale (see Appendix 4)

The quantity $A$ represents the collision frequency and is called the *frequency factor*. It can be treated as a constant for a given reacting system over a fairly wide temperature range. The Arrhenius equation shows that the rate constant is directly proportional to $A$. Therefore, as the number of collisions increase, the rate increases.

The minus sign associated with the exponent $E_a/RT$ indicates that the rate constant decreases with increasing activation energy and increases with increasing temperature.

The Arrhenius equation can be expressed in a more useful form by taking the natural logarithm of both sides.

$$\ln k = \ln\left[Ae^{-E_a/RT}\right]$$

$$\ln k = \ln A - \frac{E_a}{RT} \quad (13.12, \text{text})$$

or

$$\ln k = \left(-\frac{E_a}{R}\right)\left(\frac{1}{T}\right) + \ln A \quad (13.13, \text{text})$$

Equation (13.13) of the text has the form of the linear equation

$$y = mx + b$$

where,

*m* is the slope of the line
*b* is the *y*-intercept

A plot of ln [*k*] versus $\frac{1}{T}$ (*y* vs. *x*) gives a straight line with a slope of $\frac{-E_a}{R}$ (*m*). Thus, we can calculate the activation energy ($E_a$) from the slope of the plot.

## A. Applying the Arrhenius Equation

**EXAMPLE 13.9**
**Variation of the rate constant with temperature for the reaction**

$$NO + O_3 \rightarrow NO_2 + O_2$$

**is given in the following table. Determine graphically the activation energy for the reaction.**

| Temperature (K) | $k$ ($M^{-1}s^{-1}$) |
|---|---|
| 283 | $9.30 \times 10^6$ |
| 293 | $1.08 \times 10^7$ |
| 303 | $1.25 \times 10^7$ |
| 313 | $1.43 \times 10^7$ |
| 323 | $1.62 \times 10^7$ |

**Strategy:** A plot of ln [*k*] versus $\frac{1}{T}$ should give a straight line with a slope of $\frac{-E_a}{R}$. Thus, we can calculate the activation energy ($E_a$) from the slope of the plot.

**Solution:** From the given data we obtain:

| $1/T$ ($K^{-1}$) | ln *k* |
|---|---|
| $3.53 \times 10^{-3}$ | 16.05 |
| $3.41 \times 10^{-3}$ | 16.20 |
| $3.30 \times 10^{-3}$ | 16.34 |
| $3.19 \times 10^{-3}$ | 16.48 |
| $3.10 \times 10^{-3}$ | 16.60 |

These data, when plotted, yield the graph shown below.

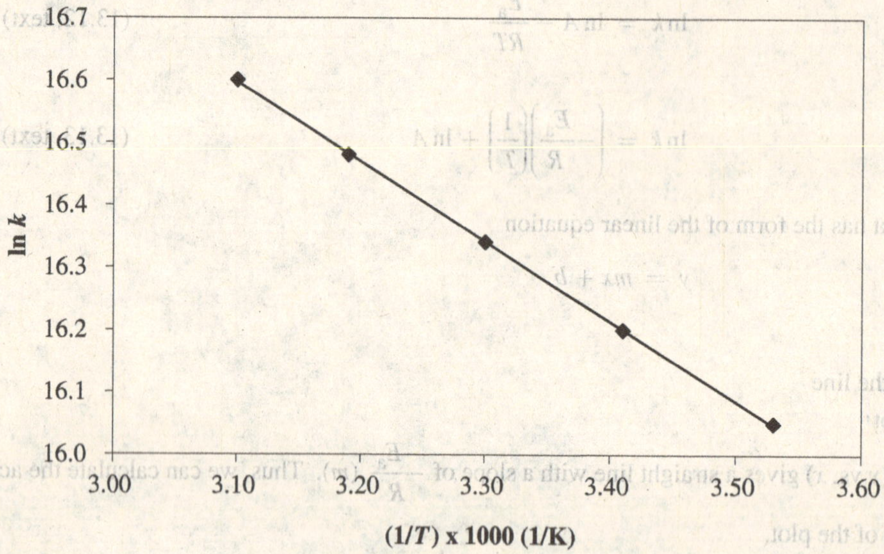


Calculating the slope from the first and last points gives:

$$\text{slope} = \frac{16.05 - 16.60}{(3.53 \times 10^{-3}) - (3.10 \times 10^{-3})} = -1.28 \times 10^3\ \text{K}$$

The slope is equal to $-E_a/R$ or

$$E_a = -\text{slope} \times R$$

$$E_a = -(-1.28 \times 10^3\ \cancel{\text{K}})(8.314\ \text{J/mol}\cdot\cancel{\text{K}}) = \mathbf{1.06 \times 10^4\ J/mol} = \mathbf{10.6\ kJ/mol}$$

## B. Applying a modified form of the Arrhenius equation that relates the rate constants at two different temperatures

Starting with Equation (13.12) of the text, we can write,

$$\ln k_1 = \ln A - \frac{E_a}{RT_1}$$

and

$$\ln k_2 = \ln A - \frac{E_a}{RT_2}$$

Subtracting $\ln k_2$ from $\ln k_1$ gives:

$$\ln k_1 - \ln k_2 = -\frac{E_a}{RT_1} + \frac{E_a}{RT_2}$$

$$\ln\frac{k_1}{k_2} = \frac{E_a}{R}\left(\frac{1}{T_2} - \frac{1}{T_1}\right)$$

or

$$\ln\frac{k_1}{k_2} = \frac{E_a}{R}\left(\frac{T_1 - T_2}{T_1T_2}\right) \quad (13.14, \text{text})$$

**Using Equation (13.14) of the text, we can calculate the activation energy if we know the rate constants at two temperatures or find the rate constant at another temperature if the activation energy is known.**

**EXAMPLE 13.10**

**For the reaction**

$$\mathbf{NO + O_3 \longrightarrow NO_2 + O_2}$$

**the following rate constants have been obtained. Calculate the activation energy for this reaction.**

| Temperature (°C) | $k$ ($M^{-1}s^{-1}$) |
|---|---|
| 10.0 | $9.3 \times 10^6$ |
| 30.0 | $1.25 \times 10^7$ |

**Strategy:** A modified form of the Arrhenius equation relates two rate constants at two different temperatures [see Equation (13.14) of the text]. The activation energy can be calculated using this equation.

**Solution:** The data are: $T_1 = 10°C = 283$ K, $k_1 = 9.3 \times 10^6\ M^{-1}s^{-1}$, $T_2 = 30°C = 303$ K, $k_2 = 1.25 \times 10^7\ M^{-1}s^{-1}$. Recall that $R = 8.314$ J/mol·K. Substituting into Equation (13.14) of the text,

$$\ln\frac{k_1}{k_2} = \frac{E_a}{R}\left(\frac{T_1 - T_2}{T_1T_2}\right) \quad (13.14, \text{text})$$

$$\ln\left(\frac{9.3 \times 10^6\ M^{-1}s^{-1}}{1.25 \times 10^7\ M^{-1}s^{-1}}\right) = \frac{E_a}{8.314\ \text{J/mol}\cdot\text{K}}\left(\frac{283\ \text{K} - 303\ \text{K}}{(283\ \text{K})(303\ \text{K})}\right)$$

$$\ln(0.744)(8.314\ \text{J/mol}\cdot\text{K}) = E_a\left(\frac{-20\ \text{K}}{8.57 \times 10^4\ \text{K}^2}\right)$$

$$(-0.296)\left(8.314\frac{\text{J}}{\text{mol}\cdot\text{K}}\right) = E_a\left(-2.33 \times 10^{-4}\frac{1}{\text{K}}\right)$$

$$\mathbf{E_a = 1.06 \times 10^4\ J/mol = 10.6\ kJ/mol}$$

Compare the answer in this example with that obtained in Example 13.9.

**PRACTICE EXERCISE**

9. At 300 K, the rate constant is $1.5 \times 10^{-5}\ M^{-1}s^{-1}$ for the following reaction:

$$2\text{NOCl} \longrightarrow 2\text{NO} + \text{Cl}_2$$

The activation energy is 90.2 kJ/mol. Calculate the value of the rate constant at 310 K.

**Text Problems: 13.38**, 13.40, 13.42

## PROBLEM TYPE 6: STUDYING REACTION MECHANISMS

For a complete discussion of reaction mechanisms, see Section 13.5 of the text. Experimental studies of reaction mechanisms begin with the collection of data (rate measurements). Next the data are analyzed to determine the rate constant and the order of reaction, so that the rate law can be written. Finally, a plausible mechanism is suggested for the reaction in terms of elementary steps. This sequence of steps is summarized in Figure 13.20 of the text.

The elementary steps of the proposed mechanism must satisfy two requirements:

- The sum of the elementary steps must give the overall balanced equation for the reaction.
- The **rate-determining step**, which is the slowest step in the sequence of steps leading to product formation, should predict the same rate law as is determined experimentally.

### EXAMPLE 13.11

**The rate law for the substitution of $NH_3$ for $H_2O$ in the following reaction is first order in $Ni(H_2O)_6^{2+}$ and zero order in $NH_3$.**

$$Ni(H_2O)_6^{2+}(aq) + NH_3(aq) \longrightarrow Ni(H_2O)_5(NH_3)^{2+}(aq) + H_2O(l)$$

**Show that the following mechanism is consistent with the experimental rate law.**

$$Ni(H_2O)_6^{2+}(aq) \longrightarrow Ni(H_2O)_5^{2+}(aq) + H_2O(l) \quad \text{(slow)}$$

$$Ni(H_2O)_5^{2+}(aq) + NH_3(aq) \longrightarrow Ni(H_2O)_5(NH_3)^{2+}(aq) \quad \text{(fast)}$$

**Strategy:** Do the elementary steps add to give the overall balanced equation? Does the rate law written from the rate-determining step match the experimentally determined rate law?

**Solution:** First, does the sum of the elementary steps give the overall balanced equation. Yes, the sum of the steps gives:

$$Ni(H_2O)_6^{2+}(aq) + NH_3(aq) \longrightarrow Ni(H_2O)_5(NH_3)^{2+}(aq) + H_2O(l)$$

Next, the reaction was experimentally determined to be first order in $Ni(H_2O)_6^{2+}$ and zero order in $NH_3$. We can write the rate law for the reaction.

$$\text{rate} = k[Ni(H_2O)_6^{2+}]$$

Does the rate law match the rate law of the proposed mechanism? We can write the rate law from the rate determining step of the proposed mechanism. Step 1 is the slow step, so we can write:

$$\text{rate} = k[Ni(H_2O)_6^{2+}]$$

which matches the experimentally determined rate law.

The elementary steps of the proposed mechanism satisfy the two requirements outlined above; therefore, it is a valid mechanism.

**PRACTICE EXERCISE**

**10.** The reaction of nitric oxide and chlorine,

$$2NO(g) + Cl_2(g) \rightarrow 2NOCl(g)$$

has been proposed to proceed by the following two-step mechanism:

$$NO(g) + Cl_2(g) \rightarrow NOCl_2(g)$$

$$NO(g) + NOCl_2(g) \rightarrow 2NOCl(g)$$

What rate law is predicted if the first step of the proposed mechanism is the rate-determining step?

**Text Problems: 13.56c**, 13.58, 13.66

## ANSWERS TO PRACTICE EXERCISES

**1.** Rate of change of NO $= -0.11$ *M*/s

**2.** Rate $= k[B]^2$

**3.** [sucrose] $= 0.083$ *M* after 24 h

**4.** $t_{\frac{1}{2}} = 3.3 \times 10^5$ s

**5.** The slope of the straight line plot equals $-k$. You should find $k = \mathbf{0.0504\ min^{-1}}$.

**6.** 442 s

**7.** 1.05 *M*

**8.** (b)

**9.** $k = 4.8 \times 10^{-5}\ M^{-1}s^{-1}$

**10.** rate $= k[NO][Cl_2]$

## SOLUTIONS TO SELECTED TEXT PROBLEMS

**13.6** (a) $$\text{rate} = -\frac{1}{2}\frac{\Delta[H_2]}{\Delta t} = -\frac{\Delta[O_2]}{\Delta t} = \frac{1}{2}\frac{\Delta[H_2O]}{\Delta t}$$

(b) $$\text{rate} = -\frac{1}{4}\frac{\Delta[NH_3]}{\Delta t} = -\frac{1}{5}\frac{\Delta[O_2]}{\Delta t} = \frac{1}{4}\frac{\Delta[NO]}{\Delta t} = \frac{1}{6}\frac{\Delta[H_2O]}{\Delta t}$$

**13.8** **Strategy:** The rate is defined as the change in concentration of a reactant or product with time. Each "change in concentration" term is divided by the corresponding stoichiometric coefficient. Terms involving reactants are preceded by a minus sign.

$$\text{rate} = -\frac{\Delta[N_2]}{\Delta t} = -\frac{1}{3}\frac{\Delta[H_2]}{\Delta t} = \frac{1}{2}\frac{\Delta[NH_3]}{\Delta t}$$

**Solution:**

(a) If hydrogen is reacting at the rate of $-0.074$ *M*/s, the rate at which ammonia is being formed is

$$\frac{1}{2}\frac{\Delta[NH_3]}{\Delta t} = -\frac{1}{3}\frac{\Delta[H_2]}{\Delta t}$$

or

$$\frac{\Delta[NH_3]}{\Delta t} = -\frac{2}{3}\frac{\Delta[H_2]}{\Delta t}$$

$$\frac{\Delta[NH_3]}{\Delta t} = -\frac{2}{3}(-0.074\ M/s) = \mathbf{0.049\ M/s}$$

**(b)** The rate at which nitrogen is reacting must be:

$$\frac{\Delta[N_2]}{\Delta t} = \frac{1}{3}\frac{\Delta[H_2]}{\Delta t} = \frac{1}{3}(-0.074\ M/s) = \mathbf{-0.025\ M/s}$$

Will the rate at which ammonia forms always be twice the rate of reaction of nitrogen, or is this true only at the instant described in this problem?

**13.14** Assume the rate law has the form:

$$\text{rate} = k[F_2]^x[ClO_2]^y$$

To determine the order of the reaction with respect to $F_2$, find two experiments in which the $[ClO_2]$ is held constant. Compare the data from experiments 1 and 3. When the concentration of $F_2$ is doubled, the reaction rate doubles. Thus, the reaction is *first-order* in $F_2$.

To determine the order with respect to $ClO_2$, compare experiments 1 and 2. When the $ClO_2$ concentration is quadrupled, the reaction rate quadruples. Thus, the reaction is *first-order* in $ClO_2$.

The rate law is:

$$\text{rate} = k[F_2][ClO_2]$$

The value of $k$ can be found using the data from any of the experiments. If we take the numbers from the second experiment we have:

$$k = \frac{\text{rate}}{[F_2][ClO_2]} = \frac{4.8\times10^{-3}\ M/s}{(0.10\ M)(0.040\ M)} = 1.2\ M^{-1}s^{-1}$$

Verify that the same value of $k$ can be obtained from the other sets of data.

Since we now know the rate law and the value of the rate constant, we can calculate the rate at any concentration of reactants.

$$\textbf{rate} = k[F_2][ClO_2] = (1.2\ M^{-1}s^{-1})(0.010\ M)(0.020\ M) = \mathbf{2.4\times10^{-4}\ M/s}$$

**13.16** **Strategy:** We are given a set of concentrations and rate data and asked to determine the order of the reaction and the initial rate for specific concentrations of X and Y. To determine the order of the reaction, we need to find the rate law for the reaction. We assume that the rate law takes the form

$$\text{rate} = k[X]^x[Y]^y$$

How do we use the data to determine $x$ and $y$? Once the orders of the reactants are known, we can calculate $k$ for any set of rate and concentrations. Finally, the rate law enables us to calculate the rate at any concentrations of X and Y.

**Solution:**

**(a)** Experiments 2 and 5 show that when we double the concentration of X at constant concentration of Y, the rate quadruples. Taking the ratio of the rates from these two experiments

$$\frac{\text{rate}_5}{\text{rate}_2} = \frac{0.509\ M/\text{s}}{0.127\ M/\text{s}} \approx 4 = \frac{k(0.40)^x(0.30)^y}{k(0.20)^x(0.30)^y}$$

Therefore,

$$\frac{(0.40)^x}{(0.20)^x} = 2^x = 4$$

or, $x = 2$. That is, the reaction is second order in X. Experiments 2 and 4 indicate that doubling [Y] at constant [X] doubles the rate. Here we write the ratio as

$$\frac{\text{rate}_4}{\text{rate}_2} = \frac{0.254\ M/\text{s}}{0.127\ M/\text{s}} = 2 = \frac{k(0.20)^x(0.60)^y}{k(0.20)^x(0.30)^y}$$

Therefore,

$$\frac{(0.60)^y}{(0.30)^y} = 2^y = 2$$

or, $y = 1$. That is, the reaction is first order in Y. Hence, the rate law is given by:

$$\text{rate} = k[\text{X}]^2[\text{Y}]$$

The order of the reaction is $(2 + 1) = 3$. The reaction is **3rd order**.

**(b)** The rate constant $k$ can be calculated using the data from any one of the experiments. Rearranging the rate law and using the first set of data, we find:

$$k = \frac{\text{rate}}{[\text{X}]^2[\text{Y}]} = \frac{0.053\ \cancel{M}/\text{s}}{(0.10\ M)^2(0.50\ \cancel{M})} = 10.6\ M^{-2}\text{s}^{-1}$$

Next, using the known rate constant and substituting the concentrations of X and Y into the rate law, we can calculate the initial rate of disappearance of X.

$$\textbf{rate} = (10.6\ M^{-2}\text{s}^{-1})(0.30\ M)^2(0.40\ M) = \mathbf{0.38\ M/s}$$

**13.18** **(a)** For a reaction first-order in A,

$$\text{Rate} = k[\text{A}]$$

$$1.6 \times 10^{-2}\ \cancel{M}/\text{s} = k(0.35\ \cancel{M})$$

$$\boldsymbol{k = 0.046\ \text{s}^{-1}}$$

**(b)** For a reaction second-order in A,

$$\text{Rate} = k[\text{A}]^2$$

$$1.6 \times 10^{-2}\ M/\text{s} = k(0.35\ M)^2$$

$$\boldsymbol{k = 0.13\ M^{-1}\cdot\text{s}^{-1}}$$

**13.20** Let $P_0$ be the pressure of $ClCO_2CCl_3$ at $t = 0$, and let $x$ be the decrease in pressure after time $t$. Note that from the coefficients in the balanced equation that the loss of 1 atmosphere of $ClCO_2CCl_3$ results in the formation of two atmospheres of $COCl_2$. We write:

$$ClCO_2CCl_3 \rightarrow 2COCl_2$$

| Time | $[ClCO_2CCl_3]$ | $[COCl_2]$ |
|---|---|---|
| $t = 0$ | $P_0$ | 0 |
| $t = t$ | $P_0 - x$ | $2x$ |

Thus the change (increase) in pressure ($\Delta P$) is $2x - x = x$. We have:

| $t$(s) | $P$ (mmHg) | $\Delta P = x$ | $P_{ClCO_2CCl_3}$ | $\ln P_{ClCO_2CCl_3}$ | $\frac{1}{P_{ClCO_2CCl_3}}$ |
|---|---|---|---|---|---|
| 0 | 15.76 | 0.00 | 15.76 | 2.757 | 0.0635 |
| 181 | 18.88 | 3.12 | 12.64 | 2.537 | 0.0791 |
| 513 | 22.79 | 7.03 | 8.73 | 2.167 | 0.115 |
| 1164 | 27.08 | 11.32 | 4.44 | 1.491 | 0.225 |

If the reaction is first order, then a plot of $\ln P_{ClCO_2CCl_3}$ vs. $t$ would be linear. If the reaction is second order, a plot of $1/P_{ClCO_2CCl_3}$ vs. $t$ would be linear. The two plots are shown below.

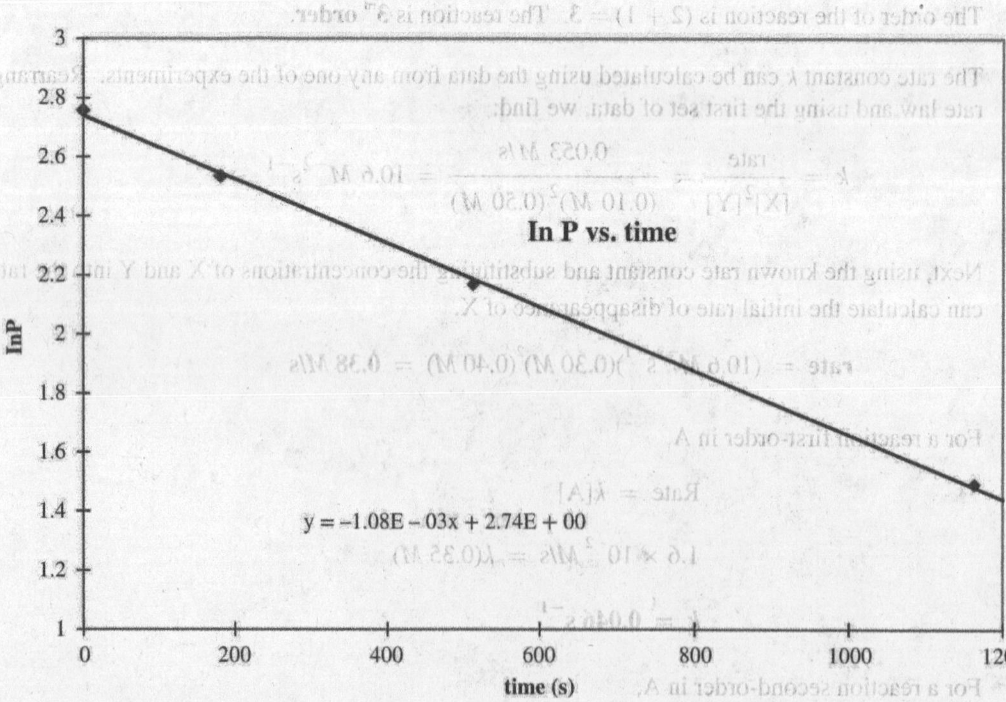

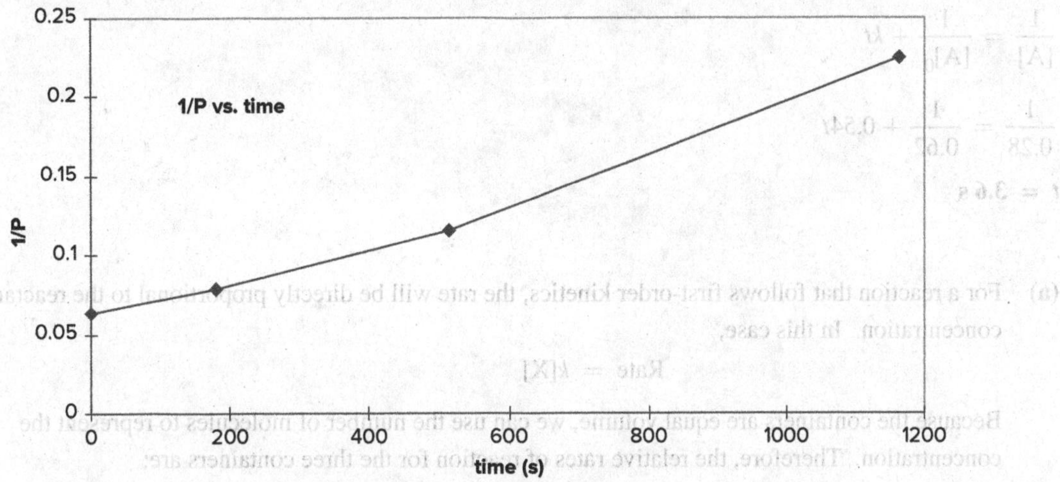


From the graphs we see that the reaction must be **first-order**. For a first-order reaction, the slope is equal to $-k$. The equation of the line is given on the graph. The rate constant is: $\boldsymbol{k = 1.08 \times 10^{-3}\ s^{-1}}$.

**13.26** **(a)**

**Strategy:** To calculate the rate constant, $k$, from the half-life of a first-order reaction, we use Equation (13.6) of the text.

**Solution:** For a first-order reaction, we only need the half-life to calculate the rate constant. From Equation (13.6)

$$k = \frac{0.693}{t_{\frac{1}{2}}}$$

$$k = \frac{0.693}{35.0\text{ s}} = \mathbf{0.0198\ s^{-1}}$$

**(b)**

**Strategy:** The relationship between the concentration of a reactant at different times in a first-order reaction is given by Equations (13.3) and (13.4) of the text. We are asked to determine the time required for 95% of the phosphine to decompose. If we initially have 100% of the compound and 95% has reacted, then what is left must be (100% − 95%), or 5%. Thus, the ratio of the percentages will be equal to the ratio of the actual concentrations; that is, $[A]_t/[A]_0 = 5\%/100\%$, or 0.05/1.00.

**Solution:** The time required for 95% of the phosphine to decompose can be found using Equation (13.3) of the text.

$$\ln\frac{[A]_t}{[A]_0} = -kt$$

$$\ln\frac{(0.05)}{(1.00)} = -(0.0198\text{ s}^{-1})t$$

$$t = -\frac{\ln(0.0500)}{0.0198\text{ s}^{-1}} = \mathbf{151\ s}$$

**13.28** $$\frac{1}{[A]} = \frac{1}{[A]_0} + kt$$

$$\frac{1}{0.28} = \frac{1}{0.62} + 0.54t$$

$t$ = **3.6 s**

**13.30** **(a)** For a reaction that follows first-order kinetics, the rate will be directly proportional to the reactant concentration. In this case,

$$\text{Rate} = k[X]$$

Because the containers are equal volume, we can use the number of molecules to represent the concentration. Therefore, the relative rates of reaction for the three containers are:

(i) Rate = $8k$

(ii) Rate = $6k$

(iii) Rate = $12k$

We can divide each rate by $2k$ to show that,

**Ratio of rates = 4 : 3 : 6**

**(b)** Doubling the volume of each container will have **no effect** on the relative rates of reaction compared to part (a). Doubling the volume would halve each of the concentrations, but the ratio of the concentrations for containers (i) – (iii) would still be 4 : 3 : 6. Therefore, the relative rates between the three containers would remain the same. The actual (absolute) rate would decrease by 50%.

**(c)** The reaction follows first-order kinetics. For a first-order reaction, the half-life is independent of the initial concentration of the reactant. Therefore, the half-lives for containers (i), (ii), and (iii), will be the **same.**

**13.38** **Strategy:** A modified form of the Arrhenius equation relates two rate constants at two different temperatures [see Equation (13.14) of the text]. Make sure the units of $R$ and $E_a$ are consistent. Since the rate of the reaction at 250°C is $1.50 \times 10^3$ times faster than the rate at 150°C, the ratio of the rate constants, $k$, is also $1.50 \times 10^3 : 1$, because rate and rate constant are directly proportional.

**Solution:** The data are: $T_1 = 250°C = 523$ K, $T_2 = 150°C = 423$ K, and $k_1/k_2 = 1.50 \times 10^3$. Substituting into Equation (13.14) of the text,

$$\ln\frac{k_1}{k_2} = \frac{E_a}{R}\left(\frac{T_1 - T_2}{T_1T_2}\right)$$

$$\ln(1.50 \times 10^3) = \frac{E_a}{8.314 \text{ J/mol}\cdot\text{K}}\left(\frac{523 \text{ K} - 423 \text{ K}}{(523 \text{ K})(423 \text{ K})}\right)$$

$$7.31 = \frac{E_a}{8.314 \frac{\text{J}}{\text{mol}\cdot\cancel{\text{K}}}}\left(4.52 \times 10^{-4} \frac{1}{\cancel{\text{K}}}\right)$$

$E_a$ = **$1.35 \times 10^5$ J/mol = 135 kJ/mol**

**13.40** Graphing Equation (13.13) of the text requires plotting ln $k$ versus $1/T$. The graph is shown below.

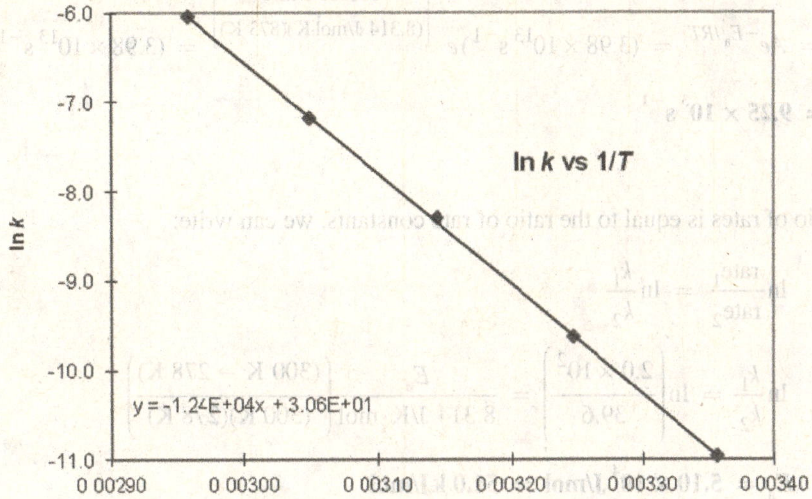


The slope of the line is $-1.24 \times 10^4$ K, which is $-E_a/R$. The activation energy is:

$$-E_a = \text{slope} \times R = (-1.24 \times 10^4\ \text{K}) \times (8.314\ \text{J/K·mol})$$

$$\mathbf{E_a = 1.03 \times 10^5\ J/mol = 103\ kJ/mol}$$

**Do you need to know the order of the reaction to find the activation energy? Is it possible to have a negative activation energy? What would a potential energy versus reaction coordinate diagram look like in such a case?**

**13.42** Use a modified form of the Arrhenius equation to calculate the temperature at which the rate constant is $8.80 \times 10^{-4}\ \text{s}^{-1}$. We carry an extra significant figure throughout this calculation to minimize rounding errors.

$$\ln\frac{k_1}{k_2} = \frac{E_a}{R}\left(\frac{1}{T_2} - \frac{1}{T_1}\right)$$

$$\ln\left(\frac{4.60 \times 10^{-4}\ \text{s}^{-1}}{8.80 \times 10^{-4}\ \text{s}^{-1}}\right) = \frac{1.04 \times 10^5\ \text{J/mol}}{8.314\ \text{J/mol}\cdot\text{K}}\left(\frac{1}{T_2} - \frac{1}{623\ \text{K}}\right)$$

$$\ln(0.5227) = (1.251 \times 10^4\ \text{K})\left(\frac{1}{T_2} - \frac{1}{623\ \text{K}}\right)$$

$$-0.6487 + 20.08 = \frac{1.251 \times 10^4\ \text{K}}{T_2}$$

$$19.43T_2 = 1.251 \times 10^4\ \text{K}$$

$$\mathbf{T_2 = 644\ K = 371°C}$$

**13.44** We use the Arrhenius equation to calculate the rate constant at a temperature of 600°C.

$$k = Ae^{-E_a/RT} = (3.98 \times 10^{13}\ \text{s}^{-1})e^{-\left[\frac{161000\ \text{J/mol}}{(8.314\ \text{J/mol·K})(873\ \text{K})}\right]} = (3.98 \times 10^{13}\ \text{s}^{-1})(2.325 \times 10^{-10})$$

$$\mathbf{k = 9.25 \times 10^3\ s^{-1}}$$

**13.46** Since the ratio of rates is equal to the ratio of rate constants, we can write:

$$\ln\frac{\text{rate}_1}{\text{rate}_2} = \ln\frac{k_1}{k_2}$$

$$\ln\frac{k_1}{k_2} = \ln\left(\frac{2.0 \times 10^2}{39.6}\right) = \frac{E_a}{8.314\ \text{J/K}\cdot\text{mol}}\left(\frac{(300\ \text{K} - 278\ \text{K})}{(300\ \text{K})(278\ \text{K})}\right)$$

$$\mathbf{E_a = 5.10 \times 10^4\ J/mol = 51.0\ kJ/mol}$$

**13.56** **(a)**

**Strategy:** We are given information as to how the concentrations of $X_2$, Y, and Z affect the rate of the reaction and are asked to determine the rate law. We assume that the rate law takes the form

$$\text{rate} = k[X_2]^x[Y]^y[Z]^z$$

How do we use the information to determine *x*, *y*, and *z*?

**Solution:** Since the reaction rate doubles when the $X_2$ concentration is doubled, the reaction is first-order in X. The reaction rate triples when the concentration of Y is tripled, so the reaction is also first-order in Y. The concentration of Z has no effect on the rate, so the reaction is zero-order in Z.

The rate law is:

$$\mathbf{rate = k[X_2][Y]}$$

**(b)** If a change in the concentration of Z has no effect on the rate, the concentration of Z is not a term in the rate law. This implies that Z does not participate in the rate-determining step of the reaction mechanism.

**(c)**

**Strategy:** The rate law, determined in part (a), shows that the slow step involves reaction of a molecule of $X_2$ with a molecule of Y. Since Z is not present in the rate law, it does not take part in the slow step and must appear in a fast step at a later time. (If the fast step involving Z happened before the rate-determining step, the rate law would involve Z in a more complex way.)

**Solution:** A mechanism that is consistent with the rate law could be:

$$X_2 + Y \longrightarrow XY + X \quad \text{(slow)}$$

$$X + Z \longrightarrow XZ \quad \text{(fast)}$$

The rate law only tells us about the slow step. Other mechanisms with different subsequent fast steps are possible. Try to invent one.

**Check:** The rate law written from the rate-determining step in the proposed mechanism matches the rate law determined in part (a). Also, the two elementary steps add to the overall balanced equation given in the problem.

**13.58** The experimentally determined rate law is first order in $H_2$ and second order in NO. In Mechanism I the slow step is bimolecular and the rate law would be:

$$\text{rate} = k[H_2][NO]$$

**Mechanism I** can be discarded.

The rate-determining step in Mechanism II involves the simultaneous collision of two NO molecules with one $H_2$ molecule. The rate law would be:

$$\text{rate} = k[H_2][NO]^2$$

Mechanism II is a possibility.

In Mechanism III we assume the forward and reverse reactions in the first fast step are in dynamic equilibrium, so their rates are equal:

$$k_f[NO]^2 = k_r[N_2O_2]$$

The slow step is bimolecular and involves collision of a hydrogen molecule with a molecule of $N_2O_2$. The rate would be:

$$\text{rate} = k_2[H_2][N_2O_2]$$

If we solve the dynamic equilibrium equation of the first step for $[N_2O_2]$ and substitute into the above equation, we have the rate law:

$$\text{rate} = \frac{k_2 k_f}{k_r}[H_2][NO]^2 = k[H_2][NO]^2$$

Mechanism III is also a possibility. Can you suggest an experiment that might help to decide between the two mechanisms?

**13.66** The rate-determining step involves the breakdown of ES to E and P. The rate law for this step is:

$$\text{rate} = k_2[ES]$$

In the first elementary step, the intermediate ES is in equilibrium with E and S. The equilibrium relationship is:

$$\frac{[ES]}{[E][S]} = \frac{k_1}{k_{-1}}$$

or

$$[ES] = \frac{k_1}{k_{-1}}[E][S]$$

Substitute [ES] into the rate law expression.

$$\textbf{rate} = k_2[ES] = \frac{k_1 k_2}{k_{-1}}[E][S]$$

**13.68** Let's count the number of molecules present at times of 0 min, 15 min, and 30 min.

0 min, 16 A atoms

15 min, 8 A atoms, 4 $A_2$ molecules

30 min, 4 A atoms, 6 $A_2$ molecules

Note that the concentration of A atoms is halved at $t = 15$ min and is halved again at $t = 30$ min. We notice that the half-life is independent of the concentration of the reactant, A, and hence the reaction is *first-order* in A. The rate constant, $k$, can now be calculated using the equation for the half-life of a first-order reaction.

$$t_{\frac{1}{2}} = \frac{0.693}{k}$$

$$k = \frac{0.693}{t_{\frac{1}{2}}} = \frac{0.693}{15\text{ min}} = \mathbf{0.046\ min^{-1}}$$

**13.70** Temperature, energy of activation, concentration of reactants, and a catalyst.

**13.72** First, calculate the radius of the 10.0 $cm^3$ sphere.

$$V = \frac{4}{3}\pi r^3$$

$$10.0\text{ cm}^3 = \frac{4}{3}\pi r^3$$

$$r = 1.34\text{ cm}$$

The surface area of the sphere is:

$$\textbf{area} = 4\pi r^2 = 4\pi(1.34\text{ cm})^2 = \mathbf{22.6\ cm^2}$$

Next, calculate the radius of the 1.25 $cm^3$ sphere.

$$V = \frac{4}{3}\pi r^3$$

$$1.25\text{ cm}^3 = \frac{4}{3}\pi r^3$$

$$r = 0.668\text{ cm}$$

The surface area of one sphere is:

$$\text{area} = 4\pi r^2 = 4\pi(0.668\text{ cm})^2 = 5.61\text{ cm}^2$$

$$\textbf{The total area of 8 spheres} = 5.61\text{ cm}^2 \times 8 = \mathbf{44.9\ cm^2}$$

Obviously, the surface area of the eight spheres (44.9 $cm^2$) is greater than that of one larger sphere (22.6 $cm^2$). A greater surface area promotes the catalyzed reaction more effectively.

It can be dangerous to work in grain elevators, because the large surface area of the grain dust can result in a violent explosion.

**13.74** The overall rate law is of the general form: rate $= k[H_2]^x[NO]^y$

**(a)** Comparing Experiment #1 and Experiment #2, we see that the concentration of NO is constant and the concentration of $H_2$ has decreased by one-half. The initial rate has also decreased by one-half. Therefore, the initial rate is directly proportional to the concentration of $H_2$; $x = 1$.

Comparing Experiment #1 and Experiment #3, we see that the concentration of $H_2$ is constant and the concentration of NO has decreased by one-half. The initial rate has decreased by one-fourth. Therefore, the initial rate is proportional to the squared concentration of NO; $y = 2$.

The overall rate law is: rate $= k[H_2][NO]^2$, and the order of the reaction is **1 + 2 = 3.**

**(b)** Using Experiment #1 to calculate the rate constant,

$$\text{rate} = k[H_2][NO]^2$$

$$k = \frac{\text{rate}}{[H_2][NO]^2}$$

$$k = \frac{2.4 \times 10^{-6}\ M/\text{s}}{(0.010\ M)(0.025\ M)^2} = \mathbf{0.38\ M^{-2}s^{-1}}$$

**(c)** Consulting the rate law, we assume that the slow step in the reaction mechanism will probably involve one $H_2$ molecule and two NO molecules. Additionally the hint tells us that O atoms are an intermediate.

$$H_2 + 2NO \rightarrow N_2 + H_2O + O \qquad \text{slow step}$$

$$O + H_2 \rightarrow H_2O \qquad \text{fast step}$$

$$2H_2 + 2NO \rightarrow N_2 + 2H_2O$$

**13.76** If water is also the solvent in this reaction, it is present in vast excess over the other reactants and products. Throughout the course of the reaction, the concentration of the water will not change by a measurable amount. As a result, the reaction rate will not appear to depend on the concentration of water.

**13.78** Since the reaction is first order in both A and B, then we can write the rate law expression:

$$\text{rate} = k[A][B]$$

Substituting in the values for the rate, [A], and [B]:

$$4.1 \times 10^{-4}\ M/\text{s} = k(1.6 \times 10^{-2})(2.4 \times 10^{-3})$$

$$\mathbf{k = 10.7/M^{-1}\cdot s^{-1}}$$

Knowing that the overall reaction was second order, could you have predicted the units for $k$?

**13.80** Recall that the pressure of a gas is directly proportional to the number of moles of gas. This comes from the ideal gas equation.

$$P = \frac{nRT}{V}$$

The balanced equation is:

$$2N_2O(g) \longrightarrow 2N_2(g) + O_2(g)$$

From the stoichiometry of the balanced equation, for every one mole of $N_2O$ that decomposes, one mole of $N_2$ and 0.5 moles of $O_2$ will be formed. Let's assume that we had 2 moles of $N_2O$ at $t = 0$. After one half-life there will be one mole of $N_2O$ remaining and one mole of $N_2$ and 0.5 moles of $O_2$ will be formed. The total number of moles of gas after one half-life will be:

$$n_T = n_{N_2O} + n_{N_2} + n_{O_2} = 1\text{ mol} + 1\text{ mol} + 0.5\text{ mol} = 2.5\text{ mol}$$

At $t = 0$, there were 2 mol of gas. Now, at $t_{\frac{1}{2}}$, there are 2.5 mol of gas. Since the pressure of a gas is directly proportional to the number of moles of gas, we can write:

$$\frac{2.10\text{ atm}}{2\text{ mol gas }(t = 0)} \times 2.5\text{ mol gas}\left(\text{at } t_{\frac{1}{2}}\right) = \textbf{2.63 atm after one half-life}$$

**13.82** The rate expression for a third order reaction is:

$$\text{rate} = -\frac{\Delta[A]}{\Delta t} = k[A]^3$$

The units for the rate law are:

$$\frac{M}{s} = kM^3$$

$$\mathbf{k = M^{-2}s^{-1}}$$

**13.84** Both compounds, A and B, decompose by first-order kinetics. Therefore, we can write a first-order rate equation for A and also one for B.

$$\ln\frac{[A]_t}{[A]_0} = -k_A t \qquad \ln\frac{[B]_t}{[B]_0} = -k_B t$$

$$\frac{[A]_t}{[A]_0} = e^{-k_A t} \qquad \frac{[B]_t}{[B]_0} = e^{-k_B t}$$

$$[A]_t = [A]_0 e^{-k_A t} \qquad [B]_t = [B]_0 e^{-k_B t}$$

We can calculate each of the rate constants, $k_A$ and $k_B$, from their respective half-lives.

$$k_A = \frac{0.693}{50.0\text{ min}} = 0.0139\text{ min}^{-1} \qquad k_B = \frac{0.693}{18.0\text{ min}} = 0.0385\text{ min}^{-1}$$

The initial concentration of A and B are equal. $[A]_0 = [B]_0$. Therefore, from the first-order rate equations, we can write:

$$\frac{[A]_t}{[B]_t} = 4 = \frac{[A]_0 e^{-k_A t}}{[B]_0 e^{-k_B t}} = \frac{e^{-k_A t}}{e^{-k_B t}} = e^{(k_B - k_A)t} = e^{(0.0385 - 0.0139)t}$$

$$4 = e^{0.0246t}$$

$$\ln 4 = 0.0246t$$

$$\mathbf{t = 56.4\ min}$$

**13.86** In comparing Reaction II to Reaction I, when the concentration of B (green spheres) is doubled from II to I, the rate increases by a factor of 4 (1 blue sphere to 4 blue spheres). The reaction is **2$^{nd}$ order** with respect to B. In comparing Reaction III to Reaction I, when the concentration of A (red spheres) is doubled from III to I, the rate increases by a factor of 2 (2 blue spheres to 4 blue spheres). The reaction is **1st order** with respect to A. The rate law for the reaction is:

$$\textbf{Rate} = k[\textbf{A}][\textbf{B}]^2$$

**13.88** **(a)** Changing the concentration of a reactant has no effect on *k*.

**(b)** If a reaction is run in a solvent other than in the gas phase, then the reaction mechanism will probably change and will thus change *k*.

**(c)** Doubling the pressure simply changes the concentration. No effect on *k*, as in (a).

**(d)** The rate constant *k* changes with temperature.

**(e)** A catalyst changes the reaction mechanism and therefore changes *k*.

**13.90** Mathematically, the amount left after ten half−lives is:

$$\left(\frac{1}{2}\right)^{10} = \mathbf{9.8 \times 10^{-4}}$$

**13.92** The net ionic equation is:

$$Zn(s) + 2H^+(aq) \longrightarrow Zn^{2+}(aq) + H_2(g)$$

**(a)** Changing from the same mass of granulated zinc to powdered zinc **increases** the rate because the surface area of the zinc (and thus its concentration) has increased.

**(b)** Decreasing the mass of zinc (in the same granulated form) will **decrease** the rate because the total surface area of zinc has decreased.

**(c)** The concentration of protons has decreased in changing from the strong acid (hydrochloric) to the weak acid (acetic); the rate will **decrease**.

**(d)** An increase in temperature will **increase** the rate constant *k*; therefore, the rate of reaction increases.

**13.94** If the reaction is 35.5% complete, the amount of A remaining is 64.5%. The ratio of $[A]_t/[A]_0$ is 64.5%/100% or 0.645/1.00. Using the first-order integrated rate law, Equation (13.3) of the text, we have

$$\ln\frac{[A]_t}{[A]_0} = -kt$$

$$\ln\frac{0.645}{1.00} = -k(4.90\text{ min})$$

$$-0.439 = -k(4.90\text{ min})$$

$$\mathbf{k = 0.0896\ min^{-1}}$$

**13.96** The first-order rate equation can be arranged to take the form of a straight line.

$$\ln[A] = -kt + \ln[A]_0$$

If a reaction obeys first-order kinetics, a plot of ln[A] vs. *t* will be a straight line with a slope of $-k$.

The slope of a plot of $\ln[N_2O_5]$ vs. $t$ is $-6.18 \times 10^{-4}\ \text{min}^{-1}$. Thus,

$$k = 6.18 \times 10^{-4}\ \text{min}^{-1}$$

The equation for the half-life of a first-order reaction is:

$$t_{\frac{1}{2}} = \frac{0.693}{k}$$

$$t_{\frac{1}{2}} = \frac{0.693}{6.18 \times 10^{-4}\ \text{min}^{-1}} = \mathbf{1.12 \times 10^3\ min}$$

**13.98** (a) In the two-step mechanism the rate-determining step is the collision of a hydrogen molecule with two iodine atoms. If visible light increases the concentration of iodine atoms, then the rate must increase. If the true rate-determining step were the collision of a hydrogen molecule with an iodine molecule (the one-step mechanism), then the visible light would have no effect (it might even slow the reaction by depleting the number of available iodine molecules).

(b) To split hydrogen molecules into atoms, one needs ultraviolet light of much higher energy.

**13.100** (a) We can write the rate law for an elementary step directly from the stoichiometry of the balanced reaction. In this rate-determining elementary step three molecules must collide simultaneously (one X and two Y's). This makes the reaction termolecular, and consequently the rate law must be third order: first order in X and second order in Y.

The rate law is:

$$\textbf{rate} = k[\mathbf{X}][\mathbf{Y}]^2$$

(b) The value of the rate constant can be found by solving algebraically for $k$.

$$k = \frac{\text{rate}}{[X][Y]^2} = \frac{3.8 \times 10^{-3}\ \cancel{M}/\text{s}}{(0.26\ \cancel{M})(0.88\ M)^2} = \mathbf{1.9 \times 10^{-2}\ M^{-2}s^{-1}}$$

Could you write the rate law if the reaction shown were the overall balanced equation and not an elementary step?

**13.102**

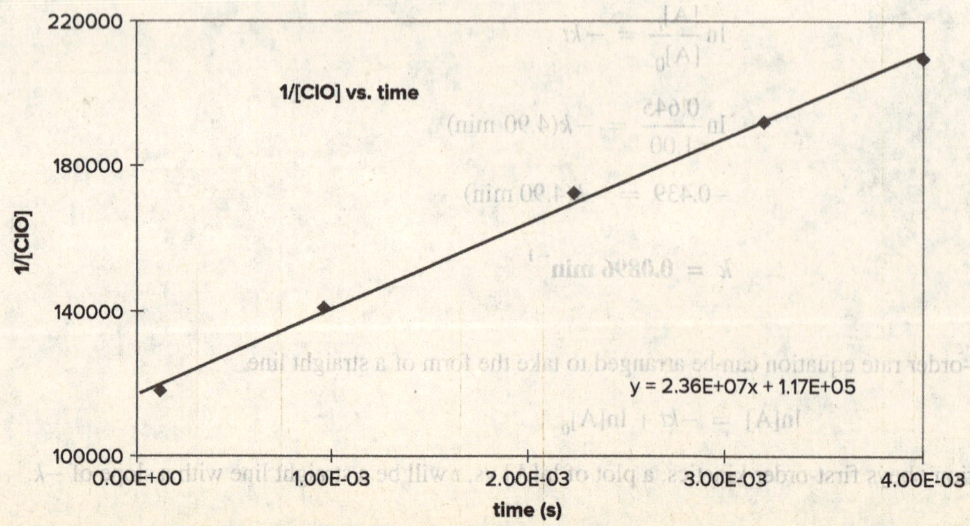

Reaction is **second-order** because a plot of 1/[ClO] vs. time is a straight line. The slope of the line equals the rate constant, $k$.

$$k = \text{Slope} = \mathbf{2.4 \times 10^7}\ \boldsymbol{M^{-1} \cdot s^{-1}}$$

**13.104** During the first five minutes or so the engine is relatively cold, so the exhaust gases will not fully react with the components of the catalytic converter. Remember, for almost all reactions, the rate of reaction increases with temperature.

**13.106** The rate law, Rate = $k[H_2][ICl]$, indicates that one molecule of $H_2$ and one molecule of ICl collide in the rate-determining step of the reaction. A possible mechanism for this reaction is:

Step 1: $H_2(g) + ICl(g) \rightarrow HCl(g) + \cancel{HI}(g)$ (slow)

Step 2: $\cancel{HI}(g) + ICl(g) \rightarrow HCl(g) + I_2(g)$ (fast)

$$H_2(g) + 2ICl(g) \rightarrow 2HCl(g) + I_2(g)$$

**13.108** First, solve for the rate constant, $k$, from the half-life of the decay.

$$t_{\frac{1}{2}} = 2.44 \times 10^5\ \text{yr} = \frac{0.693}{k}$$

$$k = \frac{0.693}{2.44 \times 10^5\ \text{yr}} = 2.84 \times 10^{-6}\ \text{yr}^{-1}$$

Now, we can calculate the time for the plutonium to decay from $5.0 \times 10^2$ g to $1.0 \times 10^2$ g using the equation for a first-order reaction relating concentration and time.

$$\ln\frac{[A]_t}{[A]_0} = -kt$$

$$\ln\frac{1.0 \times 10^2}{5.0 \times 10^2} = -(2.84 \times 10^{-6}\ \text{yr}^{-1})t$$

$$-1.61 = -(2.84 \times 10^{-6}\ \text{yr}^{-1})t$$

$$\boldsymbol{t = 5.7 \times 10^5\ \text{yr}}$$

**13.110** **(a)** Catalyst: $Mn^{2+}$; intermediate: $Mn^{3+}$

First step is rate-determining.

**(b)** Without the catalyst, the reaction would be a termolecular one involving 3 cations! ($Tl^+$ and two $Ce^{4+}$). The reaction would be slow.

**(c)** The catalyst is a homogeneous catalyst because it has the same phase (aqueous) as the reactants.

**13.112** Initially, the number of moles of gas in terms of the volume is:

$$n = \frac{PV}{RT} = \frac{(0.350\ \cancel{\text{atm}})V}{\left(0.0821\frac{\text{L}\cdot\cancel{\text{atm}}}{\text{mol}\cdot\cancel{\text{K}}}\right)(450 + 273)\cancel{\text{K}}} = 5.90 \times 10^{-3}\ V$$

We can calculate the concentration of dimethyl ether from the following equation.

$$\ln\frac{[(CH_3)_2O]_t}{[(CH_3)_2O]_0} = -kt$$

$$\frac{[(CH_3)_2O]_t}{[(CH_3)_2O]_0} = e^{-kt}$$

**Since, the volume is held constant, it will cancel out of the equation. The concentration of dimethyl ether after 8.0 minutes (480 s) is:**

$$[(CH_3)_2O]_t = \left(\frac{5.90 \times 10^{-3}\,\cancel{V}}{\cancel{V}}\right) e^{-\left(3.2 \times 10^{-4}\frac{1}{\cancel{s}}\right)(480\,\cancel{s})}$$

$$[(CH_3)_2O]_t = 5.06 \times 10^{-3}\,M$$

**After 8.0 min, the concentration of $(CH_3)_2O$ has decreased by $(5.90 \times 10^{-3} - 5.06 \times 10^{-3})M$ or $8.4 \times 10^{-4}\,M$. Since three moles of product form for each mole of dimethyl ether that reacts, the concentrations of the products are $(3)(8.4 \times 10^{-4}\,M) = 2.5 \times 10^{-3}\,M$.**

**The pressure of the system after 8.0 minutes is:**

$$P = \frac{nRT}{V} = \left(\frac{n}{V}\right)RT = MRT$$

$$P = [(5.06 \times 10^{-3}) + (2.5 \times 10^{-3})]\cancel{M} \times (0.0821\,\cancel{L}\cdot\text{atm}/\cancel{\text{mol}}\cdot\cancel{K})(723\,\cancel{K})$$

$$\boldsymbol{P = 0.45\ \text{atm}}$$

**13.114 (a)** $\dfrac{\Delta[B]}{\Delta t} = k_1[A] - k_2[B]$

**(b)** If, $\dfrac{\Delta[B]}{\Delta t} = 0$

Then, from part (a) of this problem:

$$k_1[A] = k_2[B]$$

$$[B] = \frac{k_1}{k_2}[A]$$

**13.116 (a)** The first-order rate constant can be determined from the half-life.

$$t_{\frac{1}{2}} = \frac{0.693}{k}$$

$$k = \frac{0.693}{t_{\frac{1}{2}}} = \frac{0.693}{28.1\ \text{yr}} = \mathbf{0.0247\ yr^{-1}}$$

**(b)** See Problem 13.90. Mathematically, the amount left after ten half–lives is:

$$\left(\frac{1}{2}\right)^{10} = \mathbf{9.8 \times 10^{-4}}$$

**(c)** If 99.0% has disappeared, then 1.0% remains. The ratio of $[A]_t/[A]_0$ is 1.0%/100% or 0.010/1.00. Substitute into the first-order integrated rate law, Equation (13.3) of the text, to determine the time.

$$\ln\frac{[A]_t}{[A]_0} = -kt$$

$$\ln\frac{0.010}{1.0} = -(0.0247\ \text{yr}^{-1})t$$

$$-4.6 = -(0.0247\ \text{yr}^{-1})t$$

$$t = \mathbf{186\ yr}$$

**13.118** **(a)** There are three elementary steps: A → B, B → C, and C → D.

**(b)** There are two intermediates: B and C.

**(c)** The third step, C → D, is rate determining because it has the largest activation energy.

**(d)** The overall reaction is exothermic.

**13.120** Let $k_{cat} = k_{uncat}$

Then,

$$Ae^{\frac{-E_a(\text{cat})}{RT_1}} = Ae^{\frac{-E_a(\text{uncat})}{RT_2}}$$

Since the frequency factor is the same, we can write:

$$e^{\frac{-E_a(\text{cat})}{RT_1}} = e^{\frac{-E_a(\text{uncat})}{RT_2}}$$

Taking the natural log (*ln*) of both sides of the equation gives:

$$\frac{-E_a(\text{cat})}{RT_1} = \frac{-E_a(\text{uncat})}{RT_2}$$

or,

$$\frac{E_a(\text{cat})}{T_1} = \frac{E_a(\text{uncat})}{T_2}$$

Substituting in the given values:

$$\frac{7.0\ \text{kJ/mol}}{293\ \text{K}} = \frac{42\ \text{kJ/mol}}{T_2}$$

$$T_2 = \mathbf{1.8 \times 10^3\ K}$$

This temperature is much too high to be practical.

**13.122** **(a)** The rate law for the reaction is:

$$\text{rate} = k[\text{Hb}][\text{O}_2]$$

We are given the rate constant and the concentration of Hb and $O_2$, so we can substitute in these quantities to solve for rate.

$$\text{rate} = (2.1 \times 10^{6}\ M{\cdot}\text{s})(8.0 \times 10^{-6}\ M)(1.5 \times 10^{-6}\ M)$$

$$\textbf{rate} = \mathbf{2.5 \times 10^{-5}\ \mathit{M}/s}$$

**(b)** If $HbO_2$ is being formed at the rate of $2.5 \times 10^{-5}$ $M$/s, then $O_2$ is being consumed at the same rate, **$2.5 \times 10^{-5}$ $M$/s**. Note the 1:1 mole ratio between $O_2$ and $HbO_2$.

**(c)** The rate of formation of $HbO_2$ increases, but the concentration of Hb remains the same. Assuming that temperature is constant, we can use the same rate constant as in part (a). We substitute rate, [Hb], and the rate constant into the rate law to solve for $O_2$ concentration.

$$\text{rate} = k[\text{Hb}][\text{O}_2]$$

$$1.4 \times 10^{-4}\ M/\text{s} = (2.1 \times 10^{6}\ M^{-1}{\cdot}\text{s}^{-1})(8.0 \times 10^{-6}\ M)[\text{O}_2]$$

$$\mathbf{[O_2] = 8.3 \times 10^{-6}\ \mathit{M}}$$

**13.124** $t_{\frac{1}{2}} \propto \dfrac{1}{[\text{A}]_0^{n-1}}$

$t_{\frac{1}{2}} = C\dfrac{1}{[\text{A}]_0^{n-1}}$, where $C$ is a proportionality constant.

Substituting in for zero, first, and second-order reactions gives:

$n = 0 \qquad t_{\frac{1}{2}} = C\dfrac{1}{[\text{A}]_0^{-1}} = C[\text{A}]_0$

$n = 1 \qquad t_{\frac{1}{2}} = C\dfrac{1}{[\text{A}]_0^{0}} = C$

$n = 2 \qquad t_{\frac{1}{2}} = C\dfrac{1}{[\text{A}]_0}$

Compare these results with those in Table 13.3 of the text. What is $C$ in each case?

**13.126 (a)** The units of the rate constant show the reaction to be second-order, meaning the rate law is most likely:

$$\text{Rate} = k[\text{H}_2][\text{I}_2]$$

We can use the ideal gas equation to solve for the concentrations of $H_2$ and $I_2$. We can then solve for the initial rate in terms of $H_2$ and $I_2$ and then convert to the initial rate of formation of HI. We carry an extra significant figure throughout this calculation to minimize rounding errors.

$$n = \frac{PV}{RT}$$

$$\frac{n}{V} = M = \frac{P}{RT}$$

Since the total pressure is 1658 mmHg and there are equimolar amounts of $H_2$ and $I_2$ in the vessel, the partial pressure of each gas is 829 mmHg.

$$[H_2] = [I_2] = \frac{\left(829\ \text{mmHg} \times \dfrac{1\ \text{atm}}{760\ \text{mmHg}}\right)}{\left(0.0821\dfrac{\text{L}\cdot\text{atm}}{\text{mol}\cdot\text{K}}\right)(400 + 273)\text{K}} = 0.01974\ M$$

Let's convert the units of the rate constant to $M^{-1}\cdot\text{min}^{-1}$, and then we can substitute into the rate law to solve for rate.

$$k = 2.42 \times 10^{-2}\frac{1}{M\cdot\text{s}} \times \frac{60\ \text{s}}{1\ \text{min}} = 1.452\frac{1}{M\cdot\text{min}}$$

$$\text{Rate} = k[H_2][I_2]$$

$$\text{Rate} = \left(1.452\frac{1}{M\cdot\text{min}}\right)(0.01974\ M)(0.01974\ M) = 5.658 \times 10^{-4}\ M/\text{min}$$

We know that,

$$\text{Rate} = \frac{1}{2}\frac{\Delta[\text{HI}]}{\Delta t}$$

or

$$\frac{\Delta[\text{HI}]}{\Delta t} = 2 \times \text{Rate} = (2)(5.658 \times 10^{-4}\ M/\text{min}) = \mathbf{1.13 \times 10^{-3}}\ \boldsymbol{M/\text{min}}$$

**(b)** We can use the second-order integrated rate law to calculate the concentration of $H_2$ after 10.0 minutes. We can then substitute this concentration back into the rate law to solve for rate.

$$\frac{1}{[H_2]_t} = kt + \frac{1}{[H_2]_0}$$

$$\frac{1}{[H_2]_t} = \left(1.452\frac{1}{M\cdot\text{min}}\right)(10.0\ \text{min}) + \frac{1}{0.01974\ M}$$

$$[H_2]_t = 0.01534\ M$$

We can now substitute this concentration back into the rate law to solve for rate. The concentration of $I_2$ after 10.0 minutes will also equal 0.01534 $M$.

$$\text{Rate} = k[H_2][I_2]$$

$$\text{Rate} = \left(1.452\frac{1}{M\cdot\text{min}}\right)(0.01534\ M)(0.01534\ M) = 3.417 \times 10^{-4}\ M/\text{min}$$

We know that,

$$\text{Rate} = \frac{1}{2}\frac{\Delta[\text{HI}]}{\Delta t}$$

or

$$\frac{\Delta[\text{HI}]}{\Delta t} = 2 \times \text{Rate} = (2)(3.417 \times 10^{-4}\ M/\text{min}) = \mathbf{6.83 \times 10^{-4}}\ \boldsymbol{M/\text{min}}$$

The concentration of HI after 10.0 minutes is:

$$[\text{HI}]_t = ([H_2]_0 - [H_2]_t) \times 2$$

$$\mathbf{[HI]_t} = (0.01974\ M - 0.01534\ M) \times 2 = \mathbf{8.8 \times 10^{-3}}\ \boldsymbol{M}$$

**13.128** The half-life is related to the initial concentration of A by

$$t_{\frac{1}{2}} \propto \frac{1}{[A]_0^{n-1}}$$

According to the data given, the half-life doubled when $[A]_0$ was halved. This is only possible if the half-life is inversely proportional to $[A]_0$. Substituting $n = 2$ into the above equation gives:

$$t_{\frac{1}{2}} \propto \frac{1}{[A]_0}$$

Looking at this equation, it is clear that if $[A]_0$ is halved, the half-life would double. The reaction is **second-order.**

We use Equation (13.8) of the text to calculate the rate constant.

$$t_{\frac{1}{2}} = \frac{1}{k[A]_0}$$

$$k = \frac{1}{[A]_0 t_{\frac{1}{2}}} = \frac{1}{(1.20\ M)(2.0\ \text{min})} = \mathbf{0.42/M^{-1}min^{-1}}$$

**13.130** From Equation (13.14) of the text,

$$\ln\frac{k_1}{k_2} = \frac{E_a}{R}\left(\frac{1}{T_2} - \frac{1}{T_1}\right)$$

$$\ln\left(\frac{k_1}{k_2}\right) = \frac{2.4 \times 10^5\ \text{J/mol}}{8.314\ \text{J/mol}\cdot\text{K}}\left(\frac{1}{606\ \text{K}} - \frac{1}{600\ \text{K}}\right)$$

$$\ln\left(\frac{k_1}{k_2}\right) = -0.48$$

$$\frac{k_2}{k_1} = e^{0.48} = 1.6$$

The rate constant at 606 K is 1.6 times greater than that at 600 K. This is a **60%** increase in the rate constant for a 1% increase in temperature! The result shows the profound effect of an exponential dependence. In general, the larger the $E_a$, the greater the influence of $T$ on $k$.

**13.132** The rate law can be written directly from an elementary reaction.

$$\text{Rate} = k[CH_3][C_2H_6]$$

The rate constant, $k$, is given. If the concentrations of $CH_3$ and $C_2H_6$ can be determined, the initial rate of the reaction can be calculated. The partial pressures of $CH_3$ and $C_2H_6$ can be calculated from the respective mole fractions and the total pressure. Once the partial pressures are known, the molar concentrations can be calculated using the ideal gas equation.

$$P_{CH_3} = X_{CH_3}P_T = (0.00093)(5.42\ \text{atm}) = 0.0050\ \text{atm}$$

$$P_{C_2H_6} = X_{C_2H_6}P_T = (0.00077)(5.42\ \text{atm}) = 0.0042\ \text{atm}$$

The ideal gas equation can be rearranged to solve for molar concentration.

$$\frac{n}{V} = \frac{P}{RT}$$

$$M_{CH_3} = \frac{P_{CH_3}}{RT} = \frac{(0.0050 \text{ atm})}{(0.0821 \text{ L}\cdot\text{atm/mol}\cdot\text{K})(600 \text{ K})} = 1.0 \times 10^{-4}\ M$$

$$M_{C_2H_6} = \frac{P_{C_2H_6}}{RT} = \frac{(0.0042 \text{ atm})}{(0.0821 \text{ L}\cdot\text{atm/mol}\cdot\text{K})(600 \text{ K})} = 8.5 \times 10^{-5}\ M$$

Substitute the concentrations and the rate constant into the rate law to solve for the initial rate of the reaction.

$$\text{Rate} = k[CH_3][C_2H_6]$$

$$\text{Rate} = (3.0 \times 10^4\ M^{-1}\text{s}^{-1})(1.0 \times 10^{-4}\ M)(8.5 \times 10^{-5}\ M)$$

$$\textbf{Rate} = \mathbf{2.6 \times 10^{-4}}\ \boldsymbol{M/\text{s}}$$

**13.134** See Figure 13.17(a) of the text. This diagram represents an exothermic reaction in the forward direction. For the reaction given in the problem, $E_a = 240$ kJ/mol and $\Delta H = -164$ kJ/mol for the reaction in the forward direction. The $\Delta H$ value on this diagram would be represented by the difference in energy between the reactants (A + B) and the products (C + D). The activation energy for the reverse reaction would be the energy needed to go from the products (C + D) to the activated complex. This energy difference includes $\Delta H$ for the reverse reaction (+164 kJ/mol) and the activation energy for the forward reaction.

$$E_a\textbf{(reverse)} = (+164 \text{ kJ/mol}) + (240 \text{ kJ/mol}) = \textbf{404 kJ/mol}$$

**13.136** **(a)** Since this is an elementary reaction, the rate law is:

$$\text{Rate} = k[NO]^2[O_2]$$

**(b)** Since $[O_2]$ is very large compared to [NO], then the reaction is a pseudo second-order reaction and the rate law can be simplified to:

$$\text{Rate} = k_{obs}[NO]^2$$

where $k_{obs} = k[O_2]$

**(c)** Since for a second-order reaction

$$t_{\frac{1}{2}} = \frac{1}{k[A]_0}$$

then,

$$\frac{\left(t_{\frac{1}{2}}\right)_1}{\left(t_{\frac{1}{2}}\right)_2} = \frac{\dfrac{1}{k[(A_0)_1]}}{\dfrac{1}{k[(A_0)_2]}} = \frac{[(A_0)_2]}{[(A_0)_1]}$$

$$\frac{\mathbf{6.4 \times 10^3\ min}}{\left(t_{\frac{1}{2}}\right)_2} = \frac{\mathbf{10\ ppm}}{\mathbf{2\ ppm}}$$

**Solving, the for new half-life gives:**

$$\left(t_{\frac{1}{2}}\right)_2 = \mathbf{1.3 \times 10^3\ min}$$

**You could also solve for $k$ using the half-life and concentration (2 ppm). Then substitute $k$ and the new concentration (10 ppm) into the half-life equation to solve for the new half-life. Try it.**

# CHAPTER 14
# CHEMICAL EQUILIBRIUM

## PROBLEM-SOLVING STRATEGIES AND TUTORIAL SOLUTIONS

### TYPES OF PROBLEMS

**Problem Type 1:** Homogeneous Equilibria.

**(a)** Writing expressions for $K_c$ and $K_P$.

**(b)** Calculating equilibrium partial pressures.

**(c)** Converting between $K_c$ and $K_P$.

**Problem Type 2:** Heterogeneous Equilibria, Calculating $K_P$ and $K_c$.

**Problem Type 3:** The Form of $K$ and the Equilibrium Equation.

**Problem Type 4:** Using the Reaction Quotient ($Q_c$) to Predict the Direction of a Reaction.

**Problem Type 5:** Calculating Equilibrium Concentrations.

**Problem Type 6:** Factors that Affect Chemical Equilibrium.

**(a)** Changes in concentration.

**(b)** Changes in pressure and volume.

**(c)** Changes in temperature.

**(d)** The effect of a catalyst.

## PROBLEM TYPE 1: HOMOGENEOUS EQUILIBRIA

Let's start by considering the following reversible reaction.

$$a\text{A} + b\text{B} \rightleftharpoons c\text{C} + d\text{D}$$

where *a, b, c,* and *d* are the stoichiometric coefficients for the reacting species A, B, C, and D. The equilibrium constant for the reaction at a particular temperature is:

$$K = \frac{[\text{C}]^c[\text{D}]^d}{[\text{A}]^a[\text{B}]^b}$$

This equation is the mathematical form of the **law of mass action**. It relates the concentrations of reactants and products at equilibrium in terms of a quantity called the **equilibrium constant (*K*)**.

The magnitude of the equilibrium constant, *K*, is important. If the equilibrium constant is much greater than one ($K >> 1$), we say that the equilibrium lies to the right and favors the products. Conversely, if $K << 1$, the equilibrium lies to the left and favors the reactants.

## A. Writing expressions for $K_c$ and $K_P$

The term **homogeneous equilibrium** applies to reactions in which all reacting species are in the same phase. An example of a homogeneous gas-phase equilibrium is the reaction between sulfur dioxide and oxygen to form sulfur trioxide.

$$2SO_2(g) + O_2(g) \rightleftharpoons 2SO_3(g)$$

The equilibrium constant as given in Equation (14.2) of the text is:

$$K_c = \frac{[SO_3]^2}{[SO_2]^2[O_2]}$$

The subscript c of $K_c$ denotes that the concentrations of all species are expressed in mol/L.

The concentrations of reactants and products in gas-phase reactions can also be expressed in terms of their partial pressures. Starting with the ideal gas equation, we can write:

$$P = \left(\frac{n}{V}\right)RT$$

At constant temperature, the pressure ($P$) of a gas is directly related to the concentration of the gas in mol/L, $\left(\frac{n}{V}\right)$.

Thus, for the equilibrium shown above, we can write:

$$K_P = \frac{P_{SO_3}^2}{P_{SO_2}^2 P_{O_2}}$$

where $P_{SO_3}$, $P_{SO_2}$, and $P_{O_2}$ are the equilibrium partial pressures (in atmospheres) of $SO_3$, $SO_2$, and $O_2$, respectively.

The subscript p of $K_P$ indicates the equilibrium concentrations are expressed in terms of pressure.

**EXAMPLE 14.1**

**Write the equilibrium constant expression for the following reversible reaction:**

$$4NH_3(g) + 5O_2(g) \rightleftharpoons 4NO(g) + 6H_2O(g)$$

---

**Strategy:** Remember that the concentration of each component in the equilibrium constant expression is raised to a power equal to its coefficient in the balanced equation.

**Solution:**

$$K_c = \frac{[NO]^4[H_2O]^6}{[NH_3]^4[O_2]^5}$$

You could also write the equilibrium constant expression in terms of the partial pressures of the gaseous components.

$$K_P = \frac{P_{NO}^4 P_{H_2O}^6}{P_{NH_3}^4 P_{O_2}^5}$$

**PRACTICE EXERCISE**

1. Write the equilibrium constant expression for the following reaction.

$$2N_2O(g) + 3O_2(g) \rightleftharpoons 2N_2O_4(g)$$

| Text Problem: See Review Question 14.8 |
|---|

## B. Calculating equilibrium partial pressures

**EXAMPLE 14.2**

**At 400°C, $K_P$ = 64 for the following reaction:**

$$H_2(g) + I_2(g) \rightleftharpoons 2HI(g)$$

**At equilibrium, the partial pressures of $H_2$ and $I_2$ in a closed container are 0.20 atm and 0.50 atm respectively. What is the partial pressure of HI in the mixture?**

**Strategy:** Write $K_P$ in terms of the partial pressures of the reacting species, and then substitute the known values into the equation to solve for the unknown value.

**Solution:**

$$K_P = \frac{P_{HI}^2}{P_{H_2}P_{I_2}}$$

Solve the above equation algebraically for the unknown partial pressure. Then, substitute the known values into the equation to solve for the unknown.

$$P_{HI}^2 = K_P P_{H_2} P_{I_2}$$

$$P_{HI} = \sqrt{K_P P_{H_2} P_{I_2}}$$

$$P_{HI} = \sqrt{(64)(0.20)(0.50)} = \mathbf{2.5\ atm}$$

**PRACTICE EXERCISE**

2. Consider the following reaction at a temperature of 250°C.

$$PCl_5(g) \rightleftharpoons PCl_3(g) + Cl_2(g)$$

At equilibrium, the partial pressures of $PCl_5$, $PCl_3$, and $Cl_2$ are 0.0704 atm, 0.340 atm, and 0.218 atm, respectively. What is the value of $K_P$?

## C. Converting $K_c$ to $K_P$

In general, $K_c$ is not equal to $K_P$, since the partial pressures of reactants and products are not equal to their concentrations expressed in mol/L. A simple relationship between $K_P$ and $K_c$ can be derived. See Section 14.2 of the text for the derivation. The relationship is:

$$K_P = K_c(RT)^{\Delta n}$$

where,

$$\Delta n = \text{(mol of gaseous product in balanced equation)} - \text{(mol of gaseous reactants in balanced eq.)}$$

If we express pressure in atmospheres, the gas constant $R$ is given by 0.0821 L·atm/mol·K. We can write the relationship between $K_P$ and $K_c$ as

$$K_P = K_c(0.0821\,T)^{\Delta n} \qquad \text{(14.5, text)}$$

In general, $K_P \neq K_c$ except in the special case when $\Delta n = 0$. In this case, Equation (14.5) of the text can be written as:

$$K_P = K_c(0.0821\,T)^0$$

and

$$K_P = K_c$$

**EXAMPLE 14.3**

**In the decomposition of carbon dioxide at 2000°C,**

$$2CO_2(g) \rightleftharpoons 2CO(g) + O_2(g)$$

**$K_P = 1.2 \times 10^{-4}$. Calculate $K_c$ for this reaction.**

**Strategy:** The relationship between $K_c$ and $K_P$ is given by Equation (14.5) of the text. What is the change in the number of moles of gases from reactant to product? Recall that

$$\Delta n = \text{moles of gaseous products} - \text{moles of gaseous reactants}$$

What unit of temperature should we use?

**Solution:** The relationship between $K_c$ and $K_P$ is given by Equation (14.5) of the text.

$$K_P = K_c(0.0821\,T)^{\Delta n}$$

Rearrange the equation relating $K_P$ and $K_c$, solving for $K_c$.

$$K_c = \frac{K_P}{(0.0821\,T)^{\Delta n}}$$

Substitute the given values into the above equation to solve for $K_c$. Temperature must be in units of Kelvin. **Note:** In the balanced equation, there are a total of three moles of gaseous products and two moles of gaseous reactant, so $\Delta n = 3 - 2$.

$$K_c = \frac{K_P}{(0.0821\,T)^{\Delta n}} = \frac{1.2 \times 10^{-4}}{[(0.0821)(2273\text{K})]^{(3-2)}} = \mathbf{6.4 \times 10^{-7}}$$

**PRACTICE EXERCISE**

3. At 400°C, $K_P = 64$ for the following reaction:

$$H_2(g) + I_2(g) \rightleftharpoons 2HI(g)$$

What is the value of $K_c$ for this reaction?

**Text Problems: 14.18**, 14.22

## PROBLEM TYPE 2: HETEROGENEOUS EQUILIBRIA, CALCULATING $K_P$ AND $K_c$

Whenever a reaction involves reactants and products that exist in different phases, it is called a *heterogeneous reaction.* If the reaction is carried out in a closed container, a **heterogeneous equilibrium** will result. For example, when steam is brought into contact with charcoal in a closed container, the following equilibrium is established:

$$C(s) + H_2O(g) \rightleftharpoons H_2(g) + CO(g)$$

At equilibrium, we might write the equilibrium constant as:

$$K_c' = \frac{[H_2][CO]}{[C][H_2O]}$$

However, the "concentration" of a solid, like its density, is an intensive property and thus does not depend on the amount of substance present. For this reason, the term [C] is a constant and can be combined with the equilibrium constant. We can simplify the equation by writing:

$$[C]K_c' = K_c = \frac{[H_2][CO]}{[H_2O]}$$

Keep in mind that the value of $K_c$ does not depend on how much carbon is present, as long as some amount is present at equilibrium.

The argument presented above for solids also applies to pure liquids. Thus, if a liquid is a reactant or product, we can treat its concentration as a constant and omit it from the equilibrium constant expression.

**EXAMPLE 14.4**

**What are the values of $K_P$ and $K_c$ at 1000°C for the reaction**

$$CaCO_3(s) \rightleftharpoons CaO(s) + CO_2(g)$$

**if the pressure of $CO_2$ in equilibrium with $CaCO_3$ and CaO is 3.87 atm?**

**Strategy:** Because they are constant quantities, the concentrations of solids and liquids do not appear in the equilibrium constant expressions for heterogeneous systems. The only species that enters into the equilibrium constant expression is $CO_2$. After solving for $K_P$, we can calculate $K_c$ from $K_P$.

**Solution:** As stated above, if a solid or a liquid is a reactant or product, we can treat its concentration as a constant and omit it from the equilibrium constant expression. Thus, enough information is given to calculate $K_P$ for this heterogeneous equilibrium.

$$K_P = P_{CO_2} = 3.87$$

To calculate $K_c$, rearrange the equation relating $K_P$ and $K_c$, solving for $K_c$. Then, substitute the given values into the equation to obtain the answer.

$$K_P = K_c(0.0821\,T)^{\Delta n}$$

$$K_c = \frac{K_P}{(0.0821\,T)^{\Delta n}}$$

$$K_c = \frac{3.87}{[(0.0821)(1273\text{K})]^{(1-0)}} = \mathbf{0.0370}$$

**Text Problems:** 14.22, **14.24**

## PROBLEM TYPE 3: THE FORM OF *K* AND THE EQUILIBRIUM EQUATION

The equilibrium constant expression and its value depend on how an equation is balanced. Often an equation can be balanced with more than one set of coefficients. For example,

$$2SO_2(g) + O_2(g) \rightleftharpoons 2SO_3(g) \qquad K_c = \frac{[SO_3]^2}{[SO_2]^2[O_2]}$$

and

$$SO_2(g) + \tfrac{1}{2}O_2(g) \rightleftharpoons SO_3(g) \qquad K_c' = \frac{[SO_3]}{[SO_2][O_2]^{\frac{1}{2}}}$$

Is there a relationship between the equilibrium constants for the two reactions? Note that $K_c'$ is the square root of the equilibrium constant for the first reaction, $K_c$.

$$K_c' = \frac{[SO_3]}{[SO_2][O_2]^{\frac{1}{2}}} = \sqrt{\frac{[SO_3]^2}{[SO_2]^2[O_2]}} = \sqrt{K_c}$$

The general relationship is that you raise $K_c$ to the power by which the equation was multiplied. To come up with the second balanced equation above, we had to multiply the first equation by 1/2. Thus,

$$K_c' = (K_c)^{\frac{1}{2}} = \sqrt{K_c}$$

What if the reaction is written in the reverse direction?

$$2SO_3(g) \rightleftharpoons 2SO_2(g) + O_2(g) \qquad K_c'' = \frac{[SO_2]^2[O_2]}{[SO_3]^2}$$

Is there a relationship between $K_c''$ and $K_c$? By inspection, you should find that $K_c''$ is the reciprocal of $K_c$ for the forward reaction.

$$K_c'' = \frac{[SO_2]^2[O_2]}{[SO_3]^2} = \frac{1}{K_c}$$

Remember, always use the $K_c$ expression and value that are consistent with the way in which the balanced equation is written.

**EXAMPLE 14.5**

**For the reaction, $2HBr(g) \rightleftharpoons H_2(g) + Br_2(g)$, $K_P = 1.4 \times 10^{-5}$ at 700 K. What are the values of $K_P$ for the following reactions at the same temperature?**

(a) $4HBr(g) \rightleftharpoons 2H_2(g) + 2Br_2(g)$

(b) $H_2(g) + Br_2(g) \rightleftharpoons 2HBr(g)$

**(a)** In Equation (a), the original equation has been multiplied by *two*. The general relationship is that you raise $K$ to the power by which the equation was multiplied by. Thus,

$$K_P' = (K_P)^2 = (1.4 \times 10^{-5})^2 = \mathbf{2.0 \times 10^{-10}}$$

**(b)** Equation (b) is the reverse of the original equation. By inspection, you should find that $K_P''$ is the reciprocal of $K_P$.

$$K_P'' = \frac{1}{K_P} = \frac{1}{1.4 \times 10^{-5}} = \mathbf{7.1 \times 10^4}$$

**PRACTICE EXERCISE**

4. For the reaction, $2HBr(g) \rightleftharpoons H_2(g) + Br_2(g)$, $K_P = 1.4 \times 10^{-5}$ at 700 K. What is the value of $K_P$ for the following reaction at the same temperature?

$$HBr(g) \rightleftharpoons \tfrac{1}{2}H_2(g) + \tfrac{1}{2}Br_2(g)$$

**Text Problems:** 14.30, 14.32

## PROBLEM TYPE 4: USING THE REACTION QUOTIENT (Q) TO PREDICT THE DIRECTION OF A REACTION

Equilibrium constants provide useful information about chemical reaction systems. For instance, equilibrium constants can be used to predict the direction in which a reaction will proceed to establish equilibrium.

The reaction quotient, $Q_c$, is a useful aid in predicting whether or not a reaction system is at equilibrium. Again, consider the reaction

$$2SO_2(g) + O_2(g) \rightleftharpoons 2SO_3(g)$$

The reaction quotient is:

$$Q_c = \frac{[SO_3]_0^2}{[SO_2]_0^2[O_2]_0}$$

You should notice that $Q_c$ has the same algebraic form as $K_c$. However, the concentrations are not necessarily equilibrium concentrations. We will call them initial concentrations, represented by a subscript 0 after the square brackets, $[\ ]_0$. Substituting initial concentrations into the reaction quotient, gives a value for $Q_c$. In order to predict whether the system is at equilibrium, the magnitude of $Q_c$ must be compared with that of $K_c$.

If $Q_c = K_c$, The initial concentrations are equilibrium concentrations. The system is at equilibrium.

If $Q_c > K_c$, The ratio of initial concentrations of products to reactants is too large. To reach equilibrium, products must be converted to reactants. The system proceeds from right to left (consuming products, forming reactants) to reach equilibrium.

If $Q_c < K_c$, The ratio of initial concentrations of products to reactants is too small. To reach equilibrium, reactants must be converted to products. The system proceeds from left to right (consuming reactants, forming products) to reach equilibrium.

**EXAMPLE 14.6**

**At a certain temperature, the reaction: $CO(g) + Cl_2(g) \rightleftharpoons COCl_2(g)$, has an equilibrium constant, $K_c = 13.8$. Is the following mixture an equilibrium mixture? If not, in which direction (right or left) will the reaction proceed to reach equilibrium?**

$$[CO]_0 = 2.5\ M,\ [Cl_2]_0 = 1.2\ M, \text{ and } [COCl_2]_0 = 5.0\ M$$

**Strategy:** We are given the initial concentrations of the gases, so we can calculate the reaction quotient ($Q_c$). How does a comparison of $Q_c$ with $K_c$ enable us to determine if the system is at equilibrium or, if not, in which direction the net reaction will proceed to reach equilibrium?

**Solution:** Recall that for a system to be at equilibrium, $Q_c = K_c$. Substitute the given concentrations into the equation for the reaction quotient to calculate $Q_c$.

$$Q_c = \frac{[COCl_2]_0}{[CO]_0[Cl_2]_0} = \frac{5.0}{(2.5)(1.2)} = 1.7$$

Compare $Q_c$ to $K_c$. Since $Q_c < K_c$, the ratio of initial concentrations of products to reactants is too small. To reach equilibrium, reactants must be converted to products. The system proceeds from **left to right** (consuming reactants, forming products) to reach equilibrium.

**PRACTICE EXERCISE**

5. Given the reaction,

$$N_2(g) + O_2(g) \rightleftharpoons 2NO(g) \qquad K_c = 2.5 \times 10^{-3} \text{ at } 2130°C$$

decide whether the following mixture is at equilibrium or if a net forward or reverse reaction will occur.

$[NO] = 0.0050\ M$, $[O_2] = 0.25\ M$, and $[N_2] = 0.020\ M$

**Text Problem: 14.40**

## PROBLEM TYPE 5: CALCULATING EQUILIBRIUM CONCENTRATIONS

The expected concentrations at equilibrium can be calculated from a knowledge of the initial concentrations and the equilibrium constant. In these types of problems, it will be very helpful to recall that

$$\text{equilibrium concentration} = \text{initial concentration} \pm \text{the change due to reaction.}$$

The next *two* examples illustrate this important type of calculation.

**EXAMPLE 14.7**

**A 0.25 mole sample of $N_2O_4$ dissociates and comes to equilibrium in a 1.5 L flask at 100°C. The reaction is**

$$N_2O_4(g) \rightleftharpoons 2NO_2(g) \qquad K_c = 0.36 \text{ at } 100°C$$

**What are the equilibrium concentrations of $NO_2$ and $N_2O_4$?**

**Strategy:** We are given the initial amount of $N_2O_4$ (in moles) in a vessel of known volume (in liters), so we can calculate its molar concentration. Because initially no $NO_2$ molecules are present, the system could not be at equilibrium. Therefore, some $N_2O_4$ will dissociate to form $NO_2$ molecules until equilibrium is established.

**Solution:** We follow the procedure outlined in Section 14.4 of the text to calculate the equilibrium concentrations.

*Step 1:* The initial concentration of $N_2O_4$ is 0.25 mol/1.5 L = 0.17 *M*. The stoichiometry of the problem shows 1 mole of $N_2O_4$ dissociating to 2 moles of $NO_2$ molecules. Let $x$ be the amount (in mol/L) of $N_2O_4$ dissociated. It follows that the equilibrium concentration of $NO_2$ molecules must be $2x$. We summarize the changes in concentrations as follows:

| | $N_2O_4(g)$ | $\rightleftharpoons$ | $2NO_2(g)$ |
|---|---|---|---|
| Initial (*M*): | 0.17 | | 0 |
| Change (*M*): | $-x$ | | $+2x$ |
| Equilibrium (*M*): | $0.17 - x$ | | $2x$ |

We call this type of table an **ICE** table for **I**nitial, **C**hange, and **E**quilibrium.

*Step 2:* Write the equilibrium constant expression in terms of the equilibrium concentrations. Knowing the value of the equilibrium constant, solve for $x$.

$$K_c = \frac{[NO_2]^2}{[N_2O_4]}$$

$$0.36 = \frac{(2x)^2}{0.17 - x}$$

$$0.061 - 0.36x = 4x^2$$

$$4x^2 + 0.36x - 0.061 = 0$$

The above equation is a quadratic equation of the form $ax^2 + bx + c = 0$. The solution for a quadratic equation is:

$$x = \frac{-b \pm \sqrt{b^2 - 4ac}}{2a}$$

Here, we have a = 4, b = 0.36, and c = −0.061. Substituting into the above equation,

$$x = \frac{-0.36 \pm \sqrt{(0.36)^2 - 4(4)(-0.061)}}{2(4)}$$

$$x = \frac{-0.36 \pm 1.05}{8}$$

$$x = 0.086\,M \quad \text{or} \quad x = -0.18\,M$$

The second solution is physically impossible because you cannot have a negative concentration. The first solution is the correct answer.

> **Tip:** In solving a quadratic equation of this type, one answer is always physically impossible, so the choice of which value to use for $x$ is easy to make.

*Step 3:* Having solved for $x$, calculate the equilibrium concentrations of all species.

$$[N_2O_4] = (0.17 - 0.086)M = \mathbf{0.08\ M}$$

$$[NO_2] = 2(0.086\ M) = \mathbf{0.17\ M}$$

**EXAMPLE 14.8**

**A 1.00 L vessel initially contains 0.776 mol of $SO_3$ (*g*) at 1100 K. What is the value of $K_c$ for the following reaction if 0.520 mol of $SO_3$ remain at equilibrium?**

$$2SO_3(g) \rightleftharpoons 2SO_2(g) + O_2(g)$$

---

**Strategy:** If we can calculate the equilibrium concentrations for all species, we can calculate the equilibrium constant $K_c$. Because the initial amount of $SO_3$ and the amount of $SO_3$ at equilibrium are given, the equilibrium concentration of $SO_3$ can be calculated. The equilibrium concentrations of other species can be determined from the amount of $SO_3$ reacted.

**Solution:**

*Step 1:* In this problem, we are given both the initial and equilibrium concentrations of $SO_3$. Recalling that

equilibrium concentration = initial concentration ± the change due to reaction

we can calculate the change in concentration of $SO_3$ due to reaction. Let's call this change, $2x$, because of the coefficient of 2 for $SO_3$ in the balanced equation.

$$\text{change in concentration of } SO_3 = 2x = 0.776\,M - 0.520\,M = 0.256\,M$$

$$2x = 0.256,\ x = 0.128\,M$$

Complete a table that lists the initial concentrations, the change in concentrations, and the equilibrium concentrations.

| | $2SO_3(g)$ | $\rightleftharpoons$ $2SO_2(g)$ | $+$ $O_2(g)$ |
|---|---|---|---|
| Initial ($M$): | 0.776 | 0 | 0 |
| Change ($M$): | $-2x = -0.256$ | $+2x$ | $+x$ |
| Equilibrium ($M$): | 0.520 | $2x$ | $x$ |

So, at equilibrium,

$$[SO_2] = 2x = 0.256\ M$$

$$[O_2] = x = 0.128\ M$$

**Tip:** You probably could have come up with the equilibrium concentrations of $SO_2$ and $O_2$ without the use of a table. However, a table is a simple way to keep all the data organized.

*Step 2:* Substitute the equilibrium concentrations into the equilibrium constant expression to solve for $K_c$.

$$K_c = \frac{[SO_2]^2[O_2]}{[SO_3]^2}$$

$$K_c = \frac{(0.256)^2(0.128)}{(0.520)^2} = \mathbf{0.0310}$$

**PRACTICE EXERCISE**

6. For the reaction,

$$N_2(g) + O_2(g) \rightleftharpoons 2NO(g) \qquad K_P = 3.80 \times 10^{-4} \text{ at } 2000°C$$

what equilibrium pressures of $N_2$, $O_2$, and NO will result when a 10.0 L reactor vessel is filled with 2.00 atm of $N_2$ and 0.400 atm of $O_2$ and the reaction is allowed to come to equilibrium?

7. Initially a 1.0 L vessel contains 10.0 mol of NO and 6.0 mol of $O_2$ at a certain temperature.

$$2NO(g) + O_2(g) \rightleftharpoons 2NO_2(g)$$

At equilibrium, the vessel contains 8.8 mol of $NO_2$. Determine the value of $K_c$ at this temperature.

**Text Problems: 14.42, 14.44,** 14.46, 14.48

## PROBLEM TYPE 6: FACTORS THAT AFFECT CHEMICAL EQUILIBRIUM

What effect does a change in concentration, pressure, volume, or temperature have on a system at equilibrium? This question can be answered qualitatively by using **Le Châtelier's principle**. It states that when an external stress is applied to a system at equilibrium, the system adjusts in such a way that the stress is partially offset. The word "stress" here means a change in concentration, pressure, volume, or temperature that removes a system from the equilibrium state.

## A. Changes in concentration

**EXAMPLE 14.9**

**For the following reaction at equilibrium in a closed container,**

$$2NaHCO_3(s) \rightleftharpoons Na_2CO_3(s) + H_2O(g) + CO_2(g)$$

**state the effects (increase, decrease, or no change) of the following stresses on the number of moles of sodium carbonate, $Na_2CO_3$. Note that $Na_2CO_3$ is a solid (this is a heterogeneous equilibrium); its concentration will remain constant, but its amount can change.**

(a) Removing $CO_2(g)$.

(b) Adding $H_2O(g)$.

(c) Adding $NaHCO_3(s)$.

**Strategy:** (a) The stress is the removal of $CO_2$ gas. How will the system adjust to partially offset the stress? (b) The stress is the addition of $H_2O$ gas. How will the system adjust to partially offset the stress? (c) The stress is the addition of $NaHCO_3(s)$. Do pure solids or pure liquids enter into the equilibrium constant expression?

**Solution:** Applying Le Châtelier's principle,

**(a)** If $CO_2$ concentration is lowered, the system will react to offset the change. That is, a shift to the right will replace some of the removed $CO_2$. Moles of $Na_2CO_3$ will **increase**.

**(b)** The system will respond to the stress of added $H_2O$ by shifting to the left to remove some of the water. Moles of $Na_2CO_3$ will **decrease**.

**(c)** The position of a heterogeneous equilibrium does not depend on the amounts of pure solids or liquids present. Remember that the concentrations of pure solids and liquids are constant and thus do not enter into the equilibrium constant expression. Hence, there is no shift in the equilibrium, and the amount of $Na_2CO_3$ is **unchanged**.

## B. Changes in pressure and volume

The pressure of a system of gases in chemical equilibrium can be increased by decreasing the available volume. The ideal gas equation shows this inverse relationship between pressure and volume.

$$PV = nRT$$

A decrease in volume causes the concentrations of all *gas-phase* components to increase. Remember that pressure is directly proportional to the concentration of a gas.

$$P = \left(\frac{n}{V}\right)RT$$

The stress caused by the increased pressure will be partially offset by a net reaction that will lower the total pressure. In other words, the system will adjust by lowering the total concentration of gas molecules to reestablish equilibrium. Again, consider the following equilibrium:

$$2SO_2(g) + O_2(g) \rightleftharpoons 2SO_3(g)$$

When the molecules of the above gases are compressed into a smaller volume, the total pressure increases and hence the total concentration increases (this is a stress). A net forward reaction (shift to the right) will bring the system to a new state of equilibrium, in which the *total concentration* of all molecules will be lowered somewhat. This partially offsets the initial stress on the system. The total concentration is lowered somewhat because when 2 moles of $SO_2$ react with 1 mole of $O_2$ (a total of 3 moles), only 2 moles of $SO_3$ are produced.

In general, an increase in pressure (decrease in volume) favors the net reaction that decreases the total number of moles of gases. On the other hand, a decrease in pressure (increase in volume) favors the net reaction that increases the total number of moles of gases.

**EXAMPLE 14.10**

**For the following reaction at equilibrium in a closed container,**

$$2NaHCO_3(s) \rightleftharpoons Na_2CO_3(s) + H_2O(g) + CO_2(g)$$

**state the effect (increase, decrease, or no change) of increasing the volume of the container on the number of moles of sodium carbonate, $Na_2CO_3$. Note that $Na_2CO_3$ is a solid (this is a heterogeneous equilibrium); its concentration will remain constant, but its amount can change.**

**Strategy:** A change in pressure can affect only the volume of a gas, but not that of a solid or liquid because solids and liquids are much less compressible. The stress applied is a decrease in pressure (increase in volume). According to Le Châtelier's principle, the system will adjust to partially offset this stress. In other words, the system will adjust to increase the pressure. This can be achieved by shifting to the side of the equation that has more moles of gas. Recall that pressure is directly proportional to moles of gas: $PV = nRT$ so $P \propto n$.

**Solution:** Looking at the above reaction, there are zero moles of gas on the reactants' side, and two moles of gas on the products' side. The pressure will be increased by shifting to the right (more moles of gas).

Because the reaction shifts to the right to establish equilibrium, the amount of $Na_2CO_3$ will **increase**.

## C. Changes in temperature

A change in concentration, pressure, or volume may alter the equilibrium position, but it does not change the value of the equilibrium constant. However, a change in temperature can alter the equilibrium constant.

To decide how a temperature stress will affect a system at equilibrium, you must look at whether the reaction is endothermic or exothermic. The following reaction is endothermic (positive $\Delta H$). Endothermic reactions absorb heat from the surroundings; therefore, we can think of heat as a reactant.

$$\text{heat} + PCl_5(g) \rightleftharpoons PCl_3(g) + Cl_2(g) \qquad \Delta H^\circ = 92.9 \text{ kJ/mol}$$

What happens if we heat the above system at equilibrium? We have added heat (a stress). The system will shift in the direction that will partially offset this stress by removing some heat. Shifting to the right will remove heat. $PCl_5$ dissociates into $PCl_3$ and $Cl_2$ molecules. Consequently, the equilibrium constant for the above reaction, given by

$$K_c = \frac{[PCl_3][Cl_2]}{[PCl_5]}$$

increases with temperature.

In general, a temperature *increase* favors an *endothermic* reaction, and a temperature *decrease* favors an *exothermic* reaction.

**EXAMPLE 14.11**

**For the following reaction at equilibrium in a closed container,**

$$2NaHCO_3(s) \rightleftharpoons Na_2CO_3(s) + H_2O(g) + CO_2(g) \qquad \Delta H = 128 \text{ kJ/mol}$$

**state the effect (increase, decrease, or no change) of decreasing the temperature on the number of moles of sodium carbonate, $Na_2CO_3$. Note that $Na_2CO_3$ is a solid (this is a heterogeneous equilibrium); its concentration will remain constant, but its amount can change.**

---

**Strategy:** What does the sign of $\Delta H°$ indicate about the heat change (endothermic or exothermic) for the forward reaction?

**Solution:** The stress applied is heat removed from the system. Note that the reaction is endothermic ($\Delta H° > 0$). Endothermic reactions absorb heat from the surroundings; therefore, we can think of heat as a reactant.

$$\text{heat} + 2NaHCO_3(s) \rightleftharpoons Na_2CO_3(s) + H_2O(g) + CO_2(g)$$

As heat is removed, the system will shift to replace some of the heat by shifting to the left. Thus, the reverse reaction is favored, and the amount of $Na_2CO_3$ will **decrease**.

**PRACTICE EXERCISE**

8. For the decomposition of calcium carbonate

$$CaCO_3(s) \rightleftharpoons CaO(s) + CO_2(g) \qquad \Delta H° = 175 \text{ kJ/mol}$$

how will the amount (not concentration) of $CaCO_3(s)$ change with the following stresses to the system at equilibrium?

(a) $CO_2(g)$ is removed.
(b) $CaO(s)$ is added.
(c) The temperature is raised.
(d) The volume of the container is decreased.

**Text Problems:** 14.54, **14.56**, **14.58**, 14.60, 14.62

## D. The effect of a catalyst

A catalyst increases the rate at which a reaction occurs. For a reversible reaction, a catalyst affects the rate in the forward and reverse directions to the same extent. Therefore, the presence of a catalyst does not alter the equilibrium constant, nor does it shift the position of an equilibrium system.

# ANSWERS TO PRACTICE EXERCISES

1. If concentration is expressed in mol/L, $K_c = \dfrac{[N_2O_4]^2}{[N_2O]^2[O_2]^3}$

   or, in terms of partial pressures, $K_P = \dfrac{P^2_{N_2O_4}}{P^2_{N_2O}P^3_{O_2}}$

2. $K_P = 1.05$ **3.** $K_P = K_c = 64$ **4.** $K_P' = 3.7 \times 10^{-3}$

5. Since $Q_c > K_c$, the ratio of initial concentrations of products to reactants is too large. To reach equilibrium, products must be converted to reactants. The system proceeds from **right to left** (consuming products, forming reactants) to reach equilibrium.

6. $P_{NO_2} = 1.74 \times 10^{-2}$ atm
   $P_{N_2} = 1.99$ atm
   $P_{O_2} = 0.391$ atm

7. $K_c = 34$

8. (a) The amount of $CaCO_3$ will **decrease**.
   (b) The amount of $CaCO_3$ will **not change**.
   (c) The amount of $CaCO_3$ will **decrease**.
   (d) The amount of $CaCO_3$ will **increase**.

# SOLUTIONS TO SELECTED TEXT PROBLEMS

**14.14** Note that we are comparing similar reactions at equilibrium – two reactants producing one product, all with coefficients of one in the balanced equation.

(a) The reaction, $A + C \rightleftharpoons AC$ has the **largest** equilibrium constant. Of the three diagrams, there is the most product present at equilibrium.

(b) The reaction, $A + D \rightleftharpoons AD$ has the **smallest** equilibrium constant. Of the three diagrams, there is the least amount of product present at equilibrium.

**14.16** The problem states that the system is at equilibrium, so we simply substitute the equilibrium concentrations into the equilibrium constant expression to calculate $K_c$.

*Step 1:* Calculate the concentrations of the components in units of mol/L. The molarities can be calculated by simply dividing the number of moles by the volume of the flask.

$$[H_2] = \frac{2.50 \text{ mol}}{12.0 \text{ L}} = 0.208\ M$$

$$[S_2] = \frac{1.35 \times 10^{-5} \text{ mol}}{12.0 \text{ L}} = 1.13 \times 10^{-6}\ M$$

$$[H_2S] = \frac{8.70 \text{ mol}}{12.0 \text{ L}} = 0.725\ M$$

*Step 2:* Once the molarities are known, $K_c$ can be found by substituting the molarities into the equilibrium constant expression.

$$K_c = \frac{[H_2S]^2}{[H_2]^2[S_2]} = \frac{(0.725)^2}{(0.208)^2(1.13 \times 10^{-6})} = \mathbf{1.08 \times 10^7}$$

If you forget to convert moles to moles/liter, will you get a different answer? Under what circumstances will the two answers be the same?

**14.18** **Strategy:** The relationship between $K_c$ and $K_P$ is given by Equation (14.5) of the text. What is the change in the number of moles of gases from reactant to product? Recall that

$$\Delta n = \text{moles of gaseous products} - \text{moles of gaseous reactants}$$

What unit of temperature should we use?

**Solution:** The relationship between $K_c$ and $K_P$ is given by Equation (14.5) of the text.

$$K_P = K_c(0.0821\,T)^{\Delta n}$$

Rearrange the equation relating $K_P$ and $K_c$, solving for $K_c$.

$$K_c = \frac{K_P}{(0.0821T)^{\Delta n}}$$

Because $T = 623$ K and $\Delta n = 3 - 2 = 1$, we have:

$$K_c = \frac{K_P}{(0.0821T)^{\Delta n}} = \frac{1.8 \times 10^{-5}}{(0.0821)(623 \text{ K})} = \mathbf{3.5 \times 10^{-7}}$$

**14.20** The equilibrium constant expressions are:

**(a)** $K_c = \dfrac{[NH_3]^2}{[N_2][H_2]^3}$

**(b)** $K_c = \dfrac{[NH_3]}{[N_2]^{\frac{1}{2}}[H_2]^{\frac{3}{2}}}$

Substituting the given equilibrium concentration gives:

$$\text{(a)}\quad K_c = \frac{(0.25)^2}{(0.11)(1.91)^3} = \mathbf{0.082}$$

$$\text{(b)}\quad K_c = \frac{(0.25)}{(0.11)^{\frac{1}{2}}(1.91)^{\frac{3}{2}}} = \mathbf{0.29}$$

Is there a relationship between the $K_c$ values from parts (a) and (b)?

**14.22** Because pure solids do not enter into an equilibrium constant expression, we can calculate $K_P$ directly from the pressure that is due solely to $CO_2(g)$.

$$K_P = P_{CO_2} = \mathbf{0.105}$$

Now, we can convert $K_P$ to $K_c$ using the following equation.

$$K_P = K_c(0.0821\,T)^{\Delta n}$$

$$K_c = \frac{K_P}{(0.0821T)^{\Delta n}}$$

$$K_c = \frac{0.105}{(0.0821 \times 623)^{(1-0)}} = \mathbf{2.05 \times 10^{-3}}$$

**14.24** **Strategy:** Because they are constant quantities, the concentrations of solids and liquids do not appear in the equilibrium constant expressions for heterogeneous systems. The total pressure at equilibrium that is given is due to both $NH_3$ and $CO_2$. Note that for every 1 atm of $CO_2$ produced, 2 atm of $NH_3$ will be produced due to the stoichiometry of the balanced equation. Using this ratio, we can calculate the partial pressures of $NH_3$ and $CO_2$ at equilibrium.

**Solution:** The equilibrium constant expression for the reaction is

$$K_P = P_{NH_3}^2 P_{CO_2}$$

The total pressure in the flask (0.363 atm) is a sum of the partial pressures of $NH_3$ and $CO_2$.

$$P_T = P_{NH_3} + P_{CO_2} = 0.363 \text{ atm}$$

Let the partial pressure of $CO_2 = x$. From the stoichiometry of the balanced equation, you should find that $P_{NH_3} = 2P_{CO_2}$. Therefore, the partial pressure of $NH_3 = 2x$. Substituting into the equation for total pressure gives:

$$P_T = P_{NH_3} + P_{CO_2} = 2x + x = 3x$$

$$3x = 0.363 \text{ atm}$$

$$x = P_{CO_2} = 0.121 \text{ atm}$$

$$P_{NH_3} = 2x = 0.242 \text{ atm}$$

Substitute the equilibrium pressures into the equilibrium constant expression to solve for $K_P$.

$$K_P = P_{NH_3}^2 P_{CO_2} = (0.242)^2(0.121) = \mathbf{7.09 \times 10^{-3}}$$

**14.26** If the CO pressure at equilibrium is 0.497 atm, the balanced equation requires the chlorine pressure to have the same value. The initial pressure of phosgene gas can be found from the ideal gas equation.

$$P = \frac{nRT}{V} = \frac{(3.00 \times 10^{-2} \cancel{\text{mol}})(0.0821 \cancel{\text{L}}\cdot\text{atm}/\cancel{\text{mol}}\cdot\cancel{\text{K}})(800 \cancel{\text{K}})}{(1.50 \cancel{\text{L}})} = 1.31 \text{ atm}$$

Since there is a 1:1 mole ratio between phosgene and CO, the partial pressure of CO formed (0.497 atm) equals the partial pressure of phosgene reacted. The phosgene pressure at equilibrium is:

| | $CO(g)$ | + | $Cl_2(g)$ | $\rightleftharpoons$ | $COCl_2(g)$ |
|---|---|---|---|---|---|
| Initial (atm): | 0 | | 0 | | 1.31 |
| Change (atm): | +0.497 | | +0.497 | | −0.497 |
| Equilibrium (atm): | 0.497 | | 0.497 | | 0.81 |

The value of $K_P$ is then found by substitution.

$$K_P = \frac{P_{COCl_2}}{P_{CO}P_{Cl_2}} = \frac{0.81}{(0.497)^2} = \mathbf{3.3}$$

**14.28** In this problem, you are asked to calculate $K_c$.

*Step 1:* Calculate the initial concentration of NOCl. We carry an extra significant figure throughout this calculation to minimize rounding errors.

$$[\text{NOCl}]_0 = \frac{2.50 \text{ mol}}{1.50 \text{ L}} = 1.667\ M$$

*Step 2:* Let's represent the change in concentration of NOCl as $-2x$. Setting up a table:

| | $2NOCl(g)$ | $\rightleftharpoons$ | $2NO(g)$ | + | $Cl_2(g)$ |
|---|---|---|---|---|---|
| Initial (*M*): | 1.667 | | 0 | | 0 |
| Change (*M*): | $-2x$ | | $+2x$ | | $+x$ |
| Equilibrium (*M*): | $1.667 - 2x$ | | $2x$ | | $x$ |

If 28.0 percent of the NOCl has dissociated at equilibrium, the amount reacted is:

$$(0.280)(1.667\ M) = 0.4668\ M$$

In the table above, we have represented the amount of NOCl that reacts as $2x$. Therefore,

$$2x = 0.4668\,M$$

$$x = 0.2334\,M$$

The equilibrium concentrations of NOCl, NO, and $Cl_2$ are:

$$[NOCl] = (1.67 - 2x)M = (1.667 - 0.4668)M = 1.200\,M$$

$$[NO] = 2x = 0.4668\,M$$

$$[Cl_2] = x = 0.2334\,M$$

*Step 3:* The equilibrium constant $K_c$ can be calculated by substituting the above concentrations into the equilibrium constant expression.

$$K_c = \frac{[NO]^2[Cl_2]}{[NOCl]^2} = \frac{(0.4668)^2(0.2334)}{(1.200)^2} = \mathbf{0.0353}$$

**14.30** $K = K'K''$

$$K = (6.5 \times 10^{-2})(6.1 \times 10^{-5})$$

$$K = \mathbf{4.0 \times 10^{-6}}$$

**14.32** To obtain $2SO_2$ as a reactant in the final equation, we must reverse the first equation and multiply by two. For the equilibrium, $2SO_2(g) \rightleftharpoons 2S(s) + 2O_2(g)$

$$K_c''' = \left(\frac{1}{K_c'}\right)^2 = \left(\frac{1}{4.2 \times 10^{52}}\right)^2 = 5.7 \times 10^{-106}$$

Now we can add the above equation to the second equation to obtain the final equation. Since we add the two equations, the equilibrium constant is the product of the equilibrium constants for the two reactions.

$$2SO_2(g) \rightleftharpoons 2S(s) + 2O_2(g) \qquad K_c''' = 5.7 \times 10^{-106}$$

$$2S(s) + 3O_2(g) \rightleftharpoons 2SO_3(g) \qquad K_c'' = 9.8 \times 10^{128}$$

$$2SO_2(g) + O_2(g) \rightleftharpoons 2SO_3(g) \qquad K_c = K_c''' \times K_c'' = \mathbf{5.6 \times 10^{23}}$$

**14.36** At equilibrium, the value of $K_c$ is equal to the ratio of the forward rate constant to the rate constant for the reverse reaction.

$$K_c = \frac{k_f}{k_r} = \frac{k_f}{5.1 \times 10^{-2}} = 12.6$$

$$k_f = (12.6)(5.1 \times 10^{-2}) = 0.64$$

The forward reaction is third order, so the units of $k_f$ must be:

$$\text{rate} = k_f[A]^2[B]$$

$$k_f = \frac{\text{rate}}{(\text{concentration})^3} = \frac{M/\text{s}}{M^3} = 1/M^2 \cdot \text{s}$$

$$\mathbf{k_f = 0.64\ M^{-2} \cdot s^{-1}}$$

**14.40** **Strategy:** We are given the initial concentrations of the gases, so we can calculate the reaction quotient ($Q_c$). How does a comparison of $Q_c$ with $K_c$ enable us to determine if the system is at equilibrium or, if not, in which direction the net reaction will proceed to reach equilibrium?

**Solution:** Recall that for a system to be at equilibrium, $Q_c = K_c$. Substitute the given concentrations into the equation for the reaction quotient to calculate $Q_c$.

$$Q_c = \frac{[NH_3]_0^2}{[N_2]_0[H_2]_0^3} = \frac{[0.48]^2}{[0.60][0.76]^3} = 0.87$$

Comparing $Q_c$ to $K_c$, we find that $Q_c < K_c$ $(0.87 < 1.2)$. The ratio of initial concentrations of products to reactants is too small. To reach equilibrium, reactants must be converted to products. The system proceeds from left to right (consuming reactants, forming products) to reach equilibrium.

Therefore, **$[NH_3]$ will increase** and **$[N_2]$ and $[H_2]$ will decrease** at equilibrium.

**14.42** **Strategy:** The equilibrium constant $K_P$ is given, and we start with pure $NO_2$. The partial pressure of $O_2$ at equilibrium is 0.25 atm. From the stoichiometry of the reaction, we can determine the partial pressure of NO at equilibrium. Knowing $K_P$ and the partial pressures of both $O_2$ and NO, we can solve for the partial pressure of $NO_2$.

**Solution:** Since the reaction started with only pure $NO_2$, the equilibrium concentration of NO must be twice the equilibrium concentration of $O_2$, due to the 2:1 mole ratio of the balanced equation. Therefore, the equilibrium partial pressure of **NO** is $(2 \times 0.25\ \text{atm}) =$ **0.50 atm.**

We can find the equilibrium $NO_2$ pressure by rearranging the equilibrium constant expression, then substituting in the known values.

$$K_P = \frac{P_{NO}^2 P_{O_2}}{P_{NO_2}^2}$$

$$P_{NO_2} = \sqrt{\frac{P_{NO}^2 P_{O_2}}{K_P}} = \sqrt{\frac{(0.50)^2(0.25)}{158}} = \mathbf{0.020\ atm}$$

**14.44** **Strategy:** We are given the initial amount of $I_2$ (in moles) in a vessel of known volume (in liters), so we can calculate its molar concentration. Because initially no I atoms are present, the system could not be at equilibrium. Therefore, some $I_2$ will dissociate to form I atoms until equilibrium is established.

**Solution:** We follow the procedure outlined in Section 14.4 of the text to calculate the equilibrium concentrations.

*Step 1:* The initial concentration of $I_2$ is 0.0456 mol/2.30 L = 0.0198 *M*. The stoichiometry of the problem shows 1 mole of $I_2$ dissociating to 2 moles of I atoms. Let *x* be the amount (in mol/L) of $I_2$ dissociated. It follows that the equilibrium concentration of I atoms must be 2*x*. We summarize the changes in concentrations as follows:

| | $I_2(g)$ | $\rightleftharpoons$ | $2I(g)$ |
|---|---|---|---|
| Initial (*M*): | 0.0198 | | 0.000 |
| Change (*M*): | $-x$ | | $+2x$ |
| Equilibrium (*M*): | $(0.0198 - x)$ | | $2x$ |

*Step 2:* Write the equilibrium constant expression in terms of the equilibrium concentrations. Knowing the value of the equilibrium constant, solve for *x*.

$$K_c = \frac{[I]^2}{[I_2]} = \frac{(2x)^2}{(0.0198 - x)} = 3.80 \times 10^{-5}$$

$$4x^2 + (3.80 \times 10^{-5})x - (7.52 \times 10^{-7}) = 0$$

The above equation is a quadratic equation of the form $ax^2 + bx + c = 0$. The solution for a quadratic equation is

$$x = \frac{-b \pm \sqrt{b^2 - 4ac}}{2a}$$

Here, we have $a = 4$, $b = 3.80 \times 10^{-5}$, and $c = -7.52 \times 10^{-7}$. Substituting into the above equation,

$$x = \frac{(-3.80 \times 10^{-5}) \pm \sqrt{(3.80 \times 10^{-5})^2 - 4(4)(-7.52 \times 10^{-7})}}{2(4)}$$

$$x = \frac{(-3.80 \times 10^{-5}) \pm (3.47 \times 10^{-3})}{8}$$

$$x = 4.29 \times 10^{-4}\,M \quad \text{or} \quad x = -4.39 \times 10^{-4}\,M$$

The second solution is physically impossible because you cannot have a negative concentration. The first solution is the correct answer.

*Step 3:* Having solved for *x*, calculate the equilibrium concentrations of all species.

$$[I] = 2x = (2)(4.29 \times 10^{-4}\,M) = \mathbf{8.58 \times 10^{-4}\,M}$$

$$[I_2] = (0.0198 - x) = [0.0198 - (4.29 \times 10^{-4})]\,M = \mathbf{0.0194\,M}$$

**Tip:** We could have simplified this problem by assuming that $x$ was small compared to 0.0198. We could then assume that $0.0198 - x \approx 0.0198$. By making this assumption, we could have avoided solving a quadratic equation.

**14.46** **(a)** The equilibrium constant, $K_c$, can be found by simple substitution.

$$K_c = \frac{[H_2O][CO]}{[CO_2][H_2]} = \frac{(0.040)(0.050)}{(0.086)(0.045)} = \mathbf{0.52}$$

**(b)** The magnitude of the reaction quotient $Q_c$ for the system after the concentration of $CO_2$ becomes 0.50 mol/L, but before equilibrium is reestablished, is:

$$Q_c = \frac{(0.040)(0.050)}{(0.50)(0.045)} = 0.089$$

The value of $Q_c$ is smaller than $K_c$; therefore, the system will shift to the right, increasing the concentrations of CO and $H_2O$ and decreasing the concentrations of $CO_2$ and $H_2$. Let $x$ be the depletion in the concentration of $CO_2$ at equilibrium. The stoichiometry of the balanced equation then requires that the decrease in the concentration of $H_2$ must also be $x$, and that the concentration increases of CO and $H_2O$ be equal to $x$ as well. The changes in the original concentrations are shown in the table.

| | $CO_2$ | + | $H_2$ | $\rightleftharpoons$ | CO | + | $H_2O$ |
|---|---|---|---|---|---|---|---|
| Initial ($M$): | 0.50 | | 0.045 | | 0.050 | | 0.040 |
| Change ($M$): | $-x$ | | $-x$ | | $+x$ | | $+x$ |
| Equilibrium ($M$): | $(0.50 - x)$ | | $(0.045 - x)$ | | $(0.050 + x)$ | | $(0.040 + x)$ |

The equilibrium constant expression is:

$$K_c = \frac{[H_2O][CO]}{[CO_2][H_2]} = \frac{(0.040 + x)(0.050 + x)}{(0.50 - x)(0.045 - x)} = 0.52$$

$$0.52(x^2 - 0.545x + 0.0225) = x^2 + 0.090x + 0.0020$$

$$0.48x^2 + 0.373x - (9.7 \times 10^{-3}) = 0$$

The positive root of the equation is $x = 0.025$.

The equilibrium concentrations are:

$$[CO_2] = (0.50 - 0.025)\ M = \mathbf{0.48\ M}$$

$$[H_2] = (0.045 - 0.025)\ M = \mathbf{0.020\ M}$$

$$[CO] = (0.050 + 0.025)\ M = \mathbf{0.075\ M}$$

$$[H_2O] = (0.040 + 0.025)\ M = \mathbf{0.065\ M}$$

**14.48** The initial concentrations are $[H_2] = 0.80\ \text{mol}/5.0\ \text{L} = 0.16\ M$ and $[CO_2] = 0.80\ \text{mol}/5.0\ \text{L} = 0.16\ M$.

| | $H_2(g)$ | + | $CO_2(g)$ | $\rightleftharpoons$ | $H_2O(g)$ | + | $CO(g)$ |
|---|---|---|---|---|---|---|---|
| Initial ($M$): | 0.16 | | 0.16 | | 0.00 | | 0.00 |
| Change ($M$): | $-x$ | | $-x$ | | $+x$ | | $+x$ |
| Equilibrium ($M$): | $0.16 - x$ | | $0.16 - x$ | | $x$ | | $x$ |

$$K_c = \frac{[H_2O][CO]}{[H_2][CO_2]} = 4.2 = \frac{x^2}{(0.16 - x)^2}$$

Taking the square root of both sides, we obtain:

$$\frac{x}{0.16 - x} = 2.0$$

$$x = 0.11\ M$$

The equilibrium concentrations are:

$$[H_2] = [CO_2] = (0.16 - 0.11)\ M = \mathbf{0.05\ M}$$

$$[H_2O] = [CO] = \mathbf{0.11\ M}$$

**14.54** **(a)** Removal of $CO_2(g)$ from the system would shift the position of equilibrium to the **right**.

**(b)** Addition of more solid $Na_2CO_3$ would have **no effect**. $[Na_2CO_3]$ does not appear in the equilibrium constant expression.

**(c)** Removal of some of the solid $NaHCO_3$ would have **no effect**. Same reason as (b).

**14.56** **Strategy:** A change in pressure can affect only the volume of a gas, but not that of a solid or liquid because solids and liquids are much less compressible. The stress applied is an increase in pressure. According to Le Châtelier's principle, the system will adjust to partially offset this stress. In other words, the system will adjust to decrease the pressure. This can be achieved by shifting to the side of the equation that has fewer moles of gas. Recall that pressure is directly proportional to moles of gas: $PV = nRT$ so $P \propto n$.

**Solution:**

**(a)** Changes in pressure ordinarily do not affect the concentrations of reacting species in condensed phases because liquids and solids are virtually incompressible. Pressure change should have **no effect** on this system.

**(b)** Same situation as (a).

**(c)** Only the product is in the gas phase. An increase in pressure should favor the reaction that decreases the total number of moles of gas. The equilibrium should shift to the **left**, that is, the amount of B should decrease and that of A should increase.

**(d)** In this equation there are equal moles of gaseous reactants and products. A shift in either direction will have no effect on the total number of moles of gas present. There will be **no change** when the pressure is increased.

**(e)** A shift in the direction of the reverse reaction (**left**) will have the result of decreasing the total number of moles of gas present.

**14.58** **Strategy:** (a) What does the sign of $\Delta H°$ indicate about the heat change (endothermic or exothermic) for the forward reaction? (b) The stress is the addition of $Cl_2$ gas. How will the system adjust to partially offset the stress? (c) The stress is the removal of $PCl_3$ gas. How will the system adjust to partially offset the stress? (d) The stress is an increase in pressure. The system will adjust to decrease the pressure. Remember, pressure is directly proportional to moles of gas. (e) What is the function of a catalyst? How does it affect a reacting system not at equilibrium? at equilibrium?

**Solution:**

**(a)** The stress applied is the heat added to the system. Note that the reaction is endothermic ($\Delta H° > 0$). Endothermic reactions absorb heat from the surroundings; therefore, we can think of heat as a reactant.

$$\text{heat} + PCl_5(g) \rightleftharpoons PCl_3(g) + Cl_2(g)$$

The system will adjust to remove some of the added heat by undergoing a decomposition reaction (from **left to right**)

**(b)** The stress is the addition of $Cl_2$ gas. The system will shift in the direction to remove some of the added $Cl_2$. The system shifts from **right to left** until equilibrium is reestablished.

**(c)** The stress is the removal of $PCl_3$ gas. The system will shift to replace some of the $PCl_3$ that was removed. The system shifts from **left to right** until equilibrium is reestablished.

**(d)** The stress applied is an increase in pressure. The system will adjust to remove the stress by decreasing the pressure. Recall that pressure is directly proportional to the number of moles of gas. In the balanced equation we see 1 mole of gas on the reactants side and 2 moles of gas on the products side. The pressure can be decreased by shifting to the side with the fewer moles of gas. The system will shift from **right to left** to reestablish equilibrium.

**(e)** The function of a catalyst is to increase the rate of a reaction. If a catalyst is added to the reacting system not at equilibrium, the system will reach equilibrium faster than if left undisturbed. If a system is already at equilibrium, as in this case, the addition of a catalyst will not affect either the concentrations of reactant and product, or the equilibrium constant.

**14.60** There will be no change in the pressures. A catalyst has no effect on the position of the equilibrium.

**14.62** For this system, $K_P = [CO_2]$.

This means that to remain at equilibrium, the pressure of carbon dioxide must stay at a fixed value as long as the temperature remains the same.

**(a)** If the volume is increased, the pressure of $CO_2$ will drop (Boyle's law, pressure and volume are inversely proportional). Some $CaCO_3$ will break down to form more $CO_2$ and CaO. **(Shift right)**

**(b)** Assuming that the amount of added solid CaO is not so large that the volume of the system is altered significantly, there should be **no change** at all. If a huge amount of CaO were added, this would have the effect of reducing the volume of the container. What would happen then?

**(c)** Assuming that the amount of $CaCO_3$ removed doesn't alter the container volume significantly, there should be **no change**. Removing a huge amount of $CaCO_3$ will have the effect of increasing the container volume. The result in that case will be the same as in part (a).

**(d)** The pressure of $CO_2$ will be greater and will exceed the value of $K_P$. Some $CO_2$ will combine with CaO to form more $CaCO_3$. **(Shift left)**

**(e)** Carbon dioxide combines with aqueous NaOH according to the equation

$$CO_2(g) + NaOH(aq) \rightarrow NaHCO_3(aq)$$

This will have the effect of reducing the $CO_2$ pressure and causing more $CaCO_3$ to break down to $CO_2$ and CaO. **(Shift right)**

**(f)** Carbon dioxide does not react with hydrochloric acid, but $CaCO_3$ does.

$$CaCO_3(s) + 2HCl(aq) \rightarrow CaCl_2(aq) + CO_2(g) + H_2O(l)$$

The $CO_2$ produced by the action of the acid will combine with CaO as discussed in (d) above. **(Shift left)**

**(g)** This is a decomposition reaction. Decomposition reactions are endothermic. Increasing the temperature will favor this reaction and produce more $CO_2$ and CaO. **(Shift right)**

**14.64** **(a)** Since the total pressure is 1.00 atm, the sum of the partial pressures of NO and $Cl_2$ is

$$1.00 \text{ atm} - \text{partial pressure of NOCl} = 1.00 \text{ atm} - 0.64 \text{ atm} = 0.36 \text{ atm}$$

The stoichiometry of the reaction requires that the partial pressure of NO be twice that of $Cl_2$. Hence, the partial pressure of NO is **0.24 atm** and the partial pressure of $Cl_2$ is **0.12 atm.**

**(b)** The equilibrium constant $K_P$ is found by substituting the partial pressures calculated in part (a) into the equilibrium constant expression.

$$K_P = \frac{P_{NO}^2 P_{Cl_2}}{P_{NOCl}^2} = \frac{(0.24)^2(0.12)}{(0.64)^2} = \mathbf{0.017}$$

**14.66** Since diagram (a) represents the system at equilibrium, we can solve for the equilibrium constant. The concentration of each species at equilibrium is:

$$M_{A_2} = M_{B_2} = \frac{(4)(0.020\text{ mol})}{1.0\text{ L}} = 0.080\ M$$

$$M_{AB} = \frac{(2)(0.020\text{ mol})}{1.0\text{ L}} = 0.040\ M$$

The equilibrium constant, $K_c$, is:

$$K_c = \frac{[AB]^2}{[A_2][B_2]} = \frac{(0.040)^2}{(0.080)(0.080)} = 0.25$$

Now, we can set up an ICE table to solve for the concentrations of each species in diagram (b) when the system reaches equilibrium. The initial concentration of AB is 5 × 0.020 *M* or 0.10 *M*.

| | $A_2$ + | $B_2$ ⇌ | 2AB |
|---|---|---|---|
| Initial (*M*): | 0 | 0 | 0.10 |
| Change (*M*): | $+x$ | $+x$ | $-2x$ |
| Equilibrium (*M*): | $x$ | $x$ | $0.10 - 2x$ |

$$K_c = \frac{[AB]^2}{[A_2][B_2]}$$

$$0.25 = \frac{(0.10 - 2x)^2}{x^2}$$

Taking the square root of both sides of the equation gives:

$$0.50 = \frac{0.10 - 2x}{x}$$

$$x = 0.040\ M$$

The concentrations of each species are:

$$[A_2] = [B_2] = x = \mathbf{0.040\ M}$$

$$[AB] = 0.10\ M - 2x = \mathbf{0.020\ M}$$

Check your work by substituting the above equilibrium concentrations into the $K_c$ expression to solve for $K_c$.

**14.68** The equilibrium expression for this system is given by:

$$K_P = P_{CO_2}P_{H_2O}$$

**(a)** In a closed vessel the decomposition will stop when the product of the partial pressures of $CO_2$ and $H_2O$ equals $K_P$. Adding more sodium bicarbonate will have **no effect.**

**(b)** In an open vessel, $CO_2(g)$ and $H_2O(g)$ will escape from the vessel, and the partial pressures of $CO_2$ and $H_2O$ will never become large enough for their product to equal $K_P$. Therefore, equilibrium will never be established. Adding more sodium bicarbonate will result in the production of **more $CO_2$ and $H_2O$.**

**14.70** **(a)** The equation that relates $K_P$ and $K_c$ is:

$$K_P = K_c(0.0821\,T)^{\Delta n}$$

For this reaction, $\Delta n = 3 - 2 = 1$

$$K_c = \frac{K_P}{(0.0821T)} = \frac{2\times 10^{-42}}{(0.0821\times 298)} = \mathbf{8\times 10^{-44}}$$

**(b)** Because of a very large activation energy, the reaction of hydrogen with oxygen is infinitely slow without a catalyst or an initiator. The action of a single spark on a mixture of these gases results in the explosive formation of water.

**14.72** **(a)** Calculate the value of $K_P$ by substituting the equilibrium partial pressures into the equilibrium constant expression.

$$K_P = \frac{P_B}{P_A^2} = \frac{(0.60)}{(0.60)^2} = \mathbf{1.7}$$

**(b)** The total pressure is the sum of the partial pressures for the two gaseous components, A and B. We can write:

$$P_A + P_B = 1.5 \text{ atm}$$

and

$$P_B = 1.5 - P_A$$

Substituting into the expression for $K_P$ gives:

$$K_P = \frac{(1.5 - P_A)}{P_A^2} = 1.7$$

$$1.7P_A^2 + P_A - 1.5 = 0$$

Solving the quadratic equation, we obtain:

$$P_A = \mathbf{0.69\ atm}$$

and by difference,

$$P_B = \mathbf{0.81\ atm}$$

Check that substituting these equilibrium concentrations into the equilibrium constant expression gives the equilibrium constant calculated in part (a).

$$K_P = \frac{P_B}{P_A^2} = \frac{0.81}{(0.69)^2} = 1.7$$

**14.74** Total number of moles of gas is:

$$0.020 + 0.040 + 0.96 = 1.02 \text{ mol of gas}$$

You can calculate the partial pressure of each gaseous component from the mole fraction and the total pressure.

$$P_{NO} = X_{NO}P_T = \frac{0.040}{1.02} \times 0.20 \text{ atm} = 0.0078 \text{ atm}$$

$$P_{O_2} = X_{O_2}P_T = \frac{0.020}{1.02} \times 0.20 \text{ atm} = 0.0039 \text{ atm}$$

$$P_{NO_2} = X_{NO_2}P_T = \frac{0.96}{1.02} \times 0.20 \text{ atm} = 0.19 \text{ atm}$$

Calculate $K_P$ by substituting the partial pressures into the equilibrium constant expression.

$$K_P = \frac{P_{NO_2}^2}{P_{NO}^2 P_{O_2}} = \frac{(0.19)^2}{(0.0078)^2(0.0039)} = \mathbf{1.5 \times 10^5}$$

**14.76** Set up a table that contains the initial concentrations, the change in concentrations, and the equilibrium concentration. Assume that the vessel has a volume of 1 L.

| | $H_2$ | + | $Cl_2$ | $\rightleftharpoons$ | 2HCl |
|---|---|---|---|---|---|
| Initial (*M*): | 0.47 | | 0 | | 3.59 |
| Change (*M*): | $+x$ | | $+x$ | | $-2x$ |
| Equilibrium (*M*): | $(0.47 + x)$ | | $x$ | | $(3.59 - 2x)$ |

Substitute the equilibrium concentrations into the equilibrium constant expression, then solve for $x$. Since $\Delta n = 0$, $K_c = K_P$.

$$K_c = \frac{[HCl]^2}{[H_2][Cl_2]} = \frac{(3.59 - 2x)^2}{(0.47 + x)x} = 193$$

Solving the quadratic equation,

$$x = 0.10$$

Having solved for $x$, calculate the equilibrium concentrations of all species.

$$[H_2] = 0.57\,M \qquad [Cl_2] = 0.10\,M \qquad [HCl] = 3.39\,M$$

Since we assumed that the vessel had a volume of 1 L, the above molarities also correspond to the number of moles of each component.

From the mole fraction of each component and the total pressure, we can calculate the partial pressure of each component.

$$\text{Total number of moles} = 0.57 + 0.10 + 3.39 = 4.06 \text{ mol}$$

$$P_{H_2} = \frac{0.57}{4.06} \times 2.00 = \mathbf{0.28\ atm}$$

$$P_{Cl_2} = \frac{0.10}{4.06} \times 2.00 = \mathbf{0.049\ atm}$$

$$P_{HCl} = \frac{3.39}{4.06} \times 2.00 = \mathbf{1.67\ atm}$$

**14.78** This is a difficult problem. Express the equilibrium number of moles in terms of the initial moles and the change in number of moles ($x$). Next, calculate the mole fraction of each component. Using the mole fraction, you should come up with a relationship between partial pressure and total pressure for each component. Substitute the partial pressures into the equilibrium constant expression to solve for the total pressure, $P_T$.

The reaction is:

| | $N_2$ | + | $3H_2$ | $\rightleftharpoons$ | $2NH_3$ |
|---|---|---|---|---|---|
| Initial (mol): | 1 | | 3 | | 0 |
| Change (mol): | $-x$ | | $-3x$ | | $2x$ |
| Equilibrium (mol): | $(1 - x)$ | | $(3 - 3x)$ | | $2x$ |

$$\text{Mole fraction of } NH_3 = \frac{\text{mol of } NH_3}{\text{total number of moles}}$$

$$X_{NH_3} = \frac{2x}{(1 - x) + (3 - 3x) + 2x} = \frac{2x}{4 - 2x}$$

$$0.21 = \frac{2x}{4 - 2x}$$

$$x = 0.35 \text{ mol}$$

Substituting $x$ into the following mole fraction equations, the mole fractions of $N_2$ and $H_2$ can be calculated.

$$X_{N_2} = \frac{1 - x}{4 - 2x} = \frac{1 - 0.35}{4 - 2(0.35)} = 0.20$$

$$X_{H_2} = \frac{3 - 3x}{4 - 2x} = \frac{3 - 3(0.35)}{4 - 2(0.35)} = 0.59$$

The partial pressures of each component are equal to the mole fraction multiplied by the total pressure.

$$P_{NH_3} = 0.21P_T \qquad P_{N_2} = 0.20P_T \qquad P_{H_2} = 0.59P_T$$

Substitute the partial pressures above (in terms of $P_T$) into the equilibrium constant expression, and solve for $P_T$.

$$K_P = \frac{P_{NH_3}^2}{P_{H_2}^3 P_{N_2}}$$

$$4.31 \times 10^{-4} = \frac{(0.21)^2 P_T^2}{(0.59P_T)^3(0.20P_T)}$$

$$4.31 \times 10^{-4} = \frac{1.07}{P_T^{\ 2}}$$

$$\mathbf{P_T = 5.0 \times 10^1\ atm}$$

**14.80** We carry an additional significant figure throughout this calculation to minimize rounding errors. The initial molarity of $SO_2Cl_2$ is:

$$[SO_2Cl_2] = \frac{6.75\ \text{g}\ SO_2Cl_2 \times \dfrac{1\ \text{mol}\ SO_2Cl_2}{135.0\ \text{g}\ SO_2Cl_2}}{2.00\ \text{L}} = 0.02500\ M$$

The concentration of $SO_2$ at equilibrium is:

$$[SO_2] = \frac{0.0345\ \text{mol}}{2.00\ \text{L}} = 0.01725\ M$$

Since there is a 1:1 mole ratio between $SO_2$ and $SO_2Cl_2$, the concentration of $SO_2$ at equilibrium (0.01725 *M*) equals the concentration of $SO_2Cl_2$ reacted. The concentrations of $SO_2Cl_2$ and $Cl_2$ at equilibrium are:

| | $SO_2Cl_2(g)$ | ⇌ | $SO_2(g)$ | + | $Cl_2(g)$ |
|---|---|---|---|---|---|
| Initial (*M*): | 0.02500 | | 0 | | 0 |
| Change (*M*): | −0.01725 | | +0.01725 | | +0.01725 |
| Equilibrium (*M*): | 0.00775 | | 0.01725 | | 0.01725 |

Substitute the equilibrium concentrations into the equilibrium constant expression to calculate $K_c$.

$$K_c = \frac{[SO_2][Cl_2]}{[SO_2Cl_2]} = \frac{(0.01725)(0.01725)}{(0.00775)} = \mathbf{3.84 \times 10^{-2}}$$

**14.82** $I_2(g) \rightleftharpoons 2I(g)$

Assuming 1 mole of $I_2$ is present originally and α moles reacts, at equilibrium: $[I_2] = 1 - \alpha$, $[I] = 2\alpha$. The total number of moles present in the system $= (1 - \alpha) + 2\alpha = 1 + \alpha$. From Problem 14.117(a) in the text, we know that $K_P$ is equal to:

$$K_P = \frac{4\alpha^2}{1-\alpha^2}P \qquad (1)$$

If there were no dissociation, then the pressure would be:

$$P = \frac{nRT}{V} = \frac{\left(1.00\ \text{g} \times \frac{1\ \text{mol}}{253.8\ \text{g}}\right)\left(0.0821\frac{\text{L}\cdot\text{atm}}{\text{mol}\cdot\text{K}}\right)(1473\ \text{K})}{0.500\ \text{L}} = \mathbf{0.953\ atm}$$

$$\frac{\text{observed pressure}}{\text{calculated pressure}} = \frac{1.51\ \text{atm}}{0.953\ \text{atm}} = \frac{1+\alpha}{1}$$

$$\alpha = 0.584$$

Substituting in equation (1) above:

$$K_P = \frac{4\alpha^2}{1-\alpha^2}P = \frac{(4)(0.584)^2}{1-(0.584)^2} \times 1.51 = \mathbf{3.13}$$

**14.84** According to the ideal gas law, pressure is directly proportional to the concentration of a gas in mol/L if the reaction is at constant volume and temperature. Therefore, pressure may be used as a concentration unit. The reaction is:

| | $N_2$ | + | $3H_2$ | $\rightleftharpoons$ | $2NH_3$ |
|---|---|---|---|---|---|
| Initial (atm): | 0.862 | | 0.373 | | 0 |
| Change (atm): | $-x$ | | $-3x$ | | $+2x$ |
| Equilibrium (atm): | $(0.862 - x)$ | | $(0.373 - 3x)$ | | $2x$ |

$$K_P = \frac{P_{NH_3}^2}{P_{H_2}^3 P_{N_2}}$$

$$4.31 \times 10^{-4} = \frac{(2x)^2}{(0.373-3x)^3(0.862-x)}$$

At this point, we need to make two assumptions that $3x$ is very small compared to 0.373 and that $x$ is very small compared to 0.862. Hence,

$$0.373 - 3x \approx 0.373$$

and

$$0.862 - x \approx 0.862$$

$$4.31 \times 10^{-4} \approx \frac{(2x)^2}{(0.373)^3(0.862)}$$

Solving for $x$.

$$x = 2.20 \times 10^{-3}\ \text{atm}$$

The equilibrium pressures are:

$$P_{N_2} = [0.862 - (2.20 \times 10^{-3})]\text{atm} = \mathbf{0.860\ atm}$$

$$P_{H_2} = [0.373 - (3)(2.20 \times 10^{-3})]\text{atm} = \mathbf{0.366\ atm}$$

$$P_{NH_3} = (2)(2.20 \times 10^{-3}\text{ atm}) = \mathbf{4.40 \times 10^{-3}\ atm}$$

Was the assumption valid that we made above? Typically, the assumption is considered valid if $x$ is less than 5 percent of the number that we said it was very small compared to. Is this the case?

**14.86** **(a)** The equation is:

| | fructose | $\rightleftharpoons$ | glucose |
|---|---|---|---|
| Initial (*M*): | 0.244 | | 0 |
| Change (*M*): | −0.131 | | +0.131 |
| Equilibrium (*M*): | 0.113 | | 0.131 |

Calculating the equilibrium constant,

$$K_c = \frac{[\text{glucose}]}{[\text{fructose}]} = \frac{0.131}{0.113} = \mathbf{1.16}$$

**(b)** $$\textbf{Percent converted} = \frac{\text{amount of fructose converted}}{\text{original amount of fructose}} \times 100\%$$

$$= \frac{0.131}{0.244} \times 100\% = \mathbf{53.7\%}$$

**14.88** **(a)** There is only one gas phase component, $O_2$. The equilibrium constant is simply

$$K_P = P_{O_2} = \mathbf{0.49\ atm}$$

**(b)** From the ideal gas equation, we can calculate the moles of $O_2$ produced by the decomposition of CuO.

$$n_{O_2} = \frac{PV}{RT} = \frac{(0.49\text{ atm})(2.0\text{ L})}{(0.0821\text{ L}\cdot\text{atm/K}\cdot\text{mol})(1297\text{ K})} = 9.2 \times 10^{-3}\text{ mol } O_2$$

From the balanced equation,

$$(9.2 \times 10^{-3}\text{ mol } O_2) \times \frac{4\text{ mol CuO}}{1\text{ mol } O_2} = 3.7 \times 10^{-2}\text{ mol CuO decomposed}$$

$$\textbf{Fraction of CuO decomposed} = \frac{\text{amount of CuO lost}}{\text{original amount of CuO}}$$

$$= \frac{3.7 \times 10^{-2}\text{ mol}}{0.16\text{ mol}} = \mathbf{0.23}$$

**(c)** If a 1.0 mol sample were used, the pressure of oxygen would still be the same (0.49 atm) and it would be due to the same quantity of $O_2$. Remember, a pure solid does not affect the equilibrium position. The moles of CuO lost would still be $3.7 \times 10^{-2}$ mol. Thus the fraction decomposed would be:

$$\frac{0.037}{1.0} = \mathbf{0.037}$$

**(d)** If the number of moles of CuO were less than $3.7 \times 10^{-2}$ mol, the equilibrium could not be established because the pressure of $O_2$ would be less than 0.49 atm. Therefore, the smallest number of moles of CuO needed to establish equilibrium must be **slightly greater than $3.7 \times 10^{-2}$ mol.**

**14.90** We first must find the initial concentrations of all the species in the system.

$$[H_2]_0 = \frac{0.714 \text{ mol}}{2.40 \text{ L}} = 0.298\ M$$

$$[I_2]_0 = \frac{0.984 \text{ mol}}{2.40 \text{ L}} = 0.410\ M$$

$$[HI]_0 = \frac{0.886 \text{ mol}}{2.40 \text{ L}} = 0.369\ M$$

Calculate the reaction quotient by substituting the initial concentrations into the appropriate equation.

$$Q_c = \frac{[HI]_0^2}{[H_2]_0[I_2]_0} = \frac{(0.369)^2}{(0.298)(0.410)} = 1.11$$

We find that $Q_c$ is less than $K_c$. The equilibrium will shift to the right, decreasing the concentrations of $H_2$ and $I_2$ and increasing the concentration of HI.

We set up the usual table. Let $x$ be the decrease in concentration of $H_2$ and $I_2$.

| | $H_2$ | + | $I_2$ | $\rightleftharpoons$ | 2 HI |
|---|---|---|---|---|---|
| Initial ($M$): | 0.298 | | 0.410 | | 0.369 |
| Change ($M$): | $-x$ | | $-x$ | | $+2x$ |
| Equilibrium ($M$): | $(0.298 - x)$ | | $(0.410 - x)$ | | $(0.369 + 2x)$ |

The equilibrium constant expression is:

$$K_c = \frac{[HI]^2}{[H_2][I_2]} = \frac{(0.369 + 2x)^2}{(0.298 - x)(0.410 - x)} = 54.3$$

This becomes the quadratic equation

$$50.3x^2 - 39.9x + 6.48 = 0$$

The smaller root is $x = 0.228\ M$. (The larger root is physically impossible.)

**Having solved for $x$, calculate the equilibrium concentrations.**

$$[H_2] = (0.298 - 0.228)\ M = \mathbf{0.070\ M}$$

$$[I_2] = (0.410 - 0.228)\ M = \mathbf{0.182\ M}$$

$$[HI] = [0.369 + 2(0.228)]\ M = \mathbf{0.825\ M}$$

**14.92** The gas cannot be (a) because the color became lighter with heating. Heating (a) to 150°C would produce some HBr, which is colorless and would lighten rather than darken the gas.

The gas cannot be (b) because $Br_2$ doesn't dissociate into Br atoms at 150°C, so the color shouldn't change.

The gas must be **(c)**. From 25°C to 150°C, heating causes $N_2O_4$ to dissociate into $NO_2$, thus darkening the color ($NO_2$ is a brown gas).

$$N_2O_4(g) \rightarrow 2NO_2(g)$$

Above 150°C, the $NO_2$ breaks up into colorless NO and $O_2$.

$$2NO_2(g) \rightarrow 2NO(g) + O_2(g)$$

An increase in pressure shifts the equilibrium back to the left, forming $NO_2$, thus darkening the gas again.

$$2NO(g) + O_2(g) \rightarrow 2NO_2(g)$$

**14.94** Given the following: $K_c = \dfrac{[NH_3]^2}{[N_2][H_2]^3} = 1.2$

**(a)** Temperature must have units of Kelvin.

$$K_P = K_c(0.0821\ T)^{\Delta n}$$

$$K_P = (1.2)(0.0821 \times 648)^{(2-4)} = \mathbf{4.2 \times 10^{-4}}$$

**(b)** Recalling that,

$$K_{\text{forward}} = \frac{1}{K_{\text{reverse}}}$$

Therefore,

$$K_c' = \frac{1}{1.2} = \mathbf{0.83}$$

**(c)** Since the equation

$$\tfrac{1}{2}N_2(g) + \tfrac{3}{2}H_2(g) \rightleftharpoons NH_3(g)$$

is equivalent to

$$\tfrac{1}{2}[N_2(g) + 3H_2(g) \rightleftharpoons 2NH_3(g)]$$

then, $K_c'$ for the reaction:

$$\frac{1}{2}N_2(g) + \frac{3}{2}H_2(g) \rightleftharpoons NH_3(g)$$

equals $(K_c)^{\frac{1}{2}}$ for the reaction:

$$N_2(g) + 3H_2(g) \rightleftharpoons 2NH_3(g)$$

Thus,

$$K_c' = (K_c)^{\frac{1}{2}} = \sqrt{1.2} = \mathbf{1.1}$$

**(d)** For $K_P$ in part (b):

$$K_P = (0.83)(0.0821 \times 648)^{+2} = \mathbf{2.3 \times 10^3}$$

and for $K_P$ in part (c):

$$K_P = (1.1)(0.0821 \times 648)^{-1} = \mathbf{0.021}$$

**14.96** The vapor pressure of water is equivalent to saying the partial pressure of $H_2O(g)$.

$$K_P = P_{H_2O} = \mathbf{0.0231}$$

$$K_c = \frac{K_P}{(0.0821T)^{\Delta n}} = \frac{0.0231}{(0.0821 \times 293)^1} = \mathbf{9.60 \times 10^{-4}}$$

**14.98** We can calculate the average molar mass of the gaseous mixture from the density.

$$\mathcal{M} = \frac{dRT}{P}$$

Let $\overline{\mathcal{M}}$ be the average molar mass of $NO_2$ and $N_2O_4$. The above equation becomes:

$$\overline{\mathcal{M}} = \frac{dRT}{P} = \frac{(2.3\text{ g/L})(0.0821\text{ L}\cdot\text{atm/K}\cdot\text{mol})(347\text{ K})}{1.3\text{ atm}}$$

$$\overline{\mathcal{M}} = 50.4\text{ g/mol}$$

The average molar mass is equal to the sum of the molar masses of each component times the respective mole fractions. Setting this up, we can calculate the mole fraction of each component.

$$\overline{\mathcal{M}} = X_{NO_2}\mathcal{M}_{NO_2} + X_{N_2O_4}\mathcal{M}_{N_2O_4} = 50.4\text{ g/mol}$$

$$X_{NO_2}(46.01\text{ g/mol}) + (1 - X_{NO_2})(92.01\text{ g/mol}) = 50.4\text{ g/mol}$$

$$X_{NO_2} = 0.905$$

We can now calculate the partial pressure of $NO_2$ from the mole fraction and the total pressure.

$$P_{NO_2} = X_{NO_2} P_T$$

$$P_{NO_2} = (0.905)(1.3 \text{ atm}) = 1.18 \text{ atm} = \mathbf{1.2\ atm}$$

We can calculate the partial pressure of $N_2O_4$ by difference.

$$P_{N_2O_4} = P_T - P_{NO_2}$$

$$P_{N_2O_4} = (1.3 - 1.18)\text{atm} = \mathbf{0.12\ atm}$$

Finally, we can calculate $K_P$ for the dissociation of $N_2O_4$.

$$K_P = \frac{P_{NO_2}^2}{P_{N_2O_4}} = \frac{(1.2)^2}{0.12} = \mathbf{12}$$

**14.100** Diagram **(a)** represents the system at equilibrium.

$$K_c = \frac{[Z]^3}{[X]^4[Y]} = \frac{(4 \times 0.20)^3}{(2 \times 0.20)^4(3 \times 0.20)} = 33.3$$

The value of $Q_c$ for diagram (b) is:

$$Q_c = \frac{[Z]_0^3}{[X]_0^4[Y]_0} = \frac{(3 \times 0.20)^3}{(3 \times 0.20)^4(3 \times 0.20)} = 2.8$$

$Q_c < K_c$. The system will **shift right** (toward product) to reach equilibrium.

The value of $Q_c$ for diagram (c) is:

$$Q_c = \frac{[Z]_0^3}{[X]_0^4[Y]_0} = \frac{(3 \times 0.20)^3}{(0.20)^4(4 \times 0.20)} = 169$$

$Q_c > K_c$. The system will **shift left** (toward reactants) to reach equilibrium.

**14.102** **(a)** shifts to right **(b)** shifts to right **(c)** no change **(d)** no change
**(e)** no change **(f)** shifts to left

**14.104** The equilibrium is: $N_2O_4(g) \rightleftharpoons 2NO_2(g)$

$$K_P = \frac{(P_{NO_2})^2}{P_{N_2O_4}} = \frac{0.15^2}{0.20} = 0.113$$

Volume is doubled so pressure is halved. Let's calculate $Q_P$ and compare it to $K_P$.

$$Q_P = \frac{\left(\frac{0.15}{2}\right)^2}{\left(\frac{0.20}{2}\right)} = 0.0563 < K_P$$

Equilibrium will shift to the right. Some $N_2O_4$ will react, and some $NO_2$ will be formed. Let $x$ = amount of $N_2O_4$ reacted.

| | $N_2O_4(g)$ | $\rightleftharpoons$ | $2NO_2(g)$ |
|---|---|---|---|
| Initial (atm): | 0.10 | | 0.075 |
| Change (atm): | $-x$ | | $+2x$ |
| Equilibrium (atm): | $0.10 - x$ | | $0.075 + 2x$ |

Substitute into the $K_P$ expression to solve for $x$.

$$K_P = 0.113 = \frac{(0.075 + 2x)^2}{0.10 - x}$$

$$4x^2 + 0.413x - 5.68 \times 10^{-3} = 0$$

$$x = 0.0123$$

At equilibrium:

$$P_{NO_2} = 0.075 + 2(0.0123) = 0.0996 \approx \mathbf{0.100\ atm}$$

$$P_{N_2O_4} = 0.10 - 0.0123 = \mathbf{0.09\ atm}$$

Check:

$$K_P = \frac{(0.100)^2}{0.09} = 0.111 \qquad \text{close enough to } 0.113$$

**14.106** (a) Molar mass of $PCl_5$ = 208.2 g/mol

$$P = \frac{nRT}{V} = \frac{\left(2.50\ \text{g} \times \frac{1\ \text{mol}}{208.2\ \text{g}}\right)\left(0.0821\frac{\text{L}\cdot\text{atm}}{\text{mol}\cdot\text{K}}\right)(523\ \text{K})}{0.500\ \text{L}} = \mathbf{1.03\ atm}$$

**(b)**

| | $PCl_5$ | $\rightleftharpoons$ | $PCl_3$ | + | $Cl_2$ |
|---|---|---|---|---|---|
| Initial (atm) | 1.03 | | 0 | | 0 |
| Change (atm) | $-x$ | | $+x$ | | $+x$ |
| Equilibrium (atm) | $1.03 - x$ | | $x$ | | $x$ |

$$K_P = 1.05 = \frac{x^2}{1.03 - x}$$

$$x^2 + 1.05x - 1.08 = 0$$

$$x = 0.639$$

At equilibrium:

$$P_{PCl_5} = 1.03 - 0.639 = \mathbf{0.39\ atm}$$

**(c)** $P_T = (1.03 - x) + x + x = 1.03 + 0.639 = \mathbf{1.67\ atm}$

**(d)** $\dfrac{0.639\text{ atm}}{1.03\text{ atm}} = \mathbf{0.620}$

**14.108** **(a)** $K_P = P_{Hg} = 0.0020\text{ mmHg} = 2.6 \times 10^{-6}\text{ atm} = \mathbf{2.6 \times 10^{-6}}$ (equil. constants are expressed without units)

$$K_c = \frac{K_P}{(0.0821T)^{\Delta n}} = \frac{2.6 \times 10^{-6}}{(0.0821 \times 299)^1} = \mathbf{1.1 \times 10^{-7}}$$

**(b)** Volume of lab $= (6.1\text{ m})(5.3\text{ m})(3.1\text{ m}) = 100\text{ m}^3$

$[Hg] = K_c$

$$\textbf{Total mass of Hg vapor} = \frac{1.1 \times 10^{-7}\ \cancel{\text{mol}}}{1\ \cancel{\text{L}}} \times \frac{200.6\text{ g}}{1\ \cancel{\text{mol}}} \times \frac{1\ \cancel{\text{L}}}{1000\ \cancel{\text{cm}^3}} \times \left(\frac{1\ \cancel{\text{cm}}}{0.01\ \cancel{\text{m}}}\right)^3 \times 100\ \cancel{\text{m}^3} = \mathbf{2.2\ g}$$

The concentration of mercury vapor in the room is:

$$\frac{2.2\text{ g}}{100\text{ m}^3} = 0.022\text{ g/m}^3 = \mathbf{22\ mg/m^3}$$

**Yes!** This concentration exceeds the safety limit of 0.05 mg/m$^3$. Better clean up the spill!

**14.110** There is a temporary dynamic equilibrium between the melting ice cubes and the freezing of water between the ice cubes.

**14.112** First, let's calculate the initial concentration of ammonia.

$$[NH_3] = \frac{14.6\ \cancel{\text{g}} \times \dfrac{1\text{ mol NH}_3}{17.03\ \cancel{\text{g}}\text{ NH}_3}}{4.00\text{ L}} = 0.214\ M$$

Let's set up a table to represent the equilibrium concentrations. We represent the amount of $NH_3$ that reacts as $2x$.

| | $2NH_3(g)$ | $\rightleftharpoons$ | $N_2(g)$ + | $3H_2(g)$ |
|---|---|---|---|---|
| Initial (*M*): | 0.214 | | 0 | 0 |
| Change (*M*): | $-2x$ | | $+x$ | $+3x$ |
| Equilibrium (*M*): | $0.214 - 2x$ | | $x$ | $3x$ |

Substitute into the equilibrium constant expression to solve for $x$.

$$K_c = \frac{[N_2][H_2]^3}{[NH_3]^2}$$

$$0.83 = \frac{(x)(3x)^3}{(0.214 - 2x)^2} = \frac{27x^4}{(0.214 - 2x)^2}$$

Taking the square root of both sides of the equation gives:

$$0.91 = \frac{5.20x^2}{0.214 - 2x}$$

Rearranging,

$$5.20x^2 + 1.82x - 0.195 = 0$$

Solving the quadratic equation gives the solutions:

$$x = 0.086\ M \text{ and } x = -0.44\ M$$

The positive root is the correct answer. The equilibrium concentrations are:

$$[NH_3] = 0.214 - 2(0.086) = \mathbf{0.042\ M}$$

$$[N_2] = \mathbf{0.086\ M}$$

$$[H_2] = 3(0.086) = \mathbf{0.26\ M}$$

**14.114** The equilibrium constant expression for the given reaction is

$$K_P = \frac{P_{CO}^2}{P_{CO_2}}$$

We also know that

$$P_{CO} = X_{CO}P_{total} \quad \text{and} \quad P_{CO_2} = X_{CO_2}P_{total}$$

We can solve for the mole fractions of each gas, and then substitute into the equilibrium constant expression to solve for the total pressure. We carry an additional significant figure throughout this calculation to minimize rounding errors.

$$X_{CO} = \frac{0.025\text{ mol}}{0.037\text{ mol}} = 0.676$$

and

$$X_{CO_2} = 1 - X_{CO} = 0.324$$

Substitute into the equilibrium constant expression.

$$K_P = \frac{P_{CO}^2}{P_{CO_2}} = \frac{(X_{CO}P_T)^2}{X_{CO_2}P_T}$$

$$1.9 = \frac{(0.676P_T)^2}{0.324P_T}$$

$$0.616P_T = 0.457P_T^2$$

$$0.457P_T^2 - 0.616P_T = 0$$

Solving the quadratic equation,

$$\mathbf{P_T = 1.3\ atm}$$

**Check:**

$$P_{CO} = X_{CO}P_{total} = (0.676)(1.35\text{ atm}) = 0.913\text{ atm}$$

$$P_{CO_2} = X_{CO_2}P_{total} = (0.324)(1.35\text{ atm}) = 0.437\text{ atm}$$

$$K_P = \frac{P_{CO}^2}{P_{CO_2}} = \frac{(0.913)^2}{0.437} = 1.9$$

**14.116** $K_P = \dfrac{P_{PCl_5}}{P_{PCl_3} \cdot P_{Cl_2}}$

Note that: $P_A = X_AP_T$. The total pressure at equilibrium is 2.00 atm. We need to come up with expressions for the mole fractions of each gas. Starting with 2.00 moles of $PCl_3$ and 1.00 mole $Cl_2$, let $x$ be the number of moles of $PCl_5$ at equilibrium. Therefore, the number of moles of $PCl_3$ at equilibrium is $(2.00 - x)$, and the number of moles of $Cl_2$ is $(1.00 - x)$. The total number of moles at equilibrium is:

$$x + (2.00 - x) + (1.00 - x) = 3.00 - x$$

The partial pressure of each gas is:

$$P_{PCl_5} = X_{PCl_5}P_T = \left(\frac{x}{3.00 - x}\right)P_T$$

$$P_{PCl_3} = X_{PCl_3}P_T = \left(\frac{2.00 - x}{3.00 - x}\right)P_T$$

$$P_{Cl_2} = X_{Cl_2}P_T = \left(\frac{1.00 - x}{3.00 - x}\right)P_T$$

We substitute the expressions for the partial pressures into the $K_P$ expression to solve for $x$.

$$K_P = \frac{P_{PCl_5}}{P_{PCl_3} \cdot P_{Cl_2}}$$

$$2.93 = \frac{\left(\frac{x}{3.00 - x}\right)P_T}{\left(\frac{2.00 - x}{3.00 - x}\right)P_T \cdot \left(\frac{1.00 - x}{3.00 - x}\right)P_T}$$

$$2.93 = \frac{x(3.00 - x)}{(2.00 - x)(1.00 - x)P_T}$$

$P_T = 2.00$ atm

$$2.93 = \frac{3.00x - x^2}{2.00x^2 - 6.00x + 4.00}$$

$$6.86x^2 - 20.58x + 11.72 = 0$$

Solving the quadratic equation, $x = 0.764$. (The other solution is $x = 2.24$, which is physically impossible.) Finally, we can solve for the partial pressure of each gas.

$$P_{PCl_5} = \left(\frac{x}{3.00 - x}\right)P_T = \left(\frac{0.764}{3.00 - 0.764}\right)(2.00 \text{ atm}) = \mathbf{0.683 \text{ atm}}$$

$$P_{PCl_3} = \left(\frac{2.00 - x}{3.00 - x}\right)P_T = \left(\frac{2.00 - 0.764}{3.00 - 0.764}\right)(2.00 \text{ atm}) = \mathbf{1.11 \text{ atm}}$$

$$P_{Cl_2} = \left(\frac{1.00 - x}{3.00 - x}\right)P_T = \left(\frac{1.00 - 0.764}{3.00 - 0.764}\right)(2.00 \text{ atm}) = \mathbf{0.211 \text{ atm}}$$

Check your work by substituting the above partial pressures into the $K_P$ expression to solve for $K_P$.

**14.118** To determine $\Delta H°$, we need to plot $\ln K_P$ versus $1/T$ ($y$ vs. $x$).

| $\ln K_P$ | $1/T$ |
|---|---|
| 4.93 | 0.00167 |
| 1.63 | 0.00143 |
| 0.83 | 0.00125 |
| 2.77 | 0.00111 |
| 4.34 | 0.00100 |

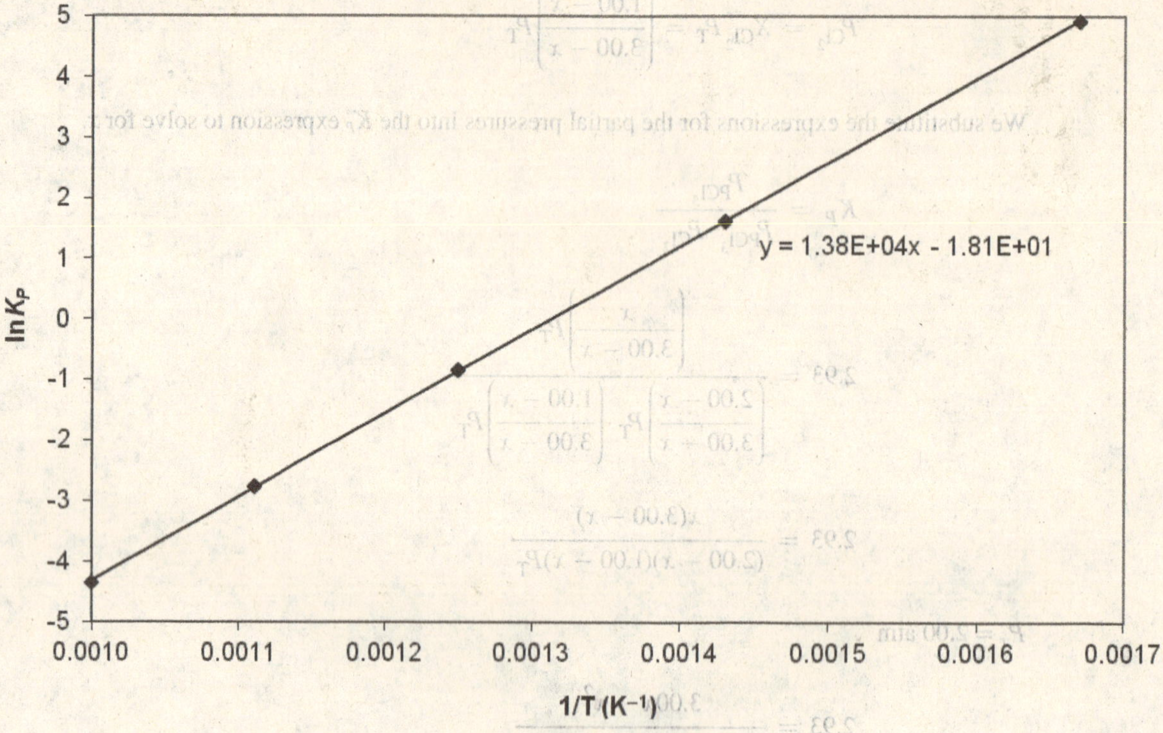


The slope of the plot equals $-\Delta H°/R$.

$$1.38 \times 10^4 \text{ K} = -\frac{\Delta H°}{8.314 \text{ J/mol} \cdot \text{K}}$$

$$\Delta H° = \mathbf{-1.15 \times 10^5 \text{ J/mol} = -115 \text{ kJ/mol}}$$

**14.120** Initially, the pressure of $SO_2Cl_2$ is 9.00 atm. The pressure is held constant, so after the reaction reaches equilibrium, $P_{SO_2Cl_2} + P_{SO_2} + P_{Cl_2} = 9.00 \text{ atm}$. The amount (pressure) of $SO_2Cl_2$ reacted must equal the pressure of $SO_2$ and $Cl_2$ produced for the pressure to remain constant. If we let $P_{SO_2} + P_{Cl_2} = x$, then the pressure of $SO_2Cl_2$ reacted must be $2x$. We set up a table showing the initial pressures, the change in pressures, and the equilibrium pressures.

| | $SO_2Cl_2(g)$ | $\rightleftharpoons$ | $SO_2(g)$ | + | $Cl_2(g)$ |
|---|---|---|---|---|---|
| Initial (atm): | 9.00 | | 0 | | 0 |
| Change (atm): | $-2x$ | | $+x$ | | $+x$ |
| Equilibrium (atm): | $9.00 - 2x$ | | $x$ | | $x$ |

Again, note that the change in pressure for $SO_2Cl_2$ ($-2x$) does not match the stoichiometry of the reaction, because we are expressing changes in pressure. The total pressure is kept at 9.00 atm throughout.

$$K_P = \frac{P_{SO_2}P_{Cl_2}}{P_{SO_2Cl_2}}$$

$$2.05 = \frac{(x)(x)}{9.00 - 2x}$$

$$x^2 + 4.10x - 18.45 = 0$$

**Solving the quadratic equation, $x = 2.71$ atm. At equilibrium,**

$$P_{SO_2} = P_{Cl_2} = x = \mathbf{2.71\ atm}$$

$$P_{SO_2Cl_2} = 9.00 - 2(2.71) = \mathbf{3.58\ atm}$$

**14.122** We start with a table.

| | $A_2$ | + | $B_2$ | $\rightleftharpoons$ | 2AB |
|---|---|---|---|---|---|
| Initial (mol): | 1 | | 3 | | 0 |
| Change (mol): | $-\frac{x}{2}$ | | $-\frac{x}{2}$ | | $+x$ |
| Equilibrium (mol): | $1-\frac{x}{2}$ | | $3-\frac{x}{2}$ | | $x$ |

After the addition of 2 moles of A,

| | $A_2$ | + | $B_2$ | $\rightleftharpoons$ | 2AB |
|---|---|---|---|---|---|
| Initial (mol): | $3-\frac{x}{2}$ | | $3-\frac{x}{2}$ | | $x$ |
| Change (mol): | $-\frac{x}{2}$ | | $-\frac{x}{2}$ | | $+x$ |
| Equilibrium (mol): | $3-x$ | | $3-x$ | | $2x$ |

We write two different equilibrium constants expressions for the two tables.

$$K = \frac{[AB]^2}{[A_2][B_2]}$$

$$K = \frac{x^2}{\left(1-\frac{x}{2}\right)\left(3-\frac{x}{2}\right)} \quad \text{and} \quad K = \frac{(2x)^2}{(3-x)(3-x)}$$

We equate the equilibrium constant expressions and solve for $x$.

$$\frac{x^2}{\left(1-\frac{x}{2}\right)\left(3-\frac{x}{2}\right)} = \frac{(2x)^2}{(3-x)(3-x)}$$

$$\frac{1}{\frac{1}{4}(x^2 - 8x + 12)} = \frac{4}{x^2 - 6x + 9}$$

$$-6x + 9 = -8x + 12$$

$$x = 1.5$$

We substitute $x$ back into one of the equilibrium constant expressions to solve for $K$.

$$K = \frac{(2x)^2}{(3-x)(3-x)} = \frac{(3)^2}{(1.5)(1.5)} = \mathbf{4.0}$$

Substitute $x$ into the other equilibrium constant expression to see if you obtain the same value for $K$. Note that we used moles rather than molarity for the concentrations, because the volume, $V$, cancels in the equilibrium constant expressions.

**14.124** **(a)** As volume of a gas is increased, pressure decreases at constant temperature (see Figure 5.7(b) of the text). As pressure decreases, LeChâtelier's principle tells us that a system will shift to increase the pressure by shifting to the side of the equation with more moles of gas. In this case, the system will shift to the right. As volume of the system is increased, more $NO_2$ will be produced increasing the molecules of gas in the system, thereby increasing the pressure of the system compared to an ideal gas. Note that a large volume corresponds to a small value of $1/V$. A plot of $P$ versus $1/V$ at constant temperature for the system and an ideal gas is shown below.

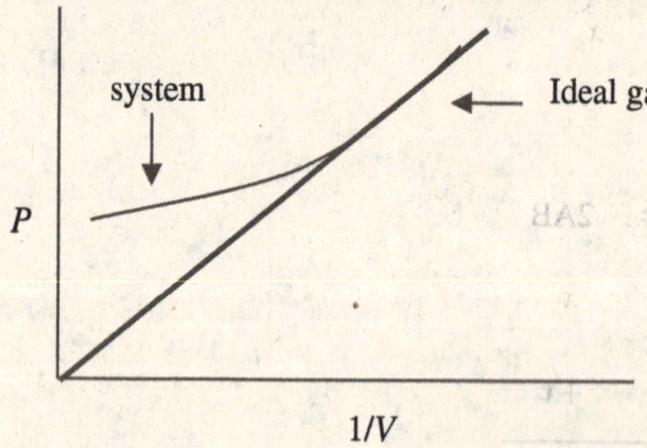


**(b)** As temperature of a gas is increased, the volume occupied by the gas increases at constant pressure (See Figure 5.9 of the text). Because this reaction is endothermic, the system will shift to the right as the temperature is increased. As the system shifts right, more $NO_2$ will be produced increasing the molecules of gas in the system, thereby increasing the volume occupied by these gases compared to an ideal gas. A plot of $V$ versus $T$ at constant pressure for the system and an ideal gas is shown below.

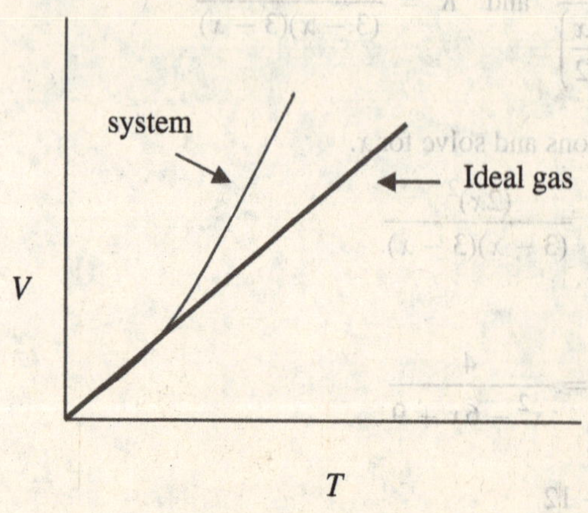

# CHAPTER 15
# ACIDS AND BASES

## PROBLEM-SOLVING STRATEGIES AND TUTORIAL SOLUTIONS

### TYPES OF PROBLEMS

**Problem Type 1:** Identifying Conjugate Acid-Base Pairs.

**Problem Type 2:** The Ion-Product Constant ($K_w$), Calculating [$H^+$] from [$OH^-$].

**Problem Type 3:** pH Calculations.

**(a)** Calculating pH from [$H^+$].

**(b)** Calculating [$H^+$] from pH.

**Problem Type 4:** Calculating the pH of a Strong Acid and a Strong Base.

**Problem Type 5:** Weak Acids.

**(a)** Ionization of a weak monoprotic acid.

**(b)** Determining $K_a$ from a pH measurement.

**Problem Type 6:** Calculating the pH of a Diprotic Acid.

**Problem Type 7:** Calculating the pH of a Weak Base.

**Problem Type 8:** Predicting the Acid-Base Properties of Salt Solutions.

## PROBLEM TYPE 1: IDENTIFYING CONJUGATE ACID-BASE PAIRS

A **conjugate acid-base pair** can be defined as *an acid and its conjugate base or a base and its conjugate acid.* The conjugate base of a Brønsted acid is the species that remains when *one* proton has been removed from the acid. Conversely, a conjugate acid results from the addition of *one* proton to a Brønsted base.

**EXAMPLE 15.1**
**Consider the reaction**

$$HSO_4^-(aq) + HCO_3^-(aq) \rightleftharpoons SO_4^{2-}(aq) + H_2CO_3(aq)$$

**(a) Identify the acids and bases for the forward and reverse reactions.**

**(b) Identify the conjugate acid-base pairs.**

**Strategy:** (a) An acid is a proton donor and a base is a proton acceptor. (b) Remember that a conjugate base always has one fewer H atom and one more negative charge (or one fewer positive charge) than the formula of the corresponding acid.

**Solution:**

**(a)** In the forward reaction, $HSO_4^-$ is the proton donor, which makes it an acid. The proton acceptor, $HCO_3^-$, is a base. In the reverse reaction, the proton donor (acid) is $H_2CO_3$, and $SO_4^{2-}$ is the proton acceptor (base).

**(b)** $HSO_4^-(aq) + HCO_3^-(aq) \rightleftharpoons SO_4^{2-}(aq) + H_2CO_3(aq)$

acid$_1$ base$_2$ base$_1$ acid$_2$

The subscripts 1 and 2 designate the two conjugate acid-base pairs.

**PRACTICE EXERCISE**

1. Identify the conjugate bases of the following acids:

(a) $CH_3COOH$ (b) $H_2S$ (c) $HSO_3^-$ (d) HClO

**Text Problems:** 15.4, 15.6, 15.8

## PROBLEM TYPE 2: THE ION-PRODUCT CONSTANT ($K_w$), CALCULATING [H$^+$] FROM [OH$^-$]

Water is a very weak electrolyte and therefore a poor conductor of electricity, but it does ionize to a small extent:

$$H_2O(l) \rightleftharpoons H^+(aq) + OH^-(aq)$$

We can write the equilibrium constant for the autoionization of water as

$$K_c = \frac{[H^+][OH^-]}{[H_2O]}$$

Since a very small fraction of water molecules are ionized, the concentration of water, $[H_2O]$, remains virtually unchanged. Therefore, we assume that the concentration of water is constant, and we write:

$$K_c[H_2O] = K_w = [H^+][OH^-] \qquad (15.3, \text{text})$$

In pure water at 25°C, the concentrations of $H^+$ and $OH^-$ ions are equal and found to be $[H^+] = [OH^-] = 1.0 \times 10^{-7}$ *M*. Substituting these concentrations into Equation (15.3) of the text,

$$K_w = [H^+][OH^-] = (1 \times 10^{-7})(1 \times 10^{-7}) = 1 \times 10^{-14}$$

Thus, knowing $K_w$ and one of the concentrations, either $[H^+]$ or $[OH^-]$, we can easily calculate the other concentration.

**EXAMPLE 15.2**

**The $OH^-$ ion concentration in a certain solution at 25°C is $5.0 \times 10^{-5}$ *M*. What is the $H^+$ concentration?**

**Strategy:** We are given the concentration of $OH^-$ ions and asked to calculate $[H^+]$. The relationship between $[H^+]$ and $[OH^-]$ in water or an aqueous solution is given by the ion-product of water, $K_w$ [Equation (15.3) of the text].

**Solution:** The ion product of water is applicable to all aqueous solutions. At 25°C,

$$K_w = 1.0 \times 10^{-14} = [H^+][OH^-]$$

Rearrange the equation to solve for $[H^+]$.

$$[H^+] = \frac{1.0\times10^{-14}}{[OH^-]} = \frac{1.0\times10^{-14}}{5.0\times10^{-5}} = \mathbf{2.0\times10^{-10}}\ \boldsymbol{M}$$

**PRACTICE EXERCISE**

2. The $H^+$ concentration in a certain solution is $6.6 \times 10^{-4}$ *M*. What is the $OH^-$ concentration?

| **Text Problems:** 15.16, 15.18, **15.20c** |
|---|

## PROBLEM TYPE 3: pH CALCULATIONS

Because the concentrations of $H^+$ and $OH^-$ ions in aqueous solutions are frequently very small numbers making them inconvenient to work with, the Danish biochemist Soren Sorensen in 1909 proposed a more practical measure called pH. The **pH** of a solution is defined as *the negative log of the hydrogen ion concentration* (in moles per liter).

$$pH = -\log[H^+]$$

The pOH of a solution is defined in a similar manner.

$$pOH = -\log[OH^-]$$

A useful relationship between pH and pOH is:

$$pH + pOH = 14.$$

See Section 15.3 of the text for a complete discussion.

### A. Calculating pH from $[H^+]$

**EXAMPLE 15.3**

**The $[H^+]$ of a solution is 0.015 *M*. What is the pH of the solution?**

**Strategy:** The relationship between $[H^+]$ and pH is: $pH = -\log[H^+]$. The hydrogen ion concentration is given in the problem.

**Solution:** Substitute the $H^+$ concentration into the above equation to calculate the pH of the solution.

$$pH = -\log[H^+] = -\log[0.015] = \mathbf{1.82}$$

**PRACTICE EXERCISE**

3. The $OH^-$ concentration of a certain ammonia solution is $7.2 \times 10^{-4}$ *M*. What is the pOH and pH?

| **Text Problem:** 15.18 |
|---|

## B. Calculating [$H^+$] from pH or pOH

**EXAMPLE 15.4**

**What is the $H^+$ concentration in a solution with a pOH of 3.9?**

**Strategy:** Here we are given the pOH of a solution and asked to calculate [$H^+$]. First, the pH can be calculated from the pOH. Because pH is defined as pH = $-\log[H^+]$, we can solve for [$H^+$] by taking the antilog of the pH; that is, $[H^+] = 10^{-pH}$.

**Solution:** Recall that pH + pOH = 14.00. Since pOH is given in the problem, we can calculate the pH.

$$pH = 14.00 - pOH$$

$$pH = 14.00 - 3.9 = 10.1$$

The $H^+$ concentration can now be determined from the pH.

$$pH = -\log[H^+]$$

or,

$$-pH = \log[H^+]$$

Taking the antilog of both sides of the equation,

$$10^{-pH} = [H^+]$$

$$[H^+] = 10^{-10.1} = \mathbf{8 \times 10^{-11}\ \mathit{M}}$$

**PRACTICE EXERCISE**

4. The pH of a solution is 4.45. What is the $H^+$ concentration?

**Text Problem: 15.20 a,b**

## PROBLEM TYPE 4: CALCULATING THE pH OF A STRONG ACID AND A STRONG BASE

**Strong acids and strong bases** are strong electrolytes that are assumed to ionize completely in water. For example, consider the strong acid, hydrochloric acid [HCl(*aq*)]. We assume that it ionizes completely in water.

$$HCl(aq) \rightarrow H^+(aq) + Cl^-(aq)$$

Since strong acids ionize completely in water, we can easily calculate the [$H^+$] in solution. For example, consider a 0.10 *M* HCl solution. Let's set up a table to determine the [$H^+$] of the solution.

| | HCl(*aq*) → | $H^+(aq)$ + | $Cl^-(aq)$ |
|---|---|---|---|
| initial conc. | 0.10 *M* | 0 | 0 |
| conc. after ionization | 0 | 0.10 *M* | 0.10 *M* |

Since there is a one:one mole ratio between HCl and $H^+$, the $H^+$ concentration after ionization equals the initial concentration of HCl.

Note that the above reaction can be written more accurately as:

$$HCl(aq) + H_2O(l) \rightarrow H_3O^+(aq) + Cl^-(aq)$$

HCl is an acid and donates a proton ($H^+$) to the weak base, $H_2O$. $H^+$ in aqueous solution is really shorthand notation for $H_3O^+$.

You need to know the six strong acids and the six strong bases. Half the battle in pH calculations is recognizing the type of species in solution (i.e., strong acid, strong base, weak acid, weak base, or salt).

**The six strong acids**

| | |
|---|---|
| $HClO_4$ | perchloric acid |
| HI | hydroiodic acid |
| HBr | hydrobromic acid |
| HCl | hydrochloric acid |
| $H_2SO_4$ | sulfuric acid |
| $HNO_3$ | nitric acid |

The six **strong bases** include the five alkali metal hydroxides (LiOH, NaOH, KOH, RbOH, and CsOH), plus barium hydroxide, $Ba(OH)_2$.

**EXAMPLE 15.5**

**Calculate the pH of a 0.10 *M* $Ba(OH)_2$ solution.**

**Strategy:** You must recognize that barium hydroxide is one of the six strong bases. Strong bases ionize completely in solution.

**Solution:** $Ba(OH)_2$ is a strong base; each $Ba(OH)_2$ unit produces two $OH^-$ ions. The changes in concentrations of all species can be represented as follows:

| | $Ba(OH)_2(aq)$ $\rightarrow$ | $Ba^{2+}(aq)$ + | $2OH^-(aq)$ |
|---|---|---|---|
| Initial (*M*): | 0.10 | 0 | 0 |
| Change (*M*): | −0.10 | +0.10 | +2(0.10) |
| Final (*M*): | 0 | 0.10 | 0.20 |

Because there is a 2:1 mole ratio between $OH^-$ and $Ba(OH)_2$, the $OH^-$ concentration is double the $Ba(OH)_2$ concentration.

Calculate the pOH from the $OH^-$ concentration.

$$pOH = -\log[OH^-]$$

$$\text{pOH} = -\log[0.20] = 0.70$$

Use the relationship, pH + pOH = 14.0 to calculate the pH of the $Ba(OH)_2$ solution.

$$\text{pH} = 14.00 - 0.70 = \mathbf{13.30}$$

**PRACTICE EXERCISE**

**5.** What is the pH of a 0.025 *M* HCl solution?

**Text Problems:** 15.24, 15.26

## PROBLEM TYPE 5: WEAK ACIDS

Most acids ionize only to a limited extent in water. Such acids are classified as **weak acids**. For example, $HNO_2$ (nitrous acid), is a weak acid. Its ionization in water is represented by

$$HNO_2(aq) + H_2O(l) \rightleftharpoons H_3O^+(aq) + NO_2^-(aq)$$

or simply

$$HNO_2(aq) \rightleftharpoons H^+(aq) + NO_2^-(aq)$$

The equilibrium constant for this acid ionization is called the **acid ionization constant, $K_a$**. It is given by:

$$K_a = \frac{[H^+][NO_2^-]}{[HNO_2]}$$

In general,

$$K_a = \frac{[H^+][A^-]}{[HA]}$$

where,

$A^-$ is the conjugate base of the weak acid, HA.

### A. Ionization of a weak monoprotic acid

This problem is very similar to equilibrium calculations in Chapter 14. Try to think of this problem as just another equilibrium calculation. The only difference is that in Chapter 14, we dealt mostly with gas phase reactions. Now, we are dealing with weak acids in aqueous solution. Some problems will ask you to calculate the equilibrium concentrations of all species, others will ask for the pH of the solution.

**EXAMPLE 15.6**

**Calculate the pH of a 0.20 *M* acetic acid, $CH_3COOH$, solution.**

**Strategy:** First, recognize that acetic acid is a weak acid. It is not one of the six strong acids, so it must be a weak acid. Recall that a weak acid only partially ionizes in water. We are given the initial quantity of a weak acid ($CH_3COOH$) and asked to calculate the pH. In order to calculate pH, we need to determine the $H^+$ concentration. In determining the $H^+$ concentration, we ignore the ionization of $H_2O$ as a source of $H^+$, so the major source of $H^+$ ions is the acid. We follow the procedure outlined in Section 15.5 of the text.

**Solution:**

*Step 1:* We ignore water's contribution to $[H^+]$. We consider $CH_3COOH$ as the only source of $H^+$ ions.

*Step 2:* Letting $x$ be the equilibrium concentration of $H^+$ and $CH_3COO^-$ ions in mol/L, we summarize:

| | $CH_3COOH(aq)$ | $\rightleftharpoons$ | $H^+(aq)$ | + | $CH_3COO^-(aq)$ |
|---|---|---|---|---|---|
| Initial ($M$): | 0.20 | | 0 | | 0 |
| Change ($M$): | $-x$ | | $+x$ | | $+x$ |
| Equilibrium ($M$): | $0.20 - x$ | | $x$ | | $x$ |

*Step 3:* Write the ionization constant expression in terms of the equilibrium concentrations. Knowing the value of the equilibrium constant ($K_a$), solve for $x$.

$$K_a = \frac{[H^+][CH_3COO^-]}{[CH_3COOH]}$$

You can look up the $K_a$ value for acetic acid in Table 15.3 of the text.

$$1.8 \times 10^{-5} = \frac{(x)(x)}{(0.20 - x)}$$

At this point, we can make an assumption that $x$ is very small compared to 0.20. Hence,

$$0.20 - x \approx 0.20$$

Oftentimes, assumptions such as these are valid if $K$ is very small. A very small value of $K$ means that a very small amount of reactants go to products. Hence, $x$ is small. If we did not make this assumption, we would have to solve a quadratic equation.

$$1.8 \times 10^{-5} \approx \frac{(x)(x)}{0.20}$$

Solving for $x$.

$$x = 1.9 \times 10^{-3}\ M = [H^+]$$

*Step 4:* Having solved for the $[H^+]$, calculate the pH of the solution.

$$\mathbf{pH} = -\log[H^+] = -\log(1.9 \times 10^{-3}) = \mathbf{2.72}$$

**Check:** Checking the validity of the assumption,

$$\frac{1.9 \times 10^{-3}}{0.20} \times 100\% = 0.95\% < 5\%$$

The assumption was valid.

**PRACTICE EXERCISE**

**6.** Calculate the pH of a 0.50 $M$ nitrous acid ($HNO_2$) solution. The $K_a$ value for $HNO_2$ is $4.5 \times 10^{-4}$.

**Text Problems: 15.44**, 15.46, 15.48, 15.50

## B. Determining $K_a$ from a pH measurement

This problem is the reverse of Example 15.6 above. From the pH of the solution, you can calculate the $[H^+]$. Then from the $[H^+]$, you can calculate the equilibrium concentrations of the other species in solution. Substitute these equilibrium concentrations into the ionization constant expression to solve for $K_a$.

**EXAMPLE 15.7**

**The pH of a 0.10 *M* solution of a weak monoprotic acid is 5.15. What is the $K_a$ of the acid?**

**Strategy:** Weak acids only partially ionize in water.

$$HA(aq) \rightleftharpoons H^+(aq) + A^-(aq)$$

Note that the concentration of the weak acid given refers to the initial concentration before ionization has started. The pH of the solution, on the other hand, refers to the situation at equilibrium. To calculate $K_a$, we need to know the concentrations of all three species, [HA], $[H^+]$, and $[A^-]$ at equilibrium. We ignore the ionization of water as a source of $H^+$ ions.

**Solution:** We proceed as follows.

*Step 1:* The major species in solution are HA, $H^+$, and the conjugate acid $A^-$.

*Step 2:* First, we need to calculate the hydrogen ion concentration from the pH value.

$$pH = -\log[H^+]$$

or,

$$-pH = \log[H^+]$$

Taking the antilog of both sides of the equation,

$$10^{-pH} = [H^+]$$

$$[H^+] = 10^{-5.15} = 7.1 \times 10^{-6}\,M$$

*Step 3:* If the concentration of $H^+$ is $7.1 \times 10^{-6}$ *M* at equilibrium, that must mean that $7.1 \times 10^{-6}$ *M* of the acid ionized. We summarize the changes.

| | HA(*aq*) | ⇌ | $H^+(aq)$ | + | $A^-(aq)$ |
|---|---|---|---|---|---|
| Initial (*M*): | 0.10 | | 0 | | 0 |
| Change (*M*): | $-7.1 \times 10^{-6}$ | | $+7.1 \times 10^{-6}$ | | $+7.1 \times 10^{-6}$ |
| Equilibrium (*M*): | $0.10 - (7.1 \times 10^{-6})$ | | $7.1 \times 10^{-6}$ | | $7.1 \times 10^{-6}$ |

*Step 4:* Substitute the equilibrium concentrations into the ionization constant expression to solve for $K_a$.

$$K_a = \frac{[H^+][A^-]}{[HA]}$$

$$K_a = \frac{(7.1 \times 10^{-6})^2}{(0.10)} = \mathbf{5.0 \times 10^{-10}}$$

**PRACTICE EXERCISE**

7. The pH of a 0.50 *M* monoprotic weak acid is 2.24. What is the $K_a$ of the acid?

**Text Problem: 15.54**

## PROBLEM TYPE 6: CALCULATING THE pH OF A DIPROTIC ACID

Diprotic acids may yield more than one hydrogen ion per molecule. These acids ionize in a stepwise manner, that is, they lose one proton at a time. An ionization constant expression can be written for each ionization stage. Consequently, two equilibrium constant expressions must be used to calculate the concentrations of species in the acid solution. For the generic diprotic acid, $H_2A$, we can write:

$$H_2A(aq) \rightleftharpoons H^+(aq) + HA^-(aq) \qquad K_{a_1} = \frac{[H^+][HA^-]}{[H_2A]}$$

$$HA^-(aq) \rightleftharpoons H^+(aq) + A^{2-}(aq) \qquad K_{a_2} = \frac{[H^+][A^{2-}]}{[HA^-]}$$

Note that the conjugate base in the first ionization becomes the acid in the second ionization.

**EXAMPLE 15.8**

**Calculate the concentrations of $H_2A$, $HA^-$, $H^+$, and $A^{2-}$ in a 1.0 *M* $H_2A$ solution. The first and second ionization constants for $H_2A$ are $1.3 \times 10^{-2}$ and $6.3 \times 10^{-8}$, respectively.**

**Strategy:** Determining the pH of a diprotic acid in aqueous solution is more involved than for a monoprotic acid. $H_2A$ is a weak acid, and the conjugate base produced in the first ionization ($HA^-$) is also a weak acid. We follow the procedure for determining the pH of a weak acid for both stages.

**Solution:** We proceed according to the following steps.

*Step 1:* Complete a table showing the concentrations for the first ionization stage.

| | $H_2A(aq)$ | $\rightleftharpoons$ | $H^+(aq)$ | + | $HA^-(aq)$ |
|---|---|---|---|---|---|
| Initial (*M*): | 1.0 | | 0 | | 0 |
| Change (*M*): | $-x$ | | $+x$ | | $+x$ |
| Equilibrium (*M*): | $1.0 - x$ | | $x$ | | $x$ |

*Step 2:* Write the ionization constant expression for $K_{a_1}$. Then, solve for $x$.

$$K_{a_1} = \frac{[H^+][HA^-]}{[H_2A]}$$

$$1.3 \times 10^{-2} = \frac{(x)(x)}{(1.0 - x)}$$

Since $K_{a_1}$ is quite large, we cannot make the assumption

$$1.0 - x \approx 1.0$$

Therefore, we must solve a quadratic equation.

$$x^2 + 0.013x - 0.013 = 0$$

The above equation is a quadratic equation of the form $ax^2 + bx + c = 0$. The solution for a quadratic equation is:

$$x = \frac{-b \pm \sqrt{b^2 - 4ac}}{2a}$$

Here, we have a = 1, b = 0.013, and c = −0.013. Substituting into the above equation,

$$x = \frac{-0.013 \pm \sqrt{(0.013)^2 - 4(1)(-0.013)}}{2(1)}$$

$$x = \frac{-0.013 \pm 0.23}{2}$$

$$x = 0.11\,M \quad \text{or} \quad x = -0.12\,M$$

The second solution is physically impossible because you cannot have a negative concentration. The first solution is the correct answer.

*Step 3:* Having solved for $x$, calculate the concentrations when the equilibrium for the first stage of ionization is reached.

Because $K_{a_1} >> K_{a_2}$, we assume that essentially all the $H^+$ comes from the first ionization stage. Hence,

$$[H^+] = [HA^-] = x = \mathbf{0.11\,M}$$

$$[H_2A] = 1.0 - x = 1.00 - 0.11 = \mathbf{0.89\,M}$$

*Step 4:* Now, consider the second stage of ionization. Set up a table showing the concentrations for the second ionization stage. Let $y$ be the change in concentration.

| | $HA^-(aq)$ | $\rightleftharpoons$ $H^+(aq)$ | + $A^{2-}(aq)$ |
|---|---|---|---|
| Initial ($M$): | 0.11 | 0.11 | 0 |
| Change ($M$): | $-y$ | $+y$ | $+y$ |
| Equilibrium ($M$): | $0.11 - y$ | $0.11 + y$ | $y$ |

*Step 5:* Write the ionization constant expression for $K_{a_2}$. Then, solve for $y$.

$$K_{a_2} = \frac{[H^+][A^{2-}]}{[HA^-]}$$

$$6.3 \times 10^{-8} = \frac{(0.11 + y)(y)}{(0.11 - y)}$$

Since $K_{a_2}$ is very small, we can make an assumption that $y$ is very small compared to 0.11.

Hence,

$$0.11 \pm y \approx 0.11$$

$$6.3 \times 10^{-8} = \frac{(0.11)(y)}{0.11}$$

Solving for $y$,

$$y = 6.3 \times 10^{-8}\ M = [A^{2-}]$$

**Check:** Checking the validity of the assumption,

$$\frac{6.3 \times 10^{-8}}{0.11} \times 100\% = 5.7 \times 10^{-5}\ \% < 5\%$$

The assumption was valid.

**PRACTICE EXERCISE**

8. The first and second ionization constants of $H_2CO_3$ are $4.2 \times 10^{-7}$ and $4.8 \times 10^{-11}$, respectively. Calculate the concentrations of $H^+$, $HCO_3^-$, $CO_3^{2-}$, and unionized $H_2CO_3$ in a 0.080 $M$ $H_2CO_3$ solution.

**Text Problems: 15.62, 15.64**

## PROBLEM TYPE 7: CALCULATING THE pH OF A WEAK BASE

The procedure used to calculate the pH of a weak base solution is essentially the same as the one used for weak acids.

**EXAMPLE 15.9**

**What is the pH of a 0.10 $M$ $C_5H_5N$ (pyridine) solution?**

**Strategy:** First, recognize that pyridine is a weak base. It is an amine. Recall that a weak base only partially ionizes in water. We are given the initial quantity of a weak base ($C_5H_5N$) and asked to calculate the pH. To calculate pH, we can first determine the pOH from the $OH^-$ concentration and then calculate pH from pOH. In determining the $OH^-$ concentration, we ignore the ionization of $H_2O$ as a source of $OH^-$, so the major source of $OH^-$ ions is the base. We follow the procedure outlined in Section 15.6 of the text.

**Solution:**

*Step 1:* We ignore water's contribution to $[OH^-]$. We consider $C_5H_5N$ as the only source of $OH^-$ ions.

*Step 2:* Letting $x$ be the equilibrium concentration of $OH^-$ and $C_5H_5NH^+$ ions in mol/L, we summarize:

| | $C_5H_5N(aq)$ + $H_2O(l)$ ⇌ | $C_5H_5NH^+(aq)$ + | $OH^-(aq)$ |
|---|---|---|---|
| Initial ($M$): | 0.10 | 0 | 0 |
| Change ($M$): | $-x$ | $+x$ | $+x$ |
| Equilibrium ($M$): | $0.10 - x$ | $x$ | $x$ |

*Step 3:* Write the ionization constant expression in terms of the equilibrium concentrations. Knowing the value of the equilibrium constant ($K_b$), solve for $x$.

$$K_b = \frac{[C_5H_5NH^+][OH^-]}{[C_5H_5N]}$$

You can look up the $K_b$ value for pyridine in Table 15.4 of the text.

$$1.7 \times 10^{-9} = \frac{(x)(x)}{(0.10 - x)}$$

At this point, we can make an assumption that $x$ is very small compared to 0.10. Hence,

$$0.10 - x \approx 0.10$$

Oftentimes, assumptions such as these are valid if $K$ is very small. A very small value of $K$ means that a very small amount of reactants go to products. Hence, $x$ is small. If we did not make this assumption, we would have to solve a quadratic equation.

$$1.7 \times 10^{-9} \approx \frac{(x)(x)}{0.10}$$

Solving for $x$,

$$x = 1.3 \times 10^{-5}\ M = [OH^-]$$

*Step 4:* Having solved for the [OH$^-$], calculate the pOH of the solution. Then use the relationship, pH + pOH = 14, to solve for the pH of the solution.

$$\text{pOH} = -\log[OH^-] = -\log(1.3 \times 10^{-5}) = 4.89$$

$$\textbf{pH} = 14.00 - \text{pOH} = 14.00 - 4.89 = \mathbf{9.11}$$

**PRACTICE EXERCISE**

**9.** What is the pH of a 2.0 $M$ $NH_3$ solution?

**Text Problem: 15.56**

## PROBLEM TYPE 8: PREDICTING THE ACID-BASE PROPERTIES OF SALT SOLUTIONS

Salts, when dissolved in water, can produce neutral, acidic, or basic solutions. We need to consider four possibilities.

1. **Salts that produce neutral solutions.** It is generally true that salts containing an alkali metal ion or an alkaline earth metal ion (except $Be^{2+}$) and the conjugate base of a strong acid (for example, $Cl^-$, $NO_3^-$, and $ClO_4^-$) do not undergo hydrolysis, and thus their solutions are neutral.

2. **Salts that produce basic solutions.** Salts that contain the conjugate base of a weak acid and an alkali metal or alkaline earth metal ion, produce basic solutions. As an example, consider potassium fluoride. The dissociation of KF in water is given by:

$$KF(s) \xrightarrow{H_2O} K^+(aq) + F^-(aq)$$

A hydrated alkali or alkaline earth metal ion has no acidic or basic properties. However, the conjugate base of a weak acid is a weak base (it has an affinity for $H^+$ ions). $F^-$ is the conjugate base of the weak acid HF. The hydrolysis reaction is given by:

$$F^-(aq) + H_2O(l) \rightleftharpoons HF(aq) + OH^-(aq)$$

Because this reaction produces $OH^-$ ions, the potassium fluoride solution will be basic.

3. **Salts that produce acidic solutions**. Salts that contain the conjugate acid of a weak base and the conjugate base of a strong acid, produce acidic solutions. As an example, consider ammonium iodide. The dissociation of $NH_4I$ in water is given by

$$NH_4I(s) \xrightarrow{H_2O} NH_4^+(aq) + I^-(aq)$$

The conjugate base of a strong acid, such as $I^-$, does not undergo hydrolysis. However, the conjugate acid of a weak base is a weak acid and does undergo hydrolysis. The reaction is

$$NH_4^+(aq) + H_2O(l) \rightleftharpoons NH_3(aq) + H_3O^+(aq)$$

Since this reaction produces $H_3O^+$ ions, the ammonium iodide solution will be acidic.

4. **Salts in which both the cation and anion hydrolyze.** In these cases, you must compare the base strength of the anion to the acid strength of the cation. We consider three situations.

- $K_b > K_a$. If $K_b$ for the anion is greater than $K_a$ for the cation, the solution is basic because the anion will hydrolyze to a greater extent than the cation. At equilibrium, there will be more $OH^-$ ions than $H^+$ ions.
- $K_b < K_a$. Conversely, if $K_b$ for the anion is smaller than $K_a$ for the cation, the solution will be acidic because cation hydrolysis will be more extensive than anion hydrolysis. At equilibrium, there will be more $H^+$ ions than $OH^-$ ions.
- $K_b \approx K_a$. If $K_b$ is approximately equal to $K_a$, the solution will be close to neutral.

**EXAMPLE 15.10**

**Is a 0.10 *M* solution of $Na_2CO_3$ acidic, basic, or neutral?**

**Strategy:** In deciding whether a salt will undergo hydrolysis, ask yourself the following questions: Is the cation a highly charged metal ion or the conjugate acid of a weak base? Is the anion the conjugate base of a weak acid? If yes to either question, then hydrolysis will occur. In cases where both the cation and the anion react with water, the pH of the solution will depend on the relative magnitudes of $K_a$ for the cation and $K_b$ for the anion (see Table 15.7 of the text).

**Solution:** We first break up the salt into its cation and anion components and then examine the possible reaction of each ion with water. The dissociation of $Na_2CO_3$ in water is given by

$$Na_2CO_3(s) \xrightarrow{H_2O} 2Na^+(aq) + CO_3^{2-}(aq)$$

A hydrated alkali or alkaline earth metal ion has no acidic or basic properties. However, the conjugate base of a weak acid is a weak base (it has an affinity for $H^+$ ions). $CO_3^{2-}$ is the conjugate base of the weak acid $HCO_3^-$. The hydrolysis reaction is given by

$$CO_3^{2-}(aq) + H_2O(l) \rightleftharpoons HCO_3^-(aq) + OH^-(aq)$$

Because this reaction produces $OH^-$ ions, the sodium carbonate solution will be **basic.**

**PRACTICE EXERCISE**

**10.** Predict whether the following aqueous solutions will be acidic, basic, or neutral.

(a) KI

(b) $NH_4Cl$

(c) $CH_3COOK$

| **Text Problems: 15.76,** 15.78, 15.80, 15.82 |
|---|

## ANSWERS TO PRACTICE EXERCISES

**1.** **(a)** $CH_3COO^-$ **(b)** $HS^-$ **(c)** $SO_3^{2-}$ **(d)** $ClO^-$

**2.** $[OH^-] = 1.5 \times 10^{-11}\ M$ **3.** $pOH = 3.14$, $pH = 10.86$ **4.** $[H^+] = 3.5 \times 10^{-5}\ M$

**5.** $pH = 1.60$ **6.** $pH = 1.82$ **7.** $K_a = 6.69 \times 10^{-5}$

**8.** $[H^+] = [HCO_3^-] = 1.8 \times 10^{-4}\ M$ $[H_2CO_3] = 0.080\ M$ $[CO_3^{2-}] = 4.8 \times 10^{-11}\ M$

**9.** $pH = 11.8$ **10.** **(a)** neutral **(b)** acidic **(c)** basic

## SOLUTIONS TO SELECTED TEXT PROBLEMS

**15.4** Recall that the conjugate base of a Brønsted acid is the species that remains when *one* proton has been removed from the acid.

**(a)** nitrite ion: $NO_2^-$

**(b)** hydrogen sulfate ion (also called bisulfate ion): $HSO_4^-$

**(c)** hydrogen sulfide ion (also called bisulfide ion): $HS^-$

**(d)** cyanide ion: $CN^-$

**(e)** formate ion: $HCOO^-$

**15.6** The conjugate acid of any base is just the base with a proton added.

(a) $H_2S$ (b) $H_2CO_3$ (c) $HCO_3^-$ (d) $H_3PO_4$ (e) $H_2PO_4^-$

(f) $HPO_4^{2-}$ (g) $H_2SO_4$ (h) $HSO_4^-$ (i) $HSO_3^-$

**15.8** The conjugate base of any acid is simply the acid minus one proton.

(a) $CH_2ClCOO^-$ (b) $IO_4^-$ (c) $H_2PO_4^-$ (d) $HPO_4^{2-}$ (e) $PO_4^{3-}$

(f) $HSO_4^-$ (g) $SO_4^{2-}$ (h) $IO_3^-$ (i) $SO_3^{2-}$ (j) $NH_3$

(k) $HS^-$ (l) $S^{2-}$ (m) $OCl^-$

**15.16** $[OH^-] = 0.62\ M$

$$[H^+] = \frac{K_w}{[OH^-]} = \frac{1.0 \times 10^{-14}}{0.62} = \mathbf{1.6 \times 10^{-14}\ M}$$

**15.18** **(a)** $Ba(OH)_2$ is ionic and fully ionized in water. The concentration of the hydroxide ion is $5.6 \times 10^{-4}\ M$ (Why? What is the concentration of $Ba^{2+}$?) We find the hydrogen ion concentration.

$$[H^+] = \frac{K_w}{[OH^-]} = \frac{1.0 \times 10^{-14}}{5.6 \times 10^{-4}} = 1.8 \times 10^{-11}\ M$$

The pH is then: $\text{pH} = -\log[H^+] = -\log(1.8 \times 10^{-11}) = \mathbf{10.74}$

**(b)** Nitric acid is a strong acid, so the concentration of hydrogen ion is also $5.2 \times 10^{-4}\ M$. The pH is:

$$\text{pH} = -\log[H^+] = -\log(5.2 \times 10^{-4}) = \mathbf{3.28}$$

**15.20** For (a) and (b) we can calculate the $H^+$ concentration using the equation representing the definition of pH.

**Strategy:** Here we are given the pH of a solution and asked to calculate $[H^+]$. Because pH is defined as $\text{pH} = -\log[H^+]$, we can solve for $[H^+]$ by taking the antilog of the pH; that is, $[H^+] = 10^{-\text{pH}}$.

**Solution:** From Equation (15.4) of the text:

**(a)** $\text{pH} = -\log[H^+] = 5.20$

$\log[H^+] = -5.20$

To calculate $[H^+]$, we need to take the antilog of $-5.20$.

$[H^+] = 10^{-5.20} = \mathbf{6.3 \times 10^{-6}\ M}$

**Check:** Because the pH is between 5 and 6, we can expect $[H^+]$ to be between $1 \times 10^{-5}$ *M* and $1 \times 10^{-6}$ *M*. Therefore, the answer is reasonable.

**(b)** $pH = -\log [H^+] = 16.00$

$$\log[H^+] = -16.00$$

$$[H^+] = 10^{-16.00} = \mathbf{1.0 \times 10^{-16}\ M}$$

**(c)** For part ©, it is probably easiest to calculate the $[H^+]$ from the ion product of water.

**Strategy:** We are given the concentration of $OH^-$ ions and asked to calculate $[H^+]$. The relationship between $[H^+]$ and $[OH^-]$ in water or an aqueous solution is given by the ion-product of water, $K_w$ [Equation (15.3) of the text].

**Solution:** The ion product of water is applicable to all aqueous solutions. At 25°C,

$$K_w = 1.0 \times 10^{-14} = [H^+][OH^-]$$

Rearranging the equation to solve for $[H^+]$, we write

$$[H^+] = \frac{1.0 \times 10^{-14}}{[OH^-]} = \frac{1.0 \times 10^{-14}}{3.7 \times 10^{-9}} = \mathbf{2.7 \times 10^{-6}\ M}$$

**Check:** Since the $[OH^-] < 1 \times 10^{-7}$ *M* we expect the $[H^+]$ to be greater than $1 \times 10^{-7}$ *M*.

**15.22** (a) acidic (b) neutral (c) basic

**15.24** $$5.50\ \text{mL} \times \frac{1\ \text{L}}{1000\ \text{mL}} \times \frac{0.360\ \text{mol}}{1\ \text{L}} = \mathbf{1.98 \times 10^{-3}\ mol\ KOH}$$

KOH is a strong base and therefore ionizes completely. The $OH^-$ concentration equals the KOH concentration, because there is a 1:1 mole ratio between KOH and $OH^-$.

$$[OH^-] = 0.360\ M$$

$$\mathbf{pOH} = -\log[OH^-] = \mathbf{0.444}$$

**15.26** Molarity of the HCl solution is: $$\frac{18.4\ \text{g HCl} \times \dfrac{1\ \text{mol HCl}}{36.46\ \text{g HCl}}}{662 \times 10^{-3}\ \text{L}} = 0.762\ M$$

$$\mathbf{pH} = -\log(0.762) = \mathbf{0.118}$$

**15.32** **(1)** The two steps in the ionization of a weak diprotic acid are:

$$H_2A(aq) + H_2O(l) \rightleftharpoons H_3O^+(aq) + HA^-(aq)$$

$$HA^-(aq) + H_2O(l) \rightleftharpoons H_3O^+(aq) + A^{2-}(aq)$$

The diagram that represents a weak diprotic acid is ©. In this diagram, we only see the first step of the ionization, because $HA^-$ is a much weaker acid than $H_2A$.

**(2)** Both **(b)** and **(d)** are chemically implausible situations. Because $HA^-$ is a much weaker acid than $H_2A$, you would not see a higher concentration of $A^{2-}$ compared to $HA^-$.

**15.34** **(a)** strong base **(b)** weak base **(c)** weak base **(d)** weak base **(e)** strong base

**15.36** **(a)** false, they are equal **(b)** true, find the value of log(1.00) on your calculator
**(c)** true **(d)** false, if the acid is strong, [HA] = 0.00 *M*

**15.38** $Cl^-$ is the conjugate base of the strong acid, HCl. It is a negligibly weak base and has no affinity for protons. Therefore, the reaction will *not* proceed from left to right to any measurable extent.

Another way to think about this problem is to consider the possible products of the reaction.

$$CH_3COOH(aq) + Cl^-(aq) \rightarrow HCl(aq) + CH_3COO^-(aq)$$

The favored reaction is the one that proceeds from right to left. HCl is a strong acid and will ionize completely, donating all its protons to the base, $CH_3COO^-$.

**15.44** **Strategy:** Recall that a weak acid only partially ionizes in water. We are given the initial quantity of a weak acid ($CH_3COOH$) and asked to calculate the concentrations of $H^+$, $CH_3COO^-$, and $CH_3COOH$ at equilibrium. First, we need to calculate the initial concentration of $CH_3COOH$. In determining the $H^+$ concentration, we ignore the ionization of $H_2O$ as a source of $H^+$, so the major source of $H^+$ ions is the acid. We follow the procedure outlined in Section 15.5 of the text.

**Solution:**
*Step 1:* Calculate the concentration of acetic acid before ionization.

$$0.0560\ \text{g acetic acid} \times \frac{1\ \text{mol acetic acid}}{60.05\ \text{g acetic acid}} = 9.33 \times 10^{-4}\ \text{mol acetic acid}$$

$$\frac{9.33 \times 10^{-4}\ \text{mol}}{0.0500\ \text{L soln}} = 0.0187\ M\ \text{acetic acid}$$

*Step 2:* We ignore water's contribution to $[H^+]$. We consider $CH_3COOH$ as the only source of $H^+$ ions.

*Step 3:* Letting $x$ be the equilibrium concentration of $H^+$ and $CH_3COO^-$ ions in mol/L, we summarize:

| | $CH_3COOH(aq)$ | $\rightleftharpoons$ $H^+(aq)$ | $+$ $CH_3COO^-(aq)$ |
|---|---|---|---|
| Initial ($M$): | 0.0187 | 0 | 0 |
| Change ($M$): | $-x$ | $+x$ | $+x$ |
| Equilibrium ($M$): | $0.0187 - x$ | $x$ | $x$ |

*Step 3:* Write the ionization constant expression in terms of the equilibrium concentrations. Knowing the value of the equilibrium constant ($K_a$), solve for $x$. You can look up the $K_a$ value in Table 15.3 of the text.

$$K_a = \frac{[H^+][CH_3COO^-]}{[CH_3COOH]}$$

$$1.8 \times 10^{-5} = \frac{(x)(x)}{(0.0187 - x)}$$

At this point, we can make an assumption that $x$ is very small compared to 0.0187. Hence,

$$0.0187 - x \approx 0.0187$$

$$1.8 \times 10^{-5} = \frac{(x)(x)}{0.0187}$$

$$\mathbf{x = 5.8 \times 10^{-4}\,M = [H^+] = [CH_3COO^-]}$$

$$[CH_3COOH] = (0.0187 - 5.8 \times 10^{-4})M = \mathbf{0.0181\,M}$$

**Check:** Testing the validity of the assumption,

$$\frac{5.8 \times 10^{-4}}{0.0187} \times 100\% = 3.1\% < 5\%$$

The assumption is valid.

**15.46** A pH of 3.26 corresponds to a $[H^+]$ of $5.5 \times 10^{-4}\,M$. Let the original concentration of formic acid be $x$. If the concentration of $[H^+]$ is $5.5 \times 10^{-4}\,M$, that means that $5.5 \times 10^{-4}\,M$ of HCOOH ionized because of the 1:1 mole ratio between HCOOH and $H^+$.

| | $HCOOH(aq)$ | $\rightleftharpoons$ $H^+(aq)$ | $+$ $HCOO^-(aq)$ |
|---|---|---|---|
| Initial ($M$): | $x$ | 0 | 0 |
| Change ($M$): | $-5.5 \times 10^{-4}$ | $+5.5 \times 10^{-4}$ | $+5.5 \times 10^{-4}$ |
| Equilibrium ($M$): | $x - (5.5 \times 10^{-4})$ | $5.5 \times 10^{-4}$ | $5.5 \times 10^{-4}$ |

Substitute $K_a$ and the equilibrium concentrations into the ionization constant expression to solve for $x$.

$$K_a = \frac{[H^+][HCOO^-]}{[HCOOH]}$$

$$1.7 \times 10^{-4} = \frac{(5.5 \times 10^{-4})^2}{x - (5.5 \times 10^{-4})}$$

$$\mathbf{x = [HCOOH] = 2.3 \times 10^{-3}\ M}$$

**15.48** Percent ionization is defined as:

$$\text{percent ionization} = \frac{\text{ionized acid concentration at equilibrium}}{\text{initial concentration of acid}} \times 100\%$$

For a monoprotic acid, HA, the concentration of acid that undergoes ionization is equal to the concentration of $H^+$ ions or the concentration of $A^-$ ions at equilibrium. Thus, we can write:

$$\text{percent ionization} = \frac{[H^+]}{[HA]_0} \times 100\%$$

**(a)** First, recognize that hydrofluoric acid is a weak acid. It is not one of the six strong acids, so it must be a weak acid.

*Step 1:* Express the equilibrium concentrations of all species in terms of initial concentrations and a single unknown $x$, that represents the change in concentration. Let $(-x)$ be the depletion in concentration (mol/L) of HF. From the stoichiometry of the reaction, it follows that the increase in concentration for both $H^+$ and $F^-$ must be $x$. Complete a table that lists the initial concentrations, the change in concentrations, and the equilibrium concentrations.

| | $HF(aq)$ | $\rightleftharpoons$ | $H^+(aq)$ | + | $F^-(aq)$ |
|---|---|---|---|---|---|
| Initial ($M$): | 0.60 | | 0 | | 0 |
| Change ($M$): | $-x$ | | $+x$ | | $+x$ |
| Equilibrium ($M$): | $0.60 - x$ | | $x$ | | $x$ |

*Step 2:* Write the ionization constant expression in terms of the equilibrium concentrations. Knowing the value of the equilibrium constant ($K_a$), solve for $x$.

$$K_a = \frac{[H^+][F^-]}{[HF]}$$

You can look up the $K_a$ value for hydrofluoric acid in Table 15.3 of your text.

$$7.1 \times 10^{-4} = \frac{(x)(x)}{(0.60 - x)}$$

At this point, we can make an assumption that $x$ is very small compared to 0.60. Hence,

$$0.60 - x \approx 0.60$$

Oftentimes, assumptions such as these are valid if $K$ is very small. A very small value of $K$ means that a very small amount of reactants go to products. Hence, $x$ is small. If we did not make this assumption, we would have to solve a quadratic equation.

$$7.1 \times 10^{-4} = \frac{(x)(x)}{0.60}$$

Solving for $x$.

$$x = 0.021\ M = [H^+]$$

*Step 3:* Having solved for the $[H^+]$, calculate the percent ionization.

$$\textbf{percent ionization} = \frac{[H^+]}{[HF]_0} \times 100\%$$

$$= \frac{0.021\ M}{0.60\ M} \times 100\% = \mathbf{3.5\%}$$

(b) – (c) are worked in a similar manner to part (a). However, as the initial concentration of HF becomes smaller, the assumption that $x$ is very small compared to this concentration will no longer be valid. You must solve a quadratic equation.

**(b)** $$K_a = \frac{[H^+][F^-]}{[HF]} = \frac{x^2}{(0.0046 - x)} = 7.1 \times 10^{-4}$$

$$x^2 + (7.1 \times 10^{-4})x - (3.3 \times 10^{-6}) = 0$$

$$x = 1.5 \times 10^{-3}\ M$$

$$\textbf{Percent ionization} = \frac{1.5 \times 10^{-3}\ M}{0.0046\ M} \times 100\% = \mathbf{33\%}$$

**(c)** $$K_a = \frac{[H^+][F^-]}{[HF]} = \frac{x^2}{(0.00028 - x)} = 7.1 \times 10^{-4}$$

$$x^2 + (7.1 \times 10^{-4})x - (2.0 \times 10^{-7}) = 0$$

$$x = 2.2 \times 10^{-4}\ M$$

$$\textbf{Percent ionization} = \frac{2.2 \times 10^{-4}\ M}{0.00028\ M} \times 100\% = \mathbf{79\%}$$

As the solution becomes more dilute, the percent ionization increases.

**15.50** The equilibrium is:

| | $C_9H_8O_4(aq)$ | $\rightleftharpoons$ $H^+(aq)$ | + $C_9H_7O_4^-(aq)$ |
|---|---|---|---|
| Initial ($M$): | 0.20 | 0 | 0 |
| Change ($M$): | $-x$ | $+x$ | $+x$ |
| Equilibrium ($M$): | $0.20 - x$ | $x$ | $x$ |

**(a)** $K_a = \dfrac{[H^+][C_9H_7O_4^-]}{[C_9H_8O_4]}$

$$3.0 \times 10^{-4} = \frac{x^2}{(0.20 - x)}$$

Assuming $(0.20 - x) \approx 0.20$

$$x = [H^+] = 7.7 \times 10^{-3}\ M$$

$$\textbf{Percent ionization} = \frac{x}{0.20} \times 100\% = \frac{7.7 \times 10^{-3}\ M}{0.20\ M} \times 100\% = \mathbf{3.9\%}$$

**(b)** At pH 1.00 the concentration of hydrogen ion is 0.10 $M$ ($[H^+] = 10^{-pH}$). The extra hydrogen ions will tend to suppress the ionization of the weak acid (LeChâtelier's principle, Section 14.5 of the text). The position of equilibrium is shifted in the direction of the un-ionized acid. Let's set up a table of concentrations with the initial concentration of $H^+$ equal to 0.10 $M$.

| | $C_9H_8O_4(aq)$ $\rightleftharpoons$ | $H^+(aq)$ + | $C_9H_7O_4^-(aq)$ |
|---|---|---|---|
| Initial ($M$): | 0.20 | 0.10 | 0 |
| Change ($M$): | $-x$ | $+x$ | $+x$ |
| Equilibrium ($M$): | $0.20 - x$ | $0.10 + x$ | $x$ |

$$K_a = \frac{[H^+][C_9H_7O_4^-]}{[C_9H_8O_4]}$$

$$3.0 \times 10^{-4} = \frac{x(0.10 + x)}{(0.20 - x)}$$

Assuming $(0.20 - x) \approx 0.20$ and $(0.10 + x) \approx 0.10$

$$x = 6.0 \times 10^{-4}\ M$$

$$\textbf{Percent ionization} = \frac{x}{0.20} \times 100\% = \frac{6.0 \times 10^{-4}\ M}{0.20\ M} \times 100\% = \mathbf{0.30\%}$$

The high acidity of the gastric juices appears to enhance the rate of absorption of unionized aspirin molecules through the stomach lining. In some cases this can irritate these tissues and cause bleeding.

**15.54** Since the solutions are of equal concentration, the one with the smallest $K_b$ value will ionize the least producing the smallest amount of hydroxide ions, and the one with the largest $K_b$ value will ionize the most producing the greatest amount of hydroxide ions. The bases listed in order of increasing $K_b$ values are:

**(c) < (a) < (b)**

**15.56** **Strategy:** Weak bases only partially ionize in water.

$$B(aq) + H_2O(l) \rightleftharpoons BH^+(aq) + OH^-(aq)$$

Note that the concentration of the weak base given refers to the initial concentration before ionization has started. The pH of the solution, on the other hand, refers to the situation at equilibrium. To calculate $K_b$, we need to know the concentrations of all three species, [B], $[BH^+]$, and $[OH^-]$ at equilibrium. We ignore the ionization of water as a source of $OH^-$ ions.

**Solution:** We proceed as follows.

*Step 1:* The major species in solution are B, $OH^-$, and the conjugate acid $BH^+$.

*Step 2:* First, we need to calculate the hydroxide ion concentration from the pH value. Calculate the pOH from the pH. Then, calculate the $OH^-$ concentration from the pOH.

$$\text{pOH} = 14.00 - \text{pH} = 14.00 - 10.66 = 3.34$$

$$\text{pOH} = -\log[OH^-]$$

$$-\text{pOH} = \log[OH^-]$$

Taking the antilog of both sides of the equation,

$$10^{-\text{pOH}} = [OH^-]$$

$$[OH^-] = 10^{-3.34} = 4.6 \times 10^{-4}\ M$$

*Step 3:* If the concentration of $OH^-$ is $4.6 \times 10^{-4}\ M$ at equilibrium, that must mean that $4.6 \times 10^{-4}\ M$ of the base ionized. We summarize the changes.

| | $B(aq)$ + $H_2O(l)$ | $\rightleftharpoons$ $BH^+(aq)$ | + $OH^-(aq)$ |
|---|---|---|---|
| Initial (*M*): | 0.30 | 0 | 0 |
| Change (*M*): | $-4.6 \times 10^{-4}$ | $+4.6 \times 10^{-4}$ | $+4.6 \times 10^{-4}$ |
| Equilibrium (*M*): | $0.30 - (4.6 \times 10^{-4})$ | $4.6 \times 10^{-4}$ | $4.6 \times 10^{-4}$ |

*Step 4:* Substitute the equilibrium concentrations into the ionization constant expression to solve for $K_b$.

$$K_b = \frac{[BH^+][OH^-]}{[B]}$$

$$\boldsymbol{K_b} = \frac{(4.6 \times 10^{-4})^2}{(0.30)} = \mathbf{7.1 \times 10^{-7}}$$

**15.58** The reaction is:

| | $NH_3(aq) + H_2O(l)$ | $\rightleftharpoons$ $NH_4^+(aq)$ | + $OH^-(aq)$ |
|---|---|---|---|
| Initial (*M*): | 0.080 | 0 | 0 |
| Change (*M*): | $-x$ | $+x$ | $+x$ |
| Equilibrium (*M*): | $0.080 - x$ | $x$ | $x$ |

At equilibrium we have:

$$K_a = \frac{[NH_4^+][OH^-]}{[NH_3]}$$

$$1.8 \times 10^{-5} = \frac{x^2}{(0.080 - x)} \approx \frac{x^2}{0.080}$$

$$x = 1.2 \times 10^{-3}\,M$$

$$\textbf{Percent } NH_3 \textbf{ present as } NH_4^+ = \frac{1.2 \times 10^{-3}}{0.080} \times 100\% = \mathbf{1.5\%}$$

**15.64** The pH of a 0.040 *M* HCl solution (strong acid) is: pH = −log(0.040) = **1.40**.

**Strategy:** Determining the pH of a diprotic acid in aqueous solution is more involved than for a monoprotic acid. The first stage of ionization for $H_2SO_4$ goes to completion. We follow the procedure for determining the pH of a strong acid for this stage. The conjugate base produced in the first ionization ($HSO_4^-$) is a weak acid. We follow the procedure for determining the pH of a weak acid for this stage.

**Solution:** We proceed according to the following steps.

*Step 1:* $H_2SO_4$ is a strong acid. The first ionization stage goes to completion. The ionization of $H_2SO_4$ is

$$H_2SO_4(aq) \rightarrow H^+(aq) + HSO_4^-(aq)$$

The concentrations of all the species ($H_2SO_4$, $H^+$, and $HSO_4^-$) before and after ionization can be represented as follows.

| | $H_2SO_4(aq)$ | $\rightarrow$ $H^+(aq)$ | + $HSO_4^-(aq)$ |
|---|---|---|---|
| Initial (*M*): | 0.040 | 0 | 0 |
| Change (*M*): | −0.040 | +0.040 | +0.040 |
| Final (*M*): | 0 | 0.040 | 0.040 |

*Step 2:* Now, consider the second stage of ionization. $HSO_4^-$ is a weak acid. Set up a table showing the concentrations for the second ionization stage. Let *x* be the change in concentration. Note that the initial concentration of $H^+$ is 0.040 *M* from the first ionization.

| | $HSO_4^-(aq)$ | $\rightleftharpoons$ $H^+(aq)$ | + $SO_4^{2-}(aq)$ |
|---|---|---|---|
| Initial (*M*): | 0.040 | 0.040 | 0 |
| Change (*M*): | $-x$ | $+x$ | $+x$ |
| Equilibrium (*M*): | $0.040 - x$ | $0.040 + x$ | $x$ |

Write the ionization constant expression for $K_a$. Then, solve for *x*. You can find the $K_a$ value in Table 15.5 of the text.

$$K_a = \frac{[H^+][SO_4^{2-}]}{[HSO_4^-]}$$

$$1.3 \times 10^{-2} = \frac{(0.040 + x)(x)}{(0.040 - x)}$$

Since $K_a$ is quite large, we cannot make the assumptions that

$$0.040 - x \approx 0.040 \quad \text{and} \quad 0.040 + x \approx 0.040$$

Therefore, we must solve a quadratic equation.

$$x^2 + 0.053x - (5.2 \times 10^{-4}) = 0$$

$$x = \frac{-0.053 \pm \sqrt{(0.053)^2 - 4(1)(-5.2 \times 10^{-4})}}{2(1)}$$

$$x = \frac{-0.053 \pm 0.070}{2}$$

$$x = 8.5 \times 10^{-3}\ M \quad \text{or} \quad x = -0.062\ M$$

The second solution is physically impossible because you cannot have a negative concentration. The first solution is the correct answer.

*Step 3:* Having solved for $x$, we can calculate the $H^+$ concentration at equilibrium. We can then calculate the pH from the $H^+$ concentration.

$$[H^+] = 0.040\ M + x = [0.040 + (8.5 \times 10^{-3})]M = 0.049\ M$$

$$\textbf{pH} = -\log(0.049) = \mathbf{1.31}$$

Without doing any calculations, could you have known that the pH of the sulfuric acid would be lower (more acidic) than that of the hydrochloric acid?

**15.66** For the first stage of ionization:

| | $H_2CO_3(aq)$ | $\rightleftharpoons$ $H^+(aq)$ | + $HCO_3^-(aq)$ |
|---|---|---|---|
| Initial ($M$): | 0.025 | 0.00 | 0.00 |
| Change ($M$): | $-x$ | $+x$ | $+x$ |
| Equilibrium ($M$): | $(0.025 - x)$ | $x$ | $x$ |

$$K_{a_1} = \frac{[H^+][HCO_3^-]}{[H_2CO_3]}$$

$$4.2 \times 10^{-7} = \frac{x^2}{(0.025 - x)} \approx \frac{x^2}{0.025}$$

$$x = 1.0 \times 10^{-4}\ M$$

For the second ionization,

| | $HCO_3^-(aq)$ | $\rightleftharpoons$ | $H^+(aq)$ | + | $CO_3^{2-}(aq)$ |
|---|---|---|---|---|---|
| Initial ($M$): | $1.0 \times 10^{-4}$ | | $1.0 \times 10^{-4}$ | | 0.00 |
| Change ($M$): | $-x$ | | $+x$ | | $+x$ |
| Equilibrium ($M$): | $(1.0 \times 10^{-4}) - x$ | | $(1.0 \times 10^{-4}) + x$ | | $x$ |

$$K_{a_2} = \frac{[H^+][CO_3^{2-}]}{[HCO_3^-]}$$

$$4.8 \times 10^{-11} = \frac{[(1.0 \times 10^{-4}) + x](x)}{(1.0 \times 10^{-4}) - x} \approx \frac{(1.0 \times 10^{-4})(x)}{(1.0 \times 10^{-4})}$$

$$x = 4.8 \times 10^{-11}\ M$$

Since $HCO_3^-$ is a very weak acid, there is little ionization at this stage. Therefore we have:

$$[H^+] = [HCO_3^-] = 1.0 \times 10^{-4}\ M \text{ and } [CO_3^{2-}] = x = 4.8 \times 10^{-11}\ M$$

**15.70** All the listed pairs are oxoacids that contain different central atoms whose elements are in the same group of the periodic table and have the same oxidation number. In this situation the acid with the most electronegative central atom will be the strongest.

(a) $H_2SO_4 > H_2SeO_4$.

(b) $H_3PO_4 > H_3AsO_4$

**15.72** The conjugate bases are $C_6H_5O^-$ from phenol and $CH_3O^-$ from methanol. The $C_6H_5O^-$ is stabilized by resonance:

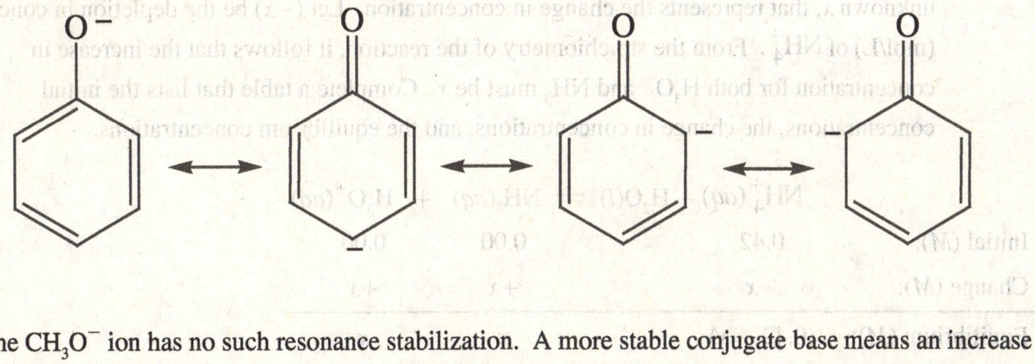

The $CH_3O^-$ ion has no such resonance stabilization. A more stable conjugate base means an increase in the strength of the acid.

**15.78** **Strategy:** In deciding whether a salt will undergo hydrolysis, ask yourself the following questions: Is the cation a highly charged metal ion or an ammonium ion? Is the anion the conjugate base of a weak acid? If yes to either question, then hydrolysis will occur. In cases where both the cation and the anion react with water, the pH of the solution will depend on the relative magnitudes of $K_a$ for the cation and $K_b$ for the anion (see Table 15.7 of the text).

**Solution:** We first break up the salt into its cation and anion components and then examine the possible reaction of each ion with water.

**(a)** The $Na^+$ cation does not hydrolyze. The $Br^-$ anion is the conjugate base of the strong acid HBr. Therefore, $Br^-$ will not hydrolyze either, and the solution is **neutral**.

**(b)** The $K^+$ cation does not hydrolyze. The $SO_3^{2-}$ anion is the conjugate base of the weak acid $HSO_3^-$ and will hydrolyze to give $HSO_3^-$ and $OH^-$. The solution will be **basic**.

**(c)** Both the $NH_4^+$ and $NO_2^-$ ions will hydrolyze. $NH_4^+$ is the conjugate acid of the weak base $NH_3$, and $NO_2^-$ is the conjugate base of the weak acid $HNO_2$. From Tables 15.3 and 15.4 of the text, we see that the $K_a$ of $NH_4^+$ ($5.6 \times 10^{-10}$) is greater than the $K_b$ of $NO_2^-$ ($2.2 \times 10^{-11}$). Therefore, the solution will be **acidic**.

**(d)** $Cr^{3+}$ is a small metal cation with a high charge, which hydrolyzes to produce $H^+$ ions. The $NO_3^-$ anion does not hydrolyze. It is the conjugate base of the strong acid, $HNO_3$. The solution will be **acidic**.

**15.80** There is an inverse relationship between acid strength and conjugate base strength. As acid strength decreases, the proton accepting power of the conjugate base increases. In general the weaker the acid, the stronger the conjugate base. All three of the potassium salts ionize completely to form the conjugate base of the respective acid. The greater the pH, the stronger the conjugate base, and therefore, the weaker the acid.

The order of increasing acid strength is **HZ < HY < HX**.

**15.82** The salt ammonium chloride completely ionizes upon dissolution, producing 0.42 *M* $[NH_4^+]$ and 0.42 *M* $[Cl^-]$ ions. $NH_4^+$ will undergo hydrolysis because it is a weak acid ($NH_4^+$ is the conjugate acid of the weak base, $NH_3$).

*Step 1:* Express the equilibrium concentrations of all species in terms of initial concentrations and a single unknown $x$, that represents the change in concentration. Let $(-x)$ be the depletion in concentration (mol/L) of $NH_4^+$. From the stoichiometry of the reaction, it follows that the increase in concentration for both $H_3O^+$ and $NH_3$ must be $x$. Complete a table that lists the initial concentrations, the change in concentrations, and the equilibrium concentrations.

| | $NH_4^+(aq) + H_2O(l)$ | $\rightleftharpoons$ $NH_3(aq)$ | + $H_3O^+(aq)$ |
|---|---|---|---|
| Initial (*M*): | 0.42 | 0.00 | 0.00 |
| Change (*M*): | $-x$ | $+x$ | $+x$ |
| Equilibrium (*M*): | $(0.42 - x)$ | $x$ | $x$ |

*Step 2:* You can calculate the $K_a$ value for $NH_4^+$ from the $K_b$ value of $NH_3$. The relationship is

$$K_a \times K_b = K_w$$

or

$$K_a = \frac{K_w}{K_b} = \frac{1.0\times10^{-14}}{1.8\times10^{-5}} = 5.6\times10^{-10}$$

*Step 3:* Write the ionization constant expression in terms of the equilibrium concentrations. Knowing the value of the equilibrium constant ($K_a$), solve for *x*.

$$K_a = \frac{[NH_3][H_3O^+]}{[NH_4^+]}$$

$$5.6\times10^{-10} = \frac{x^2}{0.42 - x} \approx \frac{x^2}{0.42}$$

$$x = [H^+] = 1.5\times10^{-5}\,M$$

$$\textbf{pH} = -\log(1.5\times10^{-5}) = \textbf{4.82}$$

Since $NH_4Cl$ is the salt of a weak base (aqueous ammonia) and a strong acid (HCl), we expect the solution to be slightly acidic, which is confirmed by the calculation.

**15.84** The acid and base reactions are:

acid: $HPO_4^{2-}(aq) \rightleftharpoons H^+(aq) + PO_4^{3-}(aq)$

base: $HPO_4^{2-}(aq) + H_2O(l) \rightleftharpoons H_2PO_4^-(aq) + OH^-(aq)$

$K_a$ for $HPO_4^{2-}$ is $4.8\times10^{-13}$. Note that $HPO_4^{2-}$ is the conjugate base of $H_2PO_4^-$, so $K_b$ is $1.6\times10^{-7}$. Comparing the two *K*'s, we conclude that the monohydrogen phosphate ion is a much stronger proton acceptor (base) than a proton donor (acid). The solution will be **basic**.

**15.88** The most basic oxides occur with metal ions having the lowest positive charges (or lowest oxidation numbers).

**(a)** $Al_2O_3 < BaO < K_2O$ **(b)** $CrO_3 < Cr_2O_3 < CrO$

**15.90** $Al(OH)_3$ is an amphoteric hydroxide. The reaction is:

$$\mathbf{Al(OH)_3(s) + OH^-(aq) \rightarrow Al(OH)_4^-(aq)}$$

This is a **Lewis acid-base reaction**. Can you identify the acid and base?

**15.94** $AlCl_3$ is a Lewis acid with an incomplete octet of electrons and $Cl^-$ is the Lewis base donating a pair of electrons.

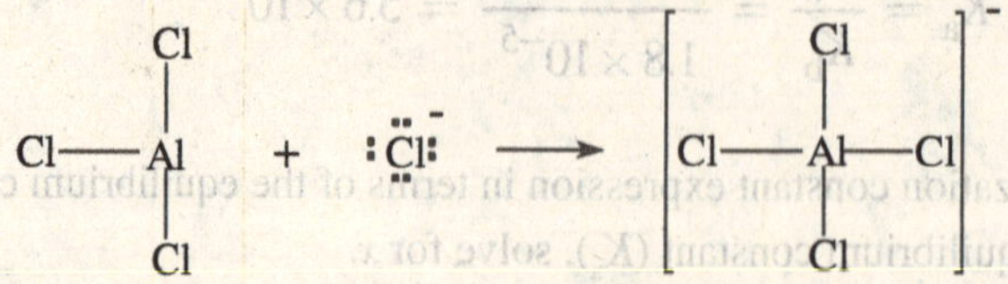

**15.96** By definition Brønsted acids are proton donors, therefore such compounds must contain at least one hydrogen atom. In Problem 15.93, Lewis acids that do not contain hydrogen, and therefore are not Brønsted acids, are $CO_2$, $SO_2$, and $BCl_3$. Can you name others?

**15.98** $NH_4^+$ is a weak acid. A pH of 5.64 corresponds to a $[H^+]$ of $2.3 \times 10^{-6}$ *M*. Let the original concentration of $NH_4^+$ be *x*. If the concentration of $[H^+]$ is $2.3 \times 10^{-6}$ *M*, that means that $2.3 \times 10^{-6}$ *M* of $NH_4^+$ ionized because of the 1:1 mole ratio between $NH_4^+$ and $H^+$.

| | $NH_4^+(aq)$ | $\rightleftharpoons$ | $H^+(aq)$ | + | $NH_3(aq)$ |
|---|---|---|---|---|---|
| Initial (*M*): | $x$ | | 0 | | 0 |
| Change (*M*): | $-2.3 \times 10^{-6}$ | | $+2.3 \times 10^{-6}$ | | $+2.3 \times 10^{-6}$ |
| Equilibrium (*M*): | $x - (2.3 \times 10^{-6})$ | | $2.3 \times 10^{-6}$ | | $2.3 \times 10^{-6}$ |

Substitute $K_a$ and the equilibrium concentrations into the ionization constant expression to solve for *I*.

$$K_a = \frac{[H^+][NH_3]}{[NH_4^+]}$$

$$5.6 \times 10^{-10} = \frac{(2.3 \times 10^{-6})^2}{x - (2.3 \times 10^{-6})}$$

$$x = [NH_4^+] = [NH_4Cl] = \mathbf{0.0094\ M}$$

Note that there is one $NH_4^+$ ion per formula unit of $NH_4Cl$; therefore, their concentrations are equal assuming the complete dissociation of the salt.

**15.100** We first find the number of moles of $CO_2$ produced in the reaction:

$$0.350\ \text{g NaHCO}_3 \times \frac{1\ \text{mol NaHCO}_3}{84.01\ \text{g NaHCO}_3} \times \frac{1\ \text{mol CO}_2}{1\ \text{mol NaHCO}_3} = 4.17 \times 10^{-3}\ \text{mol CO}_2$$

$$V_{CO_2} = \frac{n_{CO_2}RT}{P} = \frac{(4.17 \times 10^{-3}\ \text{mol})(0.0821\ \text{L}\cdot\text{atm/K}\cdot\text{mol})(37.0 + 273)\text{K}}{(1.00\ \text{atm})} = \mathbf{0.106\ L}$$

**15.102** If we assume that the unknown monoprotic acid is a strong acid that is 100% ionized, then the $[H^+]$ concentration will be 0.0642 *M*.

$$pH = -\log(0.0642) = 1.19$$

Since the actual pH of the solution is higher, the acid must be a weak acid.

**15.104** The reaction of a weak acid with a strong base is driven to completion by the formation of water. Irrespective of whether the strong base is reacting with a strong monoprotic acid or a weak monoprotic acid, the same number of moles of acid is required to react with a constant number of moles of base. Therefore **the volume of base** required to react with the same concentration of acid solutions (either both weak, both strong, or one strong and one weak) **will be the same.**

**15.106** High oxidation state leads to covalent compounds and low oxidation state leads to ionic compounds. Therefore, **CrO** is ionic and **basic** and $\mathbf{CrO_3}$ is covalent and **acidic**.

**15.108** We can write two equilibria that add up to the equilibrium in the problem.

$$CH_3COOH(aq) \rightleftharpoons H^+(aq) + CH_3COO^-(aq) \qquad K_a = \frac{[H^+][CH_3COO^-]}{[CH_3COOH]} = 1.8\times10^{-5}$$

$$H^+(aq) + NO_2^-(aq) \rightleftharpoons HNO_2(aq) \qquad K_a' = \frac{1}{K_a(HNO_2)} = \frac{1}{4.5\times10^{-4}} = 2.2\times10^3$$

$$K_a' = \frac{[HNO_2]}{[H^+][NO_2^-]}$$

$$CH_3COOH(aq) + NO_2^-(aq) \rightleftharpoons CH_3COO^-(aq) + HNO_2(aq) \qquad K = \frac{[CH_3COO^-][HNO_2]}{[CH_3COOH][NO_2^-]} = K_a \times K_a'$$

The equilibrium constant for this sum is the product of the equilibrium constants of the component reactions.

$$K = K_a \times K_a' = (1.8\times10^{-5})(2.2\times10^3) = \mathbf{4.0\times10^{-2}}$$

**15.110** In this specific case the $K_a$ of ammonium ion is the same as the $K_b$ of acetate ion $[K_a(NH_4^+) = 5.6\times10^{-10}$, $K_b(CH_3COO^-) = 5.6\times10^{-10}]$. The two are of exactly (to two significant figures) equal strength. The solution will have **pH 7.00**.

What would the pH be if the concentration were 0.1 *M* in ammonium acetate? 0.4 *M*?

**15.112** The fact that fluorine attracts electrons in a molecule more strongly than hydrogen should cause $NF_3$ to be a poor electron pair donor and a poor base. $\mathbf{NH_3}$ is the stronger base.

**15.114** The autoionization for deuterium-substituted water is: $D_2O \rightleftharpoons D^+ + OD^-$

$$[D^+][OD^-] = 1.35\times10^{-15} \qquad (1)$$

**(a)** The definition of pD is: $\mathbf{pD} = -\log[\mathrm{D}^+] = -\log\sqrt{1.35 \times 10^{-15}} = \mathbf{7.43}$

**(b)** To be acidic, the **pD** must be **< 7.43**.

**(c)** Taking −log of both sides of equation (1) above:

$$-\log[\mathrm{D}^+] + -\log[\mathrm{OD}^-] = -\log(1.35 \times 10^{-15})$$

$$\mathbf{pD + pOD = 14.87}$$

**15.116** First we must calculate the molarity of the trifluoromethane sulfonic acid. (Molar mass = 150.1 g/mol)

$$\text{Molarity} = \frac{0.616\ \cancel{\text{g}} \times \dfrac{1\ \text{mol}}{150.1\ \cancel{\text{g}}}}{0.250\ \text{L}} = 0.0164\ M$$

Since trifluoromethane sulfonic acid is a strong acid and is 100% ionized, the $[H^+]$ is 0.0165 *M*.

$$\mathbf{pH} = -\log(0.0164) = \mathbf{1.79}$$

**15.118** The reactions are $HF \rightleftharpoons H^+ + F^-$ (1)

$$F^- + HF \rightleftharpoons HF_2^- \quad (2)$$

Note that for equation (2), the equilibrium constant is relatively large with a value of 5.2. This means that the equilibrium lies to the right. Applying Le Châtelier's principle, as HF ionizes in the first step, the $F^-$ that is produced is partially removed in the second step. More HF must ionize to compensate for the removal of the $F^-$, at the same time producing more $H^+$.

**15.120** **(a)** We must consider both the complete ionization of the strong acid, and the partial ionization of water.

$$HA \longrightarrow H^+ + A^-$$

$$H_2O \rightleftharpoons H^+ + OH^-$$

From the above two equations, the $[H^+]$ in solution is:

$$[H^+] = [A^-] + [OH^-] \quad (1)$$

We can also write:

$$[H^+][OH^-] = K_w$$

$$[OH^-] = \frac{K_w}{[H^+]}$$

Substituting into Equation (1):

$$[H^+] = [A^-] + \frac{K_w}{[H^+]}$$

$$[H^+]^2 = [A^-][H^+] + K_w$$

$$[H^+]^2 - [A^-][H^+] - K_w = 0$$

Solving a quadratic equation:

$$[H^+] = \frac{[A^-] \pm \sqrt{[A^-]^2 + 4K_w}}{2}$$

**(b)** For the strong acid, HCl, with a concentration of $1.0 \times 10^{-7}$ *M*, the $[Cl^-]$ will also be $1.0 \times 10^{-7}$ *M*.

$$[H^+] = \frac{[Cl^-] \pm \sqrt{[Cl^-]^2 + 4K_w}}{2} = \frac{1 \times 10^{-7} \pm \sqrt{(1 \times 10^{-7})^2 + 4(1 \times 10^{-14})}}{2}$$

$$[H^+] = 1.6 \times 10^{-7}\ M \text{ (or } -6.0 \times 10^{-8}\ M\text{, which is impossible)}$$

$$\textbf{pH} = -\log[1.6 \times 10^{-7}] = \mathbf{6.80}$$

**15.122** The solution for the first step is standard:

$$H_3PO_4(aq) \rightleftharpoons H^+(aq) + H_2PO_4^-(aq)$$

| | $H_3PO_4(aq)$ | $H^+(aq)$ | $H_2PO_4^-(aq)$ |
|---|---|---|---|
| Initial (*M*): | 0.100 | 0.000 | 0.000 |
| Change (*M*): | $-x$ | $+x$ | $+x$ |
| Equil. (*M*): | $(0.100 - x)$ | $x$ | $x$ |

$$K_{a_1} = \frac{[H^+][H_2PO_4^-]}{[H_3PO_4]}$$

$$7.5 \times 10^{-3} = \frac{x^2}{(0.100 - x)}$$

In this case we probably cannot say that $(0.100 - x) \approx 0.100$ due to the magnitude of $K_a$. We obtain the quadratic equation:

$$x^2 + (7.5 \times 10^{-3})x - (7.5 \times 10^{-4}) = 0$$

The positive root is $x = 0.0239\ M$. We have:

$$[H^+] = [H_2PO_4^-] = 0.0239\ M$$

$$[H_3PO_4] = (0.100 - 0.0239)\ M = 0.076\ M$$

For the second ionization:

| | $H_2PO_4^-(aq)$ | $\rightleftharpoons$ $H^+(aq)$ | + $HPO_4^{2-}(aq)$ |
|---|---|---|---|
| Initial ($M$): | 0.0239 | 0.0239 | 0.000 |
| Change ($M$): | $-y$ | $+y$ | $+y$ |
| Equil ($M$): | $(0.0239 - y)$ | $(0.0239 + y)$ | $y$ |

$$K_{a_2} = \frac{[H^+][HPO_4^{2-}]}{[H_2PO_4^-]}$$

$$6.2 \times 10^{-8} = \frac{(0.0239 + y)(y)}{(0.0239 - y)} \approx \frac{(0.0239)(y)}{(0.0239)}$$

$$y = 6.2 \times 10^{-8}\ M.$$

Thus,

$$[H^+] = [H_2PO_4^-] = 0.0239\ M$$

$$[HPO_4^{2-}] = y = 6.2 \times 10^{-8}\ M$$

We set up the problem for the third ionization in the same manner.

| | $HPO_4^{2-}(aq)$ | $\rightleftharpoons$ $H^+(aq)$ | + $PO_4^{3-}(aq)$ |
|---|---|---|---|
| Initial ($M$): | $6.2 \times 10^{-8}$ | 0.0239 | 0 |
| Change ($M$): | $-z$ | $+z$ | $+z$ |
| Equil. ($M$): | $(6.2 \times 10^{-8}) - z$ | $0.0239 + z$ | $z$ |

$$K_{a_3} = \frac{[H^+][PO_4^{3-}]}{[HPO_4^{2-}]}$$

$$4.8 \times 10^{-13} = \frac{(0.0239 + z)(z)}{(6.2 \times 10^{-8}) - z} \approx \frac{(0.239)(z)}{(6.2 \times 10^{-8})}$$

$$z = 1.2 \times 10^{-18}\ M$$

The equilibrium concentrations are:

$$\mathbf{[H^+] = [H_2PO_4^-] = 0.0239\ M}$$

$$\mathbf{[H_3PO_4] = 0.076\ M}$$

$$\mathbf{[HPO_4^{2-}] = 6.2 \times 10^{-8}\ M}$$

$$\mathbf{[PO_4^{3-}] = 1.2 \times 10^{-18}\ M}$$

**15.124** Pyrex glass contains 10-25% $B_2O_3$ (See Table 11.5 of the text). $B_2O_3$ is an acidic oxide (see Figure 15.8 of the text). The strong base, NaOH, will react with the acid, $B_2O_3$, decomposing the glass.

**15.126** $0.100\ M\ Na_2CO_3 \rightarrow 0.200\ M\ Na^+ + 0.100\ M\ CO_3^{2-}$

First stage:

$$CO_3^{2-}(aq) + H_2O(l) \rightleftharpoons HCO_3^-(aq) + OH^-(aq)$$

| | $CO_3^{2-}(aq)$ | $HCO_3^-(aq)$ | $OH^-(aq)$ |
|---|---|---|---|
| Initial ($M$): | 0.100 | 0 | 0 |
| Change ($M$): | $-x$ | $+x$ | $+x$ |
| Equilibrium ($M$): | $0.100 - x$ | $x$ | $x$ |

$$K_1 = \frac{K_w}{K_2} = \frac{1.0 \times 10^{-14}}{4.8 \times 10^{-11}} = 2.1 \times 10^{-4}$$

$$K_1 = \frac{[HCO_3^-][OH^-]}{[CO_3^{2-}]}$$

$$2.1 \times 10^{-4} = \frac{x^2}{0.100 - x} \approx \frac{x^2}{0.100}$$

$$x = 4.6 \times 10^{-3}\ M = [HCO_3^-] = [OH^-]$$

Second stage:

$$HCO_3^-(aq) + H_2O(l) \rightleftharpoons H_2CO_3(aq) + OH^-(aq)$$

| | $HCO_3^-(aq)$ | $H_2CO_3(aq)$ | $OH^-(aq)$ |
|---|---|---|---|
| Initial ($M$): | $4.6 \times 10^{-3}$ | 0 | $4.6 \times 10^{-3}$ |
| Change ($M$): | $-y$ | $+y$ | $+y$ |
| Equilibrium ($M$): | $(4.6 \times 10^{-3}) - y$ | $y$ | $(4.6 \times 10^{-3}) + y$ |

$$K_2 = \frac{[H_2CO_3][OH^-]}{[HCO_3^-]}$$

$$2.4 \times 10^{-8} = \frac{y[(4.6 \times 10^{-3}) + y]}{(4.6 \times 10^{-3}) - y} \approx \frac{(y)(4.6 \times 10^{-3})}{(4.6 \times 10^{-3})}$$

$$y = 2.4 \times 10^{-8}\ M$$

At equilibrium:

$$[Na^+] = \mathbf{0.200\ M}$$

$$[CO_3^{2-}] = 0.100\ M - 4.6 \times 10^{-3}\ M = \mathbf{0.095\ M}$$

$$[HCO_3^-] = (4.6 \times 10^{-3})\ M - (2.4 \times 10^{-8})\ M \approx \mathbf{4.6 \times 10^{-3}\ M}$$

$$[H_2CO_3] = \mathbf{2.4 \times 10^{-8}\ M}$$

$$[OH^-] = (4.6 \times 10^{-3})\ M + (2.4 \times 10^{-8})\ M \approx \mathbf{4.6 \times 10^{-3}\ M}$$

$$[H^+] = \frac{1.0 \times 10^{-14}}{4.6 \times 10^{-3}} = \mathbf{2.2 \times 10^{-12}\ M}$$

**15.128** When NaCN is treated with HCl, the following reaction occurs.

$$NaCN + HCl \rightarrow NaCl + HCN$$

HCN is a very weak acid, and only partially ionizes in solution.

$$HCN(aq) \rightleftharpoons H^+(aq) + CN^-(aq)$$

The main species in solution is HCN which has a tendency to escape into the gas phase.

$$HCN(aq) \rightleftharpoons HCN(g)$$

Since the HCN(*g*) that is produced is a highly poisonous compound, it would be dangerous to treat NaCN with acids without proper ventilation.

**15.130** $pH = 2.53 = -\log[H^+]$

$[H^+] = 2.95 \times 10^{-3}\ M$

Since the concentration of $H^+$ at equilibrium is $2.95 \times 10^{-3}\ M$, that means that $2.95 \times 10^{-3}\ M$ HCOOH ionized. Let' represent the initial concentration of HCOOH as $x$. The equation representing the ionization of formic acid is:

| | $HCOOH(aq)$ | $\rightleftharpoons$ | $H^+(aq)$ | + | $HCOO^-(aq)$ |
|---|---|---|---|---|---|
| Initial (*M*): | $x$ | | 0 | | 0 |
| Change (*M*): | $-2.95 \times 10^{-3}$ | | $+2.95 \times 10^{-3}$ | | $+2.95 \times 10^{-3}$ |
| Equilibrium (*M*): | $x - (2.95 \times 10^{-3})$ | | $2.95 \times 10^{-3}$ | | $2.95 \times 10^{-3}$ |

$$K_a = \frac{[H^+][HCOO^-]}{[HCOOH]}$$

$$1.7 \times 10^{-4} = \frac{(2.95 \times 10^{-3})^2}{x - (2.95 \times 10^{-3})}$$

$$x = 0.054\ M$$

There are 0.054 moles of formic acid in 1000 mL of solution. The mass of formic acid in 100 mL is:

$$100\ \text{mL} \times \frac{0.054\ \text{mol formic acid}}{1000\ \text{mL soln}} \times \frac{46.03\ \text{g formic acid}}{1\ \text{mol formic acid}} = \mathbf{0.25\ g\ formic\ acid}$$

**15.132** The balanced equation is: $Mg + 2HCl \rightarrow MgCl_2 + H_2$

$$\text{mol of Mg} = 1.87\ \text{g Mg} \times \frac{1\ \text{mol Mg}}{24.31\ \text{g Mg}} = 0.0769\ \text{mol}$$

From the balanced equation:

$$\text{mol of HCl required for reaction} = 2 \times \text{mol Mg} = (2)(0.0769\ \text{mol}) = 0.154\ \text{mol HCl}$$

The concentration of HCl:

$pH = -0.544$, thus $[H^+] = 3.50\ M$

$$\text{initial mol HCl} = M \times \text{Vol (L)} = (3.50\ M)(0.0800\ \text{L}) = 0.280\ \text{mol HCl}$$

Moles of HCl left after reaction:

$$\text{initial mol HCl} - \text{mol HCl reacted} = 0.280\ \text{mol} - 0.154\ \text{mol} = 0.126\ \text{mol HCl}$$

Molarity of HCl left after reaction:

$$M = \text{mol/L} = 0.126\ \text{mol}/0.080\ \text{L} = 1.58\ M$$

$$\textbf{pH} = -\log(1.58) = \mathbf{-0.20}$$

**15.134** The important equation is the hydrolysis of $NO_2^-$: $NO_2^- + H_2O \rightleftharpoons HNO_2 + OH^-$

**(a)** Addition of HCl will result in the reaction of the $H^+$ from the HCl with the $OH^-$ that was present in the solution. The $OH^-$ will effectively be removed and the equilibrium will **shift to the right** to compensate (more hydrolysis).

**(b)** Addition of NaOH is effectively addition of more $OH^-$ which places stress on the right hand side of the equilibrium. The equilibrium will **shift to the left** (less hydrolysis) to compensate for the addition of $OH^-$.

**(c)** Addition of NaCl will have **no effect**.

**(d)** Recall that the percent ionization of a weak acid increases with dilution (see Figure 15.4 of the text). The same is true for weak bases. Thus dilution will cause more hydrolysis, shifting the equilibrium to the **right**.

**15.136** In Chapter 11, we found that salts with their formal electrostatic intermolecular attractions had low vapor pressures and thus high boiling points. Ammonia and its derivatives (amines) are molecules with dipole-dipole attractions; as long as the nitrogen has one direct N–H bond, the molecule will have hydrogen bonding. Even so, these molecules will have much higher vapor pressures than ionic species. Thus, if we could convert the neutral ammonia-type molecules into salts, their vapor pressures, and thus associated odors, would decrease. Lemon juice contains acids which can react with neutral ammonia-type (amine) molecules to form ammonium salts.

$$NH_3 + H^+ \rightarrow NH_4^+$$

$$RNH_2 + H^+ \rightarrow RNH_3^+$$

**15.138** $HCOOH \rightleftharpoons H^+ + HCOO^-$

| | HCOOH | $H^+$ | $HCOO^-$ |
|---|---|---|---|
| Initial (*M*): | 0.400 | 0 | 0 |
| Change (*M*): | $-x$ | $+x$ | $+x$ |
| Equilibrium (*M*): | $0.400 - x$ | $x$ | $x$ |

Total concentration of particles in solution: $(0.400 - x) + x + x = 0.400 + x$

Assuming the molarity of the solution is equal to the molality, we can write:

$$\Delta T_f = K_f m$$

$$0.758 = (1.86)(0.400 + x)$$

$$x = 0.00753 = [H^+] = [HCOO^-]$$

$$K_a = \frac{[H^+][HCOO^-]}{[HCOOH]} = \frac{(0.00753)(0.00753)}{0.400 - 0.00753} = \mathbf{1.4 \times 10^{-4}}$$

**15.140** $SO_2(g) + H_2O(l) \rightleftharpoons H^+(aq) + HSO_3^-(aq)$

Recall that 0.12 ppm $SO_2$ would mean 0.12 parts $SO_2$ per 1 million ($10^6$) parts of air by volume. The number of particles of $SO_2$ per volume will be directly related to the pressure.

$$P_{SO_2} = \frac{0.12 \text{ parts } SO_2}{10^6 \text{ parts air}} \text{ atm} = 1.2 \times 10^{-7} \text{ atm}$$

We can now calculate the $[H^+]$ from the equilibrium constant expression.

$$K = \frac{[H^+][HSO_3^-]}{P_{SO_2}}$$

$$1.3 \times 10^{-2} = \frac{x^2}{1.2 \times 10^{-7}}$$

$$x^2 = (1.3 \times 10^{-2})(1.2 \times 10^{-7})$$

$$x = 3.9 \times 10^{-5}\ M = [H^+]$$

$$\mathbf{pH} = -\log(3.9 \times 10^{-5}) = \mathbf{4.40}$$

**15.142** In inhaling the smelling salt, some of the powder dissolves in the basic solution. The ammonium ions react with the base as follows:

$$NH_4^+(aq) + OH^-(aq) \rightarrow NH_3(aq) + H_2O$$

It is the pungent odor of ammonia that prevents a person from fainting.

**15.144** (c) does not represent a Lewis acid-base reaction. In this reaction, the F–F single bond is broken and single bonds are formed between P and each F atom. For a Lewis acid-base reaction, the Lewis acid is an electron-pair acceptor and the Lewis base is an electron-pair donor.

**15.146** From the given pH's, we can calculate the $[H^+]$ in each solution.

$$\text{Solution (1): } [H^+] = 10^{-pH} = 10^{-4.12} = 7.6 \times 10^{-5}\ M$$

$$\text{Solution (2): } [H^+] = 10^{-5.76} = 1.7 \times 10^{-6}\ M$$

$$\text{Solution (3): } [H^+] = 10^{-5.34} = 4.6 \times 10^{-6}\ M$$

We are adding solutions (1) and (2) to make solution (3). The volume of solution (2) is 0.528 L. We are going to add a given volume of solution (1) to solution (2). Let's call this volume $x$. The moles of $H^+$ in solutions (1) and (2) will equal the moles of $H^+$ in solution (3).

$$\text{mol } H^+ \text{ soln (1)} + \text{mol } H^+ \text{ soln (2)} = \text{mol } H^+ \text{ soln (3)}$$

Recall that mol $= M \times$ L. We have:

$$(7.6 \times 10^{-5} \text{ mol/L})(x \text{ L}) + (1.7 \times 10^{-6} \text{ mol/L})(0.528 \text{ L}) = (4.6 \times 10^{-6} \text{ mol/L})(0.528 + x)\text{L}$$

$$(7.6 \times 10^{-5})x + (9.0 \times 10^{-7}) = (2.4 \times 10^{-6}) + (4.6 \times 10^{-6})x$$

$$(7.1 \times 10^{-5})x = 1.5 \times 10^{-6}$$

$$x = 0.021 \text{ L} = \mathbf{21\ mL}$$

**15.148** First, determine the molarity of each of the acids.

$$M\text{ (HX)} = \frac{16.9 \text{ g HX}}{1 \text{ L soln}} \times \frac{1 \text{ mol HX}}{180 \text{ g HX}} = 0.0939\ M$$

$$M\text{ (HY)} = \frac{9.05 \text{ g HY}}{1 \text{ L soln}} \times \frac{1 \text{ mol HY}}{78.0 \text{ g HY}} = 0.116\ M$$

Because both of these solutions have the same pH, they have the same concentration of $H_3O^+$ in solution. The acid with the lower concentration (**HX**) has the greater percent ionization and is therefore the stronger acid.

**15.150** The balanced equations for the two reactions are:

$$MCO_3(s) + 2HCl(aq) \longrightarrow MCl_2(aq) + CO_2(g) + H_2O(l)$$

$$HCl(aq) + NaOH(aq) \longrightarrow NaCl(aq) + H_2O(l)$$

First, let's find the number of moles of excess acid from the reaction with NaOH.

$$0.03280 \text{ L} \times \frac{0.588 \text{ mol NaOH}}{1 \text{ L soln}} \times \frac{1 \text{ mol HCl}}{1 \text{ mol NaOH}} = 0.0193 \text{ mol HCl}$$

The original number of moles of acid was:

$$0.500\,\cancel{L} \times \frac{0.100 \text{ mol HCl}}{1\,\cancel{L} \text{ soln}} = 0.0500 \text{ mol HCl}$$

The amount of hydrochloric acid that reacted with the metal carbonate is:

$$(0.0500 \text{ mol HCl}) - (0.0193 \text{ mol HCl}) = 0.0307 \text{ mol HCl}$$

The mole ratio from the balanced equation is 1 mole $MCO_3$ : 2 mole HCl. The moles of $MCO_3$ that reacted are:

$$0.0307 \,\cancel{\text{mol}} \text{ HCl} \times \frac{1 \text{ mol } MCO_3}{2 \,\cancel{\text{mol}} \text{ HCl}} = 0.01535 \text{ mol } MCO_3$$

We can now determine the molar mass of $MCO_3$, which will allow us to identify the metal.

$$\text{molar mass } MCO_3 = \frac{1.294 \text{ g } MCO_3}{0.01535 \text{ mol } MCO_3} = 84.3 \text{ g/mol}$$

We subtract off the mass of $CO_3^{2-}$ to identify the metal.

$$\text{molar mass M} = 84.3 \text{ g/mol} - 60.01 \text{ g/mol} = 24.3 \text{ g/mol}$$

The metal is **magnesium**.

**15.152** Because HF is a much stronger acid than HCN, we can assume that the pH is largely determined by the ionization of HF.

| | $HF(aq) + H_2O(l)$ | $\rightleftharpoons$ | $H_3O^+(aq)$ | + | $F^-(aq)$ |
|---|---|---|---|---|---|
| Initial (*M*): | 1.00 | | 0 | | 0 |
| Change (*M*): | $-x$ | | $+x$ | | $+x$ |
| Equilibrium (*M*): | $1.00 - x$ | | $x$ | | $x$ |

$$K_a = \frac{[H_3O^+][F^-]}{[HF]}$$

$$7.1 \times 10^{-4} = \frac{x^2}{1.00 - x} \approx \frac{x^2}{1.00}$$

$$x = 0.027\,M = [H_3O^+]$$

$$\textbf{pH} = \mathbf{1.57}$$

HCN is a very weak acid, so at equilibrium, [HCN] $\approx$ 1.00 *M*.

$$K_a = \frac{[H_3O^+][CN^-]}{[HCN]}$$

$$4.9 \times 10^{-10} = \frac{(0.027)[CN^-]}{1.00}$$

$$\mathbf{[CN^-] = 1.8 \times 10^{-8}\,M}$$

In a 1.00 *M* HCN solution, the concentration of $[CN^-]$ would be:

| | $HCN(aq) + H_2O(l) \rightleftharpoons$ | $H_3O^+(aq)$ | $+ CN^-(aq)$ |
|---|---|---|---|
| Initial (*M*): | 1.00 | 0 | 0 |
| Change (*M*): | $-x$ | $+x$ | $+x$ |
| Equilibrium (*M*): | $1.00 - x$ | $x$ | $x$ |

$$K_a = \frac{[H_3O^+][CN^-]}{[HCN]}$$

$$4.9 \times 10^{-10} = \frac{x^2}{1.00 - x} \approx \frac{x^2}{1.00}$$

$$x = \mathbf{2.2 \times 10^{-5}\ M = [CN^-]}$$

$[CN^-]$ is greater in the 1.00 *M* HCN solution compared to the 1.00 *M* HCN/1.00 *M* HF solution. According to LeChâtelier's principle, the high $[H_3O^+]$ (from HF) shifts the HCN equilibrium from right to left decreasing the ionization of HCN. The result is a smaller $[CN^-]$ in the presence of HF.

**15.154** The van't Hoff equation allows the calculation of an equilibrium constant at a different temperature if the value of the equilibrium constant at another temperature and $\Delta H°$ for the reaction are known.

$$\ln \frac{K_1}{K_2} = \frac{\Delta H°}{R}\left(\frac{1}{T_2} - \frac{1}{T_1}\right)$$

First, we calculate $\Delta H°$ for the ionization of water using data in Appendix 3 of the text.

$$H_2O(l) \rightleftharpoons H^+(aq) + OH^-(aq)$$

$$\Delta H° = [\Delta H_f°(H^+) + \Delta H_f°(OH^-)] - \Delta H_f°(H_2O)$$

$$\Delta H° = (0 - 229.94\ \text{kJ/mol}) - (-285.8\ \text{kJ/mol})$$

$$\Delta H° = 55.9\ \text{kJ/mol}$$

We substitute $\Delta H°$ and the equilibrium constant at 25°C (298 K) into the van't Hoff equation to solve for the equilibrium constant at 100°C (373 K).

$$\ln \frac{1.0 \times 10^{-14}}{K_2} = \frac{55.9 \times 10^3\ \text{J/mol}}{8.314\ \text{J/mol} \cdot \text{K}}\left(\frac{1}{373\ \text{K}} - \frac{1}{298\ \text{K}}\right)$$

$$\frac{1.0 \times 10^{-14}}{K_2} = e^{-4.537}$$

$$K_2 = 9.3 \times 10^{-13}$$

We substitute into the equilibrium constant expression for the ionization of water to solve for $[H^+]$ and then pH.

$$K_2 = [H^+][OH^-]$$

$$9.3 \times 10^{-13} = x^2$$

$$x = [H^+] = 9.6 \times 10^{-7}\ M$$

$$\textbf{pH} = -\log(9.6 \times 10^{-7}) = \textbf{6.02}$$

Note that the water is **not** acidic at 100°C because $[H^+] = [OH^-]$.

**15.156** The reactions are:

$$P_4(s) + 5O_2(g) \rightarrow P_4O_{10}(s)$$

$$P_4O_{10}(s) + 6H_2O(l) \rightarrow 4H_3PO_4(aq)$$

First, we calculate the moles of $H_3PO_4$ produced. Next, we can calculate the molarity of the phosphoric acid solution. Finally, we can determine the pH of the $H_3PO_4$ solution (a weak acid).

$$10.0\ \text{g P}_4 \times \frac{1\ \text{mol P}_4}{123.9\ \text{g P}_4} \times \frac{1\ \text{mol P}_4\text{O}_{10}}{1\ \text{mol P}_4} \times \frac{4\ \text{mol H}_3\text{PO}_4}{1\ \text{mol P}_4\text{O}_{10}} = 0.323\ \text{mol H}_3\text{PO}_4$$

$$\text{Molarity} = \frac{0.323\ \text{mol}}{0.500\ \text{L}} = 0.646\ M$$

We set up the ionization of the weak acid, $H_3PO_4$. The $K_a$ value for $H_3PO_4$ can be found in Table 15.5 of the text.

| | $H_3PO_4(aq)$ | $\rightleftharpoons$ | $H^+(aq)$ | + | $H_2PO_4^-(aq)$ |
|---|---|---|---|---|---|
| Initial ($M$): | 0.646 | | 0 | | 0 |
| Change ($M$): | $-x$ | | $+x$ | | $+x$ |
| Equilibrium ($M$): | $0.646 - x$ | | $x$ | | $x$ |

$$K_a = \frac{[H^+][H_2PO_4^-]}{[H_3PO_4]}$$

$$7.5 \times 10^{-3} = \frac{(x)(x)}{(0.646 - x)}$$

$$x^2 + 7.5 \times 10^{-3}x - 4.85 \times 10^{-3} = 0$$

Solving the quadratic equation,

$$x = 0.066\ M = [H^+]$$

Following the procedure in Problem 15.122 and the discussion in Section 15.8 of the text, we can neglect the contribution to the hydronium ion concentration from the second and third ionization steps. Thus,

$$\textbf{pH} = -\log(0.066) = \textbf{1.18}$$

**15.158** **(a)** In the diagram are 12 $OH^-$ ions and 4 $H_3O^+$ ions or there are 3 $OH^-$ ions for every one $H_3O^+$ ion.

$$[OH^-] = 3[H_3O^+]$$

$$K_w = [H_3O^+][OH^-]$$

$$1 \times 10^{-14} = [H_3O^+](3)[H_3O^+]$$

$$[H_3O^+] = 5.8 \times 10^{-8}\ M$$

$$\mathbf{pH} = -\log[H_3O^+] = -\log(5.8 \times 10^{-8}) = \mathbf{7.24}$$

**(b)** First, let's calculate $[H_3O^+]$ from the pH.

$$[H_3O^+] = 10^{-pH} = 10^{-5.0} = 1 \times 10^{-5}\ M$$

We can calculate $[OH^-]$ from $[H_3O^+]$ and then calculate the ratio of the ions.

$$K_w = [H_3O^+][OH^-]$$

$$1.0 \times 10^{-14} = (1 \times 10^{-5}\ M)[OH^-]$$

$$[OH^-] = 1 \times 10^{-9}\ M$$

$$\frac{[H_3O^+]}{[OH^-]} = \frac{1 \times 10^{-5}\ M}{1 \times 10^{-9}\ M} = 1 \times 10^4$$

To represent a pH = 5.0 solution, we would need to draw **ten thousand ($10^4$) $H_3O^+$ ions** for every $OH^-$ ion.

**15.160** In reaction (a), sodium ions ($Na^+$) and chloride ions ($Cl^-$) are spectator ions. In reaction (b), potassium ions ($K^+$) and nitrate ions ($NO_3^-$) are spectator ions. The net ionic equation for both reactions is:

$$H^+(aq) + OH^-(aq) \rightarrow H_2O(l)$$

$$\Delta H° = \Delta H_f°(H_2O) - [\Delta H_f°(H^+) + \Delta H_f°(OH^-)]$$

$$\mathbf{\Delta H°_{rxn}} = (-285.8\ \text{kJ/mol}) - [0 + (-229.94\ \text{kJ/mol})] = \mathbf{-55.9\ kJ/mol}$$

Because both reactions have the same net ionic equation, the value of $\Delta H°_{rxn}$ will be the same for both reactions. This value (−55.9 kJ/mol) is the heat of neutralization for the reaction of one mole of strong acid with one mole of strong base.

# CHAPTER 16
# ACID-BASE EQUILIBRIA AND SOLUBILITY EQUILIBRIA

## PROBLEM-SOLVING STRATEGIES AND TUTORIAL SOLUTIONS

### TYPES OF PROBLEMS

**Problem Type 1:** Buffers.

- **(a)** Identifying buffer systems.
- **(b)** Calculating the pH of a buffer system.
- **(c)** Preparing a buffer solution with a specific pH.

**Problem Type 2:** Titrations.

- **(a)** Strong acid–strong base titrations.
- **(b)** Weak acid–strong base titrations.
- **(c)** Strong acid–weak base titrations.

**Problem Type 3:** Choosing Suitable Acid-Base Indicators.

**Problem Type 4:** Solubility Equilibria.

- **(a)** Calculating $K_{sp}$ from molar solubility.
- **(b)** Calculating solubility from $K_{sp}$.
- **(c)** Predicting a precipitation reaction.
- **(d)** The effect of a common ion on solubility.

**Problem Type 5:** Complex Ion Equilibria and Solubility.

## PROBLEM TYPE 1: BUFFERS

A **buffer solution** is a solution of (1) a weak acid or a weak base and (2) its salt; both components must be present. The solution has the ability to resist change in pH upon the addition of small amounts of either acid or base. A buffer resists change in pH, because the weak acid component reacts with small amounts of added base. The weak base component of the buffer reacts with small amounts of added acid.

$$\underset{\text{weak acid of buffer}}{HA(aq)} + \underset{\text{added base}}{OH^-(aq)} \longrightarrow A^-(aq) + H_2O(l)$$

$$\underset{\text{weak base of buffer}}{A^-(aq)} + \underset{\text{added acid}}{H_3O^+(aq)} \longrightarrow HA(aq) + H_2O(l)$$

## A. Identifying buffer systems

To identify a buffer system, look for a weak acid and its conjugate base (usually the anion in a soluble salt) or a weak base and its conjugate acid (usually the cation in a soluble salt).

**EXAMPLE 16.1**

**Which of the following solutions are buffer systems: (a) $CH_3COONa/CH_3COOH$, (b) $KNO_3/HNO_3$, and (c) $NH_3/NH_4Cl$?**

**Strategy:** What constitutes a buffer system? Which of the preceding solutions contains a weak acid and its salt (containing the weak conjugate base)? Which of the preceding solutions contains a weak base and its salt (containing the weak conjugate acid)? Why is the conjugate base of a strong acid not able to neutralize an added acid?

**Solution:** The criteria for a buffer system are that we must have a weak acid and its salt (containing the weak conjugate base) or a weak base and its salt (containing the weak conjugate acid).

(a) $CH_3COOH$ (acetic acid) is a weak acid, and its conjugate base, $CH_3COO^-$ (acetate ion, the anion of the salt $CH_3COONa$), is a weak base. Therefore, this is a buffer system.

(b) Because $HNO_3$ is a strong acid, its conjugate base, $NO_3^-$, is an extremely weak base. This means that $NO_3^-$ will not combine with $H^+$ in solution to form $HNO_3$. Thus, the system cannot act as a buffer system.

(c) $NH_3$ (ammonia) is a weak base and its conjugate acid, $NH_4^+$ (ammonium ion, the cation of the salt $NH_4Cl$), is a weak acid. Therefore, this is a buffer system.

**PRACTICE EXERCISE**

1. Which of the following solutions are buffer systems: (a) KCN/HCN, (b) NaCl/HCl, and (c) $KNO_2/HNO_2$?

**Text Problem: 16.10**

## B. Calculating the pH of a buffer system

Calculating the pH of a buffer system is similar to calculating the pH of a weak acid solution or the pH of a weak base solution (see Chapter 15). The difference is that the initial concentration of the conjugate of the weak acid or weak base is *not zero*. The initial concentration of the conjugate comes from the salt component of the buffer.

For example, consider a solution that is 1.0 *M* in $CH_3COOH$ (acetic acid) and 1.0 *M* in $CH_3COONa$ (sodium acetate). If this were only a 1.0 *M* acetic acid solution, the initial concentration of the conjugate base ($CH_3COO^-$) would be *zero*. However, $CH_3COONa$ is a soluble salt that ionizes completely.

| | $CH_3COONa(aq) \longrightarrow$ | $Na^+(aq)$ + | $CH_3COO^-(aq)$ |
|---|---|---|---|
| Initial | 1.0 *M* | 0 | 0 |
| After dissociation | 0 | 1.0 *M* | 1.0 *M* |

The $CH_3COO^-$ ion is present initially and enters into the weak acid equilibrium of acetic acid. Thus, if we look at the weak acid equilibrium for acetic acid, we have:

| | $CH_3COOH(aq)$ | $\rightleftharpoons$ | $H^+(aq)$ | + | $CH_3COO^-(aq)$ |
|---|---|---|---|---|---|
| Initial ($M$): | 1.0 | | 0 | | 1.0 |
| Change ($M$): | $-x$ | | $+x$ | | $+x$ |
| Equilibrium ($M$): | $1.0 - x$ | | $x$ | | $1.0 + x$ |

You should notice that the only difference between this equilibrium, and a weak acid equilibrium (see Chapter 15), is that the initial concentration of the conjugate base, $CH_3COO^-$, is *not zero*, as it would be for a weak acid equilibrium. Remember, a buffer contains both a weak acid and its conjugate base or a weak base and its conjugate acid. The above system is a buffer because it initially contains both the weak acid ($CH_3COOH$) and its conjugate base ($CH_3COO^-$). Example 16.2 illustrates how to calculate the pH of a buffer system.

**EXAMPLE 16.2**

**(a) Calculate the pH of a buffer system containing 0.25 *M* HF and 0.50 *M* NaF.**

**(b) What is the pH of the buffer system after the addition of 0.060 mol of gaseous HCl to 1.00 L of the solution? Assume that the volume of the solution does not change upon addition of the HCl.**

**Strategy:** (a) The pH of a buffer system can be calculated in a similar manner to a weak acid equilibrium problem. The difference is that a common-ion is present in solution. The $K_a$ of HF is $7.1 \times 10^{-4}$ (see Table 15.3 of the text). (b) Added acid will react with the base component of the buffer. The concentrations of both buffer components will change. The weak base component will decrease in concentration, and the weak acid component will increase in concentration. Set up a new equilibrium calculation with these concentrations.

**Solution:**

**(a)**

*Step 1:* Recognize that NaF is a soluble salt that will completely dissociate into ions. From the dissociation of NaF, we can calculate the initial concentration of the weak base, $F^-$.

| | $NaF(aq)$ | $\longrightarrow$ | $Na^+(aq)$ | + | $F^-(aq)$ |
|---|---|---|---|---|---|
| Initial | 0.50 $M$ | | 0 | | 0 |
| After dissociation | 0 | | 0.50 $M$ | | 0.50 $M$ |

*Step 2:* The initial concentration of $F^-$ is 0.50 $M$. We summarize the concentrations of the species at equilibrium as follows:

| | $HF(aq)$ | $\rightleftharpoons$ | $H^+(aq)$ | + | $F^-(aq)$ |
|---|---|---|---|---|---|
| Initial ($M$): | 0.25 | | 0 | | 0.50 |
| Change ($M$): | $-x$ | | $+x$ | | $+x$ |
| Equilibrium ($M$): | $0.25 - x$ | | $x$ | | $0.50 + x$ |

*Step 3:* Write the ionization constant expression in terms of the equilibrium concentrations. Knowing the value of the equilibrium constant ($K_a$), solve for $x$.

$$K_a = \frac{[H^+][F^-]}{[HF]}$$

You can look up the $K_a$ value for hydrofluoric acid in Table 15.3 of the text.

$$7.1\times10^{-4} = \frac{(x)(0.50+x)}{(0.25-x)}$$

At this point, we will make two assumptions.

$$0.25 - x \approx 0.25$$

and

$$0.50 + x \approx 0.50$$

Oftentimes, assumptions such as these are valid if $K$ is very small. A very small value of $K$ means that a very small amount of reactants go to products. Hence, $x$ is small. If we did not make this assumption, we would have to solve a quadratic equation.

$$7.1\times10^{-4} \approx \frac{(x)(0.50)}{0.25}$$

Solving for $x$,

$$x = 3.6\times10^{-4}\,M = [H^+]$$

*Step 4:* Having solved for the $[H^+]$, calculate the pH of the solution.

$$\mathbf{pH} = -\log[H^+] = -\log(3.6\times10^{-4}) = \mathbf{3.44}$$

**Check:** Checking the validity of the assumption,

$$\frac{3.6\times10^{-4}}{0.25}\times100\% = 0.14\% < 5\%$$

The assumption is valid.

**(b)** We have added 0.060 mol of HCl (strong acid), which will react completely with the weak base component of the buffer. This reaction will change the equilibrium concentrations of both the acid and base components of the equilibrium. Therefore, after the reaction between HCl and the weak base of the buffer, the equilibrium must be reestablished.

*Step 1:* Write the reaction that occurs between the strong acid, HCl, and the weak base component of the buffer, $F^-$. After addition of HCl, complete ionization of HCl occurs.

| | $HCl(aq)$ | $\longrightarrow$ | $H^+(aq)$ | + | $Cl^-(aq)$ |
|---|---|---|---|---|---|
| Initial | 0.060 mol | | 0 | | 0 |
| After ionization | 0 | | 0.060 mol | | 0.060 mol |

Next, $H^+$ will react with the weak base component of the buffer, $F^-$. Since $H^+$ is the strongest acid that can exist in water, this reaction will be driven to completion.

$$H^+(aq) + F^-(aq) \rightarrow HF(aq)$$

Since we have 1.00 L of buffer solution, the number of moles of $F^-$ in solution is:

$$\text{mol } F^- = \frac{0.50\text{ mol}}{1\text{ L}}\times1.00\text{ L} = 0.50\text{ mol}$$

Similarly, the number of moles of HF is 0.25 mol.

We can now calculate the moles of $F^-$ and HF that remain after $F^-$ reacts with the strong acid HCl.

| | $H^+(aq)$ + | $F^-(aq)$ ⟶ | $HF(aq)$ |
|---|---|---|---|
| Initial (mol): | 0.060 | 0.50 | 0.25 |
| Change (mol): | −0.060 | −0.060 | +0.060 |
| Final (mol): | 0 | 0.44 | 0.31 |

*Step 2:* After the reaction above, HF and $F^-$ are no longer in equilibrium. Equilibrium must be reestablished. The concentrations of HF and $F^-$ are:

$$[HF] = \frac{0.31\ \text{mol}}{1.00\ \text{L}} = 0.31\ M$$

$$[F^-] = \frac{0.44\ \text{mol}}{1.00\ \text{L}} = 0.44\ M$$

Reestablishing equilibrium between HF and $F^-$, we have:

| | $HF(aq)$ ⇌ | $H^+(aq)$ + | $F^-(aq)$ |
|---|---|---|---|
| Initial (*M*): | 0.31 | 0 | 0.44 |
| Change (*M*): | $-x$ | $+x$ | $+x$ |
| Equilibrium (*M*): | $0.31 - x$ | $x$ | $0.44 + x$ |

*Step 3:* Write the ionization constant expression in terms of the equilibrium concentrations. Knowing the value of the equilibrium constant ($K_a$), solve for $x$.

$$K_a = \frac{[H^+][F^-]}{[HF]}$$

$$7.1 \times 10^{-4} = \frac{(x)(0.44 + x)}{(0.31 - x)}$$

$$7.1 \times 10^{-4} \approx \frac{(x)(0.44)}{0.31}$$

$$x = [H^+] = 5.0 \times 10^{-4}\ M$$

*Step 4:* Having solved for the $[H^+]$, calculate the pH of the solution.

$$\mathbf{pH} = -\log[H^+] = -\log(5.0 \times 10^{-4}) = \mathbf{3.30}$$

The pH of the solution dropped only 0.15 pH units upon addition of 0.060 mol of the strong acid, HCl. Thus, this buffer solution resisted change in pH.

**Check:** Does it make sense that the pH of the solution decreased upon the addition of a strong acid to the buffer?

**PRACTICE EXERCISE**

2. A buffer solution is prepared by mixing 500 mL of 0.600 *M* $CH_3COOH$ with 500 mL of a 1.00 *M* $CH_3COONa$ solution. What is the pH of the solution?

3. (a) Calculate the pH of a buffer system that is 0.0600 *M* $HNO_2$ and 0.160 *M* $NaNO_2$.
   (b) What is the pH after 2.00 mL of 2.00 *M* NaOH are added to 1.00 L of this buffer?

**Text Problems: 16.12, 16.14, 16.18**

## C. Preparing a buffer solution with a specific pH

To prepare a buffer solution with a specific pH, we need to consider the Henderson-Hasselbalch equation. For a derivation of this equation, see Section 16.2 of the text.

$$\text{pH} = \text{p}K_a + \log\frac{[\text{conjugate base}]}{[\text{acid}]}$$

where,

$$\text{p}K_a = -\log K_a$$

If the molar concentrations of the acid and its conjugate base are approximately equal, that is,

$$[\text{acid}] \approx [\text{conjugate base}]$$

then,

$$\log\frac{[\text{conjugate base}]}{[\text{acid}]} \approx 0$$

Substituting into the Henderson-Hasselbalch equation gives:

$$\text{pH} \approx \text{p}K_a$$

Thus, to prepare a buffer solution, we should choose a weak acid with a p$K_a$ value close to the desired pH. This choice not only gives the desired pH value for the buffer system, but also ensures that we have comparable amounts of the weak acid and its conjugate base present. Having pH ≈ p$K_a$ is a prerequisite for the buffer system to function effectively.

**EXAMPLE 16.3**
**Which of the following mixtures is suitable for making a buffer solution with an optimum pH of about 9.2?**

(a) $CH_3COONa/CH_3COOH$ (b) $NH_3/NH_4Cl$ (c) NaF/HF
(d) $NaNO_2/HNO_2$ (e) NaCl/HCl

**Strategy:** For a buffer to function effectively, the concentration of the acid component must be roughly equal to the conjugate base component. According to Equation (16.4) of the text, when the desired pH is close to the p$K_a$ of the acid, that is, when pH ≈ p$K_a$,

$$\log\frac{[\text{conjugate base}]}{[\text{acid}]} \approx 0$$

or

$$\frac{[\text{conjugate base}]}{[\text{acid}]} \approx 1$$

**Solution:** To prepare a solution of a desired pH, we should choose a weak acid with a p$K_a$ value close to the desired pH. We can rule out choice (e) immediately, because it contains a strong acid and its conjugate base. This solution is not a buffer. Calculating the p$K_a$ for each acid:

(a) $\mathrm{p}K_a = 4.74$

(b) $\mathrm{p}K_a = 9.26$

(c) $\mathrm{p}K_a = 3.15$

(d) $\mathrm{p}K_a = 3.35$

Thus, the only solution that would make an effective buffer at pH = 9.2 is choice (b), $NH_4Cl/NH_3$.

> **Tip:** We could have saved some time by recognizing that if the $K_a$ value is less than $1 \times 10^{-7}$, the p$K_a$ value will be greater than 7. The only choice that had a $K_a$ value less than $1 \times 10^{-7}$ was choice (b). Thus, it was the only choice that has a p$K_a$ value greater than 7.

**PRACTICE EXERCISE**

4. How would you prepare a liter of an $HF/F^-$ buffer at a pH of 2.85?

**Text Problem: 16.20**

## PROBLEM TYPE 2: TITRATIONS

### A. Strong acid–strong base titrations

The reaction between a strong acid ($HNO_3$) and a strong base (KOH) can be written as:

$$HNO_3(aq) + KOH(aq) \longrightarrow KNO_3(aq) + H_2O(l)$$

or in terms of the net ionic equation

$$H^+(aq) + OH^-(aq) \longrightarrow H_2O(l)$$

At the equivalence point of a titration, the $[H^+] = [OH^-]$. Thus, the only species present in solution other than water at the equivalence point is $KNO_3(aq)$. Potassium nitrate is a salt that does not undergo hydrolysis (see Chapter 15); therefore, the pH at the equivalence point of a strong acid–strong base titration is 7.

A typical strong acid–strong base titration problem involves calculating the pH at various points in the titration. Example 16.4 below illustrates this type of problem.

**EXAMPLE 16.4**

**A 20.0 mL sample of 0.0200 *M* $HNO_3$ is titrated with 0.0100 *M* KOH.**

**(a) What is the pH at the equivalence point?**

**(b) How many mL of KOH are required to reach the equivalence point?**

**(c) What is the pH before any KOH is added?**

**(d) What will be the pH after 10.0 mL of KOH are added?**

**(e) What will be the pH after 45.0 mL of KOH are added?**

**(a)** Since this is a strong acid–strong base titration, the pH at the equivalence point is 7.

**(b)** We worked this type of problem in Chapter 4.

*Step 1:* In order to have the correct mole ratio to solve the problem, you must start with a balanced chemical equation.

$$HNO_3(aq) + KOH(aq) \longrightarrow KNO_3(aq) + H_2O(l)$$

*Step 2:* From the molarity and volume of the $HNO_3$ solution, you can calculate moles of $HNO_3$. Then, using the mole ratio from the balanced equation above, you can calculate moles of KOH.

$$\text{mol KOH} = 20.0\ \text{mL soln} \times \frac{0.0200\ \text{mol HNO}_3}{1000\ \text{mL soln}} \times \frac{1\ \text{mol KOH}}{1\ \text{mol HNO}_3} = 4.00 \times 10^{-4}\ \text{mol KOH}$$

*Step 3:* Solve the molarity equation algebraically for liters of solution. Then, substitute in the moles of KOH and molarity of KOH to solve for volume of KOH.

$$M = \frac{\text{moles of solute}}{\text{liters of solution}}$$

$$\text{liters of solution} = \frac{\text{moles of solute}}{M}$$

$$\textbf{volume of KOH} = \frac{4.00 \times 10^{-4}\ \text{mol KOH}}{0.0100\ \text{mol/L}} = \mathbf{0.0400\ L = 40.0\ mL}$$

**(c)** We only have the strong acid, $HNO_3$, in solution before any KOH is added. Thus, we need to calculate the pH of a strong acid (see Chapter 15).

*Step 1:* Strong acids ionize completely in solution. Let's write the reaction and set up a table to calculate the $H^+$ concentration in solution.

| | $HNO_3(aq)$ | $\longrightarrow$ | $H^+(aq)$ | + | $NO_3^-(aq)$ |
|---|---|---|---|---|---|
| initial conc. | 0.0200 *M* | | 0 | | 0 |
| conc. after ionization | 0 | | 0.0200 *M* | | 0.0200 *M* |

Since there is a 1:1 mole ratio between $H^+$ and $HNO_3$, the $H^+$ concentration equals the $HNO_3$ concentration.

*Step 2:* Calculate the pH from the $H^+$ concentration.

$$\textbf{pH} = -\log[H^+] = -\log[0.0200] = \mathbf{1.70}$$

**(d)** Any strong base added will react completely with the strong acid in solution. The reaction is:

$$HNO_3(aq) + KOH(aq) \longrightarrow KNO_3(aq) + H_2O(l)$$

*Step 1:* Calculate the number of moles of $HNO_3$ in 20.0 mL, and calculate the number of moles of KOH in 10.0 mL.

$$\text{mol HNO}_3 = 20.0\ \text{mL soln} \times \frac{1.00\ \text{L}}{1000\ \text{mL}} \times \frac{0.0200\ \text{mol HNO}_3}{1.00\ \text{L soln}} = 4.00 \times 10^{-4}\ \text{mol}$$

$$\text{mol KOH} = 10.0\ \text{mL soln} \times \frac{1.00\ \text{L}}{1000\ \text{mL}} \times \frac{0.0100\ \text{mol KOH}}{1.00\ \text{L soln}} = 1.00 \times 10^{-4}\ \text{mol}$$

*Step 2:* Set up a table showing the number of moles of $HNO_3$ and KOH before and after the reaction.

| | $HNO_3(aq)$ + | $KOH(aq)$ ⟶ | $KNO_3(aq)$ + $H_2O(l)$ |
|---|---|---|---|
| Initial (mol): | $4.00 \times 10^{-4}$ | $1.00 \times 10^{-4}$ | 0 |
| Change (mol): | $-1.00 \times 10^{-4}$ | $-1.00 \times 10^{-4}$ | $+1.00 \times 10^{-4}$ |
| Final (mol): | $3.00 \times 10^{-4}$ | 0 | $1.00 \times 10^{-4}$ |

Thus, $3.00 \times 10^{-4}$ mol of the strong acid $HNO_3$ remain after the addition of 10.0 mL of KOH.

*Step 3:* The total volume of solution is now the sum of the volume of the acid solution and the volume of the base solution.

$$V_{\text{soln}} = V_{\text{acid}} + V_{\text{base}} = 20.0\ \text{mL} + 10.0\ \text{mL} = 30.0\ \text{mL}$$

We can now calculate the concentration of $H^+$ in solution. Since $HNO_3$ ionizes completely, the number of moles of $H^+$ in solution is $3.00 \times 10^{-4}$ mol. Calculate the molarity by dividing the moles by liters of solution.

$$30.0\ \text{mL} = 0.0300\ \text{L}$$

$$[H^+] = \frac{3.00 \times 10^{-4}\ \text{mol}}{0.0300\ \text{L}} = 0.0100\ M$$

*Step 4:* Calculate the pH from the $H^+$ concentration.

$$\textbf{pH} = -\log[H^+] = -\log[0.0100] = \mathbf{2.00}$$

**(e)** Start this part by following the same procedure as in part (d).

*Step 1:* Calculate the number of moles of KOH in 45.0 mL.

$$\text{mol KOH} = 45.0\ \text{mL soln} \times \frac{1.00\ \text{L}}{1000\ \text{mL}} \times \frac{0.0100\ \text{mol KOH}}{1.00\ \text{L soln}} = 4.50 \times 10^{-4}\ \text{mol}$$

*Step 2:* Set up a table showing the number of moles of $HNO_3$ and KOH before and after the reaction.

| | $HNO_3(aq)$ + | $KOH(aq)$ ⟶ | $KNO_3(aq)$ + $H_2O(l)$ |
|---|---|---|---|
| Initial (mol): | $4.00 \times 10^{-4}$ | $4.50 \times 10^{-4}$ | 0 |
| Change (mol): | $-4.00 \times 10^{-4}$ | $-4.00 \times 10^{-4}$ | $+4.00 \times 10^{-4}$ |
| Final (mol): | 0 | $5.00 \times 10^{-5}$ | $4.00 \times 10^{-4}$ |

We have passed the equivalence point of the titration. The only species of significance that remains in solution is the strong base, KOH. $5.00 \times 10^{-5}$ mol of the strong base KOH remain after the addition of 45.0 mL of KOH.

*Step 3:* The total volume of solution is now the sum of the volume of the acid solution and the volume of the base solution.

$$V_{soln} = V_{acid} + V_{base} = 20.0 \text{ mL} + 45.0 \text{ mL} = 65.0 \text{ mL}$$

We can now calculate the concentration of $OH^-$ in solution. Since KOH ionizes completely, the number of moles of $OH^-$ in solution is $5.00 \times 10^{-5}$ mol. Calculate the molarity by dividing the moles by liters of solution.

$$65.0 \text{ mL} = 0.0650 \text{ L}$$

$$[OH^-] = \frac{5.00 \times 10^{-5} \text{ mol}}{0.0650 \text{ L}} = 7.69 \times 10^{-4}\ M$$

*Step 4:* Calculate the pOH from the $OH^-$ concentration.

$$\text{pOH} = -\log[OH^-] = -\log[7.69 \times 10^{-4}] = 3.11$$

*Step 5:* Use the relationship, pH + pOH = 14.00, to calculate the pH of the solution.

$$\textbf{pH} = 14.00 - 3.11 = \textbf{10.89}$$

**PRACTICE EXERCISE**

5. Consider the titration of 25.0 mL of 0.250 *M* KOH with 0.100 *M* HCl.

   (a) What is the pH at the equivalence point?
   (b) How many mL of HCl are required to reach the equivalence point?
   (c) What is the pH before any HCl is added?
   (d) What will be the pH after 15.0 mL of HCl are added?
   (e) What will be the pH after 75.0 mL of HCl are added?

**Text Problem: 16.28**

## B. Weak acid–strong base titrations

Consider the neutralization between nitrous acid (a weak acid) and sodium hydroxide (a strong base):

$$HNO_2(aq) + NaOH(aq) \longrightarrow NaNO_2(aq) + H_2O(l)$$

The net ionic equation is:

$$HNO_2(aq) + OH^-(aq) \longrightarrow NO_2^-(aq) + H_2O(l)$$

At the equivalence point of a titration, the $[HNO_2] = [OH^-]$. Thus, the major species present in solution other than water at the equivalence point are $NO_2^-(aq)$ and $Na^+(aq)$. $NO_2^-$ is a weak base; it is the conjugate base of the weak acid, $HNO_2$ (see Chapter 15). The hydrolysis reaction is given by:

$$NO_2^-(aq) + H_2O(l) \rightleftharpoons HNO_2(aq) + OH^-(aq)$$

Therefore, at the equivalence point, the pH will be *greater than 7* as a result of the excess $OH^-$ formed.

A typical weak acid–strong base titration problem involves calculating the pH at various points in the titration. Example 16.5 below illustrates this type of problem.

**EXAMPLE 16.5**

**Consider the titration of 50.0 mL of 0.100 *M* $CH_3COOH$ with 0.100 *M* NaOH.**

**(a) How many mL of NaOH are required to reach the equivalence point?**

**(b) What is the pH before any NaOH is added?**

**(c) What will be the pH after 25.0 mL of NaOH are added?**

**(d) What is the pH at the equivalence point?**

**(e) What will be the pH after 60.0 mL of KOH are added?**

---

(a) We worked this type of problem in Chapter 4 and in Example 16.4 above.

*Step 1:* In order to have the correct mole ratio to solve the problem, you must start with a balanced chemical equation.

$$CH_3COOH(aq) + NaOH(aq) \longrightarrow CH_3COONa(aq) + H_2O(l)$$

*Step 2:* We can take a shortcut on this problem by recognizing that both $CH_3COOH$ and NaOH have the same concentration, 0.100 *M*. Since the mole ratio between $CH_3COOH$ and NaOH is 1:1, it will require the same volume of each component to neutralize the other.

Thus, the volume of NaOH required to reach the equivalence point is **50.0 mL.**

(b) We only have the weak acid, $CH_3COOH$, in solution before any NaOH is added. Thus, we need to calculate the pH of a weak acid (see Chapter 15).

*Step 1:* Letting $x$ be the equilibrium concentration of $H^+$ and $CH_3COO^-$ ions in mol/L, we summarize:

| | $CH_3COOH(aq)$ | $\rightleftharpoons$ $H^+(aq)$ | + $CH_3COO^-(aq)$ |
|---|---|---|---|
| Initial (*M*): | 0.100 | 0 | 0 |
| Change (*M*): | $-x$ | $+x$ | $+x$ |
| Equilibrium (*M*): | $0.100 - x$ | $x$ | $x$ |

*Step 2:* Write the ionization constant expression in terms of the equilibrium concentrations. Knowing the value of the equilibrium constant ($K_a$), solve for $x$.

$$K_a = \frac{[H^+][CH_3COO^-]}{[CH_3COOH]}$$

You can look up the $K_a$ value for acetic acid in Table 15.3 of the text.

$$1.8 \times 10^{-5} = \frac{(x)(x)}{(0.100 - x)}$$

At this point, we can make an assumption that $x$ is very small compared to 0.100. Hence,

$$0.100 - x \approx 0.100$$

and,

$$1.8 \times 10^{-5} = \frac{(x)(x)}{0.100}$$

Solving for $x$.

$$x = 1.34 \times 10^{-3}\ M = [H^+]$$

*Step 3:* Having solved for the $[H^+]$, calculate the pH of the solution.

$$\mathbf{pH} = -\log[H^+] = -\log(1.34 \times 10^{-3}) = \mathbf{2.87}$$

**Check:** Checking the validity of the assumption,

$$\frac{1.34 \times 10^{-3}}{0.100} \times 100\% = 1.34\% < 5\%$$

The assumption was valid.

**(c)** You should recognize that when 25.0 mL of NaOH are added, we are half-way to the equivalence point. For a weak acid-strong base titration or a weak base-strong acid titration, the pH at the half-way point equals $pK_a$. This relation can be derived from the Henderson-Hasselbalch equation. At the half-way point, the concentration of the acid equals the concentration of its conjugate base, because half of the acid ($CH_3COOH$) has been neutralized, forming an equal amount of its conjugate base ($CH_3COO^-$).

$$\text{pH} = pK_a + \log\frac{[\text{conjugate base}]}{[\text{acid}]}$$

At the halfway point,

$$[\text{acid}] = [\text{conjugate base}]$$

and,

$$\log\frac{[\text{conjugate base}]}{[\text{acid}]} = 0$$

Substituting into the Henderson-Hasselbalch equation gives:

$$\text{pH} = pK_a$$

$$\mathbf{pH} = pK_a = -\log(1.8 \times 10^{-5}) = \mathbf{4.74}$$

**(d)** At the equivalence point the strong base has completely neutralized the weak acid in solution. The reaction is:

$$CH_3COOH(aq) + NaOH(aq) \longrightarrow CH_3COONa(aq) + H_2O(l)$$

*Step 1:* The number of moles of acetic acid equals the number of moles of NaOH at the equivalence point. Calculate the number of moles of acetic acid and sodium hydroxide.

$$\text{mol } CH_3COOH = 50.0 \text{ mL soln} \times \frac{1.00 \text{ L}}{1000 \text{ mL}} \times \frac{0.100 \text{ mol } CH_3COOH}{1.00 \text{ L soln}} = 5.00 \times 10^{-3} \text{ mol}$$

$$\text{mol NaOH} = \text{mol } CH_3COOH$$

*Step 2:* Set up a table showing the number of moles of $CH_3COOH$ and NaOH before and after the reaction.

| | $CH_3COOH(aq)$ + | $NaOH(aq)$ ⟶ | $CH_3COONa(aq)$ + $H_2O(l)$ |
|---|---|---|---|
| Initial (mol): | $5.00 \times 10^{-3}$ | $5.00 \times 10^{-3}$ | 0 |
| Change (mol): | $-5.00 \times 10^{-3}$ | $-5.00 \times 10^{-3}$ | $+5.00 \times 10^{-3}$ |
| Final (mol): | 0 | 0 | $5.00 \times 10^{-3}$ |

The only species in solution of significance at the equivalence point is the salt $CH_3COONa$. This salt contains the conjugate base ($CH_3COO^-$) of the weak acid, $CH_3COOH$. The conjugate base is a weak base. The concentration of $CH_3COO^-$ is

$$[CH_3COO^-] = \frac{5.00 \times 10^{-3} \text{ mol } CH_3COO^-}{0.100 \text{ L soln}} = 0.0500\ M$$

Remember to calculate the total volume of solution by adding the volume of the acid to the volume of the base.

$$50.0 \text{ mL acid} + 50.0 \text{ mL base} = 100.0 \text{ mL soln} = 0.100 \text{ L soln}$$

At this point, this problem just becomes a weak base calculation (see Chapter 15).

*Step 3:* Letting $x$ be the equilibrium concentration of $OH^-$ and $CH_3COOH$ ions in mol/L, we summarize:

| | $CH_3COO^-(aq)$ + $H_2O(l)$ ⇌ | $CH_3COOH(aq)$ + | $OH^-(aq)$ |
|---|---|---|---|
| Initial ($M$): | 0.0500 | 0 | 0 |
| Change ($M$): | $-x$ | $+x$ | $+x$ |
| Equilibrium ($M$): | $0.0500 - x$ | $x$ | $x$ |

*Step 4:* Write the ionization constant expression in terms of the equilibrium concentrations. Knowing the value of the equilibrium constant ($K_b$), solve for $x$.

We can calculate $K_b$ from the $K_a$ value of acetic acid.

$$K_b = \frac{K_w}{K_a} = \frac{1.0 \times 10^{-14}}{1.8 \times 10^{-5}} = 5.6 \times 10^{-10}$$

$$K_b = \frac{[CH_3COOH][OH^-]}{[CH_3COO^-]}$$

$$5.6 \times 10^{-10} = \frac{(x)(x)}{(0.0500 - x)}$$

$$5.6 \times 10^{-10} \approx \frac{(x)(x)}{0.0500}$$

Solving for $x$.

$$x = 5.3 \times 10^{-6}\, M = [OH^-]$$

*Step 5:* Having solved for the $[OH^-]$, calculate the pOH of the solution. Then use the relationship, pH + pOH = 14, to solve for the pH of the solution.

$$pOH = -\log[OH^-] = -\log(5.3 \times 10^{-6}) = 5.28$$

$$\mathbf{pH} = 14.00 - pOH = 14.00 - 5.28 = \mathbf{8.72}$$

Does it make sense that the pH at the equivalence point is *greater than* 7?

(e) After 60.0 mL of NaOH have been added, we have passed the equivalence point. The reaction is:

$$CH_3COOH(aq) + NaOH(aq) \rightarrow CH_3COONa(aq) + H_2O(l)$$

*Step 1:* Calculate the number of moles of sodium hydroxide in 60.0 mL of solution.

$$\text{mol NaOH} = 60.0\ \text{mL soln} \times \frac{1.00\ \text{L}}{1000\ \text{mL}} \times \frac{0.100\ \text{mol NaOH}}{1.00\ \text{L soln}} = 6.00 \times 10^{-3}\ \text{mol}$$

*Step 2:* Set up a table showing the number of moles of $CH_3COOH$ and NaOH before and after the reaction.

| | $CH_3COOH(aq)$ + | $NaOH(aq)$ → | $CH_3COONa(aq)$ + | $H_2O(l)$ |
|---|---|---|---|---|
| Initial (mol): | $5.00 \times 10^{-3}$ | $6.00 \times 10^{-3}$ | 0 | |
| Change (mol): | $-5.00 \times 10^{-3}$ | $-5.00 \times 10^{-3}$ | $+5.00 \times 10^{-3}$ | |
| Final (mol): | 0 | $1.00 \times 10^{-3}$ | $5.00 \times 10^{-3}$ | |

The only species in solution of significance past the equivalence point is the strong base NaOH. The salt does contain a weak base, but because NaOH is a strong base, we can assume that all the $OH^-$ comes from NaOH. The concentration of $OH^-$ is:

$$[OH^-] = \frac{1.00 \times 10^{-3}\ \text{mol OH}^-}{0.110\ \text{L soln}} = 9.09 \times 10^{-3}\, M$$

*Step 3:* Having solved for the $[OH^-]$, calculate the pOH of the solution. Then use the relationship, pH + pOH = 14, to solve for the pH of the solution.

$$pOH = -\log[OH^-] = -\log(9.09 \times 10^{-3}) = 2.04$$

$$\mathbf{pH} = 14.00 - pOH = 14.00 - 2.04 = \mathbf{11.96}$$

**PRACTICE EXERCISE**

**6.** Consider the titration of 20.0 mL of 0.200 *M* $HNO_2$ with 0.100 *M* NaOH.

(a) How many mL of NaOH are required to reach the equivalence point?
(b) What is the pH before any NaOH is added?
(c) What will be the pH after 20.0 mL of NaOH are added?
(d) What is the pH at the equivalence point?
(e) What will be the pH after 50.0 mL of KOH are added?

**Text Problems:** 16.30, 16.32, 16.34

## C. Strong acid–weak base titrations

Consider the neutralization between ammonia (a weak base) and nitric acid (a strong acid):

$$NH_3(aq) + HNO_3(aq) \rightarrow NH_4NO_3(aq)$$

The net ionic equation is:

$$NH_3(aq) + H^+(aq) \rightarrow NH_4^+(aq)$$

At the equivalence point of a titration, the $[NH_3] = [H^+]$. Thus, the major species present in solution other than water at the equivalence point are $NH_4^+(aq)$ and $NO_3^-(aq)$. $NH_4^+$ is a weak acid; it is the conjugate acid of the weak base, $NH_3$ (see Chapter 15). The hydrolysis reaction is given by:

$$NH_4^+(aq) + H_2O(l) \rightleftharpoons NH_3(aq) + H_3O^+(aq)$$

Therefore, at the equivalence point, the pH will be *less than* 7 as a result of the excess $H_3O^+$ ions formed.

A weak base–strong acid titration problem is worked similarly to a weak acid–strong base titration problem (see Example 16.5).

**Text Problems:** 16.36, 16.38

# PROBLEM TYPE 3: CHOOSING SUITABLE ACID-BASE INDICATORS

The criterion for choosing an appropriate indicator for a given titration is whether the pH range over which the indicator changes color corresponds with the steep portion of the titration curve. If the indicator does not change color during this portion of the curve, it will not accurately identify the equivalence point. Table 16.1 of the text lists some common acid-base indicators and the pH ranges in which they change color.

Typically, the pH range for an indicator is equal to the $pK_a$ of the indicator, plus or minus one pH unit.

$$\text{pH range} = pK_a \pm 1$$

**EXAMPLE 16.6**

**Will bromophenol blue be a good choice as an indicator for a weak acid–strong base titration?**

**Strategy:** The choice of an indicator for a particular titration is based on the fact that its pH range for color change must overlap the steep portion of the titration curve. Otherwise we cannot use the color change to locate the equivalence point.

**Solution:** A typical titration curve for a weak acid–strong base titration is shown in Figure 16.5 of the text. The pH at the equivalence point is greater than 7. Checking Table 16.1 of the text, bromophenol blue changes color over the pH range, 3.0–4.6. This indicator would change color before the steep portion of the titration curve for a typical weak acid–strong base titration. Thus, bromophenol blue would *not* be a good choice as an indicator for a weak acid–strong base titration.

**PRACTICE EXERCISE**

7. Would methyl red be a suitable indicator for a titration that has a pH of 5.0 at the equivalence point?

**Text Problems:** 16.44, 16.46

## PROBLEM TYPE 4: SOLUBILITY EQUILIBRIA

Solubility equilibria typically involve salts of low solubility. For example, calcium phosphate, $[Ca_3(PO_4)_2]$, is practically insoluble in water. However, a very small amount of calcium phosphate will dissolve and dissociate completely into $Ca^{2+}$ and $PO_4^{3-}$ ions. An equilibrium will then be established between undissolved calcium phosphate and the ions in solution. Consider a saturated solution of calcium phosphate that is in contact with solid calcium phosphate. The solubility equilibrium can be represented as

$$Ca_3(PO_4)_2(s) \rightleftharpoons 3Ca^{2+}(aq) + 2PO_4^{3-}(aq)$$

We know from Chapter 14 that for heterogeneous equilibria, the concentration of a solid is a constant. Thus, we can write the equilibrium constant for the dissolution of $Ca_3(PO_4)_2$ as

$$K_{sp} = [Ca^{2+}]^3[PO_4^{3-}]^2$$

where $K_{sp}$ is called the solubility product constant or simply the solubility product. In general, the **solubility product** of a compound is the product of the molar concentrations of the constituent ions, each raised to the power of its stoichiometric coefficient in the balanced equilibrium equation.

### A. Calculating $K_{sp}$ from molar solubility

**Molar solubility** is the number of moles of solute in 1 L of a saturated solution (moles per liter). The concentrations of the ions in the solubility product expression are also molar concentrations. From the molar solubility, we can calculate the molar concentrations of the ions in solution from the stoichiometry of the balanced equilibrium equation.

**EXAMPLE 16.7**

**If the solubility of $Fe(OH)_2$ in water is $7.7 \times 10^{-6}$ mol/L at a certain temperature, what is its $K_{sp}$ value at that temperature?**

**Strategy:** From the molar solubility, the concentrations of $Fe^{2+}$ and $OH^-$ can be calculated. Then, from these concentrations, we can determine $K_{sp}$.

**Solution:** Consider the dissociation of $Fe(OH)_2$ in water. Let *s* be the molar solubility of $Fe(OH)_2$.

$$Fe(OH)_2(s) \rightleftharpoons Fe^{2+}(aq) + 2OH^-(aq)$$

| | | $Fe^{2+}$ | $OH^-$ |
|---|---|---|---|
| Initial (*M*): | | 0 | 0 |
| Change (*M*): | $-s$ | $+s$ | $+2s$ |
| Equilibrium (*M*): | | $s$ | $2s$ |

$$K_{sp} = [Fe^{2+}][OH^-]^2 = (s)(2s)^2 = 4s^3$$

The molar solubility (*s*) is given in the problem. Substitute into the equilibrium constant expression to solve for $K_{sp}$.

$$K_{sp} = [Fe^{2+}][OH^-]^2 = 4s^3 = 4(7.7 \times 10^{-6})^3 = \mathbf{1.8 \times 10^{-15}}$$

**PRACTICE EXERCISE**

8. At a certain temperature, the solubility of barium chromate ($BaCrO_4$) is $1.8 \times 10^{-5}$ mol/L. What is the $K_{sp}$ value at this temperature?

**Text Problems: 16.54**, 16.56, 16.60

## B. Calculating solubility from $K_{sp}$

As in other equilibrium calculations, the expected concentrations at equilibrium can be calculated from a knowledge of the initial concentrations and the equilibrium constant (see Chapter 14). In these types of problems, it will be very helpful to recall that

equilibrium concentration = initial concentration ± the change due to reaction.

Example 16.8 illustrates this important type of calculation.

**EXAMPLE 16.8**

**What is the molar solubility of silver phosphate ($Ag_3PO_4$) in water? $K_{sp} = 1.8 \times 10^{-18}$.**

**Strategy:** The $K_{sp}$ value for $Ag_3PO_4$ is given in the problem. Setting up the dissociation equilibrium of $Ag_3PO_4$ in water, we can solve for the molar solubility, *s*.

**Solution:** Consider the dissociation of $Ag_3PO_4$ in water.

$$Ag_3PO_4(s) \rightleftharpoons 3Ag^+(aq) + PO_4^{3-}(aq)$$

| | | $Ag^+$ | $PO_4^{3-}$ |
|---|---|---|---|
| Initial (*M*): | | 0 | 0 |
| Change (*M*): | $-s$ | $+3s$ | $+s$ |
| Equilibrium (*M*): | | $3s$ | $s$ |

Recall, that the concentration of a pure solid does not enter into an equilibrium constant expression. Therefore, the concentration of $Ag_3PO_4$ is not important.

Substitute the value of $K_{sp}$ and the concentrations of $Ag^+$ and $PO_4^{3-}$ in terms of *s* into the solubility product expression to solve for *s*, the molar solubility.

$$K_{sp} = [Ag^+]^3[PO_4^{3-}]$$

$$1.8 \times 10^{-18} = (3s)^3(s)$$

$$1.8 \times 10^{-18} = 27s^4$$

$$s = \text{molar solubility} = \mathbf{1.6 \times 10^{-5}\ mol/L}$$

The molar solubility indicates that $1.6 \times 10^{-5}$ mole of $Ag_3PO_4$ will dissolve in 1 L of an aqueous solution.

**PRACTICE EXERCISE**

9. The $K_{sp}$ value of silver carbonate ($Ag_2CO_3$) is $8.1 \times 10^{-12}$ at 25°C. What is the solubility in g/L of silver carbonate in water at 25°C?

**Text Problems: 16.58,** 16.62

## C. Predicting a precipitation reaction

To predict whether a precipitate will form, we must calculate the **ion product, *Q*.** We will follow the same procedure outlined in Section 14.4 of the text. The ion product represents the molar concentrations of the ions raised to the power of their stoichiometric coefficients. Thus, for an aqueous solution containing $Pb^{2+}$ and $S^{2-}$ ions at 25°C,

$$Q = [Pb^{2+}]_0[S^{2-}]_0$$

The subscript 0 indicates that these are initial concentrations and do not necessarily correspond to equilibrium concentrations.

There are three possible relationships between $Q$ and $K_{sp}$.

- $Q = K_{sp}$ — $[Pb^{2+}]_0[S^{2-}]_0 = 3.4 \times 10^{-28}$ — Since $Q = K_{sp}$, the reaction is already at equilibrium. The solution is saturated.

- $Q < K_{sp}$ — $[Pb^{2+}]_0[S^{2-}]_0 < 3.4 \times 10^{-28}$ — With $Q < K_{sp}$, the solution is not saturated. More solid could dissolve to produce more ions in solution. No precipitate will form.

- $Q > K_{sp}$ — $[Pb^{2+}]_0[S^{2-}]_0 > 3.4 \times 10^{-28}$ — With $Q > K_{sp}$, the solution is supersaturated. There are too many ions in solution. Some ions will combine to form PbS, which will precipitate out until the product of the ion concentrations is equal to $3.4 \times 10^{-28}$.

**EXAMPLE 16.9**

**Predict whether a precipitate of $PbI_2$ will form when 200 mL of 0.015 *M* $Pb(NO_3)_2$ and 300 mL of 0.050 *M* NaI are mixed together. $K_{sp}$ ($PbI_2$) = $1.4 \times 10^{-8}$.**

**Strategy:** Under what condition will an ionic compound precipitate from solution? The ions in solution are $Pb^{2+}$, $NO_3^-$, $Na^+$, and $I^-$. According to the solubility rules listed in Table 4.2 of the text, the only precipitate that can form is $PbI_2$. From the information given, we can calculate $[Pb^{2+}]_0$ and $[I^-]_0$ because we know the number of moles of ions in the original solutions and the volume of the combined solution. Next, we calculate the reaction quotient $Q$ ($Q = [Pb^{2+}]_0[I^-]_0^2$) and compare the value of $Q$ with $K_{sp}$ of $PbI_2$ to see if a precipitate will form, that is, if the solution is supersaturated.

**Solution:** The number of moles of $Pb^{2+}$ present in the original 200 mL of solution is:

$$200 \text{ mL} \times \frac{0.015 \text{ mol Pb}^{2+}}{1000 \text{ mL soln}} = 3.00 \times 10^{-3} \text{ mol Pb}^{2+}$$

The total volume after combining the solutions is 500 mL (0.500 L). The concentration of $Pb^{2+}$ in the 0.500 L volume is:

$$[Pb^{2+}]_0 = \frac{3.00 \times 10^{-3} \text{ mol}}{0.500 \text{ L}} = 6.00 \times 10^{-3}\ M$$

The number of moles of $I^-$ present in the original 300 mL of solution is:

$$300 \text{ mL} \times \frac{0.050 \text{ mol I}^-}{1000 \text{ mL soln}} = 1.50 \times 10^{-2} \text{ mol I}^-$$

The concentration of $I^-$ in the 0.500 L volume is:

$$[I^-]_0 = \frac{1.50 \times 10^{-2} \text{ mol}}{0.500 \text{ L}} = 3.00 \times 10^{-2}\ M$$

Now, we must compare $Q$ and $K_{sp}$ from Table 16.2 of the text.

$$PbI_2(s) \rightleftharpoons Pb^{2+}(aq) + 2I^-(aq) \qquad K_{sp} = 1.4 \times 10^{-8}$$

$$Q = [Pb^{2+}]_0[I^-]_0^2$$

$$Q = (6.0 \times 10^{-3})(3.00 \times 10^{-2})^2 = 5.4 \times 10^{-6}$$

Comparing $Q$ to $K_{sp}$, we find the $Q > K_{sp}$. This means that the solution is supersaturated. There are too many ions in solution. Some ions will combine to form $PbI_2$, which will precipitate out until the product of the ion concentrations is equal to $K_{sp}$, $1.4 \times 10^{-8}$.

**PRACTICE EXERCISE**

10. Will a precipitate of $MgF_2$ form when $6.00 \times 10^2$ mL of a solution that is $2.0 \times 10^{-4}\ M$ in $MgCl_2$ is added to $3.00 \times 10^2$ mL of a $1.1 \times 10^{-2}\ M$ NaF solution? $K_{sp}(MgF_2) = 6.6 \times 10^{-9}$.

**Text Problem:** 16.76

## D. The effect of a common ion on solubility

Consider the slightly soluble salt copper (I) iodide. We can write its equilibrium reaction in water as follows:

$$CuI(s) \rightleftharpoons Cu^{+}(aq) + I^{-}(aq)$$

Now, suppose that instead of dissolving CuI in water, we attempted to dissolve a certain quantity of CuI in a potassium iodide (KI) solution. Would the presence of KI affect the solubility of the copper(I) iodide?

Remember, that KI is a soluble salt, so it will dissociate completely into $K^+$ and $I^-$ ions.

$$KI(aq) \rightarrow K^{+}(aq) + I^{-}(aq)$$

The $I^-$ ions from KI *will affect* the copper(I) iodide equilibrium. See Chapter 14 for a discussion of Le Châtelier's principle. Le Châtelier's principle states that when an external stress is applied to a system at equilibrium, the system adjusts in such a way that the stress is partially offset. In this system, the stress is additional $I^-$ ions in solution (from KI). The CuI equilibrium will shift to the left to remove some of the additional $I^-$ ions, decreasing the solubility of CuI.

In summary, the effect of adding a **common ion**, in this case $I^-$, is to **decrease** the solubility of the salt (CuI) in solution.

Calculating the solubility of a salt when a common ion is present is very similar to calculating the solubility from $K_{sp}$ discussed earlier in this chapter. The only difference is that the initial concentration of the common ion is *not* zero. The initial concentration of the common ion comes from the soluble salt that is present in solution. Example 16.10 below illustrates this type of problem.

**EXAMPLE 16.10**

**What is the molar solubility of $PbCl_2$ in a 0.50 *M* NaCl solution? $K_{sp}$ ($PbCl_2$) = $2.4 \times 10^{-4}$.**

This problem is worked in a similar manner to Example 16.8.

**Strategy:** This is a common-ion problem. The common ion is $Cl^-$, which is supplied by both $PbCl_2$ and NaCl. Remember that the presence of a common ion will affect only the solubility of $PbCl_2$, but not the $K_{sp}$ value because it is an equilibrium constant.

**Solution:** Set up a table to find the equilibrium concentrations in 0.50 *M* NaCl. NaCl is a soluble salt that ionizes completely giving an initial concentration of $Cl^- = 0.50$ *M*.

| | $PbCl_2(s)$ $\rightleftharpoons$ | $Pb^{2+}(aq)$ + | $2Cl^{-}(aq)$ |
|---|---|---|---|
| Initial (*M*): | | 0 | 0.50 |
| Change (*M*): | $-s$ | $+s$ | $+2s$ |
| Equilibrium (*M*): | | $s$ | $0.50 + 2s$ |

Recall that the concentration of a pure solid does not enter into an equilibrium constant expression. Therefore, the concentration of $PbCl_2$ is not important.

$$K_{sp} = [Pb^{2+}][Cl^{-}]^2$$

$$2.4 \times 10^{-4} = (s)(0.50 + 2s)^2$$

Let's assume that 0.50 >> 2*s*. This is a valid assumption because $K_{sp}$ is small.

$$2.4 \times 10^{-4} \approx (s)(0.50)^2$$

$$s = \text{molar solubility} = \mathbf{9.6 \times 10^{-4}\ mol/L}$$

The molar solubility of lead(II) chloride in pure water is $3.9 \times 10^{-2}\ M$. Does it make sense that the molar solubility decreased in 0.50 *M* NaCl?

**Check:** Checking the validity of the assumption,

$$\frac{2(9.6 \times 10^{-4})}{0.50} \times 100\% = 0.38\% < 5\%$$

**PRACTICE EXERCISE**

11. What is the molar solubility of $Ag_3PO_4$ in 0.20 *M* $AgNO_3$? $K_{sp}$ $(Ag_3PO_4) = 1.8 \times 10^{-18}$.

**Text Problems: 16.68**, 16.70

## PROBLEM TYPE 5: COMPLEX ION EQUILIBRIA AND SOLUBILITY

We can define a **complex ion** as an ion containing a central metal cation bonded to one or more molecules or ions. Transition metals have a particular tendency to form complex ions. For example, silver chloride (AgCl) is insoluble in water. However, when aqueous ammonia is added to AgCl in water, the silver chloride dissolves to form $Ag(NH_3)_2^+(aq)$ and $Cl^-(aq)$. The equilibrium equation is:

$$Ag^+(aq) + 2NH_3(aq) \rightleftharpoons Ag(NH_3)_2^+(aq)$$

A measure of the tendency of a metal ion to form a particular complex ion is given by the **formation constant** $K_f$ (also called the stability constant). The formation constant is simply the equilibrium constant for complex ion formation. The larger the $K_f$ value, the more stable the complex ion. Table 16.4 of the text lists the formation constants of a number of complex ions.

The formation constant expression for the formation of $Ag(NH_3)_2^+$ can be written as:

$$K_f = \frac{[Ag(NH_3)_2^+]}{[Ag^+][NH_3]^2} = 1.5 \times 10^7$$

The very large value of $K_f$ in this case indicates the great stability of the complex ion $[Ag(NH_3)_2^+]$ in solution. Thus, the insoluble silver chloride dissolves to form the stable complex ion, $Ag(NH_3)_2^+$.

A typical complex ion equilibrium problem involves using the formation constant to calculate the equilibrium concentrations of species in solution. See Example 16.11 below.

**EXAMPLE 16.11**

**Calculate the concentration of free $Ag^+$ ions in a solution formed by adding 0.20 mol of $AgNO_3$ to 1.0 L of 1.0 *M* NaCN. Assume no volume change upon addition of $AgNO_3$.**

**Strategy:** The addition of $AgNO_3$ to the NaCN solution results in complex ion formation. In solution, $Ag^+$ ions will complex with $CN^-$ ions. The concentration of $Ag^+$ will be determined by the following equilibrium

$$Ag^+(aq) + 2CN^-(aq) \rightleftharpoons Ag(CN)_2^-$$

From Table 16.4 of the text, we see that the formation constant ($K_f$) for this reaction is very large ($K_f = 1.0 \times 10^{21}$). Because $K_f$ is so large, the reaction lies mostly to the right. At equilibrium, the concentration of $Ag^+$ will be very small. As a good approximation, we can assume that essentially all the dissolved $Ag^+$ ions end up as $Ag(CN)_2^-$ ions. What is the initial concentration of $Ag^+$ ions? A very small amount of $Ag^+$ will be present at equilibrium. Set up the $K_f$ expression for the above equilibrium to solve for $[Ag^+]$.

**Solution:** The initial concentration of $Ag^+$ is 0.20 *M*. If we assume that the above equilibrium goes to completion, we can write:

| | $Ag^+(aq)$ | + $2CN^-(aq)$ | $\rightleftharpoons$ $Ag(CN)_2^-(aq)$ |
|---|---|---|---|
| Initial (*M*): | 0.20 | 1.0 | 0 |
| Change (*M*): | −0.20 | −2(0.20) | +0.20 |
| Final (*M*): | 0 | 0.60 | 0.20 |

To find the concentration of free $Ag^+$ at equilibrium, use the formation constant expression.

$$K_f = \frac{[Ag(CN)_2^-]}{[Ag^+][CN^-]^2}$$

Rearranging,

$$[Ag^+] = \frac{[Ag(CN)_2^-]}{K_f[CN^-]^2}$$

Substitute the equilibrium concentrations calculated above into the formation constant expression to calculate the equilibrium concentration of $Ag^+$.

$$[Ag^+] = \frac{[Ag(CN)_2^-]}{K_f[CN^-]^2} = \frac{0.20}{(1.0 \times 10^{21})(0.60)^2} = \mathbf{5.6 \times 10^{-22}}\ M$$

This concentration corresponds to only *three* $Ag^+$ ions per 10 mL of solution!

**PRACTICE EXERCISE**

12. If 0.0100 mol of $Cu(NO_3)_2$ are dissolved in 1.00 L of 1.00 *M* $NH_3$, what are the concentrations of $Cu^{2+}$, $Cu(NH_3)_4^{2+}$, and $NH_3$ at equilibrium? Assume no volume change upon addition of the $Cu(NO_3)_2$ to the ammonia solution.

**Text Problems: 16.80**, 16.82, 16.84

## ANSWERS TO PRACTICE EXERCISES

**1.** Both (a) and (c). **2.** pH = 4.96 **3.** **(a)** pH = 3.77 **(b)** pH = 3.81

**4.** To prepare a pH = 2.85 buffer you would need to use a ratio of $[F^-]$ to [HF] of 0.50, or 1 to 2.

One way to prepare this buffer would be to add 0.50 mol of NaF (21 g) to 1.0 L of 1.0 *M* HF (assuming no change in volume).

**5.** **(a)** pH = 7 **(b)** 62.5 mL of HCl **(c)** pH = 13.40
**(d)** pH = 13.08 **(e)** pH = 1.90

**6.** **(a)** 40.0 mL of NaOH **(b)** pH = 2.02 **(c)** Halfway point, pH = $pK_a$ = 3.35
**(d)** pH = 8.09 **(e)** pH = 12.16

**7.** Yes, methyl red would be a suitable indicator.

**8.** $K_{sp} = 3.2 \times 10^{-10}$ **9.** solubility = 0.035 g/L **10.** No, a precipitate will *not* form.

**11.** $s$ = molar solubility = $2.3 \times 10^{-16}$ *M* **12.** $[NH_3] = 0.96\ M$

$[Cu(NH_3)_4^{2+}] = 0.0100\ M$

$[Cu^{2+}] = 2.4 \times 10^{-16}\ M$

## SOLUTIONS TO SELECTED TEXT PROBLEMS

**16.6** **(a)** This is a weak base calculation.

| | $NH_3(aq) + H_2O(l)$ | $\rightleftharpoons$ | $NH_4^+(aq)$ | $+\ OH^-(aq)$ |
|---|---|---|---|---|
| Initial (*M*): | 0.20 | | 0 | 0 |
| Change (*M*): | $-x$ | | $+x$ | $+x$ |
| Equilibrium (*M*): | $0.20 - x$ | | $x$ | $x$ |

$$K_b = \frac{[NH_4^+][OH^-]}{[NH_3]}$$

$$1.8 \times 10^{-5} = \frac{(x)(x)}{0.20 - x} \approx \frac{x^2}{0.20}$$

$$x = 1.9 \times 10^{-3}\ M = [OH^-]$$

$$\text{pOH} = 2.72$$

$$\mathbf{pH = 11.28}$$

**(b)** The initial concentration of $NH_4^+$ is 0.30 *M* from the salt $NH_4Cl$. We set up a table as in part (a).

| | $NH_3(aq) + H_2O(l)$ | $\rightleftharpoons$ | $NH_4^+(aq)$ | $+ OH^-(aq)$ |
|---|---|---|---|---|
| Initial (*M*): | 0.20 | | 0.30 | 0 |
| Change (*M*): | $-x$ | | $+x$ | $+x$ |
| Equilibrium (*M*): | $0.20 - x$ | | $0.30 + x$ | $x$ |

$$K_b = \frac{[NH_4^+][OH^-]}{[NH_3]}$$

$$1.8 \times 10^{-5} = \frac{(x)(0.30 + x)}{0.20 - x} \approx \frac{x(0.30)}{0.20}$$

$$x = 1.2 \times 10^{-5}\ M = [OH^-]$$

$$\text{pOH} = 4.92$$

$$\mathbf{pH = 9.08}$$

Alternatively, we could use the Henderson-Hasselbalch equation to solve this problem. Table 15.4 gives the value of $K_a$ for the ammonium ion. Substituting into the Henderson-Hasselbalch equation gives:

$$\text{pH} = \text{p}K_a + \log\frac{[\text{conjugate base}]}{\text{acid}} = -\log(5.6 \times 10^{-10}) + \log\frac{(0.20)}{(0.30)}$$

$$\text{pH} = 9.25 - 0.18 = \mathbf{9.07}$$

Is there any difference in the Henderson-Hasselbalch equation in the cases of a weak acid and its conjugate base and a weak base and its conjugate acid?

**16.10** **Strategy:** What constitutes a buffer system? Which of the preceding solutions contains a weak acid and its salt (containing the weak conjugate base)? Which of the preceding solutions contains a weak base and its salt (containing the weak conjugate acid)? Why is the conjugate base of a strong acid not able to neutralize an added acid?

**Solution:** The criteria for a buffer system are that we must have a weak acid and its salt (containing the weak conjugate base) or a weak base and its salt (containing the weak conjugate acid).

**(a)** HCN is a weak acid, and its conjugate base, $CN^-$, is a weak base. Therefore, this is a buffer system.

**(b)** $HSO_4^-$ is a weak acid, and its conjugate base, $SO_4^{2-}$ is a weak base (see Table 15.5 of the text). Therefore, this is a buffer system.

**(c)** $NH_3$ (ammonia) is a weak base, and its conjugate acid, $NH_4^+$ is a weak acid. Therefore, this is a buffer system.

**(d)** Because HI is a strong acid, its conjugate base, $I^-$, is an extremely weak base. This means that the $I^-$ ion will not combine with a $H^+$ ion in solution to form HI. Thus, this system cannot act as a buffer system.

**16.12** **Strategy:** The pH of a buffer system can be calculated in a similar manner to a weak acid equilibrium problem. The difference is that a common-ion is present in solution. The $K_a$ of $CH_3COOH$ is $1.8 \times 10^{-5}$ (see Table 15.3 of the text).

**Solution:**

**(a)** We summarize the concentrations of the species at equilibrium as follows:

| | $CH_3COOH(aq)$ | $\rightleftharpoons$ | $H^+(aq)$ | + | $CH_3COO^-(aq)$ |
|---|---|---|---|---|---|
| Initial ($M$): | 2.0 | | 0 | | 2.0 |
| Change ($M$): | $-x$ | | $+x$ | | $+x$ |
| Equilibrium ($M$): | $2.0 - x$ | | $x$ | | $2.0 + x$ |

$$K_a = \frac{[H^+][CH_3COO^-]}{[CH_3COOH]}$$

$$K_a = \frac{[H^+](2.0 + x)}{(2.0 - x)} \approx \frac{[H^+](2.0)}{2.0}$$

$$K_a = [H^+]$$

Taking the −log of both sides,

$$pK_a = pH$$

Thus, for a buffer system in which the [weak acid] = [weak base],

$$pH = pK_a$$

$$\mathbf{pH} = -\log(1.8 \times 10^{-5}) = \mathbf{4.74}$$

**(b)** Similar to part (a),

$$\mathbf{pH} = pK_a = \mathbf{4.74}$$

**Buffer (a)** will be a more effective buffer because the concentrations of acid and base components are ten times higher than those in (b). Thus, buffer (a) can neutralize 10 times more added acid or base compared to buffer (b).

**16.14** *Step 1:* Write the equilibrium that occurs between $H_2PO_4^-$ and $HPO_4^{2-}$. Set up a table relating the initial concentrations, the change in concentration to reach equilibrium, and the equilibrium concentrations.

| | $H_2PO_4^-(aq)$ | $\rightleftharpoons$ | $H^+(aq)$ | + | $HPO_4^{2-}(aq)$ |
|---|---|---|---|---|---|
| Initial ($M$): | 0.15 | | 0 | | 0.10 |
| Change ($M$): | $-x$ | | $+x$ | | $+x$ |
| Equilibrium ($M$): | $0.15 - x$ | | $x$ | | $0.10 + x$ |

*Step 2:* Write the ionization constant expression in terms of the equilibrium concentrations. Knowing the value of the equilibrium constant ($K_a$), solve for $x$.

$$K_a = \frac{[H^+][HPO_4^{2-}]}{[H_2PO_4^-]}$$

You can look up the $K_a$ value for dihydrogen phosphate in Table 15.5 of your text.

$$6.2 \times 10^{-8} = \frac{(x)(0.10 + x)}{(0.15 - x)}$$

$$6.2 \times 10^{-8} \approx \frac{(x)(0.10)}{(0.15)}$$

$$x = [H^+] = 9.3 \times 10^{-8}\ M$$

*Step 3:* Having solved for the $[H^+]$, calculate the pH of the solution.

$$\mathbf{pH} = -\log[H^+] = -\log(9.3 \times 10^{-8}) = \mathbf{7.03}$$

**16.16** We can use the Henderson-Hasselbalch equation to calculate the ratio $[HCO_3^-]/[H_2CO_3]$. The Henderson-Hasselbalch equation is:

$$pH = pK_a + \log\frac{[\text{conjugate base}]}{[\text{acid}]}$$

For the buffer system of interest, $HCO_3^-$ is the conjugate base of the acid, $H_2CO_3$. We can write:

$$pH = 7.40 = -\log(4.2 \times 10^{-7}) + \log\frac{[HCO_3^-]}{[H_2CO_3]}$$

$$7.40 = 6.38 + \log\frac{[HCO_3^-]}{[H_2CO_3]}$$

The [conjugate base]/[acid] ratio is:

$$\log\frac{[HCO_3^-]}{[H_2CO_3]} = 7.40 - 6.38 = 1.02$$

$$\frac{[\mathbf{HCO_3^-}]}{[\mathbf{H_2CO_3}]} = 10^{1.02} = \mathbf{1.0 \times 10^1}$$

The buffer should be more effective against an added acid because ten times more base is present compared to acid. Note that a pH of 7.40 is only a two significant figure number (Why?); the final result should only have two significant figures.

**16.18** As calculated in Problem 16.12, the pH of this buffer system is equal to $pK_a$.

$$pH = pK_a = -\log(1.8 \times 10^{-5}) = 4.74$$

**(a)** The added NaOH will react completely with the acid component of the buffer, $CH_3COOH$. NaOH ionizes completely; therefore, 0.080 mol of $OH^-$ are added to the buffer.

*Step 1:* The neutralization reaction is:

| | $CH_3COOH(aq)$ + | $OH^-(aq)$ ⟶ | $CH_3COO^-(aq) + H_2O(l)$ |
|---|---|---|---|
| Initial (mol): | 1.00 | 0.080 | 1.00 |
| Change (mol): | −0.080 | −0.080 | +0.080 |
| Final (mol): | 0.92 | 0 | 1.08 |

*Step 2:* Now, the acetic acid equilibrium is reestablished. Since the volume of the solution is 1.00 L, we can convert directly from moles to molar concentration.

| | $CH_3COOH(aq)$ ⇌ | $H^+(aq)$ + | $CH_3COO^-(aq)$ |
|---|---|---|---|
| Initial (*M*): | 0.92 | 0 | 1.08 |
| Change (*M*): | $-x$ | $+x$ | $+x$ |
| Equilibrium (*M*): | $0.92 - x$ | $x$ | $1.08 + x$ |

Write the $K_a$ expression, then solve for $x$.

$$K_a = \frac{[H^+][CH_3COO^-]}{[CH_3COOH]}$$

$$1.8 \times 10^{-5} = \frac{(x)(1.08 + x)}{(0.92 - x)} \approx \frac{x(1.08)}{0.92}$$

$$x = [H^+] = 1.5 \times 10^{-5}\,M$$

*Step 3:* Having solved for the $[H^+]$, calculate the pH of the solution.

$$\mathbf{pH} = -\log[H^+] = -\log(1.5 \times 10^{-5}) = \mathbf{4.82}$$

The pH of the buffer increased from 4.74 to 4.82 upon addition of 0.080 mol of strong base.

**(b)** The added acid will react completely with the base component of the buffer, $CH_3COO^-$. HCl ionizes completely; therefore, 0.12 mol of $H^+$ ion are added to the buffer

*Step 1:* The neutralization reaction is:

| | $CH_3COO^-(aq)$ + | $H^+(aq)$ ⟶ | $CH_3COOH(aq)$ |
|---|---|---|---|
| Initial (mol): | 1.00 | 0.12 | 1.00 |
| Change (mol): | −0.12 | −0.12 | +0.12 |
| Final (mol): | 0.88 | 0 | 1.12 |

*Step 2:* Now, the acetic acid equilibrium is reestablished. Since the volume of the solution is 1.00 L, we can convert directly from moles to molar concentration.

| | $CH_3COOH(aq)$ | $\rightleftharpoons$ | $H^+(aq)$ | + | $CH_3COO^-(aq)$ |
|---|---|---|---|---|---|
| Initial (*M*): | 1.12 | | 0 | | 0.88 |
| Change (*M*): | $-x$ | | $+x$ | | $+x$ |
| Equilibrium (*M*): | $1.12 - x$ | | $x$ | | $0.88 + x$ |

Write the $K_a$ expression, then solve for $x$.

$$K_a = \frac{[H^+][CH_3COO^-]}{[CH_3COOH]}$$

$$1.8 \times 10^{-5} = \frac{(x)(0.88 + x)}{(1.12 - x)} \approx \frac{x(0.88)}{1.12}$$

$$x = [H^+] = 2.3 \times 10^{-5}\ M$$

*Step 3:* Having solved for the $[H^+]$, calculate the pH of the solution.

$$\textbf{pH} = -\log[H^+] = -\log(2.3 \times 10^{-5}) = \mathbf{4.64}$$

The pH of the buffer decreased from 4.74 to 4.64 upon addition of 0.12 mol of strong acid.

**16.20** **Strategy:** For a buffer to function effectively, the concentration of the acid component must be roughly equal to the conjugate base component. According to Equation (16.4) of the text, when the desired pH is close to the $pK_a$ of the acid, that is, when $pH \approx pK_a$,

$$\log\frac{[\text{conjugate base}]}{[\text{acid}]} \approx 0$$

or

$$\frac{[\text{conjugate base}]}{[\text{acid}]} \approx 1$$

**Solution:** To prepare a solution of a desired pH, we should choose a weak acid with a $pK_a$ value close to the desired pH. Calculating the $pK_a$ for each acid:

For HA, $pK_a = -\log(2.7 \times 10^{-3}) = 2.57$

For HB, $pK_a = -\log(4.4 \times 10^{-6}) = 5.36$

For HC, $pK_a = -\log(2.6 \times 10^{-9}) = 8.59$

The buffer solution with a $pK_a$ closest to the desired pH is HC. Thus, **HC** is the best choice to prepare a buffer solution with pH = 8.60.

**16.22** (1) The solutions contain a weak acid and a weak base that are a conjugate acid/base pair. These are buffer solutions. The Henderson-Hasselbalch equation can be used to calculate the pH of each solution. The problem states to treat each sphere as 0.1 mole. Because HA and $A^-$ are contained in the same volume, we can plug in moles into the Henderson-Hasselbalch equation to solve for the pH of each solution.

(a) $\text{pH} = \text{p}K_\text{a} + \log\frac{[\text{A}^-]}{[\text{HA}]}$

$$\textbf{pH} = 5.00 + \log\left(\frac{0.5\text{ mol}}{0.4\text{ mol}}\right) = \textbf{5.10}$$

(b) $$\textbf{pH} = 5.00 + \log\left(\frac{0.4\text{ mol}}{0.6\text{ mol}}\right) = \textbf{4.82}$$

(c) $$\textbf{pH} = 5.00 + \log\left(\frac{0.5\text{ mol}}{0.3\text{ mol}}\right) = \textbf{5.22}$$

(d) $$\text{pH} = 5.00 + \log\left(\frac{0.4\text{ mol}}{0.4\text{ mol}}\right)$$

$$\text{pH} = \text{p}K_\text{a} = \textbf{5.00}$$

**(2)** **The added acid reacts with the base component of the buffer ($A^-$). We write out the acid-base reaction to find the number of moles of $A^-$ and HA after addition of $H^+$.**

| | $A^-(aq)$ + | $H^+(aq)$ → | $HA(aq)$ |
|---|---|---|---|
| Initial (mol): | 0.5 | 0.1 | 0.4 |
| Change (mol): | −0.1 | −0.1 | +0.1 |
| Final (mol): | 0.4 | 0 | 0.5 |

**Because the concentrations of the two buffer components are equal, the pH of this buffer equals its $\text{p}K_\text{a}$ value.**

**We use the Henderson-Hasselbalch equation to calculate the pH of this buffer.**

$$\textbf{pH} = 5.00 + \log\left(\frac{0.4\text{ mol}}{0.5\text{ mol}}\right) = \textbf{4.90}$$

**(3)** **The added base reacts with the acid component of the buffer (HA). We write out the acid-base reaction to find the number of moles of HA and $A^-$ after addition of $OH^-$.**

| | $HA(aq)$ + | $OH^-(aq)$ → | $A^-(aq)$ + $H_2O(l)$ |
|---|---|---|---|
| Initial (mol): | 0.4 | 0.1 | 0.4 |
| Change (mol): | −0.1 | −0.1 | +0.1 |
| Final (mol): | 0.3 | 0 | 0.5 |

**We use the Henderson-Hasselbalch equation to calculate the pH of this buffer.**

$$\textbf{pH} = 5.00 + \log\left(\frac{0.5\text{ mol}}{0.3\text{ mol}}\right) = \textbf{5.22}$$

**16.24** The hydrochloric acid reacts with $NH_3$ to produce $NH_4^+$.

| | $NH_3(aq)$ + | $HCl(aq)$ → | $NH_4^+(aq) + Cl^-(aq)$ |
|---|---|---|---|
| Initial ($M$): | 0.84 | $x$ | 0.96 |
| Change ($M$): | $-x$ | $-x$ | $+x$ |
| Final ($M$): | $0.84 - x$ | 0 | $0.96 + x$ |

Next, we use the Henderson-Hasselbalch equation to solve for $x$.

$$NH_4^+(aq) \rightleftharpoons H^+(aq) + NH_3(aq)$$

$$K_a = 5.6 \times 10^{-10},\ pK_a = 9.25$$

$$pH = pK_a + \log\frac{[NH_3]}{[NH_4^+]}$$

$$8.56 = 9.25 + \log\left(\frac{0.84 - x}{0.96 + x}\right)$$

$$10^{-0.69} = \frac{0.84 - x}{0.96 + x}$$

$$x = 0.53\ M$$

Because the solution volume is 1.0 L, the **moles of HCl** that must be added is **0.53 moles.**

**16.28** We want to calculate the molar mass of the diprotic acid. The mass of the acid is given in the problem, so we need to find moles of acid in order to calculate its molar mass.

want to calculate → molar mass of $H_2A$; given → g $H_2A$; need to find → mol $H_2A$

$$\text{molar mass of } H_2A = \frac{\text{g } H_2A}{\text{mol } H_2A}$$

The neutralization reaction is:

$$2KOH(aq) + H_2A(aq) \longrightarrow K_2A(aq) + 2H_2O(l)$$

From the volume and molarity of the base needed to neutralize the acid, we can calculate the number of moles of $H_2A$ reacted.

$$11.1\ \text{mL KOH} \times \frac{1.00\ \text{mol KOH}}{1000\ \text{mL}} \times \frac{1\ \text{mol } H_2A}{2\ \text{mol KOH}} = 5.55 \times 10^{-3}\ \text{mol } H_2A$$

We know that 0.500 g of the diprotic acid were reacted (1/10 of the 250 mL was tested). Divide the number of grams by the number of moles to calculate the molar mass.

$$\mathcal{M}(H_2A) = \frac{0.500\ \text{g } H_2A}{5.55 \times 10^{-3}\ \text{mol } H_2A} = \mathbf{90.1\ g/mol}$$

**16.30** We want to calculate the molarity of the $Ba(OH)_2$ solution. The volume of the solution is given (19.3 mL), so we need to find the moles of $Ba(OH)_2$ to calculate the molarity.

want to calculate → $M$ of $Ba(OH)_2$; need to find → mol $Ba(OH)_2$; given → L of $Ba(OH)_2$ soln

$$M \text{ of } Ba(OH)_2 = \frac{\text{mol } Ba(OH)_2}{\text{L of } Ba(OH)_2 \text{ soln}}$$

The neutralization reaction is:

$$2HCOOH + Ba(OH)_2 \rightarrow (HCOO)_2Ba + 2H_2O$$

From the volume and molarity of HCOOH needed to neutralize $Ba(OH)_2$, we can determine the moles of $Ba(OH)_2$ reacted.

$$20.4 \text{ mL HCOOH} \times \frac{0.883 \text{ mol HCOOH}}{1000 \text{ mL}} \times \frac{1 \text{ mol } Ba(OH)_2}{2 \text{ mol HCOOH}} = 9.01 \times 10^{-3} \text{ mol } Ba(OH)_2$$

The molarity of the $Ba(OH)_2$ solution is:

$$\frac{9.01 \times 10^{-3} \text{ mol } Ba(OH)_2}{19.3 \times 10^{-3} \text{ L}} = \mathbf{0.467\ \mathit{M}}$$

**16.32** The resulting solution is not a buffer system. There is excess NaOH and the neutralization is well past the equivalence point..

$$\text{Moles NaOH} = 0.500 \text{ L} \times \frac{0.167 \text{ mol}}{1 \text{ L}} = 0.0835 \text{ mol}$$

$$\text{Moles } CH_3COOH = 0.500 \text{ L} \times \frac{0.100 \text{ mol}}{1 \text{ L}} = 0.0500 \text{ mol}$$

| | $CH_3COOH(aq)$ + | $NaOH(aq)$ → | $CH_3COONa(aq) + H_2O(l)$ |
|---|---|---|---|
| Initial (mol): | 0.0500 | 0.0835 | 0 |
| Change (mol): | −0.0500 | −0.0500 | +0.0500 |
| Final (mol): | 0 | 0.0335 | 0.0500 |

The volume of the resulting solution is 1.00 L (500 mL + 500 mL = 1000 mL).

$$[OH^-] = \frac{0.0335 \text{ mol}}{1.00 \text{ L}} = \mathbf{0.0335\ \mathit{M}}$$

$$[Na^+] = \frac{(0.0335 + 0.0500) \text{ mol}}{1.00 \text{ L}} = \mathbf{0.0835\ \mathit{M}}$$

$$[H^+] = \frac{K_w}{[OH^-]} = \frac{1.0 \times 10^{-14}}{0.0335} = \mathbf{3.0 \times 10^{-13}}\ \boldsymbol{M}$$

$$[CH_3COO^-] = \frac{0.0500 \text{ mol}}{1.00 \text{ L}} = \mathbf{0.0500}\ \boldsymbol{M}$$

| | $CH_3COO^-(aq) + H_2O(l)$ | $\rightleftharpoons$ $CH_3COOH(aq)$ | $+ OH^-(aq)$ |
|---|---|---|---|
| Initial ($M$): | 0.0500 | 0 | 0.0335 |
| Change ($M$): | $-x$ | $+x$ | $+x$ |
| Equilibrium ($M$): | $0.0500 - x$ | $x$ | $0.0335 + x$ |

$$K_b = \frac{[CH_3COOH][OH^-]}{[CH_3COO^-]}$$

$$5.6 \times 10^{-10} = \frac{(x)(0.0335 + x)}{(0.0500 - x)} \approx \frac{(x)(0.0335)}{(0.0500)}$$

$$x = \mathbf{[CH_3COOH] = 8.4 \times 10^{-10}}\ \boldsymbol{M}$$

**16.34** Let's assume we react 1 L of HCOOH with 1 L of NaOH.

| | $HCOOH(aq)$ | $+ NaOH(aq)$ | $\rightarrow HCOONa(aq) + H_2O(l)$ |
|---|---|---|---|
| Initial (mol): | 0.10 | 0.10 | 0 |
| Change (mol): | −0.10 | −0.10 | +0.10 |
| Final (mol): | 0 | 0 | 0.10 |

The solution volume has doubled (1 L + 1 L = 2 L). The concentration of HCOONa is:

$$M\ (\text{HCOONa}) = \frac{0.10 \text{ mol}}{2 \text{ L}} = 0.050\ M$$

$HCOO^-(aq)$ is a weak base. The hydrolysis is:

| | $HCOO^-(aq) + H_2O(l)$ | $\rightleftharpoons$ $HCOOH(aq)$ | $+ OH^-(aq)$ |
|---|---|---|---|
| Initial ($M$): | 0.050 | 0 | 0 |
| Change ($M$): | $-x$ | $+x$ | $+x$ |
| Equilibrium ($M$): | $0.050 - x$ | $x$ | $x$ |

$$K_b = \frac{[HCOOH][OH^-]}{[HCOO^-]}$$

$$5.9 \times 10^{-11} = \frac{x^2}{0.050 - x} \approx \frac{x^2}{0.050}$$

$$x = 1.7 \times 10^{-6}\ M = [OH^-]$$

pOH = 5.77

**pH = 8.23**

**16.36** The reaction between $NH_3$ and HCl is:

$$NH_3(aq) + HCl(aq) \rightarrow NH_4Cl(aq)$$

We see that 1 mole $NH_3 \simeq$ 1 mol HCl. Therefore, at every stage of titration, we can calculate the number of moles of base reacting with acid, and the pH of the solution is determined by the excess base or acid left over. At the equivalence point, however, the neutralization is complete, and the pH of the solution will depend on the extent of the hydrolysis of the salt formed, which is $NH_4Cl$.

**(a)** No HCl has been added. This is a weak base calculation.

| | $NH_3(aq) + H_2O(l)$ | $\rightleftharpoons$ | $NH_4^+(aq)$ | + | $OH^-(aq)$ |
|---|---|---|---|---|---|
| Initial (*M*): | 0.300 | | 0 | | 0 |
| Change (*M*): | $-x$ | | $+x$ | | $+x$ |
| Equilibrium (*M*): | $0.300 - x$ | | $x$ | | $x$ |

$$K_b = \frac{[NH_4^+][OH^-]}{[NH_3]}$$

$$1.8 \times 10^{-5} = \frac{(x)(x)}{0.300 - x} \approx \frac{x^2}{0.300}$$

$$x = 2.3 \times 10^{-3}\ M = [OH^-]$$

pOH = 2.64

**pH = 11.36**

**(b)** The number of moles of $NH_3$ originally present in 10.0 mL of solution is:

$$10.0\ \text{mL} \times \frac{0.300\ \text{mol NH}_3}{1000\ \text{mL NH}_3\ \text{soln}} = 3.00 \times 10^{-3}\ \text{mol}$$

The number of moles of HCl in 10.0 mL is:

$$10.0\ \text{mL} \times \frac{0.100\ \text{mol HCl}}{1000\ \text{mL HCl soln}} = 1.00 \times 10^{-3}\ \text{mol}$$

We work with moles at this point because when two solutions are mixed, the solution volume increases. As the solution volume increases, molarity will change, but the number of moles will remain the same. The changes in number of moles are summarized.

| | $NH_3(aq)$ | + | $HCl(aq)$ | $\rightarrow$ | $NH_4Cl(aq)$ |
|---|---|---|---|---|---|
| Initial (mol): | $3.00 \times 10^{-3}$ | | $1.00 \times 10^{-3}$ | | 0 |
| Change (mol): | $-1.00 \times 10^{-3}$ | | $-1.00 \times 10^{-3}$ | | $+1.00 \times 10^{-3}$ |
| Final (mol): | $2.00 \times 10^{-3}$ | | 0 | | $1.00 \times 10^{-3}$ |

At this stage, we have a buffer system made up of $NH_3$ and $NH_4^+$ (from the salt, $NH_4Cl$). We use the Henderson-Hasselbalch equation to calculate the pH.

$$\text{pH} = \text{p}K_\text{a} + \log\frac{[\text{conjugate base}]}{[\text{acid}]}$$

$$\text{pH} = -\log(5.6\times10^{-10}) + \log\left(\frac{2.00\times10^{-3}}{1.00\times10^{-3}}\right)$$

$$\textbf{pH} = \textbf{9.55}$$

**(c)** This part is solved similarly to part (b).

The number of moles of HCl in 20.0 mL is:

$$20.0\ \text{mL}\times\frac{0.100\ \text{mol HCl}}{1000\ \text{mL HCl soln}} = 2.00\times10^{-3}\ \text{mol}$$

The changes in number of moles are summarized.

| | $NH_3(aq)$ | + $HCl(aq)$ | → $NH_4Cl(aq)$ |
|---|---|---|---|
| Initial (mol): | $3.00\times10^{-3}$ | $2.00\times10^{-3}$ | 0 |
| Change (mol): | $-2.00\times10^{-3}$ | $-2.00\times10^{-3}$ | $+2.00\times10^{-3}$ |
| Final (mol): | $1.00\times10^{-3}$ | 0 | $2.00\times10^{-3}$ |

At this stage, we have a buffer system made up of $NH_3$ and $NH_4^+$ (from the salt, $NH_4Cl$). We use the Henderson-Hasselbalch equation to calculate the pH.

$$\text{pH} = \text{p}K_\text{a} + \log\frac{[\text{conjugate base}]}{[\text{acid}]}$$

$$\text{pH} = -\log(5.6\times10^{-10}) + \log\left(\frac{1.00\times10^{-3}}{2.00\times10^{-3}}\right)$$

$$\textbf{pH} = \textbf{8.95}$$

**(d)** We have reached the equivalence point of the titration. $3.00\times10^{-3}$ mole of $NH_3$ reacts with $3.00\times10^{-3}$ mole HCl to produce $3.00\times10^{-3}$ mole of $NH_4Cl$. The only major species present in solution at the equivalence point is the salt, $NH_4Cl$, which contains the conjugate acid, $NH_4^+$. Let's calculate the molarity of $NH_4^+$. The volume of the solution is: (10.0 mL + 30.0 mL = 40.0 mL = 0.0400 L).

$$M\,(NH_4^+) = \frac{3.00\times10^{-3}\ \text{mol}}{0.0400\ \text{L}} = 0.0750\ M$$

We set up the hydrolysis of $NH_4^+$, which is a weak acid.

| | $NH_4^+(aq) + H_2O(l)$ | ⇌ $H_3O^+(aq)$ | + $NH_3(aq)$ |
|---|---|---|---|
| Initial ($M$): | 0.0750 | 0 | 0 |
| Change ($M$): | $-x$ | $+x$ | $+x$ |
| Equilibrium ($M$): | $0.0750 - x$ | $x$ | $x$ |

$$K_a = \frac{[H_3O^+][NH_3]}{[NH_4^+]}$$

$$5.6 \times 10^{-10} = \frac{(x)(x)}{0.0750 - x} \approx \frac{x^2}{0.0750}$$

$$x = 6.5 \times 10^{-6}\ M = [H_3O^+]$$

$$\textbf{pH} = \textbf{5.19}$$

**(e)** We have passed the equivalence point of the titration. The excess strong acid, HCl, will determine the pH at this point. The moles of HCl in 40.0 mL are:

$$40.0\ \cancel{mL} \times \frac{0.100\text{ mol HCl}}{1000\ \cancel{mL}\text{ HCl soln}} = 4.00 \times 10^{-3}\text{ mol}$$

The changes in number of moles are summarized.

| | $NH_3(aq)$ + | $HCl(aq)$ → | $NH_4Cl(aq)$ |
|---|---|---|---|
| Initial (mol): | $3.00 \times 10^{-3}$ | $4.00 \times 10^{-3}$ | 0 |
| Change (mol): | $-3.00 \times 10^{-3}$ | $-3.00 \times 10^{-3}$ | $+3.00 \times 10^{-3}$ |
| Final (mol): | 0 | $1.00 \times 10^{-3}$ | $3.00 \times 10^{-3}$ |

Let's calculate the molarity of the HCl in solution. The volume of the solution is now 50.0 mL = 0.0500 L.

$$M\text{ (HCl)} = \frac{1.00 \times 10^{-3}\text{ mol}}{0.0500\text{ L}} = 0.0200\ M$$

HCl is a strong acid. The pH is:

$$\textbf{pH} = -\log(0.0200) = \textbf{1.70}$$

**16.38** (1) Before any HCl is added, there would only be base molecules in solution – diagram **(c)**.

(2) At the halfway-point, there would be equal amounts of base and its conjugate acid – diagram **(a)**.

(3) At the equivalence point, there is only salt dissolved in water. In the diagram, $Cl^-$ and $H_2O$ are not shown, so the only species present would be $BH^+$ – diagram **(d)**.

(4) Beyond the equivalence point, excess hydronium ions would be present in solution – diagram **(b)**.

The pH is **less than 7** at the equivalence point of a titration of a weak base with a strong acid like HCl.

**16.40** The pH at the half-way point of a weak acid/strong base titration equals the $pK_a$ value of the weak acid. The 12.35 mL point of the titration is the half-way point. Hence, pH = $pK_a$ = 5.22. The $K_a$ value of acid is:

$$\mathbf{K_a} = 10^{-pK_a} = 10^{-5.22} = \mathbf{6.0 \times 10^{-6}}$$

**16.44** $CO_2$ in the air dissolves in the solution:

$$CO_2 + H_2O \rightleftharpoons H_2CO_3$$

The carbonic acid neutralizes the NaOH.

**16.46** According to Section 16.5 of the text, when [HIn] ≈ [In⁻] the indicator color is a mixture of the colors of HIn and In⁻. In other words, the indicator color changes at this point. When [HIn] ≈ [In⁻] we can write:

$$\frac{[In^-]}{[HIn]} = \frac{K_a}{[H^+]} = 1$$

$$[H^+] = K_a = 2.0 \times 10^{-6}$$

$$\mathbf{pH = 5.70}$$

**16.54** **Strategy:** In each part, we can calculate the number of moles of compound dissolved in one liter of solution (the molar solubility). Then, from the molar solubility, $s$, we can determine $K_{sp}$.

**Solution:**

**(a)** $$\frac{7.3 \times 10^{-2}\ \text{g SrF}_2}{1\ \text{L soln}} \times \frac{1\ \text{mol SrF}_2}{125.6\ \text{g SrF}_2} = 5.8 \times 10^{-4}\ \text{mol/L} = s$$

Consider the dissociation of $SrF_2$ in water. Let $s$ be the molar solubility of $SrF_2$.

$$SrF_2(s) \rightleftharpoons Sr^{2+}(aq) + 2F^-(aq)$$

| | | | |
|---|---|---|---|
| Initial (*M*): | | 0 | 0 |
| Change (*M*): | $-s$ | $+s$ | $+2s$ |
| Equilibrium (*M*): | | $s$ | $2s$ |

$$K_{sp} = [Sr^{2+}][F^-]^2 = (s)(2s)^2 = 4s^3$$

The molar solubility ($s$) was calculated above. Substitute into the equilibrium constant expression to solve for $K_{sp}$.

$$K_{sp} = [Sr^{2+}][F^-]^2 = 4s^3 = 4(5.8 \times 10^{-4})^3 = \mathbf{7.8 \times 10^{-10}}$$

**(b)** $$\frac{6.7 \times 10^{-3}\ \text{g Ag}_3\text{PO}_4}{1\ \text{L soln}} \times \frac{1\ \text{mol Ag}_3\text{PO}_4}{418.7\ \text{g Ag}_3\text{PO}_4} = 1.6 \times 10^{-5}\ \text{mol/L} = s$$

(b) is solved in a similar manner to (a)

The equilibrium equation is:

$$Ag_3PO_4(s) \rightleftharpoons 3Ag^+(aq) + PO_4^{3-}(aq)$$

| | | | |
|---|---|---|---|
| Initial (*M*): | | 0 | 0 |
| Change (*M*): | $-s$ | $+3s$ | $+s$ |
| Equilibrium (*M*): | | $3s$ | $s$ |

$$K_{sp} = [Ag^+]^3[PO_4^{3-}] = (3s)^3(s) = 27s^4 = 27(1.6 \times 10^{-5})^4 = \mathbf{1.8 \times 10^{-18}}$$

**16.56** First, we can convert the solubility of MX in g/L to mol/L.

$$\frac{4.63 \times 10^{-3} \cancel{g} \text{ MX}}{1 \text{ L soln}} \times \frac{1 \text{ mol MX}}{346 \cancel{g} \text{ MX}} = 1.34 \times 10^{-5} \text{ mol/L} = s \text{ (molar solubility)}$$

The equilibrium reaction is:

| | $MX(s) \rightleftharpoons$ | $M^{n+}(aq)$ | $+ X^{n-}(aq)$ |
|---|---|---|---|
| Initial ($M$): | | 0 | 0 |
| Change ($M$): | $-s$ | $+s$ | $+s$ |
| Equilibrium ($M$): | | $s$ | $s$ |

$$K_{sp} = [M^{n+}][X^{n-}] = s^2 = (1.34 \times 10^{-5})^2 = \mathbf{1.80 \times 10^{-10}}$$

**16.58** **Strategy:** We can look up the $K_{sp}$ value of $CaF_2$ in Table 16.2 of the text. Then, setting up the dissociation equilibrium of $CaF_2$ in water, we can solve for the molar solubility, $s$.

**Solution:** Consider the dissociation of $CaF_2$ in water.

| | $CaF_2(s) \rightleftharpoons$ | $Ca^{2+}(aq)$ | $+ 2F^-(aq)$ |
|---|---|---|---|
| Initial ($M$): | | 0 | 0 |
| Change ($M$): | $-s$ | $+s$ | $+2s$ |
| Equilibrium ($M$): | | $s$ | $2s$ |

Recall, that the concentration of a pure solid does not enter into an equilibrium constant expression. Therefore, the concentration of $CaF_2$ is not important.

Substitute the value of $K_{sp}$ and the concentrations of $Ca^{2+}$ and $F^-$ in terms of $s$ into the solubility product expression to solve for $s$, the molar solubility.

$$K_{sp} = [Ca^{2+}][F^-]^2$$

$$4.0 \times 10^{-11} = (s)(2s)^2$$

$$4.0 \times 10^{-11} = 4s^3$$

$$s = \text{molar solubility} = \mathbf{2.2 \times 10^{-4} \text{ mol/L}}$$

The molar solubility indicates that $2.2 \times 10^{-4}$ mol of $CaF_2$ will dissolve in 1 L of an aqueous solution.

**16.60** First we can calculate the $OH^-$ concentration from the pH.

$$\text{pOH} = 14.00 - \text{pH}$$

$$\text{pOH} = 14.00 - 9.68 = 4.32$$

$$[OH^-] = 10^{-\text{pOH}} = 10^{-4.32} = 4.8 \times 10^{-5} M$$

The equilibrium equation is:

$$MOH(s) \rightleftharpoons M^{+}(aq) + OH^{-}(aq)$$

From the balanced equation we know that $[M^{+}] = [OH^{-}]$

$$K_{sp} = [M^{+}][OH^{-}] = (4.8 \times 10^{-5})^2 = \mathbf{2.3 \times 10^{-9}}$$

**16.62** The net ionic equation is:

$$Sr^{2+}(aq) + 2F^{-}(aq) \longrightarrow SrF_2(s)$$

Let's find the limiting reagent in the precipitation reaction.

$$\text{Moles } F^{-} = 75\ \text{mL} \times \frac{0.060\ \text{mol}}{1000\ \text{mL soln}} = 0.0045\ \text{mol}$$

$$\text{Moles } Sr^{2+} = 25\ \text{mL} \times \frac{0.15\ \text{mol}}{1000\ \text{mL soln}} = 0.0038\ \text{mol}$$

From the stoichiometry of the balanced equation, twice as many moles of $F^{-}$ are required to react with $Sr^{2+}$. This would require 0.0076 mol of $F^{-}$, but we only have 0.0045 mol. Thus, $F^{-}$ is the limiting reagent.

Let's assume that the above reaction goes to completion. Then, we will consider the equilibrium that is established when $SrF_2$ partially dissociates into ions.

| | $Sr^{2+}(aq)$ + | $2F^{-}(aq)$ ⟶ | $SrF_2(s)$ |
|---|---|---|---|
| Initial (mol): | 0.0038 | 0.0045 | 0 |
| Change (mol): | −0.00225 | −0.0045 | +0.00225 |
| Final (mol): | 0.00155 | 0 | 0.00225 |

Now, let's establish the equilibrium reaction. The total volume of the solution is 100 mL = 0.100 L. Divide the above moles by 0.100 L to convert to molar concentration.

| | $SrF_2(s)$ ⇌ | $Sr^{2+}(aq)$ + | $2F^{-}(aq)$ |
|---|---|---|---|
| Initial (*M*): | 0.0225 | 0.0155 | 0 |
| Change (*M*): | $-s$ | $+s$ | $+2s$ |
| Equilibrium (*M*): | $0.0225 - s$ | $0.0155 + s$ | $2s$ |

Write the solubility product expression, then solve for $s$.

$$K_{sp} = [Sr^{2+}][F^{-}]^2$$

$$2.0 \times 10^{-10} = (0.0155 + s)(2s)^2 \approx (0.0155)(2s)^2$$

$$s = 5.7 \times 10^{-5}\ M$$

$$[F^{-}] = 2s = \mathbf{1.1 \times 10^{-4}\ M}$$

$$[Sr^{2+}] = 0.0155 + s = \mathbf{0.016\ M}$$

Both sodium ions and nitrate ions are spectator ions and therefore do not enter into the precipitation reaction.

$$[NO_3^-] = \frac{2(0.0038)\text{ mol}}{0.10\text{ L}} = \mathbf{0.076}\ \boldsymbol{M}$$

$$[Na^+] = \frac{0.0045\text{ mol}}{0.10\text{ L}} = \mathbf{0.045}\ \boldsymbol{M}$$

**16.64** For $Fe(OH)_3$, $K_{sp} = 1.1 \times 10^{-36}$. When $[Fe^{3+}] = 0.010\ M$, the $[OH^-]$ value is:

$$K_{sp} = [Fe^{3+}][OH^-]^3$$

or

$$[OH^-] = \left(\frac{K_{sp}}{[Fe^{3+}]}\right)^{\frac{1}{3}}$$

$$[OH^-] = \left(\frac{1.1 \times 10^{-36}}{0.010}\right)^{\frac{1}{3}} = 4.8 \times 10^{-12}\ M$$

This $[OH^-]$ corresponds to a pH of 2.68. In other words, $Fe(OH)_3$ will begin to precipitate from this solution at pH of 2.68.

For $Zn(OH)_2$, $K_{sp} = 1.8 \times 10^{-14}$. When $[Zn^{2+}] = 0.010\ M$, the $[OH^-]$ value is:

$$[OH^-] = \left(\frac{K_{sp}}{[Zn^{2+}]}\right)^{\frac{1}{2}}$$

$$[OH^-] = \left(\frac{1.8 \times 10^{-14}}{0.010}\right)^{\frac{1}{2}} = 1.3 \times 10^{-6}\ M$$

This corresponds to a pH of 8.11. In other words $Zn(OH)_2$ will begin to precipitate from the solution at pH = 8.11.

We assume a 99% precipitation of $Fe(OH)_3$ before $Zn(OH)_2$ begins to precipitate. The $Fe^{3+}$ ions remaining in solution have a concentration given by:

$$(1.00 - 0.99) \times 0.010\ M = 1.0 \times 10^{-4}\ M$$

$$K_{sp} = [Fe^{3+}][OH^-]^3$$

$$1.1 \times 10^{-36} = (1.0 \times 10^{-4})[OH^-]^3$$

$$[OH^-] = 2.2 \times 10^{-11}\ M$$

$$\text{pOH} = 10.66 \qquad \text{pH} = 3.34$$

So, the pH range over which $Fe(OH)_3$ is practically all removed and $Zn(OH)_2$ has not begun to precipitate is:

**3.34 < pH < 8.11**

**16.68** **Strategy:** In parts (b) and (c), this is a common-ion problem. In part (b), the common ion is $Br^-$, which is supplied by both $PbBr_2$ and KBr. Remember that the presence of a common ion will affect only the solubility of $PbBr_2$, but not the $K_{sp}$ value because it is an equilibrium constant. In part (c), the common ion is $Pb^{2+}$, which is supplied by both $PbBr_2$ and $Pb(NO_3)_2$.

**Solution:**

**(a)** Set up a table to find the equilibrium concentrations in pure water.

| | $PbBr_2(s)$ $\rightleftharpoons$ | $Pb^{2+}(aq)$ + | $2Br^-(aq)$ |
|---|---|---|---|
| Initial (*M*) | | 0 | 0 |
| Change (*M*) | $-s$ | $+s$ | $+2s$ |
| Equilibrium (*M*) | | $s$ | $2s$ |

$$K_{sp} = [Pb^{2+}][Br^-]^2$$

$$8.9 \times 10^{-6} = (s)(2s)^2$$

$$s = \text{molar solubility} = \mathbf{0.013\ \mathit{M}}$$

**(b)** Set up a table to find the equilibrium concentrations in 0.20 *M* KBr. KBr is a soluble salt that ionizes completely giving an initial concentration of $Br^- = 0.20\ M$.

| | $PbBr_2(s)$ $\rightleftharpoons$ | $Pb^{2+}(aq)$ + | $2Br^-(aq)$ |
|---|---|---|---|
| Initial (*M*) | | 0 | 0.20 |
| Change (*M*) | $-s$ | $+s$ | $+2s$ |
| Equilibrium (*M*) | | $s$ | $0.20 + 2s$ |

$$K_{sp} = [Pb^{2+}][Br^-]^2$$

$$8.9 \times 10^{-6} = (s)(0.20 + 2s)^2$$

$$8.9 \times 10^{-6} \approx (s)(0.20)^2$$

$$s = \text{molar solubility} = \mathbf{2.2 \times 10^{-4}\ \mathit{M}}$$

Thus, the molar solubility of $PbBr_2$ is reduced from 0.013 *M* to $2.2 \times 10^{-4}$ *M* as a result of the common ion ($Br^-$) effect.

**(c)** Set up a table to find the equilibrium concentrations in 0.20 *M* $Pb(NO_3)_2$. $Pb(NO_3)_2$ is a soluble salt that dissociates completely giving an initial concentration of $[Pb^{2+}] = 0.20\ M$.

| | $PbBr_2(s)$ $\rightleftharpoons$ | $Pb^{2+}(aq)$ + | $2Br^-(aq)$ |
|---|---|---|---|
| Initial (*M*): | | 0.20 | 0 |
| Change (*M*): | $-s$ | $+s$ | $+2s$ |
| Equilibrium (*M*): | | $0.20 + s$ | $2s$ |

$$K_{sp} = [Pb^{2+}][Br^-]^2$$

$$8.9 \times 10^{-6} = (0.20 + s)(2s)^2$$

$$8.9 \times 10^{-6} \approx (0.20)(2s)^2$$

$$s = \text{molar solubility} = \mathbf{3.3 \times 10^{-3}}\ \boldsymbol{M}$$

Thus, the molar solubility of $PbBr_2$ is reduced from 0.013 $M$ to $3.3 \times 10^{-3}$ $M$ as a result of the common ion ($Pb^{2+}$) effect.

**Check:** You should also be able to predict the decrease in solubility due to a common-ion using Le Châtelier's principle. Adding $Br^-$ or $Pb^{2+}$ ions shifts the system to the left, thus decreasing the solubility of $PbBr_2$.

**16.70** (a) The equilibrium reaction is:

| | $BaSO_4(s)$ | $\rightleftharpoons$ $Ba^{2+}(aq)$ | + $SO_4^{2-}(aq)$ |
|---|---|---|---|
| Initial ($M$): | | 0 | 0 |
| Change ($M$): | $-s$ | $+s$ | $+s$ |
| Equilibrium ($M$): | | $s$ | $s$ |

$$K_{sp} = [Ba^{2+}][SO_4^{2-}]$$

$$1.1 \times 10^{-10} = s^2$$

$$s = \mathbf{1.0 \times 10^{-5}}\ \boldsymbol{M}$$

The molar solubility of $BaSO_4$ in pure water is $1.0 \times 10^{-5}$ mol/L.

(b) The initial concentration of $SO_4^{2-}$ is 1.0 $M$.

| | $BaSO_4(s)$ | $\rightleftharpoons$ $Ba^{2+}(aq)$ | + $SO_4^{2-}(aq)$ |
|---|---|---|---|
| Initial ($M$): | | 0 | 1.0 |
| Change ($M$): | $-s$ | $+s$ | $+s$ |
| Equilibrium ($M$): | | $s$ | $1.0 + s$ |

$$K_{sp} = [Ba^{2+}][SO_4^{2-}]$$

$$1.1 \times 10^{-10} = (s)(1.0 + s) \approx (s)(1.0)$$

$$s = \mathbf{1.1 \times 10^{-10}}\ \boldsymbol{M}$$

Due to the common ion effect, the molar solubility of $BaSO_4$ decreases to $1.1 \times 10^{-10}$ mol/L in 1.0 $M$ $SO_4^{2-}(aq)$ compared to $1.0 \times 10^{-5}$ mol/L in pure water.

**16.72** **(b)** $SO_4^{2-}(aq)$ is a weak base

**(c)** $OH^-(aq)$ is a strong base

**(d)** $C_2O_4^{2-}(aq)$ is a weak base

**(e)** $PO_4^{3-}(aq)$ is a weak base.

The solubilities of the above will increase in acidic solution. Only (a), which contains an extremely weak base ($I^-$ is the conjugate base of the strong acid HI) is unaffected by the acid solution.

**16.74** From Table 16.2, the value of $K_{sp}$ for iron(II) is $1.6 \times 10^{-14}$.

**(a)** At pH = 8.00, pOH = 14.00 − 8.00 = 6.00, and $[OH^-] = 1.0 \times 10^{-6}\ M$

$$[Fe^{2+}] = \frac{K_{sp}}{[OH^-]^2} = \frac{1.6 \times 10^{-14}}{(1.0 \times 10^{-6})^2} = 0.016\ M$$

The **molar solubility** of iron(II) hydroxide at pH = 8.00 is **0.016 *M***

**(b)** At pH = 10.00, pOH = 14.00 − 10.00 = 4.00, and $[OH^-] = 1.0 \times 10^{-4}\ M$

$$[Fe^{2+}] = \frac{K_{sp}}{[OH^-]^2} = \frac{1.6 \times 10^{-14}}{(1.0 \times 10^{-4})^2} = 1.6 \times 10^{-6}\ M$$

The **molar solubility** of iron(II) hydroxide at pH = 10.00 is **$1.6 \times 10^{-6}$ *M*.**

**16.76** We first determine the effect of the added ammonia. Let's calculate the concentration of $NH_3$. This is a dilution problem.

$$M_iV_i = M_fV_f$$

$$(0.60\ M)(2.00\ \text{mL}) = M_f(1002\ \text{mL})$$

$$M_f = 0.0012\ M\ NH_3$$

Ammonia is a weak base ($K_b = 1.8 \times 10^{-5}$).

| | $NH_3$ + $H_2O$ | ⇌ | $NH_4^+$ | + $OH^-$ |
|---|---|---|---|---|
| Initial (*M*): | 0.0012 | | 0 | 0 |
| Change (*M*): | $-x$ | | $+x$ | $+x$ |
| Equil. (*M*): | $0.0012 - x$ | | $x$ | $x$ |

$$K_b = \frac{[NH_4^+][OH^-]}{[NH_3]}$$

$$1.8 \times 10^{-5} = \frac{x^2}{(0.0012 - x)}$$

Solving the resulting quadratic equation gives $x = 0.00014$, or $[OH^-] = 0.00014\ M$

This is a solution of iron(II) sulfate, which contains $Fe^{2+}$ ions. These $Fe^{2+}$ ions could combine with $OH^-$ to precipitate $Fe(OH)_2$. Therefore, we must use $K_{sp}$ for iron(II) hydroxide. We compute the value of $Q_c$ for this solution.

$$Fe(OH)_2(s) \rightleftharpoons Fe^{2+}(aq) + 2OH^-(aq)$$

$$Q = [Fe^{2+}]_0[OH^-]_0^2 = (1.0 \times 10^{-3})(0.00014)^2 = 2.0 \times 10^{-11}$$

Note that when adding 2.00 mL of $NH_3$ to 1.0 L of $FeSO_4$, the concentration of $FeSO_4$ will decrease slightly. However, rounding off to 2 significant figures, the concentration of $1.0 \times 10^{-3}\ M$ does not change. $Q$ is larger than $K_{sp}\ [Fe(OH)_2] = 1.6 \times 10^{-14}$. The concentrations of the ions in solution are greater than the equilibrium concentrations; the solution is saturated. The system will shift left to reestablish equilibrium; therefore, **a precipitate of $Fe(OH)_2$ will form.**

**16.80** **Strategy:** The addition of $Cd(NO_3)_2$ to the NaCN solution results in complex ion formation. In solution, $Cd^{2+}$ ions will complex with $CN^-$ ions. The concentration of $Cd^{2+}$ will be determined by the following equilibrium

$$Cd^{2+}(aq) + 4CN^-(aq) \rightleftharpoons Cd(CN)_4^{2-}$$

From Table 16.4 of the text, we see that the formation constant ($K_f$) for this reaction is very large ($K_f = 7.1 \times 10^{16}$). Because $K_f$ is so large, the reaction lies mostly to the right. At equilibrium, the concentration of $Cd^{2+}$ will be very small. As a good approximation, we can assume that essentially all the dissolved $Cd^{2+}$ ions end up as $Cd(CN)_4^{2-}$ ions. What is the initial concentration of $Cd^{2+}$ ions? A very small amount of $Cd^{2+}$ will be present at equilibrium. Set up the $K_f$ expression for the above equilibrium to solve for $[Cd^{2+}]$.

**Solution:** Calculate the initial concentration of $Cd^{2+}$ ions.

$$[Cd^{2+}]_0 = \frac{0.50\ \text{g} \times \dfrac{1\ \text{mol Cd(NO}_3)_2}{236.42\ \text{g Cd(NO}_3)_2} \times \dfrac{1\ \text{mol Cd}^{2+}}{1\ \text{mol Cd(NO}_3)_2}}{0.50\ \text{L}} = 4.2 \times 10^{-3}\ M$$

If we assume that the above equilibrium goes to completion, we can write

| | $Cd^{2+}(aq)$ | + | $4CN^-(aq)$ | $\longrightarrow$ | $Cd(CN)_4^{2-}(aq)$ |
|---|---|---|---|---|---|
| Initial ($M$): | $4.2 \times 10^{-3}$ | | 0.50 | | 0 |
| Change ($M$): | $-4.2 \times 10^{-3}$ | | $-4(4.2 \times 10^{-3})$ | | $+4.2 \times 10^{-3}$ |
| Final ($M$): | 0 | | 0.48 | | $4.2 \times 10^{-3}$ |

To find the concentration of free $Cd^{2+}$ at equilibrium, use the formation constant expression.

$$K_f = \frac{[Cd(CN)_4^{2-}]}{[Cd^{2+}][CN^-]^4}$$

Rearranging,

$$[Cd^{2+}] = \frac{[Cd(CN)_4^{2-}]}{K_f[CN^-]^4}$$

Substitute the equilibrium concentrations calculated above into the formation constant expression to calculate the equilibrium concentration of $Cd^{2+}$.

$$\mathbf{[Cd^{2+}]} = \frac{[Cd(CN)_4^{2-}]}{K_f[CN^-]^4} = \frac{4.2 \times 10^{-3}}{(7.1 \times 10^{16})(0.48)^4} = \mathbf{1.1 \times 10^{-18}}\ \boldsymbol{M}$$

$$[CN^-] = 0.48\ M + 4(1.1 \times 10^{-18}\ M) = \mathbf{0.48}\ \boldsymbol{M}$$

$$[Cd(CN)_4^{2-}] = (4.2 \times 10^{-3}\ M) - (1.1 \times 10^{-18}) = \mathbf{4.2 \times 10^{-3}}\ \boldsymbol{M}$$

**Check:** Substitute the equilibrium concentrations calculated into the formation constant expression to calculate $K_f$. Also, the small value of $[Cd^{2+}]$ at equilibrium, compared to its initial concentration of $4.2 \times 10^{-3}$ $M$, certainly justifies our approximation that almost all the $Cd^{2+}$ ions react.

**16.82** Silver iodide is only slightly soluble. It dissociates to form a small amount of $Ag^+$ and $I^-$ ions. The $Ag^+$ ions then complex with $NH_3$ in solution to form the complex ion $Ag(NH_3)_2^+$. The balanced equations are:

$$AgI(s) \rightleftharpoons Ag^+(aq) + I^-(aq) \qquad K_{sp} = [Ag^+][I^-] = 8.3 \times 10^{-17}$$

$$Ag^+(aq) + 2NH_3(aq) \rightleftharpoons Ag(NH_3)_2^+(aq) \qquad K_f = \frac{[Ag(NH_3)_2^+]}{[Ag^+][NH_3]^2} = 1.5 \times 10^7$$

$$\text{Overall:} \quad AgI(s) + 2NH_3(aq) \rightleftharpoons Ag(NH_3)_2^+(aq) + I^-(aq) \qquad K = K_{sp} \times K_f = 1.2 \times 10^{-9}$$

If $s$ is the molar solubility of AgI then,

| | $AgI(s)$ + | $2NH_3(aq)$ | $\rightleftharpoons$ $Ag(NH_3)_2^+(aq)$ | + $I^-(aq)$ |
|---|---|---|---|---|
| Initial ($M$): | | 1.0 | 0.0 | 0.0 |
| Change ($M$): | $-s$ | $-2s$ | $+s$ | $+s$ |
| Equilibrium ($M$): | | $(1.0 - 2s)$ | $s$ | $s$ |

Because $K_f$ is large, we can assume all of the silver ions exist as $Ag(NH_3)_2^+$. Thus,

$$[Ag(NH_3)_2^+] = [I^-] = s$$

We can write the equilibrium constant expression for the above reaction, then solve for $s$.

$$K = 1.2 \times 10^{-9} = \frac{(s)(s)}{(1.0 - 2s)^2} \approx \frac{(s)(s)}{(1.0)^2}$$

$$s = 3.5 \times 10^{-5}\ M$$

At equilibrium, $3.5 \times 10^{-5}$ moles of AgI dissolves in 1 L of 1.0 $M$ $NH_3$ solution.

**16.84** **(a)** The equations are as follows:

$$CuI_2(s) \rightleftharpoons Cu^{2+}(aq) + 2I^-(aq)$$

$$Cu^{2+}(aq) + 4NH_3(aq) \rightleftharpoons Cu(NH_3)_4^{2+}(aq)$$

The ammonia combines with the $Cu^{2+}$ ions formed in the first step to form the complex ion, $Cu(NH_3)_4^{2+}$, effectively removing the $Cu^{2+}$ ions, causing the first equilibrium to shift to the right (resulting in more $CuI_2$ dissolving).

**(b)** Similar to part (a):

$$AgBr(s) \rightleftharpoons Ag^+(aq) + Br^-(aq)$$

$$Ag^+(aq) + 2CN^-(aq) \rightleftharpoons Ag(CN)_2^-(aq)$$

**(c)** Similar to parts (a) and (b).

$$HgCl_2(s) \rightleftharpoons Hg^{2+}(aq) + 2Cl^-(aq)$$

$$Hg^{2+}(aq) + 4Cl^-(aq) \rightleftharpoons HgCl_4^{2-}(aq)$$

**16.88** Since some $PbCl_2$ precipitates, the solution is saturated. From Table 16.2, the value of $K_{sp}$ for lead(II) chloride is $2.4 \times 10^{-4}$. The equilibrium is:

$$PbCl_2(aq) \rightleftharpoons Pb^{2+}(aq) + 2Cl^-(aq)$$

We can write the solubility product expression for the equilibrium.

$$K_{sp} = [Pb^{2+}][Cl^-]^2$$

$K_{sp}$ and $[Cl^-]$ are known. Solving for the $Pb^{2+}$ concentration,

$$[Pb^{2+}] = \frac{K_{sp}}{[Cl^-]^2} = \frac{2.4 \times 10^{-4}}{(0.15)^2} = 0.011\ M$$

**16.90** Chloride ion will precipitate $Ag^+$ but not $Cu^{2+}$. So, dissolve some solid in $H_2O$ and add HCl. If a precipitate forms, the salt was $AgNO_3$. A flame test will also work. $Cu^{2+}$ gives a green flame test.

**16.92** We can use the Henderson-Hasselbalch equation to solve for the pH when the indicator is 90% acid / 10% conjugate base and when the indicator is 10% acid / 90% conjugate base.

$$\mathrm{pH} = \mathrm{p}K_\mathrm{a} + \log\frac{[\text{conjugate base}]}{[\text{acid}]}$$

Solving for the pH with 90% of the indicator in the HIn form:

$$\mathrm{pH} = 3.46 + \log\frac{[10]}{[90]} = 3.46 - 0.95 = 2.51$$

Next, solving for the pH with 90% of the indicator in the $In^-$ form:

$$\mathrm{pH} = 3.46 + \log\frac{[90]}{[10]} = 3.46 + 0.95 = 4.41$$

Thus the pH range varies from **2.51 to 4.41** as the [HIn] varies from 90% to 10%.

**16.94** The hydrochloric acid reacts with $CH_3COO^-$ to produce $CH_3COOH$. Note that the final solution volume is 1.0 L; therefore, the concentration of $CH_3COO^-$ is 0.010 $M$.

| | $CH_3COO^-(aq)$ + | $HCl(aq)$ → | $CH_3COOH(aq)$ + $Cl^-(aq)$ |
|---|---|---|---|
| Initial ($M$): | 0.010 | $x$ | 0 |
| Change ($M$): | $-x$ | $-x$ | $+x$ |
| Final ($M$): | $0.010 - x$ | 0 | $x$ |

Next, we use the Henderson-Hasselbalch equation to solve for $x$.

$$CH_3COOH(aq) \rightleftharpoons H^+(aq) + CH_3COO^-(aq)$$

$K_a = 1.8 \times 10^{-5}$, $\mathrm{p}K_a = 4.74$

$$\mathrm{pH} = \mathrm{p}K_\mathrm{a} + \log\frac{[CH_3COO^-]}{[CH_3COOH]}$$

$$5.00 = 4.74 + \log\left(\frac{0.010 - x}{x}\right)$$

$$10^{0.26} = \frac{0.010 - x}{x}$$

$$x = 0.0035\ M$$

Because the mixed solution volume is 1.0 L, the moles of HCl needed is 0.0035 mole. The volume of HCl needed to prepare this buffer with a pH = 5.00 is:

$$0.0035\ \cancel{\text{mol}}\ \text{HCl} \times \frac{1\ \text{L}}{0.020\ \cancel{\text{mol}}} = \mathbf{0.18\ L = 1.8 \times 10^2\ mL}$$

**16.96** First, calculate the pH of the 2.00 *M* weak acid ($HNO_2$) solution before any NaOH is added.

| | $HNO_2(aq)$ | $\rightleftharpoons$ | $H^+(aq)$ | + | $NO_2^-(aq)$ |
|---|---|---|---|---|---|
| Initial (*M*): | 2.00 | | 0 | | 0 |
| Change (*M*): | $-x$ | | $+x$ | | $+x$ |
| Equilibrium (*M*): | $2.00 - x$ | | $x$ | | $x$ |

$$K_a = \frac{[H^+][NO_2^-]}{[HNO_2]}$$

$$4.5 \times 10^{-4} = \frac{x^2}{2.00 - x} \approx \frac{x^2}{2.00}$$

$$x = [H^+] = 0.030\ M$$

$$pH = -\log(0.030) = 1.52$$

Since the pH after the addition is 1.5 pH units greater, the new pH = 1.52 + 1.50 = 3.02.

From this new pH, we can calculate the $[H^+]$ in solution.

$$[H^+] = 10^{-pH} = 10^{-3.02} = 9.55 \times 10^{-4}\ M$$

When the NaOH is added, we dilute our original 2.00 *M* $HNO_2$ solution to:

$$M_iV_i = M_fV_f$$

$$(2.00\ M)(400\ mL) = M_f(600\ mL)$$

$$M_f = 1.33\ M$$

Since we have not reached the equivalence point, we have a buffer solution. The reaction between $HNO_2$ and NaOH is:

$$HNO_2(aq) + NaOH(aq) \longrightarrow NaNO_2(aq) + H_2O(l)$$

Since the mole ratio between $HNO_2$ and NaOH is 1:1, the decrease in $[HNO_2]$ is the same as the decrease in [NaOH].

We can calculate the decrease in $[HNO_2]$ by setting up the weak acid equilibrium. From the pH of the solution, we know that the $[H^+]$ at equilibrium is $9.55 \times 10^{-4}$ *M*.

| | $HNO_2(aq)$ | $\rightleftharpoons$ | $H^+(aq)$ | + | $NO_2^-(aq)$ |
|---|---|---|---|---|---|
| Initial (*M*): | 1.33 | | 0 | | 0 |
| Change (*M*): | $-x$ | | | | $+x$ |
| Equilibrium (*M*): | $1.33 - x$ | | $9.55 \times 10^{-4}$ | | $x$ |

We can calculate $x$ from the equilibrium constant expression.

$$K_a = \frac{[H^+][NO_2^-]}{[HNO_2]}$$

$$4.5 \times 10^{-4} = \frac{(9.55 \times 10^{-4})(x)}{1.33 - x}$$

$$x = 0.426\ M$$

Thus, $x$ is the decrease in $[HNO_2]$ which equals the concentration of added $OH^-$. However, this is the concentration of NaOH after it has been diluted to 600 mL. We need to correct for the dilution from 200 mL to 600 mL to calculate the concentration of the original NaOH solution.

$$M_iV_i = M_fV_f$$

$$M_i(200\text{ mL}) = (0.426\ M)(600\text{ mL})$$

$$\mathbf{[NaOH]} = M_i = \mathbf{1.28\ M}$$

**16.98** The resulting solution is not a buffer system. There is excess NaOH and the neutralization is well past the equivalence point.

$$\text{Moles NaOH} = 0.500\ \cancel{L} \times \frac{0.167\text{ mol}}{1\ \cancel{L}} = 0.0835\text{ mol}$$

$$\text{Moles HCOOH} = 0.500\ \cancel{L} \times \frac{0.100\text{ mol}}{1\ \cancel{L}} = 0.0500\text{ mol}$$

| | $HCOOH(aq)$ + | $NaOH(aq)$ → | $HCOONa(aq) + H_2O(l)$ |
|---|---|---|---|
| Initial (mol): | 0.0500 | 0.0835 | 0 |
| Change (mol): | −0.0500 | −0.0500 | +0.0500 |
| Final (mol): | 0 | 0.0335 | 0.0500 |

The volume of the resulting solution is 1.00 L (500 mL + 500 mL = 1000 mL).

$$[OH^-] = \frac{0.0335\text{ mol}}{1.00\text{ L}} = \mathbf{0.0335\ M}$$

$$[Na^+] = \frac{(0.0335 + 0.0500)\text{ mol}}{1.00\text{ L}} = \mathbf{0.0835\ M}$$

$$[H^+] = \frac{K_w}{[OH^-]} = \frac{1.0 \times 10^{-14}}{0.0335} = \mathbf{3.0 \times 10^{-13}\ M}$$

$$[HCOO^-] = \frac{0.0500\text{ mol}}{1.00\text{ L}} = \mathbf{0.0500\ M}$$

| | $HCOO^-(aq) + H_2O(l)$ ⇌ | $HCOOH(aq)$ + | $OH^-(aq)$ |
|---|---|---|---|
| Initial ($M$): | 0.0500 | 0 | 0.0335 |
| Change ($M$): | $-x$ | $+x$ | $+x$ |
| Equilibrium ($M$): | $0.0500 - x$ | $x$ | $0.0335 + x$ |

$$K_b = \frac{[HCOOH][OH^-]}{[HCOO^-]}$$

$$5.9 \times 10^{-11} = \frac{(x)(0.0335 + x)}{(0.0500 - x)} \approx \frac{(x)(0.0335)}{(0.0500)}$$

$$x = [\text{HCOOH}] = \mathbf{8.8 \times 10^{-11}}\ \boldsymbol{M}$$

**16.100** The number of moles of $Ba(OH)_2$ present in the original 50.0 mL of solution is:

$$50.0\ \cancel{\text{mL}} \times \frac{1.00\ \text{mol Ba(OH)}_2}{1000\ \cancel{\text{mL}}\ \text{soln}} = 0.0500\ \text{mol Ba(OH)}_2$$

The number of moles of $H_2SO_4$ present in the original 86.4 mL of solution, assuming complete dissociation, is:

$$86.4\ \cancel{\text{mL}} \times \frac{0.494\ \text{mol H}_2\text{SO}_4}{1000\ \cancel{\text{mL}}\ \text{soln}} = 0.0427\ \text{mol H}_2\text{SO}_4$$

The reaction is:

| | $Ba(OH)_2(aq)$ | + $H_2SO_4(aq)$ | → $BaSO_4(s)$ | + $2H_2O(l)$ |
|---|---|---|---|---|
| Initial (mol): | 0.0500 | 0.0427 | 0 | |
| Change (mol): | −0.0427 | −0.0427 | +0.0427 | |
| Final (mol): | 0.0073 | 0 | 0.0427 | |

Thus the mass of $BaSO_4$ formed is:

$$0.0427\ \cancel{\text{mol}}\ \text{BaSO}_4 \times \frac{233.4\ \text{g BaSO}_4}{1\ \cancel{\text{mol}}\ \text{BaSO}_4} = \mathbf{9.97\ g\ BaSO_4}$$

The pH can be calculated from the excess $OH^-$ in solution. First, calculate the molar concentration of $OH^-$. The total volume of solution is 136.4 mL = 0.1364 L.

$$[\text{OH}^-] = \frac{0.0073\ \cancel{\text{mol}}\ \text{Ba(OH)}_2 \times \dfrac{2\ \text{mol OH}^-}{1\ \cancel{\text{mol}}\ \text{Ba(OH)}_2}}{0.1364\ \text{L}} = 0.11\ M$$

$$\text{pOH} = -\log(0.11) = 0.96$$

$$\mathbf{pH} = 14.00 - \text{pOH} = 14.00 - 0.96 = \mathbf{13.04}$$

**16.102** First, we calculate the molar solubility of $CaCO_3$.

| | $CaCO_3(s)$ | ⇌ $Ca^{2+}(aq)$ | + $CO_3^{2-}(aq)$ |
|---|---|---|---|
| Initial ($M$): | | 0 | 0 |
| Change ($M$): | $-s$ | $+s$ | $+s$ |
| Equil. ($M$): | | $s$ | $s$ |

$$K_{sp} = [Ca^{2+}][CO_3^{2-}] = s^2 = 8.7 \times 10^{-9}$$

$$s = 9.3 \times 10^{-5}\ M = 9.3 \times 10^{-5}\ \text{mol/L}$$

The moles of $CaCO_3$ in the kettle are:

$$116\ \cancel{g} \times \frac{1\ \text{mol } CaCO_3}{100.1\ \cancel{g}\ CaCO_3} = 1.16\ \text{mol } CaCO_3$$

The volume of distilled water needed to dissolve 1.16 moles of $CaCO_3$ is:

$$1.16\ \cancel{\text{mol}}\ CaCO_3 \times \frac{1\ \text{L}}{9.3 \times 10^{-5}\ \cancel{\text{mol}}\ CaCO_3} = 1.2 \times 10^4\ \text{L}$$

The number of times the kettle would have to be filled is:

$$(1.2 \times 10^4\ \cancel{L}) \times \frac{1\ \text{filling}}{2.0\ \cancel{L}} = \mathbf{6.0 \times 10^3\ fillings}$$

Note that the very important assumption is made that each time the kettle is filled, the calcium carbonate is allowed to reach equilibrium before the kettle is emptied.

**16.104** First we find the molar solubility and then convert moles to grams. The solubility equilibrium for silver carbonate is:

| | $Ag_2CO_3(s)$ | $\rightleftharpoons$ | $2Ag^+(aq)$ | + | $CO_3^{2-}(aq)$ |
|---|---|---|---|---|---|
| Initial ($M$): | | | 0 | | 0 |
| Change ($M$): | $-s$ | | $+2s$ | | $+s$ |
| Equilibrium ($M$): | | | $2s$ | | $s$ |

$$K_{sp} = [Ag^+]^2[CO_3^{2-}] = (2s)^2(s) = 4s^3 = 8.1 \times 10^{-12}$$

$$s = \left(\frac{8.1 \times 10^{-12}}{4}\right)^{\frac{1}{3}} = 1.3 \times 10^{-4}\ M$$

Converting from mol/L to g/L:

$$\frac{1.3 \times 10^{-4}\ \cancel{\text{mol}}}{1\ \text{L soln}} \times \frac{275.8\ \text{g}}{1\ \cancel{\text{mol}}} = \mathbf{0.036\ g/L}$$

**16.106** **(a)** To $2.50 \times 10^{-3}$ mole HCl (that is, 0.0250 L of 0.100 $M$ solution) is added $1.00 \times 10^{-3}$ mole $CH_3NH_2$ (that is, 0.0100 L of 0.100 $M$ solution).

| | $HCl(aq)$ + | $CH_3NH_2(aq)$ → | $CH_3NH_3Cl(aq)$ |
|---|---|---|---|
| Initial (mol): | $2.50 \times 10^{-3}$ | $1.00 \times 10^{-3}$ | 0 |
| Change (mol): | $-1.00 \times 10^{-3}$ | $-1.00 \times 10^{-3}$ | $+1.00 \times 10^{-3}$ |
| Equilibrium (mol): | $1.50 \times 10^{-3}$ | 0 | $1.00 \times 10^{-3}$ |

After the acid-base reaction, we have $1.50 \times 10^{-3}$ mol of HCl remaining. Since HCl is a strong acid, the $[H^+]$ will come from the HCl. The total solution volume is 35.0 mL = 0.0350 L.

$$[H^+] = \frac{1.50 \times 10^{-3}\ \text{mol}}{0.0350\ \text{L}} = 0.0429\ M$$

**pH = 1.37**

**(b)** When a total of 25.0 mL of $CH_3NH_2$ is added, we reach the equivalence point. That is, $2.50 \times 10^{-3}$ mol HCl reacts with $2.50 \times 10^{-3}$ mol $CH_3NH_2$ to form $2.50 \times 10^{-3}$ mol $CH_3NH_3Cl$. Since there is a total of 50.0 mL of solution, the concentration of $CH_3NH_3^+$ is:

$$[CH_3NH_3^+] = \frac{2.50 \times 10^{-3}\ \text{mol}}{0.0500\ \text{L}} = 5.00 \times 10^{-2}\ M$$

This is a problem involving the hydrolysis of the weak acid $CH_3NH_3^+$.

| | $CH_3NH_3^+(aq)$ ⇌ | $H^+(aq)$ + | $CH_3NH_2(aq)$ |
|---|---|---|---|
| Initial (*M*): | $5.00 \times 10^{-2}$ | 0 | 0 |
| Change (*M*): | $-x$ | $+x$ | $+x$ |
| Equilibrium (*M*): | $(5.00 \times 10^{-2}) - x$ | $x$ | $x$ |

$$K_a = \frac{[CH_3NH_2][H^+]}{[CH_3NH_3^+]}$$

$$2.3 \times 10^{-11} = \frac{x^2}{(5.00 \times 10^{-2}) - x} \approx \frac{x^2}{5.00 \times 10^{-2}}$$

$$1.15 \times 10^{-12} = x^2$$

$$x = 1.07 \times 10^{-6}\ M = [H^+]$$

**pH = 5.97**

**(c)** 35.0 mL of 0.100 *M* $CH_3NH_2$ ($3.50 \times 10^{-3}$ mol) is added to the 25 mL of 0.100 *M* HCl ($2.50 \times 10^{-3}$ mol).

| | $HCl(aq)$ + | $CH_3NH_2(aq)$ → | $CH_3NH_3Cl(aq)$ |
|---|---|---|---|
| Initial (mol): | $2.50 \times 10^{-3}$ | $3.50 \times 10^{-3}$ | 0 |
| Change (mol): | $-2.50 \times 10^{-3}$ | $-2.50 \times 10^{-3}$ | $+2.50 \times 10^{-3}$ |
| Equilibrium (mol): | 0 | $1.00 \times 10^{-3}$ | $2.50 \times 10^{-3}$ |

This is a buffer solution. Using the Henderson-Hasselbalch equation:

$$\text{pH} = \text{p}K_\text{a} + \log\frac{[\text{conjugate base}]}{[\text{acid}]}$$

$$\textbf{pH} = -\log(2.3\times10^{-11}) + \log\frac{(1.00\times10^{-3})}{(2.50\times10^{-3})} = \mathbf{10.24}$$

**16.108** The precipitate is $HgI_2$.

$$Hg^{2+}(aq) + 2I^-(aq) \longrightarrow HgI_2(s)$$

With further addition of $I^-$, a soluble complex ion is formed and the precipitate redissolves.

$$HgI_2(s) + 2I^-(aq) \longrightarrow HgI_4^{2-}(aq)$$

**16.110** We can use the Henderson-Hasselbalch equation to solve for the pH when the indicator is 95% acid / 5% conjugate base and when the indicator is 5% acid / 95% conjugate base.

$$\text{pH} = \text{p}K_\text{a} + \log\frac{[\text{conjugate base}]}{[\text{acid}]}$$

Solving for the pH with 95% of the indicator in the HIn form:

$$\text{pH} = 9.10 + \log\frac{[5]}{[95]} = 9.10 - 1.28 = 7.82$$

Next, solving for the pH with 95% of the indicator in the $In^-$ form:

$$\text{pH} = 9.10 + \log\frac{[95]}{[5]} = 9.10 + 1.28 = 10.38$$

Thus the pH range varies from **7.82 to 10.38** as the [HIn] varies from 95% to 5%.

**16.112** **(a)** We abbreviate the name of cacodylic acid to CacH. We set up the usual table.

| | $CacH(aq)$ ⇌ | $Cac^-(aq)$ + | $H^+(aq)$ |
|---|---|---|---|
| Initial ($M$): | 0.10 | 0 | 0 |
| Change ($M$): | $-x$ | $+x$ | $+x$ |
| Equilibrium ($M$): | $0.10 - x$ | $x$ | $x$ |

$$K_\text{a} = \frac{[\text{H}^+][\text{Cac}^-]}{[\text{CacH}]}$$

$$6.4\times10^{-7} = \frac{x^2}{0.10 - x} \approx \frac{x^2}{0.10}$$

$$x = 2.5 \times 10^{-4}\,M = [H^+]$$

$$\mathbf{pH} = -\log(2.5 \times 10^{-4}) = \mathbf{3.60}$$

**(b)** We set up a table for the hydrolysis of the anion:

| | $Cac^-(aq)$ + $H_2O(l)$ | $\rightleftharpoons$ $CacH(aq)$ + | $OH^-(aq)$ |
|---|---|---|---|
| Initial ($M$): | 0.15 | 0 | 0 |
| Change ($M$): | $-x$ | $+x$ | $+x$ |
| Equilibrium ($M$): | $0.15 - x$ | $x$ | $x$ |

The ionization constant, $K_b$, for $Cac^-$ is:

$$K_b = \frac{K_w}{K_a} = \frac{1.0 \times 10^{-14}}{6.4 \times 10^{-7}} = 1.6 \times 10^{-8}$$

$$K_b = \frac{[CacH][OH^-]}{[Cac^-]}$$

$$1.6 \times 10^{-8} = \frac{x^2}{0.15 - x} \approx \frac{x^2}{0.15}$$

$$x = 4.9 \times 10^{-5}\,M$$

$$pOH = -\log(4.9 \times 10^{-5}) = 4.31$$

$$\mathbf{pH} = 14.00 - 4.31 = \mathbf{9.69}$$

**(c)** Number of moles of CacH from (a) is:

$$50.0\ \cancel{mL}\ CacH \times \frac{0.10\text{ mol CacH}}{1000\ \cancel{mL}} = 5.0 \times 10^{-3}\text{ mol CacH}$$

Number of moles of $Cac^-$ from (b) is:

$$25.0\ \cancel{mL}\ CacNa \times \frac{0.15\text{ mol CacNa}}{1000\ \cancel{mL}} = 3.8 \times 10^{-3}\text{ mol CacNa}$$

At this point we have a buffer solution.

$$\mathbf{pH} = pK_a + \log\frac{[Cac^-]}{[CacH]} = -\log(6.4 \times 10^{-7}) + \log\frac{3.8 \times 10^{-3}}{5.0 \times 10^{-3}} = \mathbf{6.07}$$

**16.114** **(a)** $MCO_3 + 2HCl \rightarrow MCl_2 + H_2O + CO_2$

$HCl + NaOH \rightarrow NaCl + H_2O$

**(b)** We are given the mass of the metal carbonate, so we need to find moles of the metal carbonate to calculate its molar mass. We can find moles of $MCO_3$ from the moles of HCl reacted.

$$\text{Moles of HCl reacted with } MCO_3 = \text{Total moles of HCl} - \text{Moles of excess HCl}$$

$$\text{Total moles of HCl} = 18.68 \text{ mL} \times \frac{5.653 \text{ mol}}{1000 \text{ mL soln}} = 0.1056 \text{ mol HCl}$$

$$\text{Moles of excess HCl} = 12.06 \text{ mL} \times \frac{1.789 \text{ mol}}{1000 \text{ mL soln}} = 0.02158 \text{ mol HCl}$$

$$\text{Moles of HCl reacted with } MCO_3 = 0.1056 \text{ mol} - 0.02158 \text{ mol} = 0.0840 \text{ mol HCl}$$

$$\text{Moles of } MCO_3 \text{ reacted} = 0.0840 \text{ mol HCl} \times \frac{1 \text{ mol } MCO_3}{2 \text{ mol HCl}} = 0.0420 \text{ mol } MCO_3$$

$$\textbf{Molar mass of } \mathbf{MCO_3} = \frac{3.542 \text{ g}}{0.0420 \text{ mol}} = \mathbf{84.3 \text{ g/mol}}$$

$$\text{Molar mass of } CO_3 = 60.01 \text{ g}$$

$$\text{Molar mass of M} = 84.3 \text{ g/mol} - 60.01 \text{ g/mol} = 24.3 \text{ g/mol}$$

The metal, M, is **Mg**!

**16.116** The number of moles of NaOH reacted is:

$$15.9 \text{ mL NaOH} \times \frac{0.500 \text{ mol NaOH}}{1000 \text{ mL soln}} = 7.95 \times 10^{-3} \text{ mol NaOH}$$

Since two moles of NaOH combine with one mole of oxalic acid, the number of moles of oxalic acid reacted is $3.98 \times 10^{-3}$ mol. This is the number of moles of oxalic acid hydrate in 25.0 mL of solution. In 250 mL, the number of moles present is $3.98 \times 10^{-2}$ mol. Thus the molar mass is:

$$\frac{5.00 \text{ g}}{3.98 \times 10^{-2} \text{ mol}} = 126 \text{ g/mol}$$

From the molecular formula we can write:

$$2(1.008)\text{g} + 2(12.01)\text{g} + 4(16.00)\text{g} + x(18.02)\text{g} = 126 \text{ g}$$

Solving for $x$:

$$\mathbf{x = 2}$$

**16.118** **(a)** $$\text{pH} = \text{p}K_\text{a} + \log\frac{[\text{conjugate base}]}{[\text{acid}]}$$

$$8.00 = 9.10 + \log\frac{[\text{ionized}]}{[\text{un-ionized}]}$$

$$\frac{[\text{un-ionized}]}{[\text{ionized}]} = \mathbf{12.6} \qquad (1)$$

**(b)** First, let's calculate the total concentration of the indicator. 2 drops of the indicator are added and each drop is 0.050 mL.

$$2\text{ drops} \times \frac{0.050\text{ mL phenolphthalein}}{1\text{ drop}} = 0.10\text{ mL phenolphthalein}$$

This 0.10 mL of phenolphthalein of concentration 0.060 $M$ is diluted to 50.0 mL.

$$M_iV_i = M_fV_f$$

$$(0.060\ M)(0.10\text{ mL}) = M_f(50.0\text{ mL})$$

$$M_f = 1.2 \times 10^{-4}\ M$$

Using equation (1) above and letting $y$ = [ionized], then [un-ionized] = $(1.2 \times 10^{-4}) - y$.

$$\frac{(1.2 \times 10^{-4}) - y}{y} = 12.6$$

$$y = \mathbf{8.8 \times 10^{-6}}\ M$$

**16.120** **(a)** Add sulfate. $Na_2SO_4$ is soluble, $BaSO_4$ is not.

**(b)** Add sulfide. $K_2S$ is soluble, PbS is not

**(c)** Add iodide. $ZnI_2$ is soluble, $HgI_2$ is not.

**16.122** The amphoteric oxides cannot be used to prepare buffer solutions because they are insoluble in water.

**16.124** The ionized polyphenols have a dark color. In the presence of citric acid from lemon juice, the anions are converted to the lighter-colored acids.

**16.126** Assuming the density of water to be 1.00 g/mL, 0.05 g $Pb^{2+}$ per $10^6$ g water is equivalent to $5 \times 10^{-5}$ g $Pb^{2+}$/L

$$\frac{0.05\text{ g Pb}^{2+}}{1 \times 10^6\text{ g H}_2\text{O}} \times \frac{1\text{ g H}_2\text{O}}{1\text{ mL H}_2\text{O}} \times \frac{1000\text{ mL H}_2\text{O}}{1\text{ L H}_2\text{O}} = 5 \times 10^{-5}\text{ g Pb}^{2+}/\text{L}$$

| | $PbSO_4$ | $\rightleftharpoons$ $Pb^{2+}$ | + $SO_4^{2-}$ |
|---|---|---|---|
| Initial ($M$): | | 0 | 0 |
| Change ($M$): | $-s$ | $+s$ | $+s$ |
| Equilibrium ($M$): | | $s$ | $s$ |

$$K_{sp} = [Pb^{2+}][SO_4^{2-}]$$

$$1.6 \times 10^{-8} = s^2$$

$$s = 1.3 \times 10^{-4}\ M$$

The solubility of $PbSO_4$ in g/L is:

$$\frac{1.3 \times 10^{-4}\ \text{mol}}{1\ \text{L}} \times \frac{303.3\ \text{g}}{1\ \text{mol}} = 4.0 \times 10^{-2}\ \text{g/L}$$

**Yes.** The $[Pb^{2+}]$ exceeds the safety limit of $5 \times 10^{-5}$ g $Pb^{2+}$/L.

**16.128** (c) has the highest $[H^+]$

$$F^- + SbF_5 \rightarrow SbF_6^-$$

Removal of $F^-$ promotes further ionization of HF.

**16.130** (a) This is a common ion ($CO_3^{2-}$) problem.

The dissociation of $Na_2CO_3$ is:

$$Na_2CO_3(s) \xrightarrow{H_2O} 2Na^+(aq) + CO_3^{2-}(aq)$$

$$2(0.050\ M) \qquad 0.050\ M$$

Let $s$ be the molar solubility of $CaCO_3$ in $Na_2CO_3$ solution. We summarize the changes as:

| | $CaCO_3(s) \rightleftharpoons$ | $Ca^{2+}(aq)$ + | $CO_3^{2-}(aq)$ |
|---|---|---|---|
| Initial ($M$): | | 0.00 | 0.050 |
| Change ($M$): | | $+s$ | $+s$ |
| Equil. ($M$): | | $+s$ | $0.050 + s$ |

$$K_{sp} = [Ca^{2+}][CO_3^{2-}]$$

$$8.7 \times 10^{-9} = s(0.050 + s)$$

Since $s$ is small, we can assume that $0.050 + s \approx 0.050$

$$8.7 \times 10^{-9} = 0.050s$$

$$\mathbf{s = 1.7 \times 10^{-7}\ M}$$

Thus, the addition of washing soda to permanent hard water removes most of the $Ca^{2+}$ ions as a result of the common ion effect.

**(b)** $Mg^{2+}$ is not removed by this procedure, because $MgCO_3$ is fairly soluble ($K_{sp} = 4.0 \times 10^{-5}$).

**(c)** The $K_{sp}$ for $Ca(OH)_2$ is $8.0 \times 10^{-6}$.

$$Ca(OH)_2 \rightleftharpoons Ca^{2+} + 2OH^-$$

At equil.: $s$ $\quad 2s$

$$K_{sp} = 8.0 \times 10^{-6} = [Ca^{2+}][OH^-]^2$$

$$4s^3 = 8.0 \times 10^{-6}$$

$$s = 0.0126\ M$$

$$[OH^-] = 2s = 0.0252\ M$$

$$pOH = -\log(0.0252) = 1.60$$

$$\mathbf{pH = 12.40}$$

**(d)** The $[OH^-]$ calculated above is 0.0252 *M*. At this rather high concentration of $OH^-$, most of the $Mg^{2+}$ will be removed as $Mg(OH)_2$. The small amount of $Mg^{2+}$ remaining in solution is due to the following equilibrium:

$$Mg(OH)_2(s) \rightleftharpoons Mg^{2+}(aq) + 2OH^-(aq)$$

$$K_{sp} = [Mg^{2+}][OH^-]^2$$

$$1.2 \times 10^{-11} = [Mg^{2+}](0.0252)^2$$

$$\mathbf{[Mg^{2+}] = 1.9 \times 10^{-8}\ M}$$

**(e)** Remove $Ca^{2+}$ first because it is present in larger amounts.

**16.132** $pH = pK_a + \log\dfrac{[\text{conjugate base}]}{[\text{acid}]}$

**At pH = 1.0,**

$-COOH$ $\quad 1.0 = 2.3 + \log\dfrac{[-COO^-]}{[-COOH]}$

$$\frac{[-COOH]}{[-COO^-]} = 20$$

$-NH_3^+$ $\quad 1.0 = 9.6 + \log\dfrac{[-NH_2]}{[-NH_3^+]}$

$$\frac{[-NH_3^+]}{[-NH_2]} = 4 \times 10^8$$

Therefore the **predominant species** is: $^{+}NH_3 - CH_2 - COOH$

**At pH = 7.0,**

$-COOH$ $$7.0 = 2.3 + \log\frac{[-COO^-]}{[-COOH]}$$

$$\frac{[-COO^-]}{[-COOH]} = 5 \times 10^4$$

$-NH_3^+$ $$7.0 = 9.6 + \log\frac{[-NH_2]}{[-NH_3^+]}$$

$$\frac{[-NH_3^+]}{[-NH_2]} = 4 \times 10^2$$

**Predominant species:** $^{+}NH_3 - CH_2 - COO^-$

**At pH = 12.0,**

$-COOH$ $$12.0 = 2.3 + \log\frac{[-COO^-]}{[-COOH]}$$

$$\frac{[-COO^-]}{[-COOH]} = 5 \times 10^9$$

$-NH_3^+$ $$12.0 = 9.6 + \log\frac{[-NH_2]}{[-NH_3^+]}$$

$$\frac{[-NH_2]}{[-NH_3^+]} = 2.5 \times 10^2$$

**Predominant species:** $NH_2 - CH_2 - COO^-$

**16.134** **(a)** Let $x$ equal the moles of NaOH added to the solution. The acid-base reaction is:

| | $HF(aq)$ + | $NaOH(aq)$ → | $NaF(aq) + H_2O(l)$ |
|---|---|---|---|
| Initial (mol): | 0.0050 | $x$ | 0 |
| Change (mol): | $-x$ | $-x$ | $+x$ |
| Final (mol) | $0.0050 - x$ | 0 | $x$ |

At pH = 2.85, we are in the buffer region of the titration. The pH of a 0.20 *M* HF solution is 1.92, and clearly the equivalence point of the titration has not been reached. NaOH is the limiting reagent. The Henderson-Hasselbalch equation can be used to solve for $x$. Because HF and $F^-$ are contained in the same volume of solution, we can plug in moles into the Henderson-Hasselbalch equation to solve for $x$. The $K_a$ value for HF is $7.1 \times 10^{-4}$.

$$pH = pK_a + \log\frac{[F^-]}{[HF]}$$

$$2.85 = 3.15 + \log\left(\frac{x}{0.0050 - x}\right)$$

$$10^{-0.30} = \frac{x}{0.0050 - x}$$

$$0.0025 - 0.50x = x$$

$$x = 0.00167 \text{ mol}$$

The volume of NaOH added can be calculated from the molarity of NaOH and moles of NaOH.

$$\frac{1 \text{ L}}{0.20 \text{ mol NaOH}} \times 0.00167 \text{ mol NaOH} = \mathbf{0.0084 \text{ L}} = \mathbf{8.4 \text{ mL}}$$

**(b)** The p$K_a$ value of HF is 3.15. This is halfway to the equivalence point of the titration, where the concentration of HF equals the concentration of $F^-$. For the concentrations of HF and $F^-$ to be equal, **12.5 mL** of 0.20 *M* NaOH must be added to 25.0 mL of 0.20 *M* HF.

| | $HF(aq)$ + | $NaOH(aq)$ → | $NaF(aq) + H_2O(l)$ |
|---|---|---|---|
| Initial (mol): | 0.0050 | 0.0025 | 0 |
| Change (mol): | −0.0025 | −0.0025 | +0.0025 |
| Final (mol) | 0.0025 | 0 | 0.0025 |

**(c)** At pH = 11.89, the equivalence point of the titration has been passed. Excess $OH^-$ ions are present in solution. First, let's calculate the $[OH^-]$ at pH = 11.89.

$$\text{pOH} = 2.11$$

$$[OH^-] = 10^{-\text{pOH}} = 0.0078\ M$$

Because the initial concentrations of HF and NaOH are equal, it would take 25.0 mL of NaOH to reach the equivalence point. Therefore, the total solution volume at the equivalence point is 50.0 mL. At the equivalence point, there are no hydroxide ions present in solution. After the equivalence point, excess NaOH is added. Let *x* equal the volume of NaOH added after the equivalence point. A NaOH solution with a concentration of 0.20 *M* is being added, and a pH = 11.89 solution has a hydroxide concentration of 0.0078 *M*. A dilution calculation can be set up to solve for *x*.

$$M_1V_1 = M_2V_2$$

$$(0.20\ M)(x) = (0.0078\ M)(50 \text{ mL} + x)$$

$$0.192x = 0.39$$

$$x = 2.0 \text{ mL}$$

This volume of 2.0 mL NaOH solution is the volume added after the equivalence point. The total volume of NaOH added to reach a pH of 11.89 is:

$$\textbf{volume NaOH} = 25.0 \text{ mL} + 2.0 \text{ mL} = \mathbf{27.0 \text{ mL}}$$

**16.136** **(a)** Before dilution:

$$pH = pK_a + \log\frac{[CH_3COO^-]}{[CH_3COOH]}$$

$$\mathbf{pH} = 4.74 + \log\frac{[0.500]}{[0.500]} = \mathbf{4.74}$$

After a 10-fold dilution:

$$\mathbf{pH} = 4.74 + \log\frac{[0.0500]}{[0.0500]} = \mathbf{4.74}$$

There is no change in the pH of a buffer upon dilution.

**(b)** Before dilution:

| | $CH_3COOH(aq) + H_2O(l) \rightleftharpoons$ | $H_3O^+(aq)$ | $+\ CH_3COO^-(aq)$ |
|---|---|---|---|
| Initial ($M$): | 0.500 | 0 | 0 |
| Change ($M$): | $-x$ | $+x$ | $+x$ |
| Equilibrium ($M$): | $0.500 - x$ | $x$ | $x$ |

$$K_a = \frac{[H_3O^+][CH_3COO^-]}{[CH_3COOH]}$$

$$1.8 \times 10^{-5} = \frac{x^2}{0.500 - x} \approx \frac{x^2}{0.500}$$

$$x = 3.0 \times 10^{-3}\ M = [H_3O^+]$$

$$\mathbf{pH} = -\log(3.0 \times 10^{-3}) = \mathbf{2.52}$$

After dilution:

$$1.8 \times 10^{-5} = \frac{x^2}{0.0500 - x} \approx \frac{x^2}{0.0500}$$

$$x = 9.5 \times 10^{-4}\ M = [H_3O^+]$$

$$\mathbf{pH} = -\log(9.5 \times 10^{-4}) = \mathbf{3.02}$$

**16.138** The reaction is:

$$NH_3 + HCl \rightarrow NH_4Cl$$

First, we calculate moles of HCl and $NH_3$.

$$n_{HCl} = \frac{PV}{RT} = \frac{\left(372\ \text{mmHg} \times \dfrac{1\ \text{atm}}{760\ \text{mmHg}}\right)(0.96\ \text{L})}{\left(0.0821\dfrac{\text{L}\cdot\text{atm}}{\text{mol}\cdot\text{K}}\right)(295\ \text{K})} = 0.0194\ \text{mol}$$

$$n_{NH_3} = \frac{0.57\ \text{mol NH}_3}{1\ \text{L soln}} \times 0.034\ \text{L} = 0.0194\ \text{mol}$$

The mole ratio between $NH_3$ and HCl is 1:1, so we have complete neutralization.

| | $NH_3$ | + HCl | → $NH_4Cl$ |
|---|---|---|---|
| Initial (mol): | 0.0194 | 0.0194 | 0 |
| Change (mol): | −0.0194 | −0.0194 | +0.0194 |
| Final (mol): | 0 | 0 | 0.0194 |

$NH_4^+$ is a weak acid. We set up the reaction representing the hydrolysis of $NH_4^+$.

| | $NH_4^+(aq)$ + $H_2O(l)$ ⇌ | $H_3O^+(aq)$ + | $NH_3(aq)$ |
|---|---|---|---|
| Initial ($M$): | 0.0194 mol/0.034 L | 0 | 0 |
| Change ($M$): | $-x$ | $+x$ | $+x$ |
| Equilibrium ($M$): | $0.57 - x$ | $x$ | $x$ |

$$K_a = \frac{[H_3O^+][NH_3]}{[NH_4^+]}$$

$$5.6 \times 10^{-10} = \frac{x^2}{0.57 - x} \approx \frac{x^2}{0.57}$$

$$x = 1.79 \times 10^{-5}\ M = [H_3O^+]$$

$$\mathbf{pH} = -\log(1.79 \times 10^{-5}) = \mathbf{4.75}$$

**16.140** **(a)** The potential precipitate is $MgCO_3$. We solve for $Q = K_{sp}$ to determine the maximum mass of $Na_2CO_3$ that can be added without $MgCO_3$ precipitating. The concentration of $Mg^{2+}$ ions is 0.100 $M$.

$$Q = [Mg^{2+}][CO_3^{2-}]$$

$$4.0 \times 10^{-5} = (0.100\ M)\ [CO_3^{2-}]$$

$$[CO_3^{2-}] = 4.0 \times 10^{-4}\ M$$

$$\textbf{mass Na}_2\textbf{CO}_3 = 0.200\ \text{L} \times \frac{4.0 \times 10^{-4}\ \text{mol CO}_3^{2-}}{1\ \text{L soln}} \times \frac{1\ \text{mol Na}_2\text{CO}_3}{1\ \text{mol CO}_3^{2-}} \times \frac{105.99\ \text{g Na}_2\text{CO}_3}{1\ \text{mol Na}_2\text{CO}_3} = \mathbf{0.0085\ g}$$

**(b)** The potential precipitate is AgCl. We solve for $Q = K_{sp}$ to determine the maximum mass of $AgNO_3$ that can be added without AgCl precipitating. The concentration of $Cl^-$ ions is 0.200 $M$ because 1 mole of $MgCl_2$ yields 2 moles of $Cl^-$.

$$Q = [Ag^+][Cl^-]$$

$$1.6 \times 10^{-10} = [Ag^+](0.200\ M)$$

$$[Ag^+] = 8.0 \times 10^{-10}\ M$$

$$\textbf{mass AgNO}_\mathbf{3} = 0.200\ \cancel{L} \times \frac{8.00 \times 10^{-10}\ \cancel{mol}\ Ag^+}{1\ \cancel{L}\ soln} \times \frac{1\ \cancel{mol}\ AgNO_3}{1\ \cancel{mol}\ Ag^+} \times \frac{169.9\ g\ AgNO_3}{1\ \cancel{mol}\ AgNO_3} = \mathbf{2.7 \times 10^{-8}\ g}$$

**(c)** The potential precipitate is $Mg(OH)_2$. We solve for $Q = K_{sp}$ to determine the maximum mass of KOH that can be added without $Mg(OH)_2$ precipitating. The concentration of $Mg^{2+}$ ions is 0.100 $M$.

$$Q = [Mg^{2+}][OH^-]^2$$

$$1.2 \times 10^{-11} = (0.100\ M)[OH^-]^2$$

$$[OH^-] = 1.1 \times 10^{-5}\ M$$

$$\textbf{mass KOH} = 0.200\ \cancel{L} \times \frac{1.1 \times 10^{-5}\ \cancel{mol}\ OH^-}{1\ \cancel{L}\ soln} \times \frac{1\ \cancel{mol}\ KOH}{1\ \cancel{mol}\ OH^-} \times \frac{56.11\ g\ KOH}{1\ \cancel{mol}\ KOH} = \mathbf{1.2 \times 10^{-4}\ g}$$

**16.142** (1) The initial pH of acid (a) is lower.
(2) The pH at the half-way to the equivalence point is lower for acid (a). At the half-way point, pH = $pK_a$. The $K_a$ value of the first acid is therefore larger.
(3) The pH at the equivalence point is lower for acid (a). At this point, the pH is determined by the hydrolysis of the conjugate base of the weak acid. A lower pH indicates a weaker conjugate base and hence a stronger acid.

Based on these three observations, acid **(a)** is the stronger acid.

**16.144** When the $CuSO_4$ solution is mixed with the $Ba(OH)_2$ solution, both $Cu(OH)_2$ and $BaSO_4$ precipitate from solution. The solutions are mixed in the correct stoichiometric ratio. If we calculate $Q$ (the reaction quotient) for each of these salts, we would find that $Q > K_{sp}$, indicating that both salts precipitate.

| | $CuSO_4(aq)$ + | $Ba(OH)_2(aq)$ → | $Cu(OH)_2(s)$ + | $BaSO_4(s)$ |
|---|---|---|---|---|
| Initial (mol): | 0.0100 | 0.0100 | 0.00 | 0.00 |
| Change (mol): | −0.0100 | −0.0100 | +0.0100 | +0.0100 |
| Final (mol): | 0 | 0 | 0.0100 | 0.0100 |

We can now consider the dissociation of each precipitate in water to determine the concentration of ions in the combined solution.

| | $Cu(OH)_2(s)$ | $\rightleftharpoons$ | $Cu^{2+}(aq)$ | + | $2OH^-(aq)$ |
|---|---|---|---|---|---|
| Initial (*M*): | | | 0 | | 0 |
| Change (*M*): | $-s$ | | $+s$ | | $+2s$ |
| Equilibrium (*M*): | | | $s$ | | $2s$ |

Recall that the concentration of a pure solid does not enter into an equilibrium constant expression. Therefore, the amount of $Cu(OH)_2$ is unimportant as long as precipitate is present.

$$K_{sp} = [Cu^{2+}][OH^-]^2$$

$$2.2 \times 10^{-20} = (s)(2s)^2$$

$$2.2 \times 10^{-20} = 4s^3$$

$$s = 1.8 \times 10^{-7}\ M$$

$$\mathbf{[Cu^{2+}] = s = 1.8 \times 10^{-7}\ M}$$

$$\mathbf{[OH^-] = 2s = 3.6 \times 10^{-7}\ M}$$

| | $BaSO_4(s)$ | $\rightleftharpoons$ | $Ba^{2+}(aq)$ | + | $SO_4^{2-}(aq)$ |
|---|---|---|---|---|---|
| Initial (*M*): | | | 0 | | 0 |
| Change (*M*): | $-s$ | | $+s$ | | $+s$ |
| Equilibrium (*M*): | | | $s$ | | $s$ |

$$K_{sp} = [Ba^{2+}][SO_4^{2-}]$$

$$1.1 \times 10^{-10} = s^2$$

$$s = 1.0 \times 10^{-5}\ M$$

$$\mathbf{[Ba^{2+}] = [SO_4^{2-}] = s = 1.0 \times 10^{-5}\ M}$$

# CHAPTER 17
# ENTROPY, FREE ENERGY, AND EQUILIBRIUM

## PROBLEM-SOLVING STRATEGIES AND TUTORIAL SOLUTIONS

### TYPES OF PROBLEMS

**Problem Type 1:** Entropy.

- **(a)** Predicting entropy changes.
- **(b)** Calculating entropy changes of a system.

**Problem Type 2:** Free Energy.

- **(a)** Calculating standard free energy changes.
- **(b)** Calculating standard free energy change from enthalpy and entropy changes.
- **(c)** Entropy changes due to phase transitions.

**Problem Type 3:** Free Energy and Chemical Equilibrium.

- **(a)** Using the standard free energy change to calculate the equilibrium constant.
- **(b)** Using the free energy change to predict the direction of a reaction.

## PROBLEM TYPE 1: ENTROPY

**Entropy (*S*)** is a direct measure of the randomness or disorder of a system. The greater the disorder of a system, the greater its entropy. Conversely, the more ordered a system, the lower its entropy.

### A. Predicting entropy changes

For any substance, the particles in the solid are more ordered than those in the liquid state, which in turn are more ordered than those in the gaseous state. See Figure 17.3 of the text. Thus, for any substance, the entropy for the same molar amount always increases in the following order.

$$S_{\text{solid}} < S_{\text{liquid}} < S_{\text{gas}}$$

There are a number of other factors that you need to consider to predict entropy changes.

- Heating increases the entropy of a system because it increases the random motion of atoms and molecules.
- When a solid dissolves in water, the highly ordered structure of the solid and part of the order of the water are destroyed. Consequently, the solution possesses greater disorder than the pure solute and pure solvent.
- If a reaction produces more gas molecules than it consumes, the entropy of the system increases. If the total number of gas molecules diminishes during a reaction, the entropy of the system decreases. The dependence of the entropy of reaction on the number of gas molecules is due to the fact that gas molecules possess much greater entropy than either solid or liquid molecules.
- If there is no net change in the total number of gas molecules in a reaction, then the entropy of the system may increase or decrease, but it will be a relatively small change.

As discussed, the entropy is often described as a measure of disorder or randomness. While useful, these terms must be used with caution because they are subjective concepts. In general, it is preferable to view a change in entropy of a system in terms of the change in the number of microstates of the systems. If the number of microstates increases, entropy increases. Conversely, if the number of microstates decreases, the entropy decreases. See Section 17.3 of the text for a complete discussion.

**EXAMPLE 17.1**

**Predict whether the entropy increases, decreases, or remains essentially unchanged for the following reactions.**

(a) $H_2O_2(l) \longrightarrow H_2O(l) + \frac{1}{2}O_2(g)$

(b) $H^+(aq) + OH^-(aq) \longrightarrow H_2O(l)$

(c) $Ca(OH)_2(s) + CO_2(g) \longrightarrow CaCO_3(s) + H_2O(g)$

**Strategy:** To determine the entropy change in each case, we examine whether the number of microstates of the system increases or decreases. The sign of $\Delta S$ will be positive if there is an increase in the number of microstates and negative if the number of microstates decreases.

**Solution:**

**(a)** The number of moles of gaseous compounds in the products is greater than in the reactant. $\Delta S > 0$; increase in number of moles of gas (an increase in microstates).

**(b)** Two reactants combine into one product in this reaction. The number of microstates decreases and so entropy decreases ($\Delta S < 0$).

**(c)** The number of moles of gas phase products is the same as the reactants. The entropy remains essentially unchanged as the number of microstates is relatively constant.

**PRACTICE EXERCISE**

**1.** Predict whether the entropy increases, decreases, or remains essentially unchanged for the following reactions:

(a) $CaO(s) + CO_2(g) \longrightarrow CaCO_3(s)$

(b) $CuSO_4(s) \longrightarrow Cu^{2+}(aq) + SO_4^{\ 2-}(aq)$

(c) $2HCl(g) + Br_2(l) \longrightarrow 2HBr(g) + Cl_2(g)$

**Text Problems:** 17.10, 17.14

## B. Calculating entropy changes of a system

The universe is made up of the system and the surroundings. The entropy change in the universe ($\Delta S_{univ}$) for any process is the *sum* of the entropy changes in the system ($\Delta S_{sys}$) and in the surroundings ($\Delta S_{surr}$).

$$\Delta S_{univ} = \Delta S_{sys} + \Delta S_{surr}$$

For a spontaneous process, the entropy of the universe increases.

$$\Delta S_{univ} = \Delta S_{sys} + \Delta S_{surr} > 0$$

In chemistry, we typically focus on the entropy of the system. Let's suppose that the system is represented by the following reaction:

$$a\text{A} + b\text{B} \longrightarrow c\text{C} + d\text{D}$$

As is the case for the enthalpy of a reaction [see Equation (6.18) in the text], the standard entropy change $\Delta S^\circ$ is given by:

$$\Delta S^\circ_{\text{rxn}} = [cS^\circ(\text{C}) + dS^\circ(\text{D})] - [aS^\circ(\text{A}) + bS^\circ(\text{B})]$$

or, using $\Sigma$ to represent summation and *m* and *n* for the stoichiometric coefficients in the reaction,

$$\Delta S^\circ_{\text{rxn}} = \Sigma nS^\circ(\text{products}) - \Sigma mS^\circ(\text{reactants}) \qquad (17.7, \text{text})$$

Using the standard entropy values ($S^\circ$) listed in Appendix 3 of the text, we can calculate $\Delta S^\circ_{\text{rxn}}$, which corresponds to $\Delta S_{\text{sys}}$.

**EXAMPLE 17.2**

**Use standard entropy values to calculate the standard entropy change ($\Delta S^\circ_{\text{rxn}}$) for the reaction**

$$H_2(g) + \tfrac{1}{2}O_2(g) \longrightarrow H_2O(l)$$

---

**Strategy:** To calculate the standard entropy change of a reaction, we look up the standard entropies of reactants and products in Appendix 3 of the text and apply Equation (17.7) of the text. As in the calculation of enthalpy of reaction, the stoichiometric coefficients have no units, so $\Delta S^\circ_{\text{rxn}}$ is expressed in units of J/K·mol.

**Solution:** The standard entropy change for a reaction can be calculated using the following equation.

$$\Delta S^\circ_{\text{rxn}} = \Sigma nS^\circ(\text{products}) - \Sigma mS^\circ(\text{reactants})$$

$$\Delta S^\circ_{\text{rxn}} = S^\circ[H_2O(l)] - \{S^\circ[H_2(g)] + \left(\tfrac{1}{2}\right)S^\circ[O_2(g)]\}$$

$$\Delta S^\circ_{\text{rxn}} = (1)(69.9\ \text{J/K}\cdot\text{mol}) - [(1)(131.0\ \text{J/K}\cdot\text{mol}) + \left(\tfrac{1}{2}\right)(205.0\ \text{J/K}\cdot\text{mol})]$$

$$\Delta S^\circ_{\text{rxn}} = 69.9\ \text{J/K}\cdot\text{mol} - 233.5\ \text{J/K}\cdot\text{mol} = \mathbf{-163.6\ J/K\cdot mol}$$

**Tip:** This reaction is known to be spontaneous, so you would think that $\Delta S$ should be positive. Remember, that the value of $\Delta S^\circ_{\text{rxn}}$ applies only to the system; $\Delta S_{\text{univ}}$ will have a positive value.

**PRACTICE EXERCISE**

2. Hydrate lime or slaked lime, $Ca(OH)_2$, can be reformed into quicklime, CaO, by heating:

$$Ca(OH)_2(s) \xrightarrow{\text{heat}} CaO(s) + H_2O(g)$$

Use standard entropy values to calculate the standard entropy change ($\Delta S^\circ_{\text{rxn}}$) for this reaction.

**Text Problem: 17.12**

## PROBLEM TYPE 2: FREE ENERGY

The second law of thermodynamics tells us that a spontaneous reaction increases the entropy of the universe; that is, $\Delta S_{univ} > 0$. To calculate $\Delta S_{univ}$, we must calculate both $\Delta S_{surr}$ and $\Delta S_{sys}$. However, it can be difficult to calculate $\Delta S_{surr}$, and typically we are only concerned with what happens in a particular system.

Therefore, considering only the system, we use another thermodynamic function to determine if a reaction will occur spontaneously. This function, called **Gibbs free energy (*G*)**, or simply free energy is given by the following equation:

$$G = H - TS$$

where,

$H$ is the enthalpy
$S$ is the entropy
$T$ is the temperature (in K)

The change in free energy ($\Delta G$) of a system for a reaction at constant temperature is:

$$\Delta G = \Delta H - T\Delta S \qquad (17.10, \text{text})$$

The sign of $\Delta G$ will allow us to predict whether a reaction is spontaneous. At constant temperature and pressure, we can summarize the following conditions in terms of $\Delta G$.

- $\Delta G < 0$ A spontaneous process in the forward direction.
- $\Delta G > 0$ A nonspontaneous reaction as written. However, the reaction is spontaneous in the reverse direction.
- $\Delta G = 0$ The system is at equilibrium. There is no net change in the system.

### A. Calculating standard free energy changes

Let's again suppose that the system is represented by the following reaction:

$$a\text{A} + b\text{B} \longrightarrow c\text{C} + d\text{D}$$

As is the case for the entropy of a reaction the standard free energy change $\Delta G°$ is given by:

$$\Delta G^\circ_{rxn} = [c\,\Delta G^\circ_f\,(\text{C}) + d\,\Delta G^\circ_f\,(\text{D})] - [a\,\Delta G^\circ_f\,(\text{A}) + b\,\Delta G^\circ_f\,(\text{B})]$$

or,

$$\Delta G^\circ_{rxn} = \Sigma n\Delta G^\circ_f\,(\text{products}) - \Sigma m\Delta G^\circ_f\,(\text{reactants}) \qquad (17.12, \text{text})$$

where $m$ and $n$ are stoichiometric coefficients. The term $\Delta G^\circ_f$ is the **standard free energy of formation** of a compound. It is the free energy change that occurs when 1 mole of the compound is synthesized from its elements in their standard states (see Table 17.2 of the text for conventions used for standard states). By definition, the standard free energy of formation of any element in its stable form is *zero*.

For example,

$$\Delta G^\circ_f\,[O_2(g)] = 0 \quad \text{and} \quad \Delta G^\circ_f\,[\text{Na}(s)] = 0$$

Other standard free energies of formation can be found in Appendix 3 of the text.

**EXAMPLE 17.3**

**Calculate $\Delta G^\circ_{rxn}$ at 25°C for the following reaction given that $\Delta G^\circ_f(Fe_2O_3) = -741.0$ kJ/mol.**

$$2Al(s) + Fe_2O_3(s) \longrightarrow Al_2O_3(s) + 2Fe(s)$$

---

**Strategy:** To calculate the standard free-energy change of a reaction, we look up the standard free energies of formation of reactants and products in Appendix 3 of the text and apply Equation (17.12) of the text. Note that all the stoichiometric coefficients have no units so $\Delta G^\circ_{rxn}$ is expressed in units of kJ/mol. The standard free energy of formation of any element in its stable allotropic form at 1 atm and 25°C is zero.

**Solution:** The standard free energy change for a reaction can be calculated using the following equation.

$$\Delta G^\circ_{rxn} = \Sigma n\Delta G^\circ_f(\text{products}) - \Sigma m\Delta G^\circ_f(\text{reactants})$$

$$\Delta G^\circ_{rxn} = \Delta G^\circ_f[Al_2O_3(s)] + 2\Delta G^\circ_f[Fe(s)] - \{2\Delta G^\circ_f[Al(s)] + \Delta G^\circ_f[Fe_2O_3(s)]\}$$

$$\Delta G^\circ_{rxn} = [(1)(-1576.41 \text{ kJ/mol}) + (2)(0)] - [(2)(0) + (1)(-741.0 \text{ kJ/mol})]$$

$$\Delta G^\circ_{rxn} = -1576.41 \text{ kJ/mol} + 741.0 \text{ kJ/mol} = \mathbf{-835.4 \text{ kJ/mol}}$$

**Tip:** Remember that the standard free energy of formation of any element in its stable form is *zero*. Therefore, the standard free energy of formation values for both Al(*s*) and Fe(*s*) are *zero*.

**PRACTICE EXERCISE**

3. Using Appendix 3 of the text, calculate $\Delta G^\circ_{rxn}$ values for the following reactions:

(a) $3CaO(s) + 2Al(s) \longrightarrow 3Ca(s) + Al_2O_3(s)$

(b) $ZnO(s) \longrightarrow Zn(s) + \frac{1}{2}O_2(g)$

**Text Problems: 17.18,** 17.32

## B. Calculating standard free energy change from enthalpy and entropy changes

We can determine the sign of $\Delta G$ if we know the signs of both $\Delta H$ and $\Delta S$. A negative $\Delta H$ (an exothermic reaction) and a positive $\Delta S$ (increase in disorder), give a negative $\Delta G$. In addition, temperature may influence the direction of a spontaneous reaction. There are four possible outcomes for the relationship

$$\Delta G = \Delta H - T\Delta S$$

- If both $\Delta H$ and $\Delta S$ are positive, $\Delta G$ will be negative only when $T\Delta S$ is greater in magnitude than $\Delta H$. This condition is met when $T$ is large (high temperature).
- If $\Delta H$ is positive and $\Delta S$ is negative, $\Delta G$ will always be positive, regardless of temperature. The reaction is nonspontaneous.
- If $\Delta H$ is negative and $\Delta S$ is positive, $\Delta G$ will always be negative, regardless of temperature. The reaction is spontaneous.
- If $\Delta H$ is negative and $\Delta S$ is negative, $\Delta G$ will be negative only when $\Delta H$ is greater in magnitude than $T\Delta S$. This condition is met when $T$ is small (low temperature).

We can also calculate the value of $\Delta G$ if we know the values of $\Delta H$, $\Delta S$, and the temperature. Typically, $\Delta H°$ is calculated from standard enthalpies of formation.

$$\Delta H^\circ_{rxn} = \sum n\Delta H^\circ_f\text{(products)} - \sum m\Delta H^\circ_f\text{(reactants)}$$

Also, $\Delta S°$ is calculated from standard entropy values (see Problem Type 1B).

$$\Delta S^\circ_{rxn} = \sum nS^\circ\text{(products)} - \sum mS^\circ\text{(reactants)}$$

Then, for reactions carried out under standard-state conditions, we can substitute $\Delta H°$, $\Delta S°$, and the temperature into the following equation to calculate $\Delta G°$.

$$\Delta G^\circ = \Delta H^\circ - T\Delta S^\circ \quad \text{(17.10, text)}$$

**EXAMPLE 17.4**

**Calculate $\Delta G°$ for the following reaction at 298 K:**

$$2H_2(g) + CO(g) \rightleftharpoons CH_3OH(g)$$

**given that $\Delta H° = -90.7$ kJ/mol and $\Delta S° = -221.5$ J/K·mol for this process.**

---

**Strategy:** The standard free energy change is given by, $\Delta G° = \Delta H° - T\Delta S°$ [Equation (17.10) of the text]. Make sure that the units of $\Delta H°$ and $\Delta S°$ are consistent. What unit of temperature should be used?

**Solution:** Let's convert $\Delta S°$ to units of kJ/K, so that we have consistent units.

$$-221.5\frac{\cancel{J}}{K\cdot mol}\times\frac{1\ kJ}{1000\ \cancel{J}} = -0.2215\ kJ/K\cdot mol$$

Substitute $\Delta H°$, $\Delta S°$, and the temperature (in K) into the Equation (17.10) to calculate $\Delta G°$.

$$\Delta G^\circ = \Delta H^\circ - T\Delta S^\circ$$

$$\Delta G^\circ = -90.7\ kJ/mol - [(298\ \cancel{K})(-0.2215\ kJ/\cancel{K}\cdot mol)]$$

$$\Delta G^\circ = -90.7\ kJ/mol + 66.0\ kJ/mol = \mathbf{-24.7\ kJ/mol}$$

**PRACTICE EXERCISE**

4. For the following reaction, $\Delta H° = -1204$ kJ/mol and $\Delta S° = -216.4$ J/K·mol.

$$2Mg(s) + O_2(g) \longrightarrow 2MgO(s)$$

At 25°C, are reactants or products favored at equilibrium under standard conditions?

**Text Problem: 17.20**

## C. Entropy changes due to phase transitions

At the transition temperature, the melting or boiling point, a system is at equilibrium and $\Delta G = 0$. Thus, we can write,

$$\Delta G = \Delta H - T\Delta S$$

or

$$0 = \Delta H - T\Delta S$$

and

$$\Delta S = \frac{\Delta H}{T}$$

$\Delta S$ is the entropy change due to the phase transition.

**EXAMPLE 17.5**

**The heat of fusion of water, $\Delta H_{fus}$, at 0°C is 6.01 kJ/mol. What is $\Delta S_{fus}$ for 1 mole of $H_2O$ at the melting point?**

**Strategy:** At the melting point, solid and liquid phase water are at equilibrium, so $\Delta G = 0$. From equation (17.10) of the text, we have $\Delta G = 0 = \Delta H - T\Delta S$ or $\Delta S = \Delta H/T$. To calculate the entropy change for the solid water → liquid water transition, we write $\Delta S_{fus} = \Delta H_{fus}/T$. What temperature unit should we use?

**Solution:** The entropy change due to the phase transition (the melting of water), can be calculated using the following equation. Recall that the temperature must be in units of Kelvin (0°C = 273 K).

$$\Delta S = \frac{\Delta H}{T}$$

$$\Delta S = \frac{6.01 \text{ kJ/mol}}{273 \text{ K}} = \mathbf{+0.0220 \ kJ/mol \cdot K = +22.0 \ J/mol \cdot K}$$

**Check:** The increase in entropy upon melting the solid corresponds to the increase in molecular disorder (increase in microstates) in the liquid state compared to the solid state.

**PRACTICE EXERCISE**

5. The enthalpy of vaporization of mercury is 58.5 kJ/mol and the normal boiling point is 630 K. What is the entropy of vaporization of mercury?

**Text Problem: 17.64**

## PROBLEM TYPE 3: FREE ENERGY AND CHEMICAL EQUILIBRIUM

### A. Using the standard free energy change to calculate the equilibrium constant

There is a relationship between the free energy change ($\Delta G$) and the standard free energy change.

$$\Delta G = \Delta G° + RT \ln Q \qquad (17.13, \text{text})$$

where,

*R* is the gas constant (8.314 J/mol·K)

*T* is the absolute temperature of the reaction (in K)

*Q* is the reaction quotient (see Chapter 14)

At equilibrium, by definition, $\Delta G = 0$ and $Q = K$, where *K* is the equilibrium constant. Thus,

$$0 = \Delta G° + RT \ln K$$

or

$$\Delta G° = -RT \ln K \qquad (17.14, \text{text})$$

$K$ can be either $K_P$, used for gases, or $K_c$ used for reactions in solution. Note that the larger the value of $K$, the more negative the value of $\Delta G°$. This should make sense because a large value of $K$ means that the equilibrium lies far to the right (toward products), and a negative value for $\Delta G°$ also means that products are favored at equilibrium. See Table 17.4 of the text for a summary of the relation between $K$ and $\Delta G°$.

**EXAMPLE 17.6**

**The standard free energy change for the reaction**

$$\tfrac{1}{2}N_2(g) + \tfrac{3}{2}H_2(g) \rightleftharpoons NH_3(g)$$

**is $\Delta G° = 26.9$ kJ/mol at 700 K. Calculate the equilibrium constant at this temperature.**

**Strategy:** According to Equation (17.14) of the text, the equilibrium constant for the reaction is related to the standard free energy change; that is, $\Delta G° = -RT\ln K$. Since we are given the free energy change in the problem, we can solve for equilibrium constant. What temperature unit should be used?

**Solution:** The equilibrium constant is related to the standard free energy change by the following equation.

$$\Delta G° = -RT\ln K$$

$R$ has units of J/mol·K, so we must convert $\Delta G°$ from units of kJ/mol to J/mol.

$$\frac{26.9\ \cancel{\text{kJ}}}{1\ \text{mol}} \times \frac{1000\ \text{J}}{1\ \cancel{\text{kJ}}} = 2.69 \times 10^4\ \text{J/mol}$$

Substitute $\Delta G°$, $R$, and $T$ into Equation (17.14) to calculate the equilibrium constant, $K_P$. We are calculating $K_P$ in this problem since this a gas-phase reaction.

$$\Delta G° = -RT\ln K_P$$

$$2.69 \times 10^4\ \cancel{\text{J}}/\cancel{\text{mol}} = -(8.314\ \cancel{\text{J}}/\cancel{\text{mol}}\cdot\cancel{\text{K}})(700\ \cancel{\text{K}})\ln K_P$$

$$-4.62 = \ln K_P$$

Taking the antilog of both sides gives:

$$e^{-4.62} = K_P$$

$$K_P = 9.9 \times 10^{-3}$$

**Check:** A positive $\Delta G°$ value indicates that reactants are favored at equilibrium. A small value for $K$ also indicates that reactants are favored at equilibrium.

**PRACTICE EXERCISE**

6. In Chapter 14, we saw that for the reaction

$$H_2(g) + I_2(g) \longrightarrow 2HI(g)$$

the equilibrium constant at 400°C is $K_P = 64$. Calculate the value of $\Delta G°$ at this temperature.

**Text Problems: 17.24**, 17.26, 17.28, 17.30

## B. Using the free energy change to predict the direction of a reaction

To predict the direction of a reaction, we can calculate $\Delta G$ from Equation (17.13) of the text.

$$\Delta G = \Delta G^\circ + RT \ln Q \qquad (17.13, \text{text})$$

To calculate the free energy change ($\Delta G$), you must be given or you must calculate $\Delta G^\circ$, and you must calculate the reaction quotient $Q$. Substitute the values of $\Delta G^\circ$ and $Q$ into Equation (17.13) to solve for $\Delta G$. Recall the meaning of the sign of $\Delta G$:

- $\Delta G < 0$ — A spontaneous process in the forward direction.
- $\Delta G > 0$ — A nonspontaneous reaction as written. However, the reaction is spontaneous in the reverse direction.
- $\Delta G = 0$ — The system is at equilibrium. There is no net change in the system.

**EXAMPLE 17.7**

**Using the reaction and data given in Example 17.6, calculate $\Delta G$ at 700 K if the reaction mixture consists of 30.0 atm of $H_2$, 20.0 atm of $N_2$, and 0.500 atm of $NH_3$.**

**Strategy:** From the information given we see that neither the reactants nor products are at their standard state of 1 atm. We use Equation (17.13) of the text to calculate the free-energy change under non-standard-state conditions. Note that the partial pressures are expressed as dimensionless quantities in the reaction quotient $Q_P$.

**Solution:** Under non-standard-state conditions, $\Delta G$ is related to the reaction quotient $Q$ by the following equation.

$$\Delta G = \Delta G^\circ + RT \ln Q_P$$

We are using $Q_P$ in the equation because this is a gas-phase reaction.

$\Delta G^\circ$ is given in Example 17.6. We must calculate $Q_P$.

$$\tfrac{1}{2}N_2(g) + \tfrac{3}{2}H_2(g) \rightleftharpoons NH_3(g)$$

$$Q_P = \frac{P_{NH_3}}{P_{N_2}^{\frac{1}{2}} \cdot P_{H_2}^{\frac{3}{2}}} = \frac{0.50}{(20.0)^{\frac{1}{2}}(30.0)^{\frac{3}{2}}} = 6.80 \times 10^{-4}$$

Substitute $\Delta G^\circ = 2.69 \times 10^4$ J/mol and $Q_P$ into Equation (17.13) to calculate $\Delta G$.

$$\Delta G = \Delta G^\circ + RT \ln Q_P$$

$$\Delta G = (2.69 \times 10^4 \text{ J/mol}) + (8.314 \text{ J/mol}\cdot\text{K})(700 \text{ K}) \ln(6.80 \times 10^{-4})$$

$$\boldsymbol{\Delta G} = (2.69 \times 10^4 \text{ J/mol}) - (4.24 \times 10^4 \text{ J/mol}) = \mathbf{-1.55 \times 10^4\ J/mol = -15.5\ kJ/mol}$$

**Comment:** By making the partial pressures of $N_2$ and $H_2$ high and that of $NH_3$ low, the reaction is spontaneous in the forward direction. This condition corresponds to $Q_P < K_P$, and so the reaction proceeds in the forward direction until $Q_P = K_P$ (see Chapter 14).

**PRACTICE EXERCISE**

**7.** Calculate $\Delta G$ for the following reaction at 25°C when the pressure of $CO_2$ is 0.0010 atm, given that $\Delta H° = 177.8$ kJ/mol and $\Delta S° = 160.5$ J/K·mol.

$$CaCO_3(s) \longrightarrow CaO(s) + CO_2(g)$$

**Text Problem: 17.28**

## ANSWERS TO PRACTICE EXERCISES

**1.** **(a)** entropy decreases **(b)** entropy increases **(c)** entropy increases

**2.** $\Delta S°_{rxn} = +145.1$ J/K·mol **3.** **(a)** $\Delta G°_{rxn} = 236$ kJ/mol **(b)** $\Delta G°_{rxn} = 318.2$ kJ/mol

**4.** $\Delta G° = -1.14 \times 10^3$ kJ/mol. Since $\Delta G° < 0$, products are favored at equilibrium.

**5.** $\Delta S_{vap} = 92.9$ J/mol·K **6.** $\Delta G° = -23.2$ kJ/mol **7.** $\Delta G = 113$ kJ/mol

## SOLUTIONS TO SELECTED TEXT PROBLEMS

**17.6** The probability ($P$) of finding all the molecules in the same flask becomes progressively smaller as the number of molecules increases. An equation that relates the probability to the number of molecules is given by:

$$P = \left(\frac{1}{2}\right)^N$$

where, $N$ is the total number of molecules present.

Using the above equation, we find:

**(a)** $P = 0.25$ **(b)** $P = 8 \times 10^{-31}$ **(c)** $P \approx 0$

Extending the calculation to a macroscopic system would result in such a small number for the probability (similar to the calculation with Avogadro's number above) that for all practical purposes, there is zero probability that all molecules would be found in the same bulb.

**17.10** In order of increasing entropy per mole at 25°C:

**(c) < (d) < (e) < (a) < (b)**

(c) Na($s$): ordered, crystalline material.

(d) NaCl($s$): ordered crystalline material, but with more particles per mole than Na(s).

(e) $H_2$: a diatomic gas, hence of higher entropy than a solid.

(a) Ne($g$): a monatomic gas of higher molar mass than $H_2$.

(b) $SO_2(g)$: a polyatomic gas of higher molar mass than Ne.

**17.12** **Strategy:** To calculate the standard entropy change of a reaction, we look up the standard entropies of reactants and products in Appendix 3 of the text and apply Equation (17.7). As in the calculation of enthalpy of reaction, the stoichiometric coefficients have no units, so $\Delta S^\circ_{rxn}$ is expressed in units of J/K·mol.

**Solution:** The standard entropy change for a reaction can be calculated using the following equation.

$$\Delta S^\circ_{rxn} = \Sigma nS^\circ(\text{products}) - \Sigma mS^\circ(\text{reactants})$$

**(a)** $\Delta S^\circ_{rxn} = S^\circ(Cu) + S^\circ(H_2O) - [S^\circ(H_2) + S^\circ(CuO)]$

$= (1)(33.3\ \text{J/K·mol}) + (1)(188.7\ \text{J/K·mol}) - [(1)(131.0\ \text{J/K·mol}) + (1)(43.5\ \text{J/K·mol})]$

$= \mathbf{47.5\ J/K{\cdot}mol}$

**(b)** $\Delta S^\circ_{rxn} = S^\circ(Al_2O_3) + 3S^\circ(Zn) - [2S^\circ(Al) + 3S^\circ(ZnO)]$

$= (1)(50.99\ \text{J/K·mol}) + (3)(41.6\ \text{J/K·mol}) - [(2)(28.3\ \text{J/K·mol}) + (3)(43.9\ \text{J/K·mol})]$

$= \mathbf{-12.5\ J/K{\cdot}mol}$

**(c)** $\Delta S^\circ_{rxn} = S^\circ(CO_2) + 2S^\circ(H_2O) - [S^\circ(CH_4) + 2S^\circ(O_2)]$

$= (1)(213.6\ \text{J/K·mol}) + (2)(69.9\ \text{J/K·mol}) - [(1)(186.2\ \text{J/K·mol}) + (2)(205.0\ \text{J/K·mol})]$

$= \mathbf{-242.8\ J/K{\cdot}mol}$

Why was the entropy value for water different in parts (a) and (c)?

**17.14** (a) $\Delta S < 0$; gas reacting with a liquid to form a solid (decrease in number of moles of gas, hence a decrease in microstates).

(b) $\Delta S > 0$; solid decomposing to give a liquid and a gas (an increase in microstates).

(c) $\Delta S > 0$; increase in number of moles of gas (an increase in microstates).

(d) $\Delta S < 0$; gas reacting with a solid to form a solid (decrease in number of moles of gas, hence a decrease in microstates).

**17.18** **Strategy:** To calculate the standard free-energy change of a reaction, we look up the standard free energies of formation of reactants and products in Appendix 3 of the text and apply Equation (17.12). Note that all the stoichiometric coefficients have no units so $\Delta G^\circ_{rxn}$ is expressed in units of kJ/mol. The standard free energy of formation of any element in its stable allotropic form at 1 atm and 25°C is zero.

**Solution:** The standard free energy change for a reaction can be calculated using the following equation.

$$\Delta G^\circ_{rxn} = \Sigma n\Delta G^\circ_f(\text{products}) - \Sigma m\Delta G^\circ_f(\text{reactants})$$

**(a)** $\Delta G^\circ_{rxn} = 2\Delta G^\circ_f(MgO) - [2\Delta G^\circ_f(Mg) + \Delta G^\circ_f(O_2)]$

$\Delta G^\circ_{rxn} = (2)(-569.6\ \text{kJ/mol}) - [(2)(0) + (1)(0)] = \mathbf{-1139\ kJ/mol}$

**(b)** $\Delta G^\circ_{rxn} = 2\Delta G^\circ_f(SO_3) - [2\Delta G^\circ_f(SO_2) + \Delta G^\circ_f(O_2)]$

$$\Delta G^\circ_{rxn} = (2)(-370.4 \text{ kJ/mol}) - [(2)(-300.4 \text{ kJ/mol}) + (1)(0)] = \mathbf{-140.0 \text{ kJ/mol}}$$

**(c)** $\Delta G^\circ_{rxn} = 4\Delta G^\circ_f[CO_2(g)] + 6\Delta G^\circ_f[H_2O(l)] - \{2\Delta G^\circ_f[C_2H_6(g)] + 7\Delta G^\circ_f[O_2(g)]\}$

$$\Delta G^\circ_{rxn} = (4)(-394.4 \text{ kJ/mol}) + (6)(-237.2 \text{ kJ/mol}) - [(2)(-32.89 \text{ kJ/mol}) + (7)(0)] = \mathbf{-2935.0 \text{ kJ/mol}}$$

**17.20** **Reaction A:** Calculate $\Delta G$ from $\Delta H$ and $\Delta S$.

$$\Delta G = \Delta H - T\Delta S = -126{,}000 \text{ J/mol} - (298 \text{ K})(84 \text{ J/K·mol}) = -151{,}000 \text{ J/mol}$$

The free energy change is negative so the reaction is spontaneous at 298 K. Since $\Delta H$ is negative and $\Delta S$ is positive, **the reaction is spontaneous at all temperatures.**

**Reaction B:** Calculate $\Delta G$.

$$\Delta G = \Delta H - T\Delta S = -11{,}700 \text{ J/mol} - (298 \text{ K})(-105 \text{ J/K·mol}) = +19{,}600 \text{ J}$$

The free energy change is positive at 298 K which means the reaction is not spontaneous at that temperature. The positive sign of $\Delta G$ results from the large negative value of $\Delta S$. At lower temperatures, the $-T\Delta S$ term will be smaller thus allowing the free energy change to be negative.

$\Delta G$ will equal zero when $\Delta H = T\Delta S$.

Rearranging,

$$T = \frac{\Delta H}{\Delta S} = \frac{-11700 \text{ J/mol}}{-105 \text{ J/K·mol}} = \mathbf{111 \text{ K}}$$

At temperatures **below 111 K**, $\Delta G$ will be negative and the reaction will be spontaneous.

**17.24** **Strategy:** According to Equation (17.14) of the text, the equilibrium constant for the reaction is related to the standard free energy change; that is, $\Delta G^\circ = -RT\ln K$. Since we are given the equilibrium constant in the problem, we can solve for $\Delta G^\circ$. What temperature unit should be used?

**Solution:** The equilibrium constant is related to the standard free energy change by the following equation.

$$\Delta G^\circ = -RT\ln K$$

Substitute $K_w$, $R$, and $T$ into the above equation to calculate the standard free energy change, $\Delta G^\circ$. The temperature at which $K_w = 1.0 \times 10^{-14}$ is 25°C = 298 K.

$$\Delta G^\circ = -RT\ln K_w$$

$$\Delta G^\circ = -(8.314 \text{ J/mol·K})(298 \text{ K})\ln(1.0 \times 10^{-14}) = \mathbf{8.0 \times 10^4 \text{ J/mol} = 8.0 \times 10^1 \text{ kJ/mol}}$$

**17.26** Use standard free energies of formation from Appendix 3 to find the standard free energy difference.

$$\Delta G^\circ_{\text{rxn}} = 2\Delta G^\circ_f[H_2(g)] + \Delta G^\circ_f[O_2(g)] - 2\Delta G^\circ_f[H_2O(g)]$$

$$\Delta G^\circ_{\text{rxn}} = (2)(0) + (1)(0) - (2)(-228.6\ \text{kJ/mol})$$

$$\boldsymbol{\Delta G^\circ_{\text{rxn}} = 457.2\ \text{kJ/mol} = 4.572 \times 10^5\ \text{J/mol}}$$

We can calculate $K_p$ using the following equation. We carry additional significant figures in the calculation to minimize rounding errors when calculating $K_p$.

$$\Delta G^\circ = -RT\ln K_p$$

$$4.572 \times 10^5\ \text{J/mol} = -(8.314\ \text{J/mol·K})(298\ \text{K})\ln K_p$$

$$-184.54 = \ln K_p$$

Taking the antiln of both sides,

$$e^{-184.54} = K_p$$

$$\boldsymbol{K_p = 7.2 \times 10^{-81}}$$

**17.28** **(a)** The equilibrium constant is related to the standard free energy change by the following equation.

$$\Delta G^\circ = -RT\ln K$$

Substitute $K_p$, $R$, and $T$ into the above equation to the standard free energy change, $\Delta G^\circ$.

$$\Delta G^\circ = -RT\ln K_p$$

$$\boldsymbol{\Delta G^\circ} = -(8.314\ \text{J/mol·K})(2000\ \text{K})\ln(4.40) = -2.464 \times 10^4\ \text{J/mol} = \boldsymbol{-24.6\ \text{kJ/mol}}$$

**(b)**

**Strategy:** From the information given we see that neither the reactants nor products are at their standard state of 1 atm. We use Equation (17.13) of the text to calculate the free-energy change under non-standard-state conditions. Note that the partial pressures are expressed as dimensionless quantities in the reaction quotient $Q_P$.

**Solution:** Under non-standard-state conditions, $\Delta G$ is related to the reaction quotient $Q$ by the following equation.

$$\Delta G = \Delta G^\circ + RT\ln Q_P$$

We are using $Q_p$ in the equation because this is a gas-phase reaction.

*Step 1:* $\Delta G^\circ$ was calculated in part (a). We must calculate $Q_P$. We carry additional significant figures in this calculation to minimize rounding errors.

$$Q_P = \frac{P_{H_2O} \cdot P_{CO}}{P_{H_2} \cdot P_{CO_2}} = \frac{(0.66)(1.20)}{(0.25)(0.78)} = 4.062$$

*Step 2:* Substitute $\Delta G° = -2.46 \times 10^4$ J/mol and $Q_P$ into the following equation to calculate $\Delta G$.

$$\Delta G = \Delta G° + RT\ln Q_P$$

$$\Delta G = -2.464 \times 10^4 \text{ J/mol} + (8.314 \text{ J/mol·K})(2000 \text{ K})\ln(4.062)$$

$$\Delta G = (-2.464 \times 10^4 \text{ J/mol}) + (2.331 \times 10^4 \text{ J/mol})$$

$$\mathbf{\Delta G = -1.33 \times 10^3 \text{ J/mol} = -1.33 \text{ kJ/mol}}$$

**17.30** We use the given $K_P$ to find the standard free energy change.

$$\Delta G° = -RT\ln K$$

$$\Delta G° = -(8.314 \text{ J/K·mol})(298 \text{ K})\ln(5.62 \times 10^{35}) = 2.04 \times 10^5 \text{ J/mol} = -204 \text{ kJ/mol}$$

The standard free energy of formation of one mole of $COCl_2$ can now be found using the standard free energy of reaction calculated above and the standard free energies of formation of $CO(g)$ and $Cl_2(g)$.

$$\Delta G°_{rxn} = \Sigma n\Delta G°_f(\text{products}) - \Sigma m\Delta G°_f(\text{reactants})$$

$$\Delta G°_{rxn} = \Delta G°_f[COCl_2(g)] - \{\Delta G°_f[CO(g)] + \Delta G°_f[Cl_2(g)]\}$$

$$-204 \text{ kJ/mol} = (1)\Delta G°_f[COCl_2(g)] - [(1)(-137.3 \text{ kJ/mol}) + (1)(0)]$$

$$\mathbf{\Delta G°_f[COCl_2(g)] = -341 \text{ kJ/mol}}$$

**17.32** The standard free energy change is given by:

$$\Delta G°_{rxn} = \Delta G°_f(\text{graphite}) - \Delta G°_f(\text{diamond})$$

You can look up the standard free energy of formation values in Appendix 3 of the text.

$$\mathbf{\Delta G°_{rxn}} = (1)(0) - (1)(2.87 \text{ kJ/mol}) = \mathbf{-2.87 \text{ kJ/mol}}$$

Thus, the formation of graphite from diamond is **favored** under standard-state conditions at 25°C. However, the rate of the diamond to graphite conversion is very slow (due to a high activation energy) so that it will take millions of years before the process is complete.

**17.36** The equation for the coupled reaction is:

**glucose + ATP → glucose 6−phosphate + ADP**

$$\Delta G° = 13.4 \text{ kJ/mol} - 31 \text{ kJ/mol} = -18 \text{ kJ/mol}$$

As an estimate:

$$\ln K = \frac{-\Delta G°}{RT}$$

$$\ln K = \frac{-(-18 \times 10^3 \text{ J/mol})}{(8.314 \text{ J/K}\cdot\text{mol})(298 \text{ K})} = 7.3$$

$$\mathbf{K = 1 \times 10^3}$$

**17.38** In each part of this problem we can use the following equation to calculate $\Delta G$.

$$\Delta G = \Delta G° + RT\ln Q$$

or,

$$\Delta G = \Delta G° + RT\ln[\text{H}^+][\text{OH}^-]$$

**(a)** In this case, the given concentrations are equilibrium concentrations at 25°C. Since the reaction is at equilibrium, $\boldsymbol{\Delta G = 0}$. This is advantageous, because it allows us to calculate $\Delta G°$. Also recall that at equilibrium, $Q = K$. We can write:

$$\Delta G° = -RT\ln K_w$$

$$\Delta G° = -(8.314 \text{ J/K}\cdot\text{mol})(298 \text{ K})\ln(1.0 \times 10^{-14}) = 8.0 \times 10^4 \text{ J/mol}$$

**(b)** $\Delta G = \Delta G° + RT\ln Q = \Delta G° + RT\ln[\text{H}^+][\text{OH}^-]$

$$\boldsymbol{\Delta G} = (8.0 \times 10^4 \text{ J/mol}) + (8.314 \text{ J/K}\cdot\text{mol})(298 \text{ K})\ln[(1.0 \times 10^{-3})(1.0 \times 10^{-4})] = \mathbf{4.0 \times 10^4 \text{ J/mol}}$$

**(c)** $\Delta G = \Delta G° + RT\ln Q = \Delta G° + RT\ln[\text{H}^+][\text{OH}^-]$

$$\boldsymbol{\Delta G} = (8.0 \times 10^4 \text{ J/mol}) + (8.314 \text{ J/K}\cdot\text{mol})(298 \text{ K})\ln[(1.0 \times 10^{-12})(2.0 \times 10^{-8})] = \mathbf{-3.2 \times 10^4 \text{ J/mol}}$$

**(d)** $\Delta G = \Delta G° + RT\ln Q = \Delta G° + RT\ln[\text{H}^+][\text{OH}^-]$

$$\boldsymbol{\Delta G} = (8.0 \times 10^4 \text{ J/mol}) + (8.314 \text{ J/K}\cdot\text{mol})(298 \text{ K})\ln[(3.5)(4.8 \times 10^{-4})] = \mathbf{6.4 \times 10^4 \text{ J/mol}}$$

**17.39** We use Equation (17.7) of the text to calculate $\Delta S°_{\text{soln}}$.

**(a)** $\Delta S°_{\text{soln}} = [S°(\text{NH}_4^+) + S°(\text{NO}_3^-)] - S°(\text{NH}_4\text{NO}_3)$

$$\Delta S°_{\text{soln}} = [(1)(112.8 \text{ J/K}\cdot\text{mol}) + (1)(146.4 \text{ J/K}\cdot\text{mol})] - (1)(151 \text{ J/K}\cdot\text{mol}) = \mathbf{108 \text{ J/K}\cdot\text{mol}}$$

**(b)** $\Delta S°_{\text{soln}} = [S°(\text{Fe}^{3+}) + 3S°(\text{Cl}^-)] - S°(\text{FeCl}_3)$

$$\Delta S°_{\text{soln}} = [(1)(-293.3 \text{ J/K}\cdot\text{mol}) + (3)(56.5 \text{ J/K}\cdot\text{mol})] - (1)(142.3 \text{ J/K}\cdot\text{mol}) = \mathbf{-266.1 \text{ J/K}\cdot\text{mol}}$$

Usually, the solution process leads to an increase in entropy, as in part (a). However, we must also consider hydration. For small, highly charged ions such as $Fe^{3+}$, the decrease in entropy due to hydration outweighs the increase in entropy due to mixing and dissociation. The overall entropy change is negative.

**17.40** Because the entropy of the system decreases, the entropy of the surroundings *must increase* for the reaction to be spontaneous. $\Delta S_{surr}$ is **positive.**

**17.42** One possible explanation is simply that no reaction is possible, namely that there is an unfavorable free energy difference between products and reactants ($\Delta G > 0$).

A second possibility is that the potential for spontaneous change is there ($\Delta G < 0$), but that the reaction is extremely slow (very large activation energy).

A remote third choice is that the student accidentally prepared a mixture in which the components were already at their equilibrium concentrations.

Which of the above situations would be altered by the addition of a catalyst?

**17.44** For a solid to liquid phase transition (melting) the entropy always increases ($\Delta S > 0$) and the reaction is always endothermic ($\Delta H > 0$).

**(a)** Melting is always spontaneous above the melting point, so $\Delta G < 0$.

**(b)** At the melting point (−77.7°C), solid and liquid are in equilibrium, so $\Delta G = 0$.

**(c)** Melting is not spontaneous below the melting point, so $\Delta G > 0$.

**17.46** If the process is *spontaneous* as well as *endothermic*, the signs of $\Delta G$ and $\Delta H$ must be negative and positive, respectively. Since $\Delta G = \Delta H - T\Delta S$, the sign of **$\Delta S$ must be positive ($\Delta S > 0$)** for $\Delta G$ to be negative.

**17.48** **(a)** Using the relationship:

$$\frac{\Delta H_{vap}}{T_{b.p.}} = \Delta S_{vap} \approx 90 \text{ J/K·mol}$$

benzene $\Delta S_{vap} = 87.8$ J/K·mol

hexane $\Delta S_{vap} = 90.1$ J/K·mol

mercury $\Delta S_{vap} = 93.7$ J/K·mol

toluene $\Delta S_{vap} = 91.8$ J/K·mol

Most liquids have $\Delta S_{vap}$ approximately equal to a constant value because the order of the molecules in the liquid state is similar. The order of most gases is totally random; thus, $\Delta S$ for liquid → vapor should be similar for most liquids.

**(b)** Using the data in Table 11.6 of the text, we find:

ethanol $\Delta S_{vap} = 111.9$ J/K·mol

water $\Delta S_{vap} = 109.4$ J/K·mol

Both water and ethanol have a larger $\Delta S_{vap}$ because the liquid molecules are more ordered due to hydrogen bonding (there are fewer microstates in these liquids).

**17.50** **(a)** $2CO + 2NO \rightarrow 2CO_2 + N_2$

**(b)** The oxidizing agent is NO; the reducing agent is CO.

**(c)** $\Delta G° = 2\Delta G_f^\circ(CO_2) + \Delta G_f^\circ(N_2) - 2\Delta G_f^\circ(CO) - 2\Delta G_f^\circ(NO)$

$$\Delta G° = (2)(-394.4 \text{ kJ/mol}) + (0) - (2)(-137.3 \text{ kJ/mol}) - (2)(86.7 \text{ kJ/mol}) = -687.6 \text{ kJ/mol}$$

$$\Delta G° = -RT\ln K_P$$

$$\ln K_P = \frac{6.876 \times 10^5 \text{ J/mol}}{(8.314 \text{ J/K}\cdot\text{mol})(298 \text{ K})} = 277.5$$

$$\mathbf{K_P = 3 \times 10^{120}}$$

**(d)** $$Q_P = \frac{P_{N_2}P_{CO_2}^2}{P_{CO}^2P_{NO}^2} = \frac{(0.80)(0.030)^2}{(5.0 \times 10^{-5})^2(5.0 \times 10^{-7})^2} = \mathbf{1.2 \times 10^{18}}$$

Since $Q_P << K_P$, the reaction will proceed from **left to right**.

**(e)** $\Delta H° = 2\Delta H_f^\circ(CO_2) + \Delta H_f^\circ(N_2) - 2\Delta H_f^\circ(CO) - 2\Delta H_f^\circ(NO)$

$$\Delta H° = (2)(-393.5 \text{ kJ/mol}) + (0) - (2)(-110.5 \text{ kJ/mol}) - (2)(90.4 \text{ kJ/mol}) = -746.8 \text{ kJ/mol}$$

Since $\Delta H°$ is negative, raising the temperature will decrease $K_P$, thereby increasing the amount of reactants and decreasing the amount of products. **No**, the formation of $N_2$ and $CO_2$ is not favored by raising the temperature.

**17.52** We can use data in Appendix 3 of the text to calculate the standard free energy change for the reaction. Then, we can use Equation (17.14) of the text to calculate the equilibrium constant, $K_{sp}$.

$$AgCl(s) \rightarrow Ag^+(aq) + Cl^-(aq)$$

$$\Delta G° = \Delta G_f^\circ(Ag^+) + \Delta G_f^\circ(Cl^-) - \Delta G_f^\circ(AgCl)$$

$$\Delta G° = [(1)(77.1 \text{ kJ/mol}) + (1)(-131.2 \text{ kJ/mol})] - (1)(-109.7 \text{ kJ/mol}) = 55.6 \text{ kJ/mol}$$

$$\Delta G° = -RT\ln K_{sp}$$

$$\ln K_{sp} = -\frac{55.6 \times 10^3 \text{ J/mol}}{(8.314 \text{ J/K}\cdot\text{mol})(298 \text{ K})} = -22.4$$

$$\mathbf{K_{sp} = 2 \times 10^{-10}}$$

This value of $K_{sp}$ matches the $K_{sp}$ value in Table 16.2 of the text.

**17.54** The equilibrium reaction is:

$$AgCl(s) \rightleftharpoons Ag^+(aq) + Cl^-(aq)$$

$$K_{sp} = [Ag^+][Cl^-] = 1.6 \times 10^{-10}$$

We can calculate the standard enthalpy of reaction from the standard enthalpies of formation in Appendix 3 of the text.

$$\Delta H° = \Delta H_f°(Ag^+) + \Delta H_f°(Cl^-) - \Delta H_f°(AgCl)$$

$$\Delta H° = (1)(105.9\ \text{kJ/mol}) + (1)(-167.2\ \text{kJ/mol}) - (1)(-127.0\ \text{kJ/mol}) = 65.7\ \text{kJ/mol}$$

From Problem 17.51(a):

$$\ln\frac{K_2}{K_1} = \frac{\Delta H°}{R}\left(\frac{T_2 - T_1}{T_1T_2}\right)$$

$K_1 = 1.6 \times 10^{-10}$ $T_1 = 298$ K

$K_2 = ?$ $T_2 = 333$ K

$$\ln\frac{K_2}{1.6 \times 10^{-10}} = \frac{6.57 \times 10^4\ \text{J}}{8.314\ \text{J/K}\cdot\text{mol}}\left(\frac{333\ \text{K} - 298\ \text{K}}{(333\ \text{K})(298\ \text{K})}\right)$$

$$\ln\frac{K_2}{1.6 \times 10^{-10}} = 2.79$$

$$\frac{K_2}{1.6 \times 10^{-10}} = e^{2.79}$$

$$\mathbf{K_2 = 2.6 \times 10^{-9}}$$

The increase in $K$ indicates that the solubility increases with temperature.

**17.56** Assuming that both $\Delta H°$ and $\Delta S°$ are temperature independent, we can calculate both $\Delta H°$ and $\Delta S°$.

$$\Delta H° = \Delta H_f°(CO) + \Delta H_f°(H_2) - [\Delta H_f°(H_2O) + \Delta H_f°(C)]$$

$$\Delta H° = (1)(-110.5\ \text{kJ/mol}) + (1)(0)] - [(1)(-241.8\ \text{kJ/mol}) + (1)(0)]$$

$$\Delta H° = 131.3\ \text{kJ/mol}$$

$$\Delta S° = S°(CO) + S°(H_2) - [S°(H_2O) + S°(C)]$$

$$\Delta S° = [(1)(197.9\ \text{J/K}\cdot\text{mol}) + (1)(131.0\ \text{J/K}\cdot\text{mol})] - [(1)(188.7\ \text{J/K}\cdot\text{mol}) + (1)(5.69\ \text{J/K}\cdot\text{mol})]$$

$$\Delta S° = 134.5\ \text{J/K}\cdot\text{mol}$$

It is obvious from the given conditions that the reaction must take place at a fairly high temperature (in order to have red–hot coke). Setting $\Delta G^\circ = 0$

$$0 = \Delta H^\circ - T\Delta S^\circ$$

$$T = \frac{\Delta H^\circ}{\Delta S^\circ} = \frac{131.3\ \text{kJ/mol} \times \dfrac{1000\ \text{J}}{1\ \text{kJ}}}{134.5\ \text{J/K}\cdot\text{mol}} = \mathbf{976\ K = 703^\circ C}$$

The temperature must be greater than 703°C for the reaction to be spontaneous.

**17.58** For a reaction to be spontaneous at constant temperature and pressure, $\Delta G < 0$. The process of crystallization proceeds with more order (less disorder), so $\boldsymbol{\Delta S < 0}$. We also know that

$$\Delta G = \Delta H - T\Delta S$$

Since $\Delta G$ must be negative, and since the entropy term will be positive ($-T\Delta S$, where $\Delta S$ is negative), then $\Delta H$ must be negative ($\boldsymbol{\Delta H < 0}$). The reaction will be exothermic.

**17.60** For the reaction to be spontaneous, $\Delta G$ must be negative.

$$\Delta G = \Delta H - T\Delta S$$

Given that $\Delta H = 19$ kJ/mol = 19,000 J/mol, then

$$\Delta G = 19{,}000\ \text{J/mol} - (273\ \text{K} + 72\ \text{K})(\Delta S)$$

Solving the equation with the value of $\Delta G = 0$

$$0 = 19{,}000\ \text{J/mol} - (273\ \text{K} + 72\ \text{K})(\Delta S)$$

$$\boldsymbol{\Delta S = 55\ \text{J/K}\cdot\text{mol}}$$

This value of $\Delta S$ which we solved for is the value needed to produce a $\Delta G$ value of zero. The *minimum* value of $\Delta S$ that will produce a spontaneous reaction will be any value of entropy *greater than* 55 J/K·mol.

**17.62** The second law states that the entropy of the universe must increase in a spontaneous process. But the entropy of the universe is the sum of two terms: the entropy of the system plus the entropy of the surroundings. One of the entropies can decrease, but not both. In this case, the decrease in system entropy is offset by an increase in the entropy of the surroundings. The reaction in question is exothermic, and the heat released raises the temperature (and the entropy) of the surroundings.

Could this process be spontaneous if the reaction were endothermic?

**17.64** **Strategy:** At the boiling point, liquid and gas phase ethanol are at equilibrium, so $\Delta G = 0$. From Equation (17.10) of the text, we have $\Delta G = 0 = \Delta H - T\Delta S$ or $\Delta S = \Delta H/T$. To calculate the entropy change for the liquid ethanol $\rightarrow$ gas ethanol transition, we write $\Delta S_{vap} = \Delta H_{vap}/T$. What temperature unit should we use?

**Solution:** The entropy change due to the phase transition (the vaporization of ethanol), can be calculated using the following equation. Recall that the temperature must be in units of Kelvin (78.3°C = 351 K).

$$\Delta S_{\text{vap}} = \frac{\Delta H_{\text{vap}}}{T_{\text{b.p.}}}$$

$$\Delta S_{\text{vap}} = \frac{39.3 \text{ kJ/mol}}{351 \text{ K}} = 0.112 \text{ kJ/mol}\cdot\text{K} = 112 \text{ J/mol}\cdot\text{K}$$

The problem asks for the change in entropy for the vaporization of 0.50 moles of ethanol. The $\Delta S$ calculated above is for 1 mole of ethanol.

$$\mathbf{\Delta S \text{ for } 0.50 \text{ mol}} = (112 \text{ J/}\cancel{\text{mol}}\cdot\text{K})(0.50 \cancel{\text{mol}}) = \mathbf{56 \text{ J/K}}$$

**17.66** For the given reaction we can calculate the standard free energy change from the standard free energies of formation (see Appendix 3 of the text). Then, we can calculate the equilibrium constant, $K_P$, from the standard free energy change.

$$\Delta G^\circ = \Delta G^\circ_f[\text{Ni(CO)}_4] - [4\Delta G^\circ_f(\text{CO}) + \Delta G^\circ_f(\text{Ni})]$$

$$\Delta G^\circ = (1)(-587.4 \text{ kJ/mol}) - [(4)(-137.3 \text{ kJ/mol}) + (1)(0)] = -38.2 \text{ kJ/mol} = -3.82 \times 10^4 \text{ J/mol}$$

Substitute $\Delta G^\circ$, $R$, and $T$ (in K) into the following equation to solve for $K_P$.

$$\Delta G^\circ = -RT\ln K_P$$

$$\ln K_P = \frac{-\Delta G^\circ}{RT} = \frac{-(-3.82 \times 10^4 \text{ J/mol})}{(8.314 \text{ J/K}\cdot\text{mol})(353 \text{ K})}$$

$$\mathbf{K_P = 4.5 \times 10^5}$$

**17.68** We carry additional significant figures throughout this calculation to minimize rounding errors. The equilibrium constant is related to the standard free energy change by the following equation:

$$\Delta G^\circ = -RT\ln K_P$$

$$2.12 \times 10^5 \cancel{\text{J/mol}} = -(8.314 \cancel{\text{J/mol}\cdot\text{K}})(298 \cancel{\text{K}})\ln K_P$$

$$K_P = 6.894 \times 10^{-38}$$

We can write the equilibrium constant expression for the reaction.

$$K_P = \sqrt{P_{O_2}}$$

$$P_{O_2} = (K_P)^2$$

$$P_{O_2} = (6.894 \times 10^{-38})^2 = \mathbf{4.8 \times 10^{-75} \text{ atm}}$$

This pressure is far too small to measure.

**17.70** Both (a) and (b) apply to a reaction with a negative $\Delta G°$ value. Statement (c) is not always true. An endothermic reaction that has a positive $\Delta S°$ (increase in entropy) will have a negative $\Delta G°$ value at high temperatures.

**17.72** We write the two equations as follows. The standard free energy change for the overall reaction will be the sum of the two steps.

$$\mathrm{CuO}(s) \rightleftharpoons \mathrm{Cu}(s) + \tfrac{1}{2}\mathrm{O_2}(g) \qquad \Delta G° = 127.2 \text{ kJ/mol}$$

$$\mathrm{C(graphite)} + \tfrac{1}{2}\mathrm{O_2}(g) \rightleftharpoons \mathrm{CO}(g) \qquad \Delta G° = -137.3 \text{ kJ/mol}$$

$$\mathbf{CuO + C(graphite) \rightleftharpoons Cu(s) + CO(g)} \qquad \Delta G° = -10.1 \text{ kJ/mol}$$

We can now calculate the equilibrium constant from the standard free energy change, $\Delta G°$.

$$\ln K = \frac{-\Delta G°}{RT} = \frac{-(-10.1\times 10^3 \text{ J/mol})}{(8.314 \text{ J/K}\cdot\text{mol})(673 \text{ K})}$$

$$\ln K = 1.81$$

$$\boldsymbol{K = 6.1}$$

**17.74** As discussed in Chapter 17 of the text for the decomposition of calcium carbonate, a reaction favors the formation of products at equilibrium when

$$\Delta G° = \Delta H° - T\Delta S° < 0$$

If we can calculate $\Delta H°$ and $\Delta S°$, we can solve for the temperature at which decomposition begins to favor products. We use data in Appendix 3 of the text to solve for $\Delta H°$ and $\Delta S°$.

$$\Delta H° = \Delta H_f°[\mathrm{MgO}(s)] + \Delta H_f°[\mathrm{CO_2}(g)] - \Delta H_f°[\mathrm{MgCO_3}(s)]$$

$$\Delta H° = -601.8 \text{ kJ/mol} + (-393.5 \text{ kJ/mol}) - (-1112.9 \text{ kJ/mol}) = 117.6 \text{ kJ/mol}$$

$$\Delta S° = S°[\mathrm{MgO}(s)] + S°[\mathrm{CO_2}(g)] - S°[\mathrm{MgCO_3}(s)]$$

$$\Delta S° = 26.78 \text{ J/K·mol} + 213.6 \text{ J/K·mol} - 65.69 \text{ J/K·mol} = 174.7 \text{ J/K·mol}$$

For the reaction to begin to favor products,

$$\Delta H° - T\Delta S° < 0$$

or

$$T > \frac{\Delta H°}{\Delta S°}$$

$$T > \frac{117.6\times 10^3 \text{ J/mol}}{174.7 \text{ J/K}\cdot\text{mol}}$$

$$\boldsymbol{T > 673.2 \text{ K}}$$

**17.76** **(a)** $\Delta G° = \Delta G_f°(H_2) + \Delta G_f°(Fe^{2+}) - [\Delta G_f°(Fe) + 2\Delta G_f°(H^+)]$

$\Delta G° = (1)(0) + (1)(-84.9\text{ kJ/mol}) - [(1)(0) + (2)(0)]$

$\Delta G° = -84.9\text{ kJ/mol}$

$\Delta G° = -RT\ln K$

$-84.9 \times 10^3\text{ J/mol} = -(8.314\text{ J/mol·K})(298\text{ K})\ln K$

$\mathbf{K = 7.6 \times 10^{14}}$

**(b)** $\Delta G° = \Delta G_f°(H_2) + \Delta G_f°(Cu^{2+}) - [\Delta G_f°(Cu) + 2\Delta G_f°(H^+)]$

$\Delta G° = 64.98\text{ kJ/mol}$

$\Delta G° = -RT\ln K$

$64.98 \times 10^3\text{ J/mol} = -(8.314\text{ J/mol·K})(298\text{ K})\ln K$

$\mathbf{K = 4.1 \times 10^{-12}}$

The activity series is correct. The very large value of $K$ for reaction (a) indicates that *products* are highly favored; whereas, the very small value of $K$ for reaction (b) indicates that *reactants* are highly favored.

**17.78** **(a)** It is a "reverse" disproportionation redox reaction.

**(b)** $\Delta G° = (2)(-228.6\text{ kJ/mol}) - [(2)(-33.0\text{ kJ/mol}) + (1)(-300.4\text{ kJ/mol})]$

$\Delta G° = -90.8\text{ kJ/mol}$

$-90.8 \times 10^3\text{ J/mol} = -(8.314\text{ J/mol·K})(298\text{ K})\ln K$

$\mathbf{K = 8.2 \times 10^{15}}$

Because of the large value of $K$, this method is efficient for removing $SO_2$.

**(c)** $\Delta H° = [(2)(-241.8\text{ kJ/mol}) + (3)(0)] - [(2)(-20.15\text{ kJ/mol}) + (1)(-296.1\text{ kJ/mol})]$

$\Delta H° = -147.2\text{ kJ/mol}$

$\Delta S° = [(2)(188.7\text{ J/K·mol}) + (3)(31.88\text{ J/K·mol})] - [(2)(205.64\text{ J/K·mol}) + (1)(248.5\text{ J/K·mol})]$

$\Delta S° = -186.7\text{ J/K·mol}$

$\Delta G° = \Delta H° - T\Delta S°$

Due to the negative entropy change, $\Delta S°$, the free energy change, $\Delta G°$, will become positive at higher temperatures. Therefore, the reaction will be **less effective** at high temperatures.

**17.80** $2O_3 \rightleftharpoons 3O_2$

$$\Delta G° = 3\Delta G_f°(O_2) - 2\Delta G_f°(O_3) = 0 - (2)(163.4\text{ kJ/mol})$$

$$\Delta G° = -326.8\text{ kJ/mol}$$

$$-326.8 \times 10^3\text{ J/mol} = -(8.314\text{ J/mol·K})(243\text{ K})\ln K_P$$

$$\mathbf{K_P = 1.8 \times 10^{70}}$$

Due to the large magnitude of $K$, you would expect this reaction to be spontaneous in the forward direction. However, this reaction has a **large activation energy**, so the rate of reaction is extremely slow.

**17.82** Heating the ore alone is not a feasible process. Looking at the coupled process:

| | |
|---|---|
| $Cu_2S \rightarrow 2Cu + S$ | $\Delta G° = 86.1\text{ kJ/mol}$ |
| $S + O_2 \rightarrow SO_2$ | $\Delta G° = -300.4\text{ kJ/mol}$ |
| $Cu_2S + O_2 \rightarrow 2Cu + SO_2$ | $\mathbf{\Delta G° = -214.3\text{ kJ/mol}}$ |

Since $\Delta G°$ is a large negative quantity, the coupled reaction is feasible for extracting sulfur.

**17.84** First, we need to calculate $\Delta H°$ and $\Delta S°$ for the reaction in order to calculate $\Delta G°$.

$$\Delta H° = -41.2\text{ kJ/mol} \qquad \Delta S° = -42.0\text{ J/K·mol}$$

Next, we calculate $\Delta G°$ at 300°C or 573 K, assuming that $\Delta H°$ and $\Delta S°$ are temperature independent.

$$\Delta G° = \Delta H° - T\Delta S°$$

$$\Delta G° = -41.2 \times 10^3\text{ J/mol} - (573\text{ K})(-42.0\text{ J/K·mol})$$

$$\Delta G° = -1.71 \times 10^4\text{ J/mol}$$

Having solved for $\Delta G°$, we can calculate $K_P$.

$$\Delta G° = -RT\ln K_P$$

$$-1.71 \times 10^4\text{ J/mol} = -(8.314\text{ J/K·mol})(573\text{ K})\ln K_P$$

$$\ln K_P = 3.59$$

$$\mathbf{K_P = 36}$$

Due to the negative entropy change calculated above, we expect that $\Delta G°$ will become positive at some temperature higher than 300°C. We need to find the temperature at which $\Delta G°$ becomes zero. This is the temperature at which reactants and products are equally favored ($K_P = 1$).

$$\Delta G° = \Delta H° - T\Delta S°$$

$$0 = \Delta H° - T\Delta S°$$

$$T = \frac{\Delta H°}{\Delta S°} = \frac{-41.2 \times 10^3 \text{ J/mol}}{-42.0 \text{ J/K} \cdot \text{mol}}$$

$$\mathbf{T = 981\ K = 708°C}$$

This calculation shows that at 708°C, $\Delta G° = 0$ and the equilibrium constant $K_p = 1$. Above 708°C, $\Delta G°$ is positive and $K_p$ will be smaller than 1, meaning that reactants will be favored over products. Note that the temperature 708°C is only an estimate, as we have assumed that both $\Delta H°$ and $\Delta S°$ are independent of temperature.

Using a more efficient catalyst will **not** increase $K_p$ at a given temperature, because the catalyst will speed up both the forward and reverse reactions. The value of $K_P$ will stay the same.

**17.86** Energy must be supplied from the surroundings to dissociate the reactant molecules into atoms. The kinetic energy of the surrounding molecules decreases and therefore the entropy of the surrounding molecules decreases. $\Delta S_{surr}$ is **negative.**

**17.88** butane → isobutane

$$\Delta G° = \Delta G_f°(\text{isobutane}) - \Delta G_f°(\text{butane})$$

$$\Delta G° = (1)(-18.0 \text{ kJ/mol}) - (1)(-15.9 \text{ kJ/mol})$$

$$\Delta G° = -2.1 \text{ kJ/mol}$$

For a mixture at equilibrium at 25°C:

$$\Delta G° = -RT\ln K_P$$

$$-2.1 \times 10^3 \text{ J/mol} = -(8.314 \text{ J/mol}\cdot\text{K})(298 \text{ K})\ln K_P$$

$$K_P = 2.3$$

$$K_P = \frac{P_{\text{isobutane}}}{P_{\text{butane}}} \propto \frac{\text{mol isobutane}}{\text{mol butane}}$$

$$2.3 = \frac{\text{mol isobutane}}{\text{mol butane}}$$

This shows that there are 2.3 times as many moles of isobutane as moles of butane. Or, we can say for every one mole of butane, there are 2.3 moles of isobutane.

$$\textbf{mol \% isobutane} = \frac{2.3 \text{ mol}}{2.3 \text{ mol} + 1.0 \text{ mol}} \times 100\% = \mathbf{70\%}$$

By difference, the mole % of butane is **30%**.

**Yes,** this result supports the notion that straight-chain hydrocarbons like butane are less stable than branched-chain hydrocarbons like isobutane.

**17.90** We use Equation (17.7) of the text to determine $S°$ for the one species not listed in each reaction.

(a) $$\Delta S^\circ_{rxn} = S°[Na(l)] - S°[Na(s)]$$

$$48.64 \text{ J/K}\cdot\text{mol} = (1)S°[Na(l)] - (1)(51.05 \text{ J/K}\cdot\text{mol})$$

$$\mathbf{S°[Na(l)] = 99.69 \text{ J/K•mol}}$$

(b) $$\Delta S^\circ_{rxn} = S°[S_2Cl_2(g)] - \{2S°[S(\text{monoclinic})] + S°[Cl_2(g)]\}$$

$$43.4 \text{ J/K}\cdot\text{mol} = (1)S°[S_2Cl_2(g)] - [(2)(32.55 \text{ J/K}\cdot\text{mol}) + (1)(223.0 \text{ J/K}\cdot\text{mol})]$$

$$\mathbf{S°[S_2Cl_2(g)] = 331.5 \text{ J/K•mol}}$$

(c) $$\Delta S^\circ_{rxn} = \{S°[Fe^{2+}(aq)] + 2S°[Cl^-(aq)]\} - S°[FeCl_2(s)]$$

$$-118.3 \text{ J/K}\cdot\text{mol} = (1)(-113.39 \text{ J/K}\cdot\text{mol}) + (2)(56.5 \text{ J/K}\cdot\text{mol}) - (1)S°[FeCl_2(s)]$$

$$\mathbf{S°[FeCl_2(s)] = 117.9 \text{ J/K•mol}}$$

**17.92** We can calculate $K_P$ from $\Delta G°$.

$$\Delta G° = [(1)(-394.4 \text{ kJ/mol}) + (0)] - [(1)(-137.3 \text{ kJ/mol}) + (1)(-255.2 \text{ kJ/mol})]$$

$$\Delta G° = -1.9 \text{ kJ/mol}$$

$$-1.9 \times 10^3 \text{ J/mol} = -(8.314 \text{ J/mol}\cdot\text{K})(1173 \text{ K})\ln K_P$$

$$K_P = 1.2$$

Now, from $K_P$, we can calculate the mole fractions of CO and $CO_2$.

$$K_P = \frac{P_{CO_2}}{P_{CO}} = 1.2 \qquad P_{CO_2} = 1.2P_{CO}$$

$$X_{CO} = \frac{P_{CO}}{P_{CO} + P_{CO_2}} = \frac{P_{CO}}{P_{CO} + 1.2P_{CO}} = \frac{1}{2.2} = \mathbf{0.45}$$

$$X_{CO_2} = 1 - 0.45 = \mathbf{0.55}$$

We assumed that $\Delta G°$ calculated from $\Delta G^\circ_f$ values was temperature independent. The $\Delta G^\circ_f$ values in Appendix 3 of the text are measured at 25°C, but the temperature of the reaction is 900°C.

**17.94** For a phase transition, $\Delta G = 0$. We write:

$$\Delta G = \Delta H - T\Delta S$$

$$0 = \Delta H - T\Delta S$$

$$\Delta S_{\text{sub}} = \frac{\Delta H_{\text{sub}}}{T}$$

Substituting $\Delta H$ and the temperature, $(-78° + 273°)\text{K} = 195\text{ K}$, gives

$$\Delta S_{\text{sub}} = \frac{\Delta H_{\text{sub}}}{T} = \frac{62.4 \times 10^3\text{ J/mol}}{195\text{ K}} = 3.20 \times 10^2\text{ J/K}\cdot\text{mol}$$

This value of $\Delta S_{\text{sub}}$ is for the sublimation of 1 mole of $CO_2$. We convert to the $\Delta S$ value for the sublimation of 84.8 g of $CO_2$.

$$84.8\text{ g CO}_2 \times \frac{1\text{ mol CO}_2}{44.01\text{ g CO}_2} \times \frac{3.20 \times 10^2\text{ J}}{\text{K}\cdot\text{mol}} = \mathbf{617\text{ J/K}}$$

**17.96** First, let's convert the age of the universe from units of years to units of seconds.

$$(13 \times 10^9\text{ yr}) \times \frac{365\text{ days}}{1\text{ yr}} \times \frac{24\text{ h}}{1\text{ day}} \times \frac{3600\text{ s}}{1\text{ h}} = 4.1 \times 10^{17}\text{ s}$$

The probability of finding all 100 molecules in the same flask is $8 \times 10^{-31}$. Multiplying by the number of seconds gives:

$$(8 \times 10^{-31})(4.1 \times 10^{17}\text{ s}) = \mathbf{3 \times 10^{-13}\text{ s}}$$

**17.98** We can calculate $\Delta S_{\text{sys}}$ from standard entropy values in Appendix 3 of the text. We can calculate $\Delta S_{\text{surr}}$ from the $\Delta H_{\text{sys}}$ value given in the problem. Finally, we can calculate $\Delta S_{\text{univ}}$ from the $\Delta S_{\text{sys}}$ and $\Delta S_{\text{surr}}$ values.

$$\Delta S_{\text{sys}} = (2)(69.9\text{ J/K}\cdot\text{mol}) - [(2)(131.0\text{ J/K}\cdot\text{mol}) + (1)(205.0\text{ J/K}\cdot\text{mol})] = \mathbf{-327\text{ J/K}\cdot\text{mol}}$$

$$\Delta S_{\text{surr}} = \frac{-\Delta H_{\text{sys}}}{T} = \frac{-(-571.6 \times 10^3\text{ J/mol})}{298\text{ K}} = \mathbf{1918\text{ J/K}\cdot\text{mol}}$$

$$\Delta S_{\text{univ}} = \Delta S_{\text{sys}} + \Delta S_{\text{surr}} = (-327 + 1918)\text{J/K}\cdot\text{mol} = \mathbf{1591\text{ J/K}\cdot\text{mol}}$$

**17.100** ***q*** and ***w*** are *not* state functions. Recall that state functions represent properties that are determined by the state of the system, regardless of how that condition is achieved. Heat and work are not state functions because they are not properties of the system. They manifest themselves only during a process (during a change). Thus their values depend on the path of the process and vary accordingly.

**17.102** Since the adsorption is spontaneous, $\Delta G$ must be negative ($\mathbf{\Delta G < 0}$). When hydrogen bonds to the surface of the catalyst, the system becomes more ordered ($\mathbf{\Delta S < 0}$). Since there is a decrease in entropy, the adsorption must be exothermic for the process to be spontaneous ($\mathbf{\Delta H < 0}$).

**17.104** **(a)** Each CO molecule has two possible orientations in the crystal,

CO or OC

If there is no preferred orientation, then for one molecule there are two, or $2^1$, choices of orientation. Two molecules have four or $2^2$ choices, and for 1 mole of CO there are $2^{N_A}$ choices. From Equation (17.1) of the text:

$$S = k\ln W$$

$$S = (1.38\times10^{-23}\text{ J/K})\ln 2^{6.022\times10^{23}}$$

$$S = (1.38\times10^{-23}\text{ J/K})(6.022\times10^{23}\text{ /mol})\ln 2$$

$$S = \mathbf{5.76\ J/K{\cdot}mol}$$

**(b)** The fact that the actual residual entropy is 4.2 J/K·mol means that the orientation is not totally random.

**17.106** We use data in Appendix 3 of the text to calculate $\Delta H°$ and $\Delta S°$.

$$\Delta H° = \Delta H_{vap} = \Delta H_f°[C_6H_6(g)] - \Delta H_f°[C_6H_6(l)]$$

$$\boldsymbol{\Delta H°} = 82.93\text{ kJ/mol} - 49.04\text{ kJ/mol} = \mathbf{33.89\ kJ/mol}$$

$$\Delta S° = S°[C_6H_6(g)] - S°[C_6H_6(l)]$$

$$\boldsymbol{\Delta S°} = 269.2\text{ J/K·mol} - 172.8\text{ J/K·mol} = \mathbf{96.4\ J/K{\cdot}mol}$$

We can now calculate $\Delta G°$ at 298 K.

$$\Delta G° = \Delta H° - T\Delta S°$$

$$\Delta G° = 33.89\text{ kJ/mol} - (298\text{ K})(96.4\text{ J/K·mol})\times\frac{1\text{ kJ}}{1000\text{ J}}$$

$$\boldsymbol{\Delta G°} = \mathbf{5.2\ kJ/mol}$$

$\Delta H°$ is positive because this is an endothermic process. We also expect $\Delta S°$ to be positive because this is a liquid → vapor phase change. $\Delta G°$ is positive because we are at a temperature that is below the boiling point of benzene (80.1°C).

**17.108** We can calculate $\Delta G°$ at 872 K from the equilibrium constant, $K_1$.

$$\Delta G° = -RT\ln K$$

$$\Delta G° = -(8.314\text{ J/mol·K})(872\text{ K})\ln(1.80\times10^{-4})$$

$$\boldsymbol{\Delta G°} = 6.25\times10^4\text{ J/mol} = \mathbf{62.5\ kJ/mol}$$

We use the equation derived in Problem 17.51 to calculate $\Delta H°$.

$$\ln\frac{K_2}{K_1} = \frac{\Delta H°}{R}\left(\frac{1}{T_1} - \frac{1}{T_2}\right)$$

$$\ln\frac{0.0480}{1.80 \times 10^{-4}} = \frac{\Delta H°}{8.314 \text{ J/mol}\cdot\text{K}}\left(\frac{1}{872 \text{ K}} - \frac{1}{1173 \text{ K}}\right)$$

$$\Delta H° = \mathbf{157.8 \text{ kJ/mol}}$$

Now that both $\Delta G°$ and $\Delta H°$ are known, we can calculate $\Delta S°$ at 872 K.

$$\Delta G° = \Delta H° - T\Delta S°$$

$$62.5 \times 10^3 \text{ J/mol} = (157.8 \times 10^3 \text{ J/mol}) - (872 \text{ K})\Delta S°$$

$$\Delta S° = \mathbf{109 \text{ J/K}\cdot\text{mol}}$$

**17.110** We can solve for the minimum partial pressure of $N_2$ by first assuming that the system is at equilibrium. From the given value of $K_p$ we write:

$$K_P = \frac{P_{NH_3}^2}{P_{N_2} \cdot P_{H_2}^3}$$

$$2.4 \times 10^{-3} = \frac{(2.1 \times 10^{-2})^2}{P_{N_2}(1.52)^3}$$

$$P_{N_2} = \mathbf{0.052 \text{ atm}}$$

For the reaction to proceed spontaneously in the forward direction, the nitrogen gas pressure must be slightly larger than 0.052 atm.

An alternative way to solve this problem is to use Equation (17.13) of the text. We first set $\Delta G = 0$, calculate $\Delta G°$ from the $K_p$ value, and then solve for pressure of nitrogen gas. We again obtain 0.052 atm for the pressure of nitrogen gas. A slight increase in the nitrogen pressure (from 0.052 atm) will result in $\Delta G < 0$ so the reaction is spontaneous in the forward direction.

# CHAPTER 18
# ELECTROCHEMISTRY

## PROBLEM-SOLVING STRATEGIES AND TUTORIAL SOLUTIONS

### TYPES OF PROBLEMS

**Problem Type 1:** Balancing Redox Equations.

**Problem Type 2:** Standard Reduction Potentials.
- (a) Comparing strengths of oxidizing agents.
- (b) Calculating the standard emf ($E°$) of a galvanic cell.

**Problem Type 3:** Spontaneity of Redox Reactions.
- (a) Predicting whether a redox reaction is spontaneous.
- (b) Calculating $\Delta G°$ and $K$ from $E°$.

**Problem Type 4:** The Nernst Equation.
- (a) Using the Nernst equation to predict the spontaneity of a redox reaction.
- (b) Using the Nernst equation to calculate concentration.

**Problem Type 5:** Electrolysis.
- (a) Predicting the products of electrolysis.
- (b) Calculating the quantity of products in electrolysis.

## PROBLEM TYPE 1: BALANCING REDOX EQUATIONS

We will use the **ion-electron method** to balance redox reactions. In this approach, the overall reaction is divided into two half-reactions, one for oxidation and one for reduction. The two equations are balanced separately and then added together to give the overall balanced equation. Example 18.1 demonstrates how to balance a redox reaction by the ion-electron method.

**EXAMPLE 18.1**

**Balance the following redox reaction in an acidic solution.**

$$\mathbf{Sn + NO_3^- \longrightarrow Sn^{2+} + NO}$$

**Strategy:** We follow the procedure for balancing redox reactions presented in Section 18.1 of the text.

**Solution:**

*Step 1:* Write the unbalanced equation for the reaction in ionic form.

The equation given in the problem is already in ionic form.

*Step 2:* Separate the equation into two half-reactions.

$$\overset{0}{\text{Sn}} \xrightarrow{\text{oxidation}} \text{Sn}^{2+}$$

$$\overset{+5}{\text{N}}\text{O}_3^- \xrightarrow{\text{reduction}} \overset{+2}{\text{N}}\text{O}$$

*Step 3:* We balance each half-reaction for number and type of atoms and charges.

The *oxidation half-reaction* is already balanced for Sn atoms. There are two net positive charges on the right, so we add two electrons to the same side to balance the charge.

$$Sn \longrightarrow Sn^{2+} + 2e^-$$

The *reduction half-reaction* is balanced for nitrogen atoms. To balance the O atoms, we add two water molecules on the right side.

$$NO_3^- \longrightarrow NO + 2H_2O$$

To balance the H atoms, we add 4 $H^+$ to the left-hand side.

$$4H^+ + NO_3^- \longrightarrow NO + 2H_2O$$

There are three net positive charges on the left, so we add three electrons to the same side to balance the charge.

$$4H^+ + NO_3^- + 3e^- \longrightarrow NO + 2H_2O$$

*Step 4:* We now add the oxidation and reduction half-reactions to give the overall reaction. In order to equalize the number of electrons, we need to multiply the oxidation half-reaction by 3 and the reduction half-reaction by 2.

$$3Sn \longrightarrow 3Sn^{2+} + 6e^-$$

$$8H^+ + 2NO_3^- + 6e^- \longrightarrow 2NO + 4H_2O$$

---

$$8H^+ + 2NO_3^- + 3Sn + 6e^- \longrightarrow 2NO + 3Sn^{2+} + 4H_2O + 6e^-$$

The electrons on both sides cancel, and we are left with the balanced net ionic equation:

$$\mathbf{8H^+ + 2NO_3^- + 3Sn \longrightarrow 2NO + 3Sn^{2+} + 4H_2O}$$

*Step 5:* Check to see that the equation is balanced by verifying that the equation has the same types and numbers of atoms and the same charges on both sides of the equation.

The equation is "atomically" balanced. There are 8 H, 2 N, 6 O, and 3 Sn atoms on each side of the equation. The equation is also "electrically" balanced. The net charge is +6 on each side of the equation.

**PRACTICE EXERCISE**

1. The redox reaction between permanganate ion and iron(II) ions in acidic solution can be used to analyze iron ore for its iron content. Balance this redox reaction.

$$MnO_4^-(aq) + Fe^{2+}(aq) \longrightarrow Fe^{3+}(aq) + Mn^{2+}(aq)$$

**Text Problem: 18.2**

## PROBLEM TYPE 2: STANDARD REDUCTION POTENTIALS

### A. Comparing strengths of oxidizing agents

For a reduction reaction at an electrode when all solutes are 1 *M* and all gases are at 1 atm, the voltage is called the **standard reduction potential**. Table 18.1 of the text lists the standard reduction potentials for a number of half-reactions. The more positive the value of $E°$, the greater the tendency for the substance to be reduced, and therefore, the stronger its tendency to act as an oxidizing agent. Looking at Table 18.1, we see that $F_2$ is the strongest oxidizing agent and $Li^+$ the weakest.

**EXAMPLE 18.2**

**Arrange the following species in order of increasing strength as oxidizing agents under standard-state conditions: $Zn^{2+}$, $MnO_4^-$ (in acid solution), and $Ag^+$.**

**Strategy:** The greater the tendency for the substance to be reduced, the stronger its tendency to act as an oxidizing agent. The species that has a stronger tendency to be reduced will have a larger reduction potential.

**Solution:** Consulting Table 18.1 of the text, we write the half-reactions in the order of decreasing standard reduction potentials, $E°$.

$$MnO_4^-(aq) + 8H^+(aq) + 5e^- \longrightarrow Mn^{2+}(aq) + 4H_2O(l) \qquad E° = +1.51\text{ V}$$

$$Ag^+(aq) + e^- \longrightarrow Ag(s) \qquad E° = +0.80\text{ V}$$

$$Zn^{2+}(aq) + 2e^- \longrightarrow Zn(s) \qquad E° = -0.76\text{ V}$$

Since the reduction of $MnO_4^-$ has the highest reduction potential and $Zn^{2+}$ has the lowest reduction potential, the order of increasing strength of oxidizing agents is:

$$Zn^{2+} < Ag^+ < MnO_4^-$$

The large positive reduction potential for $MnO_4^-$ indicates the strong tendency for permanganate ion to be reduced, and therefore the stronger its tendency to act as an oxidizing agent.

**PRACTICE EXERCISE**

2. Arrange the following species in order of increasing strength as oxidizing agents: $Ce^{4+}$, $O_2$, $H_2O_2$, and $SO_4^{2-}$.

| **Text Problems:** 18.14, **18.18** |
|---|

## B. Calculating the standard emf ($E°$) of a galvanic cell

In a galvanic cell, electrons flow from one electrode to the other. This indicates that there is a voltage difference between the two electrodes. This voltage difference is called the **electromotive force**, or **emf (*E*)**, and it can be measured by connecting a voltmeter to both electrodes. The electromotive force is also called the cell voltage or cell potential. It is usually measured in volts. The electrode at which reduction occurs is called the ***cathode***, and the electrode at which oxidation occurs is called the ***anode***.

If all solutes have a concentration of 1 *M* and all gases have a pressure of 1 atm (standard conditions), the voltage difference between the two electrodes of the cell is called the **standard emf ($E°_{cell}$)**. The ***standard emf*** which is composed of a contribution from the *anode* and a contribution from the *cathode*, is given by:

$$E°_{cell} = E°_{cathode} - E°_{anode}$$

where $E°_{cathode}$ and $E°_{anode}$ are the standard reduction potentials of the cathode and anode, respectively.

Under standard-state conditions, a positive $E°_{cell}$ means the redox reaction will favor the formation of products at equilibrium. Conversely, a negative standard cell emf, means that more reactants than products will be formed at equilibrium.

We can calculate the standard emf of the cell using a table of standard reduction potentials (Table 18.1 of the text). As an example, consider the galvanic cell represented by the following reaction:

$$Cu(s) + 2Ag^+(aq) \longrightarrow Cu^{2+}(aq) + 2Ag(s)$$

Let's break this reaction down into its two half-reactions.

$$Cu(s) \xrightarrow{\text{oxidation (anode)}} Cu^{2+}(aq) + 2e^-$$

$$2Ag^+(aq) + 2e^- \xrightarrow{\text{reduction (cathode)}} 2Ag(s)$$

We can calculate the standard cell emf by subtracting $E^\circ_{anode}$ from $E^\circ_{cathode}$.

$$E^\circ_{cell} = E^\circ_{cathode} - E^\circ_{anode} = E^\circ_{Ag^+/Ag} - E^\circ_{Cu^{2+}/Cu}$$

$$E^\circ_{cell} = 0.80\text{ V} - 0.34\text{ V} = 0.46\text{ V}$$

**EXAMPLE 18.3**

**Calculate the standard cell potential for a galvanic cell in which the following reaction takes place:**

$$\mathbf{Cl_2(g) + 2Br^-(aq) \longrightarrow Br_2(l) + 2Cl^-(aq)}$$

**Strategy:** Separate the reaction into half-reactions. Then look up reduction potentials in Table 18.1 of the text.

**Solution:**

$$Cl_2(g) + 2e^- \xrightarrow{\text{reduction (cathode)}} 2Cl^-(aq)$$

$$2Br^-(aq) \xrightarrow{\text{oxidation (anode)}} Br_2(l) + 2e^-$$

We can look up standard reduction potentials in Table 18.1 of the text.

$$E^\circ_{cathode} = E^\circ_{Cl_2/Cl^-} = +1.36\text{ V}$$

$$E^\circ_{anode} = E^\circ_{Br_2/Br^-} = +1.07\text{ V}$$

The standard cell emf is given by

$$E^\circ_{cell} = E^\circ_{cathode} - E^\circ_{anode} = E^\circ_{Cl_2/Cl^-} - E^\circ_{Br_2/Br^-}$$

$$E^\circ_{cell} = +1.36\text{ V} - 1.07\text{ V} = \mathbf{+0.29\text{ V}}$$

**PRACTICE EXERCISE**

3. Consider a uranium-bromine galvanic cell in which U is oxidized and $Br_2$ is reduced. The reduction half-reactions are:

$$U^{3+}(aq) + 3e^- \longrightarrow U(s) \qquad E^\circ_{U^{3+}/U} = ?$$

$$Br_2(l) + 2e^- \longrightarrow 2Br^-(aq) \qquad E^\circ_{Br_2/Br^-} = 1.07\text{ V}$$

If the standard cell emf is 2.91 V, what is the standard reduction potential for uranium?

**Text Problem: 18.12**

## PROBLEM TYPE 3: SPONTANEITY OF REDOX REACTIONS

### A. Predicting whether a redox reaction is spontaneous

There is a relationship between free energy change and cell emf.

$$\Delta G = -nFE_{\text{cell}} \qquad (18.2, \text{text})$$

where,

$n$ is the number of moles of electrons transferred during the redox reaction
$F$ is the Faraday constant, which is the electrical charge contained in 1 mole of electrons

$$1\ F = 96{,}500\ \text{C/mol} = 96{,}500\ \text{J/V·mol}$$

For a derivation of this relationship, see Section 18.4 of the text. Both $n$ and $F$ are positive quantities, and we know from Chapter 17 that $\Delta G$ is *negative* for a spontaneous process. Therefore, $E_{\text{cell}}$ is *positive* for a spontaneous process.

For reactions in which reactants and products are in their standard states, Equation (18.2) of the text becomes:

$$\Delta G^\circ = -nFE^\circ_{\text{cell}} \qquad (18.3, \text{text})$$

A *negative* $\Delta G^\circ$ and a *positive* $E^\circ_{\text{cell}}$ mean that formation of products is favored at equilibrium. A *positive* $\Delta G^\circ$ and a *negative* $E^\circ_{\text{cell}}$ mean that formation of reactants is favored at equilibrium. Table 18.2 of the text summarizes the relationships between $\Delta G^\circ$ and $E^\circ_{\text{cell}}$.

**EXAMPLE 18.4**
**Predict whether products or reactants are favored at equilibrium when the following reaction is run under standard-state conditions.**

$$Fe^{2+} + Cr_2O_7^{2-} \longrightarrow Fe^{3+} + Cr^{3+}$$

**Strategy:** A positive $E^\circ_{\text{cell}}$ means the reaction will favor the formation of products at equilibrium. Calculate the standard cell emf from the potentials for the two half-reactions.

$$E^\circ_{\text{cell}} = E^\circ_{\text{cathode}} - E^\circ_{\text{anode}}$$

**Solution:** Separate the reaction into half-reactions to calculate the standard cell emf.

$$Fe^{2+}(aq) \xrightarrow{\text{oxidation (anode)}} Fe^{3+}(aq) + e^- \qquad E^\circ_{\text{anode}} = +0.77\ \text{V}$$

At this point, we could balance the reduction half-reaction and then come up with the overall balanced equation. But, we are not asked to do that. We can save time by looking at Table 18.1 of the text and finding a reaction that contains both $Cr_2O_7^{2-}$ and $Cr^{3+}$.

$$Cr_2O_7^{2-}(aq) + 14H^+(aq) + 6e^- \xrightarrow{\text{reduction (cathode)}} 2Cr^{3+}(aq) + 7H_2O(l)$$

$$E^\circ_{\text{cathode}} = +1.33\ \text{V}$$

We have not come up with a balanced equation, but we do not need a balanced equation to calculate $E^\circ_{\text{cell}}$.

$$E^\circ_{\text{cell}} = E^\circ_{\text{cathode}} - E^\circ_{\text{anode}}$$

$$E^\circ_{\text{cell}} = 1.33\ \text{V} - 0.77\ \text{V} = \mathbf{+0.56\ V}$$

Since $E^\circ_{\text{cell}}$ is positive, products are favored at equilibrium.

**PRACTICE EXERCISE**

4. Predict whether a spontaneous reaction will occur when the following reactants and products are in their standard states.

(a) $2Fe^{3+}(aq) + 2I^-(aq) \longrightarrow 2Fe^{2+}(aq) + I_2(s)$

(b) $Cu(s) + 2H^+(aq) \longrightarrow Cu^{2+}(aq) + H_2(g)$

**Text Problem: 18.16**

## B. Calculating $\Delta G°$ and $K$ from $E°$

Equation (18.3) of the text shows the relationship between standard free energy change and standard emf.

$$\Delta G° = -nFE°_{cell} \qquad (18.3, text)$$

Also recall that in Section 18.6 of the text, we saw that the standard free energy change for a reaction is related to its equilibrium constant $K$ by the following equation:

$$\Delta G° = -RT \ln K$$

Therefore, from the two equations we obtain:

$$-nFE°_{cell} = -RT \ln K$$

Solving for $E°_{cell}$, we obtain:

$$E°_{cell} = \frac{RT}{nF} \ln K \qquad (18.4, text)$$

At 298 K, we can simplify Equation (18.4) of the text by substituting for $R$ and $F$.

$$E°_{cell} = \frac{(8.314 \text{ J/mol}\cdot\text{K})(298 \text{ K})}{n(96{,}500 \text{ J/V}\cdot\text{mol})} \ln K$$

$$E°_{cell} = \frac{0.0257 \text{ V}}{n} \ln K \qquad (18.5, text)$$

We now have relationships between $\Delta G°$ and $E°_{cell}$, between $\Delta G°$ and $K$, and between $E°_{cell}$ and $K$. Figure 18.5 of the text summarizes the relationships among $\Delta G°$, $K$, and $E°_{cell}$.

**EXAMPLE 18.5**

**Calculate $\Delta G°$ and the equilibrium constant at 25°C for the reaction**

$$2Br^-(aq) + I_2(s) \longrightarrow Br_2(l) + 2I^-(aq)$$

**Strategy:** The relationship between the standard free energy change and the standard emf of the cell is given by Equation (18.3) of the text: $\Delta G° = -nFE°_{cell}$. The relationship between the equilibrium constant, $K$, and the standard emf is given by Equation (18.5) of the text: $E°_{cell} = (0.0257 \text{ V}/n)\ln K$. Thus, if we can determine $E°_{cell}$, we can calculate $\Delta G°$ and $K$. We can determine the $E°_{cell}$ of a hypothetical galvanic cell made up of two couples ($Br_2/Br^-$ and $I_2/I^-$) from the standard reduction potentials in Table 18.1 of the text.

**Solution:** The half-cell reactions are:

$$I_2(s) + 2e^- \xrightarrow{\text{reduction (cathode)}} 2I^-(aq) \qquad E°_{cathode} = +0.53\ V$$

$$2Br^-(aq) \xrightarrow{\text{oxidation (anode)}} Br_2(l) + 2e^- \qquad E°_{anode} = +1.07\ V$$

$$2Br^-(aq) + I_2(s) \longrightarrow Br_2(l) + 2I^-(aq) \qquad E°_{cell} = E°_{cathode} - E°_{anode} = -0.54\ V$$

Substitute $E°_{cell}$ into Equation (18.3) of the text to calculate the standard free energy change. In the balanced equation above, 2 moles of electrons are transferred. Therefore, $n = 2$.

$$\Delta G° = -nFE°_{cell}$$

$$\Delta G° = -(2)(96{,}500\ J/V{\cdot}mol)(-0.54\ V)$$

$$\boldsymbol{\Delta G° = 1.04 \times 10^5\ J/mol = 104\ kJ/mol}$$

Both the positive value of $\Delta G°$ and the negative value for $E°_{cell}$ indicate that the reactants are favored at equilibrium under standard-state conditions.

Rearrange Equation (18.5) of the text to solve for the equilibrium constant, $K$.

$$\ln K = \frac{nE°}{0.0257\ V}$$

$$\ln K = \frac{(2)(-0.54\ V)}{0.0257\ V} = -42.0$$

Taking the anti-ln of both sides of the equation,

$$\boldsymbol{K = e^{-42.0} = 6 \times 10^{-19}}$$

You should notice that the magnitude of $K$ relates to what we have learned from $\Delta G°$ and $E°_{cell}$. A very small value for the equilibrium constant indicates that reactants are highly favored at equilibrium. This agrees with the signs of $\Delta G°$ and $E°_{cell}$, which indicate that more reactants than products will be formed at equilibrium.

**PRACTICE EXERCISE**

5. Calculate the equilibrium constant for the following redox reaction at 25°C.

$$2Fe^{2+}(aq) + Ni^{2+}(aq) \longrightarrow 2Fe^{3+}(aq) + Ni(s)$$

**Text Problems: 18.24**, 18.26, **18.28**, 18.40

## PROBLEM TYPE 4: THE NERNST EQUATION

Thus far, we have only focused on redox reactions in which the reactants and products are in their standard states. However, standard-state conditions are often difficult and sometimes impossible to maintain. Therefore, we need a relationship between cell emf and the concentrations of reactants and products under *nonstandard-state* conditions.

In Chapter 17, we encountered a relationship between the free energy change ($\Delta G$) and the standard free energy change ($\Delta G°$).

$$\Delta G = \Delta G° + RT\ln Q$$

where,

$Q$ is the reaction quotient.

We also know that:

$$\Delta G = -nFE_{\text{cell}}$$

and

$$\Delta G° = -nFE°_{\text{cell}}$$

Substituting for $\Delta G$ and $\Delta G°$ in the first equation, we find:

$$-nFE_{\text{cell}} = -nFE°_{\text{cell}} + RT\ln Q$$

Dividing both sides of the equation by $-nF$ and omitting "cell" for simplicity gives:

$$E = E° - \frac{RT}{nF}\ln Q \qquad (18.7, \text{text})$$

Equation (18.7) is known as the **Nernst equation**. At 298 K, Equation (18.7) of the text can be simplified by substituting $R$, $T$, and $F$ into the equation.

$$E = E° - \frac{0.0257\ \text{V}}{n}\ln Q \qquad (18.8, \text{text})$$

At equilibrium, there is no net transfer of electrons, so $E = 0$ and $Q = K$, where $K$ is the equilibrium constant of the redox reaction. Substituting into Equation (18.8) gives

$$E° = \frac{0.0257\ \text{V}}{n}\ln K \qquad (18.5, \text{text})$$

This is Equation (18.5) of the text derived in Problem Type 3B.

### A. Using the Nernst equation to predict the spontaneity of a redox reaction

To predict the spontaneity of a redox reaction, we can use the Nernst equation to calculate the cell emf, $E$. If the cell emf is *positive*, the redox reaction is *spontaneous*. Conversely, if $E$ is *negative*, the reaction is *not spontaneous* in the direction written.

**EXAMPLE 18.6**

**Calculate the cell emf and predict whether the following reaction is spontaneous at 25°C.**

$$\mathbf{Zn(s) + 2H^+(1 \times 10^{-4}\ M) \longrightarrow Zn^{2+}(1.5\ M) + H_2(1\ atm)}$$

**Strategy:** The standard emf ($E°$) can be calculated using the standard reduction potentials in Table 18.1 of the text. Because the reactions are not run under standard-state conditions (concentrations are not 1 *M*), we need Nernst's equation [Equation (18.8) of the text] to calculate the emf ($E$) of a hypothetical galvanic cell. Remember that solids do not appear in the reaction quotient ($Q$) term in the Nernst equation.

**Solution:** Calculate the standard cell emf, $E°$, from standard reduction potentials (Table 18.1). Separate the reaction into its half-reactions.

$$2H^+(aq) + 2e^- \xrightarrow{\text{reduction (cathode)}} H_2(g) \qquad E°_{\text{cathode}} = 0.00\text{ V}$$

$$Zn(s) \xrightarrow{\text{oxidation (anode)}} Zn^{2+}(aq) + 2e^- \qquad E°_{\text{anode}} = -0.76\text{ V}$$

$$Zn(s) + 2H^+(aq) \longrightarrow Zn^{2+}(aq) + H_2(g) \qquad E°_{\text{cell}} = E°_{\text{cathode}} - E°_{\text{anode}} = +0.76\text{ V}$$

Calculate the cell emf, $E$, from the Nernst equation. Two moles of electrons were transferred during the reaction, so $n = 2$.

$$E = E° - \frac{0.0257\text{ V}}{n}\ln\frac{[Zn^{2+}]P_{H_2}}{[H^+]^2}$$

$$E = 0.76\text{ V} - \frac{0.0257\text{ V}}{2}\ln\frac{(1.5)(1)}{(1 \times 10^{-4})^2}$$

$$E = 0.76\text{ V} - 0.24\text{ V} = \mathbf{0.52\text{ V}}$$

The reaction is spontaneous because the cell emf, $E$, is positive.

**PRACTICE EXERCISE**

6. Calculate the cell emf for the following reaction:

$$2Ag^+(0.10\,M) + H_2(1.0\text{ atm}) \longrightarrow 2Ag(s) + 2H^+\ (\text{pH} = 8.00)$$

**Text Problems: 18.32**, 18.34, 18.36

## B. Using the Nernst equation to calculate concentration

Recall that the reaction quotient $Q$ equals the concentrations of the products raised to the power of their stoichiometric coefficients divided by the concentrations of the reactants raised to the power of their stoichiometric coefficients.

$$Q = \frac{[\text{products}]^x}{[\text{reactants}]^y}$$

Since $Q$ is in the Nernst equation, if the cell emf is measured and the standard cell emf is calculated, we can determine the concentration of one of the components if the concentrations of the other components are known. See Example 18.7 below.

**EXAMPLE 18.7**

**A galvanic cell is constructed from a silver half-cell and a copper half-cell. The copper half-cell contains 0.10 *M* $Cu(NO_3)_2$ and the concentration of silver ions in the other half-cell is unknown. If the Ag electrode is the cathode and the cell emf is measured and found to be 0.10 V at 25°C, what is the $Ag^+$ ion concentration?**

**Strategy:** We are given the cell emf, $E = 0.10$ V, in the problem. If we can write the reaction occurring in the cell, we can calculate the standard cell emf, $E°$. We are also given the $Cu^{2+}$ concentration in the problem. Given all this information, we can calculate the $Ag^+$ ion concentration using the Nernst equation.

**Solution:** Write the half-reactions to calculate the standard cell emf, $E°$. We are told that the Ag electrode is the cathode (reduction occurs at the cathode). We can write:

$$Ag^+(aq) + e^- \xrightarrow{\text{reduction (cathode)}} Ag(s) \qquad E^\circ_{\text{cathode}} = +0.80 \text{ V}$$

If $Ag^+$ is reduced, Cu must be oxidized.

$$Cu(s) \xrightarrow{\text{oxidation (anode)}} Cu^{2+}(aq) + 2e^- \qquad E^\circ_{\text{anode}} = +0.34 \text{ V}$$

$$E^\circ_{\text{cell}} = E^\circ_{\text{cathode}} - E^\circ_{\text{anode}}$$

$$E^\circ_{\text{cell}} = 0.80 \text{ V} - 0.34 \text{ V} = \mathbf{+0.46 \text{ V}}$$

We must multiply the Ag half-reaction by two so that the number of electrons gained equals the number of electrons lost. The balanced reaction is:

$$2Ag^+(aq) + Cu(s) \longrightarrow 2Ag(s) + Cu^{2+}(aq)$$

Substitute $E$, $E°$, and $[Cu^{2+}]$ into the Nernst equation to calculate the $[Ag^+]$. Also, $n = 2$ since two moles of electrons are transferred during the reaction.

$$E = E^\circ - \frac{0.0257 \text{ V}}{n}\ln Q$$

$$E = E^\circ - \frac{0.0257 \text{ V}}{n}\ln\frac{[Cu^{2+}]}{[Ag^+]^2}$$

$$0.10 \text{ V} = 0.46 \text{ V} - \frac{0.0257 \text{ V}}{2}\ln\frac{[0.10]}{[Ag^+]^2}$$

$$\frac{2(0.10 \text{ V} - 0.46 \text{ V})}{-0.0257 \text{ V}} = \ln\frac{0.10}{[Ag^+]^2}$$

$$28.0 = \ln\frac{0.10}{[Ag^+]^2}$$

Taking the anti-ln of both sides of the equation,

$$e^{28.0} = \frac{0.10}{[Ag^+]^2}$$

$$1.45 \times 10^{12} = \frac{0.10}{[Ag^+]^2}$$

$$[Ag^+] = \sqrt{\frac{0.10}{1.45 \times 10^{12}}}$$

$$\mathbf{[Ag^+] = 2.6 \times 10^{-7}\ \mathit{M}}$$

**PRACTICE EXERCISE**

7. When the concentration of $Zn^{2+}$ is 0.15 *M*, the measured voltage of a Zn-Cu galvanic cell is 0.40 V. What is the $Cu^{2+}$ ion concentration?

| Text Problem: 18.122 |
|---|

## PROBLEM TYPE 5: ELECTROLYSIS

**Electrolysis** is the process in which electrical energy is used to cause a nonspontaneous chemical reaction to occur. The same principles underlie electrolysis and the processes that take place in galvanic cells.

### A. Predicting the products of electrolysis

**EXAMPLE 18.8**

**Predict the products of the electrolysis of an aqueous $MgCl_2$ solution.**

This aqueous solution contains several species that could be oxidized or reduced. Two species that could be reduced are the metal ion ($Mg^{2+}$) and $H_2O$. The reduction half-reactions that might occur at the cathode are

(1) $Mg^{2+}(aq) + 2e^- \xrightarrow{\text{reduction (cathode)}} Mg(s) \qquad E^\circ_{\text{cathode}} = -2.37\text{ V}$

(2) $2H_2O(l) + 2e^- \xrightarrow{\text{reduction (cathode)}} H_2(g) + 2OH^-(aq) \qquad E^\circ_{\text{cathode}} = -0.83\text{ V}$

(3) $2H^+(aq) + 2e^- \xrightarrow{\text{reduction (cathode)}} H_2(g) \qquad E^\circ_{\text{cathode}} = 0.00\text{ V}$

We can rule out Reaction (1) immediately. The very negative standard reduction potential indicates that $Mg^{2+}$ has essentially no tendency to undergo reduction. Reaction (3) is preferred over Reaction (2) under standard-state conditions. However, at a pH of 7 (as is the case for a $MgCl_2$ solution), they are equally probable. We generally use Reaction (2) to describe the cathode reaction because the concentration of $H^+$ ions is too low (about $1 \times 10^{-7}$ *M*) to make (3) a reasonable choice.

The oxidation reactions that might occur at the anode are:

(4) $2Cl^-(aq) \xrightarrow{\text{oxidation (anode)}} Cl_2(g) + 2e^- \qquad E^\circ_{\text{anode}} = +1.36\text{ V}$

(5) $2H_2O(l) \xrightarrow{\text{oxidation (anode)}} O_2(g) + 4H^+(aq) + 2e^- \qquad E^\circ_{\text{anode}} = +1.23\text{ V}$

The standard reduction potentials of (4) and (5) are not very different, but the values do suggest that $H_2O$ should be preferentially oxidized at the anode. However, we find by experiment that the gas liberated at the anode is $Cl_2$, not $O_2$. The large overvoltage for $O_2$ formation prevents its production when the $Cl^-$ ion is there to compete (see Section 18.8 of the text).

The overall reaction is:

$$2H_2O(l) + 2e^- \longrightarrow H_2(g) + 2OH^-(aq)$$

$$2Cl^-(aq) \longrightarrow Cl_2(g) + 2e^-$$

$$\overline{2H_2O(l) + 2Cl^-(aq) \longrightarrow H_2(g) + 2OH^-(aq) + Cl_2(g)}$$

## B. Calculating the quantity of products in electrolysis

During electrolysis, the mass of product formed (or reactant consumed) at an electrode is proportional to both the amount of electricity transferred at the electrode and the molar mass of the substance in question.

The steps involved in calculating the quantities of substances produced in electrolysis are shown below.

| | | | | | | | | |
|---|---|---|---|---|---|---|---|---|
| Current | | Charge | | Number | | Moles of | | Grams of |
| (amperes) | → | in | → | of mol of | → | substance reduced | → | substance reduced |
| and time | | coulombs | | electrons | | or oxidized | | or oxidized |

To convert to coulombs, recall that

$$1\text{ C} = 1\text{ A} \times 1\text{ s}$$

To convert from coulombs to moles of electrons, use the conversion factor

$$1\text{ mol } e^- = 96{,}500\text{ C}$$

Finally, we need to consider the number of moles of electrons transferred during the reaction. For example, to reduce $Al^{3+}$ ions to Al metal, 3 moles of electrons are needed to reduce 1 mole of $Al^{3+}$ ions.

$$Al^{3+} + 3e^- \longrightarrow Al$$

**EXAMPLE 18.9**

**How many moles of electrons are transferred in an electrolytic cell when a current of 12 amps flows for 16 hours?**

**Strategy:** According to Figure 18.20 of the text, we can carry out the following conversion steps to calculate the moles of electrons transferred.

$$\text{current} \times \text{time} \rightarrow \text{coulombs} \rightarrow \text{mol } e^-$$

This is a large number of steps, so let's break it down into two parts. First, we calculate the coulombs of electricity that pass through the cell. Then, we will continue on to moles of electrons.

**Solution:** First, we can calculate the number of coulombs passing through the cell in 16 hours. Recall that

$$1\text{ C} = 1\text{ A·s}$$

Converting from amps to coulombs:

$$12\,\cancel{\text{A}} \times \frac{1\text{ C}}{1\,\cancel{\text{A·s}}} \times \frac{3600\,\cancel{\text{s}}}{1\,\cancel{\text{h}}} \times 16\,\cancel{\text{h}} = 6.9 \times 10^5\text{ C}$$

Next, we convert from coulombs to moles of electrons using the following conversion factor.

$$1\text{ mol } e^- = 96{,}500\text{ C}$$

$$\textbf{? mol } e^- = (6.9 \times 10^5\,\cancel{\text{C}}) \times \frac{1\text{ mol e}^-}{96{,}500\,\cancel{\text{C}}} = \textbf{7.2 mol } e^-$$

**EXAMPLE 18.10**

**How many grams of copper metal would be deposited from a solution of $CuSO_4$ by the passage of 3.0 A of electrical current through an electrolytic cell for 2.0 h?**

**Strategy:** According to Figure 18.20 of the text, we can carry out the following conversion steps to calculate the quantity of Cu in grams.

$$\text{current} \times \text{time} \rightarrow \text{coulombs} \rightarrow \text{mol } e^- \rightarrow \text{mol Cu} \rightarrow \text{g Cu}$$

This is a large number of steps, so let's break it down into two parts. First, we calculate the coulombs of electricity that pass through the cell. Then, we will continue on to calculate grams of Cu.

**Solution:** Let's start by writing the half-reaction to determine the moles of electrons required to produce 1 mole of Cu. Start by writing the half-reaction for the reduction of $Cu^{2+}$ to Cu metal.

$$Cu^{2+}(aq) + 2e^- \longrightarrow Cu(s)$$

This half-reaction tells us that 2 mol of $e^-$ are required to produce 1 mol of Cu ($s$).

Next, we calculate the coulombs of electricity that pass through the cell.

$$3.0\ \cancel{A} \times \frac{1\ \text{C}}{1\ \cancel{A}\cdot\cancel{s}} \times \frac{3600\ \cancel{s}}{1\ \cancel{h}} \times 2.0\ \cancel{h} = 2.2 \times 10^4\ \text{C}$$

The grams of Cu produced at the cathode are:

$$\textbf{? g Cu} = (2.2 \times 10^4\ \cancel{C}) \times \frac{1\ \cancel{\text{mol } e^-}}{96{,}500\ \cancel{C}} \times \frac{1\ \cancel{\text{mol}}\ \text{Cu}}{2\ \cancel{\text{mol } e^-}} \times \frac{63.55\ \text{g Cu}}{1\ \cancel{\text{mol}}\ \text{Cu}} = \textbf{7.3 g Cu}$$

**Note:** This predicted amount of Cu is based on a process that is 100 percent efficient. Any side reactions or any oxidation or reduction of impurities will cause the actual yield to be less than the theoretical yield.

**EXAMPLE 18.11**

**How long will it take to electrodeposit (plate out) 1.0 g of Ni from a $NiSO_4$ solution using a current of 2.5 A?**

**Strategy:** This calculation is opposite to Example 18.10. We start with grams of Ni and are asked to calculate the time required to electrodeposit 1.0 g. We follow the strategy:

$$\text{g Ni} \rightarrow \text{mol Ni} \rightarrow \text{mol } e^- \rightarrow \text{coulombs} \rightarrow \text{time}$$

Let's break the calculation down into two steps, first converting grams of Ni to mole of electrons and then converting from moles of electrons to time.

**Solution:** First, find the number of moles of electrons required to electrodeposit 1.0 g of Ni from solution. The half-reaction for the reduction of $Ni^{2+}$ is:

$$Ni^{2+}(aq) + 2e^- \longrightarrow Ni(s)$$

2 moles of electrons are required to reduce 1 mol of $Ni^{2+}$ ions to Ni metal. But, we are electrodepositing less than 1 mole of Ni($s$). We need to complete the following conversions:

$$\text{g Ni} \rightarrow \text{mol Ni} \rightarrow \text{mol of } e^-$$

$$? \text{ mol of } e^- = 1.0 \text{ g Ni} \times \frac{1 \text{ mol Ni}}{58.7 \text{ g Ni}} \times \frac{2 \text{ mol } e^-}{1 \text{ mol Ni}} = 0.034 \text{ mol } e^-$$

Determine how long it will take for 0.034 moles of electrons to flow through the cell when the current is 2.5 C/s. We need to complete the following conversions:

$$\text{mol of } e^- \rightarrow \text{coulombs} \rightarrow \text{seconds}$$

$$? \textbf{ seconds} = 0.034 \text{ mol } e^- \times \frac{96{,}500 \text{ C}}{1 \text{ mol } e^-} \times \frac{1 \text{ s}}{2.5 \text{ C}} = \mathbf{1.3 \times 10^3 \text{ s}} \text{ (22 min)}$$

**PRACTICE EXERCISES**

**8.** How many moles of electrons are transferred in an electrolytic cell when a current of 2.0 amps flows for 6.0 hours?

**9.** How many grams of cobalt can be electroplated by passing a constant current of 5.2 A through a solution of $CoCl_3$ for 60.0 min?

**10.** How long will it take to produce 54 kg of Al metal by the reduction of $Al^{3+}$ in an electrolytic cell using a current of $5.0 \times 10^2$ amps?

**Text Problems: 18.48**, 18.50, 18.54, 18.56, 18.58, 18.60, 18.62

## ANSWERS TO PRACTICE EXERCISES

**1.** $5Fe^{2+}(aq) + MnO_4^-(aq) + 8H^+(aq) \longrightarrow 5Fe^{3+}(aq) + Mn^{2+}(aq) + 4H_2O(l)$

**2.** $SO_4^{2-} > O_2 > Ce^{4+} > H_2O_2$

**3.** $E^\circ_{U^{3+}/U} = -1.84$ V

**4.** (a) $E^\circ_{cell} = +0.24$ V. The positive value of the standard cell emf indicates that the reaction is spontaneous under standard-state conditions.

(b) $E^\circ_{cell} = -0.34$ V. A negative standard cell emf indicates that this reaction is not spontaneous under standard-state conditions. Copper will not dissolve in 1 *M* HCl.

**5.** $K = 3.0 \times 10^{-35}$

**6.** $E^\circ = 1.21$ V

**7.** $[Cu^{2+}] = 3.1 \times 10^{-25}$ *M*

**8.** Number of mol $e^-$ transferred = 0.45 mol $e^-$

**9.** Grams of Co electroplated = 3.81 g Co

**10.** Time = 322 h

## SOLUTIONS TO SELECTED TEXT PROBLEMS

**18.2** **Strategy:** We follow the procedure for balancing redox reactions presented in Section 18.1 of the text.

**Solution:**

**(a)**

*Step 1:* The unbalanced equation is given in the problem.

$$Mn^{2+} + H_2O_2 \longrightarrow MnO_2 + H_2O$$

*Step 2:* The two half-reactions are:

$$Mn^{2+} \xrightarrow{\text{oxidation}} MnO_2$$

$$H_2O_2 \xrightarrow{\text{reduction}} H_2O$$

*Step 3:* We balance each half-reaction for number and type of atoms and charges.

The *oxidation half-reaction* is already balanced for Mn atoms. To balance the O atoms, we add two water molecules on the left side.

$$Mn^{2+} + 2H_2O \longrightarrow MnO_2$$

To balance the H atoms, we add 4 $H^+$ to the right-hand side.

$$Mn^{2+} + 2H_2O \longrightarrow MnO_2 + 4H^+$$

There are four net positive charges on the right and two net positive charge on the left, we add two electrons to the right side to balance the charge.

$$Mn^{2+} + 2H_2O \longrightarrow MnO_2 + 4H^+ + 2e^-$$

*Reduction half-reaction*: we add one $H_2O$ to the right-hand side of the equation to balance the O atoms.

$$H_2O_2 \longrightarrow 2H_2O$$

To balance the H atoms, we add $2H^+$ to the left-hand side.

$$H_2O_2 + 2H^+ \longrightarrow 2H_2O$$

There are two net positive charges on the left, so we add two electrons to the same side to balance the charge.

$$H_2O_2 + 2H^+ + 2e^- \longrightarrow 2H_2O$$

*Step 4:* We now add the oxidation and reduction half-reactions to give the overall reaction. Note that the number of electrons gained and lost is equal.

$$Mn^{2+} + 2H_2O \longrightarrow MnO_2 + 4H^+ + 2e^-$$

$$H_2O_2 + 2H^+ + 2e^- \longrightarrow 2H_2O$$

$$\overline{Mn^{2+} + H_2O_2 + 2e^- \longrightarrow MnO_2 + 2H^+ + 2e^-}$$

The electrons on both sides cancel, and we are left with the balanced net ionic equation in acidic medium.

$$Mn^{2+} + H_2O_2 \longrightarrow MnO_2 + 2H^+$$

Because the problem asks to balance the equation in basic medium, we add one $OH^-$ to both sides for each $H^+$ and combine pairs of $H^+$ and $OH^-$ on the same side of the arrow to form $H_2O$.

$$Mn^{2+} + H_2O_2 + 2OH^- \longrightarrow MnO_2 + 2H^+ + 2OH^-$$

Combining the $H^+$ and $OH^-$ to form water we obtain:

$$\mathbf{Mn^{2+} + H_2O_2 + 2OH^- \longrightarrow MnO_2 + 2H_2O}$$

*Step 5:* Check to see that the equation is balanced by verifying that the equation has the same types and numbers of atoms and the same charges on both sides of the equation.

**(b)** This problem can be solved by the same methods used in part (a).

$$\mathbf{2Bi(OH)_3 + 3SnO_2^{2-} \longrightarrow 2Bi + 3H_2O + 3SnO_3^{2-}}$$

**(c)**

*Step 1:* The unbalanced equation is given in the problem.

$$Cr_2O_7^{2-} + C_2O_4^{2-} \longrightarrow Cr^{3+} + CO_2$$

*Step 2:* The two half-reactions are:

$$C_2O_4^{2-} \xrightarrow{\text{oxidation}} CO_2$$

$$Cr_2O_7^{2-} \xrightarrow{\text{reduction}} Cr^{3+}$$

*Step 3:* We balance each half-reaction for number and type of atoms and charges.

In the *oxidation half-reaction*, we first need to balance the C atoms.

$$C_2O_4^{2-} \longrightarrow 2CO_2$$

The O atoms are already balanced. There are two net negative charges on the left, so we add two electrons to the right to balance the charge.

$$C_2O_4^{2-} \longrightarrow 2CO_2 + 2e^-$$

In the *reduction half-reaction*, we first need to balance the Cr atoms.

$$Cr_2O_7^{2-} \longrightarrow 2Cr^{3+}$$

We add seven $H_2O$ molecules on the right to balance the O atoms.

$$Cr_2O_7^{2-} \longrightarrow 2Cr^{3+} + 7H_2O$$

To balance the H atoms, we add $14H^+$ to the left-hand side.

$$Cr_2O_7^{2-} + 14H^+ \longrightarrow 2Cr^{3+} + 7H_2O$$

There are twelve net positive charges on the left and six net positive charges on the right. We add six electrons on the left to balance the charge.

$$Cr_2O_7^{2-} + 14H^+ + 6e^- \longrightarrow 2Cr^{3+} + 7H_2O$$

*Step 4:* We now add the oxidation and reduction half-reactions to give the overall reaction. In order to equalize the number of electrons, we need to multiply the oxidation half-reaction by 3.

$$\begin{array}{l}3(C_2O_4^{2-} \longrightarrow 2CO_2 + 2e^-) \\ Cr_2O_7^{2-} + 14H^+ + 6e^- \longrightarrow 2Cr^{3+} + 7H_2O \\ \hline 3C_2O_4^{2-} + Cr_2O_7^{2-} + 14H^+ + 6e^- \longrightarrow 6CO_2 + 2Cr^{3+} + 7H_2O + 6e^-\end{array}$$

The electrons on both sides cancel, and we are left with the balanced net ionic equation in acidic medium.

$$\mathbf{3C_2O_4^{2-} + Cr_2O_7^{2-} + 14H^+ \longrightarrow 6CO_2 + 2Cr^{3+} + 7H_2O}$$

*Step 5:* Check to see that the equation is balanced by verifying that the equation has the same types and numbers of atoms and the same charges on both sides of the equation.

**(d)** This problem can be solved by the same methods used in part (c).

$$\mathbf{2Cl^- + 2ClO_3^- + 4H^+ \longrightarrow Cl_2 + 2ClO_2 + 2H_2O}$$

**18.12** **Strategy:** At first, it may not be clear how to assign the electrodes in the galvanic cell. From Table 18.1 of the text, we write the standard reduction potentials of Al and Ag and apply the diagonal rule to determine which is the anode and which is the cathode.

**Solution:** The standard reduction potentials are:

$$Ag^+(1.0\ M) + e^- \rightarrow Ag(s) \qquad E^\circ = 0.80\ V$$

$$Al^{3+}(1.0\ M) + 3e^- \rightarrow Al(s) \qquad E^\circ = -1.66\ V$$

Applying the diagonal rule, we see that $Ag^+$ will oxidize Al.

| | |
|---|---|
| Anode (oxidation): | $Al(s) \rightarrow Al^{3+}(1.0\ M) + 3e^-$ |
| Cathode (reduction): | $3Ag^+(1.0\ M) + 3e^- \rightarrow 3Ag(s)$ |
| Overall: | $\mathbf{Al(s) + 3Ag^+(1.0\ M) \rightarrow Al^{3+}(1.0\ M) + 3Ag(s)}$ |

Note that in order to balance the overall equation, we multiplied the reduction of $Ag^+$ by 3. We can do so because, as an intensive property, $E^\circ$ is not affected by this procedure. We find the emf of the cell using Equation (18.1) and Table 18.1 of the text.

$$E^\circ_{cell} = E^\circ_{cathode} - E^\circ_{anode} = E^\circ_{Ag^+/Ag} - E^\circ_{Al^{3+}/Al}$$

$$E^\circ_{cell} = 0.80\ V - (-1.66\ V) = \mathbf{+2.46\ V}$$

**Check:** The positive value of $E^\circ$ shows that the forward reaction is favored.

**18.14** The half-reaction for oxidation is:

$$\text{Anode (oxidation):} \quad 2H_2O(l) \rightarrow O_2(g) + 4H^+(aq) + 4e^- \qquad E^\circ_{anode} = +1.23\ V$$

The species that can oxidize water to molecular oxygen must have an $E^\circ_{red}$ more positive than +1.23 V. From Table 18.1 of the text we see that only $\mathbf{Cl_2(g)}$ and $\mathbf{MnO_4^-(aq)}$ in acid solution can oxidize water to oxygen.

**18.16** **Strategy:** $E^\circ_{cell}$ is *positive* for a spontaneous reaction. In each case, we can calculate the standard cell emf from the potentials for the two half-reactions.

$$E^\circ_{cell} = E^\circ_{cathode} - E^\circ_{anode}$$

**Solution:**

(a) $E^\circ = -0.40\text{ V} - (-2.87\text{ V}) = \mathbf{2.47\text{ V}}$. The reaction is **spontaneous.**

(b) $E^\circ = -0.14\text{ V} - 1.07\text{ V} = \mathbf{-1.21\text{ V}}$. The reaction is **not spontaneous.**

(c) $E^\circ = -0.25\text{ V} - 0.80\text{ V} = \mathbf{-1.05\text{ V}}$. The reaction is **not spontaneous.**

(d) $E^\circ = 0.77\text{ V} - 0.15\text{ V} = \mathbf{0.62\text{ V}}$. The reaction is **spontaneous.**

**18.18** **Strategy:** The greater the tendency for the substance to be oxidized, the stronger its tendency to act as a reducing agent. The species that has a stronger tendency to be oxidized will have a smaller reduction potential.

**Solution:** In each pair, look for the one with the smaller reduction potential. This indicates a greater tendency for the substance to be oxidized.

(a) Li (b) $H_2$ (c) $Fe^{2+}$ (d) $Br^-$

**18.20** In cell diagrams, the anode is written first, to the left of the double lines. See Table 18.1 of the text for the reduction potential of $Ni^{2+}$.

$$E^\circ_{cell} = E^\circ_{cathode} - E^\circ_{anode}$$

$$1.54\text{ V} = -0.25\text{ V} - E^\circ_{anode}$$

$$E^\circ_{anode} = \mathbf{-1.79\text{ V}} = E^\circ_{U^{3+}/U}$$

**18.24** **Strategy:** The relationship between the equilibrium constant, *K*, and the standard emf is given by Equation (18.5) of the text: $E^\circ_{cell} = (0.0257\text{ V}/n)\ln K$. Thus, knowing *n* (the moles of electrons transferred) and the equilibrium constant, we can determine $E^\circ_{cell}$.

**Solution:** The equation that relates *K* and the standard cell emf is:

$$E^\circ_{cell} = \frac{0.0257\text{ V}}{n}\ln K$$

We see in the reaction that Sr goes to $Sr^{2+}$ and $Mg^{2+}$ goes to Mg. Therefore, two moles of electrons are transferred during the redox reaction. Substitute the equilibrium constant and the moles of $e^-$ transferred ($n = 2$) into the above equation to calculate $E^\circ$.

$$E^\circ = \frac{(0.0257\text{ V})\ln K}{n} = \frac{(0.0257\text{ V})\ln(2.69\times10^{12})}{2} = \mathbf{0.368\text{ V}}$$

**18.26** **(a)** We break the equation into two half–reactions:

| | | |
|---|---|---|
| Anode (oxidation): | $Mg(s) \rightarrow Mg^{2+}(aq) + 2e^-$ | $E^\circ_{anode} = -2.37$ V |
| Cathode (reduction): | $Pb^{2+}(aq) + 2e^- \rightarrow Pb(s)$ | $E^\circ_{cathode} = -0.13$ V |

The standard emf is given by

$$E^\circ_{cell} = E^\circ_{cathode} - E^\circ_{anode} = -0.13\text{ V} - (-2.37\text{ V}) = 2.24\text{ V}$$

We can calculate $\Delta G^\circ$ from the standard emf.

$$\Delta G^\circ = -nFE^\circ_{cell}$$

$$\boldsymbol{\Delta G^\circ = -(2)(96500\text{ J/}\cancel{\text{V}}\cdot\text{mol})(2.24\ \cancel{\text{V}}) = -432\text{ kJ/mol}}$$

Next, we can calculate $K$ using Equation (18.5) of the text.

$$E^\circ_{cell} = \frac{0.0257\text{ V}}{n}\ln K$$

or

$$\ln K = \frac{nE^\circ_{cell}}{0.0257\text{ V}}$$

and

$$K = e^{\frac{nE^\circ}{0.0257}}$$

$$\boldsymbol{K = e^{\frac{(2)(2.24)}{0.0257}} = 5 \times 10^{75}}$$

**Tip:** You could also calculate $K_c$ from the standard free energy change, $\Delta G^\circ$, using the equation: $\Delta G^\circ = -RT\ln K_c$.

**(b)** We break the equation into two half–reactions:

$$Br_2(l) + 2e^- \xrightarrow{\text{reduction (cathode)}} 2Br^-(aq) \qquad E^\circ_{cathode} = 1.07\text{ V}$$

$$2I^-(aq) \xrightarrow{\text{oxidation (anode)}} I_2(s) + 2e^- \qquad E^\circ_{anode} = 0.53\text{ V}$$

The standard emf is

$$E^\circ_{cell} = E^\circ_{cathode} - E^\circ_{anode} = 1.07\text{ V} - 0.53\text{ V} = 0.54\text{ V}$$

We can calculate $\Delta G^\circ$ from the standard emf.

$$\Delta G^\circ = -nFE^\circ_{cell}$$

$$\boldsymbol{\Delta G^\circ = -(2)(96500\text{ J/}\cancel{\text{V}}\cdot\text{mol})(0.54\ \cancel{\text{V}}) = -104\text{ kJ/mol}}$$

Next, we can calculate $K$ using Equation (18.5) of the text.

$$K = e^{\frac{nE^\circ}{0.0257}}$$

$$\mathbf{K = e^{\frac{(2)(0.54)}{0.0257}} = 2 \times 10^{18}}$$

**(c)** This is worked in an analogous manner to parts (a) and (b).

$$E^\circ_{\text{cell}} = E^\circ_{\text{cathode}} - E^\circ_{\text{anode}} = 1.23\text{ V} - 0.77\text{ V} = 0.46\text{ V}$$

$$\Delta G^\circ = -nFE^\circ_{\text{cell}}$$

$$\mathbf{\Delta G^\circ = -(4)(96500\text{ J/}\cancel{\text{V}}\cdot\text{mol})(0.46\ \cancel{\text{V}}) = -178\text{ kJ/mol}}$$

$$K = e^{\frac{nE^\circ}{0.0257}}$$

$$\mathbf{K = e^{\frac{(4)(0.46)}{0.0257}} = 1 \times 10^{31}}$$

**(d)** This is worked in an analogous manner to parts (a), (b), and (c).

$$E^\circ_{\text{cell}} = E^\circ_{\text{cathode}} - E^\circ_{\text{anode}} = 0.53\text{ V} - (-1.66\text{ V}) = 2.19\text{ V}$$

$$\Delta G^\circ = -nFE^\circ_{\text{cell}}$$

$$\mathbf{\Delta G^\circ = -(6)(96500\text{ J/}\cancel{\text{V}}\cdot\text{mol})(2.19\ \cancel{\text{V}}) = -1.27 \times 10^3\text{ kJ/mol}}$$

$$K = e^{\frac{nE^\circ}{0.0257}}$$

$$\mathbf{K = e^{\frac{(6)(2.19)}{0.0257}} = 8 \times 10^{211}}$$

**18.28** **Strategy:** The relationship between the standard free energy change and the standard emf of the cell is given by Equation (18.3) of the text: $\Delta G^\circ = -nFE^\circ_{\text{cell}}$. The relationship between the equilibrium constant, $K$, and the standard emf is given by Equation (18.5) of the text: $E^\circ_{\text{cell}} = (0.0257\text{ V}/n)\ln K$. Thus, if we can determine $E^\circ_{\text{cell}}$, we can calculate $\Delta G^\circ$ and $K$. We can determine the $E^\circ_{\text{cell}}$ of a hypothetical galvanic cell made up of two couples ($Cu^{2+}/Cu^+$ and $Cu^+/Cu$) from the standard reduction potentials in Table 18.1 of the text.

**Solution:** The half-cell reactions are:

| | |
|---|---|
| Anode (oxidation): | $Cu^+(1.0\ M) \rightarrow Cu^{2+}(1.0\ M) + e^-$ |
| Cathode (reduction): | $Cu^+(1.0\ M) + e^- \rightarrow Cu(s)$ |
| Overall: | $2Cu^+(1.0\ M) \rightarrow Cu^{2+}(1.0\ M) + Cu(s)$ |

$$E^\circ_{\text{cell}} = E^\circ_{\text{cathode}} - E^\circ_{\text{anode}} = E^\circ_{\text{Cu}^+/\text{Cu}} - E^\circ_{\text{Cu}^{2+}/\text{Cu}^+}$$

$$E^\circ_{\textbf{cell}} = 0.52\text{ V} - 0.15\text{ V} = \mathbf{0.37\ V}$$

Now, we use Equation (18.3) of the text. The overall reaction shows that $n = 1$.

$$\Delta G^\circ = -nFE^\circ_{\text{cell}}$$

$$\Delta G^\circ = -(1)(96500\text{ J/V}\cdot\text{mol})(0.37\text{ V}) = \mathbf{-36\ kJ/mol}$$

Next, we can calculate $K$ using Equation (18.5) of the text.

$$E^\circ_{\text{cell}} = \frac{0.0257\text{ V}}{n}\ln K$$

or

$$\ln K = \frac{nE^\circ_{\text{cell}}}{0.0257\text{ V}}$$

and

$$K = e^{\frac{nE^\circ}{0.0257}}$$

$$K = e^{\frac{(1)(0.37)}{0.0257}} = e^{14.4} = \mathbf{2 \times 10^6}$$

**Check:** The negative value of $\Delta G^\circ$ and the large positive value of $K$, both indicate that the reaction favors products at equilibrium. The result is consistent with the fact that $E^\circ$ for the galvanic cell is positive.

**18.32** **Strategy:** The standard emf ($E^\circ$) can be calculated using the standard reduction potentials in Table 18.1 of the text. Because the reactions are not run under standard-state conditions (concentrations are not 1 $M$), we need Nernst's equation [Equation (18.8) of the text] to calculate the emf ($E$) of a hypothetical galvanic cell. Remember that solids do not appear in the reaction quotient ($Q$) term in the Nernst equation. We can calculate $\Delta G$ from $E$ using Equation (18.2) of the text: $\Delta G = -nFE_{\text{cell}}$.

**Solution:**

(a) The half-cell reactions are:

| | |
|---|---|
| Anode (oxidation): | $Mg(s) \rightarrow Mg^{2+}(1.0\ M) + 2e^-$ |
| Cathode (reduction): | $Sn^{2+}(1.0\ M) + 2e^- \rightarrow Sn(s)$ |
| Overall: | $Mg(s) + Sn^{2+}(1.0\ M) \rightarrow Mg^{2+}(1.0\ M) + Sn(s)$ |

$$E^\circ_{\text{cell}} = E^\circ_{\text{cathode}} - E^\circ_{\text{anode}} = E^\circ_{\text{Sn}^{2+}/\text{Sn}} - E^\circ_{\text{Mg}^{2+}/\text{Mg}}$$

$$E^\circ_{\textbf{cell}} = -0.14\text{ V} - (-2.37\text{ V}) = \mathbf{2.23\ V}$$

From Equation (18.8) of the text, we write:

$$E = E^\circ - \frac{0.0257\text{ V}}{n}\ln Q$$

$$E = E^\circ - \frac{0.0257\text{ V}}{n}\ln\frac{[\text{Mg}^{2+}]}{[\text{Sn}^{2+}]}$$

$$E = 2.23\text{ V} - \frac{0.0257\text{ V}}{2}\ln\frac{0.045}{0.035} = \mathbf{2.23\text{ V}}$$

We can now find the free energy change at the given concentrations using Equation (18.2) of the text. Note that in this reaction, $n = 2$.

$$\Delta G = -nFE_{\text{cell}}$$

$$\Delta G = -(2)(96500\text{ J/V}\cdot\text{mol})(2.23\text{ V}) = \mathbf{-430\text{ kJ/mol}}$$

**(b)** The half-cell reactions are:

| | |
|---|---|
| Anode (oxidation): | $3[\text{Zn}(s) \rightarrow \text{Zn}^{2+}(1.0\,M) + 2e^-]$ |
| Cathode (reduction): | $2[\text{Cr}^{3+}(1.0\,M) + 3e^- \rightarrow \text{Cr}(s)]$ |
| Overall: | $3\text{Zn}(s) + 2\text{Cr}^{3+}(1.0\,M) \rightarrow 3\text{Zn}^{2+}(1.0\,M) + 2\text{Cr}(s)$ |

$$E^\circ_{\text{cell}} = E^\circ_{\text{cathode}} - E^\circ_{\text{anode}} = E^\circ_{\text{Cr}^{3+}/\text{Cr}} - E^\circ_{\text{Zn}^{2+}/\text{Zn}}$$

$$E^\circ_{\text{cell}} = -0.74\text{ V} - (-0.76\text{ V}) = \mathbf{0.02\text{ V}}$$

From Equation (18.8) of the text, we write:

$$E = E^\circ - \frac{0.0257\text{ V}}{n}\ln Q$$

$$E = E^\circ - \frac{0.0257\text{ V}}{n}\ln\frac{[\text{Zn}^{2+}]^3}{[\text{Cr}^{3+}]^2}$$

$$E = 0.02\text{ V} - \frac{0.0257\text{ V}}{6}\ln\frac{(0.0085)^3}{(0.010)^2} = \mathbf{0.04\text{ V}}$$

We can now find the free energy change at the given concentrations using Equation (18.2) of the text. Note that in this reaction, $n = 6$.

$$\Delta G = -nFE_{\text{cell}}$$

$$\Delta G = -(6)(96500\text{ J/V}\cdot\text{mol})(0.04\text{ V}) = \mathbf{-23\text{ kJ/mol}}$$

**18.34** Let's write the two half-reactions to calculate the standard cell emf. (Oxidation occurs at the Pb electrode.)

| | | |
|---|---|---|
| Anode (oxidation): | $\text{Pb}(s) \rightarrow \text{Pb}^{2+}(aq) + 2e^-$ | $E^\circ_{\text{anode}} = -0.13\text{ V}$ |
| Cathode (reduction): | $2\text{H}^+(aq) + 2e^- \rightarrow \text{H}_2(g)$ | $E^\circ_{\text{cathode}} = 0.00\text{ V}$ |
| | $2\text{H}^+(aq) + \text{Pb}(s) \rightarrow \text{H}_2(g) + \text{Pb}^{2+}(aq)$ | |

$$E^\circ_{\text{cell}} = E^\circ_{\text{cathode}} - E^\circ_{\text{anode}} = 0.00\text{ V} - (-0.13\text{ V}) = 0.13\text{ V}$$

Using the Nernst equation, we can calculate the cell emf, $E$.

$$E = E^\circ - \frac{0.0257\text{ V}}{n}\ln\frac{[\text{Pb}^{2+}]P_{\text{H}_2}}{[\text{H}^+]^2}$$

$$E = 0.13\text{ V} - \frac{0.0257\text{ V}}{2}\ln\frac{(0.10)(1.0)}{(0.050)^2} = \mathbf{0.083\text{ V}}$$

18.36 All concentration cells have the same standard emf: *zero* volts.

$$\text{Anode (oxidation):}\quad \text{Mg}(s) \rightarrow \text{Mg}^{2+}(aq) + 2e^- \qquad E^\circ_{\text{anode}} = -2.37\text{ V}$$

$$\text{Cathode (reduction):}\quad \text{Mg}^{2+}(aq) + 2e^- \rightarrow \text{Mg}(s) \qquad E^\circ_{\text{cathode}} = -2.37\text{ V}$$

$$E^\circ_{\text{cell}} = E^\circ_{\text{cathode}} - E^\circ_{\text{anode}} = -2.37\text{ V} - (-2.37\text{ V}) = 0.00\text{ V}$$

We use the Nernst equation to compute the emf. There are two moles of electrons transferred from the reducing agent to the oxidizing agent in this reaction, so $n = 2$.

$$E = E^\circ - \frac{0.0257\text{ V}}{n}\ln Q$$

$$E = E^\circ - \frac{0.0257\text{ V}}{n}\ln\frac{[\text{Mg}^{2+}]_{\text{ox}}}{[\text{Mg}^{2+}]_{\text{red}}}$$

$$E = 0\text{ V} - \frac{0.0257\text{ V}}{2}\ln\frac{0.24}{0.53} = \mathbf{0.010\text{ V}}$$

What is the direction of spontaneous change in all concentration cells?

18.40 We can calculate the standard free energy change, $\Delta G^\circ$, from the standard free energies of formation, $\Delta G^\circ_f$ using Equation (17.12) of the text. Then, we can calculate the standard cell emf, $E^\circ_{\text{cell}}$, from $\Delta G^\circ$.

The overall reaction is:

$$\text{C}_3\text{H}_8(g) + 5\text{O}_2(g) \longrightarrow 3\text{CO}_2(g) + 4\text{H}_2\text{O}(l)$$

$$\Delta G^\circ_{\text{rxn}} = 3\Delta G^\circ_f[\text{CO}_2(g)] + 4\Delta G^\circ_f[\text{H}_2\text{O}(l)] - \{\Delta G^\circ_f[\text{C}_3\text{H}_8(g)] + 5\Delta G^\circ_f[\text{O}_2(g)]\}$$

$$\Delta G^\circ_{\text{rxn}} = (3)(-394.4\text{ kJ/mol}) + (4)(-237.2\text{ kJ/mol}) - [(1)(-23.5\text{ kJ/mol}) + (5)(0)] = -2108.5\text{ kJ/mol}$$

We can now calculate the standard emf using the following equation:

$$\Delta G^\circ = -nFE^\circ_{\text{cell}}$$

or

$$E^\circ_{\text{cell}} = \frac{-\Delta G^\circ}{nF}$$

Check the half-reactions on the page of the text listed in the problem to determine that 20 moles of electrons are transferred during this redox reaction.

$$E^\circ_{\text{cell}} = \frac{-(-2108.5 \times 10^3 \text{ J/mol})}{(20)(96500 \text{ J/V} \cdot \text{mol})} = \mathbf{1.09\ V}$$

Does this suggest that, in theory, it should be possible to construct a galvanic cell (battery) based on any conceivable spontaneous reaction?

**18.48** **(a)** The only ions present in molten $BaCl_2$ are $Ba^{2+}$ and $Cl^-$. The electrode reactions are:

$$\text{anode:} \quad 2Cl^-(aq) \longrightarrow Cl_2(g) + 2e^-$$

$$\text{cathode:} \quad Ba^{2+}(aq) + 2e^- \longrightarrow Ba(s)$$

This cathode half-reaction tells us that 2 moles of $e^-$ are required to produce 1 mole of $Ba(s)$.

**(b)**

**Strategy:** According to Figure 18.20 of the text, we can carry out the following conversion steps to calculate the quantity of Ba in grams.

$$\text{current} \times \text{time} \rightarrow \text{coulombs} \rightarrow \text{mol } e^- \rightarrow \text{mol Ba} \rightarrow \text{g Ba}$$

This is a large number of steps, so let's break it down into two parts. First, we calculate the coulombs of electricity that pass through the cell. Then, we will continue on to calculate grams of Ba.

**Solution:** First, we calculate the coulombs of electricity that pass through the cell.

$$0.50 \text{ A} \times \frac{1 \text{ C}}{1 \text{ A} \cdot \text{s}} \times \frac{60 \text{ s}}{1 \text{ min}} \times 30 \text{ min} = 9.0 \times 10^2 \text{ C}$$

We see that for every mole of Ba formed at the cathode, 2 moles of electrons are needed. The grams of Ba produced at the cathode are:

$$? \text{ g Ba} = (9.0 \times 10^2 \text{ C}) \times \frac{1 \text{ mol } e^-}{96{,}500 \text{ C}} \times \frac{1 \text{ mol Ba}}{2 \text{ mol } e^-} \times \frac{137.3 \text{ g Ba}}{1 \text{ mol Ba}} = \mathbf{0.64\ g\ Ba}$$

**18.50** The cost for producing various metals is determined by the moles of electrons needed to produce a given amount of metal. For each reduction, let's first calculate the number of tons of metal produced per 1 mole of electrons (1 ton = $9.072 \times 10^5$ g). The reductions are:

$$Mg^{2+} + 2e^- \longrightarrow Mg \qquad \frac{1 \text{ mol Mg}}{2 \text{ mol } e^-} \times \frac{24.31 \text{ g Mg}}{1 \text{ mol Mg}} \times \frac{1 \text{ ton}}{9.072 \times 10^5 \text{ g}} = 1.340 \times 10^{-5} \text{ ton Mg/mol } e^-$$

$$Al^{3+} + 3e^- \longrightarrow Al \qquad \frac{1\text{ mol Al}}{3\text{ mol }e^-} \times \frac{26.98\text{ g Al}}{1\text{ mol Al}} \times \frac{1\text{ ton}}{9.072 \times 10^5\text{ g}} = 9.913 \times 10^{-6}\text{ ton Al/mol }e^-$$

$$Na^+ + e^- \longrightarrow Na \qquad \frac{1\text{ mol Na}}{1\text{ mol }e^-} \times \frac{22.99\text{ g Na}}{1\text{ mol Na}} \times \frac{1\text{ ton}}{9.072 \times 10^5\text{ g}} = 2.534 \times 10^{-5}\text{ ton Na/mol }e^-$$

$$Ca^{2+} + 2e^- \longrightarrow Ca \qquad \frac{1\text{ mol Ca}}{2\text{ mol }e^-} \times \frac{40.08\text{ g Ca}}{1\text{ mol Ca}} \times \frac{1\text{ ton}}{9.072 \times 10^5\text{ g}} = 2.209 \times 10^{-5}\text{ ton Ca/mol }e^-$$

Now that we know the tons of each metal produced per mole of electrons, we can convert from \$155/ton Mg to the cost to produce the given amount of each metal.

**(a)** For aluminum :

$$\frac{\$155}{1\text{ ton Mg}} \times \frac{1.340 \times 10^{-5}\text{ ton Mg}}{1\text{ mol }e^-} \times \frac{1\text{ mol }e^-}{9.913 \times 10^{-6}\text{ ton Al}} \times 10.0\text{ tons Al} = \mathbf{\$2.10 \times 10^3}$$

**(b)** For sodium:

$$\frac{\$155}{1\text{ ton Mg}} \times \frac{1.340 \times 10^{-5}\text{ ton Mg}}{1\text{ mol }e^-} \times \frac{1\text{ mol }e^-}{2.534 \times 10^{-5}\text{ ton Na}} \times 30.0\text{ tons Na} = \mathbf{\$2.46 \times 10^3}$$

**(c)** For calcium:

$$\frac{\$155}{1\text{ ton Mg}} \times \frac{1.340 \times 10^{-5}\text{ ton Mg}}{1\text{ mol }e^-} \times \frac{1\text{ mol }e^-}{2.209 \times 10^{-5}\text{ ton Ca}} \times 50.0\text{ tons Ca} = \mathbf{\$4.70 \times 10^3}$$

**18.52** **(a)** The half–reaction is:

$$2H_2O(l) \longrightarrow O_2(g) + 4H^+(aq) + 4e^-$$

First, we can calculate the number of moles of oxygen produced using the ideal gas equation.

$$n_{O_2} = \frac{PV}{RT}$$

$$n_{O_2} = \frac{(1.0\text{ atm})(0.84\text{ L})}{(0.0821\text{ L}\cdot\text{atm/mol}\cdot\text{K})(298\text{ K})} = 0.034\text{ mol }O_2$$

From the half-reaction, we see that 1 mol $O_2 \simeq 4$ mol $e^-$.

$$\textbf{? mol } e^- = 0.034\text{ mol }O_2 \times \frac{4\text{ mol }e^-}{1\text{ mol }O_2} = \mathbf{0.14\text{ mol }e^-}$$

**(b)** The half–reaction is:

$$2Cl^-(aq) \longrightarrow Cl_2(g) + 2e^-$$

The number of moles of chlorine produced is:

$$n_{Cl_2} = \frac{PV}{RT}$$

$$n_{Cl_2} = \frac{\left(750\ \text{mmHg} \times \dfrac{1\ \text{atm}}{760\ \text{mmHg}}\right)(1.50\ \text{L})}{(0.0821\ \text{L}\cdot\text{atm/mol}\cdot\text{K})(298\ \text{K})} = 0.0605\ \text{mol } Cl_2$$

From the half-reaction, we see that 1 mol $Cl_2 \simeq 2$ mol $e^-$.

$$? \textbf{ mol } e^- = 0.0605\ \text{mol } Cl_2 \times \frac{2\ \text{mol } e^-}{1\ \text{mol } Cl_2} = \mathbf{0.121\ mol\ } e^-$$

(c) The half–reaction is:

$$Sn^{2+}(aq) + 2e^- \longrightarrow Sn(s)$$

The number of moles of Sn(*s*) produced is

$$? \text{ mol Sn} = 6.0\ \text{g Sn} \times \frac{1\ \text{mol Sn}}{118.7\ \text{g Sn}} = 0.051\ \text{mol Sn}$$

From the half-reaction, we see that 1 mol Sn $\simeq 2$ mol $e^-$.

$$? \textbf{ mol } e^- = 0.051\ \text{mol Sn} \times \frac{2\ \text{mol } e^-}{1\ \text{mol Sn}} = \mathbf{0.10\ mol\ } e^-$$

**18.54** (a) The half–reaction is:

$$\mathbf{Ag^+(aq) + e^- \longrightarrow Ag(s)}$$

(b) Since this reaction is taking place in an aqueous solution, the probable oxidation is the oxidation of water. (Neither $Ag^+$ nor $NO_3^-$ can be further oxidized.)

$$\mathbf{2H_2O(l) \longrightarrow O_2(g) + 4H^+(aq) + 4e^-}$$

(c) The half-reaction tells us that 1 mole of electrons is needed to reduce 1 mol of $Ag^+$ to Ag metal. We can set up the following strategy to calculate the quantity of electricity (in C) needed to deposit 0.67 g of Ag.

grams Ag → mol Ag → mol $e^-$ → coulombs

$$0.67\ \text{g Ag} \times \frac{1\ \text{mol Ag}}{107.9\ \text{g Ag}} \times \frac{1\ \text{mol } e^-}{1\ \text{mol Ag}} \times \frac{96500\ \text{C}}{1\ \text{mol } e^-} = \mathbf{6.0 \times 10^2\ C}$$

**18.56** **(a)** First find the amount of charge needed to produce 2.00 g of silver according to the half–reaction:

$$Ag^{+}(aq) + e^{-} \longrightarrow Ag(s)$$

$$2.00\ \cancel{g}\ Ag \times \frac{1\ \cancel{mol}\ Ag}{107.9\ \cancel{g}\ Ag} \times \frac{1\ \cancel{mol}\ e^{-}}{1\ \cancel{mol}\ Ag} \times \frac{96500\ C}{1\ \cancel{mol}\ e^{-}} = 1.79 \times 10^{3}\ C$$

The half–reaction for the reduction of copper(II) is:

$$Cu^{2+}(aq) + 2e^{-} \longrightarrow Cu(s)$$

From the amount of charge calculated above, we can calculate the mass of copper deposited in the second cell.

$$(1.79 \times 10^{3}\ \cancel{C}) \times \frac{1\ \cancel{mol}\ e^{-}}{96500\ \cancel{C}} \times \frac{1\ \cancel{mol}\ Cu}{2\ \cancel{mol}\ e^{-}} \times \frac{63.55\ g\ Cu}{1\ \cancel{mol}\ Cu} = \mathbf{0.589\ g\ Cu}$$

**(b)** We can calculate the current flowing through the cells using the following strategy.

Coulombs → Coulombs/hour → Coulombs/second

Recall that 1 C = 1 A·s

The current flowing through the cells is:

$$(1.79 \times 10^{3}\ A\cdot\cancel{s}) \times \frac{1\ \cancel{h}}{3600\ \cancel{s}} \times \frac{1}{3.75\ \cancel{h}} = \mathbf{0.133\ A}$$

**18.58** *Step 1:* Balance the half–reaction.

$$Cr_2O_7^{2-}(aq) + 14H^{+}(aq) + 12e^{-} \longrightarrow 2Cr(s) + 7H_2O(l)$$

*Step 2:* Calculate the quantity of chromium metal by calculating the volume and converting this to mass using the given density.

Volume Cr = thickness × surface area

$$\text{Volume Cr} = (1.0 \times 10^{-2}\ \cancel{mm}) \times \frac{1\ m}{1000\ \cancel{mm}} \times 0.25\ m^{2} = 2.5 \times 10^{-6}\ m^{3}$$

Converting to cm$^3$,

$$(2.5 \times 10^{-6}\ \cancel{m^{3}}) \times \left(\frac{1\ cm}{0.01\ \cancel{m}}\right)^{3} = 2.5\ cm^{3}$$

Next, calculate the mass of Cr.

Mass = density × volume

$$\text{Mass Cr} = 2.5\ \text{cm}^3 \times \frac{7.19\ \text{g}}{1\ \text{cm}^3} = 18\ \text{g Cr}$$

*Step 3:* Find the number of moles of electrons required to electrodeposit 18 g of Cr from solution. The half-reaction is:

$$Cr_2O_7^{2-}(aq) + 14H^+(aq) + 12e^- \longrightarrow 2Cr(s) + 7H_2O(l)$$

Six moles of electrons are required to reduce 1 mol of Cr metal. But, we are electrodepositing less than 1 mole of Cr(*s*). We need to complete the following conversions:

$$\text{g Cr} \rightarrow \text{mol Cr} \rightarrow \text{mol } e^-$$

$$?\ \text{faradays} = 18\ \text{g Cr} \times \frac{1\ \text{mol Cr}}{52.00\ \text{g Cr}} \times \frac{6\ \text{mol } e^-}{1\ \text{mol Cr}} = 2.1\ \text{mol } e^-$$

*Step 4:* Determine how long it will take for 2.1 moles of electrons to flow through the cell when the current is 25.0 C/s. We need to complete the following conversions:

$$\text{mol } e^- \rightarrow \text{coulombs} \rightarrow \text{seconds} \rightarrow \text{hours}$$

$$?\ \text{h} = 2.1\ \text{mol } e^- \times \frac{96{,}500\ \text{C}}{1\ \text{mol } e^-} \times \frac{1\ \text{s}}{25.0\ \text{C}} \times \frac{1\ \text{h}}{3600\ \text{s}} = \mathbf{2.3\ h}$$

Would any time be saved by connecting several bumpers together in a series?

**18.60** Based on the half-reaction, we know that one faraday will produce half a mole of copper.

$$Cu^{2+}(aq) + 2e^- \longrightarrow Cu(s)$$

First, let's calculate the charge (in C) needed to deposit 0.300 g of Cu.

$$(3.00\ \text{A})(304\ \text{s}) \times \frac{1\ \text{C}}{1\ \text{A}\cdot\text{s}} = 912\ \text{C}$$

We know that one faraday will produce half a mole of copper, but we don't have a half a mole of copper. We have:

$$0.300\ \text{g Cu} \times \frac{1\ \text{mol Cu}}{63.55\ \text{g Cu}} = 4.72 \times 10^{-3}\ \text{mol}$$

We calculated the number of coulombs (912 C) needed to produce $4.72 \times 10^{-3}$ mol of Cu. How many coulombs will it take to produce 0.500 moles of Cu? This will be Faraday's constant.

$$\frac{912\ \text{C}}{4.72 \times 10^{-3}\ \text{mol Cu}} \times 0.500\ \text{mol Cu} = \mathbf{9.66 \times 10^4\ C = 1\ \mathit{F}}$$

**18.62** First we can calculate the number of moles of hydrogen produced using the ideal gas equation.

$$n_{H_2} = \frac{PV}{RT}$$

$$n_{H_2} = \frac{\left(782\ \text{mmHg} \times \dfrac{1\ \text{atm}}{760\ \text{mmHg}}\right)(0.845\ \text{L})}{(0.0821\ \text{L}\cdot\text{atm/K}\cdot\text{mol})(298\ \text{K})} = 0.0355\ \text{mol}$$

The half-reaction in the problem shows that 2 moles of electrons are required to produce 1 mole of $H_2$.

$$0.0355\ \text{mol}\ H_2 \times \frac{2\ \text{mol}\ e^-}{1\ \text{mol}\ H_2} = \mathbf{0.0710\ mol\ e^-}$$

**18.64** (a) Anode: $Cu(s) \longrightarrow Cu^{2+}(aq) + 2e^-$

Cathode: $Cu^{2+}(aq) + 2e^- \longrightarrow Cu(s)$

(b) The quantity of charge passing through the solution is:

$$20\ \text{A} \times \frac{1\ \text{C}}{1\ \text{A}\cdot\text{s}} \times \frac{60\ \text{s}}{1\ \text{min}} \times \frac{1\ \text{mol}\ e^-}{96500\ \text{C}} \times \frac{60\ \text{min}}{1\ \text{h}} \times 10\ \text{h} = 7.5\ \text{mol}\ e^-$$

Since the charge of the copper ion is +2, the mass of copper purified is:

$$7.5\ \text{mol}\ e^- \times \frac{1\ \text{mol Cu}}{2\ \text{mol}\ e^-} \times \frac{63.55\ \text{g Cu}}{1\ \text{mol Cu}} = \mathbf{2.4 \times 10^2\ g\ Cu}$$

(c) Copper metal is more easily oxidized than Ag and Au. Copper ions, $Cu^{2+}$, are more easily reduced than $Fe^{2+}$ and $Zn^{2+}$ ions. See Table 18.1 of the text.

**18.66** All concentration cells have the same standard emf: *zero* volts. First, we calculate the concentrations of each solution from the osmotic pressures. Then, we use the Nernst equation to compute the cell emf. There are two moles of electrons transferred from the reducing agent to the oxidizing agent in this reaction, so $n = 2$.

$$\pi = iMRT$$

Soln A: $\pi = iMRT$

$$48.9\ \text{atm} = (2)M(0.0821\ \text{L}\cdot\text{atm/mol}\cdot\text{K})(298\ \text{K})$$

$$M = 0.999\ \text{mol/L}$$

Since the osmotic pressure of solution B is one-tenth that of solution A, the concentration is also one-tenth that of solution A. ($M_{\text{soln B}} = 0.0999$ mol/L). We substitute into the Nernst equation to calculate $E_{\text{cell}}$.

$$E = E^\circ - \frac{0.0257\ \text{V}}{n}\ln Q$$

$$E = E^\circ - \frac{0.0257\text{ V}}{n}\ln\frac{[Cu^{2+}]_{dil}}{[Cu^{2+}]_{conc}}$$

$$E = 0\text{ V} - \frac{0.0257\text{ V}}{2}\ln\frac{0.0999}{0.999} = \mathbf{0.0296\text{ V}}$$

**18.68** The balanced equation is:

$$\mathbf{Cr_2O_7^{2-} + 6\,Fe^{2+} + 14H^+ \longrightarrow 2Cr^{3+} + 6Fe^{3+} + 7H_2O}$$

The remainder of this problem is a solution stoichiometry problem.

The number of moles of potassium dichromate in 26.0 mL of the solution is:

$$26.0\text{ mL} \times \frac{0.0250\text{ mol}}{1000\text{ mL soln}} = 6.50 \times 10^{-4}\text{ mol }K_2Cr_2O_7$$

From the balanced equation it can be seen that 1 mole of dichromate is stoichiometrically equivalent to 6 moles of iron(II). The number of moles of iron(II) oxidized is therefore

$$(6.50 \times 10^{-4}\text{ mol }Cr_2O_7^{2-}) \times \frac{6\text{ mol }Fe^{2+}}{1\text{ mol }Cr_2O_7^{2-}} = 3.90 \times 10^{-3}\text{ mol }Fe^{2+}$$

Finally, the molar concentration of $Fe^{2+}$ is:

$$\frac{3.90 \times 10^{-3}\text{ mol}}{25.0 \times 10^{-3}\text{ L}} = 0.156\text{ mol/L} = \mathbf{0.156\ \mathit{M}\ Fe^{2+}}$$

**18.70** The balanced equation is:

$$MnO_4^- + 5Fe^{2+} + 8H^+ \longrightarrow Mn^{2+} + 5Fe^{3+} + 4H_2O$$

First, let's calculate the number of moles of potassium permanganate in 23.30 mL of solution.

$$23.30\text{ mL} \times \frac{0.0194\text{ mol}}{1000\text{ mL soln}} = 4.52 \times 10^{-4}\text{ mol }KMnO_4$$

From the balanced equation it can be seen that 1 mole of permanganate is stoichiometrically equivalent to 5 moles of iron(II). The number of moles of iron(II) oxidized is therefore

$$(4.52 \times 10^{-4}\text{ mol }MnO_4^-) \times \frac{5\text{ mol }Fe^{2+}}{1\text{ mol }MnO_4^-} = 2.26 \times 10^{-3}\text{ mol }Fe^{2+}$$

The mass of $Fe^{2+}$ oxidized is:

$$\text{mass }Fe^{2+} = (2.26 \times 10^{-3}\text{ mol }Fe^{2+}) \times \frac{55.85\text{ g }Fe^{2+}}{1\text{ mol }Fe^{2+}} = 0.126\text{ g }Fe^{2+}$$

Finally, the mass percent of iron in the ore can be calculated.

$$\text{mass \% Fe} = \frac{\text{mass of iron}}{\text{total mass of sample}} \times 100\%$$

$$\%\text{Fe} = \frac{0.126\text{ g}}{0.2792\text{ g}} \times 100\% = \mathbf{45.1\%}$$

**18.72** **(a)** The half–reactions are:

(i) $MnO_4^-(aq) + 8H^+(aq) + 5e^- \longrightarrow Mn^{2+}(aq) + 4H_2O(l)$

(ii) $C_2O_4^{2-}(aq) \longrightarrow 2CO_2(g) + 2e^-$

We combine the half-reactions to cancel electrons, that is, [2 × equation (i)] + [5 × equation (ii)]

$$\mathbf{2MnO_4^-(aq) + 16H^+(aq) + 5C_2O_4^{2-}(aq) \longrightarrow 2Mn^{2+}(aq) + 10CO_2(g) + 8H_2O(l)}$$

**(b)** We can calculate the moles of $KMnO_4$ from the molarity and volume of solution.

$$24.0\text{ mL } KMnO_4 \times \frac{0.0100\text{ mol } KMnO_4}{1000\text{ mL soln}} = 2.40 \times 10^{-4}\text{ mol } KMnO_4$$

We can calculate the mass of oxalic acid from the stoichiometry of the balanced equation. The mole ratio between oxalate ion and permanganate ion is 5:2.

$$(2.40 \times 10^{-4}\text{ mol } KMnO_4) \times \frac{5\text{ mol } H_2C_2O_4}{2\text{ mol } KMnO_4} \times \frac{90.04\text{ g } H_2C_2O_4}{1\text{ mol } H_2C_2O_4} = 0.0540\text{ g } H_2C_2O_4$$

Finally, the percent by mass of oxalic acid in the sample is:

$$\textbf{\% oxalic acid} = \frac{0.0540\text{ g}}{1.00\text{ g}} \times 100\% = \mathbf{5.40\%}$$

**18.74** The balanced equation is:

$$2MnO_4^- + 5C_2O_4^{2-} + 16H^+ \longrightarrow 2Mn^{2+} + 10CO_2 + 8H_2O$$

Therefore, 2 mol $MnO_4^-$ reacts with 5 mol $C_2O_4^{2-}$.

$$\text{Moles of } MnO_4^- \text{ reacted} = 24.2\text{ mL} \times \frac{9.56 \times 10^{-4}\text{ mol } MnO_4^-}{1000\text{ mL soln}} = 2.31 \times 10^{-5}\text{ mol } MnO_4^-$$

Recognize that the mole ratio of $Ca^{2+}$ to $C_2O_4^{2-}$ is 1:1 in $CaC_2O_4$. The mass of $Ca^{2+}$ in 10.0 mL is:

$$(2.31 \times 10^{-5}\text{ mol } MnO_4^-) \times \frac{5\text{ mol } Ca^{2+}}{2\text{ mol } MnO_4^-} \times \frac{40.08\text{ g } Ca^{2+}}{1\text{ mol } Ca^{2+}} = 2.31 \times 10^{-3}\text{ g } Ca^{2+}$$

Finally, converting to mg/mL, we have:

$$\frac{2.31\times10^{-3}\ \text{g}\ Ca^{2+}}{10.0\ \text{mL}}\times\frac{1000\ \text{mg}}{1\ \text{g}} = \mathbf{0.231\ mg\ Ca^{2+}/mL\ blood}$$

**18.76** (a) The half−reactions are:

$$2H^+(aq) + 2e^- \longrightarrow H_2(g) \qquad E^\circ_{\text{anode}} = 0.00\ \text{V}$$

$$Ag^+(aq) + e^- \longrightarrow Ag(s) \qquad E^\circ_{\text{cathode}} = 0.80\ \text{V}$$

$$E^\circ_{\text{cell}} = E^\circ_{\text{cathode}} - E^\circ_{\text{anode}} = 0.80\ \text{V} - 0.00\ \text{V} = \mathbf{0.80\ V}$$

**(b)** The spontaneous cell reaction under standard-state conditions is:

$$\mathbf{2Ag^+(aq) + H_2(g) \longrightarrow 2Ag(s) + 2H^+(aq)}$$

(c) Using the Nernst equation we can calculate the cell potential under nonstandard-state conditions.

$$E = E^\circ - \frac{0.0257\ \text{V}}{n}\ln\frac{[H^+]^2}{[Ag^+]^2 P_{H_2}}$$

(i) The potential is:

$$E = 0.80\ \text{V} - \frac{0.0257\ \text{V}}{2}\ln\frac{(1.0\times10^{-2})^2}{(1.0)^2(1.0)} = \mathbf{0.92\ V}$$

(ii) The potential is:

$$E = 0.80\ \text{V} - \frac{0.0257\ \text{V}}{2}\ln\frac{(1.0\times10^{-5})^2}{(1.0)^2(1.0)} = \mathbf{1.10\ V}$$

(d) From the results in part (c), we deduce that this cell is a pH meter; its potential is a sensitive function of the hydrogen ion concentration. Each 1 unit increase in pH causes a voltage increase of 0.060 V.

**18.78** The overvoltage of oxygen is not large enough to prevent its formation at the anode. Applying the diagonal rule, we see that water is oxidized before fluoride ion.

$$F_2(g) + 2e^- \longrightarrow 2F^-(aq) \qquad E^\circ = 2.87\ \text{V}$$

$$O_2(g) + 4H^+(aq) + 4e^- \longrightarrow 2H_2O(l) \qquad E^\circ = 1.23\ \text{V}$$

The very positive standard reduction potential indicates that $F^-$ has essentially no tendency to undergo oxidation. The oxidation potential of chloride ion is much smaller (−1.36 V), and hence $Cl_2(g)$ can be prepared by electrolyzing a solution of NaCl.

This fact was one of the major obstacles preventing the discovery of fluorine for many years. HF was usually chosen as the substance for electrolysis, but two problems interfered with the experiment. First, any water in the HF was oxidized before the fluoride ion. Second, pure HF without any water in it is a nonconductor of electricity (HF is a weak acid!). The problem was finally solved by dissolving KF in liquid HF to give a conducting solution.

**18.80** We can calculate the amount of charge that 4.0 g of $MnO_2$ can produce.

$$4.0\ \text{g MnO}_2 \times \frac{1\ \text{mol}}{86.94\ \text{g}} \times \frac{2\ \text{mol}\ e^-}{2\ \text{mol MnO}_2} \times \frac{96500\ \text{C}}{1\ \text{mol}\ e^-} = 4.44 \times 10^3\ \text{C}$$

Since a current of one ampere represents a flow of one coulomb per second, we can find the time it takes for this amount of charge to pass.

$0.0050\ \text{A} = 0.0050\ \text{C/s}$

$$(4.44 \times 10^3\ \text{C}) \times \frac{1\ \text{s}}{0.0050\ \text{C}} \times \frac{1\ \text{h}}{3600\ \text{s}} = \mathbf{2.5 \times 10^2\ h}$$

**18.82** Since this is a concentration cell, the standard emf is zero. (Why?) Using the Nernst equation, we can write equations to calculate the cell voltage for the two cells.

$$(1) \quad E_{\text{cell}} = -\frac{RT}{nF}\ln Q = -\frac{RT}{2F}\ln\frac{[\text{Hg}_2^{2+}]\text{soln A}}{[\text{Hg}_2^{2+}]\text{soln B}}$$

$$(2) \quad E_{\text{cell}} = -\frac{RT}{nF}\ln Q = -\frac{RT}{1F}\ln\frac{[\text{Hg}^{+}]\text{soln A}}{[\text{Hg}^{+}]\text{soln B}}$$

In the first case, two electrons are transferred per mercury ion ($n = 2$), while in the second only one is transferred ($n = 1$). Note that the concentration ratio will be 1:10 in both cases. The voltages calculated at 18°C are:

$$(1) \quad E_{\text{cell}} = \frac{-(8.314\ \text{J/K}\cdot\text{mol})(291\ \text{K})}{2(96500\ \text{J}\cdot\text{V}^{-1}\text{mol}^{-1})}\ln 10^{-1} = 0.0289\ \text{V}$$

$$(2) \quad E_{\text{cell}} = \frac{-(8.314\ \text{J/K}\cdot\text{mol})(291\ \text{K})}{1(96500\ \text{J}\cdot\text{V}^{-1}\text{mol}^{-1})}\ln 10^{-1} = 0.0577\ \text{V}$$

Since the calculated cell potential for cell (1) agrees with the measured cell emf, we conclude that the mercury(I) ion exists as $\mathbf{Hg_2^{2+}}$ in solution.

**18.84** We begin by treating this like an ordinary stoichiometry problem (see Chapter 3).

*Step 1:* Calculate the number of moles of Mg and $Ag^+$.

The number of moles of magnesium is:

$$1.56\ \text{g Mg} \times \frac{1\ \text{mol Mg}}{24.31\ \text{g Mg}} = 0.0642\ \text{mol Mg}$$

The number of moles of silver ion in the solution is:

$$\frac{0.100\ \text{mol Ag}^+}{1\ \text{L}} \times 0.1000\ \text{L} = 0.0100\ \text{mol Ag}^+$$

*Step 2:* Calculate the mass of Mg remaining by determining how much Mg reacts with $Ag^+$.

The balanced equation for the reaction is:

$$2\text{Ag}^+(aq) + \text{Mg}(s) \longrightarrow 2\text{Ag}(s) + \text{Mg}^{2+}(aq)$$

Since you need twice as much $Ag^+$ compared to Mg for complete reaction, $Ag^+$ is the limiting reagent. The amount of Mg consumed is:

$$0.0100\ \text{mol Ag}^+ \times \frac{1\ \text{mol Mg}}{2\ \text{mol Ag}^+} = 0.00500\ \text{mol Mg}$$

The amount of magnesium remaining is:

$$(0.0642 - 0.00500)\ \text{mol Mg} \times \frac{24.31\ \text{g Mg}}{1\ \text{mol Mg}} = \mathbf{1.44\ g\ Mg}$$

*Step 3:* Assuming complete reaction, calculate the concentration of $Mg^{2+}$ ions produced.

Since the mole ratio between Mg and $Mg^{2+}$ is 1:1, the mol of $Mg^{2+}$ formed will equal the mol of Mg reacted. The concentration of $Mg^{2+}$ is:

$$[\text{Mg}^{2+}]_0 = \frac{0.00500\ \text{mol}}{0.100\ \text{L}} = 0.0500\ M$$

*Step 4:* We can calculate the equilibrium constant for the reaction from the standard cell emf.

$$E^\circ_{\text{cell}} = E^\circ_{\text{cathode}} - E^\circ_{\text{anode}} = 0.80\ \text{V} - (-2.37\ \text{V}) = 3.17\ \text{V}$$

We can then compute the equilibrium constant.

$$K = e^{\frac{nE^\circ_{\text{cell}}}{0.0257}}$$

$$K = e^{\frac{(2)(3.17)}{0.0257}} = 1 \times 10^{107}$$

*Step 5:* To find equilibrium concentrations of $Mg^{2+}$ and $Ag^+$, we have to solve an equilibrium problem.

Let $x$ be the small amount of $Mg^{2+}$ that reacts to achieve equilibrium. The concentration of $Ag^+$ will be $2x$ at equilibrium. Assume that essentially all $Ag^+$ has been reduced so that the initial concentration of $Ag^+$ is zero.

| | $2Ag^+(aq)$ + $Mg(s)$ | $\rightleftharpoons$ | $2Ag(s)$ + $Mg^{2+}(aq)$ |
|---|---|---|---|
| Initial ($M$): | 0.0000 | | 0.0500 |
| Change ($M$): | $+2x$ | | $-x$ |
| Equilibrium ($M$): | $2x$ | | $(0.0500 - x)$ |

$$K = \frac{[Mg^{2+}]}{[Ag^+]^2}$$

$$1 \times 10^{107} = \frac{(0.0500 - x)}{(2x)^2}$$

We can assume $0.0500 - x \approx 0.0500$.

$$1 \times 10^{107} \approx \frac{0.0500}{(2x)^2}$$

$$(2x)^2 = \frac{0.0500}{1 \times 10^{107}} = 0.0500 \times 10^{-107}$$

$$(2x)^2 = 5.00 \times 10^{-109} = 50.0 \times 10^{-110}$$

$$2x = 7 \times 10^{-55}\ M$$

$$[Ag^+] = 2x = \mathbf{7 \times 10^{-55}}\ \boldsymbol{M}$$

$$[Mg^{2+}] = 0.0500 - x = \mathbf{0.0500}\ \boldsymbol{M}$$

**18.86** (a) Since this is an acidic solution, the gas must be hydrogen gas from the reduction of hydrogen ion. The two electrode reactions and the overall cell reaction are:

anode: $Cu(s) \longrightarrow Cu^{2+}(aq) + 2e^-$

cathode: $2H^+(aq) + 2e^- \longrightarrow H_2(g)$

$$Cu(s) + 2H^+(aq) \longrightarrow Cu^{2+}(aq) + H_2(g)$$

Since 0.584 g of copper was consumed, the amount of hydrogen gas produced is:

$$0.584\ \text{g Cu} \times \frac{1\ \text{mol Cu}}{63.55\ \text{g Cu}} \times \frac{1\ \text{mol } H_2}{1\ \text{mol Cu}} = 9.20 \times 10^{-3}\ \text{mol } H_2$$

At STP, 1 mole of an ideal gas occupies a volume of 22.41 L. Thus, the volume of $H_2$ at STP is:

$$V_{H_2} = (9.20 \times 10^{-3}\ \text{mol } H_2) \times \frac{22.41\ \text{L}}{1\ \text{mol}} = \mathbf{0.206\ L}$$

**(b)** From the current and the time, we can calculate the amount of charge:

$$1.18\,\cancel{A} \times \frac{1\text{ C}}{1\,\cancel{A \cdot s}} \times (1.52 \times 10^3\,\cancel{s}) = 1.79 \times 10^3\text{ C}$$

Since we know the charge of an electron, we can compute the number of electrons.

$$(1.79 \times 10^3\,\cancel{C}) \times \frac{1\,e^-}{1.6022 \times 10^{-19}\,\cancel{C}} = 1.12 \times 10^{22}\,e^-$$

Using the amount of copper consumed in the reaction and the fact that 2 mol of $e^-$ are produced for every 1 mole of copper consumed, we can calculate Avogadro's number.

$$\frac{1.12 \times 10^{22}\,e^-}{9.20 \times 10^{-3}\,\cancel{\text{mol}}\text{ Cu}} \times \frac{1\,\cancel{\text{mol}}\text{ Cu}}{2\text{ mol }e^-} = \mathbf{6.09 \times 10^{23}\,/\,mol}\;\boldsymbol{e^-}$$

In practice, Avogadro's number can be determined by electrochemical experiments like this. The charge of the electron can be found independently by Millikan's experiment.

**18.88** **(a)** We can calculate $\Delta G°$ from standard free energies of formation.

$$\Delta G° = 2\Delta G_f^\circ(N_2) + 6\Delta G_f^\circ(H_2O) - [4\Delta G_f^\circ(NH_3) + 3\Delta G_f^\circ(O_2)]$$

$$\Delta G = 0 + (6)(-237.2\text{ kJ/mol}) - [(4)(-16.6\text{ kJ/mol}) + 0]$$

$$\boldsymbol{\Delta G = -1356.8\text{ kJ/mol}}$$

**(b)** The half-reactions are:

$$4NH_3(g) \longrightarrow 2N_2(g) + 12H^+(aq) + 12e^-$$

$$3O_2(g) + 12H^+(aq) + 12e^- \longrightarrow 6H_2O(l)$$

The overall reaction is a 12-electron process. We can calculate the standard cell emf from the standard free energy change, $\Delta G°$.

$$\Delta G° = -nFE^\circ_{cell}$$

$$E^\circ_{cell} = \frac{-\Delta G°}{nF} = \frac{-\left(\dfrac{-1356.8\,\cancel{kJ}}{1\,\cancel{mol}} \times \dfrac{1000\,\cancel{J}}{1\,\cancel{kJ}}\right)}{(12)(96500\,\cancel{J}/V \cdot \cancel{mol})} = \mathbf{1.17\ V}$$

**18.90** The reduction of $Ag^+$ to Ag metal is:

$$Ag^+(aq) + e^- \longrightarrow Ag$$

We can calculate both the moles of Ag deposited and the moles of Au deposited.

$$? \text{ mol Ag} = 2.64 \not{g} \text{ Ag} \times \frac{1 \text{ mol Ag}}{107.9 \not{g} \text{ Ag}} = 2.45 \times 10^{-2} \text{ mol Ag}$$

$$? \text{ mol Au} = 1.61 \not{g} \text{ Au} \times \frac{1 \text{ mol Au}}{197.0 \not{g} \text{ Au}} = 8.17 \times 10^{-3} \text{ mol Au}$$

We do not know the oxidation state of Au ions, so we will represent the ions as $Au^{n+}$. If we divide the mol of Ag by the mol of Au, we can determine the ratio of $Ag^+$ reduced compared to $Au^{n+}$ reduced.

$$\frac{2.45 \times 10^{-2} \text{ mol Ag}}{8.17 \times 10^{-3} \text{ mol Au}} = 3$$

That is, the same number of electrons that reduced the $Ag^+$ ions to Ag reduced only one-third the number of moles of the $Au^{n+}$ ions to Au. Thus, each $Au^{n+}$ required three electrons per ion for every one electron for $Ag^+$. The oxidation state for the gold ion is **+3**; the ion is $Au^{3+}$.

$$Au^{3+}(aq) + 3e^- \longrightarrow Au$$

**18.92** We reverse the first half–reaction and add it to the second to come up with the overall balanced equation

$$Hg_2^{2+} \longrightarrow 2Hg^{2+} + 2e^- \qquad E^\circ_{\text{anode}} = +0.92 \text{ V}$$

$$Hg_2^{2+} + 2e^- \longrightarrow 2Hg \qquad E^\circ_{\text{cathode}} = +0.85 \text{ V}$$

$$2Hg_2^{2+} \longrightarrow 2Hg^{2+} + 2Hg \qquad E^\circ_{\text{cell}} = 0.85 \text{ V} - 0.92 \text{ V} = -0.07 \text{ V}$$

Since the standard cell potential is an intensive property,

$$Hg_2^{2+}(aq) \longrightarrow Hg^{2+}(aq) + Hg(l) \qquad E^\circ_{\text{cell}} = -0.07 \text{ V}$$

We calculate $\Delta G^\circ$ from $E^\circ$.

$$\boldsymbol{\Delta G^\circ} = -nFE^\circ = -(1)(96500 \text{ J/}\not{V}\cdot\text{mol})(-0.07 \not{V}) = \mathbf{6.8 \text{ kJ/mol}}$$

The corresponding equilibrium constant is:

$$K = \frac{[Hg^{2+}]}{[Hg_2^{2+}]}$$

We calculate $K$ from $\Delta G^\circ$.

$$\Delta G^\circ = -RT\ln K$$

$$\ln K = \frac{-6.8 \times 10^3 \not{J}/\not{mol}}{(8.314 \not{J}/\not{K}\cdot\not{mol})(298 \not{K})}$$

$$\mathbf{K = 0.064}$$

**18.94** In both cells, the anode is on the left and the cathode is on the right, but the signs of the anodes (and those of the cathodes) are opposite to each other. In the galvanic cell, the anode is negatively charged and the cathode is positively charged. In an electrolytic cell, an external voltage source (either from a battery or a galvanic cell) causes the anode to be positively charged with respect to the cathode. Thus, electrons flow from the anode in the galvanic cell to the cathode in the electrolytic cell and electrons flow from the anode in the electrolytic cell to the cathode in the galvanic cell. In the solutions, the anions migrate toward the anode and the cations migrate toward the cathode.

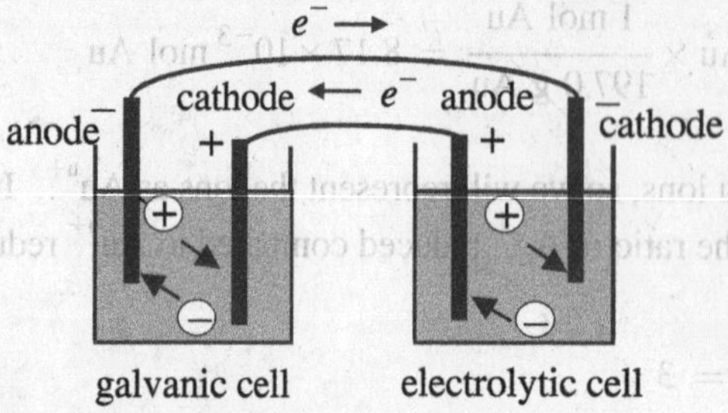


**18.96** The reactions for the electrolysis of NaCl(*aq*) are:

| | |
|---|---|
| Anode: | $2Cl^-(aq) \longrightarrow Cl_2(g) + 2e^-$ |
| Cathode: | $2H_2O(l) + 2e^- \longrightarrow H_2(g) + 2OH^-(aq)$ |
| Overall: | $2H_2O(l) + 2Cl^-(aq) \longrightarrow H_2(g) + Cl_2(g) + 2OH^-(aq)$ |

From the pH of the solution, we can calculate the $OH^-$ concentration. From the $[OH^-]$, we can calculate the moles of $OH^-$ produced. Then, from the moles of $OH^-$ we can calculate the average current used.

$$pH = 12.24$$

$$pOH = 14.00 - 12.24 = 1.76$$

$$[OH^-] = 1.74 \times 10^{-2}\,M$$

The moles of $OH^-$ produced are:

$$\frac{1.74 \times 10^{-2}\text{ mol}}{1\text{ L}} \times 0.300\text{ L} = 5.22 \times 10^{-3}\text{ mol OH}^-$$

From the balanced equation, it takes 1 mole of $e^-$ to produce 1 mole of $OH^-$ ions.

$$(5.22 \times 10^{-3}\text{ mol OH}^-) \times \frac{1\text{ mol }e^-}{1\text{ mol OH}^-} \times \frac{96500\text{ C}}{1\text{ mol }e^-} = 504\text{ C}$$

Recall that 1 C = 1 A·s

$$504\text{ C} \times \frac{1\text{ A}\cdot\text{s}}{1\text{ C}} \times \frac{1\text{ min}}{60\text{ s}} \times \frac{1}{6.00\text{ min}} = \mathbf{1.4\text{ A}}$$

**18.98** The reaction is:

$$Pt^{n+} + ne^- \longrightarrow Pt$$

Thus, we can calculate the charge of the platinum ions by realizing that *n* mol of $e^-$ are required per mol of Pt formed.

The moles of Pt formed are:

$$9.09\ \text{g Pt} \times \frac{1\ \text{mol Pt}}{195.1\ \text{g Pt}} = 0.0466\ \text{mol Pt}$$

Next, calculate the charge passed in C.

$$C = 2.00\ \text{h} \times \frac{3600\ \text{s}}{1\ \text{h}} \times \frac{2.50\ \text{C}}{1\ \text{s}} = 1.80 \times 10^4\ \text{C}$$

Convert to moles of electrons.

$$?\ \text{mol}\ e^- = (1.80 \times 10^4\ \text{C}) \times \frac{1\ \text{mol}\ e^-}{96500\ \text{C}} = 0.187\ \text{mol}\ e^-$$

We now know the number of moles of electrons (0.187 mol $e^-$) needed to produce 0.0466 mol of Pt metal. We can calculate the number of moles of electrons needed to produce 1 mole of Pt metal.

$$\frac{0.187\ \text{mol}\ e^-}{0.0466\ \text{mol Pt}} = 4.01\ \text{mol}\ e^-/\text{mol Pt}$$

Since we need 4 moles of electrons to reduce 1 mole of Pt ions, the charge on the Pt ions must be **+4**.

**18.100** The half–reaction for the oxidation of water to oxygen is:

$$2H_2O(l) \xrightarrow{\text{oxidation (anode)}} O_2(g) + 4H^+(aq) + 4e^-$$

Knowing that one mole of any gas at STP occupies a volume of 22.41 L, we find the number of moles of oxygen.

$$4.26\ \text{L}\ O_2 \times \frac{1\ \text{mol}}{22.41\ \text{L}} = 0.190\ \text{mol}\ O_2$$

Since four electrons are required to form one oxygen molecule, the number of electrons must be:

$$0.190\ \text{mol}\ O_2 \times \frac{4\ \text{mol e}^-}{1\ \text{mol}\ O_2} \times \frac{6.022 \times 10^{23}\ e^-}{1\ \text{mol}} = 4.58 \times 10^{23}\ e^-$$

The amount of charge passing through the solution is:

$$6.00\ \text{A} \times \frac{1\ \text{C}}{1\ \text{A}\cdot\text{s}} \times \frac{3600\ \text{s}}{1\ \text{h}} \times 3.40\ \text{h} = 7.34 \times 10^4\ \text{C}$$

We find the electron charge by dividing the amount of charge by the number of electrons.

$$\frac{7.34 \times 10^4\ \text{C}}{4.58 \times 10^{23}\ e^-} = \mathbf{1.60 \times 10^{-19}\ C/e^-}$$

In actual fact, this sort of calculation can be used to find Avogadro's number, not the electron charge. The latter can be measured independently, and one can use this charge together with electrolytic data like the above to calculate the number of objects in one mole. See also Problem 18.86.

**18.102** Cells of higher voltage require very reactive oxidizing and reducing agents, which are difficult to handle. (From Table 18.1 of the text, we see that 5.92 V is the theoretical limit of a cell made up of $Li^+/Li$ and $F_2/F^-$ electrodes under standard-state conditions.) Batteries made up of several cells in series are easier to use.

**18.104** In the compartment with the lower pressure of $Cl_2$ (0.20 atm), $Cl^-$ will be oxidized to $Cl_2$. This is the anode. In the second compartment with the higher pressure of $Cl_2$ (2.00 atm), $Cl_2$ will be reduced to $Cl^-$. This is the cathode. The cell diagram is:

$$\mathbf{Pt(\mathit{s})|Cl_2(0.20\ atm)|Cl^-(1.0\ \mathit{M})||Cl^-(1.0\ \mathit{M})|Cl_2(2.0\ atm)|Pt(\mathit{s})}$$

Anode: $2Cl^-(aq) \longrightarrow Cl_2(g, 0.20\text{ atm}) + 2e^-$ $\qquad E^\circ_{\text{anode}} = 1.36\text{ V}$

Cathode: $Cl_2(g, 2.0\text{ atm}) + 2e^- \longrightarrow 2Cl^-(aq)$ $\qquad E^\circ_{\text{cathode}} = 1.36\text{ V}$

$Cl_2(2.0\text{ atm}) \rightarrow Cl_2(0.20\text{ atm})$ $\qquad E^\circ_{\text{cell}} = 0$

As for all concentration cells, the standard emf is zero. We use the Nernst equation to calculate the cell emf ($E_{\text{cell}}$).

$$E = E^\circ - \frac{0.0257\text{ V}}{n}\ln Q$$

$$E = E^\circ - \frac{0.0257\text{ V}}{n}\ln\frac{P_{Cl_2}}{P_{Cl_2}}$$

$$E = 0\text{ V} - \frac{0.0257\text{ V}}{2}\ln\frac{0.20}{2.0} = \mathbf{0.030\ V}$$

**18.106** The half-reactions are:

$Zn(s) + 4OH^-(aq) \rightarrow Zn(OH)_4^{2-}(aq) + 2e^-$ $\qquad E^\circ_{\text{anode}} = -1.36\text{ V}$

$Zn^{2+}(aq) + 2e^- \rightarrow Zn(s)$ $\qquad E^\circ_{\text{cathode}} = -0.76\text{ V}$

$Zn^{2+}(aq) + 4OH^-(aq) \rightarrow Zn(OH)_4^{2-}(aq)$ $\qquad E^\circ_{\text{cell}} = -0.76\text{ V} - (-1.36\text{ V}) = 0.60\text{ V}$

$$E^\circ_{\text{cell}} = \frac{0.0257\text{ V}}{n}\ln K_f$$

$$K_f = e^{\frac{nE^\circ}{0.0257}} = e^{\frac{(2)(0.60)}{0.0257}} = \mathbf{2 \times 10^{20}}$$

**18.108** **(a)** Since electrons flow from X to SHE, $E^\circ$ for X must be negative. Thus $E^\circ$ for Y must be positive.

**(b)** $Y^{2+} + 2e^- \rightarrow Y$ $\qquad E^\circ_{\text{cathode}} = 0.34\text{ V}$

$X \rightarrow X^{2+} + 2e^-$ $\qquad E^\circ_{\text{anode}} = -0.25\text{ V}$

$X + Y^{2+} \rightarrow X^{2+} + Y$ $\qquad E^\circ_{\text{cell}} = 0.34\text{ V} - (-0.25\text{ V}) = \mathbf{0.59\ V}$

**18.110** (a) Gold does not tarnish in air because the reduction potential for oxygen is insufficient to result in the oxidation of gold.

$$O_2 + 4H^+ + 4e^- \rightarrow 2H_2O \qquad E^\circ_{cathode} = 1.23\text{ V}$$

That is, $E^\circ_{cell} = E^\circ_{cathode} - E^\circ_{anode} < 0$, for either oxidation by $O_2$ to $Au^+$ or $Au^{3+}$.

$$E^\circ_{cell} = 1.23\text{ V} - 1.50\text{ V} < 0$$

or

$$E^\circ_{cell} = 1.23\text{ V} - 1.69\text{ V} < 0$$

(b) $3(Au^+ + e^- \rightarrow Au) \qquad E^\circ_{cathode} = 1.69\text{ V}$

$Au \rightarrow Au^{3+} + 3e^- \qquad E^\circ_{anode} = 1.50\text{ V}$

$3Au^+ \rightarrow 2Au + Au^{3+} \qquad E^\circ_{cell} = 1.69\text{ V} - 1.50\text{ V} = 0.19\text{ V}$

Calculating $\Delta G$,

$$\Delta G^\circ = -nFE^\circ = -(3)(96{,}500\text{ J/V}\cdot\text{mol})(0.19\text{ V}) = -55.0\text{ kJ/mol}$$

For spontaneous electrochemical equations, $\Delta G^\circ$ must be negative. Thus, **the disproportionation occurs spontaneously**.

(c) Since the most stable oxidation state for gold is $Au^{3+}$, the predicted reaction is:

$$\mathbf{2Au + 3F_2 \rightarrow 2AuF_3}$$

**18.112** The balanced equation is: $5Fe^{2+} + MnO_4^- + 8H^+ \longrightarrow Mn^{2+} + 5Fe^{3+} + 4H_2O$

Calculate the amount of iron(II) in the original solution using the mole ratio from the balanced equation.

$$23.0\text{ mL} \times \frac{0.0200\text{ mol KMnO}_4}{1000\text{ mL soln}} \times \frac{5\text{ mol Fe}^{2+}}{1\text{ mol KMnO}_4} = 0.00230\text{ mol Fe}^{2+}$$

The concentration of iron(II) must be:

$$[Fe^{2+}] = \frac{0.00230\text{ mol}}{0.0250\text{ L}} = \mathbf{0.0920\ \mathit{M}}$$

The total iron concentration can be found by simple proportion because the same sample volume (25.0 mL) and the same $KMnO_4$ solution were used.

$$[Fe]_{total} = \frac{40.0\text{ mL KMnO}_4}{23.0\text{ mL KMnO}_4} \times 0.0920\ M = 0.160\ M$$

$$[Fe^{3+}] = [Fe]_{total} - [Fe^{2+}] = \mathbf{0.0680\ \mathit{M}}$$

Why are the two titrations with permanganate necessary in this problem?

**18.114** From Table 18.1 of the text.

$$H_2O_2(aq) + 2H^+(aq) + 2e^- \rightarrow 2H_2O(l) \qquad E^\circ_{cathode} = 1.77\ V$$

$$H_2O_2(aq) \rightarrow O_2(g) + 2H^+(aq) + 2e^- \qquad E^\circ_{anode} = 0.68\ V$$

$$2H_2O_2(aq) \rightarrow 2H_2O(l) + O_2(g) \qquad E^\circ_{cell} = E^\circ_{cathode} - E^\circ_{anode} = 1.77\ V - (0.68\ V) = \mathbf{1.09\ V}$$

Because $E^\circ$ is positive, the decomposition is **spontaneous.**

**18.116** **(a)** Anode: $Mg(s) \longrightarrow Mg^{2+}(aq) + 2e^- \qquad E^\circ_{anode} = -2.37\ V$

Cathode: $X^{2+}(aq) + 2e^- \longrightarrow X(s) \qquad E^\circ_{cathode} = ?$

We solve for $E^\circ_{cathode}$ to identify X.

$$E^\circ_{cell} = E^\circ_{cathode} - E^\circ_{anode}$$

$$2.12\ V = E^\circ_{cathode} - (-2.37\ V)$$

$$E^\circ_{cathode} = -0.25\ V$$

Checking Table 18.1 of the text, **X is Ni.**

**(b)** We solve for $E^\circ_{cathode}$ to identify X.

$$E^\circ_{cell} = E^\circ_{cathode} - E^\circ_{anode}$$

$$2.24\ V = E^\circ_{cathode} - (-2.37\ V)$$

$$E^\circ_{cathode} = -0.13\ V$$

Checking Table 18.1 of the text, **X is Pb.**

**(c)** We solve for $E^\circ_{cathode}$ to identify X.

$$E^\circ_{cell} = E^\circ_{cathode} - E^\circ_{anode}$$

$$1.61\ V = E^\circ_{cathode} - (-2.37\ V)$$

$$E^\circ_{cathode} = -0.76\ V$$

Checking Table 18.1 of the text, **X is Zn.**

**(d)** We solve for $E^\circ_{cathode}$ to identify X.

$$E^\circ_{cell} = E^\circ_{cathode} - E^\circ_{anode}$$

$$1.93\text{ V} = E^{\circ}_{\text{cathode}} - (-2.37\text{ V})$$

$$E^{\circ}_{\text{cathode}} = -0.44\text{ V}$$

Checking Table 18.1 of the text, **X** is **Fe.**

**18.118** **(a)** unchanged **(b)** unchanged **(c)** squared **(d)** doubled **(e)** doubled

**18.120** $F_2\,(g) + 2H^+\,(aq) + 2e^- \rightarrow 2HF\,(g)$

$$E = E^{\circ} - \frac{RT}{2F}\ln\frac{P_{HF}^2}{P_{F_2}[H^+]^2}$$

With increasing $[H^+]$, $E$ will be larger. $F_2$ will become a **stronger oxidizing agent.**

**18.122**

$$Pb \rightarrow Pb^{2+} + 2e^- \qquad E^{\circ}_{\text{anode}} = -0.13\text{ V}$$

$$2H^+ + 2e^- \rightarrow H_2 \qquad E^{\circ}_{\text{cathode}} = 0.00\text{ V}$$

$$Pb + 2H^+ \rightarrow Pb^{2+} + H_2 \qquad E^{\circ}_{\text{cell}} = 0.00\text{ V} - (-0.13\text{ V}) = 0.13\text{ V}$$

$$\text{pH} = 1.60$$

$$[H^+] = 10^{-1.60} = 0.025\ M$$

$$E = E^{\circ} - \frac{RT}{nF}\ln\frac{[Pb^{2+}]P_{H_2}}{[H^+]^2}$$

$$0 = 0.13 - \frac{0.0257\text{ V}}{2}\ln\frac{(0.035)P_{H_2}}{0.025^2}$$

$$\frac{0.26}{0.0257} = \ln\frac{(0.035)P_{H_2}}{0.025^2}$$

$$P_{H_2} = \mathbf{4.4 \times 10^2\ atm}$$

**18.124** **(a)** The half-reactions are:

Anode: $Zn \rightarrow Zn^{2+} + 2e^-$

Cathode: $\frac{1}{2}O_2 + 2e^- \rightarrow O^{2-}$

Overall: $Zn + \frac{1}{2}O_2 \rightarrow ZnO$

To calculate the standard emf, we first need to calculate $\Delta G°$ for the reaction. From Appendix 3 of the text we write:

$$\Delta G° = \Delta G_f°(\text{ZnO}) - [\Delta G_f°(\text{Zn}) + \tfrac{1}{2}\Delta G_f°(\text{O}_2)]$$

$$\Delta G° = -318.2 \text{ kJ/mol} - [0 + 0]$$

$$\Delta G° = -318.2 \text{ kJ/mol}$$

$$\Delta G° = -nFE°$$

$$-318.2 \times 10^3 \text{ J/mol} = -(2)(96{,}500 \text{ J/V·mol})E°$$

$$E° = \mathbf{1.65\ V}$$

**(b)** We use the following equation:

$$E = E° - \frac{RT}{nF}\ln Q$$

$$E = 1.65 \text{ V} - \frac{0.0257 \text{ V}}{2}\ln\frac{1}{P_{O_2}}$$

$$E = 1.65 \text{ V} - \frac{0.0257 \text{ V}}{2}\ln\frac{1}{0.21}$$

$$E = 1.65 \text{ V} - 0.020 \text{ V}$$

$$E = \mathbf{1.63\ V}$$

**(c)** Since the free energy change represents the maximum work that can be extracted from the overall reaction, the maximum amount of energy that can be obtained from this reaction is the free energy change. To calculate the energy density, we multiply the free energy change by the number of moles of Zn present in 1 kg of Zn.

$$\textbf{energy density} = \frac{318.2 \text{ kJ}}{1 \text{ mol Zn}} \times \frac{1 \text{ mol Zn}}{65.39 \text{ g Zn}} \times \frac{1000 \text{ g Zn}}{1 \text{ kg Zn}} = \mathbf{4.87 \times 10^3\ kJ/kg\ Zn}$$

**(d)** One ampere is 1 C/s. The charge drawn every second is given by $nF$.

$$\text{charge} = nF$$

$$2.1 \times 10^5 \text{ C} = n(96{,}500 \text{ C/mol } e^-)$$

$$n = 2.2 \text{ mol } e^-$$

From the overall balanced reaction, we see that 4 moles of electrons will reduce 1 mole of $O_2$; therefore, the number of moles of $O_2$ reduced by 2.2 moles of electrons is:

$$\text{mol } O_2 = 2.2 \text{ mol } e^- \times \frac{1 \text{ mol } O_2}{4 \text{ mol } e^-} = 0.55 \text{ mol } O_2$$

The volume of oxygen at 1.0 atm partial pressure can be obtained by using the ideal gas equation.

$$V_{O_2} = \frac{nRT}{P} = \frac{(0.55 \text{ mol})(0.0821 \text{ L·atm/mol·K})(298 \text{ K})}{(1.0 \text{ atm})} = 13 \text{ L}$$

Since air is 21 percent oxygen by volume, the volume of air required every second is:

$$V_{air} = 13\text{ L }O_2 \times \frac{100\%\text{ air}}{21\%\ O_2} = \mathbf{62\text{ L of air}}$$

**18.126** We can calculate $\Delta G^\circ_{rxn}$ using the following equation.

$$\Delta G^\circ_{rxn} = \Sigma n\Delta G^\circ_f(\text{products}) - \Sigma m\Delta G^\circ_f(\text{reactants})$$

$$\Delta G^\circ_{rxn} = 0 + 0 - [(1)(-293.8\text{ kJ/mol}) + 0] = 293.8\text{ kJ/mol}$$

Next, we can calculate $E^\circ$ using the equation

$$\Delta G^\circ = -nFE^\circ$$

We use a more accurate value for Faraday's constant.

$$293.8 \times 10^3\text{ J/mol} = -(1)(96485.3\text{ J/V}\cdot\text{mol})E^\circ$$

$$\mathbf{E^\circ = -3.05\text{ V}}$$

**18.128** First, we need to calculate $E^\circ_{cell}$, then we can calculate $K$ from the cell potential.

$$H_2(g) \rightarrow 2H^+(aq) + 2e^- \qquad E^\circ_{anode} = 0.00\text{ V}$$

$$2H_2O(l) + 2e^- \rightarrow H_2(g) + 2OH^- \qquad E^\circ_{cathode} = -0.83\text{ V}$$

$$2H_2O(l) \rightarrow 2H^+(aq) + 2OH^-(aq) \qquad E^\circ_{cell} = -0.83\text{ V} - 0.00\text{ V} = -0.83\text{ V}$$

We want to calculate $K$ for the reaction: $H_2O(l) \rightarrow H^+(aq) + OH^-(aq)$. The cell potential for this reaction will be the same as the above reaction, but the moles of electrons transferred, $n$, will equal one.

$$E^\circ_{cell} = \frac{0.0257\text{ V}}{n}\ln K_w$$

$$\ln K_w = \frac{nE^\circ_{cell}}{0.0257\text{ V}}$$

$$K_w = e^{\frac{nE^\circ}{0.0257}}$$

$$\mathbf{K_w} = e^{\frac{(1)(-0.83)}{0.0257}} = e^{-32} = \mathbf{1 \times 10^{-14}}$$

**18.130** **(a)** $1\text{A}\cdot\text{h} = 1\text{A} \times 3600\text{s} = 3600\text{ C}$

**(b)** Anode: $Pb + SO_4^{2-} \rightarrow PbSO_4 + 2e^-$

Two moles of electrons are produced by 1 mole of Pb. Recall that the charge of 1 mol $e^-$ is 96,500 C. We can set up the following conversions to calculate the capacity of the battery.

mol Pb $\rightarrow$ mol $e^-$ $\rightarrow$ coulombs $\rightarrow$ ampere hour

$$406\text{ g Pb} \times \frac{1\text{ mol Pb}}{207.2\text{ g Pb}} \times \frac{2\text{ mol }e^-}{1\text{ mol Pb}} \times \frac{96500\text{ C}}{1\text{ mol }e^-} = (3.74 \times 10^5\text{ C}) \times \frac{1\text{ h}}{3600\text{ s}} = \mathbf{104\ A \cdot h}$$

This ampere·hour cannot be fully realized because the concentration of $H_2SO_4$ keeps decreasing.

(c) $E^\circ_{\text{cell}} = 1.70\text{ V} - (-0.31\text{ V}) = \mathbf{2.01\ V}$ (From Table 18.1 of the text)

$$\Delta G^\circ = -nFE^\circ$$

$$\Delta G^\circ = -(2)(96500\text{ J/V}\cdot\text{mol})(2.01\text{ V}) = \mathbf{-3.88 \times 10^5\ J/mol}$$

Spontaneous as expected.

**18.132** The surface area of an open cylinder is $2\pi rh$. The surface area of the culvert is

$$2\pi(0.900\text{ m})(40.0\text{ m}) \times 2\text{ (for both sides of the iron sheet)} = 452\text{ m}^2$$

Converting to units of $\text{cm}^2$,

$$452\text{ m}^2 \times \left(\frac{100\text{ cm}}{1\text{ m}}\right)^2 = 4.52 \times 10^6\text{ cm}^2$$

The volume of the Zn layer is

$$0.200\text{ mm} \times \frac{1\text{ cm}}{10\text{ mm}} \times (4.52 \times 10^6\text{ cm}^2) = 9.04 \times 10^4\text{ cm}^3$$

The mass of Zn needed is

$$(9.04 \times 10^4\text{ cm}^3) \times \frac{7.14\text{ g}}{1\text{ cm}^3} = 6.45 \times 10^5\text{ g Zn}$$

$$Zn^{2+} + 2e^- \rightarrow Zn$$

$$Q = (6.45 \times 10^5\text{ g Zn}) \times \frac{1\text{ mol Zn}}{65.39\text{ g Zn}} \times \frac{2\text{ mol }e^-}{1\text{ mol Zn}} \times \frac{96500\text{ C}}{1\text{ mol }e^-} = 1.90 \times 10^9\text{ C}$$

$$1\text{ J} = 1\text{ C} \times 1\text{ V}$$

$$\text{Total energy} = \frac{(1.90 \times 10^9\text{ C})(3.26\text{ V})}{0.95 \leftarrow \text{(efficiency)}} = 6.52 \times 10^9\text{ J}$$

$$\text{Cost} = (6.52 \times 10^9\text{ J}) \times \frac{1\text{ kw}}{1000\frac{\text{J}}{\text{s}}} \times \frac{1\text{ h}}{3600\text{ s}} \times \frac{\$0.12}{1\text{ kwh}} = \mathbf{\$217}$$

**18.134** It might appear that because the sum of the first two half-reactions gives Equation (3), $E_3^\circ$ is given by $E_1^\circ + E_2^\circ = 0.33$ V. This is not the case, however, because emf is not an extensive property. We cannot set $E_3^\circ = E_1^\circ + E_2^\circ$. On the other hand, the Gibbs energy is an extensive property, so we can add the separate Gibbs energy changes to obtain the overall Gibbs energy change.

$$\Delta G_3^\circ = \Delta G_1^\circ + \Delta G_2^\circ$$

Substituting the relationship $\Delta G^\circ = -nFE^\circ$, we obtain

$$n_3FE_3^\circ = n_1FE_1^\circ + n_2FE_2^\circ$$

$$E_3^\circ = \frac{n_1E_1^\circ + n_2E_2^\circ}{n_3}$$

$n_1 = 2$, $n_2 = 1$, and $n_3 = 3$.

$$E_3^\circ = \frac{(2)(-0.44\text{ V}) + (1)(0.77\text{ V})}{3} = \mathbf{-0.037\text{ V}}$$

**18.136** First, calculate the standard emf of the cell from the standard reduction potentials in Table 18.1 of the text. Then, calculate the equilibrium constant from the standard emf using Equation (18.5) of the text.

$$E_{\text{cell}}^\circ = E_{\text{cathode}}^\circ - E_{\text{anode}}^\circ = 0.34\text{ V} - (-0.76\text{ V}) = 1.10\text{ V}$$

$$\ln K = \frac{nE_{\text{cell}}^\circ}{0.0257\text{ V}}$$

$$K = e^{\frac{nE_{\text{cell}}^\circ}{0.0257\text{ V}}} = e^{\frac{(2)(1.10\text{ V})}{0.0257\text{ V}}}$$

$$\mathbf{K = 2 \times 10^{37}}$$

The very large equilibrium constant means that the oxidation of Zn by $Cu^{2+}$ is virtually complete.

**18.138** The standard free-energy change, $\Delta G^\circ$, can be calculated from the cell potential.

$$E_{\text{cell}}^\circ = E_{\text{cathode}}^\circ - E_{\text{anode}}^\circ = 1.23\text{ V} - 0.42\text{ V} = 0.81\text{ V}$$

$$\Delta G^\circ = -nFE_{\text{cell}}^\circ$$

$$\Delta G^\circ = -(4)(96{,}500\text{ J/V}\cdot\text{mol})(0.81\text{ V})$$

$$\Delta G^\circ = -3.13 \times 10^5\text{ J/mol} = -313\text{ kJ/mol}$$

This is the free-energy change for the oxidation of 2 moles of nitrite ($NO_2^-$)

Looking at Section 17.7 of the text, we find that it takes 31 kJ of free energy to synthesize 1 mole of ATP from ADP. The yield of ATP synthesis per mole of nitrite oxidized is:

$$156.5\ \text{kJ} \times \frac{1\ \text{mol ATP}}{31\ \text{kJ}} = \mathbf{5.0\ mol\ ATP}$$

**18.140** We follow the procedure shown in Problem 18.126.

$$F_2(g) + H_2(g) \rightarrow 2H^+(aq) + 2F^-(aq)$$

We can calculate $\Delta G^\circ_{rxn}$ using the following equation and data in Appendix 3 of the text.

$$\Delta G^\circ_{rxn} = \Sigma n\Delta G^\circ_f(\text{products}) - \Sigma m\Delta G^\circ_f(\text{reactants})$$

$$\Delta G^\circ_{rxn} = [0 + (2)(-276.48\ \text{kJ/mol})] - [0 + 0] = -552.96\ \text{kJ/mol}$$

Next, we can calculate $E^\circ$ using the equation

$$\Delta G^\circ = -nFE^\circ$$

We use a more accurate value for Faraday's constant.

$$-552.96 \times 10^3\ \text{J/mol} = -(2)(96485.3\ \text{J/V}\cdot\text{mol})E^\circ$$

$$E^\circ = \mathbf{2.87\ V}$$

# CHAPTER 19
# NUCLEAR CHEMISTRY

## PROBLEM-SOLVING STRATEGIES AND TUTORIAL SOLUTIONS

### TYPES OF PROBLEMS

**Problem Type 1:** Balancing Nuclear Equations.

**Problem Type 2:** Nuclear Stability

**Problem Type 3:** Calculating Nuclear Binding Energy.

**Problem Type 4:** Kinetics of Radioactive Decay.

**Problem Type 5:** Balancing Nuclear Transmutation Equations.

## PROBLEM TYPE 1: BALANCING NUCLEAR EQUATIONS

Writing a nuclear equation differs slightly from writing equations for chemical reactions. In addition to writing the symbols for various chemical elements, we must also explicitly indicate protons, neutrons, and electrons. In fact, we must show the numbers of protons and neutrons present in *every* species in the reaction.

You must know the symbols for elementary particles in order to balance a nuclear equation.

| $^{1}_{1}\text{p}$ or $^{1}_{1}\text{H}$ | $^{1}_{0}\text{n}$ | $^{0}_{-1}e$ or $^{0}_{-1}\beta$ | $^{0}_{+1}e$ or $^{0}_{+1}\beta$ | $^{4}_{2}\text{He}$ or $^{4}_{2}\alpha$ |
|---|---|---|---|---|
| proton | neutron | electron | positron | α particle |

In balancing any nuclear equation, we observe the following rules:

- The total number of protons plus neutrons in the products and in the reactants must be the same (conservation of mass number).
- The total number of nuclear charges in the products and in the reactants must be the same (conservation of atomic number).

If the atomic numbers and mass numbers of all the species but one in a nuclear equation are known, the unknown species can be identified by applying the above rules.

**EXAMPLE 19.1**
**Identify X in the following nuclear reactions:**

(a) $$^{14}_{7}\text{N} + ^{1}_{0}\text{n} \longrightarrow ^{14}_{6}\text{C} + \text{X}$$

(b) $$^{226}\text{Ra} \longrightarrow ^{4}_{2}\alpha + \text{X}$$

**Strategy:** In balancing nuclear equations, note that the sum of atomic numbers and that of mass numbers must match on both sides of the equation.

**Solution:**

**(a)** Because the sum of the mass numbers must be conserved, the unknown product will have a mass number of 1. Because the sum of the atomic numbers must be conserved, the nuclear charge of the unknown product must be 1.

The particle is a proton.

$$^{14}_{7}N + ^{1}_{0}n \longrightarrow ^{14}_{6}C + ^{1}_{1}\mathbf{p}$$

**(b)** Note that the atomic number of radium is missing. Look at a periodic table to find that Ra is element number 88. Balancing the mass numbers first, we find that the unknown product must have a mass of 222. Balancing the nuclear charges, we find that the nuclear charge of the unknown must be 86. Element number 86 is radon (Rn).

$$^{226}_{88}Ra \longrightarrow ^{4}_{2}\alpha + ^{222}_{86}\mathbf{Rn}$$

**PRACTICE EXERCISE**

1. Identify X in the following nuclear reactions:

(a) $^{239}_{94}Pu \longrightarrow ^{235}_{92}U + X$

(b) $^{90}_{38}Sr \longrightarrow ^{90}_{37}Rb + X$

**Text Problems: 19.8**, 19.30

## PROBLEM TYPE 2: NUCLEAR STABILITY

The principal factor for determining the stability of a nucleus is the *neutron-to-proton ratio* (*n*/*p*). For stable elements of low atomic number, the *n*/*p* ratio is close to 1. As the atomic number increases, the *n*/*p* ratios of stable nuclei become greater than 1. The deviation in the *n*/*p* ratio at higher atomic numbers arises because a larger number of neutrons is needed to stabilize the nucleus by counteracting the strong repulsion among the large number of protons. Figure 19.1 of the text shows a plot of the number of neutrons versus the number of protons in various isotopes. The stable nuclei are located in an area of the graph called the *belt of stability*. Most radioactive nuclei lie outside the belt.

The following rules are useful in predicting nuclear stability.

- Nuclei that contain 2, 8, 20, 50, 82, or 126 protons or neutrons are generally more stable than nuclei that do not possess these numbers. These numbers are called *magic numbers*.
- Nuclei with even numbers of both protons and neutrons are generally more stable than those with odd numbers of these particles (see Table 19.2 of the text).
- All isotopes of the elements starting with polonium (Po, Z = 84) are radioactive. All isotopes of technetium (Tc, Z = 43) and promethium (Pm, Z = 61) are also radioactive.

**EXAMPLE 19.2**

**Rank the following nuclei in order of increasing nuclear stability:**

$^{40}_{20}\text{Ca}$ $^{39}_{20}\text{Ca}$ $^{11}_{5}\text{B}$

**Strategy:** The principal factor for determining the stability of a nucleus is the *neutron-to-proton ratio* (*n*/*p*). For stable elements of low atomic number, the *n*/*p* ratio is close to 1. As the atomic number increases, the *n*/*p* ratios of stable nuclei become greater than 1. The following rules are useful in predicting nuclear stability.

1) Nuclei that contain 2, 8, 20, 50, 82, or 126 protons or neutrons are generally more stable than nuclei that do not possess these numbers. These numbers are called *magic numbers*.

2) Nuclei with even numbers of both protons and neutrons are generally more stable than those with odd numbers of these particles (see Table 19.2 of the text).

**Solution:**

Boron-11 has both an odd number of protons and an odd number of neutrons; therefore, it should be the least stable of the three nuclei. Calcium-39 has a magic number of protons (20), but an odd number of neutrons. Calcium-40 has both a magic number of protons and neutrons (20), so it should be more stable than calcium-39. The order of increasing nuclear stability is

$$^{11}_{5}\text{B} < {}^{39}_{20}\text{Ca} < {}^{40}_{20}\text{Ca}$$

**PRACTICE EXERCISE**

2. Rank the following isotopes in order of increasing nuclear stability:

$^{39}_{20}\text{Ca}$ $^{222}_{86}\text{Rn}$ $^{98}_{43}\text{Tc}$

**Text Problems: 19.16**, 19.18

## PROBLEM TYPE 3: CALCULATING NUCLEAR BINDING ENERGY

A quantitative measure of nuclear stability is the **nuclear binding energy**, which is the energy required to break up a nucleus into its component protons and neutrons. The concept of nuclear binding energy evolved from studies showing that the masses of nuclei are always less than the sum of the masses of the **nucleons** (the protons and neutrons in a nucleus). The difference between the mass of an atom and the sum of the masses of its protons, neutrons, and electrons is called the **mass defect**. According to Einstein's mass-energy equivalence relationship

$$E = mc^2$$

where,

$E$ is energy
$m$ is mass
$c$ is the velocity of light

the loss in mass shows up as energy (heat) given off to the surroundings. We can calculate the amount of energy released by writing:

$$\Delta E = (\Delta m)c^2$$

where,

$$\Delta E = \text{energy of products} - \text{energy of reactants}$$

$$\Delta m = \text{mass of products} - \text{mass of reactants}$$

See Example 19.3 below for a detailed calculation.

**EXAMPLE 19.3**

**Calculate the nuclear binding energy of the light isotope of helium, $^3$He. The atomic mass of $^3_2$He is 3.01603 amu.**

---

**Strategy:** To calculate the nuclear binding energy, we first determine the difference between the mass of the nucleus and the mass of all the protons and neutrons, which gives us the mass defect. Next, we apply Einstein's mass-energy relationship [$\Delta E = (\Delta m)c^2$].

**Solution:** The binding energy is the energy required for the process

$$^3_2\text{He} \longrightarrow 2\ ^1_1\text{p} + ^1_0\text{n}$$

There are 2 protons and 1 neutron in the helium nucleus. The mass of 2 protons is

$$(2)(1.007825 \text{ amu}) = 2.015650 \text{ amu}$$

and the mass of a neutron is 1.008665 amu

Therefore, the predicted mass of $^3_2$He is 2.015650 + 1.008665 = 3.024315 amu, and the mass defect is

$$\Delta m = 3.024315 \text{ amu} - 3.01603 \text{ amu} = 0.00829 \text{ amu}$$

The energy change ($\Delta E$) for the process is

$$\Delta E = (\Delta m)c^2$$

$$\Delta E = (0.00829 \text{ amu})(3.00 \times 10^8 \text{ m/s})^2$$

$$\Delta E = 7.46 \times 10^{14} \frac{\text{amu} \cdot \text{m}^2}{\text{s}^2}$$

Let's convert to more familiar energy units (J/He atom).

$$7.46 \times 10^{14} \frac{\text{amu} \cdot \text{m}^2}{\text{s}^2} \times \frac{1.00 \text{ g}}{6.022 \times 10^{23} \text{ amu}} \times \frac{1 \text{ kg}}{1000 \text{ g}} \times \frac{1 \text{ J}}{\frac{1 \text{ kg} \cdot \text{m}^2}{\text{s}^2}} = \mathbf{1.24 \times 10^{-12} \text{ J/atom}}$$

This is the nuclear binding energy. It's the energy required to break up one helium-3 nucleus into 2 protons and 1 neutron.

When comparing the stability of any two nuclei we must account for the fact that they have different numbers of nucleons. For this reason, it is more meaningful to use the *nuclear binding energy per nucleon*, defined as

$$\text{nuclear binding energy per nucleon} = \frac{\text{nuclear binding energy}}{\text{number of nucleons}}$$

For the helium-3 nucleus,

$$\text{nuclear binding energy per nucleon} = \frac{1.24 \times 10^{-12}\ \text{J/He}\ \cancel{\text{atom}}}{3\ \text{nucleons/He}\ \cancel{\text{atom}}} = \mathbf{4.13 \times 10^{-13}\ J/nucleon}$$

**PRACTICE EXERCISE**

**3.** (a) Calculate the binding energy and the binding energy per nucleon of $^{27}_{13}\text{Al}$. The atomic mass of $^{27}_{13}\text{Al}$ is 26.98154 amu.

(b) Compare the result from part (a) to the binding energy of $^{28}_{14}\text{Si}$, which has an even number of protons and neutrons. The atomic mass of $^{28}_{14}\text{Si}$ is 27.976928 amu.

**Text Problem: 19.22**

## PROBLEM TYPE 4: KINETICS OF RADIOACTIVE DECAY

All radioactive decays obey first-order kinetics. To review first-order reactions, see Sections 13.2 and 13.3 of the text. The decay rate at any time *t* is given by

$$\text{rate of decay at time } t = \lambda N$$

where,

$\lambda$ is the first-order rate constant
$N$ is the number of radioactive nuclei present at time $t$

The relationship between the number of radioactive nuclei present at time zero ($N_0$) and the number of nuclei remaining at a later time $t$ ($N_t$) is given by:

$$\ln\frac{N_t}{N_0} = -\lambda t$$

The corresponding half-life for a first-order reaction is given by:

$$t_{\frac{1}{2}} = \frac{0.693}{\lambda}$$

Unlike ordinary chemical reactions, the rate constants for nuclear reactions are unaffected by changes in environmental conditions such as temperature and pressure.

**EXAMPLE 19.4**

**Cobalt-60 is used in radiation therapy. It has a half-life of 5.26 years.**

**(a) Calculate the rate constant for radioactive decay.**

**(b) What fraction of a certain sample will remain after 12 years?**

---

**Strategy:** (a) According to Equation (13.6) of the text, the half-life of first-order decay is:

$$t_{\frac{1}{2}} = \frac{0.693}{\lambda}$$

We can solve for the first-order rate constant (λ) given the half-life.

(b) According to Equation (13.3) of the text, the number of radioactive nuclei at time zero ($N_0$) and time $t$ ($N_t$) is:

$$\ln\frac{N_t}{N_0} = -\lambda t$$

We want to calculate the fraction of the sample remaining, which will equal $\frac{N_t}{N_0}$.

**Solution:**

**(a)** The rate constant for the decay can be calculated from the half-life.

$$t_{\frac{1}{2}} = \frac{0.693}{\lambda}$$

Rearrange the equation to solve for the rate constant (λ).

$$\lambda = \frac{0.693}{t_{\frac{1}{2}}} = \frac{0.693}{5.26\text{ yr}} = \mathbf{0.132\ yr^{-1}}$$

**(b)** The fraction of a certain sample that will remain after 12 years is

$$\frac{N_t}{N_0}$$

where $t = 12$ yr.

Rearrange the equation, $\ln\frac{N_t}{N_0} = -\lambda t$, to solve for $\frac{N_t}{N_0}$.

$$\ln\frac{N_t}{N_0} = -\lambda t$$

$$\frac{N_t}{N_0} = e^{-\lambda t}$$

$$\frac{N_t}{N_0} = e^{-(0.132\ \cancel{\text{yr}}^{-1})(12\ \cancel{\text{yr}})} = \mathbf{0.205}$$

In other words, 20.5 percent of the original sample will remain after 12 years.

**PRACTICE EXERCISE**

**4.** Estimate the age of a bottle of wine that has a tritium $\left({}_{1}^{3}\text{H}\right)$ content 75.0 percent of the tritium content of environmental water obtained from the area where the grapes were grown. $t_{\frac{1}{2}} = 12.3$ yr.

**Text Problems: 19.26,** 19.28, 19.32

## PROBLEM TYPE 5: BALANCING NUCLEAR TRANSMUTATION EQUATIONS

Nuclei can undergo change as a result of bombardment by neutrons, protons, or other nuclei. This process is called **nuclear transmutation**. Unlike radioactive decay, nuclear transmutation is *not* a spontaneous process; consequently, nuclear transmutation reactions have more than one reactant. As an example, consider the synthesis of neptunium (Np), which was the first transuranium element to be synthesized by scientists.

First, uranium-238 is bombarded with neutrons to produce uranium-239.

$$^{238}_{92}\text{U} + {}^{1}_{0}\text{n} \longrightarrow {}^{239}_{92}\text{U}$$

This is a nuclear transmutation reaction. Uranium-239 is unstable and decays spontaneously to neptunium-239 by emitting a β particle.

$$^{239}_{92}\text{U} \longrightarrow {}^{239}_{93}\text{Np} + {}^{0}_{-1}\beta$$

To balance a nuclear transmutation reaction, follow the same rules used to balance nuclear equations, Problem Type 1.

- The total number of protons plus neutrons in the products and in the reactants must be the same (conservation of mass number).
- The total number of nuclear charges in the products and in the reactants must be the same (conservation of atomic number).

**EXAMPLE 19.5**

**Write and balance the following reactions. (a) When aluminum-27 is bombarded with α particles, phosphorus-30 and one other particle are produced. (b) Phosphorus-30 has a low *n/p* ratio and decays spontaneously by positron emission.**

**Strategy:** (a) The first reaction is a nuclear transmutation. You are given both reactants and one of the two products in the problem. To balance the equation, remember that both mass number and atomic number must be conserved. (b) The second reaction is spontaneous so there is only one reactant, phosphorus-30. The problem indicates that phosphorus-30 decays by positron emission.

**Solution:**

(a) Let's start by writing what is given in the problem.

$$^{27}_{13}\text{Al} + {}^{4}_{2}\alpha \longrightarrow {}^{30}_{15}\text{P} + \text{X}$$

To balance the mass number, the missing particle (X) must have a mass number of 1. To balance the atomic number, X must have an atomic number of 0. X must be a neutron.

$$^{27}_{13}\text{Al} + ^{4}_{2}\alpha \longrightarrow ^{30}_{15}\text{P} + ^{1}_{0}\text{n}$$

(b) The second reaction is spontaneous so there is only one reactant, phosphorus-30. The problem indicates that phosphorus-30 decays by positron emission. Let's write down what we know so far.

$$^{30}_{15}\text{P} \longrightarrow \text{X} + ^{0}_{+1}\beta$$

To balance the mass number, the missing element (X) must have a mass number of 30. To balance the atomic number, X must have an atomic number of 14. X must be silicon-30.

$$^{30}_{15}\text{P} \longrightarrow ^{30}_{14}\text{Si} + ^{0}_{+1}\beta$$

**PRACTICE EXERCISE**

5. Write and balance the following reactions. When chlorine-37 is bombarded with neutrons, only one product is produced. The product is unstable and spontaneously decays by beta emission.

**Text Problems:** 19.38, 19.40

## ANSWERS TO PRACTICE EXERCISES

1. (a) $X = ^{4}_{2}\text{He}$

   (b) $X = ^{0}_{+1}\beta$

2. Technetium-98 should be unstable because it has an odd number of both protons and neutrons. In fact, all isotopes of technetium are radioactive. Radon-222 will perhaps be slightly more stable than technetium-98, because it has an even number of both protons and neutrons. Remember, however, that all isotopes of the elements starting with polonium (Po, Z = 84) are radioactive. Calcium-39 has 20 protons (a "magic number") and should be the most stable of the three isotopes. The correct order of increasing nuclear stability is

$$^{98}_{43}\text{Tc} \quad < \quad ^{222}_{86}\text{Rn} \quad < \quad ^{39}_{20}\text{Ca}$$

3. (a) Binding energy $= 3.6 \times 10^{-11}$ J/Al atom $= 1.3 \times 10^{-12}$ J/nucleon

   (b) Binding energy $= 3.8 \times 10^{-11}$ J/Si atom $= 1.4 \times 10^{-12}$ J/nucleon

   When comparing the stability of any two nuclei, it is best to compare the binding energy in units of J/nucleon. Silicon has the greater binding energy per nucleon and hence is more stable than Al. You should have expected this result, because Si has an even number of protons and neutrons; whereas, Al has an odd number of protons and an even number of neutrons. Nuclei with even numbers of both protons and neutrons are generally more stable than those with odd numbers of these particles.

**4.** You can calculate the rate constant (λ) from the half-life.

$$\lambda = 0.0563\ \text{yr}^{-1}$$

Next, $\dfrac{N_t}{N_0} = 0.750$. The age ($t$) of the bottle of wine can then be calculated from the equation,

$$\ln\frac{N_t}{N_0} = -\lambda t$$

$$t = 5.07\ \text{yr}$$

**5.** $^{1}_{0}\text{n} + {}^{37}_{17}\text{Cl} \longrightarrow {}^{38}_{17}\text{Cl}$

$^{38}_{17}\text{Cl} \longrightarrow {}^{38}_{18}\text{Ar} + {}^{0}_{-1}\beta$

## SOLUTIONS TO SELECTED TEXT PROBLEMS

**19.8** **Strategy:** In balancing nuclear equations, note that the sum of atomic numbers and that of mass numbers must match on both sides of the equation.

**Solution:**

**(a)** The sum of the mass numbers must be conserved. Thus, the unknown product will have a mass number of 0. The atomic number must be conserved. Thus, the nuclear charge of the unknown product must be −1. The particle is a β particle.

$$^{135}_{53}\text{I} \longrightarrow {}^{135}_{54}\text{Xe} + {}^{0}_{-1}\beta$$

**(b)** Balancing the mass numbers first, we find that the unknown product must have a mass of 40. Balancing the nuclear charges, we find that the atomic number of the unknown must be 20. Element number 20 is calcium (Ca).

$$^{40}_{19}\text{K} \longrightarrow {}^{0}_{-1}\beta + {}^{40}_{20}\text{Ca}$$

**(c)** Balancing the mass numbers, we find that the unknown product must have a mass of 4. Balancing the nuclear charges, we find that the nuclear charge of the unknown must be 2. The unknown particle is an alpha (α) particle.

$$^{59}_{27}\text{Co} + {}^{1}_{0}\text{n} \longrightarrow {}^{56}_{25}\text{Mn} + {}^{4}_{2}\alpha$$

**(d)** Balancing the mass numbers, we find that the unknown products must have a combined mass of 2. Balancing the nuclear charges, we find that the combined nuclear charge of the two unknown particles must be 0. The unknown particles are neutrons.

$$^{235}_{92}\text{U} + {}^{1}_{0}\text{n} \longrightarrow {}^{99}_{40}\text{Zr} + {}^{135}_{52}\text{Te} + 2\,{}^{1}_{0}\text{n}$$

**19.16** **Strategy:** The principal factor for determining the stability of a nucleus is the *neutron-to-proton ratio* (*n*/*p*). For stable elements of low atomic number, the *n*/*p* ratio is close to 1. As the atomic number increases, the *n*/*p* ratios of stable nuclei become greater than 1. The following rules are useful in predicting nuclear stability.

1) Nuclei that contain 2, 8, 20, 50, 82, or 126 protons or neutrons are generally more stable than nuclei that do not possess these numbers. These numbers are called *magic numbers*.

2) Nuclei with even numbers of both protons and neutrons are generally more stable than those with odd numbers of these particles (see Table 19.2 of the text).

**Solution:**

(a) **Lithium-9** should be less stable. The neutron-to-proton ratio is too high. For small atoms, the *n*/*p* ratio will be close to 1:1.

(b) **Sodium-25** is less stable. Its neutron-to-proton ratio is probably too high.

(c) **Scandium-48** is less stable because of odd numbers of protons and neutrons. We would not expect calcium-48 to be stable even though it has a magic number of protons. Its *n*/*p* ratio is too high.

**19.18** (a) **Neon-17** should be radioactive. It falls below the belt of stability (low *n*/*p* ratio).

(b) **Calcium-45** should be radioactive. It falls above the belt of stability (high *n*/*p* ratio).

(c) All **technetium** isotopes are radioactive.

(d) **Mercury-195** should be radioactive. Mercury-196 has an even number of both neutrons and protons.

(e) All **curium** isotopes are unstable.

**19.20** We can use the equation, $\Delta E = \Delta mc^2$, to solve the problem. Recall the following conversion factor:

$$1\text{ J} = \frac{1\text{ kg}\cdot\text{m}^2}{\text{s}^2}$$

The energy loss in one second is:

$$\Delta m = \frac{\Delta E}{c^2} = \frac{\dfrac{5\times 10^{26}\text{ kg}\cdot\cancel{\text{m}^2}}{1\,\cancel{\text{s}^2}}}{\left(3.00\times 10^8\,\dfrac{\cancel{\text{m}}}{\cancel{\text{s}}}\right)^2} = 6\times 10^9\text{ kg}$$

Therefore the rate of mass loss is $\mathbf{6 \times 10^9}$ **kg/s**.

**19.22** **Strategy:** To calculate the nuclear binding energy, we first determine the difference between the mass of the nucleus and the mass of all the protons and neutrons, which gives us the mass defect. Next, we apply Einstein's mass-energy relationship [$\Delta E = (\Delta m)c^2$].

**Solution:**

**(a)** The binding energy is the energy required for the process

$$^{4}_{2}\text{He} \rightarrow 2\,^{1}_{1}\text{p} + 2\,^{1}_{0}\text{n}$$

There are 2 protons and 2 neutrons in the helium nucleus. The mass of 2 protons is

$$(2)(1.007825 \text{ amu}) = 2.015650 \text{ amu}$$

and the mass of 2 neutrons is

$$(2)(1.008665 \text{ amu}) = 2.017330 \text{ amu}$$

Therefore, the predicted mass of $^{4}_{2}\text{He}$ is 2.015650 + 2.017330 = 4.032980 amu, and the mass defect is

$$\Delta m = 4.0026 \text{ amu} - 4.032980 \text{ amu} = -0.0304 \text{ amu}$$

The energy change ($\Delta E$) for the process is

$$\Delta E = (\Delta m)c^2$$

$$= (-0.0304 \text{ amu})(3.00 \times 10^8 \text{ m/s})^2$$

$$= -2.74 \times 10^{15} \frac{\text{amu}\cdot\text{m}^2}{\text{s}^2}$$

Let's convert to more familiar energy units (J/He atom).

$$\frac{-2.74 \times 10^{15} \text{ amu}\cdot\text{m}^2}{1 \text{ s}^2} \times \frac{1.00 \text{ g}}{6.022 \times 10^{23} \text{ amu}} \times \frac{1 \text{ kg}}{1000 \text{ g}} \times \frac{1 \text{ J}}{\frac{1 \text{ kg}\cdot\text{m}^2}{\text{s}^2}} = -4.55 \times 10^{-12} \text{ J}$$

The nuclear binding energy is **$4.55 \times 10^{-12}$ J**. It's the energy required to break up one helium-4 nucleus into 2 protons and 2 neutrons.

When comparing the stability of any two nuclei we must account for the fact that they have different numbers of nucleons. For this reason, it is more meaningful to use the *nuclear binding energy per nucleon*, defined as

$$\text{nuclear binding energy per nucleon} = \frac{\text{nuclear binding energy}}{\text{number of nucleons}}$$

For the helium-4 nucleus,

$$\text{nuclear binding energy per nucleon} = \frac{4.55 \times 10^{-12} \text{ J/He atom}}{4 \text{ nucleons/He atom}} = \mathbf{1.14 \times 10^{-12} \text{ J/nucleon}}$$

**(b)** The binding energy is the energy required for the process

$$^{184}_{74}\text{W} \rightarrow 74\ ^{1}_{1}\text{p} + 110\ ^{1}_{0}\text{n}$$

There are 74 protons and 110 neutrons in the tungsten nucleus. The mass of 74 protons is

$$(74)(1.007825 \text{ amu}) = 74.57905 \text{ amu}$$

and the mass of 110 neutrons is

$$(110)(1.008665 \text{ amu}) = 110.9532 \text{ amu}$$

Therefore, the predicted mass of $^{184}_{74}\text{W}$ is 74.57905 + 110.9532 = 185.5323 amu, and the mass defect is

$$\Delta m = 183.9510 \text{ amu} - 185.5323 \text{ amu} = -1.5813 \text{ amu}$$

The energy change ($\Delta E$) for the process is

$$\Delta E = (\Delta m)c^2$$

$$= (-1.5813 \text{ amu})(3.00 \times 10^8 \text{ m/s})^2$$

$$= -1.42 \times 10^{17} \frac{\text{amu}\cdot\text{m}^2}{\text{s}^2}$$

Let's convert to more familiar energy units (J/W atom).

$$\frac{-1.42 \times 10^{17}\ \cancel{\text{amu}}\cdot\cancel{\text{m}^2}}{1\ \cancel{\text{s}^2}} \times \frac{1.00\ \cancel{\text{g}}}{6.022 \times 10^{23}\ \cancel{\text{amu}}} \times \frac{1\ \cancel{\text{kg}}}{1000\ \cancel{\text{g}}} \times \frac{1 \text{ J}}{\dfrac{1\ \cancel{\text{kg}}\cdot\cancel{\text{m}^2}}{\cancel{\text{s}^2}}} = -2.36 \times 10^{-10} \text{ J}$$

The nuclear binding energy is **2.36 × 10⁻¹⁰ J**. It's the energy required to break up one tungsten-184 nucleus into 74 protons and 110 neutrons.

When comparing the stability of any two nuclei we must account for the fact that they have different numbers of nucleons. For this reason, it is more meaningful to use the *nuclear binding energy per nucleon*, defined as

$$\text{nuclear binding energy per nucleon} = \frac{\text{nuclear binding energy}}{\text{number of nucleons}}$$

For the tungsten-184 nucleus,

$$\text{nuclear binding energy per nucleon} = \frac{2.36 \times 10^{-10} \text{ J/W}\ \cancel{\text{atom}}}{184 \text{ nucleons/W}\ \cancel{\text{atom}}} = \mathbf{1.28 \times 10^{-12}\ J/nucleon}$$

**19.26** **Strategy:** According to Equation (13.3) of the text, the number of radioactive nuclei at time zero ($N_0$) and time $t$ ($N_t$) is

$$\ln\frac{N_t}{N_0} = -\lambda t$$

and the corresponding half-life of the reaction is given by Equation (13.6) of the text:

$$t_{\frac{1}{2}} = \frac{0.693}{\lambda}$$

Using the information given in the problem and the first equation above, we can calculate the rate constant, λ. Then, the half-life can be calculated from the rate constant.

**Solution:** We can use the following equation to calculate the rate constant λ for each point.

$$\ln\frac{N_t}{N_0} = -\lambda t$$

From day 0 to day 1, we have

$$\ln\frac{389}{500} = -\lambda(1\text{ d})$$

$$\lambda = 0.251\text{ d}^{-1}$$

Following the same procedure for the other days,

| *t* (d) | mass (g) | λ ($d^{-1}$) |
|---|---|---|
| 0 | 500 | |
| 1 | 389 | 0.251 |
| 2 | 303 | 0.250 |
| 3 | 236 | 0.250 |
| 4 | 184 | 0.250 |
| 5 | 143 | 0.250 |
| 6 | 112 | 0.249 |

The average value of λ is **0.250 d$^{-1}$**.

We use the average value of λ to calculate the half-life.

$$t_{\frac{1}{2}} = \frac{0.693}{\lambda} = \frac{0.693}{0.250\text{ d}^{-1}} = \mathbf{2.77\text{ d}}$$

**19.28** Since all radioactive decay processes have first–order rate laws, the decay rate is proportional to the amount of radioisotope at any time. The half-life is given by the following equation:

$$t_{\frac{1}{2}} = \frac{0.693}{\lambda} \qquad (1)$$

There is also an equation that relates the number of nuclei at time zero ($N_0$) and time $t$ ($N_t$).

$$\ln\frac{N_t}{N_0} = -\lambda t$$

We can use this equation to solve for the rate constant, λ. Then, we can substitute λ into Equation (1) to calculate the half-life.

The time interval is:

(2:15 p.m., 12/17/92) – (1:00 p.m., 12/3/92) = 14 d + 1 hr + 15 min = 20,235 min

$$\ln\left(\frac{2.6 \times 10^4 \text{ dis/min}}{9.8 \times 10^5 \text{ dis/min}}\right) = -\lambda(20{,}235 \text{ min})$$

$$\lambda = 1.8 \times 10^{-4} \text{ min}^{-1}$$

Substitute λ into equation (1) to calculate the half-life.

$$t_{\frac{1}{2}} = \frac{0.693}{\lambda} = \frac{0.693}{1.8 \times 10^{-4} \text{ min}^{-1}} = \mathbf{3.9 \times 10^3 \text{ min or } 2.7 \text{ d}}$$

**19.30** The equation for the overall process is:

$$^{232}_{90}\text{Th} \longrightarrow 6\,^{4}_{2}\text{He} + 4\,^{0}_{-1}\beta + \text{X}$$

The final product isotope must be $\mathbf{^{208}_{82}Pb}$.

**19.32** Let's consider the decay of A first.

$$\lambda = \frac{0.693}{t_{\frac{1}{2}}} = \frac{0.693}{4.50 \text{ s}} = 0.154 \text{ s}^{-1}$$

Let's convert λ to units of day$^{-1}$.

$$0.154\frac{1}{\cancel{s}} \times \frac{3600\,\cancel{s}}{1\,\cancel{h}} \times \frac{24\,\cancel{h}}{1 \text{ d}} = 1.33 \times 10^4 \text{ d}^{-1}$$

Next, use the first-order rate equation to calculate the amount of A left after 30 days.

$$\ln\frac{[A]_t}{[A]_0} = -\lambda t$$

Let $x$ be the amount of A left after 30 days.

$$\ln\frac{x}{100} = -(1.33 \times 10^4\ \mathrm{d}^{-1})(30\ \mathrm{d}) = -3.99 \times 10^5$$

$$\frac{x}{100} = e^{(-3.99 \times 10^5)}$$

$$x \approx 0$$

Thus, **no A remains.**

For B: As calculated above, all of A is converted to B in less than 30 days. In fact, essentially all of A is gone in less than 1 day! This means that at the beginning of the 30 day period, there is 1.00 mol of B present. The half life of B is 15 days, so that after two half-lives (30 days), there should be **0.25 mole of B** left.

For C: As in the case of A, the half-life of C is also very short. Therefore, at the end of the 30–day period, **no C is left.**

For D: D is not radioactive. 0.75 mol of B reacted in 30 days; therefore, due to a 1:1 mole ratio between B and D, there should be **0.75 mole of D** present after 30 days.

**19.38** (a) $^{80}_{34}\mathrm{Se} + ^{2}_{1}\mathrm{H} \longrightarrow ^{81}_{34}\mathrm{Se} + ^{1}_{1}\mathrm{p}$

(b) $^{9}_{4}\mathrm{Be} + ^{2}_{1}\mathrm{H} \longrightarrow ^{9}_{3}\mathrm{Li} + 2^{1}_{1}\mathrm{p}$

(c) $^{10}_{5}\mathrm{B} + ^{1}_{0}\mathrm{n} \longrightarrow ^{7}_{3}\mathrm{Li} + ^{4}_{2}\alpha$

**19.40** Upon bombardment with neutrons, mercury–198 is first converted to mercury–199, which then emits a proton. The reaction is:

$$^{198}_{80}\mathrm{Hg} + ^{1}_{0}\mathrm{n} \longrightarrow ^{199}_{80}\mathrm{Hg} \longrightarrow ^{198}_{79}\mathrm{Au} + ^{1}_{1}\mathrm{p}$$

**19.52** The fact that the radioisotope appears only in the $I_2$ shows that the $IO_3^-$ is formed only from the $IO_4^-$. Does this result rule out the possibility that $I_2$ could be formed from $IO_4^-$ as well? Can you suggest an experiment to answer the question?

**19.54** Add iron-59 to the person's diet, and allow a few days for the iron-59 isotope to be incorporated into the person's body. Isolate red blood cells from a blood sample and monitor radioactivity from the hemoglobin molecules present in the red blood cells.

**19.56** **(a)** $^{50}_{25}\text{Mn} \longrightarrow {}^{50}_{24}\text{Cr} + {}^{0}_{+1}\beta$

**(b)** There are 32 total spheres in the diagram (4 red, 28 green). Starting with 32 red spheres of $^{50}$Mn, after one half-life there would be 16 red spheres ($^{50}$Mn) and 16 green spheres ($^{50}$Cr). After two half-lives, there would be 8 red spheres ($^{50}$Mn) and 24 green spheres ($^{50}$Cr). After 3 half-lives, there would be 4 red spheres ($^{50}$Mn) and 28 green spheres ($^{50}$Cr). Based on the diagram, **three half-lives** have elapsed.

**19.58** Apparently there is a sort of Pauli exclusion principle for nucleons as well as for electrons. When neutrons pair with neutrons and when protons pair with protons, their spins cancel. Even–even nuclei are the only ones with no net spin.

**19.60** **(a)** One millicurie represents $3.70 \times 10^7$ disintegrations/s. The rate of decay of the isotope is given by the rate law: rate = $\lambda N$, where $N$ is the number of atoms in the sample. We find the value of $\lambda$ in units of $s^{-1}$:

$$\lambda = \frac{0.693}{t_{\frac{1}{2}}} = \frac{0.693}{2.20 \times 10^6 \text{ yr}} \times \frac{1 \text{ yr}}{365 \text{ d}} \times \frac{1 \text{ d}}{24 \text{ h}} \times \frac{1 \text{ h}}{3600 \text{ s}} = 9.99 \times 10^{-15} \text{ s}^{-1}$$

The number of atoms ($N$) in a 0.500 g sample of neptunium–237 is:

$$0.500 \text{ g} \times \frac{1 \text{ mol}}{237.0 \text{ g}} \times \frac{6.022 \times 10^{23} \text{ atoms}}{1 \text{ mol}} = 1.27 \times 10^{21} \text{ atoms}$$

$$\text{rate of decay} = \lambda N$$

$$= (9.99 \times 10^{-15} \text{ s}^{-1})(1.27 \times 10^{21} \text{ atoms}) = 1.27 \times 10^7 \text{ atoms/s}$$

We can also say that:

$$\text{rate of decay} = 1.27 \times 10^7 \text{ disintegrations/s}$$

The activity in millicuries is:

$$(1.27 \times 10^7 \text{ disintegrations/s}) \times \frac{1 \text{ millicurie}}{3.70 \times 10^7 \text{ disintegrations/s}} = \mathbf{0.343\ millicuries}$$

**(b)** The decay equation is:

$$^{237}_{93}\text{Np} \longrightarrow {}^{4}_{2}\alpha + {}^{233}_{91}\text{Pa}$$

**19.62** We use the same procedure as in Problem 19.22.

| | Isotope | Atomic Mass (amu) | Nuclear Binding Energy (J/nucleon) |
|---|---|---|---|
| (a) | $^{10}B$ | 10.0129 | $1.040 \times 10^{-12}$ |
| (b) | $^{11}B$ | 11.00931 | $1.111 \times 10^{-12}$ |
| (c) | $^{14}N$ | 14.00307 | $1.199 \times 10^{-12}$ |
| (d) | $^{56}Fe$ | 55.9349 | $1.410 \times 10^{-12}$ |

**19.64** When an isotope is above the belt of stability, the neutron/proton ratio is too high. The only mechanism to correct this situation is beta emission; the process turns a neutron into a proton. Direct neutron emission does not occur.

$$^{18}_{7}\text{N} \longrightarrow {}^{18}_{8}\text{O} + {}^{0}_{-1}\beta$$

Oxygen-18 is a stable isotope.

**19.66** The age of the fossil can be determined by radioactively dating the age of the deposit that contains the fossil.

**19.68** (a) $^{209}_{83}\text{Bi} + {}^{4}_{2}\alpha \longrightarrow {}^{211}_{85}\text{At} + 2\,{}^{1}_{0}\text{n}$

(b) $^{209}_{83}\text{Bi}(\alpha, 2\text{n})^{211}_{85}\text{At}$

**19.70** Because of the relative masses, the force of gravity on the sun is much greater than it is on Earth. Thus the nuclear particles on the sun are already held much closer together than the equivalent nuclear particles on the earth. Less energy (lower temperature) is required on the sun to force fusion collisions between the nuclear particles.

**19.72** ***Step 1:*** The half-life of carbon-14 is 5730 years. From the half-life, we can calculate the rate constant, λ.

$$\lambda = \frac{0.693}{t_{\frac{1}{2}}} = \frac{0.693}{5730\text{ yr}} = 1.21 \times 10^{-4}\text{ yr}^{-1}$$

***Step 2:*** The age of the object can now be calculated using the following equation.

$$\ln\frac{N_t}{N_0} = -\lambda t$$

$N =$ the number of radioactive nuclei. In the problem, we are given disintegrations per second per gram. The number of disintegrations is directly proportional to the number of radioactive nuclei. We can write,

$$\ln\frac{\text{decay rate of old sample}}{\text{decay rate of fresh sample}} = -\lambda t$$

$$\ln\frac{0.186\text{ dps/g C}}{0.260\text{ dps/g C}} = -(1.21\times10^{-4}\text{ yr}^{-1})t$$

$$t = \mathbf{2.77\times10^3\text{ yr}}$$

**19.74** **(a)** The balanced equation is:

$$^{40}_{19}\text{K} \longrightarrow {}^{40}_{18}\text{Ar} + {}^{0}_{+1}\beta$$

**(b)** First, calculate the rate constant λ.

$$\lambda = \frac{0.693}{t_{\frac{1}{2}}} = \frac{0.693}{1.2\times10^9\text{ yr}} = 5.8\times10^{-10}\text{ yr}^{-1}$$

Then, calculate the age of the rock by substituting λ into the following equation. ($N_t = 0.18N_0$)

$$\ln\frac{N_t}{N_0} = -\lambda t$$

$$\ln\frac{0.18}{1.00} = -(5.8\times10^{-10}\text{ yr}^{-1})t$$

$$t = \mathbf{3.0\times10^9\text{ yr}}$$

**19.76** **(a)** In the $^{90}$Sr decay, the mass defect is:

$$\Delta m = (\text{mass }^{90}\text{Y} + \text{mass e}^-) - \text{mass }^{90}\text{Sr}$$

$$= [(89.907152\text{ amu} + 5.4857\times10^{-4}\text{ amu}) - 89.907738\text{ amu}] = -3.743\times10^{-5}\text{ amu}$$

$$= (-3.743\times10^{-5}\text{ amu})\times\frac{1\text{ g}}{6.022\times10^{23}\text{ amu}} = -6.216\times10^{-29}\text{ g} = -6.216\times10^{-32}\text{ kg}$$

The energy change is given by:

$$\Delta E = (\Delta m)c^2$$

$$= (-6.126\times10^{-32}\text{ kg})(3.00\times10^8\text{ m/s})^2$$

$$= -5.59\times10^{-15}\text{ kg m}^2\text{/s}^2 = -5.59\times10^{-15}\text{ J}$$

Similarly, for the $^{90}$Y decay, we have

$$\Delta m = (\text{mass }^{90}\text{Zr} + \text{mass e}^-) - \text{mass }^{90}\text{Y}$$

$$= [(89.904703 \text{ amu} + 5.4857 \times 10^{-4} \text{ amu}) - 89.907152 \text{ amu}] = -1.900 \times 10^{-3} \text{ amu}$$

$$= (-1.900 \times 10^{-3} \cancel{\text{amu}}) \times \frac{1 \text{ g}}{6.022 \times 10^{23} \cancel{\text{amu}}} = -3.156 \times 10^{-27} \text{ g} = -3.156 \times 10^{-30} \text{ kg}$$

and the energy change is:

$$\Delta E = (-3.156 \times 10^{-30} \text{ kg})(3.00 \times 10^{8} \text{ m/s})^2 = -2.84 \times 10^{-13} \text{ J}$$

The energy released in the above two decays is **$5.59 \times 10^{-15}$ J** and **$2.84 \times 10^{-13}$ J**. The total amount of energy released is:

$$(5.59 \times 10^{-15} \text{ J}) + (2.84 \times 10^{-13} \text{ J}) = 2.90 \times 10^{-13} \text{ J}.$$

**(b)** This calculation requires that we know the rate constant for the decay. From the half-life, we can calculate $\lambda$.

$$\lambda = \frac{0.693}{t_{\frac{1}{2}}} = \frac{0.693}{28.1 \text{ yr}} = 0.0247 \text{ yr}^{-1}$$

To calculate the number of moles of $^{90}$Sr decaying in a year, we apply the following equation:

$$\ln \frac{N_t}{N_0} = -\lambda t$$

$$\ln \frac{x}{1.00} = -(0.0247 \cancel{\text{yr}}^{-1})(1.00 \cancel{\text{yr}})$$

where $x$ is the number of moles of $^{90}$Sr nuclei left over. Solving, we obtain:

$$x = 0.9756 \text{ mol } ^{90}\text{Sr}$$

Thus the number of moles of nuclei which decay in a year is

$$(1.00 - 0.9756) \text{ mol} = 0.0244 \text{ mol} = \mathbf{0.024 \text{ mol}}$$

This is a reasonable number since it takes 28.1 years for 0.5 mole of $^{90}$Sr to decay.

**(c)** Since the half–life of $^{90}$Y is much shorter than that of $^{90}$Sr, we can safely assume that *all* the $^{90}$Y formed from $^{90}$Sr will be converted to $^{90}$Zr. The energy changes calculated in part (a) refer to the decay of individual nuclei. In 0.0244 mole, the number of nuclei that have decayed is:

$$0.0244 \cancel{\text{mol}} \times \frac{6.022 \times 10^{23} \text{ nuclei}}{1 \cancel{\text{mol}}} = 1.47 \times 10^{22} \text{ nuclei}$$

Realize that there are two decay processes occurring, so we need to add the energy released for each process calculated in part (a). Thus, the heat released from 1 mole of $^{90}Sr$ waste in a year is given by:

$$\textbf{heat released} = (1.47 \times 10^{22} \text{ nuclei}) \times \frac{2.90 \times 10^{-13} \text{ J}}{1 \text{ nucleus}} = \mathbf{4.26 \times 10^9 \text{ J} = 4.26 \times 10^6 \text{ kJ}}$$

This amount is roughly equivalent to the heat generated by burning 50 tons of coal! Although the heat is released slowly during the course of a year, effective ways must be devised to prevent heat damage to the storage containers and subsequent leakage of radioactive material to the surroundings.

**19.78** First, let's calculate the number of disintegrations/s to which 7.4 mCi corresponds.

$$7.4 \text{ mCi} \times \frac{1 \text{ Ci}}{1000 \text{ mCi}} \times \frac{3.7 \times 10^{10} \text{ disintegrations/s}}{1 \text{ Ci}} = 2.7 \times 10^8 \text{ disintegrations/s}$$

This is the rate of decay. We can now calculate the number of iodine-131 atoms to which this radioactivity corresponds. First, we calculate the half-life in seconds:

$$t_{\frac{1}{2}} = 8.1 \text{ d} \times \frac{24 \text{ h}}{1 \text{ d}} \times \frac{3600 \text{ s}}{1 \text{ h}} = 7.0 \times 10^5 \text{ s}$$

$$\lambda = \frac{0.693}{t_{\frac{1}{2}}} \qquad \text{Therefore, } \lambda = \frac{0.693}{7.0 \times 10^5 \text{ s}} = 9.9 \times 10^{-7} \text{ s}^{-1}$$

$$\text{rate} = \lambda N$$

$$2.7 \times 10^8 \text{ disintegrations/s} = (9.9 \times 10^{-7} \text{ s}^{-1})N$$

$$\mathbf{N = 2.7 \times 10^{14} \text{ iodine-131 atoms}}$$

**19.80** One curie represents $3.70 \times 10^{10}$ disintegrations/s. The rate of decay of the isotope is given by the rate law: rate $= \lambda N$, where $N$ is the number of atoms in the sample and $\lambda$ is the first-order rate constant. We find the value of $\lambda$ in units of $s^{-1}$:

$$\lambda = \frac{0.693}{t_{\frac{1}{2}}} = \frac{0.693}{1.6 \times 10^3 \text{ yr}} = 4.3 \times 10^{-4} \text{ yr}^{-1}$$

$$\frac{4.3 \times 10^{-4}}{1 \text{ yr}} \times \frac{1 \text{ yr}}{365 \text{ d}} \times \frac{1 \text{ d}}{24 \text{ h}} \times \frac{1 \text{ h}}{3600 \text{ s}} = 1.4 \times 10^{-11} \text{ s}^{-1}$$

Now, we can calculate $N$, the number of Ra atoms in the sample.

$$\text{rate} = \lambda N$$

$$3.7 \times 10^{10} \text{ disintegrations/s} = (1.4 \times 10^{-11} \text{ s}^{-1})N$$

$$N = 2.6 \times 10^{21} \text{ Ra atoms}$$

By definition, 1 curie corresponds to exactly $3.7 \times 10^{10}$ nuclear disintegrations per second which is the decay rate equivalent to that of *1 g of radium.* Thus, the mass of $2.6 \times 10^{21}$ Ra atoms is 1 g.

$$\frac{2.6 \times 10^{21}\ \text{Ra atoms}}{1.0\ \cancel{\text{g}}\ \text{Ra}} \times \frac{226.03\ \cancel{\text{g}}\ \text{Ra}}{1\ \text{mol Ra}} = \mathbf{5.9 \times 10^{23}\ atoms/mol} = \mathbf{N_A}$$

**19.82** All except gravitational have a nuclear origin.

**19.84** U-238, $t_{\frac{1}{2}} = 4.5 \times 10^9$ yr and Th-232, $t_{\frac{1}{2}} = 1.4 \times 10^{10}$ yr.

They are still present because of their long half lives.

**19.86** $E = \dfrac{hc}{\lambda}$

$$\lambda = \frac{hc}{E} = \frac{(3.00 \times 10^8\ \cancel{\text{m/s}})(6.63 \times 10^{-34}\ \cancel{\text{J}\cdot\text{s}})}{2.4 \times 10^{-13}\ \cancel{\text{J}}} = 8.3 \times 10^{-13}\ \text{m} = \mathbf{8.3 \times 10^{-4}\ nm}$$

This wavelength is clearly in the γ-ray region of the electromagnetic spectrum.

**19.88** Only $^{3}$**H** has a suitable half-life. The other half-lives are either too long or too short to accurately determine the time span of 6 years.

**19.90** Obviously, a small scale chain reaction took place. Copper played the crucial role of reflecting neutrons from the splitting uranium-235 atoms back into the uranium sphere to trigger the chain reaction. Note that a sphere has the most appropriate geometry for such a chain reaction. In fact, during the implosion process prior to an atomic explosion, fragments of uranium-235 are pressed roughly into a sphere for the chain reaction to occur (see Section 19.5 of the text).

**19.92** In this problem, we are asked to calculate the molar mass of a radioactive isotope. Grams of sample are given in the problem, so if we can find moles of sample we can calculate the molar mass. The rate constant can be calculated from the half-life. Then, from the rate of decay and the rate constant, the number of radioactive nuclei can be calculated. The number of radioactive nuclei can be converted to moles.

First, we convert the half-life to units of minutes because the rate is given in dpm (disintegrations per minute). Then, we calculate the rate constant from the half-life.

$$(1.3 \times 10^9\ \cancel{\text{yr}}) \times \frac{365\ \cancel{\text{days}}}{1\ \cancel{\text{yr}}} \times \frac{24\ \cancel{\text{h}}}{1\ \cancel{\text{day}}} \times \frac{60\ \text{min}}{1\ \cancel{\text{h}}} = 6.8 \times 10^{14}\ \text{min}$$

$$\lambda = \frac{0.693}{t_{\frac{1}{2}}} = \frac{0.693}{6.8 \times 10^{14}\ \text{min}} = 1.0 \times 10^{-15}\ \text{min}^{-1}$$

Next, we calculate the number of radioactive nuclei from the rate and the rate constant.

$$\text{rate} = \lambda N$$

$$2.9 \times 10^{4}\ \text{dpm} = (1.0 \times 10^{-15}\ \text{min}^{-1})N$$

$$N = 2.9 \times 10^{19}\ \text{nuclei}$$

Convert to moles of nuclei, and then determine the molar mass.

$$(2.9 \times 10^{19}\ \text{nuclei}) \times \frac{1\ \text{mol}}{6.022 \times 10^{23}\ \text{nuclei}} = 4.8 \times 10^{-5}\ \text{mol}$$

$$\textbf{molar mass} = \frac{\text{g of substance}}{\text{mol of substance}} = \frac{0.0100\ \text{g}}{4.8 \times 10^{-5}\ \text{mol}} = \mathbf{2.1 \times 10^{2}\ g/mol}$$

**19.94** $^{234}_{90}\text{Th} \longrightarrow {}^{234}_{91}\text{Pa} + {}^{0}_{-1}\beta$

$^{234}_{91}\text{Pa} \longrightarrow {}^{234}_{92}\text{U} + {}^{0}_{-1}\beta$

$^{234}_{92}\text{U} \longrightarrow {}^{230}_{90}\text{Th} + {}^{4}_{2}\alpha$

$^{230}_{90}\text{Th} \longrightarrow {}^{226}_{88}\text{Ra} + {}^{4}_{2}\alpha$

$^{226}_{88}\text{Ra} \longrightarrow {}^{222}_{86}\text{Rn} + {}^{4}_{2}\alpha$

**19.96** **(a)** $^{238}_{94}\text{Pu} \rightarrow {}^{4}_{2}\text{He} + {}^{234}_{92}\text{U}$

**(b)** At $t = 0$, the number of $^{238}$Pu atoms is

$$(1.0 \times 10^{-3}\ \text{g}) \times \frac{1\ \text{mol}}{238\ \text{g}} \times \frac{6.022 \times 10^{23}\ \text{atoms}}{1\ \text{mol}} = 2.53 \times 10^{18}\ \text{atoms}$$

The decay rate constant, $\lambda$, is

$$\lambda = \frac{0.693}{t_{\frac{1}{2}}} = \frac{0.693}{86\ \text{yr}} = 0.00806 \frac{1}{\text{yr}} \times \frac{1\ \text{yr}}{365\ \text{d}} \times \frac{1\ \text{d}}{24\ \text{h}} \times \frac{1\ \text{h}}{3600\ \text{s}} = 2.56 \times 10^{-10}\ \text{s}^{-1}$$

$$\text{rate} = \lambda N_0 = (2.56 \times 10^{-10}\ \text{s}^{-1})(2.53 \times 10^{18}\ \text{atoms}) = 6.48 \times 10^{8}\ \text{decays/s}$$

$$\text{Power} = (\text{decays/s}) \times (\text{energy/decay})$$

$$\text{Power} = (6.48 \times 10^8 \text{ decays/s})(9.0 \times 10^{-13} \text{ J/decay}) = 5.8 \times 10^{-4} \text{ J/s} = \mathbf{5.8 \times 10^{-4}\ W = 0.58\ mW}$$

At $t = 10$ yr,

$$\text{Power} = (0.58 \text{ mW})(0.92) = \mathbf{0.53\ mW}$$

**19.98** 1 Ci = $3.7 \times 10^{10}$ decays/s

Let $R_0$ be the activity of the injected 20.0 mCi $^{99m}$Tc.

$$R_0 = (20.0 \times 10^{-3} \text{ Ci}) \times \frac{3.70 \times 10^{10} \text{ decays/s}}{1 \text{ Ci}} = 7.4 \times 10^8 \text{ decays/s}$$

$R_0 = \lambda N_0$, where $N_0$ = number of $^{99m}$Tc nuclei present.

$$\lambda = \frac{0.693}{t_{\frac{1}{2}}} = \frac{0.693}{6.0 \text{ h}} = 0.1155 \frac{1}{\text{h}} \times \frac{1 \text{ h}}{3600 \text{ s}} = 3.208 \times 10^{-5} \text{ s}^{-1}$$

$$N_0 = \frac{R_0}{\lambda} = \frac{7.4 \times 10^8 \text{ deays/s}}{3.208 \times 10^{-5}\text{/s}} = 2.307 \times 10^{13} \text{ decays} = 2.307 \times 10^{13} \text{ nuclei}$$

Each of the nuclei emits a photon of energy $2.29 \times 10^{-14}$ J. The total energy absorbed by the patient is

$$\boldsymbol{E} = \frac{2}{3}(2.307 \times 10^{13} \text{ nuclei}) \times \left(\frac{2.29 \times 10^{-14} \text{ J}}{1 \text{ nuclei}}\right) = \mathbf{0.352\ J}$$

The rad is:

$$\frac{0.352 \text{ J}/10^{-2} \text{ J}}{70} = \mathbf{0.503\ rad}$$

Given that RBE = 0.98, the rem is:

$$(0.503)(0.98) = \mathbf{0.49\ rem}$$

**19.100** The ignition of a fission bomb requires an ample supply of neutrons. In addition to the normal neutron source placed in the bomb, the high temperature attained during the chain reaction causes a small scale nuclear fusion between deuterium and tritium.

$$^2_1\text{H} + {}^3_1\text{H} \rightarrow {}^4_2\text{He} + {}^1_0\text{n}$$

The additional neutrons produced will enhance the efficiency of the chain reaction and result in a more powerful bomb.

**19.102** The five radioactive decays that lead to the production of five α particles per $^{226}$Ra decay to $^{206}$Pb are:

$$^{226}\text{Ra} \rightarrow {}^{222}\text{Rn} + \alpha$$

$$^{222}\text{Rn} \rightarrow {}^{218}\text{Po} + \alpha$$

$$^{218}\text{Po} \rightarrow {}^{214}\text{Pb} + \alpha$$

$$^{214}\text{Po} \rightarrow {}^{210}\text{Pb} + \alpha$$

$$^{210}\text{Po} \rightarrow {}^{206}\text{Pb} + \alpha$$

Because the time frame of the experiment (100 years) is much longer than any of the half-lives following the decay of $^{226}$Ra, we assume that 5 α particles are generated per $^{226}$Ra decay to $^{206}$Pb. Additional significant figures are carried throughout the calculation to minimize rounding errors.

The rate of decay of 1.00 g of $^{226}$Ra can be calculated using the following equation.

$$\text{rate} = \lambda N = \frac{0.693}{t_{\frac{1}{2}}}N$$

The number of radium atoms in 1.00 g is:

$$1.00\ \text{g Ra} \times \frac{1\ \text{mol Ra}}{226\ \text{g Ra}} \times \frac{6.022 \times 10^{23}\ \text{Ra atoms}}{1\ \text{mol Ra}} = 2.665 \times 10^{21}\ \text{Ra atoms}$$

The number of disintegrations in 100 years will be:

$$\text{rate} = \lambda N = \frac{0.693}{t_{\frac{1}{2}}}N$$

$$\text{rate} = \left(\frac{0.693}{1.60 \times 10^{3}\ \text{yr}}\right)(2.665 \times 10^{21}\ \text{Ra atoms}) = 1.154 \times 10^{18}\ \text{Ra atoms/yr}$$

$$\frac{1.154 \times 10^{18}\ \text{Ra atoms}}{1\ \text{yr}} \times 100\ \text{yr} = 1.154 \times 10^{20}\ \text{Ra atoms} = 1.154 \times 10^{20}\ \text{disintegrations}$$

As determined above, 5 α particles are generated per $^{226}$Ra decay to $^{206}$Pb. In 100 years, the amount of α particles produced is:

$$1.154 \times 10^{20}\ \text{Ra atoms} \times \frac{5\ \alpha\ \text{particles}}{1\ \text{Ra atoms}} \times \frac{1\ \text{mol}\ \alpha}{6.022 \times 10^{23}\ \alpha\ \text{particles}} = 9.58 \times 10^{-4}\ \text{mol}\ \alpha$$

Each α particle forms a helium atom by gaining two electrons. The volume of He collected at STP is:

$$V_{He} = \frac{n_{He}RT}{P} = \frac{(9.58 \times 10^{-4}\ \cancel{mol})(0.0821\ L \cdot \cancel{atm}/\cancel{mol} \cdot \cancel{K})(273\ \cancel{K})}{1\ \cancel{atm}}$$

$$V_{He} = \mathbf{0.0215\ L = 21.5\ mL}$$

**19.104** No, this does not violate the law of conservation of mass. In this case, kinetic energy generated during the collision of the high-speed particles is converted to mass ($E = mc^2$). But energy also has mass, and the total mass, from particles and energy, in a closed system is conserved.

**19.106** **(a)** The mass defect is the difference between the mass of $^{14}_{7}N$ and the combined mass of a neutron and $^{13}_{7}N$.

$$\Delta m = 14.003074\ \text{amu} - (13.005738\ \text{amu} + 1.00866\ \text{amu}) = -0.01132\ \text{amu}$$

The binding energy for a single neutron has the opposite sign of the energy released when this mass difference is converted to energy.

$$\Delta E = \Delta mc^2 = \left(-0.01132\ \cancel{amu} \times \frac{1\ kg}{6.022 \times 10^{26}\ \cancel{amu}}\right)(3.00 \times 10^8\ m/s)^2 = -1.69 \times 10^{-12}\ J$$

The binding energy of a single neutron is $\mathbf{1.69 \times 10^{-12}\ J}$.

**(b)** In this case, the mass defect is the difference between the mass of $^{14}_{7}N$ and the combined mass of a proton and $^{13}_{6}C$.

$$\Delta m = 14.003074\ \text{amu} - (13.003355\ \text{amu} + 1.00794\ \text{amu}) = -0.00822\ \text{amu}$$

The binding energy for a single proton has the opposite sign of the energy released when this mass difference is converted to energy.

$$\Delta E = \Delta mc^2 = \left(-0.00822\ \cancel{amu} \times \frac{1\ kg}{6.022 \times 10^{26}\ \cancel{amu}}\right)(3.00 \times 10^8\ m/s)^2 = -1.23 \times 10^{-12}\ J$$

The binding energy of a single proton is $\mathbf{1.23 \times 10^{-12}\ J}$.

Because a proton feels the repulsion from other protons, it has a smaller binding energy than a neutron.

# CHAPTER 20
# CHEMISTRY IN THE ATMOSPHERE

Almost all the types of problems in this chapter have been encountered in previous chapters. Therefore, we will not repeat the problem types here, but will refer you to problem types where appropriate.

## SOLUTIONS TO SELECTED TEXT PROBLEMS

**20.6** Using the information in Table 20.1 and Problem 20.5, 0.033 percent of the volume (and therefore the pressure) of dry air is due to $CO_2$. The partial pressure of $CO_2$ is:

$$P_{CO_2} = X_{CO_2}P_T = (3.3 \times 10^{-4})(754\ \text{mmHg}) \times \frac{1\ \text{atm}}{760\ \text{mmHg}} = \mathbf{3.3 \times 10^{-4}\ atm}$$

**20.8** From Problem 5.106, the total mass of air is $5.25 \times 10^{18}$ kg. Table 20.1 lists the composition of air by volume. Under the same conditions of $P$ and $T$, $V \propto n$ (Avogadro's law).

$$\text{Total moles of gases} = (5.25 \times 10^{21}\ \text{g}) \times \frac{1\ \text{mol}}{29.0\ \text{g}} = 1.81 \times 10^{20}\ \text{mol}$$

Mass of $N_2$ (78.03%):

$$(0.7803)(1.81 \times 10^{20}\ \text{mol}) \times \frac{28.02\ \text{g}}{1\ \text{mol}} = 3.96 \times 10^{21}\ \text{g} = \mathbf{3.96 \times 10^{18}\ kg}$$

Mass of $O_2$ (20.99%):

$$(0.2099)(1.81 \times 10^{20}\ \text{mol}) \times \frac{32.00\ \text{g}}{1\ \text{mol}} = 1.22 \times 10^{21}\ \text{g} = \mathbf{1.22 \times 10^{18}\ kg}$$

Mass of $CO_2$ (0.033%):

$$(3.3 \times 10^{-4})(1.81 \times 10^{20}\ \text{mol}) \times \frac{44.01\ \text{g}}{1\ \text{mol}} = 2.63 \times 10^{18}\ \text{g} = \mathbf{2.63 \times 10^{15}\ kg}$$

**20.12** **Strategy:** We are given the wavelength of the emitted photon and asked to calculate its energy. Equation (7.2) of the text relates the energy and frequency of an electromagnetic wave.

$$E = h\nu$$

First, we calculate the frequency from the wavelength, then we can calculate the energy difference between the two levels.

**Solution:** Calculate the frequency from the wavelength.

$$\nu = \frac{c}{\lambda} = \frac{3.00 \times 10^{8}\ \text{m/s}}{558 \times 10^{-9}\ \text{m}} = 5.38 \times 10^{14}\ \text{/s}$$

Now, we can calculate the energy difference from the frequency.

$$\Delta E = h\nu = (6.63 \times 10^{-34}\ \text{J}\cdot\text{s})(5.38 \times 10^{14}\ /\text{s})$$

$$\mathbf{\Delta E = 3.57 \times 10^{-19}\ J}$$

**20.22** The quantity of ozone lost is:

$$(0.06)(3.2 \times 10^{12}\ \text{kg}) = 1.9 \times 10^{11}\ \text{kg of } O_3$$

Assuming no further deterioration, the kilograms of $O_3$ that would have to be manufactured on a daily basis are:

$$\frac{1.9 \times 10^{11}\ \text{kg } O_3}{100\ \text{yr}} \times \frac{1\ \text{yr}}{365\ \text{days}} = \mathbf{5.2 \times 10^{6}\ kg/day}$$

The standard enthalpy of formation (from Appendix 3 of the text) for ozone:

$$\tfrac{3}{2}O_2 \rightarrow O_3 \qquad \Delta H_f^\circ = 142.2\ \text{kJ/mol}$$

The *total* energy required is:

$$(1.9 \times 10^{14}\ \text{g of } O_3) \times \frac{1\ \text{mol } O_3}{48.00\ \text{g } O_3} \times \frac{142.2\ \text{kJ}}{1\ \text{mol } O_3} = \mathbf{5.6 \times 10^{14}\ kJ}$$

**20.24** The energy of the photons of UV radiation in the troposphere is insufficient (that is, the wavelength is too long and the frequency is too small) to break the bonds in CFCs.

**20.26** First, we need to calculate the energy needed to break one bond.

$$\frac{276 \times 10^{3}\ \text{J}}{1\ \text{mol}} \times \frac{1\ \text{mol}}{6.022 \times 10^{23}\ \text{molecules}} = 4.58 \times 10^{-19}\ \text{J/molecule}$$

The longest wavelength required to break this bond is:

$$\lambda = \frac{hc}{E} = \frac{(3.00 \times 10^{8}\ \text{m/s})(6.63 \times 10^{-34}\ \text{J}\cdot\text{s})}{4.58 \times 10^{-19}\ \text{J}} = 4.34 \times 10^{-7}\ \text{m} = \mathbf{434\ nm}$$

434 nm is in the visible region of the electromagnetic spectrum; therefore, $CF_3Br$ will be decomposed in **both** the troposphere and stratosphere.

**20.28** The Lewis structure of HCFC−123 is:

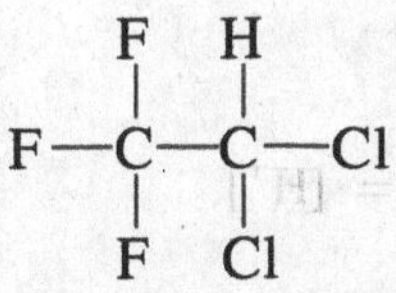

The Lewis structure for $CF_3CFH_2$ is:

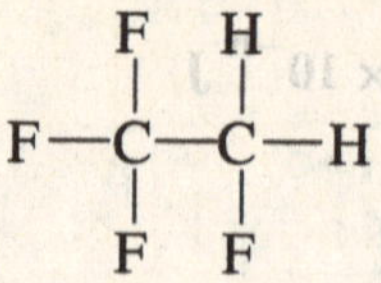

Lone pairs on the outer atoms have been omitted.

**20.40** **Strategy:** Looking at the balanced equation, how do we compare the amounts of CaO and $CO_2$? We can compare them based on the mole ratio from the balanced equation.

**Solution:** Because the balanced equation is given in the problem, the mole ratio between CaO and $CO_2$ is known: 1 mole CaO ≏ 1 mole $CO_2$. If we convert grams of CaO to moles of CaO, we can use this mole ratio to convert to moles of $CO_2$. Once moles of $CO_2$ are known, we can convert to grams $CO_2$.

$$\text{mass } CO_2 = (1.7 \times 10^{13} \text{ g CaO}) \times \frac{1 \text{ mol CaO}}{56.08 \text{ g CaO}} \times \frac{1 \text{ mol } CO_2}{1 \text{ mol CaO}} \times \frac{44.01 \text{ g}}{1 \text{ mol } CO_2}$$

$$= 1.3 \times 10^{13} \text{ g } CO_2 = \mathbf{1.3 \times 10^{10} \text{ kg } CO_2}$$

**20.42** Ethane and propane are greenhouse gases. They would contribute to global warming.

**20.50** Recall that ppm means the number of parts of substance per 1,000,000 parts. We can calculate the partial pressure of $SO_2$ in the troposphere.

$$P_{SO_2} = \frac{0.16 \text{ molecules of } SO_2}{10^6 \text{ parts of air}} \times 1 \text{ atm} = 1.6 \times 10^{-7} \text{ atm}$$

Next, we need to set up the equilibrium constant expression to calculate the concentration of $H^+$ in the rainwater. From the concentration of $H^+$, we can calculate the pH.

$$SO_2 \quad + \quad H_2O \rightleftharpoons H^+ \ + \ HSO_3^-$$

Equilibrium: $1.6 \times 10^{-7}$ atm $\qquad x \qquad x$

$$K = \frac{[H^+][HSO_3^-]}{P_{SO_2}} = 1.3 \times 10^{-2}$$

$$1.3 \times 10^{-2} = \frac{x^2}{1.6 \times 10^{-7}}$$

$$x^2 = 2.1 \times 10^{-9}$$

$$x = 4.6 \times 10^{-5}\, M = [H^+]$$

$$\textbf{pH} = -\log(4.6 \times 10^{-5}) = \mathbf{4.34}$$

**20.58** **Strategy:** This problem gives the volume, temperature, and pressure of PAN. Is the gas undergoing a change in any of its properties? What equation should we use to solve for moles of PAN? Once we have determined moles of PAN, we can convert to molarity and use the first-order rate law to solve for rate.

**Solution:** Because no changes in gas properties occur, we can use the ideal gas equation to calculate the moles of PAN. 0.55 ppm by volume means:

$$\frac{V_{\text{PAN}}}{V_{\text{T}}} = \frac{0.55\text{ L}}{1 \times 10^6\text{ L}}$$

Rearranging Equation (5.8) of the text, at STP, the number of moles of PAN in 1.0 L of air is:

$$n = \frac{PV}{RT} = \frac{(1\cancel{\text{atm}})\left(\dfrac{0.55\cancel{\text{L}}}{1 \times 10^6\cancel{\text{L}}} \times 1.0\cancel{\text{L}}\right)}{(0.0821\cancel{\text{L}}\cdot\cancel{\text{atm}}/\cancel{\text{K}}\cdot\text{mol})(273\cancel{\text{K}})} = 2.5 \times 10^{-8}\text{ mol}$$

Since the decomposition follows first-order kinetics, we can write:

$$\text{rate} = k[\text{PAN}]$$

$$\textbf{rate} = (4.9 \times 10^{-4}\text{ /s})\left(\frac{2.5 \times 10^{-8}\text{ mol}}{1.0\text{ L}}\right) = \mathbf{1.2 \times 10^{-11}}\ \boldsymbol{M/\text{s}}$$

**20.60** The Gobi desert lacks the primary pollutants (nitric oxide, carbon monoxide, hydrocarbons) to have photochemical smog. The primary pollutants are present both in New York City and in Boston. However, the sunlight that is required for the conversion of the primary pollutants to the secondary pollutants associated with smog is more likely in a July afternoon than one in January. Therefore, answer **(b)** is correct.

**20.66** **Strategy:** After writing a balanced equation, how do we compare the amounts of $CaCO_3$ and $CO_2$? We can compare them based on the mole ratio from the balanced equation. Once we have moles of $CO_2$, we can then calculate moles of air using the ideal gas equation. From the moles of $CO_2$ and the moles of air, we can calculate the percentage of $CO_2$ in the air.

**Solution:** First, we need to write a balanced equation.

$$CO_2 + Ca(OH)_2 \rightarrow CaCO_3 + H_2O$$

The mole ratio between $CaCO_3$ and $CO_2$ is: 1 mole $CaCO_3 \simeq$ 1 mole $CO_2$. If we convert grams of $CaCO_3$ to moles of $CaCO_3$, we can use this mole ratio to convert to moles of $CO_2$. Once moles of $CO_2$ are known, we can convert to grams $CO_2$.

Moles of $CO_2$ reacted:

$$0.026\cancel{\text{g}}\,CaCO_3 \times \frac{1\cancel{\text{mol}}\,CaCO_3}{100.1\cancel{\text{g}}\,CaCO_3} \times \frac{1\text{ mol }CO_2}{1\cancel{\text{mol}}\,CaCO_3} = 2.6 \times 10^{-4}\text{ mol }CO_2$$

The total number of moles of air can be calculated using the ideal gas equation.

$$n = \frac{PV}{RT} = \frac{\left(747\ \text{mmHg} \times \frac{1\ \text{atm}}{760\ \text{mmHg}}\right)(5.0\ \text{L})}{(0.0821\ \text{L}\cdot\text{atm/mol}\cdot\text{K})(291\ \text{K})} = 0.21\ \text{mol air}$$

The percentage by volume of $CO_2$ in air is:

$$\frac{V_{CO_2}}{V_{air}} \times 100\% = \frac{n_{CO_2}}{n_{air}} \times 100\% = \frac{2.6 \times 10^{-4}\ \text{mol}}{0.21\ \text{mol}} \times 100\% = \mathbf{0.12\%}$$

**20.68** An increase in temperature has shifted the system to the right; the equilibrium constant has increased with an increase in temperature. If we think of heat as a reactant (endothermic)

$$\text{heat} + N_2 + O_2 \rightleftharpoons 2\,NO$$

based on Le Châtelier's principle, adding heat would indeed shift the system to the right. Therefore, the reaction is **endothermic**.

**20.70** The concentration of $O_2$ could be monitored. Formation of $CO_2$ must deplete $O_2$.

**20.72** In Problem 20.6, we determined the partial pressure of $CO_2$ in dry air to be $3.3 \times 10^{-4}$ atm. Using Henry's law, we can calculate the concentration of $CO_2$ in water.

$$c = kP$$

$$[CO_2] = (0.032\ \text{mol/L}\cdot\text{atm})(3.3 \times 10^{-4}\ \text{atm}) = 1.06 \times 10^{-5}\ \text{mol/L}$$

We assume that all of the dissolved $CO_2$ is converted to $H_2CO_3$, thus giving us $1.06 \times 10^{-5}$ mol/L of $H_2CO_3$. $H_2CO_3$ is a weak acid. Setup the equilibrium of this acid in water and solve for $[H^+]$.

The equilibrium expression is:

| | $H_2CO_3$ | $\rightleftharpoons$ | $H^+$ | + | $HCO_3^-$ |
|---|---|---|---|---|---|
| Initial (*M*): | $1.06 \times 10^{-5}$ | | 0 | | 0 |
| Change (*M*): | $-x$ | | $+x$ | | $+x$ |
| Equilibrium (*M*): | $(1.06 \times 10^{-5}) - x$ | | $x$ | | $x$ |

$$K\ (\text{from Table 15.5}) = 4.2 \times 10^{-7} = \frac{[H^+][HCO_3^-]}{[H_2CO_3]} = \frac{x^2}{(1.06 \times 10^{-5}) - x}$$

Solving the quadratic equation:

$$x = 1.9 \times 10^{-6}\ M = [H^+]$$

$$\mathbf{pH} = -\log(1.9 \times 10^{-6}) = \mathbf{5.72}$$

**20.74** **Strategy:** From $\Delta H_f^\circ$ values given in Appendix 3 of the text, we can calculate $\Delta H^\circ$ for the reaction

$$NO_2 \rightarrow NO + O$$

Then, we can calculate $\Delta E^\circ$ from $\Delta H^\circ$. The $\Delta E^\circ$ calculated will have units of kJ/mol. If we can convert this energy to units of J/molecule, we can calculate the wavelength required to decompose $NO_2$.

**Solution:** We use the $\Delta H_f^\circ$ values in Appendix 3 and Equation (6.18) of the text.

$$\Delta H_{rxn}^\circ = \sum n\Delta H_f^\circ(\text{products}) - \sum m\Delta H_f^\circ(\text{reactants})$$

Consider reaction (1):

$$\Delta H^\circ = \Delta H_f^\circ(NO) + \Delta H_f^\circ(O) - \Delta H_f^\circ(NO_2)$$

$$\Delta H^\circ = (1)(90.4\ \text{kJ/mol}) + (1)(249.4\ \text{kJ/mol}) - (1)(33.85\ \text{kJ/mol})$$

$$\Delta H^\circ = 306.0\ \text{kJ/mol}$$

From Equation (6.10) of the text, $\Delta E^\circ = \Delta H^\circ - RT\Delta n$

$$\Delta E^\circ = (306.0 \times 10^3\ \text{J/mol}) - (8.314\ \text{J/mol·K})(298\ \text{K})(1)$$

$$\Delta E^\circ = 304 \times 10^3\ \text{J/mol}$$

This is the energy needed to dissociate 1 mole of $NO_2$. We need the energy required to dissociate *one molecule* of $NO_2$.

$$\frac{304 \times 10^3\ \text{J}}{1\ \text{mol}\ NO_2} \times \frac{1\ \text{mol}\ NO_2}{6.022 \times 10^{23}\ \text{molecules}\ NO_2} = 5.05 \times 10^{-19}\ \text{J/molecule}$$

The longest wavelength that can dissociate $NO_2$ is:

$$\lambda = \frac{hc}{E} = \frac{(6.63 \times 10^{-34}\ \text{J·s})(3.00 \times 10^8\ \text{m/s})}{5.05 \times 10^{-19}\ \text{J}} = \mathbf{3.94 \times 10^{-7}\ m = 394\ nm}$$

**20.76** This reaction has a high activation energy.

**20.78** The size of tree rings can be related to $CO_2$ content, where the number of rings indicates the age of the tree. The amount of $CO_2$ in ice can be directly measured from portions of polar ice in different layers obtained by drilling. The "age" of $CO_2$ can be determined by radiocarbon dating and other methods.

**20.80** $Cl_2 + O_2 \rightarrow 2ClO$

$$\Delta H^\circ = \Sigma BE(\text{reactants}) - \Sigma BE(\text{products})$$

$$\Delta H^\circ = (1)(242.7\ \text{kJ/mol}) + (1)(498.7\ \text{kJ/mol}) - (2)(206\ \text{kJ/mol})$$

$$\Delta H^\circ = 329\ \text{kJ/mol}$$

$$\Delta H° = 2\Delta H_f°(\text{ClO}) - 2\Delta H_f°(\text{Cl}_2) - 2\Delta H_f°(\text{O}_2)$$

$$329 \text{ kJ/mol} = 2\Delta H_f°(\text{ClO}) - 0 - 0$$

$$\mathbf{\Delta H_f°(ClO)} = \frac{329 \text{ kJ/mol}}{2} = \mathbf{165\ kJ/mol}$$

**20.82** In one second, the energy absorbed by $CO_2$ is 6.7 J. If we can calculate the energy of one photon of light with a wavelength of 14993 nm, we can then calculate the number of photons absorbed per second.

The energy of one photon with a wavelength of 14993 nm is:

$$E = \frac{hc}{\lambda} = \frac{(6.63 \times 10^{-34} \text{ J}\cdot\text{s})(3.00 \times 10^8 \text{ m/s})}{14993 \times 10^{-9} \text{ m}} = 1.3266 \times 10^{-20} \text{ J}$$

The number of photons absorbed by $CO_2$ per second is:

$$6.7 \text{ J} \times \frac{1 \text{ photon}}{1.3266 \times 10^{-20} \text{ J}} = \mathbf{5.1 \times 10^{20}\ photons}$$

**20.84** **(a)** We use Equation (13.14) of the text.

$$\ln\frac{k_1}{k_2} = \frac{E_a}{R}\left(\frac{T_1 - T_2}{T_1T_2}\right)$$

$$\ln\frac{2.6 \times 10^{-7} \text{ s}^{-1}}{3.0 \times 10^{-4} \text{ s}^{-1}} = \frac{E_a}{8.314 \text{ J/mol}\cdot\text{K}}\left(\frac{233 \text{ K} - 298 \text{ K}}{(233 \text{ K})(298 \text{ K})}\right)$$

$$\mathbf{E_a} = 6.26 \times 10^4 \text{ J/mol} = \mathbf{62.6\ kJ/mol}$$

**(b)** The unit for the rate constant indicates that the reaction is first-order. The half-life is:

$$t_{\frac{1}{2}} = \frac{0.693}{k} = \frac{0.693}{3.0 \times 10^{-4} \text{ s}^{-1}} = \mathbf{2.3 \times 10^3\ s = 38\ min}$$

**20.86** In order to end up with the desired equation, we keep the second equation as written, but we must reverse the first equation and multiply by two.

$$2\text{S}(s) + 3\text{O}_2(g) \rightleftharpoons 2\text{SO}_3(g) \qquad K_2$$

$$2\text{SO}_2(g) \rightleftharpoons 2\text{S}(s) + 2\text{O}_2(g) \qquad K_1' = \frac{1}{(K_1)^2}$$

$$2\text{SO}_2(g) + \text{O}_2(g) \rightleftharpoons 2\text{SO}_3(g) \qquad K = K_2 \times \frac{1}{(K_1)^2}$$

$$K = K_2 \times \frac{1}{(K_1)^2} = (9.8 \times 10^{128}) \times \frac{1}{(4.2 \times 10^{52})^2}$$

$$\mathbf{K = 5.6 \times 10^{23}}$$

**Thus, the reaction favors the formation of $SO_3$. But, this reaction has a high activation energy and requires a catalyst to promote it.**

**20.88** 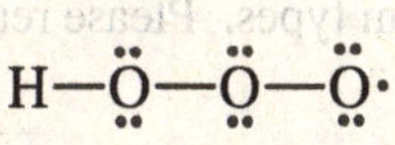

# CHAPTER 21
# METALLURGY AND THE CHEMISTRY OF METALS

This chapter is of a very descriptive nature. As such, there are essentially no defined problem types. Please read Chapter 21 of the text carefully before answering the end-of-chapter problems.

## SOLUTIONS TO SELECTED TEXT PROBLEMS

**21.12** The cathode reaction is: $Cu^{2+}(aq) + 2e^- \longrightarrow Cu(s)$

First, let's calculate the number of moles of electrons needed to reduce 5.0 kg of Cu.

$$5.00\ \text{kg Cu} \times \frac{1000\ \text{g}}{1\ \text{kg}} \times \frac{1\ \text{mol Cu}}{63.55\ \text{g Cu}} \times \frac{2\ \text{mol}\ e^-}{1\ \text{mol Cu}} = 1.57 \times 10^2\ \text{mol}\ e^-$$

Next, let's determine how long it will take for $1.57 \times 10^2$ moles of electrons to flow through the cell when the current is 37.8 C/s.

$$(1.57 \times 10^2\ \text{mol}\ e^-) \times \frac{96{,}500\ \text{C}}{1\ \text{mol}\ e^-} \times \frac{1\ \text{s}}{37.8\ \text{C}} \times \frac{1\ \text{h}}{3600\ \text{s}} = \mathbf{111\ h}$$

**21.14** The sulfide ore is first roasted in air:

$$2ZnS(s) + 3O_2(g) \longrightarrow 2ZnO(s) + 2SO_2(g)$$

The zinc oxide is then mixed with coke and limestone in a blast furnace where the following reductions occur:

$$ZnO(s) + C(s) \longrightarrow Zn(g) + CO(g)$$

$$ZnO(s) + CO(g) \longrightarrow Zn(g) + CO_2(g)$$

The zinc vapor formed distills from the furnace into an appropriate receiver.

**21.16** **(a)** We first find the mass of ore containing $2.0 \times 10^8$ kg of copper.

$$(2.0 \times 10^8\ \text{kg Cu}) \times \frac{100\%\ \text{ore}}{0.80\%\ \text{Cu}} = 2.5 \times 10^{10}\ \text{kg ore}$$

We can then compute the volume from the density of the ore.

$$(2.5 \times 10^{10}\ \text{kg}) \times \frac{1000\ \text{g}}{1\ \text{kg}} \times \frac{1\ \text{cm}^3}{2.8\ \text{g}} = \mathbf{8.9 \times 10^{12}\ cm^3}$$

**(b)** From the formula of chalcopyrite it is clear that two moles of sulfur dioxide will be formed per mole of copper. The mass of sulfur dioxide formed will be:

$$(2.0 \times 10^8\ \text{kg Cu}) \times \frac{1\ \text{mol Cu}}{0.06355\ \text{kg Cu}} \times \frac{2\ \text{mol}\ SO_2}{1\ \text{mol Cu}} \times \frac{0.06407\ \text{kg}\ SO_2}{1\ \text{mol}\ SO_2} = \mathbf{4.0 \times 10^8\ kg\ SO_2}$$

**21.18** Iron can be produced by reduction with coke in a blast furnace; whereas, aluminum is usually produced electrolytically, which is a much more expensive process.

**21.28** (a) $2Na(s) + 2H_2O(l) \longrightarrow 2NaOH(aq) + H_2(g)$

(b) $2NaOH(aq) + CO_2(g) \longrightarrow Na_2CO_3(aq) + H_2O(l)$

(c) $Na_2CO_3(s) + 2HCl(aq) \longrightarrow 2NaCl(aq) + CO_2(g) + H_2O(l)$

(d) $NaHCO_3(aq) + HCl(aq) \longrightarrow NaCl(aq) + CO_2(g) + H_2O(l)$

(e) $2NaHCO_3(s) \longrightarrow Na_2CO_3(s) + CO_2(g) + H_2O(g)$

(f) $Na_2CO_3(s) \longrightarrow$ no reaction. Unlike $CaCO_3(s)$, $Na_2CO_3(s)$ is not decomposed by moderate heating.

**21.30** The balanced equation is: $Na_2CO_3(s) + 2HCl(aq) \longrightarrow 2NaCl(aq) + CO_2(g) + H_2O(l)$

$$\text{mol } CO_2 \text{ produced} = 25.0\text{ g } Na_2CO_3 \times \frac{1 \text{ mol } Na_2CO_3}{106.0 \text{ g } Na_2CO_3} \times \frac{1 \text{ mol } CO_2}{1 \text{ mol } Na_2CO_3} = 0.236 \text{ mol } CO_2$$

$$V_{CO_2} = \frac{nRT}{P} = \frac{(0.236 \text{ mol})(0.0821 \text{ L}\cdot\text{atm/K}\cdot\text{mol})(283 \text{ K})}{\left(746 \text{ mmHg} \times \dfrac{1 \text{ atm}}{760 \text{ mmHg}}\right)} = \mathbf{5.59\ L}$$

**21.34** First magnesium is treated with concentrated nitric acid (redox reaction) to obtain magnesium nitrate.

$$3Mg(s) + 8HNO_3(aq) \longrightarrow 3Mg(NO_3)_2(aq) + 4H_2O(l) + 2NO(g)$$

The magnesium nitrate is recovered from solution by evaporation, dried, and heated in air to obtain magnesium oxide:

$$2Mg(NO_3)_2(s) \longrightarrow 2MgO(s) + 4NO_2(g) + O_2(g)$$

**21.36** The electron configuration of magnesium is $[Ne]3s^2$. The 3*s* electrons are outside the neon core (shielded), so they have relatively low ionization energies. Removing the third electron means separating an electron from the neon (closed shell) core, which requires a great deal more energy.

**21.38** Even though helium and the Group 2A metals have $ns^2$ outer electron configurations, helium has a closed shell noble gas configuration and the Group 2A metals do not. The electrons in He are much closer to and more strongly attracted by the nucleus. Hence, the electrons in He are not easily removed. Helium is inert.

**21.40** (a) quicklime: $CaO(s)$ (b) slaked lime: $Ca(OH)_2(s)$

(c) limewater: an aqueous suspension of $Ca(OH)_2$

**21.44** The reduction reaction is: $Al^{3+}(aq) + 3e^- \rightarrow Al(s)$

First, we can calculate the amount of charge needed to deposit 664 g of Al.

$$664\ \text{g Al} \times \frac{1\ \text{mol Al}}{26.98\ \text{g Al}} \times \frac{3\ \text{mol}\ e^-}{1\ \text{mol Al}} \times \frac{96{,}500\ \text{C}}{1\ \text{mol}\ e^-} = 7.12 \times 10^6\ \text{C}$$

Since a current of one ampere represents a flow of one coulomb per second, we can find the time it takes to pass this amount of charge.

$32.6\ \text{A} = 32.6\ \text{C/s}$

$$(7.12 \times 10^6\ \text{C}) \times \frac{1\ \text{s}}{32.6\ \text{C}} \times \frac{1\ \text{h}}{3600\ \text{s}} = \mathbf{60.7\ h}$$

**21.46** **(a)** The relationship between cell voltage and free energy difference is:

$$\Delta G = -nFE$$

In the given reaction $n = 6$. We write:

$$E = \frac{-\Delta G}{nF} = \frac{-594 \times 10^3\ \text{J/mol}}{(6)(96500\ \text{J/V}\cdot\text{mol})} = -1.03\ \text{V}$$

The balanced equation shows *two* moles of aluminum. Is this the voltage required to produce *one* mole of aluminum? If we divide everything in the equation by two, we obtain:

$$\tfrac{1}{2}Al_2O_3(s) + \tfrac{3}{2}C(s) \rightarrow Al(l) + \tfrac{3}{2}CO(g)$$

For the new equation $n = 3$ and $\Delta G$ is $\left(\frac{1}{2}\right)(594\ \text{kJ/mol}) = 297\ \text{kJ/mol}$. We write:

$$E = \frac{-\Delta G}{nF} = \frac{-297 \times 10^3\ \text{J/mol}}{(3)(96500\ \text{J/V}\cdot\text{mol})} = -1.03\ \text{V}$$

The minimum voltage that must be applied is **1.03 V** (a negative sign in the answers above means that 1.03 V is required to produce the Al). The voltage required to produce one mole or one thousand moles of aluminum is the same; the amount of *current* will be different in each case.

**(b)** First we convert 1.00 kg (1000 g) of Al to moles.

$$(1.00 \times 10^3\ \text{g Al}) \times \frac{1\ \text{mol Al}}{26.98\ \text{g Al}} = 37.1\ \text{mol Al}$$

The reaction in part (a) shows us that three moles of electrons are required to produce one mole of aluminum. The voltage is three times the minimum calculated above (namely, −3.09 V or −3.09 J/C). We can find the electrical energy by using the same equation with the other voltage.

$$\Delta G = -nFE = -(37.1)\left(\frac{3\ \text{mol}\ e^-}{1\ \text{mol Al}} \times \frac{96500\ \text{C}}{1\ \text{mol}\ e^-}\right)\left(\frac{-3.09\ \text{J}}{1\ \text{C}}\right) = \mathbf{3.32 \times 10^7\ J/mol = 3.32 \times 10^4\ kJ/mol}$$

This equation can be used because electrical work can be calculated by multiplying the voltage by the amount of charge transported through the circuit (joules = volts × coulombs). The $nF$ term in Equation (18.2) of the text used above represents the amount of charge.

What is the significance of the positive sign of the free energy change? Would the manufacturing of aluminum be a different process if the free energy difference were negative?

**21.48** $4Al(NO_3)_3(s) \longrightarrow 2Al_2O_3(s) + 12NO_2(g) + 3O_2(g)$

**21.50** The "bridge" bonds in $Al_2Cl_6$ break at high temperature: $Al_2Cl_6(g) \rightleftharpoons 2AlCl_3(g)$.

This increases the number of molecules in the gas phase and causes the pressure to be higher than expected for pure $Al_2Cl_6$.

If you know the equilibrium constants for the above reaction at higher temperatures, could you calculate the expected pressure of the $AlCl_3$–$Al_2Cl_6$ mixture?

**21.52** In $Al_2Cl_6$, each aluminum atom is surrounded by 4 bonding pairs of electrons ($AB_4$–type molecule), and therefore each aluminum atom is ***$sp^3$* hybridized**. VSEPR analysis shows $AlCl_3$ to be an $AB_3$–type molecule (no lone pairs on the central atom). The geometry should be trigonal planar, and the aluminum atom should therefore be ***$sp^2$* hybridized**.

**21.54** The formulas of the metal oxide and sulfide are MO and MS (why?). The balanced equation must therefore be:

$$2MS(s) + 3O_2(g) \rightarrow 2MO(s) + 2SO_2(g)$$

The number of moles of MO and MS are equal. We let $x$ be the molar mass of metal. The number of moles of metal oxide is:

$$0.972\text{ g} \times \frac{1\text{ mol}}{(x + 16.00)\text{g}}$$

The number of moles of metal sulfide is:

$$1.164\text{ g} \times \frac{1\text{ mol}}{(x + 32.07)\text{g}}$$

The moles of metal oxide equal the moles of metal sulfide.

$$\frac{0.972}{(x + 16.00)} = \frac{1.164}{(x + 32.07)}$$

We solve for $x$.

$$0.972(x + 32.07) = 1.164(x + 16.00)$$

$$\mathbf{x = 65.4\ g/mol}$$

**21.56** Copper(II) ion is more easily reduced than either water or hydrogen ion (How can you tell? See Section 18.3 of the text.) Copper metal is more easily oxidized than water. Water should not be affected by the copper purification process.

**21.58** Using Equation (17.12) from the text:

**(a)** $\Delta G^\circ_{rxn} = [4\Delta G^\circ_f(Fe) + 3\Delta G^\circ_f(O_2)] - 2\Delta G^\circ_f(Fe_2O_3)$

$\boldsymbol{\Delta G^\circ_{rxn}} = [(4)(0) + (3)(0)] - (2)(-741.0 \text{ kJ/mol}) = \mathbf{1482 \text{ kJ/mol}}$

**(b)** $\Delta G^\circ_{rxn} = [4\Delta G^\circ_f(Al) + 3\Delta G^\circ_f(O_2)] - 2\Delta G^\circ_f(Al_2O_3)$

$\boldsymbol{\Delta G^\circ_{rxn}} = [(4)(0) + (3)(0)] - (2)(-1576.4 \text{ kJ/mol}) = \mathbf{3152.8 \text{ kJ/mol}}$

**21.60** At high temperature, magnesium metal reacts with nitrogen gas to form magnesium nitride.

$$3Mg(s) + N_2(g) \longrightarrow Mg_3N_2(s)$$

Can you think of any gas other than a noble gas that could provide an inert atmosphere for processes involving magnesium at high temperature?

**21.62** **(a)** In water the aluminum(III) ion causes an increase in the concentration of hydrogen ion (lower pH). This results from the effect of the small diameter and high charge (3+) of the aluminum ion on surrounding water molecules. The aluminum ion draws electrons in the O–H bonds to itself, thus allowing easy formation of $H^+$ ions.

**(b)** $Al(OH)_3$ is an amphoteric hydroxide. It will dissolve in strong base with the formation of a complex ion.

$$Al(OH)_3(s) + OH^-(aq) \longrightarrow Al(OH)_4^-(aq)$$

The concentration of $OH^-$ in aqueous ammonia is too low for this reaction to occur.

**21.64** Calcium oxide is a base. The reaction is a neutralization.

$$CaO(s) + 2HCl(aq) \longrightarrow CaCl_2(aq) + H_2O(l)$$

**21.66** Metals have closely spaced energy levels and (referring to Figure 21.10 of the text) a very small energy gap between filled and empty levels. Consequently, many electronic transitions can take place with absorption and subsequent emission continually occurring. Some of these transitions fall in the visible region of the spectrum and give rise to the flickering appearance.

**21.68** NaF is used in toothpaste to fight tooth decay.

$Li_2CO_3$ is used to treat mental illness.

$Mg(OH)_2$ is an antacid.

$CaCO_3$ is an antacid.

$BaSO_4$ is used to enhance X ray images of the digestive system.

$Al(OH)_2NaCO_3$ is an antacid.

**21.70** Both Li and Mg form oxides ($Li_2O$ and MgO). Other Group 1A metals (Na, K, etc.) also form peroxides and superoxides. In Group 1A, only Li forms nitride ($Li_3N$), like Mg ($Mg_3N_2$).

Li resembles Mg in that its carbonate, fluoride, and phosphate have low solubilities.

**21.72** You might know that Ag, Cu, Au, and Pt are found as free elements in nature, which leaves **Zn** by process of elimination. You could also look at Table 18.1 of the text to find the metal that is easily oxidized. Looking at the table, the standard oxidation potential of Zn is +0.76 V. The positive value indicates that Zn is easily oxidized to $Zn^{2+}$ and will not exist as a free element in nature.

**21.74** Because only B and C react with 0.5 *M* HCl, they are more electropositive than A and D. The fact that when B is added to a solution containing the ions of the other metals, metallic A, C, and D are formed indicates that B is the most electropositive metal. Because A reacts with 6 *M* $HNO_3$, A is more electropositive than D. The metals arranged in increasing order as reducing agents are:

$$\mathbf{D < A < C < B}$$

Examples are: D = Au, A = Cu, C = Zn, B = Mg

**21.76** First, we calculate the density of $O_2$ in $KO_2$ using the mass percentage of $O_2$ in the compound.

$$\frac{32.00\text{ g O}_2}{71.10\text{ g KO}_2} \times \frac{2.15\text{ g}}{1\text{ cm}^3} = 0.968\text{ g/cm}^3$$

Now, we can use Equation (5.11) of the text to calculate the pressure of oxygen gas that would have the same density as that provided by $KO_2$.

$$\frac{0.968\text{ g}}{1\text{ cm}^3} \times \frac{1000\text{ cm}^3}{1\text{ L}} = 968\text{ g/L}$$

$$d = \frac{PM}{RT}$$

or

$$P = \frac{dRT}{M} = \frac{\left(\frac{968\text{ g}}{1\text{ L}}\right)\left(0.0821\frac{\text{L}\cdot\text{atm}}{\text{mol}\cdot\text{K}}\right)(293\text{ K})}{\left(\frac{32.00\text{ g}}{1\text{ mol}}\right)} = \mathbf{727\text{ atm}}$$

Obviously, using $O_2$ instead of $KO_2$ is not practical.

# CHAPTER 22
# NONMETALLIC ELEMENTS AND THEIR COMPOUNDS

This chapter is of a very descriptive nature. As such, there are essentially no defined problem types. Please read Chapter 22 of the text carefully before answering the end-of-chapter problems.

## SOLUTIONS TO SELECTED TEXT PROBLEMS

**22.12** **(a)** Hydrogen reacts with alkali metals to form ionic hydrides:

$$2\text{Na}(s) + \text{H}_2(g) \rightarrow 2\text{NaH}(s)$$

The oxidation number of hydrogen drops from 0 to −1 in this reaction.

**(b)** Hydrogen reacts with oxygen (combustion) to form water:

$$2\text{H}_2(g) + \text{O}_2(g) \rightarrow 2\text{H}_2\text{O}(l)$$

The oxidation number of hydrogen increases from 0 to +1 in this reaction.

**22.14** Hydrogen forms an interstitial hydride with palladium, which behaves almost like a solution of hydrogen atoms in the metal. At elevated temperatures hydrogen atoms can pass through solid palladium; other substances cannot.

**22.16** The number of moles of deuterium gas is:

$$n = \frac{PV}{RT} = \frac{(0.90\ \cancel{\text{atm}})(2.0\ \cancel{\text{L}})}{(0.0821\ \cancel{\text{L}}\cdot\cancel{\text{atm}}/\cancel{\text{K}}\cdot\text{mol})(298\ \cancel{\text{K}})} = 0.074\ \text{mol}$$

If the abundance of deuterium is 0.015 percent, the number of moles of water must be:

$$0.074\ \text{mol D}_2 \times \frac{100\%\ \text{H}_2\text{O}}{0.015\%\ \text{D}_2} = 4.9 \times 10^2\ \text{mol H}_2\text{O}$$

At a recovery of 80 percent the amount of water needed is:

$$\frac{4.9 \times 10^2\ \cancel{\text{mol}}\ \text{H}_2\text{O}}{0.80} \times \frac{0.01802\ \text{kg H}_2\text{O}}{1.0\ \cancel{\text{mol}}\ \text{H}_2\text{O}} = \mathbf{11\ kg\ H_2O}$$

**22.18** **(a)** $\text{H}_2 + \text{Cl}_2 \rightarrow 2\text{HCl}$

**(b)** $3\text{H}_2 + \text{N}_2 \rightarrow 2\text{NH}_3$

**(c)** $2\text{Li} + \text{H}_2 \rightarrow 2\text{LiH}$

$\text{LiH} + \text{H}_2\text{O} \rightarrow \text{LiOH} + \text{H}_2$

**22.26** The Lewis structure is:

$$[:C\equiv C:]^{2-}$$

**22.28** **(a)** The reaction is: $2NaHCO_3(s) \rightarrow Na_2CO_3(s) + H_2O(g) + CO_2(g)$

Is this an endo- or an exothermic process?

**(b)** The hint is generous. The reaction is:

$$Ca(OH)_2(aq) + CO_2(g) \rightarrow CaCO_3(s) + H_2O(l)$$

The visual proof is the formation of a white precipitate of $CaCO_3$. Why would a water solution of NaOH be unsuitable to qualitatively test for carbon dioxide?

**22.30** Heat causes bicarbonates to decompose according to the reaction:

$$2HCO_3^- \rightarrow CO_3^{2-} + H_2O + CO_2$$

Generation of carbonate ion causes precipitation of the insoluble $MgCO_3$.

Do you think there is much chance of finding natural mineral deposits of calcium or magnesium bicarbonates?

**22.32** The wet sodium hydroxide is first converted to sodium carbonate:

$$2NaOH(aq) + CO_2(g) \rightarrow Na_2CO_3(aq) + H_2O(l)$$

and then to sodium hydrogen carbonate: $Na_2CO_3(aq) + H_2O(l) + CO_2(g) \rightarrow 2NaHCO_3(aq)$

Eventually, the sodium hydrogen carbonate precipitates (the water solvent evaporates since $NaHCO_3$ is not hygroscopic). Thus, most of the white solid is $NaHCO_3$ plus some $Na_2CO_3$.

**22.34** Carbon monoxide and molecular nitrogen are isoelectronic. Both have 14 electrons. What other diatomic molecules discussed in these problems are isoelectronic with CO?

**22.40** **(a)** $2NaNO_3(s) \rightarrow 2NaNO_2(s) + O_2(g)$

**(b)** $NaNO_3(s) + C(s) \rightarrow NaNO_2(s) + CO(g)$

**22.42** The balanced equation is: $2NH_3(g) + CO_2(g) \rightarrow (NH_2)_2CO(s) + H_2O(l)$

If pressure increases, the position of equilibrium will shift in the direction with the smallest number of molecules in the gas phase, that is, to the right. Therefore, the reaction is best run at high pressure.

Write the expression for $Q_p$ for this reaction. Does increasing pressure cause $Q_p$ to increase or decrease? Is this consistent with the above prediction?

**22.44** The density of a gas depends on temperature, pressure, and the molar mass of the substance. When two gases are at the same pressure and temperature, the ratio of their densities should be the same as the ratio of their molar masses. The molar mass of ammonium chloride is 53.5 g/mol, and the ratio of this to the molar mass of molecular hydrogen (2.02 g/mol) is 26.8. The experimental value of 14.5 is roughly half this amount. Such results usually indicate breakup or dissociation into smaller molecules in the gas phase (note the temperature). The measured molar mass is the average of all the molecules in equilibrium.

$$NH_4Cl(g) \rightleftharpoons NH_3(g) + HCl(g)$$

Knowing that ammonium chloride is a stable substance at 298 K, is the above reaction exo- or endothermic?

**22.46** The highest oxidation state possible for a Group 5A element is +5. This is the oxidation state of nitrogen in nitric acid ($HNO_3$).

**22.48** Nitric acid is a strong oxidizing agent in addition to being a strong acid (see Table 18.1 of the text, $E^\circ_{red} = +0.96V$). The primary action of a good reducing agent like zinc is reduction of nitrate ion to ammonium ion.

$$4Zn(s) + NO_3^-(aq) + 10H^+(aq) \rightarrow 4Zn^{2+}(aq) + NH_4^+(aq) + 3H_2O(l)$$

**22.50** One of the best Lewis structures for nitrous oxide is:

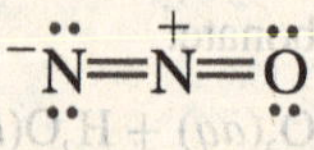

There are no lone pairs on the central nitrogen, making this an $AB_2$ VSEPR case. All such molecules are linear. Other resonance forms are:

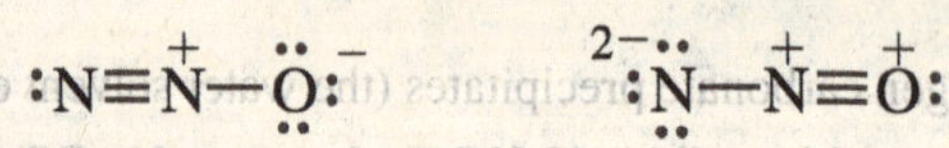

Are all the resonance forms consistent with a linear geometry?

**22.52** $\Delta H^\circ = 4\Delta H_f^\circ[NO(g)] + 6\Delta H_f^\circ[H_2O(l)] - \{4\Delta H_f^\circ[NH_3(g)] + 5\Delta H_f^\circ[O_2(g)]\}$

$\Delta H^\circ = [(4)(90.4 \text{ kJ/mol}) + (6)(-285.8 \text{ kJ/mol})] - [(4)(-46.3 \text{ kJ/mol}) + (5)(0)] = \mathbf{-1168 \text{ kJ/mol}}$

**22.54** $\Delta T_b = K_b m = 0.409°C$

$$\text{molality} = \frac{0.409\cancel{°C}}{2.34\cancel{°C}/m} = 0.175\ m$$

The number of grams of white phosphorus in 1 kg of solvent is:

$$\frac{1.645 \text{ g phosphorus}}{75.5 \cancel{\text{g}} \text{ CS}_2} \times \frac{1000 \cancel{\text{g}}}{1 \text{ kg}} = 21.8 \text{ g phosphorus/kg CS}_2$$

The molar mass of white phosphorus is:

$$\frac{21.8 \text{ g phosphorus/kg CS}_2}{0.175 \text{ mol phosphorus/kg CS}_2} = \mathbf{125 \text{ g/mol}}$$

Let the molecular formula of white phosphorus be $P_n$ so that:

$$n \times 30.97 \text{ g/mol} = 125 \text{ g/mol}$$

$$n = 4$$

The molecular formula of white phosphorus is $\mathbf{P_4}$.

**22.56** The balanced equation is:

$$\mathbf{P_4O_{10}(s) + 4HNO_3(aq) \rightarrow 2N_2O_5(g) + 4HPO_3(l)}$$

The theoretical yield of $N_2O_5$ is :

$$79.4 \text{ g } P_4O_{10} \times \frac{1 \text{ mol } P_4O_{10}}{283.9 \text{ g } P_4O_{10}} \times \frac{2 \text{ mol } N_2O_5}{1 \text{ mol } P_4O_{10}} \times \frac{108.0 \text{ g } N_2O_5}{1 \text{ mol } N_2O_5} = \mathbf{60.4 \text{ g } N_2O_5}$$

**22.58** $PH_4^+$ is similar to $NH_4^+$. The hybridization of phosphorus in $PH_4^+$ is $\boldsymbol{sp^3}$.

**22.66** $\Delta G° = [\Delta G_f°(NO_2) + \Delta G_f°(O_2)] - [\Delta G_f°(NO) + \Delta G_f°(O_3)]$

$$\Delta G° = [(1)(51.8 \text{ kJ/mol}) + (0)] - [(1)(86.7 \text{ kJ/mol}) + (1)(163.4 \text{ kJ/mol})] = \mathbf{-198.3 \text{ kJ/mol}}$$

$$\Delta G° = -RT\ln K_P$$

$$\ln K_P = \frac{-\Delta G°}{RT} = \frac{198.3 \times 10^3 \text{ J/mol}}{(8.314 \text{ J/K}\cdot\text{mol})(298 \text{ K})}$$

$$\mathbf{K_P = 6 \times 10^{34}}$$

Since there is no change in the number of moles of gases, $K_c$ is *equal* to $K_P$.

**22.68** Following the rules given in Section 4.4 of the text, we assign hydrogen an oxidation number of +1 and **fluorine** an oxidation number of **−1.** Since HFO is a neutral molecule, the oxidation number of **oxygen** is **zero.** Can you think of other compounds in which oxygen has this oxidation number?

**22.70** First, let's calculate the moles of sulfur in 48 million tons of sulfuric acid.

$$(48 \times 10^6 \text{ tons } H_2SO_4) \times \frac{2000 \text{ lb}}{1 \text{ ton}} \times \frac{453.6 \text{ g}}{1 \text{ lb}} \times \frac{1 \text{ mol } H_2SO_4}{98.09 \text{ g } H_2SO_4} \times \frac{1 \text{ mol S}}{1 \text{ mol } H_2SO_4} = \mathbf{4.4 \times 10^{11} \text{ mol S}}$$

Converting to grams of sulfur:

$$(4.4 \times 10^{11} \text{ mol S}) \times \frac{32.07 \text{ g S}}{1 \text{ mol S}} = \mathbf{1.4 \times 10^{13} \text{ g S}}$$

**22.72** There are actually several steps involved in removing sulfur dioxide from industrial emissions with calcium carbonate. First calcium carbonate is heated to form carbon dioxide and calcium oxide.

$$CaCO_3(s) \rightleftharpoons CaO(s) + CO_2(g)$$

The CaO combines with sulfur dioxide to form calcium sulfite.

$$CaO(s) + SO_2(g) \rightarrow CaSO_3(s)$$

Alternatively, calcium sulfate forms if enough oxygen is present.

$$2CaSO_3(s) + O_2(g) \rightarrow 2CaSO_4(s)$$

The amount of calcium carbonate (limestone) needed in this problem is:

$$50.6 \text{ g } SO_2 \times \frac{1 \text{ mol } SO_2}{64.07 \text{ g } SO_2} \times \frac{1 \text{ mol } CaCO_3}{1 \text{ mol } SO_2} \times \frac{100.1 \text{ g } CaCO_3}{1 \text{ mol } CaCO_3} = \mathbf{79.1 \text{ g } CaCO_3}$$

The calcium oxide–sulfur dioxide reaction is an example of a Lewis acid-base reaction (see Section 15.12 of the text) between oxide ion and sulfur dioxide. Can you draw Lewis structures showing this process? Which substance is the Lewis acid and which is the Lewis base?

**22.74** The usual explanation for the fact that no chemist has yet succeeded in making $SCl_6$, $SBr_6$ or $SI_6$ is based on the idea of excessive crowding of the six chlorine, bromine, or iodine atoms around the sulfur. Others suggest that sulfur in the +6 oxidation state would oxidize chlorine, bromine, or iodine in the −1 oxidation state to the free elements. In any case, none of these substances has been made as of the date of this writing.

It is of interest to point out that thirty years ago all textbooks confidently stated that compounds like $ClF_5$ could not be prepared.

Note that $PCl_6^-$ is a known species. How different are the sizes of S and P?

**22.76** First we convert gallons of water to grams of water.

$$(2.0 \times 10^2 \text{ gal}) \times \frac{3.785 \text{ L}}{1 \text{ gal}} \times \frac{1000 \text{ mL}}{1 \text{ L}} \times \frac{1.00 \text{ g } H_2O}{1 \text{ mL}} = 7.6 \times 10^5 \text{ g } H_2O$$

An $H_2S$ concentration of 22 ppm indicates that in 1 million grams of water, there will be 22 g of $H_2S$. First, let's calculate the number of moles of $H_2S$ in $7.6 \times 10^5$ g of $H_2O$:

$$(7.6 \times 10^5 \text{ g } H_2O) \times \frac{22 \text{ g } H_2S}{1.0 \times 10^6 \text{ g } H_2O} \times \frac{1 \text{ mol } H_2S}{34.09 \text{ g } H_2S} = 0.49 \text{ mol } H_2S$$

The mass of chlorine required to react with 0.49 mol of $H_2S$ is:

$$0.49 \text{ mol } H_2S \times \frac{1 \text{ mol } Cl_2}{1 \text{ mol } H_2S} \times \frac{70.90 \text{ g } Cl_2}{1 \text{ mol } Cl_2} = \mathbf{35 \text{ g } Cl_2}$$

**22.78** A check of Table 18.1 of the text shows that sodium ion cannot be reduced by any of the substances mentioned in this problem; it is a "spectator ion". We focus on the substances that are actually undergoing oxidation or reduction and write half-reactions for each.

$$2I^-(aq) \rightarrow I_2(s)$$

$$H_2SO_4(aq) \rightarrow H_2S(g)$$

Balancing the oxygen, hydrogen, and charge gives:

$$2I^-(aq) \rightarrow I_2(s) + 2e^-$$

$$H_2SO_4(aq) + 8H^+(aq) + 8e^- \rightarrow H_2S(g) + 4H_2O(l)$$

Multiplying the iodine half-reaction by four and combining gives the balanced redox equation.

$$H_2SO_4(aq) + 8I^-(aq) + 8H^+(aq) \rightarrow H_2S(g) + 4I_2(s) + 4H_2O(l)$$

The hydrogen ions come from extra sulfuric acid. We add one sodium ion for each iodide ion to obtain the final equation.

$$\mathbf{9H_2SO_4(aq) + 8NaI(aq) \rightarrow H_2S(g) + 4I_2(s) + 4H_2O(l) + 8NaHSO_4(aq)}$$

**22.82** Sulfuric acid is added to solid sodium chloride, not aqueous sodium chloride. Hydrogen chloride is a gas at room temperature and can escape from the reacting mixture.

$$\mathbf{H_2SO_4(l) + NaCl(s) \rightarrow HCl(g) + NaHSO_4(s)}$$

The reaction is driven to the right by the continuous loss of HCl($g$) (Le Châtelier's principle).

What happens when sulfuric acid is added to a water solution of NaCl? Could you tell the difference between this solution and the one formed by adding hydrochloric acid to aqueous sodium sulfate?

**22.84** The reaction is: $2Br^-(aq) + Cl_2(g) \rightarrow 2Cl^-(aq) + Br_2(l)$

The number of moles of chlorine needed is:

$$167 \text{ g } Br^- \times \frac{1 \text{ mol } Br^-}{79.90 \text{ g } Br^-} \times \frac{1 \text{ mol } Cl_2}{2 \text{ mol } Br^-} = 1.05 \text{ mol } Cl_2(g)$$

Use the ideal gas equation to calculate the volume of $Cl_2$ needed.

$$V_{Cl_2} = \frac{nRT}{P} = \frac{(1.05 \text{ mol})(0.0821 \text{ L}\cdot\text{atm/K}\cdot\text{mol})(293 \text{ K})}{(1 \text{ atm})} = \mathbf{25.3 \text{ L}}$$

**22.86** As with iodide salts, a redox reaction occurs between sulfuric acid and sodium bromide.

$$2H_2SO_4(aq) + 2NaBr\,(aq) \rightarrow SO_2(g) + Br_2(l) + 2H_2O(l) + Na_2SO_4(aq)$$

**22.88** The balanced equation is:

$$Cl_2(g) + 2Br^-(aq) \rightarrow 2Cl^-(aq) + Br_2(g)$$

The number of moles of bromine is the same as the number of moles of chlorine, so this problem is essentially a gas law exercise in which $P$ and $T$ are changed for some given amount of gas.

$$V_2 = \frac{P_1V_1}{T_1} \times \frac{T_2}{P_2} = \frac{(760\ \text{mmHg})(2.00\ \text{L})}{288\ \text{K}} \times \frac{373\ \text{K}}{700\ \text{mmHg}} = \mathbf{2.81\ L}$$

**22.90** The balanced equation is:

$$\mathbf{I_2O_5(s) + 5CO(g) \rightarrow I_2(s) + 5CO_2(g)}$$

The oxidation number of iodine changes from +5 to 0 and the oxidation number of carbon changes from +2 to +4. **Iodine is reduced; carbon is oxidized.**

**22.92** **(a)** $SiCl_4$ **(b)** $F^-$ **(c)** F **(d)** $CO_2$

**22.94** There is no change in oxidation number; it is zero for both compounds.

**22.96** **(a)** $2Na + 2D_2O \rightarrow 2NaOD + D_2$

**(b)** $2D_2O \xrightarrow{\text{electrolysis}} 2D_2 + O_2$

$D_2 + Cl_2 \rightarrow 2DCl$

**(c)** $Mg_3N_2 + 6D_2O \rightarrow 3Mg(OD)_2 + 2ND_3$

**(d)** $CaC_2 + 2D_2O \rightarrow C_2D_2 + Ca(OD)_2$

**(e)** $Be_2C + 4D_2O \rightarrow 2Be(OD)_2 + CD_4$

**(f)** $SO_3 + D_2O \rightarrow D_2SO_4$

**22.98** **(a)** At elevated pressures, water boils above 100°C.
**(b)** Water is sent down the outermost pipe so that it is able to melt a larger area of sulfur.
**(c)** Sulfur deposits are structurally weak. There will be a danger of the sulfur mine collapsing.

**22.100** The oxidation is probably initiated by breaking a C–H bond (the rate-determining step). The C–D bond breaks at a slower rate than the C–H bond; therefore, replacing H by D decreases the rate of oxidation.

**22.102** Organisms need a source of energy to sustain the processes of life. Respiration creates that energy. Molecular oxygen is a powerful oxidizing agent, reacting with substances such as glucose to release energy for growth and function. Molecular nitrogen (containing the nitrogen-to-nitrogen triple bond) is too unreactive at room temperature to be of any practical use.

**22.104** We know that $\Delta G^\circ = -RT\ln K$ and $\Delta G^\circ = \Delta H^\circ - T\Delta S^\circ$. We can first calculate $\Delta H^\circ$ and $\Delta S^\circ$ using data in Appendix 3 of the text. Then, we can calculate $\Delta G^\circ$ and lastly $K$.

$$\Delta H^\circ = 2\Delta H_f^\circ[CO(g)] - \{\Delta H_f^\circ[C(s)] + \Delta H_f^\circ[CO_2(g)]\}$$

$$\Delta H^\circ = (2)(-110.5\text{ kJ/mol}) - [0 + (-393.5\text{ kJ/mol})] = 172.5\text{ kJ/mol}$$

$$\Delta S^\circ = 2S^\circ[CO(g)] - \{S^\circ[C(s)] + S^\circ[CO_2(g)]\}$$

$$\Delta S^\circ = (2)(197.9\text{ J/K·mol}) - (5.69\text{ J/K·mol} + 213.6\text{ J/K·mol}) = 176.5\text{ J/K·mol}$$

At 298 K (25°C),

$$\Delta G^\circ = \Delta H^\circ - T\Delta S^\circ = (172.5\times10^3\text{ J/mol}) - (298\text{ K})(176.5\text{ J/K·mol}) = 1.199\times10^5\text{ J/mol}$$

$$\Delta G^\circ = -RT\ln K$$

$$K = e^{\frac{-\Delta G^\circ}{RT}} = e^{\frac{-(1.199\times10^5\text{ J/mol})}{(8.314\text{ J/K·mol})(298\text{ K})}} = \mathbf{9.61\times10^{-22}}$$

At 1273 K (1000°C),

$$\Delta G^\circ = (172.5\times10^3\text{ J/mol}) - (1273\text{ K})(176.5\text{ J/K·mol}) = -5.218\times10^4\text{ J/mol}$$

$$K = e^{\frac{-\Delta G^\circ}{RT}} = e^{\frac{-(-5.218\times10^4\text{ J/mol})}{(8.314\text{ J/K·mol})(1273\text{ K})}} = \mathbf{138}$$

The much larger value of $K$ at the higher temperature indicates the formation of CO is favored at higher temperatures (achieved by using a blast furnace).

**22.106** The reactions are:

$$P_4(s) + 5O_2(g) \rightarrow P_4O_{10}(s)$$
$$P_4O_{10}(s) + 6H_2O(l) \rightarrow 4H_3PO_4(aq)$$

First, we calculate the moles of $H_3PO_4$ produced. Next, we can calculate the molarity of the phosphoric acid solution. Finally, we can determine the pH of the $H_3PO_4$ solution (a weak acid).

$$10.0\text{ g P}_4 \times \frac{1\text{ mol P}_4}{123.9\text{ g P}_4} \times \frac{1\text{ mol P}_4\text{O}_{10}}{1\text{ mol P}_4} \times \frac{4\text{ mol H}_3\text{PO}_4}{1\text{ mol P}_4\text{O}_{10}} = 0.323\text{ mol H}_3\text{PO}_4$$

$$\text{Molarity} = \frac{0.323\text{ mol}}{0.500\text{ L}} = 0.646\ M$$

We set up the ionization of the weak acid, $H_3PO_4$. The $K_a$ value for $H_3PO_4$ can be found in Table 15.5 of the text.

| | $H_3PO_4(aq) + H_2O \rightleftharpoons$ | $H_3O^+(aq)$ | $+\ H_2PO_4^-(aq)$ |
|---|---|---|---|
| Initial (*M*): | 0.646 | 0 | 0 |
| Change (*M*): | $-x$ | $+x$ | $+x$ |
| Equilibrium (*M*): | $0.646 - x$ | $x$ | $x$ |

$$K_a = \frac{[H_3O^+][H_2PO_4^-]}{[H_3PO_4]}$$

$$7.5 \times 10^{-3} = \frac{(x)(x)}{(0.646 - x)}$$

$$x^2 + 7.5 \times 10^{-3}x - 4.85 \times 10^{-3} = 0$$

Solving the quadratic equation,

$$x = 0.066\ M = [H_3O^+]$$

Following the procedure in Problem 15.122 and the discussion in Section 15.8 of the text, we can neglect the contribution to the hydronium ion concentration from the second and third ionization steps. Thus,

$$\textbf{pH} = -\log(0.066) = \mathbf{1.18}$$

# CHAPTER 23
# TRANSITION METAL CHEMISTRY AND COORDINATION COMPOUNDS

## PROBLEM-SOLVING STRATEGIES AND TUTORIAL SOLUTIONS

### TYPES OF PROBLEMS

**Problem Type 1:** Assigning Oxidation Numbers to the Metal Atom in Coordination Compounds.

**Problem Type 2:** Naming Coordination Compounds.

**Problem Type 3:** Writing Formulas for Coordination Compounds.

**Problem Type 4:** Predicting the Number of Unpaired Spins in a Coordination Compound.

## PROBLEM TYPE 1: ASSIGNING OXIDATION NUMBERS TO THE METAL ATOM IN COORDINATION COMPOUNDS

Transition metals exhibit variable oxidation states in their compounds. The charge on the central metal atom and its surrounding ligands sum to zero in a neutral coordination compound. In a complex ion, the charges on the central metal atom and the surrounding ligands sum to the net charge of the ion.

**EXAMPLE 23.1**

**Specify the oxidation number of the central metal atom in each of the following compounds:**
**(a) $[Co(NH_3)_6]Cl_3$ and (b) $Co(CN)_6^{3-}$**

**Strategy:** The oxidation number of the metal atom is equal to its charge. First we look for known charges in the species. Recall that alkali metals are +1 and alkaline earth metals are +2. Also determine if the ligand is a charged or neutral species. From the known charges, we can deduce the net charge of the metal and hence its oxidation number.

**Solution:**

**(a)** $NH_3$ is a neutral species. Since each chloride ion carries a −1 charge, and there are three $Cl^-$ ions, the oxidation number of Co must be +3.

**(b)** Each cyanide ion has a charge of −1. The sum of the oxidation number of Co and the −6 charge for the six cyanide ions is −3. Therefore, the oxidation number of Co must be +3.

**PRACTICE EXERCISE**

**1.** Specify the oxidation number of the central metal atom in each of the following compounds:
(a) $[Pt(NH_3)_3Cl_3]NO_3$ and (b) $Ni(CO)_4$.

**Text Problems:** 23.12, **23.14**

## PROBLEM TYPE 2: NAMING COORDINATION COMPOUNDS

The complete set of rules for naming coordination compounds is given in Section 23.3 of the text. Presented below is an abridged version of those rules.

1. The cation is named before the anion, as is the case for other ionic compounds.
2. Within a complex ion the ligands are named first, in alphabetical order, and the metal ion is named last.
3. The names of anionic ligands end with the letter *o*, whereas a neutral ligand is usually called by the name of the molecule. Exceptions are listed in Table 23.4 of the text.
4. Greek prefixes are used to indicate the number of ligands of a particular kind present. If the ligand itself contains a Greek prefix, the prefixes *bis* (2), *tris* (3), and *tetrakis* (4) are used to indicate the number of ligands present.
5. The oxidation number of the metal is written in Roman numerals following the name of the metal.
6. If the complex is an anion, its name ends in "−ate".

### EXAMPLE 23.2

**Name the following coordination compounds and complex ions: (a) $[Ni(NH_3)_6]^{2+}$, (b) $K_2[Cu(CN)_4]$, (c) $[Pt(NH_3)_4Cl_2]Cl_2$.**

**Strategy:** We follow the procedure for naming coordination compounds outlined in Section 23.3 of the text and refer to Tables 23.4 and 23.5 of the text for names of ligands and anions containing metal atoms.

**Solution:**

**(a)** $NH_3$ is a neutral species; therefore, the Ni must have a +2 charge. Ammonia as a ligand in a coordination compound is called ammine. The complex ion is called **hexaamminenickel(II) ion.**

**(b)** Potassium is an alkali metal; it always has a +1 charge in ionic compounds. Since the two potassium ions have a total charge of +2, the charge on the complex ion $[Cu(CN)_4]$ must be −2. Cyanide ion has a −1 charge; therefore, Cu must have a +2 charge. Potassium is the cation and is named first. The compound is called **potassium tetracyanocuprate(II)**. We use an "−ate" ending because the complex is an anion.

**(c)** The complex ion has a +2 charge balanced by the total charge of −2 for the two chloride ions. Focusing on the complex ion, $NH_3$ is a neutral species and the two chloride ligands each have a −1 charge. Since the charge of the complex ion is +2, platinum must have a +4 charge. This compound is called **tetraamminedichloroplatinum(IV) chloride**.

### PRACTICE EXERCISE

2. Name the following coordination compounds and complex ions:

(a) $[Cr(OH)_4]^-$ (b) $[Pt(NH_3)_3Cl_3]NO_3$.

**Text Problem: 23.16**

## PROBLEM TYPE 3: WRITING FORMULAS FOR COORDINATION COMPOUNDS

Follow the nomenclature rules given in Problem Type 2 above. Remember that the cation is named before the anion.

### EXAMPLE 23.3

**Write the formulas for the following compounds or complex ions:**

**(a) tetrahydroxoaluminate(III) ion, (b) potassium hexachloropalladate(IV), (c) diaquodicyanocopper(II)**

**Strategy:** We follow the procedure in Section 23.3 of the text and refer to Tables 23.4 and 23.5 of the text for names of ligands and anions containing metal atoms.

**Solution:**

**(a)** Tetrahydroxo refers to four hydroxide ligands. The "–ate" of aluminate indicates that the complex is an anion. The Roman numeral (III) indicates a +3 charge on aluminum. Since each hydroxide ligand has a −1 charge, the charge on the ion is −1. The formula for the complex ion is $\mathbf{[Al(OH_4)]^-}$.

**(b)** The complex anion, hexachloropalladate(IV), has a −2 charge. Each of the six chloride ligands has a −1 charge (total of −6) and palladium has a +4 charge. To balance the −2 charge of the anion, there must be two potassium +1 ions. The formula for the compound is $\mathbf{K_2[PdCl_6]}$.

**(c)** Diaquo refers to two water ligands, and dicyano refers to two cyanide ligands. Each cyanide ligand has a −1 charge (total of −2), which balances the +2 charge on copper. The compound is electrically neutral. The formula for the compound is $\mathbf{[Cu(H_2O)_2(CN)_2]}$.

### PRACTICE EXERCISE

3. Write the formulas for the following compounds:

(a) tris(ethylenediamine)nickel(II) sulfate
(b) tetraamminediaquocobalt(III) chloride
(c) potassium hexacyanoferrate(II)

**Text Problem: 23.18**

## PROBLEM TYPE 4: PREDICTING THE NUMBER OF UNPAIRED SPINS IN A COORDINATION COMPOUND

First, we must write the electron configuration for the transition metal of the coordination compound. Remember that the *ns* shell fills before the (*n* − 1)*d* shell. When the number of *d* electrons in the transition metal is known, we must decide how to place them in the *d* orbitals.

**Crystal Field Theory** tells us that in an octahedral complex, the five *d* orbitals are *not* equivalent in energy. The $d_{xy}$, $d_{xz}$, and $d_{yz}$ are degenerate and at a lower energy than the degenerate set, $d_{z^2}$ and $d_{x^2-y^2}$. The energy difference between these two sets of d orbitals is called the **crystal field splitting (Δ)**.

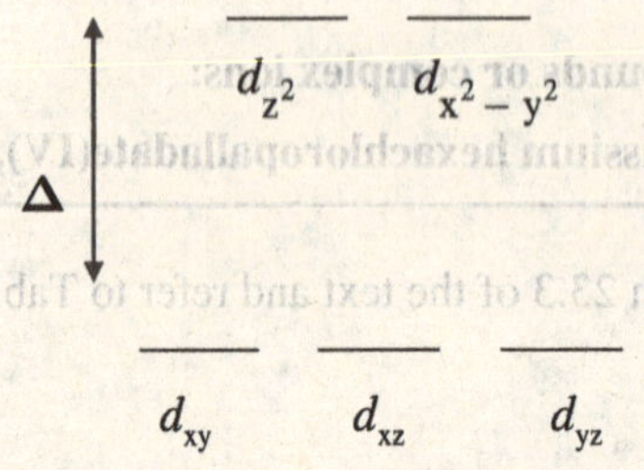


Due to the crystal field splitting, for metals with electron configurations of $d^4$, $d^5$, $d^6$, or $d^7$, there are two ways to place the electrons in the five *d* orbitals. Let's consider a $d^4$ case. According to Hund's rule (see Section 7.8 of the text), maximum stability is reached if the electrons are placed in four separate orbitals with parallel spin. But, this arrangement can be achieved only at a cost; one of the four electrons must be energetically promoted to the higher energy $d_{z^2}$ or $d_{x^2-y^2}$ orbital. This energy investment is not needed if all four electrons enter the $d_{xy}$, $d_{xz}$, and $d_{yz}$ orbitals. However, in this electron arrangement, we must pair up two of the electrons. This pairing also takes energy. The two possible electron configurations are shown below.

**high spin complex**

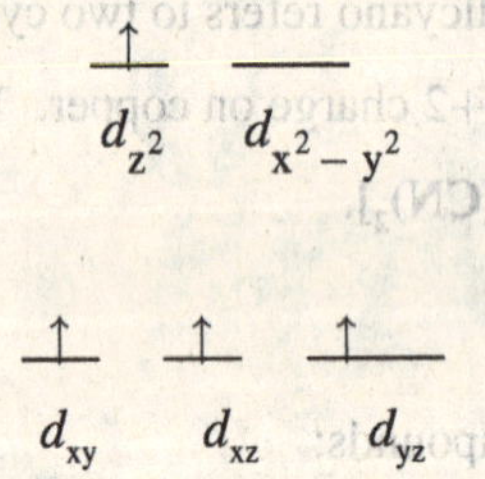


**low spin complex**

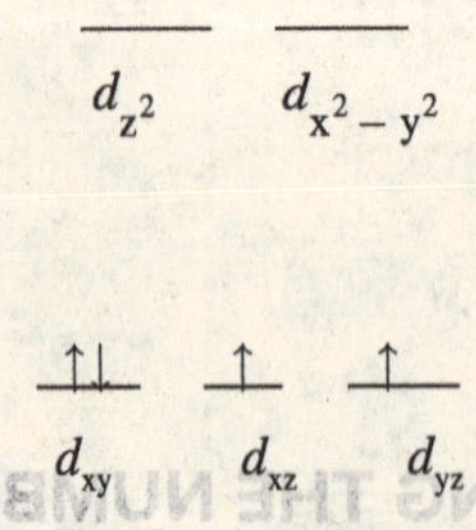


Whether a complex is high spin or low spin depends on the ligands that are bonded to the metal center. If a ligand is a weak-field ligand, the crystal field splitting (Δ) is small. Therefore, only a small energy expenditure is needed to keep all spins parallel, resulting in *high spin complexes*. On the other hand, the crystal field splitting is larger when a strong-field ligand is bound to the metal center. In these cases, it is energetically more favorable to pair up the electrons, resulting in *low spin complexes*.

In summary, the actual arrangement of the *d*-electrons is determined by the amount of stability gained by having maximum parallel spins versus the investment in energy required to promote electrons to higher energy *d* orbitals.

### EXAMPLE 23.4

**Predict the number of unpaired spins in the $[Co(CN)_6]^{3-}$ ion.**

---

**Strategy:** The electron configuration of $Co^{3+}$ is $[Ar]3d^6$. Since $CN^-$ is a strong-field ligand, we expect $[Co(CN)_6]^{3-}$ to be a low spin complex.

**Solution:** All six electrons will be placed in the lower energy *d* orbitals ($d_{xy}$, $d_{xz}$, and $d_{yz}$), and there will be no unpaired spins.

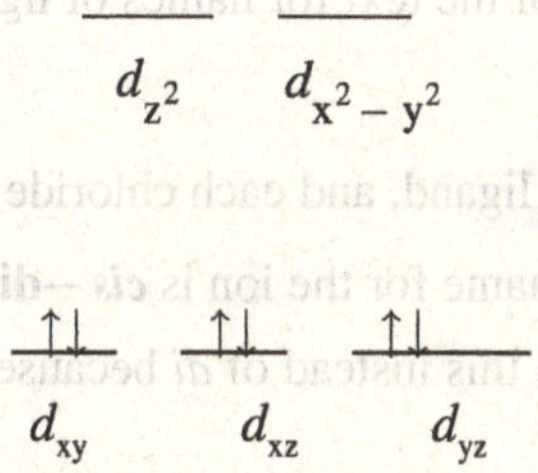


### PRACTICE EXERCISE

4. Predict the number of unpaired spins in the $[Fe(H_2O)_6]^{3+}$ ion. Water is a weak-field ligand.

| **Text Problem:** 23.60 |
|---|

## ANSWERS TO PRACTICE EXERCISES

1. (a) Pt, +4 (b) Ni, 0
2. (a) tetrahydroxochromate(III) ion (b) triamminetrichloroplatinum(IV) nitrate
3. (a) $[Ni(NH_2CH_2CH_2NH_2)_3]SO_4$ (b) $[Co(NH_3)_4(H_2O)_2]Cl_3$ (c) $K_4[Fe(CN)_6]$
4. five unpaired spins

## SOLUTIONS TO SELECTED TEXT PROBLEMS

**23.12** (a) The oxidation number of Cr is **+3**.

(b) The coordination number of Cr is **6**.

(c) **Oxalate ion** ($C_2O_4^{2-}$) is a bidentate ligand.

**23.14** **Strategy:** The oxidation number of the metal atom is equal to its charge. First we look for known charges in the species. Recall that alkali metals are +1 and alkaline earth metals are +2. Also determine if the ligand is a charged or neutral species. From the known charges, we can deduce the net charge of the metal and hence its oxidation number.

**Solution:**

(a) Since **sodium** is always **+1** and the oxygens are −2, **Mo** must have an oxidation number of **+6**.

(b) **Magnesium** is **+2** and oxygen −2; therefore **W** is **+6**.

(c) CO ligands are neutral species, so the iron atom bears no net charge. The oxidation number of **Fe** is **0**.

**23.16** **Strategy:** We follow the procedure for naming coordination compounds outlined in Section 23.3 of the text and refer to Tables 23.4 and 23.5 of the text for names of ligands and anions containing metal atoms.

**Solution:**

(a) Ethylenediamine is a neutral ligand, and each chloride has a −1 charge. Therefore, cobalt has a oxidation number of +3. The correct name for the ion is ***cis*—dichlorobis(ethylenediammine)cobalt(III)**. The prefix *bis* means two; we use this instead of *di* because *di* already appears in the name ethylenediamine.

(b) There are four chlorides each with a −1 charge; therefore, Pt has a +4 charge. The correct name for the compound is **pentaamminechloroplatinum(IV) chloride**.

(c) There are three chlorides each with a −1 charge; therefore, Co has a +3 charge. The correct name for the compound is **pentaamminechlorocobalt(III) chloride**.

**23.18** **Strategy:** We follow the procedure in Section 23.3 of the text and refer to Tables 23.4 and 23.5 of the text for names of ligands and anions containing metal atoms.

**Solution:**

(a) There are two ethylenediamine ligands and two chloride ligands. The correct formula is $\mathbf{[Cr(en)_2Cl_2]^+}$.

(b) There are five carbonyl (CO) ligands. The correct formula is $\mathbf{Fe(CO)_5}$.

(c) There are four cyanide ligands each with a −1 charge. Therefore, the complex ion has a −2 charge, and two $K^+$ ions are needed to balance the −2 charge of the anion. The correct formula is $\mathbf{K_2[Cu(CN)_4]}$.

(d) There are four $NH_3$ ligands and two $H_2O$ ligands. Two chloride ions are needed to balance the +2 charge of the complex ion. The correct formula is $\mathbf{[Co(NH_3)_4(H_2O)Cl]Cl_2}$.

**23.24** (a) In general for any $MA_2B_4$ octahedral molecule, only **two** geometric isomers are possible. The only real distinction is whether the two A−ligands are *cis* or *trans*. In Figure 23.11 of the text, (a) and (c) are the same compound (Cl atoms *cis* in both), and (b) and (d) are identical (Cl atoms *trans* in both).

(b) A model or a careful drawing is very helpful to understand the $MA_3B_3$ octahedral structure. There are only **two** possible geometric isomers. The first has all A's (and all B's) *cis*; this is called the facial isomer. The second has two A's (and two B's) at opposite ends of the molecule (*trans*). Try to make or draw other possibilities. What happens?

**23.26** **(a)** There are *cis* and *trans* geometric isomers (See Problem 23.24). No optical isomers.

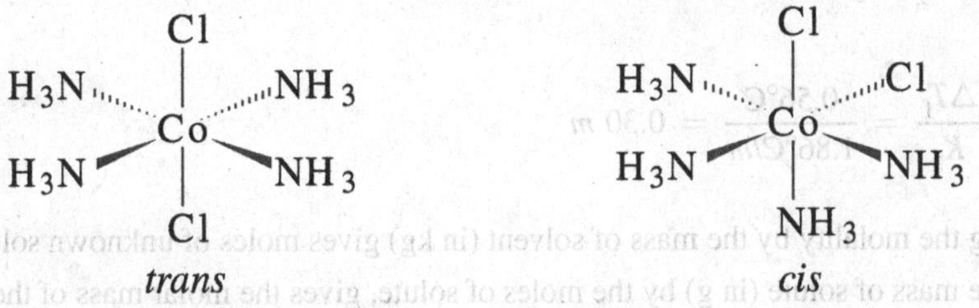


**(b)** There are two optical isomers. See Figure 23.7 of the text. The three bidentate en ligands are represented by the curved lines.

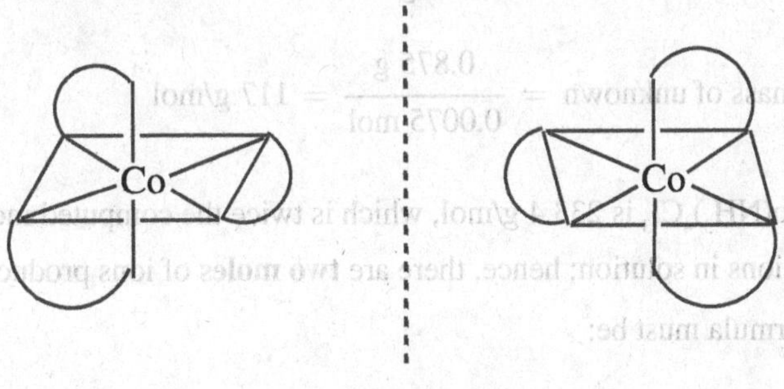


mirror

**23.34** When a substance appears to be yellow, it is absorbing light from the blue-violet, high energy end of the visible spectrum. Often this absorption is just the tail of a strong absorption in the ultraviolet. Substances that appear green or blue to the eye are absorbing light from the lower energy red or orange part of the spectrum.

Cyanide ion is a very strong field ligand. It causes a larger crystal field splitting than water, resulting in the absorption of higher energy (shorter wavelength) radiation when a *d* electron is excited to a higher energy *d* orbital.

**23.36** **(a)** Wavelengths of 470 nm fall between blue and blue-green, corresponding to an observed color in the **orange** part of the spectrum.

**(b)** We convert wavelength to photon energy using the Planck relationship.

$$\Delta E = \frac{hc}{\lambda} = \frac{(6.63 \times 10^{-34}\ \text{J}\cdot\text{s})(3.00 \times 10^{8}\ \text{m/s})}{470 \times 10^{-9}\ \text{m}} = 4.23 \times 10^{-19}\ \text{J}$$

$$\frac{4.23 \times 10^{-19}\ \text{J}}{1\ \text{photon}} \times \frac{6.022 \times 10^{23}\ \text{photons}}{1\ \text{mol}} \times \frac{1\ \text{kJ}}{1000\ \text{J}} = \mathbf{255\ kJ/mol}$$

**23.38** *Step 1:* The equation for freezing-point depression is

$$\Delta T_f = K_f m$$

Solve this equation algebraically for molality ($m$), then substitute $\Delta T_f$ and $K_f$ into the equation to calculate the molality.

$$m = \frac{\Delta T_f}{K_f} = \frac{0.56°\cancel{C}}{1.86°\cancel{C}/m} = 0.30\ m$$

*Step 2:* Multiplying the molality by the mass of solvent (in kg) gives moles of unknown solute. Then, dividing the mass of solute (in g) by the moles of solute, gives the molar mass of the unknown solute.

$$?\ \text{mol of unknown solute} = \frac{0.30\ \text{mol solute}}{1\ \cancel{\text{kg}}\ \text{water}} \times 0.0250\ \cancel{\text{kg}}\ \text{water} = 0.0075\ \text{mol solute}$$

$$\text{molar mass of unknown} = \frac{0.875\ \text{g}}{0.0075\ \text{mol}} = 117\ \text{g/mol}$$

The molar mass of $Co(NH_3)_4Cl_3$ is 233.4 g/mol, which is twice the computed molar mass. This implies dissociation into two ions in solution; hence, there are **two moles** of ions produced per one mole of $Co(NH_3)_4Cl_3$. The formula must be:

**$[Co(NH_3)_4Cl_2]Cl$**

which contains the complex ion $[Co(NH_3)_4Cl_2]^+$ and a chloride ion, $Cl^-$. Refer to Problem 23.26 (a) for a diagram of the structure of the complex ion.

**23.42** Use a radioactive label such as $^{14}CN^-$ (in NaCN). Add NaCN to a solution of $K_3Fe(CN)_6$. Isolate some of the $K_3Fe(CN)_6$ and check its radioactivity. If the complex shows radioactivity, then it must mean that the $CN^-$ ion has participated in the exchange reaction.

**23.44** The white precipitate is copper(II) cyanide.

$$Cu^{2+}(aq) + 2CN^-(aq) \rightarrow Cu(CN)_2(s)$$

This forms a soluble complex with excess cyanide.

$$Cu(CN)_2(s) + 2CN^-(aq) \rightarrow Cu(CN)_4^{2-}(aq)$$

Copper(II) sulfide is normally a very insoluble substance. In the presence of excess cyanide ion, the concentration of the copper(II) ion is so low that CuS precipitation cannot occur. In other words, the cyanide complex of copper has a very large formation constant.

**23.46** The formation constant expression is:

$$K_f = \frac{[Fe(H_2O)_5NCS^{2+}]}{[Fe(H_2O)_6^{3+}][SCN^-]}$$

Notice that the original volumes of the Fe(III) and $SCN^-$ solutions were both 1.0 mL and that the final volume is 10.0 mL. This represents a tenfold dilution, and the concentrations of Fe(III) and $SCN^-$ become 0.020 *M* and $1.0 \times 10^{-4}$ *M*, respectively. We make a table.

| | $Fe(H_2O)_6^{3+}$ | + $SCN^-$ | ⇌ $Fe(H_2O)_5NCS^{2+}$ | + $H_2O$ |
|---|---|---|---|---|
| Initial (*M*): | 0.020 | $1.0 \times 10^{-4}$ | 0 | |
| Change (*M*): | $-7.3 \times 10^{-5}$ | $-7.3 \times 10^{-5}$ | $+7.3 \times 10^{-5}$ | |
| Equilibrium (*M*): | 0.020 | $2.7 \times 10^{-5}$ | $7.3 \times 10^{-5}$ | |

$$K_f = \frac{7.3 \times 10^{-5}}{(0.020)(2.7 \times 10^{-5})} = \mathbf{1.4 \times 10^2}$$

**23.48** $Mn^{3+}$ is $3d^4$ and $Cr^{3+}$ is $3d^3$. The $3d^3$ electron configuration of $Cr^{3+}$ is stable; therefore, $\mathbf{Mn^{3+}}$ has a greater tendency to accept an electron and is a stronger oxidizing agent.

**23.50** Ti is +3 and Fe is +3.

**23.52** A 100.00 g sample of hemoglobin contains 0.34 g of iron. In moles this is:

$$0.34 \text{ g Fe} \times \frac{1 \text{ mol}}{55.85 \text{ g}} = 6.1 \times 10^{-3} \text{ mol Fe}$$

The amount of hemoglobin that contains one mole of iron must be:

$$\frac{100.00 \text{ g hemoglobin}}{6.1 \times 10^{-3} \text{ mol Fe}} = \mathbf{1.6 \times 10^4 \text{ g hemoglobin/mol Fe}}$$

We compare this to the actual molar mass of hemoglobin:

$$\frac{6.5 \times 10^4 \text{ g hemoglobin}}{1 \text{ mol hemoglobin}} \times \frac{1 \text{ mol Fe}}{1.6 \times 10^4 \text{ g hemoglobin}} = 4 \text{ mol Fe/1 mol hemoglobin}$$

The discrepancy between our minimum value and the actual value can be explained by realizing that one hemoglobin molecule contains **four** iron atoms.

**23.54** (a) $[Cr(H_2O)_6]Cl_3$, (b) $[Cr(H_2O)_5Cl]Cl_2 \cdot H_2O$, (c) $[Cr(H_2O)_4Cl_2]Cl \cdot 2H_2O$

The compounds can be identified by a conductance experiment. Compare the conductances of equal molar solutions of the three compounds with equal molar solutions of NaCl, $MgCl_2$, and $FeCl_3$. The solution that has similar conductance to the NaCl solution contains (c); the solution with the conductance similar to $MgCl_2$ contains (b); and the solution with conductance similar to $FeCl_3$ contains (a).

**23.56**

$$\text{Zn}(s) \rightarrow \text{Zn}^{2+}(aq) + 2e^- \qquad E^\circ_{\text{anode}} = -0.76\text{ V}$$

$$2[\text{Cu}^{2+}(aq) + e^- \rightarrow \text{Cu}^+(aq)] \qquad E^\circ_{\text{cathode}} = 0.15\text{ V}$$

$$\text{Zn}(s) + 2\text{Cu}^{2+}(aq) \rightarrow \text{Zn}^{2+}(aq) + 2\text{Cu}^+(aq) \qquad E^\circ_{\text{cell}} = E^\circ_{\text{cathode}} - E^\circ_{\text{anode}} = 0.15\text{ V} - (-0.76\text{ V}) = 0.91\text{ V}$$

We carry additional significant figures throughout the remainder of this calculation to minimize rounding errors.

$$\Delta G^\circ = -nFE^\circ = -(2)(96500\text{ J/V}\cdot\text{mol})(0.91\text{ V}) = -1.756 \times 10^5\text{ J/mol} = \mathbf{-1.8 \times 10^2\text{ kJ/mol}}$$

$$\Delta G^\circ = -RT\ln K$$

$$\ln K = \frac{-\Delta G^\circ}{RT} = \frac{-(-1.756 \times 10^5\text{ J/mol})}{(8.314\text{ J/K}\cdot\text{mol})(298\text{ K})}$$

$$\ln K = 70.88$$

$$K = e^{70.88} = \mathbf{6 \times 10^{30}}$$

**23.58** Iron is much more abundant than cobalt.

**23.60** Oxyhemoglobin absorbs higher energy light than deoxyhemoglobin. Oxyhemoglobin is diamagnetic (low spin), while deoxyhemoglobin is paramagnetic (high spin). These differences occur because oxygen ($O_2$) is a strong–field ligand. The crystal field splitting diagrams are:

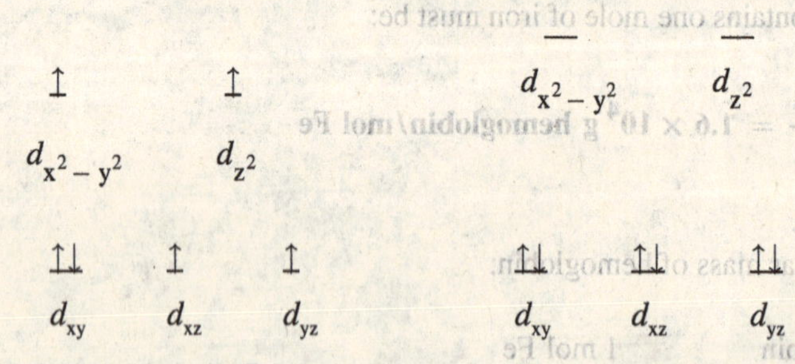


deoxyhemoglobin oxyhemoglobin

**23.62** Complexes are expected to be colored when the highest occupied orbitals have between one and nine $d$ electrons. Such complexes can therefore have $d \rightarrow d$ transitions (that are usually in the visible part of the electromagnetic radiation spectrum). The ions $V^{5+}$, $Ca^{2+}$, and $Sc^{3+}$ have $d^0$ electron configurations and $Cu^+$, $Zn^{2+}$, and $Pb^{2+}$ have $d^{10}$ electron configurations: these complexes are colorless. The other complexes have outer electron configurations of $d^1$ to $d^9$ and are therefore colored.

**23.64** Dipole moment measurement. Only the *cis* isomer has a dipole moment.

**23.66** EDTA sequesters metal ions (like $Ca^{2+}$ and $Mg^{2+}$) which are essential for growth and function, thereby depriving the bacteria to grow and multiply.

**23.68** The square planar complex shown in the problem has **3** geometric isomers. They are:

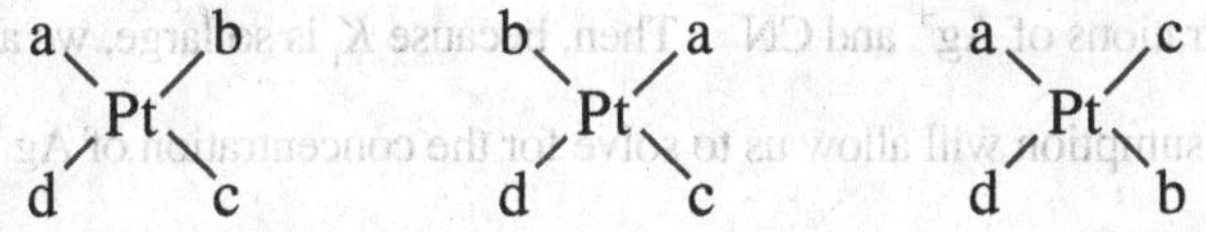


Note that in the first structure a is *trans* to c, in the second a is *trans* to d, and in the third a is *trans* to b. Make sure you realize that if we switch the positions of b and d in structure 1, we do not obtain another geometric isomer. A 180° rotation about the a–Pt–c axis gives structure 1.

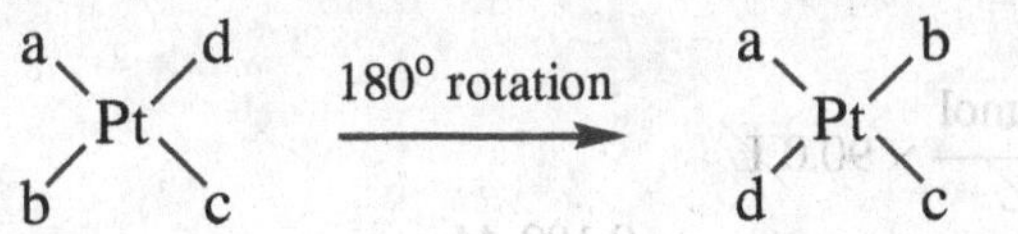


**23.70** Because the magnitude of $K_f$ is very large, we initially assume that the equilibrium goes to completion. We can write:

| | $Pb^{2+}$ | + EDTA | $\longrightarrow$ $Pb(EDTA)^{2-}$ |
|---|---|---|---|
| Initial (*M*): | $1.0 \times 10^{-3}$ | $2.0 \times 10^{-3}$ | 0 |
| Change (*M*): | $-1.0 \times 10^{-3}$ | $-1.0 \times 10^{-3}$ | $+1.0 \times 10^{-3}$ |
| Final (*M*): | 0 | $1.0 \times 10^{-3}$ | $1.0 \times 10^{-3}$ |

In actuality, there will be a very small amount of $Pb^{2+}$ in solution at equilibrium. We can find the concentration of free $Pb^{2+}$ at equilibrium using the formation constant expression. The concentrations of EDTA and $Pb(EDTA)^{2-}$ will remain essentially constant because $Pb(EDTA)^{2-}$ only dissociates to a slight extent.

$$K_f = \frac{[Pb(EDTA)^{2-}]}{[Pb^{2+}][EDTA]}$$

Rearranging,

$$[Pb^{2+}] = \frac{[Pb(EDTA)^{2-}]}{K_f[EDTA]}$$

Substitute the equilibrium concentrations of EDTA and $Pb(EDTA)^{2-}$ calculated above into the formation constant expression to calculate the equilibrium concentration of $Pb^{2+}$.

$$\mathbf{[Pb^{2+}]} = \frac{[Pb(EDTA)^{2-}]}{K_f[EDTA]} = \frac{1.0 \times 10^{-3}}{(1.0 \times 10^{18})(1.0 \times 10^{-3})} = \mathbf{1.0 \times 10^{-18}}\ \boldsymbol{M}$$

**23.72** The reaction is: $Ag^+(aq) + 2CN^-(aq) \rightleftharpoons Ag(CN)_2^-(aq)$

$$K_f = 1.0 \times 10^{21} = \frac{[Ag(CN)_2^-]}{[Ag^+][CN^-]^2}$$

First, we calculate the initial concentrations of $Ag^+$ and $CN^-$. Then, because $K_f$ is so large, we assume that the reaction goes to completion. This assumption will allow us to solve for the concentration of $Ag^+$ at equilibrium. The initial concentrations of $Ag^+$ and $CN^-$ are:

$$[CN^-] = \frac{\dfrac{5.0\text{ mol}}{1\text{ L}} \times 9.0\text{ L}}{99.0\text{ L}} = 0.455\ M$$

$$[Ag^+] = \frac{\dfrac{0.20\text{ mol}}{1\text{ L}} \times 90.0\text{ L}}{99.0\text{ L}} = 0.182\ M$$

We set up a table to determine the concentrations after complete reaction.

| | $Ag^+(aq)$ + | $2CN^-(aq)$ $\rightleftharpoons$ | $Ag(CN)_2^-(aq)$ |
|---|---|---|---|
| Initial (*M*): | 0.182 | 0.455 | 0 |
| Change (*M*): | −0.182 | −(2)(0.182) | +0.182 |
| Final (*M*): | 0 | 0.0910 | 0.182 |

$$K_f = \frac{[Ag(CN)_2^-]}{[Ag^+][CN^-]^2}$$

$$1.0 \times 10^{21} = \frac{0.182\ M}{[Ag^+](0.0910\ M)^2}$$

$$[Ag^+] = 2.2 \times 10^{-20}\ M$$

**23.74** (a) The equilibrium constant can be calculated from $\Delta G°$. We can calculate $\Delta G°$ from the cell potential.

From Table 18.1 of the text,

$Cu^{2+} + 2e^- \rightarrow Cu$ $\quad E° = 0.34\text{ V}$ and $\Delta G° = -(2)(96500\text{ J/V·mol})(0.34\text{ V}) = -6.562 \times 10^4\text{ J/mol}$

$Cu^{2+} + e^- \rightarrow Cu^+$ $\quad E° = 0.15\text{ V}$ and $\Delta G° = -(1)(96500\text{ J/V·mol})(0.15\text{ V}) = -1.448 \times 10^4\text{ J/mol}$

These two equations need to be arranged to give the disproportionation reaction in the problem. We keep the first equation as written and reverse the second equation and multiply by two.

$$Cu^{2+} + 2e^- \rightarrow Cu \qquad \Delta G° = -6.562 \times 10^4 \text{ J/mol}$$

$$2Cu^+ \rightarrow 2Cu^{2+} + 2e^- \qquad \Delta G° = +(2)(1.448 \times 10^4 \text{ J/mol})$$

$$2Cu^+ \rightarrow Cu^{2+} + Cu \qquad \Delta G° = -6.562 \times 10^4 \text{ J/mol} + 2.896 \times 10^4 \text{ J/mol} = -3.666 \times 10^4 \text{ J/mol}$$

We use Equation (17.14) of the text to calculate the equilibrium constant.

$$\Delta G° = -RT \ln K$$

$$K = e^{-\Delta G°/RT}$$

$$K = e^{\frac{-(-3.666 \times 10^4 \text{ J/mol})}{(8.314 \text{ J/mol·K})(298 \text{ K})}}$$

$$\mathbf{K = 2.7 \times 10^6}$$

**(b)** Free $Cu^+$ ions are unstable in solution [as shown in part (a)]. Therefore, the only stable compounds containing $Cu^+$ ions are insoluble.

**23.76** **(a)** $Cu^{3+}$ would not be stable in solution because it can be easily reduced to the more stable $Cu^{2+}$. From Figure 23.3 of the text, we see that the 3rd ionization energy is quite high, about 3500 kJ/mol. Therefore, $Cu^{3+}$ has a great tendency to accept an electron.

**(b)** **Potassium hexafluorocuprate(III)**. $Cu^{3+}$ is $3d^8$. $CuF_6^{3-}$ has an **octahedral** geometry. According to Figure 23.17 of the text, $CuF_6^{3-}$ should be **paramagnetic**, containing two unpaired electrons. (Because it is $3d^8$, it does not matter whether the ligand is a strong or weak-field ligand. See Figure 23.22 of the text.)

**(c)** We refer to Figure 23.24 of the text. The splitting pattern is such that all the square-planar complexes of $Cu^{3+}$ should be diamagnetic.

# CHAPTER 24
# ORGANIC CHEMISTRY

## PROBLEM-SOLVING STRATEGIES AND TUTORIAL SOLUTIONS

### TYPES OF PROBLEMS

**Problem Type 1:** Determining the Number of Structural Isomers.

**Problem Type 2:** Organic Nomenclature.
- **(a)** Alkanes.
- **(b)** Alkenes.
- **(c)** Alkynes.
- **(d)** Aromatic Compounds.

**Problem Type 3:** Addition Reactions.

**Problem Type 4:** Distinguishing between Structural and Geometric Isomers.

**Problem Type 5:** Functional Groups.

**Problem Type 6:** Chirality.

## PROBLEM TYPE 1: DETERMINING THE NUMBER OF STRUCTURAL ISOMERS

**Structural isomers** are molecules that have the same molecular formula but a different order of linking the atoms. For small hydrocarbon molecules (eight or fewer carbon atoms), it is relatively easy to determine the number of structural isomers by trial and error.

**EXAMPLE 24.1**

**How many structural isomers can be identified for hexane, $C_6H_{14}$?**

**Strategy:** For small hydrocarbon molecules (eight or fewer carbons), it is relatively easy to determine the number of structural isomers by trial and error.

**Solution:** Start by writing the straight-chain structure.

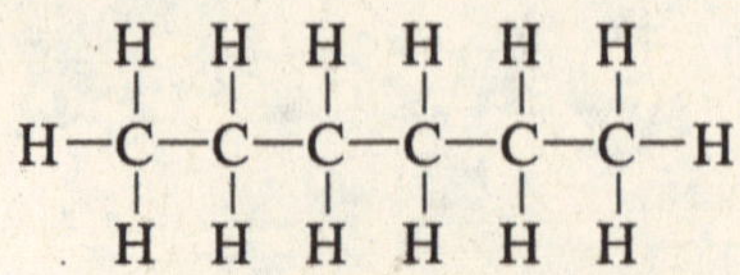

*n*-hexane

By necessity, the other structures must have branched chains. First, try single methyl substituents.

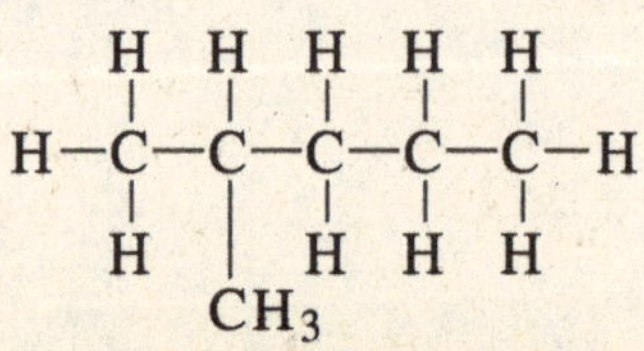

2-methylpentane

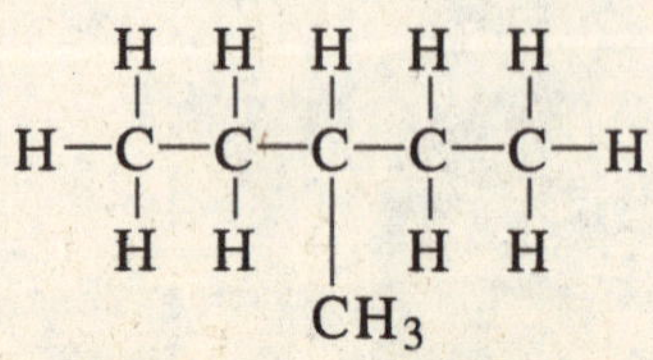

3-methylpentane

Then, try structures that have two methyl groups.

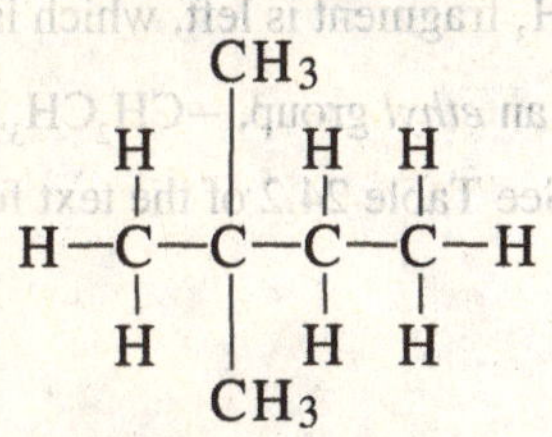

2,2-dimethylbutane

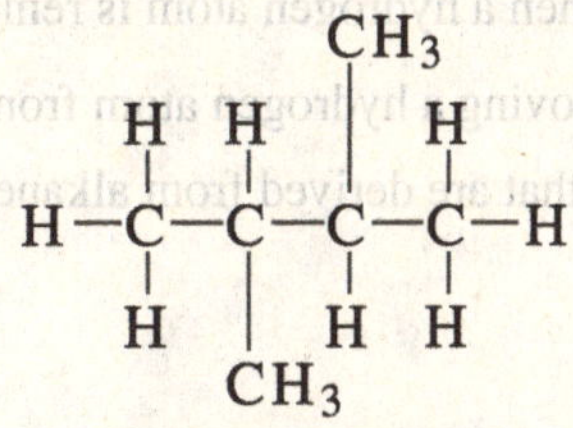

2,3-dimethylbutane

Hexane has five structural isomers, in which the numbers of carbon and hydrogen atoms remain unchanged despite the differences in structure.

**PRACTICE EXERCISE**

1. How many structural isomers are there of $C_4H_{10}O$? **Hint:** Consider both alcohols and ethers.

**Text Problems: 24.12**, 24.14

## PROBLEM TYPE 2: ORGANIC NOMENCLATURE

The first step in learning the nomenclature of hydrocarbons is to know the names of the first ten straight-chain alkanes. Except for the first four members, the number of carbon atoms in each alkane is identified by the Greek prefix. See Table 24.1 below.

**TABLE 24.1**
**The First Ten Straight-Chain Alkanes**

| Name of hydrocarbon | Molecular formula |
|---|---|
| Methane | $CH_4$ |
| Ethane | $CH_3-CH_3$ |
| Propane | $CH_3-CH_2-CH_3$ |
| Butane | $CH_3-(CH_2)_2-CH_3$ |
| Pentane | $CH_3-(CH_2)_3-CH_3$ |
| Hexane | $CH_3-(CH_2)_4-CH_3$ |
| Heptane | $CH_3-(CH_2)_5-CH_3$ |
| Octane | $CH_3-(CH_2)_6-CH_3$ |
| Nonane | $CH_3-(CH_2)_7-CH_3$ |
| Decane | $CH_3-(CH_2)_8-CH_3$ |

Next, you need to learn the names of substituents (other than hydrogen) attached to hydrocarbon chains or aromatic compounds. For example, when a hydrogen atom is removed from methane, a $-CH_3$ fragment is left, which is called a *methyl* group. Similarly, removing a hydrogen atom from an ethane molecule gives an *ethyl* group, $-CH_2CH_3$. These groups or substituents that are derived from alkanes are called *alkyl* groups. See Table 24.2 of the text for other common alkyl groups.

## A. Alkanes

Alkanes have the general formula $C_nH_{2n+2}$, where $n = 1, 2, \ldots$. Alkanes contain only single covalent bonds. The bonds are said to be saturated because no more hydrogen atoms can be added to the carbon atoms. Thus, alkanes are also called *saturated hydrocarbons*.

There are other rules to follow when naming hydrocarbons.

1. Find the longest carbon chain in the molecule. The parent name of the compound is based on the longest carbon chain. For example, if the longest carbon chain contains five C atoms, the parent name of the compound is *pentane*. See Table 24.1.

2. Next, you must specify the name and location of any groups attached to the longest carbon chain. See Table 24.2 of the text for naming *alkyl* groups. To specify the location of the group or groups, start numbering the longest chain from the end that is closer to the carbon atom bearing the substituent group.

3. If there is more than one of a particular group attached to the longest carbon chain, you must specify the number of groups with a prefix. The prefixes are di– (2 groups), tri– (3 groups), tetra– (4 groups), and so on.

4. There can be many different types of substituents other than alkyl groups. Table 24.3 of the text lists the names of some common functional groups.

**EXAMPLE 24.2**
**Give the correct name for the following structure:**

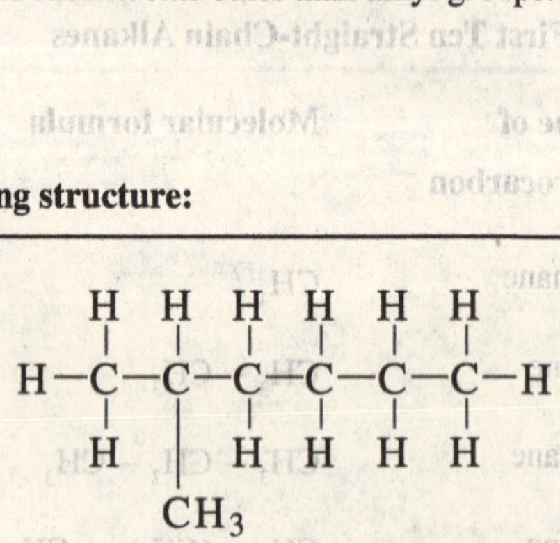


**Strategy:** We follow the IUPAC rules and use the information in Table 24.2 of the text to name the compound. How many C atoms are there in the longest chain?

**Solution:** Find the longest carbon chain. The longest chain has *six* carbons. Therefore, the parent name of the compound is *hexane*.

Next, specify the name and location of any groups attached to the longest carbon chain. You should see that there is a methyl group attached to the second carbon from the left of the chain. You should number the carbon chain from the end that is closer to the carbon atom bearing the substituent group. If you start numbering from the left, the methyl group is on the second carbon in the chain. However, if you start numbering from the right, the methyl group is on the fifth carbon in the chain. Numbering from the left is correct.

The correct name for this compound is **2-methylhexane.**

> **Tip:** You should always put a dash (–) between a number and a "word" when naming an organic compound.

**EXAMPLE 24.3**
**Give the correct name for the following structure:**

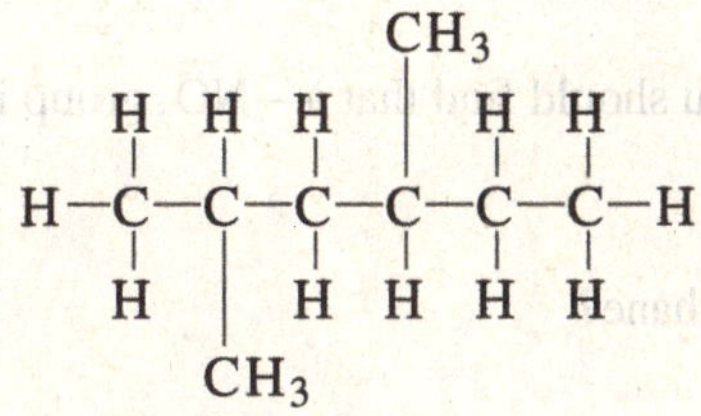

**Strategy:** We follow the IUPAC rules and use the information in Table 24.2 of the text to name the compound. How many C atoms are there in the longest chain?

**Solution:** Find the longest carbon chain. You should find that the longest carbon chain has six carbons. The parent name is *hexane*.

Numbering from the left, there are methyl groups on the second and fourth carbons. Since there are two methyl groups, we must specify this by using the prefix di–.

The correct name for the compound is **2,4-dimethylhexane.**

**Comment:** Let's examine the name of the compound if we had numbered the longest chain from right to left. The methyl groups would be on the third and fifth carbons. Hence, the name would be 3,5-dimethylhexane, which is incorrect. The correct name should always have the lowest numbering scheme as possible.

> **Tip:** Always use commas to separate numbers when naming organic compounds.

**EXAMPLE 24.4**
**Give the correct name for each of the following structures:**

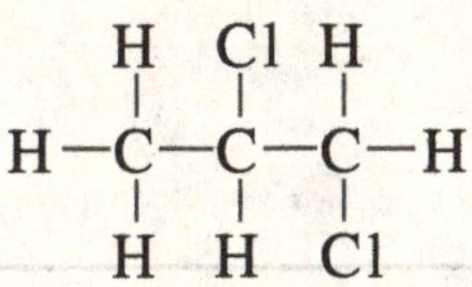

**(a)**

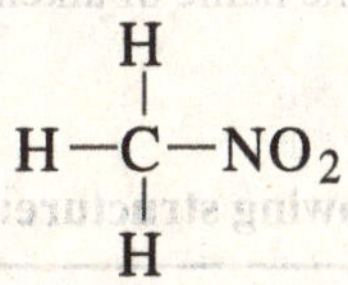

**(b)**

**Strategy:** We follow the IUPAC rules and use the information in Table 24.2 of the text to name the compound. How many C atoms are there in the longest chain? See Table 24.3 of the text for the names of common substituent groups.

**Solution:**

(a) You should find that the longest carbon chain in the molecule is *three* carbons. The parent name is *propane*.

Referring to Table 24.3 of the text, you should find that a –Cl group is called a chloro group. There are chloro groups on the 1 and 2 carbons, numbering from right to left. (Why not number from left to right?)

Hence, the correct name for the compound is **1,2-dichloropropane.**

(b) There is only one carbon in the molecule, so the parent name is *methane*.

Referring to Table 24.3 of the text, you should find that a $-NO_2$ group is called a nitro group. The correct name for the compound is **nitromethane.**

Why isn't the correct name 1-nitromethane?

**PRACTICE EXERCISE**

2. Give the systematic name for each of the following structural formulas:

(a) $CH_3(CH_2)_7CH_3$

(b) 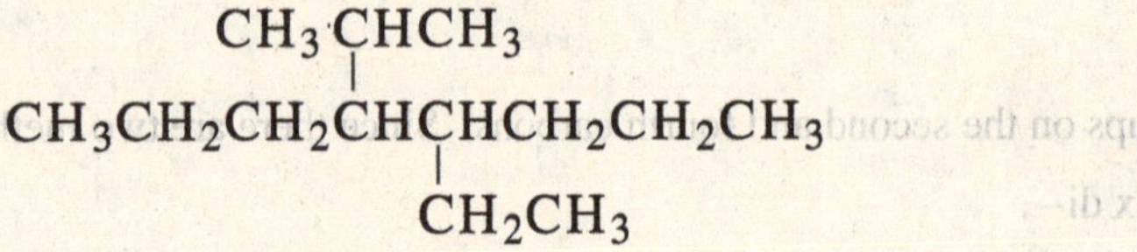

(c) $BrCH_2CH_2CHBrCH_2Br$

**Text Problem: 24.26**

## B. Alkenes

Alkenes are molecules that contain at least one carbon-carbon double bond. Alkenes have the general formula $C_nH_{2n}$, where $n = 2, 3, \ldots$.

To name alkenes, you follow the same rules as outlined for alkanes; except, you must specify the position(s) of the carbon-carbon double bond(s), and the name of alkenes ends with an "–ene" suffix.

**EXAMPLE 24.5**

**Give the correct name for the following structure:**

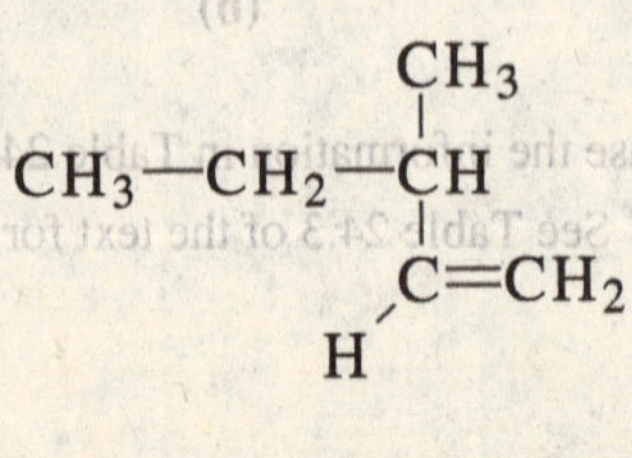

**Strategy:** We follow the IUPAC rules and use the information in Table 24.2 of the text to name the compound. How many C atoms are there in the longest chain? For alkenes, the suffix is "–ene".

**Solution:** The longest carbon chain in the molecule is five carbons. The parent name is pent*ene*, because the molecule has a double bond.

For this molecule, we number from right to left, because we want the double bond to be at the lowest number possible. In this case, the double bond starts on the first carbon if we number from right to left. There is also a methyl substituent on the third carbon.

The correct name for this compound is **3-methyl-1-pentene**. The number 1 before pentene specifies that the double bond starts on the first carbon.

**PRACTICE EXERCISE**

3. Draw the structural formula for 4-methyl-2-hexene.

**Text Problems:** 24.26, 24.28

## C. Alkynes

Alkynes contain at least one carbon-carbon triple bond. They have the general formula $C_nH_{2n-2}$, where $n = 2, 3, \ldots$.

To name alkynes, you follow the same rules as outlined above for alkanes; except, you must specify the position(s) of the carbon-carbon triple bond(s), and the name of alkynes ends with an "−yne" suffix.

**EXAMPLE 24.6**
**Give the correct name for the following structure:**

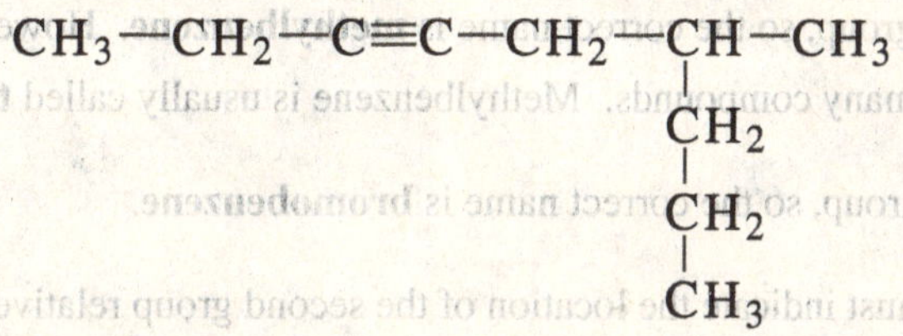

**Strategy:** We follow the IUPAC rules and use the information in Table 24.2 of the text to name the compound. How many C atoms are there in the longest chain? For alkynes the suffix is "−yne".

**Solution:** The longest carbon chain in the molecule contains nine carbons. The parent name is non*yne* because there is a triple bond.

You should number the carbon chain from left to right, placing the triple bond on the third carbon. If you numbered from right to left, the triple bond would be on the sixth carbon. You should also notice that there is a methyl substituent on the sixth carbon.

The correct name for this molecule is **6-methyl-3-nonyne**.

**PRACTICE EXERCISE**

4. Give an acceptable name for each of the following structures:

**(a)** $(CH_3)_3CC{\equiv}CCH_2CH_3$

(b) $CH_3CH_2CH_2\underset{\displaystyle CH_2CH_2CH_3}{\underset{|}{C}}HC{\equiv}CCH_3$

**Text Problems:** 24.26, 24.28

## D. Aromatic Compounds

Benzene is the parent compound of this large family of organic substances. In naming aromatic compounds, we will consider both mono- and di-substituted benzene rings.

The naming of mono-substituted benzene rings (benzenes in which one H atom has been replaced by another atom or groups of atoms) is straightforward. You simply name the substituent followed by the name "benzene".

**EXAMPLE 24.7**
**Give the correct name for each of the following structures:**

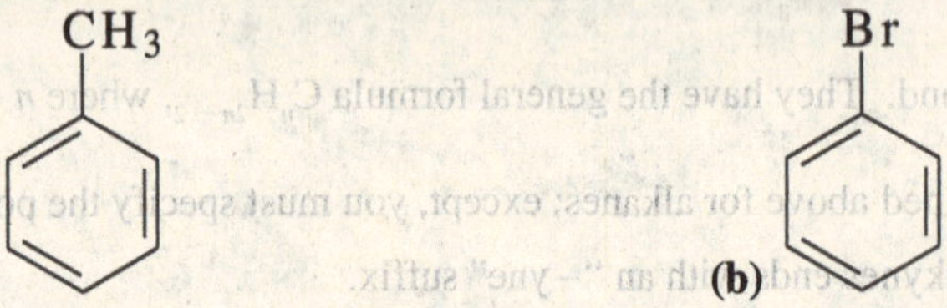


**(a)** **(b)**

**Strategy:** If a benzene ring contains only one substituent, simply name the substituent followed by "benzene".

**Solution:**

**(a)** The substituent is a methyl group, so the correct name is **methylbenzene.** However, there are older names that are still in common use for many compounds. Methylbenzene is usually called **toluene.**

**(b)** The substituent is a bromo group, so the correct name is **bromobenzene.**

For di-substituted benzenes, we must indicate the location of the second group relative to the first. For example, three different dichlorobenzenes are possible.

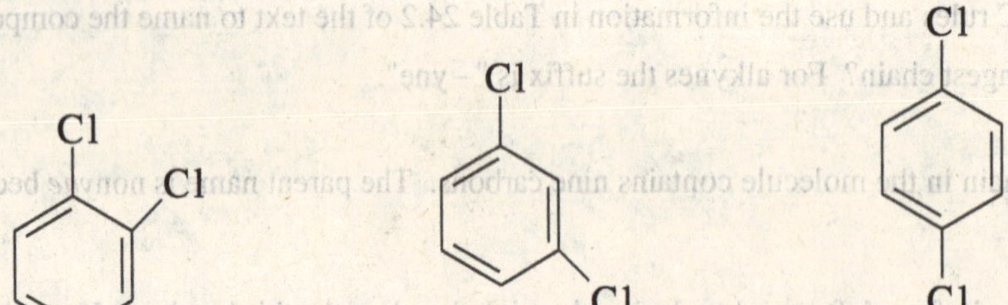


**(a)** **(b)** **(c)**

The systematic way to name these molecules is to number the carbon atoms of the benzene ring as follows:

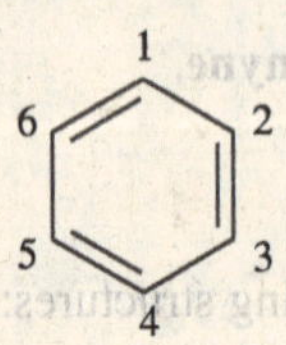

Thus, (a) would be named 1,2-dichlorobenzene, (b) 1,3-dichlorobenzene, and (c) 1,4-dichlorobenzene. However, the prefixes *o*- (*ortho*-), *m*- (*meta*), and *p*- (*para*-) are used more often to denote the relative positions of the two substituted groups. *Ortho*- designates 1,2 substituents, *meta*- designates 1,3 substituents, and *para*- designates 1,4 substituents. Thus, (a) is named *o*-dichlorobenzene, (b) *m*-dichlorobenzene, and (c) *p*-dichlorobenzene.

In compounds with two different substituted groups, the positions of the substituents can be specified with numbers or with the *o*-, *m*-, or *p*- prefixes.

**PRACTICE EXERCISE**

**5.** Draw structural formulas for:

(a) ethylbenzene (b) *m*-bromochlorobenzene (c) *p*-nitrotoluene

**Text Problems:** 24.26, **24.32**

Finally in some molecules, benzene is named as a substituent. A benzene molecule minus a hydrogen atom ($C_6H_5$) is called a *phenyl* group.

**EXAMPLE 24.8**
**Give the correct name for the following structure:**

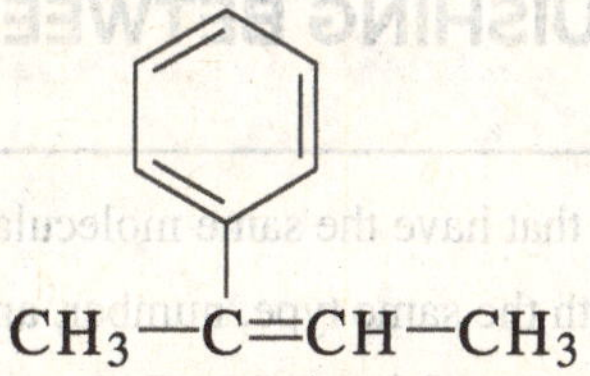

**Strategy:** For this molecule, it would be difficult to name the carbon chain as a substituent on the benzene ring. Therefore, we name the benzene ring as a substituent on the carbon chain.

**Solution:** The longest carbon chain contains four carbons. The parent name is but*ene* because the molecule contains a double bond.

You should number the carbon chain from left to right so that both the double bond and the phenyl group are on the second carbon.

The correct name for the molecule is **2-phenyl-2-butene**.

## PROBLEM TYPE 3: ADDITION REACTIONS

Alkenes, alkynes, and aromatic compounds are called *unsaturated hydrocarbons*, compounds with double or triple carbon-carbon bonds. Unsaturated hydrocarbons commonly undergo **addition reactions** in which one molecule adds to another to form a single product. An example of an addition reaction is *hydrogenation*, which is the addition of molecular hydrogen to compounds containing C=C and C≡C bonds.

$$CH_2{=}CH_2 + H_2 \longrightarrow CH_3{-}CH_3$$

$$CH{\equiv}CH + 2\,H_2 \longrightarrow CH_3{-}CH_3$$

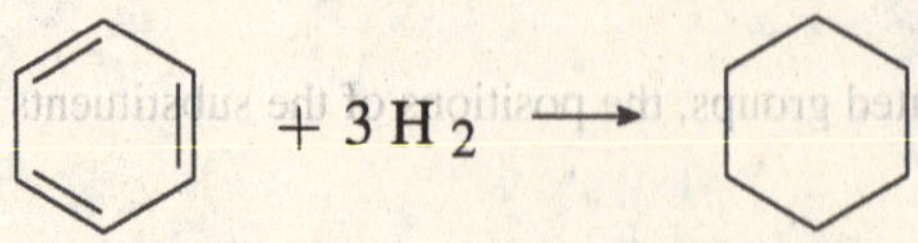

Alkanes are called *saturated hydrocarbons* because no more hydrogen can be added to the carbon atoms. The carbon atoms in a saturated hydrocarbon are already bonded to the maximum number of H atoms.

$$CH_3{-}CH_3 + H_2 \longrightarrow \text{no reaction}$$

Other addition reactions involve an addition of HX or $X_2$ to the multiple bond, where X represents a halogen (Cl, Br, or I). Examples of these addition reactions are:

$$CH_2{=}CH_2 + HCl \longrightarrow CH_3{-}CH_2Cl$$

$$CH_2{=}CH_2 + Cl_2 \longrightarrow CH_2Cl{-}CH_2Cl$$

## PROBLEM TYPE 4: DISTINGUISHING BETWEEN STRUCTURAL AND GEOMETRIC ISOMERS

Recall that **structural isomers** are molecules that have the same molecular formula but a different order of linking the atoms. **Geometric isomers** are molecules with the same type, number, and order of attachment of atoms and the same chemical bonds, but different spatial arrangements of the atoms. For example, 1,2-dichloroethene exists as two geometric isomers.

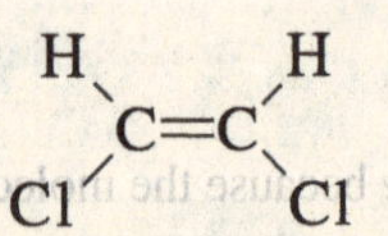

*cis*-1,2-dichloroethene

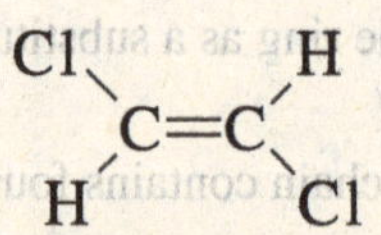

*trans*-1,2-dichloroethene

Note that in the *cis* isomer, the two Cl atoms (and the two H atoms) are adjacent to each other, whereas in the *trans* isomer, the Cl atoms are on the opposite side of the C=C bond.

**EXAMPLE 24.9**

**Draw Lewis structures for all compounds, not including cyclic compounds, with the molecular formula $C_5H_{10}$ and determine which are geometric isomers.**

**Strategy:** Alkenes have the general formula $C_nH_{2n}$. Thus the structures with molecular formula $C_5H_{10}$ should be alkenes.

**Solution:** To draw the correct Lewis structures, follow the procedure outlined in Chapter 9. We can draw six Lewis structures as follows:

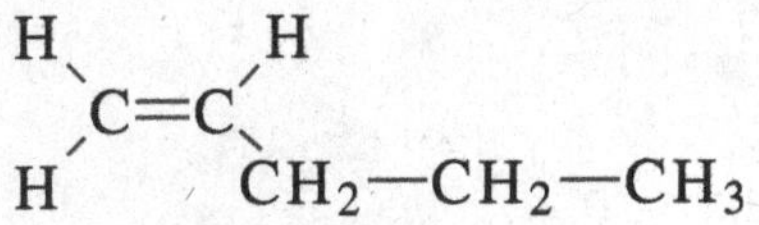

1-pentene

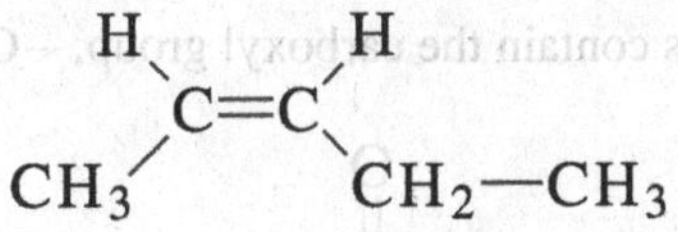

*cis*-2-pentene

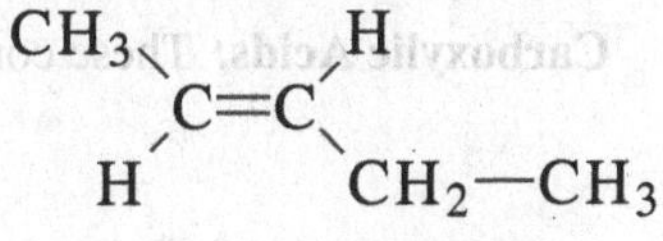

*trans*-2-pentene

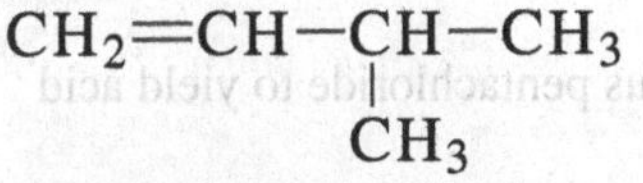

3-methyl-1-butene

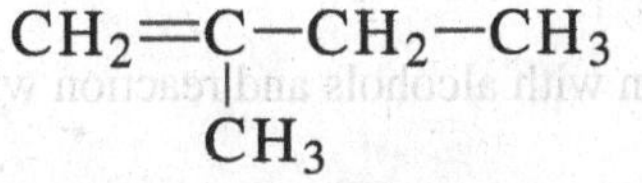

2-methyl-1-butene

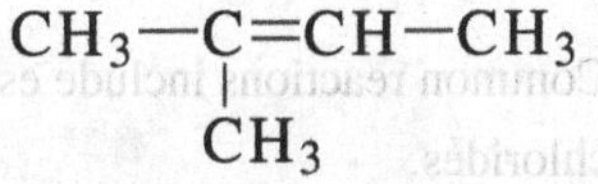

2-methyl-2-butene

The geometric isomers are *cis*- and *trans*-2-pentene. 1-pentene, 3-methyl-1-butene, 2-methyl-1-butene, and 2-methyl-2-butene are structural isomers to the geometric isomers because the atoms are attached in a different order.

Why don't 1-pentene, 3-methyl-1-butene, 2-methyl-1-butene, and 2-methyl-2-butene have geometric isomers?

**PRACTICE EXERCISE**

6. Draw the structural formula for 3-bromo-2,5-dimethyl-*trans*-3-hexene.

**Text Problem:** 24.24

## PROBLEM TYPE 5: FUNCTIONAL GROUPS

We will discuss the most common organic functional groups, with emphasis on classes of compounds in which the functional groups include oxygen or nitrogen.

1. **Alcohols.** All alcohols contain the hydroxyl group, –OH. Figure 24.8 of the text shows some common alcohols. Alcohols typically undergo esterification (formation of an ester) reactions with carboxylic acids. Oxidation to aldehydes, ketones, and carboxylic acids are also common reactions.

2. **Ethers.** Ethers contain the R–O–R' linkage, where R and R' are either an alkyl group or a group derived from an aromatic hydrocarbon.

3. **Aldehydes and Ketones.** These compounds contain the carbonyl functional group.



The difference between aldehydes and ketones is that in aldehydes at least one hydrogen atom is bonded to the carbon atom of the carbonyl group. In ketones, no hydrogen atoms are bonded to the carbonyl carbon atom. Common reactions include reduction to yield alcohols. Oxidation of aldehydes yields carboxylic acids.

4. **Carboxylic Acids.** These compounds contain the carboxyl group, −COOH.

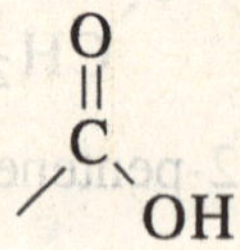

Common reactions include esterification with alcohols and reaction with phosphorus pentachloride to yield acid chlorides.

5. **Esters.** Esters have the general formula R'COOR.

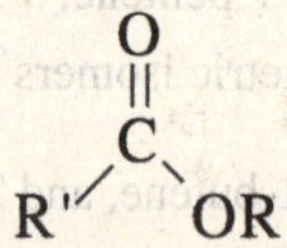

R' can be H, an alkyl, or an aromatic hydrocarbon group, and R is an alkyl or an aromatic hydrocarbon group. A common reaction is hydrolysis to yield acids and alcohols.

6. **Amines.** Amines are organic bases. They have the general formula $R_3N$, where one of the R groups must be an alkyl group or an aromatic hydrocarbon group. A common reaction is formation of ammonium salts with acids.

Table 24.4 of the text summarizes important functional groups. For more detailed discussions of reactions, see Section 24.4 of the text.

**EXAMPLE 24.10**

**Identify the functional groups in the following molecules: (a) $C_5H_{11}OH$, (b) $CH_3CHO$, (c) $C_3H_7OCH_3$, (d) $CH_3COC_2H_5$, (e) $CH_3COOCH_3$.**

**Strategy:** Learning to recognize functional groups requires memorization of their structural formulas. Table 24.4 of the text shows a number of the important functional groups.

**Solution:**

(a) $C_5H_{11}OH$ contains a *hydroxyl* group. It is an *alcohol.*

(b) $CH_3CHO$ is a way to represent

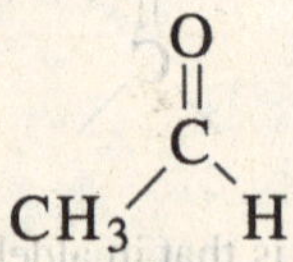

on a single line of type. C=O is a carbonyl group. Since there is a hydrogen atom bonded to the carbonyl carbon, $CH_3CHO$ is an *aldehyde.*

(c) $C_3H_7OC_2H_5$ contains a C−O−C group and is therefore an *ether.*

**(d)** $CH_3COC_2H_5$ contains a carbonyl group. Since a hydrogen atom is not bonded to the carbonyl carbon, this molecule is a *ketone*.

**(e)** $CH_3COOCH_3$ is a condensed structural formula for an *ester*.

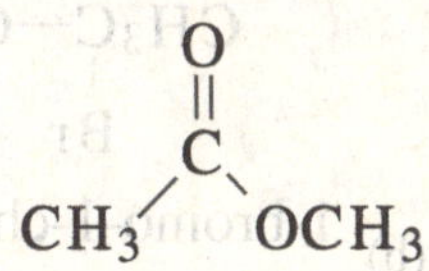


**PRACTICE EXERCISE**

**7.** Indicate the functional groups by name that are in the following molecules.

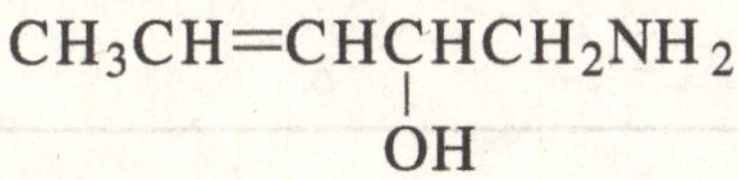


(a)

(b)

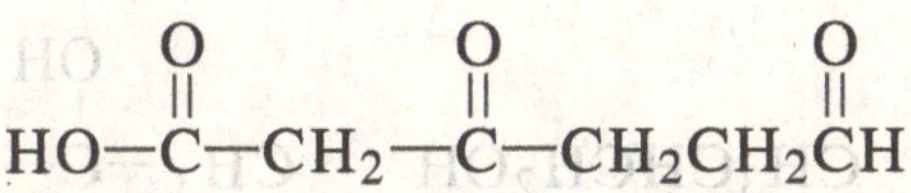


**Text Problems: 24.36**, 24.38, 24.40, 24.42

## PROBLEM TYPE 6: CHIRALITY

Compounds that come as mirror-image pairs can be compared with left-handed and right-handed gloves and are thus referred to as **chiral**, or handed, molecules. While every molecule can have a mirror image, the *chiral* mirror-image pairs are *nonsuperimposable*. Conversely, achiral (nonchiral) pairs are superimposable. See Figure 24.3 of the text for an illustration of superimposable compared to nonsuperimposable.

Observations show that most simple chiral molecules contain at least one *asymmetric* carbon atom; that is, a carbon atom bonded to four different atoms or groups of atoms.

**Tip:** Consider your hands when thinking about chirality. If you view your left hand in a mirror, the mirror image of your left hand is a right hand. However, your left and right hands are nonsuperimposable. To verify this, try putting a "left-handed" glove on your right hand.

**EXAMPLE 24.11**

**Classify the following objects as chiral or achiral:**

**(a) shoe** **(b) screw** **(c) fork** **(d) coffee cup**

A shoe and a screw are chiral. The mirror image of a right shoe is a left shoe. A left shoe is not superimposable on a right shoe. A mirror image of a screw with clockwise threads will have counterclockwise threads. The screws would not be superimposable.

A fork and a coffee cup are achiral. Their mirror images are superimposable.

**PRACTICE EXERCISE**

8. Are the following molecules chiral?

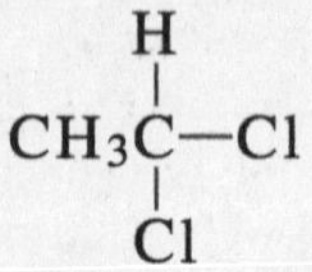

(a) 1,1-dichloroethane

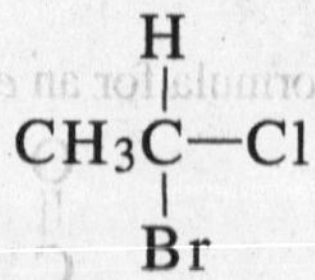

(b) 1-bromo-1-chloroethane

**Text Problem:** 24.56

## ANSWERS TO PRACTICE EXERCISES

1. There are seven structural isomers—four alcohols and three ethers. The structures are shown below.

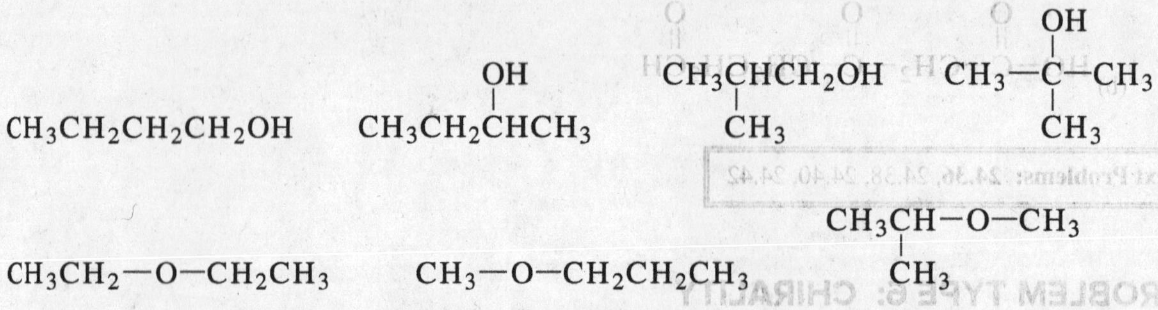

2. (a) nonane (b) 4-ethyl-5-isopropyloctane (c) 1,2,4-tribromobutane

3. 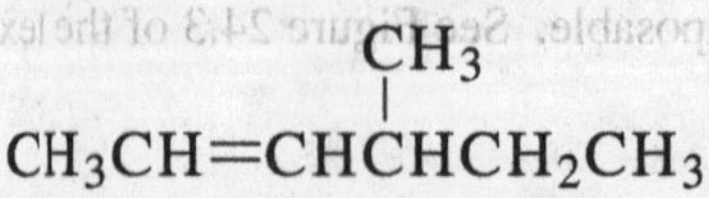

4. (a) 2,2-dimethyl-3-hexyne (b) 4-propyl-2-heptyne

5. The structures are:

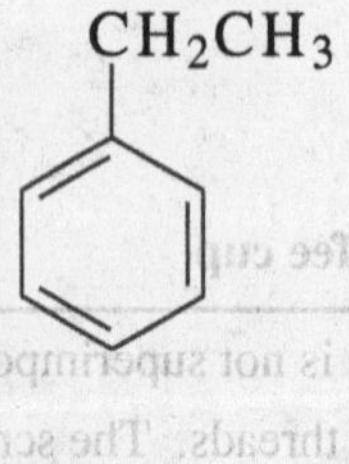

(a)

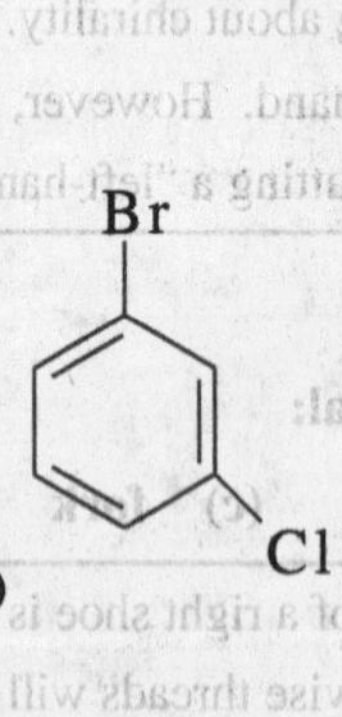

(b)

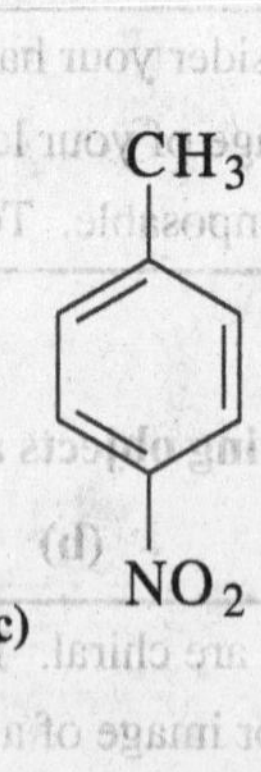

(c)

**6.** The structure of 3-bromo-2,5-dimethyl-*trans*-3-hexene is:

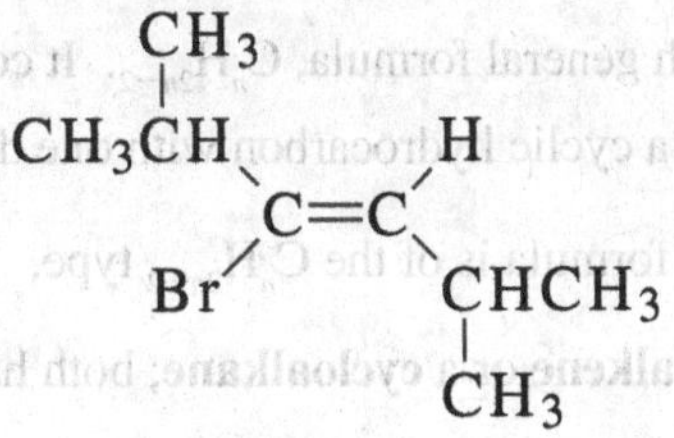

**7.** **(a)** carbon-carbon double bond, hydroxyl (–OH), and amine (R–$NH_2$).

**(b)** carboxyl (–COOH), carbonyl (ketone), and carbonyl (aldehyde).

**8.** **(a)** There are only three different groups bonded to the second carbon atom. This molecule is *achiral.*

**(b)** Replacing one of the chloro groups with a bromo group places four different groups on the second carbon atom. This molecule is *chiral.*

## SOLUTIONS TO SELECTED TEXT PROBLEMS

**24.12** **Strategy:** For small hydrocarbon molecules (eight or fewer carbons), it is relatively easy to determine the number of structural isomers by trial and error.

**Solution:** We are starting with *n*-pentane, so we do not need to worry about any branched chain structures. In the chlorination reaction, a Cl atom replaces one H atom. There are three different carbons on which the Cl atom can be placed. Hence, *three* structural isomers of chloropentane can be derived from *n*–pentane:

$CH_3CH_2CH_2CH_2CH_2Cl$ $CH_3CH_2CH_2CHClCH_3$ $CH_3CH_2CHClCH_2CH_3$

**24.14** Both alkenes and cycloalkanes have the general formula $C_nH_{2n}$. Let's start with $C_3H_6$. It could be an alkene or a cycloalkane.

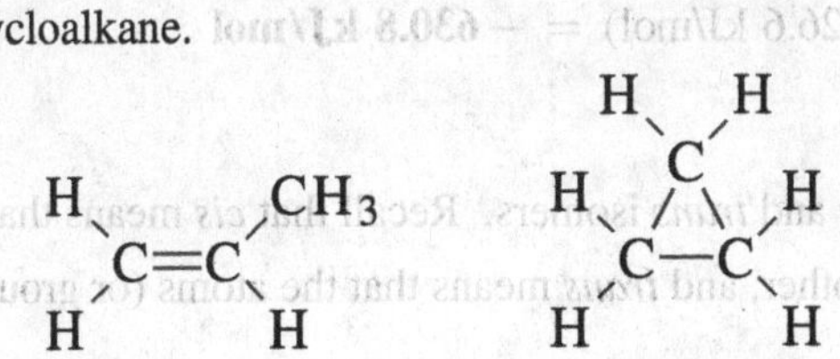

Now, let's replace one H with a Br atom to form $C_3H_5Br$. *Four* isomers are possible.

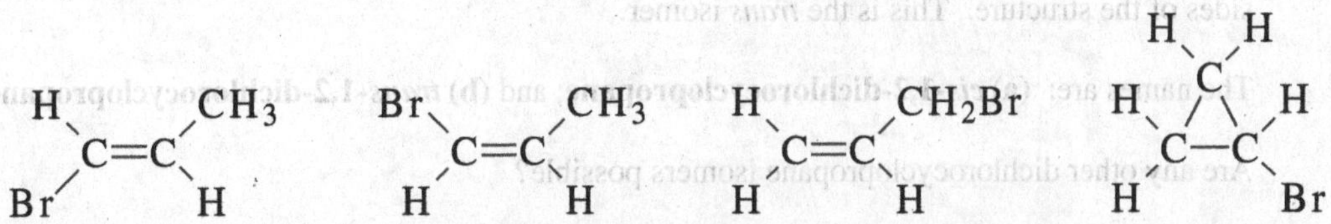

There is only one isomer for the cycloalkane. Note that all three carbons are equivalent in this structure.

**24.16** **(a)** This compound could be an **alkene** or a **cycloalkane**; both have the general formula, $C_nH_{2n}$.

**(b)** This could be an **alkyne** with general formula, $C_nH_{2n-2}$. It could also be a hydrocarbon with two double bonds (a diene). It could be a cyclic hydrocarbon with one double bond (a cycloalkene).

**(c)** This must be an **alkane**; the formula is of the $C_nH_{2n+2}$ type.

**(d)** This compound could be an **alkene** or a **cycloalkane**; both have the general formula, $C_nH_{2n}$.

**(e)** This compound could be an **alkyne** with one triple bond, or it could be a cyclic alkene (unlikely because of ring strain).

**24.18** If cyclobutadiene were square or rectangular, the C–C–C angles must be 90°. If the molecule is diamond-shaped, two of the C–C–C angles must be less than 90°. Both of these situations result in a great deal of distortion and strain in the molecule. Cyclobutadiene is very unstable for these and other reasons.

**24.20** One compound is an alkane; the other is an alkene. Alkenes characteristically undergo addition reactions with hydrogen, with halogens ($Cl_2$, $Br_2$, $I_2$) and with hydrogen halides (HCl, HBr, HI). Alkanes do not react with these substances under ordinary conditions.

**24.22** In this problem you are asked to calculate the standard enthalpy of reaction. This type of problem was covered in Chapter 6.

$$\Delta H^\circ_{\text{rxn}} = \Sigma n\Delta H^\circ_f(\text{products}) - \Sigma m\Delta H^\circ_f(\text{reactants})$$

$$\Delta H^\circ_{\text{rxn}} = \Delta H^\circ_f(C_6H_6) - 3\Delta H^\circ_f(C_2H_2)$$

You can look up $\Delta H^\circ_f$ values in Appendix 3 of your textbook.

$$\boldsymbol{\Delta H^\circ_{\text{rxn}}} = (1)(49.04\text{ kJ/mol}) - (3)(226.6\text{ kJ/mol}) = \mathbf{-630.8\ kJ/mol}$$

**24.24** In this problem you must distinguish between *cis* and *trans* isomers. Recall that *cis* means that two particular atoms (or groups of atoms) are adjacent to each other, and *trans* means that the atoms (or groups of atoms) are on opposite sides in the structural formula.

In (a), the Cl atoms are adjacent to each other. This is the *cis* isomer. In (b), the Cl atoms are on opposite sides of the structure. This is the *trans* isomer.

The names are: **(a) *cis*-1,2-dichlorocyclopropane**; and **(b) *trans*-1,2-dichlorocyclopropane**.

Are any other dichlorocyclopropane isomers possible?

**24.26** (a) This is a branched hydrocarbon. The name is based on the longest carbon chain. The name is **2−methylpentane.**

(b) This is also a branched hydrocarbon. The longest chain includes the $C_2H_5$ group; the name is based on hexane, not pentane. This is an old trick. Carbon chains are flexible and don't have to lie in a straight line. The name is **2,3,4−trimethylhexane**. Why not 3,4,5−trimethylhexane?

(c) How many carbons in the longest chain? It doesn't have to be straight! The name is **3−ethylhexane.**

(d) An alkene with two double bonds is called a diene. The name is **3−methyl−1,4−pentadiene.**

(e) The name is **2−pentyne.**

(f) The name is **3−phenyl−1−pentene.**

**24.28** The hydrogen atoms have been omitted from the skeletal structure for simplicity.

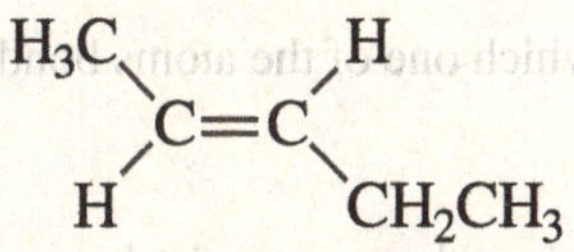
(a)

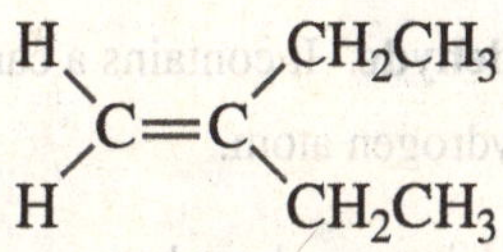
(b)

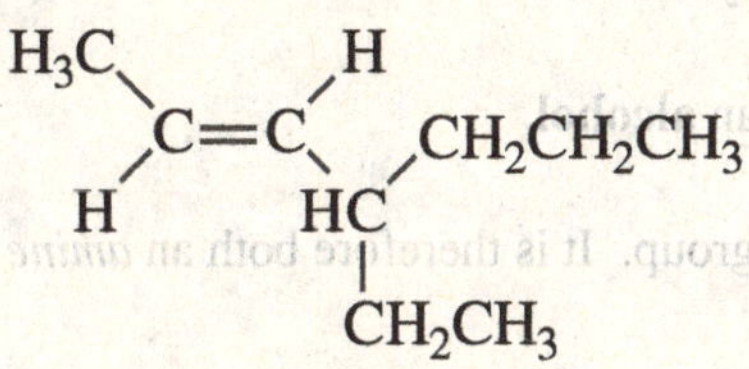
(c)

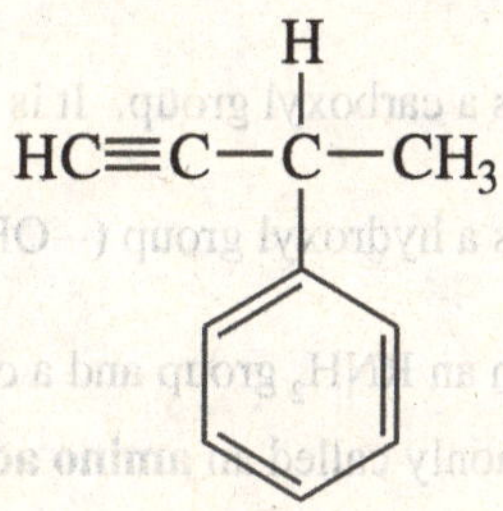
(d)

**24.32** **Strategy:** We follow the IUPAC rules and use the information in Table 24.2 of the text. When a benzene ring has more than *two* substituents, you must specify the location of the substituents with numbers. Remember to number the ring so that you end up with the lowest numbering scheme as possible, giving preference to alphabetical order.

**Solution:**

(a) Since a chloro group comes alphabetically before a methyl group, let's start by numbering the top carbon of the ring as 1. If we number clockwise, this places the second chloro group on carbon 3 and a methyl group on carbon 4.

This compound is **1,3−dichloro−4−methylbenzene.**

**(b)** If we start numbering counterclockwise from the bottom carbon of the ring, the name is 2−ethyl−1,4−dinitrobenzene. Numbering clockwise from the top carbon gives 3−ethyl−1,4−dinitrobenzene.

Numbering as low as possible, the correct name is **2−ethyl−1,4−dinitrobenzene.**

**(c)** Again, keeping the numbers as low as possible, the correct name for this compound is **1,2,4,5−tetramethylbenzene.** You should number clockwise from the top carbon of the ring.

**24.36** **Strategy:** Learning to recognize functional groups requires memorization of their structural formulas. Table 24.4 of the text shows a number of the important functional groups.

**Solution:**

**(a)** $H_3C{-}O{-}CH_2{-}CH_3$ contains a C−O−C group and is therefore an **ether**.

**(b)** This molecule contains an $RNH_2$ group and is therefore an **amine**.

**(c)** This molecule is an **aldehyde.** It contains a carbonyl group in which one of the atoms bonded to the carbonyl carbon is a hydrogen atom.

**(d)** This molecule also contains a carbonyl group. However, in this case there are no hydrogen atoms bonded to the carbonyl carbon. This molecule is a **ketone**.

**(e)** This molecule contains a carboxyl group. It is a **carboxylic acid**.

**(f)** This molecule contains a hydroxyl group (−OH). It is an **alcohol.**

**(g)** This molecule has both an $RNH_2$ group and a carboxyl group. It is therefore both an *amine* and a *carboxylic acid*, commonly called an **amino acid.**

**24.38** Alcohols react with carboxylic acids to form esters. The reaction is:

$$HCOOH + CH_3OH \longrightarrow HCOOCH_3 + H_2O$$

The structure of the product is:

$$H{-}\overset{\overset{\displaystyle O}{\|}}{C}{-}O{-}CH_3 \quad \text{(methyl formate)}$$

**24.40** The fact that the compound does not react with sodium metal eliminates the possibility that the substance is an alcohol. The only other possibility is the ether functional group. There are three ethers possible with this molecular formula:

$$CH_3{-}CH_2{-}O{-}CH_2{-}CH_3 \qquad CH_3{-}CH_2{-}CH_2{-}O{-}CH_3 \qquad (CH_3)_2CH{-}O{-}CH_3$$

Light–induced reaction with chlorine results in substitution of a chlorine atom for a hydrogen atom (the other product is HCl). For the first ether there are only two possible chloro derivatives:

$$ClCH_2{-}CH_2{-}O{-}CH_2{-}CH_3 \qquad CH_3{-}CHCl{-}O{-}CH_2{-}CH_3$$

For the second there are four possible chloro derivatives. Three are shown below. Can you draw the fourth?

$$CH_3{-}CHCl{-}CH_2{-}O{-}CH_3 \qquad CH_3{-}CH_2{-}CHCl{-}O{-}CH_3 \qquad CH_2Cl{-}CH_2{-}CH_2{-}O{-}CH_3$$

For the third there are three possible chloro derivatives:

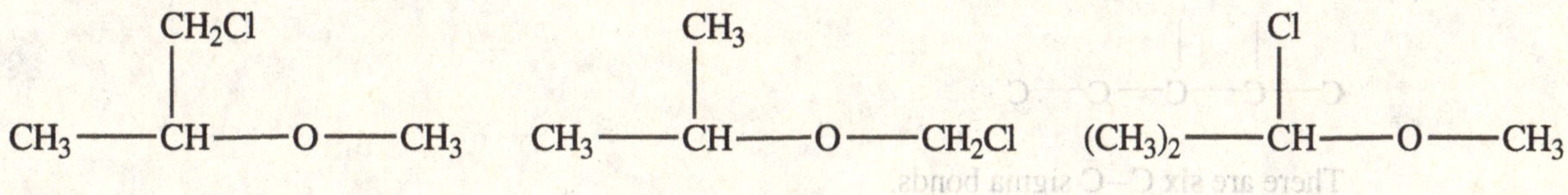

The $\mathbf{(CH_3)_2CH{-}O{-}CH_3}$ choice is the original compound.

**24.42** (a) ketone (b) ester (c) ether

**24.44** This is a Hess's Law problem. See Chapter 6.

If we rearrange the equations given and multiply times the necessary factors, we have:

$$2CO_2(g) + 2H_2O(l) \longrightarrow C_2H_4(g) + 3O_2(g) \qquad \Delta H^\circ = 1411 \text{ kJ/mol}$$

$$C_2H_2(g) + \tfrac{5}{2}O_2(g) \longrightarrow 2CO_2(g) + H_2O(l) \qquad \Delta H^\circ = -1299.5 \text{ kJ/mol}$$

$$H_2(g) + \tfrac{1}{2}O_2(g) \longrightarrow H_2O(l) \qquad \Delta H^\circ = -285.8 \text{ kJ/mol}$$

$$C_2H_2(g) + H_2(g) \longrightarrow C_2H_4(g) \qquad \mathbf{\Delta H^\circ = -174 \text{ kJ/mol}}$$

The heat of hydrogenation for acetylene is **−174 kJ/mol.**

**24.46** To form a hydrogen bond *with water* a molecule must have at least one H–F, H–O, or H–N bond, *or* must contain an O, N, or F atom. The following can form hydrogen bonds with water:

(a) carboxylic acids (c) ethers (d) aldehydes (f) amines

**24.48** (a) rubbing alcohol (b) vinegar (c) moth balls (d) organic synthesis

(e) organic synthesis (f) antifreeze (g) fuel (natural gas) (h) synthetic polymers

**24.50** (a) 2–butyne has **three** C–C sigma bonds.

(b) Anthracene is:

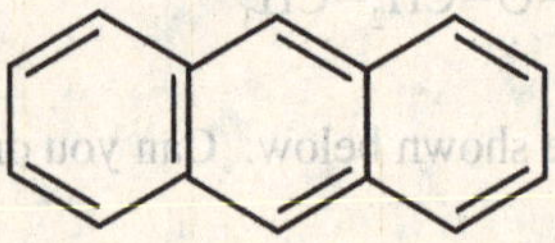

There are **sixteen** C–C sigma bonds.

(c)

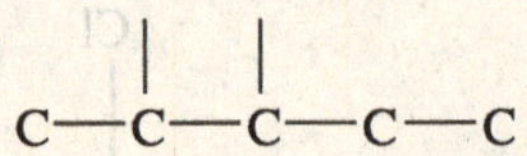


There are **six** C–C sigma bonds.

**24.52** (a) The easiest way to calculate the mg of C in $CO_2$ is by mass ratio. There are 12.01 g of C in 44.01 g $CO_2$ or 12.01 mg C in 44.01 mg $CO_2$.

$$\textbf{? mg C} = 57.94\ \text{mg CO}_2 \times \frac{12.01\ \text{mg C}}{44.01\ \text{mg CO}_2} = \textbf{15.81 mg C}$$

Similarly,

$$\textbf{? mg H} = 11.85\ \text{mg H}_2\text{O} \times \frac{2.016\ \text{mg H}}{18.02\ \text{mg H}_2\text{O}} = \textbf{1.326 mg H}$$

The mg of oxygen can be found by difference.

$$\textbf{? mg O} = 20.63\ \text{mg Y} - 15.81\ \text{mg C} - 1.326\ \text{mg H} = \textbf{3.49 mg O}$$

(b) *Step 1:* Calculate the number of moles of each element present in the sample. Use molar mass as a conversion factor.

$$?\ \text{mol C} = (15.81 \times 10^{-3}\ \text{g C}) \times \frac{1\ \text{mol C}}{12.01\ \text{g C}} = 1.316 \times 10^{-3}\ \text{mol C}$$

Similarly,

$$?\ \text{mol H} = (1.326 \times 10^{-3}\ \text{g H}) \times \frac{1\ \text{mol H}}{1.008\ \text{g H}} = 1.315 \times 10^{-3}\ \text{mol H}$$

$$?\ \text{mol O} = (3.49 \times 10^{-3}\ \text{g O}) \times \frac{1\ \text{mol O}}{16.00\ \text{g O}} = 2.18 \times 10^{-4}\ \text{mol O}$$

Thus, we arrive at the formula $C_{1.316 \times 10^{-3}}H_{1.315 \times 10^{-3}}O_{2.18 \times 10^{-4}}$, which gives the identity and the ratios of atoms present. However, chemical formulas are written with whole numbers.

*Step 2:* Try to convert to whole numbers by dividing all the subscripts by the smallest subscript.

$$\mathbf{C}: \frac{1.316 \times 10^{-3}}{2.18 \times 10^{-4}} = 6.04 \approx 6 \qquad \mathbf{H}: \frac{1.315 \times 10^{-3}}{2.18 \times 10^{-4}} = 6.03 \approx 6 \qquad \mathbf{O}: \frac{2.18 \times 10^{-4}}{2.18 \times 10^{-4}} = 1.00$$

This gives us the empirical formula, $\mathbf{C_6H_6O}$.

(c) The presence of six carbons and a corresponding number of hydrogens suggests a benzene derivative. A plausible structure is shown on the right.

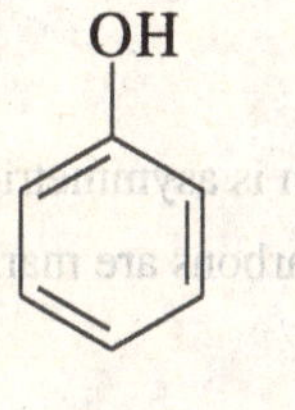


**24.54** First, calculate the moles of each element.

$$\text{C:} \quad (9.708 \times 10^{-3} \cancel{g}\ CO_2) \times \frac{1\ \cancel{mol}\ CO_2}{44.01\ \cancel{g}\ CO_2} \times \frac{1\ \text{mol C}}{1\ \cancel{mol}\ CO_2} = 2.206 \times 10^{-4}\ \text{mol C}$$

$$\text{H:} \quad (3.969 \times 10^{-3} \cancel{g}\ H_2O) \times \frac{1\ \cancel{mol}\ H_2O}{18.02\ \cancel{g}\ H_2O} \times \frac{2\ \text{mol H}}{1\ \cancel{mol}\ H_2O} = 4.405 \times 10^{-4}\ \text{mol H}$$

The mass of oxygen is found by difference:

$$3.795\ \text{mg compound} - (2.649\ \text{mg C} + 0.445\ \text{mg H}) = 0.701\ \text{mg O}$$

$$\text{O:} \quad (0.701 \times 10^{-3} \cancel{g}\ O) \times \frac{1\ \text{mol O}}{16.00\ \cancel{g}\ O} = 4.38 \times 10^{-5}\ \text{mol O}$$

This gives the formula is $C_{2.206 \times 10^{-4}}H_{4.405 \times 10^{-4}}O_{4.38 \times 10^{-5}}$. Dividing by the smallest number of moles gives the empirical formula, $\mathbf{C_5H_{10}O}$.

We calculate moles using the ideal gas equation, and then calculate the molar mass.

$$n = \frac{PV}{RT} = \frac{(1.00\ \cancel{atm})(0.0898\ \cancel{L})}{(0.0821\ \cancel{L} \cdot \cancel{atm}/\cancel{K} \cdot \text{mol})(473\ \cancel{K})} = 0.00231\ \text{mol}$$

$$\textbf{molar mass} = \frac{\text{g of substance}}{\text{mol of substance}} = \frac{0.205\ \text{g}}{0.00231\ \text{mol}} = \mathbf{88.7\ g/mol}$$

The formula mass of $C_5H_{10}O$ is 86.13 g, so this is also the molecular formula. Three possible structures are:

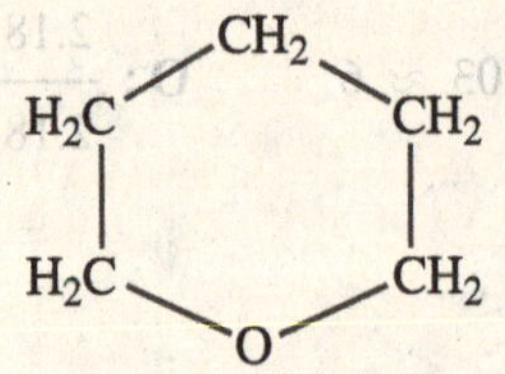

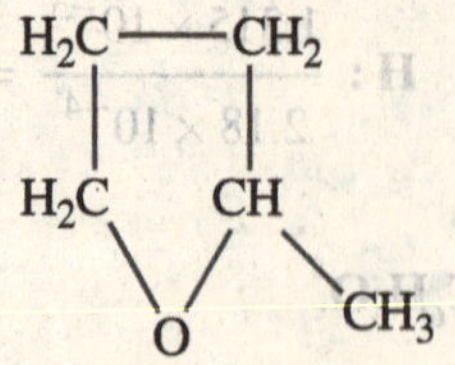

$H_2C{=}CH{-}CH_2{-}O{-}CH_2{-}CH_3$

**24.56** A carbon atom is asymmetric if it is bonded to four different atoms or groups. In the given structures the asymmetric carbons are marked with an asterisk (*).

**(a)**

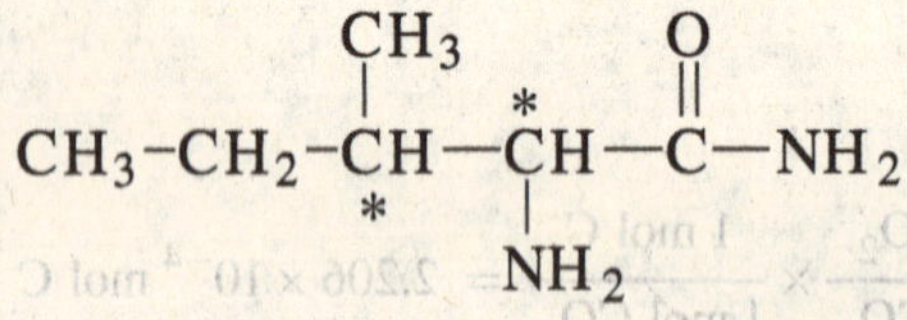

**(b)**

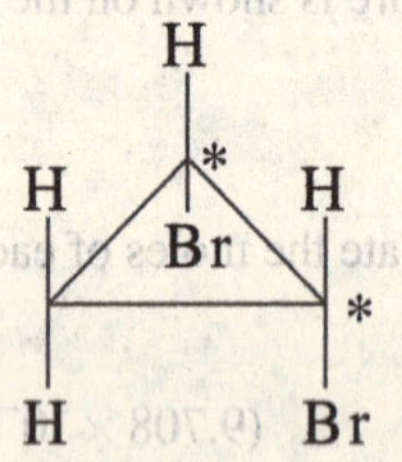

**24.58** Acetone is a ketone with the formula, $CH_3COCH_3$. We must write the structure of an aldehyde that has the same number and types of atoms ($C_3H_6O$). Removing the aldehyde functional group (–CHO) from the formula leaves $C_2H_5$. This is the formula of an ethyl group. The aldehyde that is a structural isomer of acetone is:

$$CH_3CH_2\overset{\overset{\displaystyle O}{\|}}{C}{-}H$$

**24.60** **(a)** alcohol **(b)** ether **(c)** aldehyde **(d)** carboxylic acid **(e)** amine

**24.62** In Chapter 11, we found that salts with their electrostatic intermolecular attractions had low vapor pressures and thus high boiling points. Ammonia and its derivatives (amines) are molecules with dipole–dipole attractions. If the nitrogen has one direct N–H bond, the molecule will have hydrogen bonding. Even so, these molecules will have much weaker intermolecular attractions than ionic species and hence higher vapor pressures. Thus, if we could convert the neutral ammonia–type molecules into salts, their vapor pressures, and thus associated odors, would decrease. Lemon juice contains acids which can react with ammonia–type (amine) molecules to form ammonium salts.

$$NH_3 + H^+ \longrightarrow NH_4^+ \qquad RNH_2 + H^+ \longrightarrow RNH_3^+$$

**24.64** Marsh gas (methane, $CH_4$); grain alcohol (ethanol, $C_2H_5OH$); wood alcohol (methanol, $CH_3OH$); rubbing alcohol (isopropyl alcohol, $(CH_3)_2CHOH$); antifreeze (ethylene glycol, $CH_2OHCH_2OH$); mothballs (naphthalene, $C_{10}H_8$); vinegar (acetic acid, $CH_3COOH$).

**24.66** The asymmetric carbons are shown by asterisks:

(a) 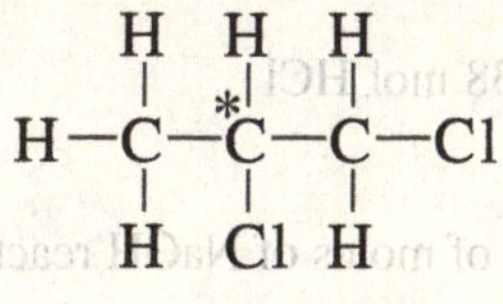

(b) 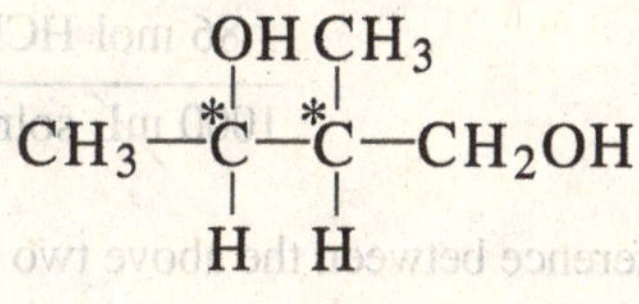

(c) All of the carbon atoms in the ring are asymmetric. Therefore there are **five** asymmetric carbon atoms.

**24.68** The red bromine vapor absorbs photons of blue light and dissociates to form bromine atoms.

$$Br_2 \rightarrow 2Br\bullet$$

The bromine atoms collide with methane molecules and abstract hydrogen atoms.

$$Br\bullet + CH_4 \rightarrow HBr + \bullet CH_3$$

The methyl radical then reacts with $Br_2$, giving the observed product and regenerating a bromine atom to start the process over again:

$$\bullet CH_3 + Br_2 \rightarrow CH_3Br + Br\bullet$$

$$Br\bullet + CH_4 \rightarrow HBr + \bullet CH_3 \quad \text{and so on...}$$

**24.70** 2–butanone is

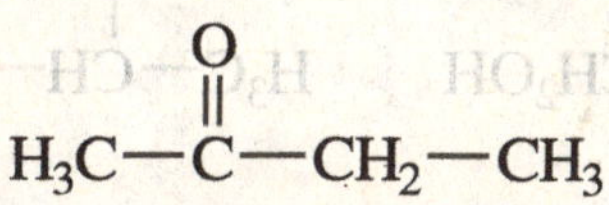

Reduction with $LiAlH_4$ produces 2-butanol.

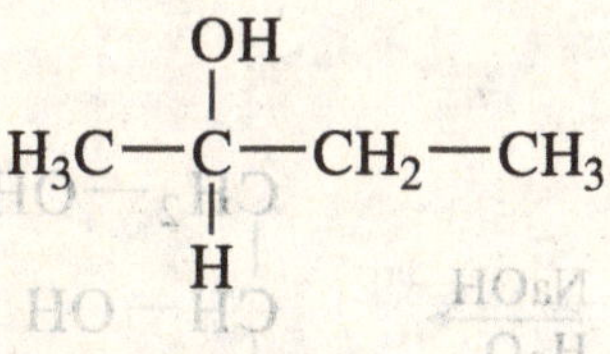

This molecule possesses an asymmetric carbon atom and should be chiral. However, the reduction produces an equimolar *d* and *l* isomers; that is, a racemic mixture (see Section 23.4 of the text). Therefore, the optical rotation as measured in a polarimeter is zero.

**24.72** To help determine the molecular formula of the alcohol, we can calculate the molar mass of the carboxylic acid, and then determine the molar mass of the alcohol from the molar mass of the acid. Grams of carboxylic acid are given (4.46 g), so we need to determine the moles of acid to calculate its molar mass.

The number of moles in 50.0 mL of 2.27 *M* NaOH is

$$\frac{2.27 \text{ mol NaOH}}{1000 \text{ mL soln}} \times 50.0 \text{ mL} = 0.1135 \text{ mol NaOH}$$

The number of moles in 28.7 mL of 1.86 $M$ HCl is

$$\frac{1.86 \text{ mol HCl}}{1000 \cancel{\text{mL}} \text{ soln}} \times 28.7 \cancel{\text{mL}} = 0.05338 \text{ mol HCl}$$

The difference between the above two numbers is the number of moles of NaOH reacted with the carboxylic acid.

$$0.1135 \text{ mol} - 0.05338 \text{ mol} = 0.06012 \text{ mol}$$

This is the number of moles present in 4.46 g of the carboxylic acid. The molar mass is

$$M = \frac{4.46 \text{ g}}{0.06012 \text{ mol}} = 74.18 \text{ g/mol}$$

A carboxylic acid contains a –COOH group and an alcohol has an –OH group. When the alcohol is oxidized to a carboxylic acid, the change is from $-CH_2OH$ to –COOH. Therefore, the molar mass of the alcohol is

$$74.18 \text{ g} - 16 \text{ g} + (2)(1.008 \text{ g}) = 60.2 \text{ g/mol}$$

With a molar mass of 60.2 g/mol for the alcohol, there can only be 1 oxygen atom and 3 carbon atoms in the molecule, so the formula must be $C_3H_8O$. The alcohol has one of the following two molecular formulas.

$CH_3CH_2CH_2OH$ $\qquad$ $H_3C-CH(OH)-CH_3$

**24.74** (a) Reaction between glycerol and carboxylic acid (formation of an ester).

(b) 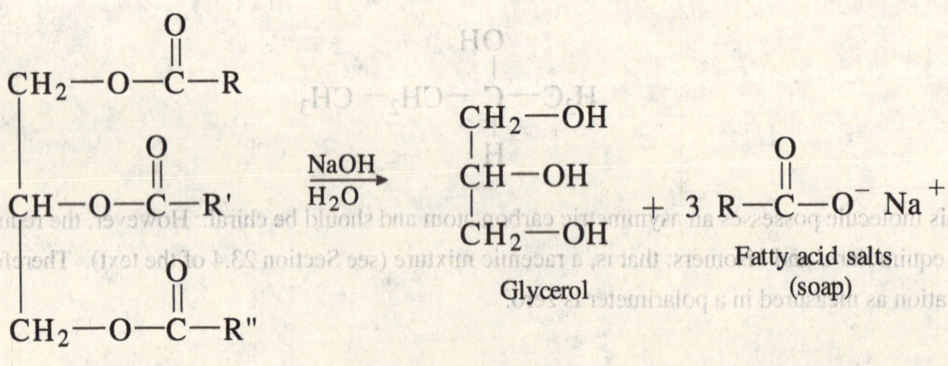


A fat or oil

(c) Molecules having more C=C bonds are harder to pack tightly together. Consequently, the compound has a lower melting point.

(d) $H_2$ gas with either a heterogeneous or homogeneous catalyst would be used. See Section 13.6 of the text.

(e) Number of moles of $Na_2S_2O_3$ reacted is:

$$20.6\ \text{mL} \times \frac{1\ \text{L}}{1000\ \text{mL}} \times \frac{0.142\ \text{mol Na}_2\text{S}_2\text{O}_3}{1\ \text{L}} = 2.93 \times 10^{-3}\ \text{mol Na}_2\text{S}_2\text{O}_3$$

The mole ratio between $I_2$ and $Na_2S_2O_3$ is 1:2. The number of grams of $I_2$ left over is:

$$(2.93 \times 10^{-3}\ \text{mol Na}_2\text{S}_2\text{O}_3) \times \frac{1\ \text{mol I}_2}{2\ \text{mol Na}_2\text{S}_2\text{O}_3} \times \frac{253.8\ \text{g I}_2}{1\ \text{mol I}_2} = 0.372\ \text{g I}_2$$

Number of grams of $I_2$ reacted is: $(43.8 - 0.372)\text{g} = 43.4\ \text{g I}_2$

The *iodine number* is the number of grams of iodine that react with 100 g of corn oil.

$$\textit{\textbf{iodine number}} = \frac{43.4\ \text{g I}_2}{35.3\ \text{g corn oil}} \times 100\ \text{g corn oil} = \mathbf{123}$$

# CHAPTER 25
# SYNTHETIC AND NATURAL ORGANIC POLYMERS

## PROBLEM-SOLVING STRATEGIES AND TUTORIAL SOLUTIONS

### TYPES OF PROBLEMS

**Problem Type 1:** Synthetic Organic Polymers.

**(a)** Addition reactions.

**(b)** Condensation reactions.

**Problem Type 2:** Proteins.

**Problem Type 3:** Nucleic Acids.

## PROBLEM TYPE 1: SYNTHETIC ORGANIC POLYMERS

The word **polymer** means "many parts". A **polymer** is a compound with an unusually high molecular mass, consisting of a large number of molecular units linked together. The small unit that is repeated many times is called a **monomer**. A typical polymer contains chains of monomers several thousand units long.

### A. Addition reactions

Addition polymers are made by adding monomer to monomer until a long chain is produced. Ethylene (ethene) and its derivatives are excellent monomers for forming addition polymers. In Chapter 24, we saw that addition reactions occur with unsaturated compounds containing C=C and C≡C bonds. In an addition reaction, the polymerization process is initiated by a radical or an ion. When ethylene is heated to 250°C under high pressure (1000–3000 atm) in the presence of a little oxygen or benzoyl peroxide (the initiator), addition polymers with masses of about 30,000 amu are obtained. This reaction is represented by

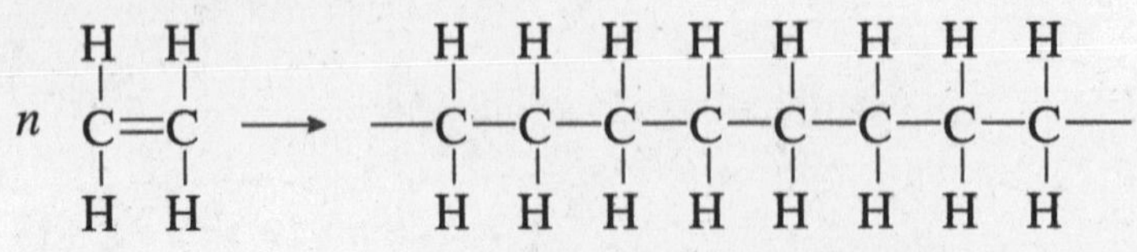

ethylene a segment of polyethylene

The general equation for addition polymerization is:

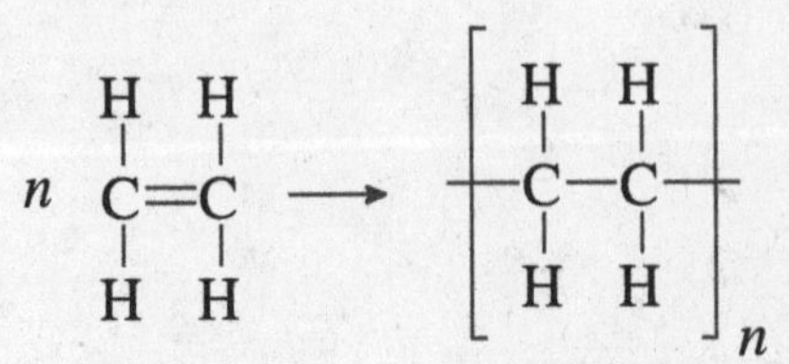

monomer repeating unit

Substitution for one or more hydrogen atoms in ethylene with Cl atoms, acetate, CN, or F provides a wide selection of monomers from which to make addition polymers with various properties. For instance, substitution of a Cl atom for a H atom in ethylene gives the monomer called vinyl chloride, $CH_2{=}CHCl$. Polymerization of vinyl chloride yields the polymer, polyvinyl chloride (PVC).

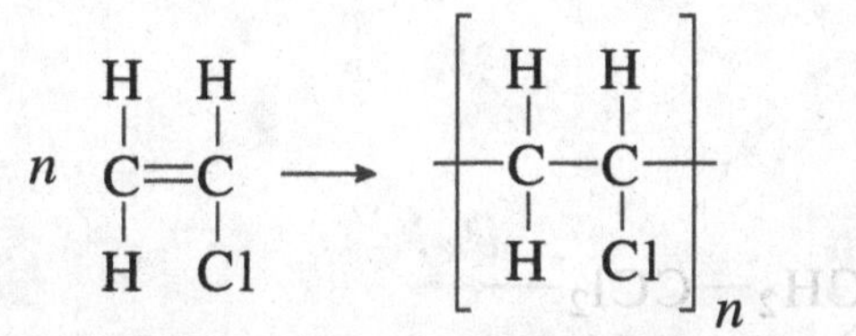

vinyl chloride monomer — polyvinyl chloride repeating unit

Polymers that are made from one type of monomer such as polyvinyl chloride are called **homopolymers.** Table 25.1 of the text gives the names, structures, and uses of a number of monomers and addition polymers.

For more detailed information on the reaction mechanism of addition polymerization, see Section 25.2 of the text.

**EXAMPLE 25.1**

**Write the formulas of the monomers used to prepare the following polymers:**

**(a) Teflon, (b) Polystyrene, and (c) PVC.**

---

Refer to Table 25.1 of the text.

**(a)** Teflon is an addition polymer with the formula

$$\left( CF_2-CF_2 \right)_n$$

It is prepared from the monomer tetrafluoroethylene ($CF_2{=}CF_2$).

**(b)** The monomer used to prepare polystyrene is styrene.

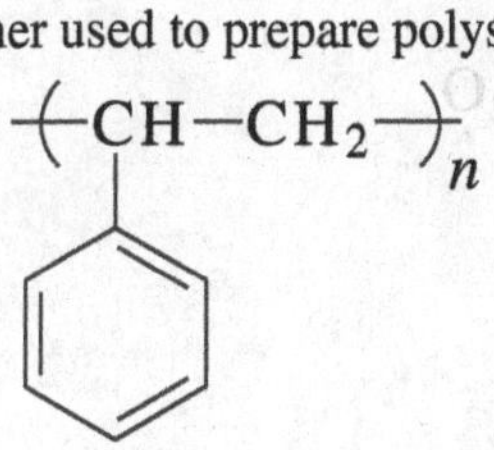

polystyrene

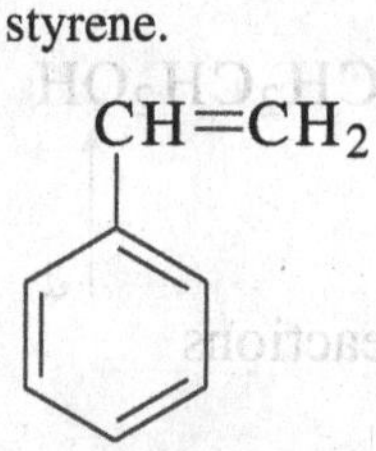

styrene

**(c)** PVC (polyvinylchloride) is prepared by the successive addition of vinyl chloride molecules ($CH_2{=}CHCl$).

**PRACTICE EXERCISE**

1. Draw the structures of the monomers from which the following polymers are formed:

(a) 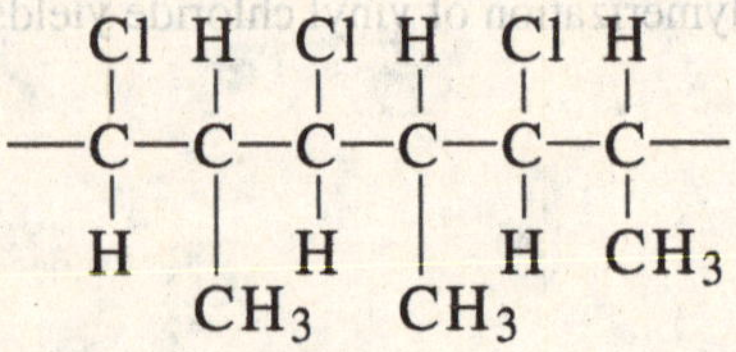

(b) $—CH_2—CCl_2—CH_2—CCl_2—CH_2—CCl_2—$

**Text Problems:** 25.8, 25.10, 25.12a

## B. Condensation reactions

**Copolymers** are polymers that contain two or more different types of monomers. Polyesters, such as the well-known Dacron, are copolymers. In Dacron, one monomer is an alcohol and the other is a carboxylic acid, which can be joined by an *esterification* reaction. The alcohol and the acid both must contain two functional groups. The monomers of Dacron are the dicarboxylic acid called terephthalic acid and the dialcohol called ethylene glycol.

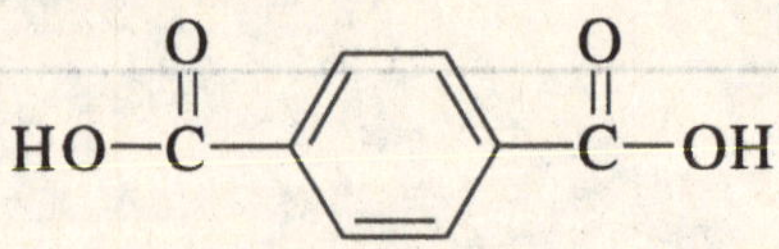

terephthalic acid

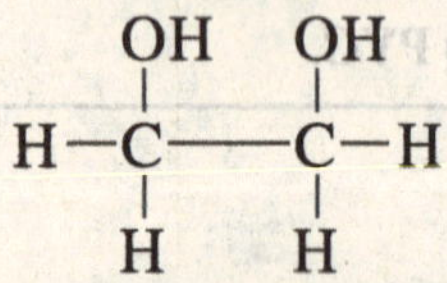

ethylene glycol

Condensation reactions differ from addition reactions in that the former always result in the formation of a small molecule such as water. When terephthalic acid and ethylene glycol react to form an ester, the first products are:

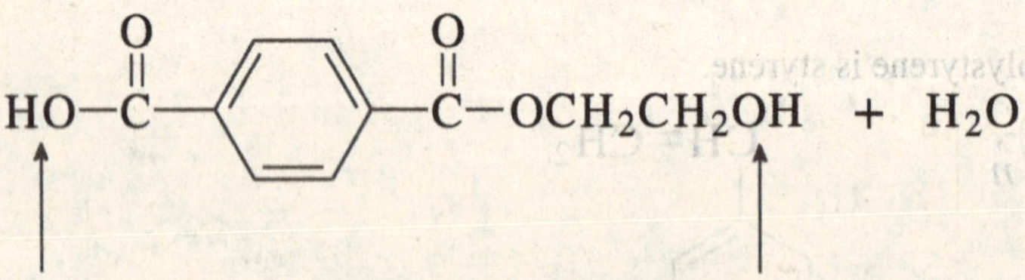

sites for further condensation reactions

When this product reacts with another molecule of the diacid, the polymer chain grows longer.

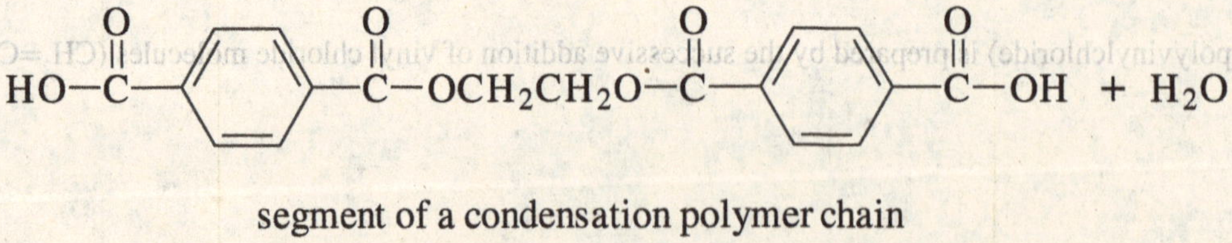

segment of a condensation polymer chain

The general formula for the polyester, Dacron, is:

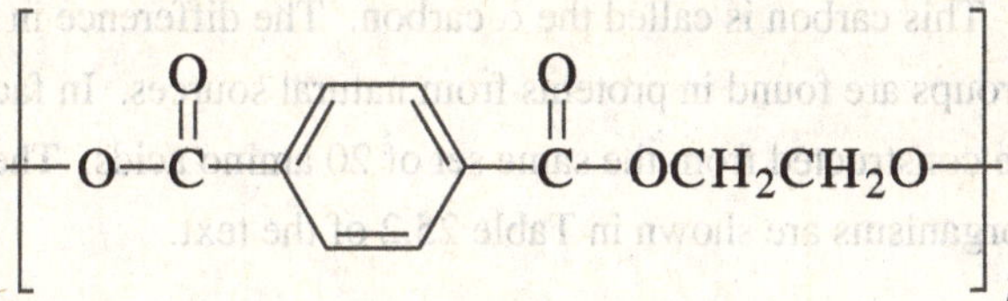

**EXAMPLE 25.2**
**Draw structures for the monomers used to make the following polyester:**

$$\left[-\overset{O}{\overset{\|}{C}}-(CH_2)_4-\overset{O}{\overset{\|}{C}}-O-(CH_2)_3-O-\right]_n$$

**Answer**

$$HO-\overset{O}{\overset{\|}{C}}-(CH_2)_4-\overset{O}{\overset{\|}{C}}-OH \qquad HO-(CH_2)_3-OH$$

**PRACTICE EXERCISE**

2. List two examples of condensation polymers discussed in the textbook.

| **Text Problem:** 25.12b |
|---|

## PROBLEM TYPE 2: PROTEINS

**Proteins** are *polymers of amino acids*. Proteins are truly giant molecules having molecular masses that range from about 10,000 to several million amu. Proteins play many roles in living organisms, where they function as catalysts (enzymes), transport molecules (hemoglobin), contractile fibers (muscle), protective agents (blood clots), hormones (chemical messengers), and structural members (feathers, horns, and nails). The word *protein* comes from the Greek work *proteios*, meaning "first". From the partial list of protein functions above, it is easy to see why proteins occupy "first place" among biomolecules in their importance to life.

Even though each protein is unique, all proteins are built from a set of only **20 amino acids**. An amino acid consists of an amino group, a carboxylic acid group, a hydrogen atom, and a distinctive R group, all bonded to the same carbon atom.

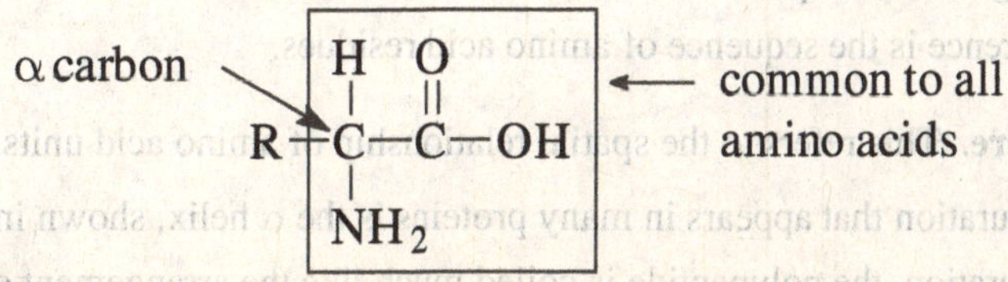

All amino acids in proteins have a common structural feature that is the attachment of the amino group ($-NH_2$) to the carbon atom adjacent to the carboxylic acid group. This carbon is called the α carbon. The difference in amino acids is due to different R groups. Twenty different R groups are found in proteins from natural sources. In fact, all proteins in all species, from bacteria to humans, are constructed from the same set of 20 amino acids. The structural formulas of the 20 amino acids essential to living organisms are shown in Table 25.2 of the text.

In proteins, the amino acid units are joined together to form a *polypeptide* chain. The carboxyl group of one amino acid is joined to the amino group of another amino acid by the formation of a peptide bond.

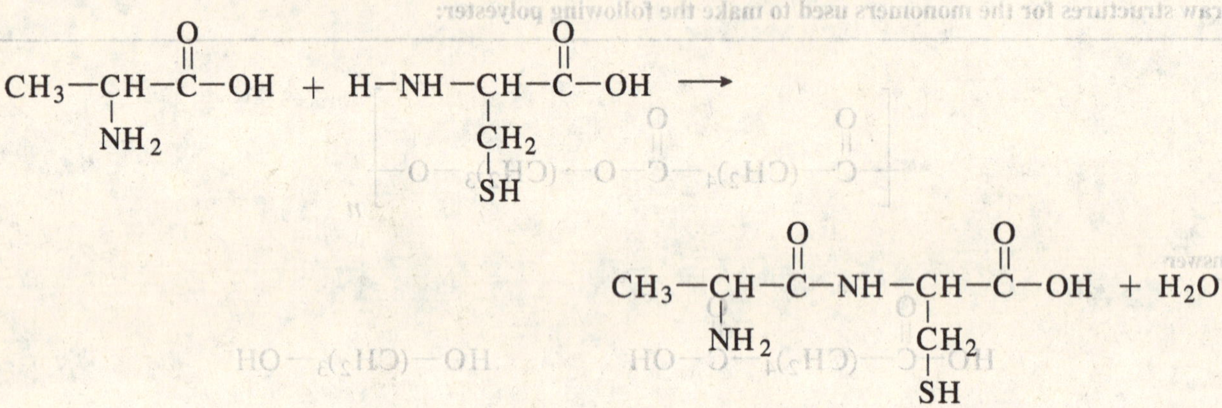

This type of reaction is another example of a condensation reaction. The new C–N covalent bond is called a *peptide* or an *amide* bond. The amide functional group present in all proteins is:

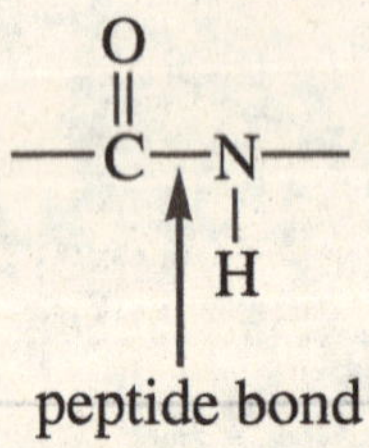


The molecule above, in which two amino acids are joined, is called a *dipeptide*. Peptides are structures intermediate in size between amino acids and proteins. The term *polypeptide* refers to long molecular chains containing many amino acid units. An amino acid unit in a polypeptide chain is called a *residue*.

Proteins are so complex that four levels of structural features have been identified. The structure of proteins is extremely important in determining how efficiently and effectively a protein will function. We will discuss the first two levels of protein structure.

1. **The Primary Structure**. Each protein has a unique amino acid sequence of its polypeptide chain. It is the amino acid sequence that distinguishes one protein from another. Proteins also differ in the numbers and types of amino acids, but the main difference is the sequence of amino acid residues.

2. **The Secondary Structure**. This refers to the spatial relationship of amino acid units that are close to one another in sequence. The configuration that appears in many proteins is the α helix, shown in Figure 25.11 of the textbook. In this configuration, the polypeptide is coiled much like the arrangement of stairs in a spiral staircase. Figure 25.11 also shows that the α helix is stabilized by the presence of intermolecular hydrogen bonds (dashed lines). The C=O group of each amino acid is hydrogen bonded to the NH group of the amino acid that is located

four amino acids ahead in the sequence. The α helix is the main structural feature of the oxygen-transport protein, hemoglobin, and of many other proteins. The structure of hemoglobin is shown in Figure 25.13 of the text.

The β-pleated sheet is another common secondary structure. In this structure, a polypeptide chain interacts strongly with adjacent chains by forming many hydrogen bonds. See Figure 25.12 of the text.

**EXAMPLE 25.3**
**What are the five chemical elements found in proteins?**

Proteins are made up of the same elements as amino acids. Therefore, proteins contain C, H, O, and N. Two amino acids, methionine and cysteine, also contain sulfur, S.

**PRACTICE EXERCISE**

3. Sketch a portion of the polypeptide chain consisting of the amino acids glycine, valine, and alanine, in that order.

**Text Problems:** 25.20, 25.22

## PROBLEM TYPE 3: NUCLEIC ACIDS

The chemical composition of the cell nucleus was first studied in the 1860s by Friedrich Miescher. He found the major components to be protein and a new material not previously isolated. This material was found to be acidic, and so it was referred to as **nucleic acid**.

Nucleic acids are polymers that store genetic information and control protein synthesis. There are two types of nucleic acids, deoxyribonucleic acid (DNA) and ribonucleic acid (RNA). ***DNA*** carries all the genetic information necessary to carry on reproduction and to sustain an organism throughout its lifetime. ***RNA*** transcribes these instructions and controls the synthesis of proteins needed to implement them.

DNA molecules are among the largest known with molar masses up to 10 billion g/mol. Hydrolysis of nucleic acids show that they are composed of only four types of building blocks: a phosphate group, a sugar, purines, and pyrimidines.

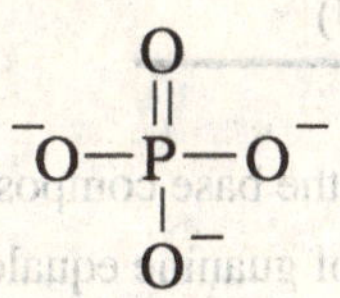

Phosphate

One of two sugars is present, ribose in RNA and deoxyribose in DNA.

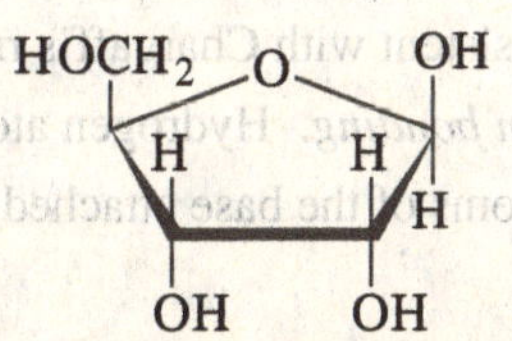

Ribose

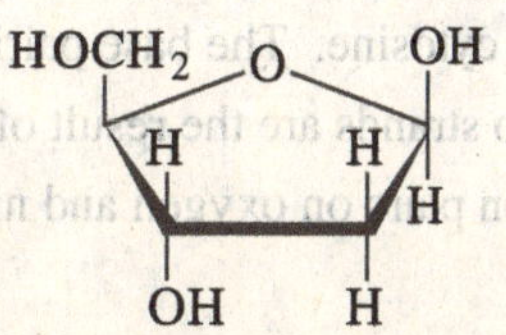

Deoxyribose

Two purines, adenine and guanine, are found in both DNA and RNA.

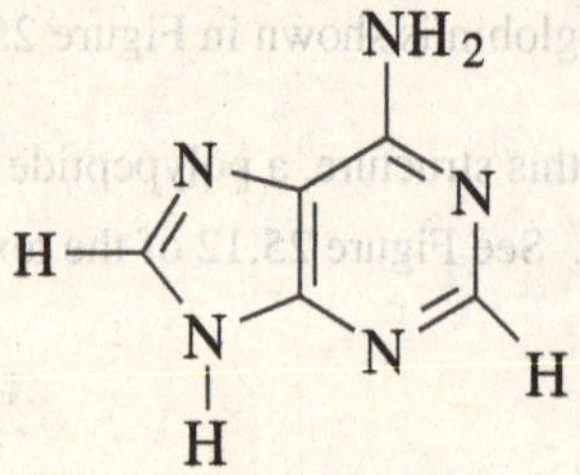

Adenine

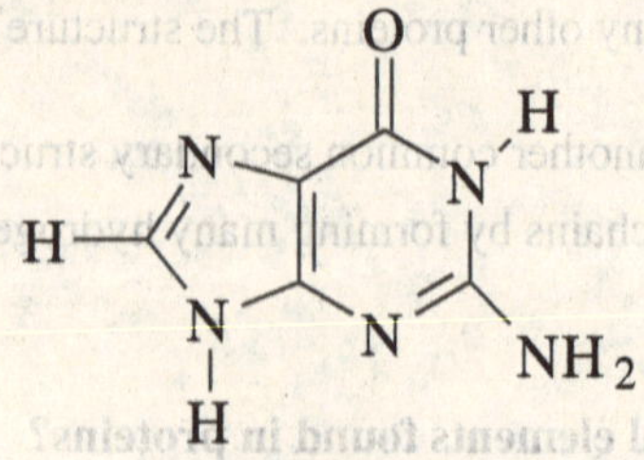

Guanine

DNA also contains the pyrimidines thymine and cytosine. RNA also contains thymine and another pyrimidine, uracil.

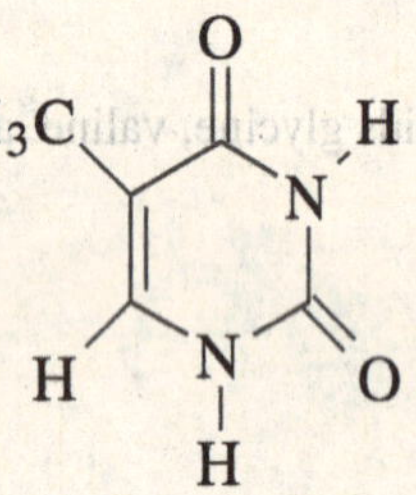

Thymine

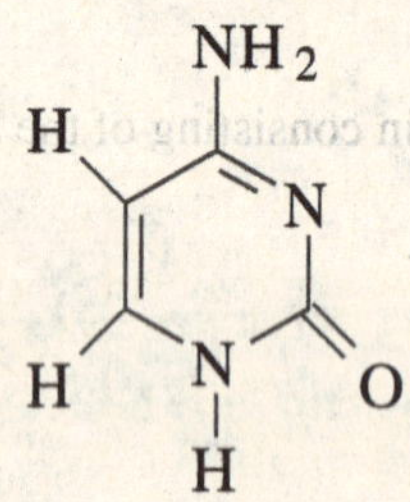

Cytosine

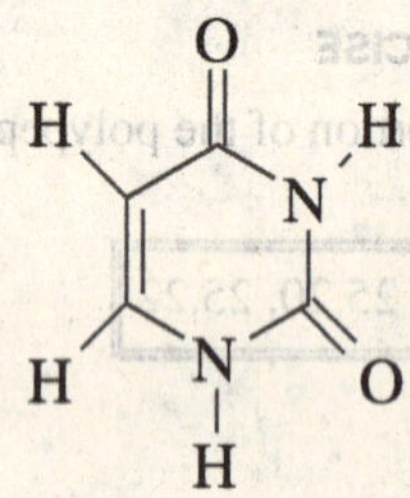

Uracil

The purines and pyrimidines are collectively called ***bases***. The table below summarizes the building blocks of DNA and RNA.

**Building Blocks of DNA and RNA**

| | DNA | RNA |
|---|---|---|
| | Phosphate (P) | Phosphate (P) |
| **Sugar** | Deoxyribose (D) | Ribose (R) |
| **Purines** | Adenine (A) | Adenine (A) |
| | Guanine (G) | Guanine (G) |
| **Pyrimidines** | Thymine (T) | Thymine (T) |
| | Cytosine (C) | Uracil (U) |

In the 1940s, Edwin Chargaff studied the composition of DNA. His analysis of the base composition of DNA showed that the amount of adenine always equaled that of thymine and that the amount of guanine equaled that of cytosine.

In 1953, Watson and Crick proposed a double-helical structure for DNA. The repeating unit in each strand is called a **nucleotide**, which consists of a base-deoxyribose-phosphate linkage (see Figure 25.18 of the text). The two strands are *not* identical, rather they are complementary. Adenine in one strand is always paired with thymine in the other strand. Guanine is always paired with cytosine. The base pairing is consistent with Chargaff's rules. Base pairing and the resulting association of the two strands are the result of *hydrogen bonding*. Hydrogen atoms in a base in one strand are attracted to unshared electron pairs on oxygen and nitrogen atoms of the base attached to the other strand (see Figure 25.19 of the text).

RNA, on the other hand, does not follow the base-pairing rules. X ray data and other evidence ruled out a double-helical structure for RNA. RNA is single-stranded.

**EXAMPLE 25.4**

**What types of forces cause base pairing in the double-stranded helical DNA molecule?**

The purine bases (adenine and guanine) in one strand of DNA form hydrogen bonds to the pyrimidine bases (thymine and cytosine) in the other DNA strand. Hydrogen atoms covalently bonded to nitrogen carry a partial positive charge. These H atoms are attracted to lone-electron pairs on oxygen and nitrogen atoms of another base. Since two complementary bases are attached to different strands, the hydrogen bonds hold the strands together.

**PRACTICE EXERCISE**

4. If the base sequence in one strand of DNA is A, T, G, C, T, what is the base sequence in the complementary strand?

**Text Problem:** 25.28

## ANSWERS TO PRACTICE EXERCISES

1. (a)

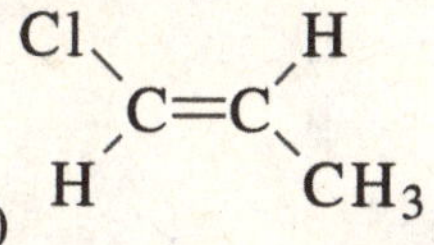

(b)

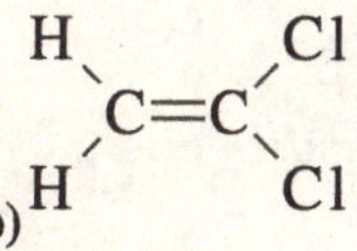

2. Nylon 66 (first prepared in 1931) and polyester.

3.

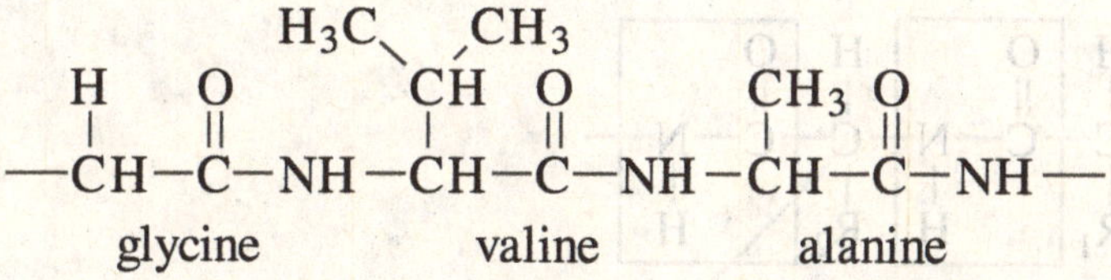

4. Recall that according to Chargaff's rules, adenine (A) is always paired with thymine (T) and guanine (G) is always paired with cytosine (C). The complementary base sequence is T, A, C, G, A.

## SOLUTIONS TO SELECTED TEXT PROBLEMS

**25.8** The repeating structural unit of the polymer is:

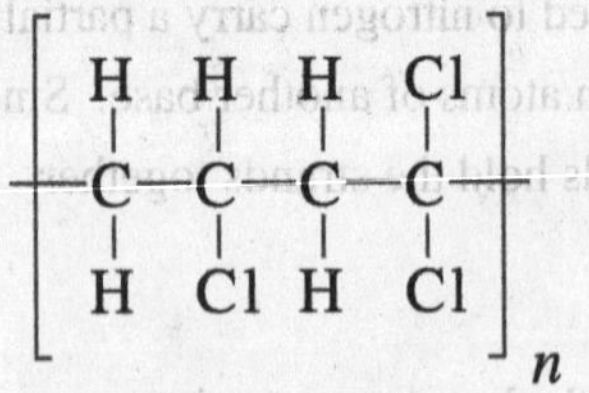

Does each carbon atom still obey the octet rule?

**25.10** Polystyrene is formed by an addition polymerization reaction with the monomer, styrene, which is a phenyl–substituted ethylene. The structures of styrene and polystyrene are shown in Table 25.1 of your text.

**25.12** The structures are shown.

**(a)** $H_2C{=}CH{-}CH{=}CH_2$

**(b)**

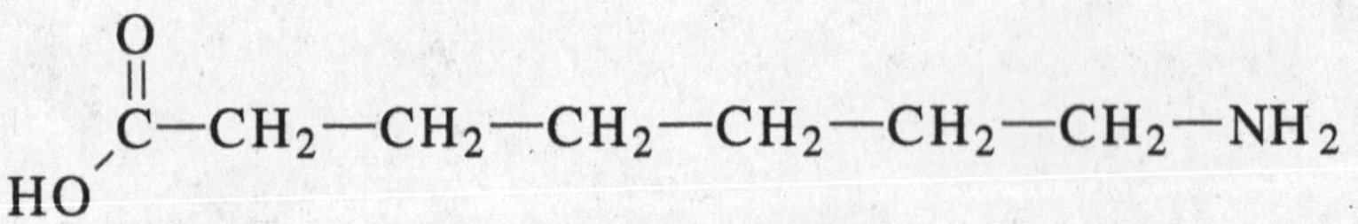

**25.20** The main backbone of a polypeptide chain is made up of the α carbon atoms and the amide group repeating alternately along the chain.

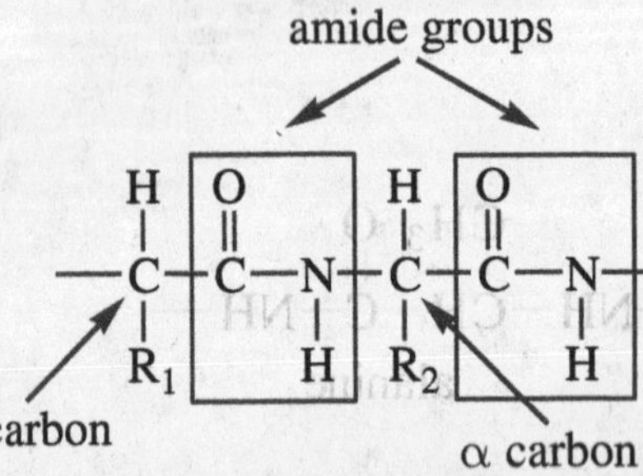


For each R group shown above, substitute the distinctive side groups of the two amino acids. There are two possible dipeptides depending on how the two amino acids are connected, either glycine–lysine or lysine–glycine. The structures of the dipeptides are:

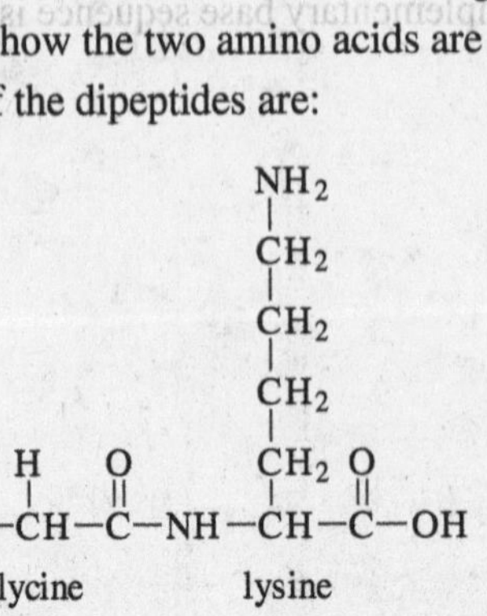

and

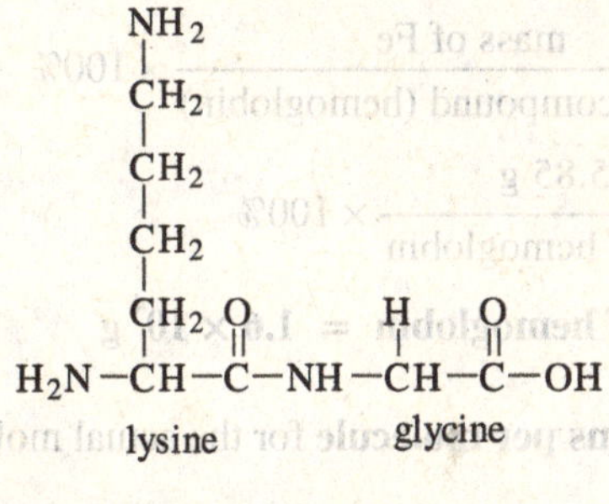


**25.22** The rate increases in an expected manner from 10°C to 30°C and then drops rapidly. The probable reason for this is the loss of catalytic activity of the enzyme because of denaturation at high temperature.

**25.28** Nucleic acids play an essential role in protein synthesis. Compared to proteins, which are made of up to 20 different amino acids, the composition of nucleic acids is considerably simpler. A DNA or RNA molecule contains only four types of building blocks: purines, pyrimidines, furanose sugars, and phosphate groups. Nucleic acids have simpler, uniform structures because they are primarily used for protein synthesis, whereas proteins have many uses.

**25.30** The sample that has the higher percentage of C–G base pairs has a higher melting point because C–G base pairs are held together by three hydrogen bonds. The A–T base pair interaction is relatively weaker because it has only two hydrogen bonds. Hydrogen bonds are represented by dashed lines in the structures below.

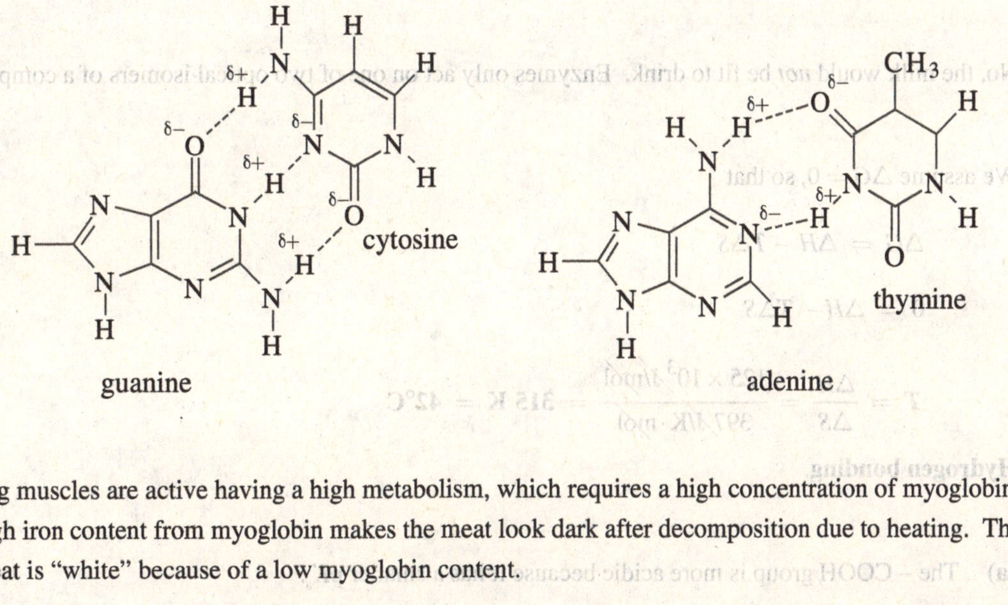


**25.32** Leg muscles are active having a high metabolism, which requires a high concentration of myoglobin. The high iron content from myoglobin makes the meat look dark after decomposition due to heating. The breast meat is "white" because of a low myoglobin content.

**25.34** Insects have blood that contains no hemoglobin. Thus, they rely on simple diffusion to supply oxygen. It is unlikely that a human-sized insect could obtain sufficient oxygen by diffusion alone to sustain its metabolic requirements.

**25.36** From the mass % Fe in hemoglobin, we can determine the mass of hemoglobin.

$$\% \text{ Fe} = \frac{\text{mass of Fe}}{\text{mass of compound (hemoglobin)}} \times 100\%$$

$$0.34\% = \frac{55.85 \text{ g}}{\text{mass of hemoglobin}} \times 100\%$$

$$\textbf{minimum mass of hemoglobin} = \mathbf{1.6 \times 10^4 \text{ g}}$$

Hemoglobin must contain **four Fe atoms per molecule** for the actual molar mass to be four times the minimum value calculated.

**25.38** The type of intermolecular attractions that occur are mostly attractions between nonpolar groups. This type of intermolecular attraction is called a **dispersion force**.

**25.40** This is as much a puzzle as it is a chemistry problem. The puzzle involves breaking up a nine-link chain in various ways and trying to deduce the original chain sequence from the various pieces. Examine the pieces and look for patterns. Remember that depending on how the chain is cut, the same link (amino acid) can show up in more than one fragment.

Since there are only seven different amino acids represented in the fragments, at least one must appear more than once. The nonapeptide is:

**Gly–Ala–Phe–Glu–His–Gly–Ala–Leu–Val**

Do you see where all the pieces come from?

**25.42** No, the milk would *not* be fit to drink. Enzymes only act on one of two optical isomers of a compound.

**25.44** We assume $\Delta G = 0$, so that

$$\Delta G = \Delta H - T\Delta S$$

$$0 = \Delta H - T\Delta S$$

$$T = \frac{\Delta H}{\Delta S} = \frac{125 \times 10^3 \cancel{\text{J}}/\cancel{\text{mol}}}{397 \cancel{\text{J}}/\text{K} \cdot \cancel{\text{mol}}} = \mathbf{315 \text{ K} = 42°\text{C}}$$

**25.46** **Hydrogen bonding.**

**25.48** **(a)** The –COOH group is more acidic because it has a smaller $pK_a$.

**(b)** We use the Henderson-Hasselbalch equation, Equation (16.4) of the text.

$$\text{pH} = \text{p}K_a + \log\frac{[\text{conjugate base}]}{[\text{acid}]}$$

**At pH = 1.0,**

$-COOH$ $$1.0 = 2.32 + \log\frac{[-COO^-]}{[-COOH]}$$

$$\frac{[-COOH]}{[-COO^-]} = 21$$

$-NH_3^+$ $$1.0 = 9.62 + \log\frac{[-NH_2]}{[-\overset{+}{N}H_3]}$$

$$\frac{[-\overset{+}{N}H_3]}{[-NH_2]} = 4.2 \times 10^8$$

Therefore the **predominant species** is:

$$\mathbf{CH(CH_3)_2 - CH(\overset{+}{N}H_3) - COOH}$$

**At pH = 7.0,**

$-COOH$ $$7.0 = 2.32 + \log\frac{[-COO^-]}{[-COOH]}$$

$$\frac{[-COO^-]}{[-COOH]} = 4.8 \times 10^4$$

$-NH_3^+$ $$7.0 = 9.62 + \log\frac{[-NH_2]}{[-\overset{+}{N}H_3]}$$

$$\frac{[-\overset{+}{N}H_3]}{[-NH_2]} = 4.2 \times 10^2$$

**Predominant species:**

$$\mathbf{CH(CH_3)_2 - CH(\overset{+}{N}H_3) - COO^-}$$

**At pH = 12.0,**

$-COOH$ $$12.0 = 2.32 + \log\frac{[-COO^-]}{[-COOH]}$$

$$\frac{[-COO^-]}{[-COOH]} = 4.8 \times 10^9$$

$-NH_3^+$ $$12.0 = 9.62 + \log\frac{[-NH_2]}{[-\overset{+}{N}H_3]}$$

$$\frac{[-NH_2]}{[-\overset{+}{N}H_3]} = 2.4 \times 10^2$$

**Predominant species:**

$$CH(CH_3)_2-CH(NH_2)-COO^-$$

**(c)** $$pI = \frac{pK_{a_1} + pK_{a_2}}{2} = \frac{2.32 + 9.62}{2} = \mathbf{5.97}$$

**25.50** **(a)** $$\bar{M}_n = \frac{\sum N_i M_i}{\sum N_i} = \frac{[(1)(1.0) + (1)(3.0) + (2)(4.0) + (1)(6.0)]\text{kg/mol}}{5} = \mathbf{3.6\ kg/mol}$$

$$\bar{M}_w = \frac{\sum N_i M_i^2}{\sum N_i M_i} = \frac{[(1)(1.0)^2 + (1)(3.0)^2 + (2)(4.0)^2 + (1)(6.0)^2](\text{kg/mol})^2}{[(1)(1.0) + (1)(3.0) + (2)(4.0) + (1)(6.0)]\text{kg/mol}} = \mathbf{4.3\ kg/mol}$$

**(b)** $$\bar{M}_n = \frac{\sum N_i M_i}{\sum N_i} = \frac{(4)(5\text{ kg/mol})}{4} = \mathbf{5\ kg/mol}$$

$$\bar{M}_w = \frac{\sum N_i M_i^2}{\sum N_i M_i} = \frac{(4)(5\text{ kg/mol})^2}{(4)(5\text{ kg/mol})} = \mathbf{5\ kg/mol}$$

**(c)** If there is a small spread in the distribution of sizes of polymer chains in the sample (i.e. the masses of each polymer chain in the sample are similar), $\bar{M}_n$ and $\bar{M}_w$ will be close in value. On the other hand, if there is a large spread in the distribution of sizes of polymer chains in the sample, there will be a larger difference in the values of $\bar{M}_n$ and $\bar{M}_w$.

**(d)** The hemoglobin molecule is made up of four subunits (see Figure 25.13 of the text). In solution, the molecules dissociate to a varying extent so there is a distribution of molar mass. Thus, $\bar{M}_n \neq \bar{M}_w$. For myoglobin or cytochrome *c*, there are no subunits so there are no dissociations and hence $\bar{M}_n = \bar{M}_w$.